Y0-AEY-658

Statistics:
Probability, Inference, and Decision

Second Edition

SERIES IN QUANTITATIVE METHODS FOR
DECISION MAKING

Robert L. Winkler, Consulting Editor

Statistics:
Probability, Inference, and Decision

Second Edition

Robert L. Winkler
Indiana University

William L. Hays
University of Georgia

HOLT, RINEHART AND WINSTON
New York · Chicago · San Francisco · Atlanta
Dallas · Montreal · Toronto · London · Sydney

Library of Congress Cataloging in Publication Data

Winkler, Robert L
Statistics—probability, inference, and decision.

(Series in quantitative methods for decision making)
In the 1st ed. W. L. Hays' name appeared first on t. p.
Bibliography: p. li
Includes index.
1. Mathematical statistics. 2. Probabilities.
I. Hays, William Lee, 1926–   joint author.
II. Title.
QA276.W55   1975   519.5   74-23848
ISBN 0-03-014011-0
Printed in the United States of America
0 038 09876

To Dorth and Palma

PREFACE

This book is intended as an introduction to probability theory, statistical inference, and decision theory (no previous knowledge of probability and statistics is assumed). The emphasis is on basic concepts and the theory underlying statistical methods rather than on a detailed exposition of all of the different methods which statisticians find useful, although most of the elementary techniques of statistical inference and decision are discussed. In other words, this book is not meant as a "cookbook"; when a new concept is introduced, every attempt is made to justify and explain the concept, both mathematically and verbally. Our intention is to give the reader a good basic knowledge of the important concepts of probability and statistics. Because of this emphasis on the conceptual framework, readers should be able to pursue more advanced work in probability and statistics upon the completion of this book, if they so desire.

It has been our experience that the mathematical backgrounds of students approaching a first course in probability and statistics are usually quite varied. Because of the current trend toward requiring some elementary calculus of students in virtually all fields of endeavor, and because the use of calculus makes it possible to discuss continuous probability models and their statistical applications in a more rigorous manner, we have used elementary calculus in some sections of this book. This will enable students who have had some calculus to gain a better appreciation of continuous probability models than is possible with college algebra alone. However, most of the material in the book does not involve calculus, and when it does, the mathematical formulations are accompanied by heuristic explanations, so that the book can also be used profitably by students with no calculus background. In this respect, we have attempted to strike a happy medium between the books using absolutely no calculus and the "mathematical statistics" books relying heavily on calculus, incorporating the advantages of both. This has resulted in a longer book, but it should prove quite valuable from a pedagogical standpoint.

The book is divided into three major parts: probability theory, statistical inference and decision, and some specific statistical techniques. A useful feature of the book is the relatively complete development of probability theory in Chapters 1–4, the first major part of the book. The set-theoretic approach is used, elementary rules of probability are discussed, and both discrete and continuous random variables are covered at some length. Also, because of the later discussion of Bayesian inference and decision theory, quite a bit of space is given to the interpretation of probability. We feel that a sound understanding of basic probability theory is a necessary prerequisite to the study of statistical inference and decision.

Statistical inference and decision is covered in Chapters 5–9, the second major part of the book. After a brief discussion of descriptive measures, the concept of a sampling distribution is introduced, and estimation and hypothesis testing are covered at some length. Next, the Bayesian approach to statistics is discussed and the relationship between Bayesian and "classical," or "sampling theory," techniques is covered. Finally, the concepts of decision theory are introduced and the distinction between inferential problems and decision-making problems is discussed. We feel that a unique feature of the book is the extensive coverage of Bayesian inference and decision theory and the integration of this material with the previous material on "classical" inferential procedures. Chapters 8 and 9 include many topics that are not generally covered in books of this nature and level.

In Chapters 10–12, the third major part of the book is presented. Regression and correlation are discussed extensively, including the theoretical regression curve, simple linear regression, least-squares curve fitting, curvilinear regression, and multiple regression. Next, sampling theory, experimental design, and several analysis of variance models are discussed. Finally, a chapter on nonparametric methods is included.

Numerous exercises are included at the end of each chapter, ranging from straightforward applications of the textual material to exercises which require considerably more thought. The exercises are an integral part of the book, serving to reinforce the reader's grasp of the concepts presented in the text and to point out possible applications and extensions of these concepts. Answers to selected exercises are presented at the end of the book, and a solutions manual containing reasonably detailed solutions of the exercises is available to instructors from the publisher.

The book is self-contained in the sense that all necessary tables are included at the end of the book. Additionally, a short appendix with some commonly encountered differentiation and integration formulas and an appendix on matrix algebra, which is used very briefly in the discussion of multiple regression in Chapter 10, are included. Finally, a list of references is provided at the end of the book for those wishing to pursue any topic further.

As the above summary indicates, there is easily enough material for a two-semester course. The chapters are divided rather closely into sections, however, so that a certain amount of flexibility is afforded. The book can be used for a one-semester or two-quarter course in probability, inference, and decision theory if some sections are deleted. Depending on the background of the students, there are various other possibilities. For instance, if the students have had some probability theory, the book could be used for a one-semester course in statistics, starting with Chapter 5.

We have been encouraged by the response to the first edition of this book, and we feel that some of the "rough spots" are smoothed out in the second edition. The basic objectives of the book are unchanged, and although we have reorganized, added, and deleted some topics within chapters, the overall organization of the book remains the same. We have attempted to reduce verbosity while still maintaining our intention to explain carefully the concepts that are introduced. Particular attention has been given to improving the examples and exercises. Many examples and exercises have been replaced or supplemented by examples and exercises involving more realistic situations, and the exercises have increased both in number (almost twice as many as in the first edition) and in variety.

We are especially grateful to all of the people who provided help and encouragement in bringing this work to completion. As noted in the preface to the first edition, we are indebted to William A. Ericson and Ingram Olkin for providing numerous valuable comments during the preparation of that edition. In the preparation of the second edition, we are grateful for many helpful suggestions from several reviewers (including Mal Golden, Stephen C. Hora, Steven A. Lippman, Ingram Olkin, Lyn D. Pankoff, and Richard E. Trueman), from numerous students and instructors who have used the first edition of the book, and from reviews of the first edition that have appeared in professional journals. In addition, we thank Dorothy M. Winkler for editorial and proofreading assistance and William McCormick, Jeannette Prietsch, and Suzanna Wong for aid in the preparation of the solutions manual. Of course, there are many people who contribute in an indirect manner to a book such as this, including contributors to the literature in probability and statistics, our colleagues, and our students; we acknowledge their contributions collectively rather than individually. Our greatest thanks, however, go to our wives, Dorth and Palma, for their seemingly endless supply of patience, encouragement, and assistance during the writing of this book.

Bloomington, Indiana
Athens, Georgia
November 1974

Robert L. Winkler
William L. Hays

CONTENTS

3 Probability Distributions 116

4   Special Probability Distributions   203

5   Frequency and Sampling Distributions   271

6   Estimation   334

7   Hypothesis Testing   402

8   Bayesian Inference   471

9   Decision Theory   550

10   Regression and Correlation   643

11   Sampling Theory, Experimental Design, and Analysis of Variance   730

12   Nonparametric Methods   815

Appendix A   i

Some Common Differentiation and Integration Formulas

Appendix B   iv

Matrix Algebra

Tables   xiii

References and Suggestions for Further Reading   li

Answers to Selected Exercises   lv

Index   lxvi

Statistics:
Probability, Inference, and Decision

Second Edition

INTRODUCTION

Applications of statistics occur in virtually all fields of endeavor—business, the social sciences, the physical sciences, the biological sciences, education, and so on, almost without end. Although the specific details of the methods differ somewhat in the different fields, the applications all rest on the same general theory of statistics. By examining what the fields have in common in their applications of statistics, we can gain a picture of the general nature of statistics.

To begin, it is convenient to identify three major branches of statistics: **descriptive statistics, inferential statistics,** and **statistical decision theory. Descriptive statistics** is a body of techniques for the effective organization, summarization, and communication of data. The everyday use of the term "statistics" usually refers to data organized by the methods of descriptive statistics. For example, consider statistics included in a newspaper's report of a sporting event such as a basketball game or a track meet, economic statistics such as the unemployment rate and the gross national product, and medical statistics such as mortality rates for different diseases. Descriptive statistics deals with the data at hand in any given situation. In contrast, **inferential statistics** is a body of methods for arriving at conclusions extending *beyond* the immediate data. Given some information regarding a small subset of a given population, the methods of inferential statistics can be used to make inferences about the entire population. For example, a political analyst may make inferences about the outcome of an election on the basis of information regarding the voting preferences of a small proportion of the electorate, a medical researcher may make inferences about the effectiveness of a new drug on the basis of an experiment involving patients at a particular hospital, and a quality-control manager may make inferences about the proportion of defective items in a large shipment of items on the basis of a careful study of a small number of items from the shipment. Finally, **statistical decision theory** goes one step further; instead of just making inferential statements about the population, the statistician uses the available information to choose among a number of alternative actions. For example, an executive might use the procedures of statistical decision theory to decide upon an investment from a list of potential investments, and a firm might use statistical decision theory to decide whether or not to market a new product.

The major concern in this book is with inferential statistics and statistical decision theory, both of which are closely related to the concept of uncertainty. Inferential statistics involves drawing conclusions or making predictions on the basis of limited information, and statistical decision theory involves making decisions on the basis of limited information. In both instances, the statistician is facing uncertainty, since only limited information is available. Thus, in one sense, mathematical statistics is a theory about uncertainty. Granted that certain conditions are fulfilled, the theory permits deductions about the likelihood of the various possible outcomes of interest. In this manner, the essential concepts in statistics are based on the theory of probability; the statistician is interested in the probabilities of particular kinds of outcomes, given that certain initial conditions are met. Therefore, it is necessary for the student to have a basic understanding of probability theory before learning statistics.

It should be emphasized that this book deals primarily with the basic concepts of statistics and with the theory underlying statistical methods rather than with a detailed exposition of all of the different methods that statisticians find useful. In other words, this is not a "cookbook" that will equip the student to meet every possible situation that might be encountered. It is true that many methods will be introduced and we will, in fact, discuss most of the elementary techniques for statistical inference and decision. In the past few years, however, there has been a proliferation of new techniques, particularly in decision theory and in problems involving several variables. This is partly due to the development of computers, which have opened up new avenues of data analysis for the statistician, making it possible for him to answer questions that were formerly unanswerable because of computational complexity. This proliferation of statistical methods can be expected to continue, with some of the current methods being replaced with newer techniques. Thus, we feel that it is better for the student to gain a basic understanding of the concepts of statistics than simply to learn a myriad of specialized methods.

Essentially, the book is divided into three parts. The first part, Chapters 1–4, deals with probability theory, thus laying a foundation for the study of statistics. Chapters 5–9 involve the basic theory of statistical inference and decision, and the third part, Chapters 10–12, concerns a number of specialized topics (specialized in comparison with the basic concepts presented in the second part). Throughout the book we have attempted, when introducing a new concept, to present both the relevant mathematics (any formulas, proofs, and so on) *and* a clear explanation in words of the concept. Those mathematical expressions that are of particular importance are denoted by an asterisk following the number of the expression. The numbering system should be self-explanatory; 2.4.1 is the first numbered expression in Section 4 of Chapter 2, for example.

1

SETS
AND
FUNCTIONS

It may seem surprising that a book about statistics starts off with a discussion of sets. Although the study of sets and functions is not always a part of a course in statistics, these topics actually provide the most fundamental concepts we will use. Set theory will be the basis for our discussion of probability theory, which will be introduced in the next chapter. The idea of a function pervades virtually all areas of application of mathematics and statistics. Even at the high-school level most of us are exposed to the notion of a mathematical function, and we know that saying that "Y is a function of X" expresses something about a relation between two things. However, without a very good background in mathematics, a student is usually somewhat vague about the precise meaning of the word "function" used in mathematical or scientific writing. One of the primary purposes of this chapter is to give a very concrete and restricted meaning to this term. We will show that the idea of a function is a very simple one, which grows out of the notion of a set.

1.1 SETS

The concept of a set is the starting point for all of modern mathematics, and yet this idea could hardly be more simple.

Any well-defined collection of objects is a set.

The individual objects making up the set are known as the "elements" or "members" of the set. The set itself is the aggregate or totality of its members. If a given object is in the set, then we say that the object is an **element** of, or a **member** of, or **belongs** to, the set.

For example, the students enrolled at a given university in a given year form a set, and any particular student is an element of that set. All certified public accountants in a given state form another set, and all integers between 10 and 10,000, inclusive, form another set. The securities on the New York Stock Exchange selling for more than $50 per share at a particular time constitute a set, while the securities selling for less than $30 constitute a different set. Another set is the set of securities on the New York Stock Exchange selling for more than $10 but less than $60, and this set has some elements in common with each of the two preceding sets.

It is important to note that in the definition of a set the qualification "well-defined" occurs. This means that *it must be possible, at least in principle, to specify the set so that we can decide whether any given object does or does not belong.* This does not mean that sets can be discussed only if their members actually exist; it is perfectly possible to speak of the set of all women presidents of the United States, for example, even though there are not any such objects at this writing. What *is* required is some procedure or rule for deciding whether an object is or is not in the set; given an object, one can decide if it meets the qualifications of a female president of the United States, and thus the set is well-defined.

The word "object" in the definition can also be interpreted quite liberally. Often sets are discussed having members that are not objects in the usual sense but rather are "phenomena," or "happenings," or possible outcomes of observation. We will have occasion to use sets of "logical possibilities," all the different ways something might happen, where each distinct possibility is thought of as one member of the set. For instance, we might be interested in the set of all possible combinations of individuals that could make up a particular committee of three, where the committee is to be selected from all of the lawyers in a given city.

In discussing sets, we will follow the practice of letting a capital letter, such as A, symbolize the set itself, with a small letter, such as a, used to indicate a particular member of the set. The symbol "$\in$" is often used to indicate "is a member of"; thus,

$$a \in A$$

is read "a is a member of A."

There are two different ways of specifying a set. The first way is by *listing* all of the members. For example,

$$A = \{1, 2, 3, 4, 5\}$$

is a complete specification of the set A, saying that it consists of the numbers 1, 2, 3, 4, and 5. The braces around the listing are used to indicate that the list makes up a set. Another set specified in a similar way is

$$B = \{\text{Roosevelt, Truman, Eisenhower, Kennedy, Johnson, Nixon}\}.$$

In each instance the respective elements of the set are simply listed. These sets are thought of as unordered, since the order of the listing is completely irrelevant as long as each member is included once and only once in the list.

The other way of specifying a set is to give a *rule* that lets us decide whether or not an object is a member of the set in question. Thus, set A may be specified by

$$A = \{a \mid a \text{ is an integer and } 1 \leq a \leq 5\}.$$

(The symbol $\mid$ is read as "such that," so that the expression above is, in words, "the set of all elements a such that a is an integer between 1 and 5 inclusive.") Given this rule, and any potential element of the set A, we can decide immediately whether or not the object actually does belong to the set. A rule for set B would be

$$B = \{b \mid b \text{ is a United States President elected between 1930 and 1972}\}.$$

It is quite possible to specify sets in terms of other sets. For example, we might specify a set C by

$$C = \{c \mid c = a + 15, \text{ for all } a \in A\}.$$

When listed, the elements of C would be

$$C = \{16, 17, 18, 19, 20\}.$$

It is usually far more convenient to specify a set by its rule rather than by listing the elements. For example, sets such as the following would be awkward or impossible to list:

$D = \{x \mid x \text{ is a U. S. citizen}\}$,

$E = \{x \mid x \text{ is a U. S. corporation with at least five employees}\}$,

$F = \{x \mid x \text{ is a positive number}\}$,

$G = \{y \mid y \in F \text{ and } y \text{ is an integer}\}$.

1.2 FINITE AND INFINITE SETS

In future work it will be necessary to distinguish between **finite** and **infinite** sets. For our purposes, a finite set has members equal in number to some specifiable positive integer or to zero. Otherwise, if the number of members is greater than any positive integer you conceivably can name, the set is considered infinite. Thus, the set of all names in the Manhattan telephone book, the set of all books printed in the nineteenth century, the set of all houses in Rhode Island, and the set of all living mammals are all finite sets, even though the number of members each includes is very large.

On the other hand, the set of all points lying on a circle, the set of all real numbers (including all rational and irrational, positive and negative numbers), the set of all positive integers, and the set of all intervals into which a straight line may be divided are examples of infinite sets.

Sets with an infinite number of members fall into two general classifications. If the members of the set can be put into one-to-one correspondence with the positive integers, so that each member of the set is associated with one and only one positive integer, then the set is said to be **countably infinite.** For example, suppose that someone decides to toss a coin until heads first appears. This might take one toss, two tosses, three tosses, or even more tosses. In theory, the coin could be tossed 10,000 times without heads appearing, although if it is an ordinary coin, this would be a surprising result. The possible numbers of tosses until heads first appears form a countably infinite set. Still another kind of set is **uncountably infinite.** Here the elements of the set are so "dense" that even the counting numbers (the positive integers) would fail to exhaust all of them. An example of such a set is the set of all possible points on a straight line. Regardless of how positive integers are assigned to the points, there will always be points left over.

The distinctions among finite sets, countably infinite sets, and uncountably infinite sets are important when probability distributions are discussed. In the next few sections, however, all sets will be treated as though they were finite. The general ideas apply to infinite sets as well, but certain qualifications must be made in some of the definitions, and we will overlook these qualifications.

1.3 UNIVERSAL SETS AND THE EMPTY SET

There are many instances in set theory when it becomes convenient to consider only objects belonging to some "large" set. Then, given that all objects to be discussed belong to this "universal" set, we proceed to talk of particular groupings of elements. A **universal set** acts to set the stage for the kinds of sets that will be introduced. For example, we may wish to deal only with U. S. students enrolled in college in the year 1974, and so we begin by specifying a universal set:

$$W = \{w \mid w \text{ was a U. S. student enrolled in college in 1974}\}.$$

Then particular subsets of W are introduced:

$$A = \{a \mid a \in W \text{ and } a \text{ was a student at Harvard}\},$$

$$B = \{b \mid b \in W \text{ and } b \text{ was classified as a sophomore}\},$$

and so on. Or perhaps the universal set is

$$W = \{w \mid w \text{ is an automobile}\},$$

with subsets such as

$$A = \{a \mid a \in W \text{ and } a \text{ was manufactured in Sweden}\}$$

and $B = \{b \mid b \in W \text{ and } b \text{ has four doors}\}.$

Up to this point, the individual member of a set has been denoted by a small letter, a, b, w, and so on, depending on the capital letter used for the set itself. This is not strictly necessary, however, as any symbol that serves as a "place holder" in the rule specifying the set would do as well. In future sections it will sometimes prove convenient to use the same symbol for a member of several sets, so that the neutral symbol x will be used, indicating any member of the universal set under discussion. Furthermore, it is redundant to state that $x \in W$ after the universal set W has been specified, and so this statement will be omitted from the rule for particular subsets of the universal set. Nevertheless, it is always understood that any member of a set is automatically a member of some universe W.

Another set of particular interest is the **empty set,** which plays a role in set theory that is similar to the role of the number zero in ordinary arithmetic. **Any set that contains no members is called the empty set.** The empty set is usually identified by the symbol $\emptyset$. It is not hard to dream up examples of the empty set: the set of all months containing 38 days, the set of all numbers not divisible by 1, and so on.

1.4 SUBSETS

Suppose that there is some set A and another set B such that any element which is in B is also in A. Then B **is a subset of** A. This is symbolized by

$$B \subseteq A.$$

More formally,

> **the set B is a subset of the set A if and only if**
> **for each $x \in B$, then $x \in A$.**

For example, consider the set A of all federal employees in the United States, and let B be the set of employees of the U. S. Department of State. Or, let A be the set of all real numbers, and let B indicate the set of all whole numbers. Once again, B is a subset of A.

If B is a subset of A *and* there is at least one element of A not in B, then B is a **proper subset** of A, symbolized by

$$B \subset A.$$

In both of the examples just given, the set B is a proper subset of A, since some federal employees are not employed by the State Department, and some numbers are not whole. On the other hand, if every single element in B is in A, *and* every single element in A is in B, so that $B \subseteq A$ *and* $A \subseteq B$, then

$$A = B;$$

the two sets are equivalent or equal. That is, two sets are **equal** if they contain precisely the same elements.

Note the similarity of the symbols "$\subseteq$," for "is a subset of," and "$\subset$," for "is a proper subset of," to the symbols "$\leq$," for "is less than or equal to," and "$<$," for "is less than." The difference between "$\subseteq$" and "$\subset$" is the fact that one set can be a subset of another even though the two sets actually are identical or are equal element by element; a set can be a proper subset of another only if the two sets are not equal. It should also be noted that the statements "$A \subseteq B$" and "$B \supseteq A$" are equivalent, as are the statements "$A \subset B$" and "$B \supset A$." In each case, the set at the *closed* end of the "horseshoe" is the subset (or proper subset), and the set at the *open* end is the larger set. This is similar to the correspondence between "$x \leq y$" and "$y \geq x$," where x and y are numbers.

The universal set and the empty set have special properties. **Every set is a subset of the universal set, and the empty set is a subset of every set: $\varnothing \subseteq A \subseteq W$ for every A.** Notice that since the empty set has no members, the definition of subset is not contradicted.

A scheme that is very useful for illustrating sets and for showing relations among them is the Venn diagram (named after the logician J. Venn). These diagrams are sometimes referred to as Euler diagrams (after the mathematician L. Euler); both men made important contributions to the theory of sets. A Venn diagram pictures a set as all points contained within a circle, square, or other closed geometrical figure.

Since there is an infinite number of possible points within any such figure, Venn diagrams actually represent infinite sets, but they are useful for representing any set. For example, a Venn diagram picturing the universal set W and the subset A would look like Figure 1.4.1. Since A is a subset of W, the area of the circle A is completely included in the area of the rectangle W.

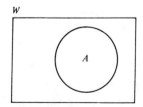

Figure 1.4.1   A Set Represented on a Venn Diagram

Now consider another set $B \subseteq W$. If $B \subset A$, the Venn diagram would be as shown by Figure 1.4.2. On the other hand, if B were not a subset of A, then Figure 1.4.3 would be the Venn diagram if A and B shared members in common, or perhaps Figure 1.4.4 would be the Venn diagram if the two sets had no members in common.

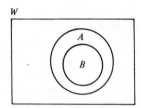

Figure 1.4.2   One Set as a Proper Subset of Another Set

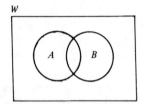

Figure 1.4.3   Two Sets with Common Members

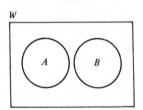

Figure 1.4.4   Two Sets without Common Members

1.5 THE UNION OF SETS

Given some sets A, B, C, and so on, contained in a universal set W, it is possible to "operate" on these sets to form new sets, each of which is also a subset of W.

The first of these operations is the **union** of two or more sets. Given W and the two subsets A and B, the union of A and B is written $A \cup B$ (a useful mnemonic device for the symbol $\cup$ is the u in *union*). By definition,

the union of sets A and B, written $A \cup B$, is the set of all elements that are members of A, or of B, or of both:

$$A \cup B = \{x \mid x \in A \text{ or } x \in B, \text{ or both}\}.$$

For example, let

   $W = \{x \mid x$ has received a degree from Indiana University$\}$,

   $A = \{x \mid x$ has received a bachelor's degree from Indiana University$\}$,

and $B = \{x \mid x$ has received a master's degree from Indiana University$\}$.

Then

$A \cup B = \{x \mid x$ has received a bachelor's degree or a master's degree or both from Indiana University$\}$.

In the Venn diagram in Figure 1.5.1, the shaded portion shows the union of A and B. Notice that we include the possibility that an element of the union could be a member of *both* A and B. Note further that in this example, $A \cup B \subset W$ (the union is a *proper* subset of the universal set), since there are some degree holders from Indiana University who belong neither to set A nor to set B (for example, individuals who received only a doctorate from Indiana University).

The idea of the union may be extended to more than two sets. For instance, given W and three sets A, B, and C, then

$$A \cup B \cup C = \{x \mid x \in A, \text{ or } x \in B, \text{ or } x \in C\},$$

as shown in Figure 1.5.2. In general, the union of K sets $A_1, A_2, \ldots, A_K$ is defined as follows:

$$\bigcup_{i=1}^{K} A_i = A_1 \cup A_2 \cup \cdots \cup A_K = \{x \mid x \in A_1, \text{ or } x \in A_2, \ldots, \text{ or } x \in A_K\}$$

$$= \{x \mid x \text{ is a member of at least one of the } K \text{ sets } A_1, A_2, \ldots, A_K\}.$$

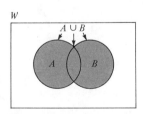

Figure 1.5.1   The Union of Two Sets

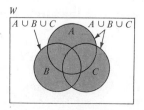

Figure 1.5.2   The Union of Three Sets

As an example of the union of three sets, let

$$W = \{x \mid x \text{ is a positive integer, } 1 \le x \le 10\},$$

$$A = \{x \mid x \text{ is a perfect square}\} = \{1, 4, 9\},$$

$$B = \{x \mid x \text{ is divisible by 3}\} = \{3, 6, 9\},$$

and

$$C = \{x \mid x \text{ is divisible by 7}\} = \{7\}.$$

Then

$$A \cup B \cup C = \{1, 3, 4, 6, 7, 9\}.$$

If the union of a number of sets equals W, then those sets are said to be **collectively exhaustive.** That is, taken together, they "exhaust" the universal set W. In the preceding example, the sets A, B, and C do not exhaust W, since there are elements of W that are not in A, B, or C. If a fourth set,

$$D = \{x \mid x \text{ is divisible by 2 or 5}\} = \{2, 4, 5, 6, 8, 10\},$$

is added, then the sets A, B, C, and D are collectively exhaustive, since

$$A \cup B \cup C \cup D = \{1, 2, 3, 4, 5, 6, 7, 8, 9, 10\} = W.$$

The union of any set with a subset of itself is simply the larger set: if $B \subseteq A$, then $A \cup B = B \cup A = A$. It follows that

$$A \cup \varnothing = A$$

and

$$W \cup A = W.$$

1.6   THE INTERSECTION OF SETS

The verbal rule for the union of two sets always involves the word "or"; the union $A \cup B$ is the set made by finding the elements that are members of A *or* of B or of both. However, what about the set of elements in *both A and B?* This set is included in the union, but we are often interested in this set by itself, and it is referred to as the **intersection** of A and B.

The intersection of sets A and B, written $A \cap B$, is the set of all members belonging to both A and B:

$$A \cap B = \{x \mid x \in A \text{ and } x \in B\}.$$

The intersection of two sets presented in a Venn diagram appears as Figure 1.6.1.

For example, let

$$W = \{x \mid x \text{ is a chemical compound}\},$$

$$A = \{x \mid x \text{ contains chlorine}\},$$

and $\qquad B = \{x \mid x \text{ contains oxygen}\}.$

Then

$$A \cap B = \{x \mid x \text{ contains chlorine } and \text{ oxygen}\}.$$

As another example, consider the sets of numbers

$$W = \{x \mid x \text{ is a positive integer}\},$$

$$A = \{1, 2, 3, 4, 5, 6, 7, 8, 9, 10\},$$

and $\qquad B = \{8, 9, 10, 11, 12, 13\}.$

Then

$$A \cap B = \{8, 9, 10\},$$

since only these elements appear in both A and B.

The intersection of two sets A and B is always a subset of their union:

$$A \cap B \subseteq A \cup B.$$

If the intersection of two sets is empty ($A \cap B = \varnothing$), then the sets are said to be **disjoint** or **mutually exclusive**.

Whenever B is a subset of A, then the intersection $A \cap B$ or $B \cap A$ is equal to the *smaller* of the two sets, or B. Thus,

$$A \cap \varnothing = \varnothing$$

and $\qquad A \cap W = A,$

since $\varnothing$ is a subset of A and A is a subset of W.

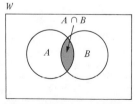

Figure 1.6.1   The Intersection of Two Sets

The intersection may be defined for any number of sets taken together. For example, the intersection of three sets A, B, and C is

$$A \cap B \cap C = \{x \mid x \in A \text{ and } x \in B \text{ and } x \in C\}.$$

Thus, if A were all women, B were all lawyers, and C were all persons with blue eyes, then $A \cap B \cap C$ would be the set of all women lawyers having blue eyes. In general, the intersection of K sets A_1, $A_2, \ldots$, A_K is defined as follows:

$$\bigcap_{i=1}^{K} A_i = A_1 \cap A_2 \cap \cdots \cap A_K$$

$$= \{x \mid x \in A_1 \text{ and } x \in A_2 \text{ and } \cdots \text{ and } x \in A_K\}$$

$$= \{x \mid x \text{ is a member of } all \text{ of the } K \text{ sets } A_1, \ldots, A_K\}.$$

The sets A_1, A_2, $\ldots$, A_K are said to be **disjoint,** or **mutually exclusive,** if *all possible pairs* of sets selected from the given K sets are disjoint. Thus, A, B, and C are disjoint if $A \cap B = \varnothing$, $A \cap C = \varnothing$, and $B \cap C = \varnothing$ (see Figure 1.6.2). Note that $A \cap B \cap C = \varnothing$ is *not* a sufficient condition for A, B, and C to be disjoint. In Figure 1.6.3, $A \cap B \cap C = \varnothing$, but the three sets are *not* disjoint, since $A \cap B \neq \varnothing$.

If the sets A_1, A_2, $\ldots$, A_K are both mutually exclusive *and* collectively exhaustive, then they are said to form a K-**fold partition** of W, or simply a **partition** of W. In terms of a Venn diagram, this means that the K sets do not overlap but that every point in the rectangle representing W is included in one of the K sets (see Figure 1.6.4). For example, let

$$W = \{1, 2, 3, 4, 5, 6\},$$

$$A = \{1, 3, 5\},$$

$$B = \{2, 4\},$$

$$C = \{1, 2, 6\},$$

and

$$D = \{6\}.$$

A, B, and C do not form a partition of W—they are collectively exhaustive, but they are not mutually exclusive, since 1 appears in both A and C and

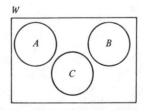

Figure 1.6.2   Three Mutually Exclusive Sets

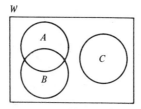

Figure 1.6.3   Three Sets that Are Not Mutually Exclusive

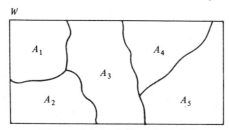

Figure 1.6.4   A Partition of W

2 appears in both B and C. However, A, B, and D are mutually exclusive and collectively exhaustive, so they form a partition (a three-fold partition) of W.

1.7 THE COMPLEMENT OF A SET

Another operation on sets is taking the **complement** of a set.

The complement of a set A, written $\bar{A}$, is made up of all members of W that are *not* in A:

$$\bar{A} = \{x \mid x \notin A\},$$

where the symbol $\notin$ is read as "is not a member of" the set. For example, if

$$W = \{x \mid x \text{ is a name in the 1974 Detroit, Michigan telephone directory}\}$$

and

$$A = \{x \mid x \text{ begins with the letter "S"}\},$$

then

$$\bar{A} = \{x \mid x \text{ does not begin with the letter "S"}\}.$$

The intersection of any set and its complement is always empty. This is,

$$A \cap \bar{A} = \varnothing,$$

since no element could be a member of both A and $\bar{A}$ simultaneously. The union of A and $\bar{A}$, however, always equals the universal set W:

$$A \cup \bar{A} = W.$$

Notice that the complement of any set is always relative to the universal set; this is why specifying the universal set is so important, since we cannot determine the complement of a set without doing so. For example, suppose that we specified the following set:

$$A = \{x \mid x \text{ has blue eyes}\}.$$

What is $\bar{A}$? That depends on what we assumed W to be. If

$$W = \{x \mid x \text{ is a woman living in the United States}\},$$

then $\bar{A}$ consists of all nonblue-eyed women living in the United States. However, if

$$W = \{x \mid x \text{ is a person living in the United States}\},$$

then $\bar{A}$ is quite a different set, including nonblue-eyed men and children as well. If

$$W = \{x \mid x \text{ is an organism}\},$$

then $\bar{A}$ would include all nonblue-eyed organisms, among them some that have no eyes at all. *The complement of a set simply has no meaning without a universal set for reference.*

The complement of a set is shown in Figure 1.7.1 as the shaded area in the Venn diagram.

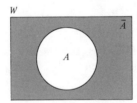

Figure 1.7.1   A Set and Its Complement

1.8   THE DIFFERENCE BETWEEN TWO SETS

The **difference** between two sets is closely allied to the idea of a complement.

The difference between sets A and B is

$$A - B = A \cap \bar{B} = \{x \mid x \in A \text{ and } x \in \bar{B}\}.$$

In other words, the difference contains all elements that are members of A *but not* members of B.

Although it makes no difference which set we write first in the symbols

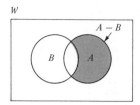

Figure 1.8.1   The Difference between Two Sets

for union and intersection, since

$$A \cup B = B \cup A$$

and $\qquad\qquad\quad A \cap B = B \cap A,$

the order *is* very important for the difference between two sets. The difference $A - B$ is *not* the same as the difference $B - A$.

As an example of the two differences, $A - B$ and $B - A$, let

$W = \{x \mid x$ was an elected official of the U. S. government in 1975$\}$,

$A = \{x \mid x$ was a member of the U. S. Congress$\}$,

and $B = \{x \mid x$ was a lawyer$\}$.

Then

$A - B = \{x \mid x$ was a member of the U. S. Congress who was not a lawyer$\}$

and

$B - A = \{x \mid x$ was a lawyer who was not a member of the U. S. Congress$\}$.

It is easy to see that $A - B$ and $B - A$ are quite different sets.

The difference $A - B$ is shown in Figure 1.8.1 as the shaded area in the Venn diagram.

1.9 THE ALGEBRA OF SETS

In elementary algebra, there are certain rules which the operations symbolized by "$+$," "$\cdot$," and so on must obey. The study of algebra is largely the study of these basic rules and of the mathematical consequences of the application of these operations to numbers. In the same way, an **algebra of sets** (often called a Boolean algebra, after the mathematician G. Boole) exists, consisting of a system for forming and manipulating sets by operations of union, intersection, complement, and so on, according to a specific set of rules or postulates. An algebra of sets has a great many similarities to, as well as some important differences from, ordinary algebra.

Although there is little point in dwelling upon set algebra at length in a course in statistics, it may be instructive to list the basic rules for an algebra of sets and to give examples of how simple theorems about sets may be proved using those rules. Just as in any formal mathematical system, the basic postulates are used to derive new conclusions by logical arguments. Since we are going to be discussing a closely related system in the next chapter, a view of the foundations of this simplest of all mathematical systems may give the reader a somewhat better idea of how postulates are used.

The postulates of set theory can be stated as follows.

1. **Closure laws:**
    (a) if A and B are sets, then $A \cup B$ is a set;
    (b) if A and B are sets, then $A \cap B$ is a set.

2. **Identity laws:**
    (a) there is one and only one set $\varnothing$, such that $A \cup \varnothing = A$ for any set A;
    (b) there is one and only one set W, such that $A \cap W = A$ for any set A.

3. **Commutative laws:**
    (a) $A \cup B = B \cup A$;
    (b) $A \cap B = B \cap A$.

4. **Associative laws:**
    (a) $(A \cup B) \cup C = A \cup (B \cup C)$;
    (b) $(A \cap B) \cap C = A \cap (B \cap C)$.

5. **Distributive laws:**
    (a) $A \cup (B \cap C) = (A \cup B) \cap (A \cup C)$;
    (b) $A \cap (B \cup C) = (A \cap B) \cup (A \cap C)$.

6. **For every set A there is one and only one set $\bar{A}$ such that**
    $$A \cup \bar{A} = W \text{ and } A \cap \bar{A} = \varnothing.$$

7. (a) $A = B$ and $C = D$ implies that $A \cup C = B \cup D$;
    (b) $A = B$ and $C = D$ implies that $A \cap C = B \cap D$;
    (c) $A = B$ implies that $\bar{A} = \bar{B}$.

8. **There are at least two distinct sets.**

Upon these statements about sets and the operations "$\cup$," "$\cap$," and "complement," an elaborate mathematical structure can be erected. It is very important to note that nowhere in these postulates is there an explicit definition of what is meant by a "set," by "$\cup$," and by "$\cap$." The postulates deal with equivalences among *undefined operations* carried out on *undefined objects*. In this sense, the set of postulates is said to be *formal*.

It is interesting to note that some of these postulates are true if we happen to be talking about the ordinary algebra of numbers rather than sets, and if "∪" is replaced by "+" and "∩" is replaced by "·." For instance, postulate 1(a) corresponds to a similar postulate about numbers: if x and y are numbers, then $x + y$ is a number. Similarly, postulates 3, 4, and 5(b) have equivalent statements in ordinary algebra:

Commutative laws:

$$x + y = y + x \quad \text{and} \quad x \cdot y = y \cdot x.$$

Associative laws:

$$(x \cdot y) \cdot z = x \cdot (y \cdot z) \quad \text{and} \quad x + (y + z) = (x + y) + z.$$

Distributive law:

$$x \cdot (y + z) = x \cdot y + x \cdot z.$$

Notice, however, that a statement corresponding to postulate 5(a) is *not*, in general, true of numbers:

$$x + (y \cdot z) \text{ is not equal to } (x + y) \cdot (x + z).$$

Thus, the algebra of sets, though similar to ordinary algebra, does differ from it in important respects.

Now an example will be given of how these postulates are used to arrive at new statements, or theorems, not specifically among these original statements. As with any mathematical system, a logical argument is used to arrive at true conclusions *given* that the postulates are true. First of all, a simple but very important theorem will be proved.

Theorem I

For any set A, $A \cap A = A$.

$$
\begin{aligned}
A \cap A &= (A \cap A) \cup \varnothing & &\text{by Postulate 2(a)}\\
&= (A \cap A) \cup (A \cap \bar{A}) & &\text{Postulate 6}\\
&= A \cap (A \cup \bar{A}) & &\text{Postulate 5(b)}\\
&= A \cap W & &\text{Postulate 6}\\
&= A & &\text{Postulate 2(b).}
\end{aligned}
$$

By a series of equivalent statements (the substitutions familiar from algebra), each justified by a postulate, we have arrived at a new statement, $A \cap A = A$, which we know must be true whenever the postulates are true.

As a slightly more complicated example of how these postulates are used, together with theorems already proved, consider the following theorem.

Theorem II

If $A \cup B = B$, then $A \cap B = A$.

$$
\begin{aligned}
B &= A \cup B &&\text{Given.} \\
A \cap B &= A \cap (A \cup B) &&\text{Postulate 7(b)} \\
&= (A \cap A) \cup (A \cap B) &&\text{Postulate 5(b)} \\
&= A \cup (A \cap B) &&\text{Theorem I} \\
&= (A \cap W) \cup (A \cap B) &&\text{Postulate 2(b)} \\
&= A \cap (W \cup B) &&\text{Postulate 5(b).}
\end{aligned}
$$

However,

$$
\begin{aligned}
(B \cup W) &= (B \cup W) \cap (B \cup \bar{B}) &&\text{Postulates 2(b) and 6} \\
&= B \cup (W \cap \bar{B}) &&\text{Postulate 5(a)} \\
&= B \cup \bar{B} &&\text{Postulate 2(b)} \\
&= W &&\text{Postulate 6,}
\end{aligned}
$$

so that

$$
\begin{aligned}
A \cap B &= A \cap W &&\text{Substitution} \\
A \cap B &= A &&\text{Postulate 2(b).}
\end{aligned}
$$

Here we have proved a much less obvious statement to be true. Anyone who has studied elementary geometry will appreciate the fact that with an accumulation of such theorems, plus the original postulates, there is virtually no end to the numbers of new theorems that can be proved in this way. Moreover, propositions about abstract sets will be true of real sets satisfying the postulates exactly as an expression in algebra will be true when the symbols are turned into numbers. Just as algebra is a mathematical model for solving problems about real quantities, and we can find characteristics of real figures using the model of Euclidean geometry, so does the algebra of sets become a useful mathematical model in some real situations. One striking example is the theory underlying large electronic computers, although very many other instances could be given.

Although this very hurried sketch of set theory may not give you any real facility with the set language, perhaps it will help when the essential ideas about sets recur in later chapters. In addition, another motive that underlies the discussion of sets is that it permits us to go on to a topic of great importance, both in statistics and in scientific enterprise in general: the study of mathematical relations.

1.10   SET PRODUCTS AND RELATIONS

One of the aims of science or of any other field of knowledge is to discover and describe relationships among things. Everybody knows what is meant by a relationship or connection among objects or phenomena; in order to use language itself we must group our experiences into classes or sets (using the nouns and adjectives) and then state relationships linking one kind of object with another (using the verbs). However, it is very important that we settle on a formal definition of what makes a **mathematical relation** before turning to other matters. It will be seen that the idea of a mathematical relation involves the notions of set and subset.

Just as a set is defined as a collection of objects, it is also possible to define a special kind of set made up of *pairs of objects*. Let a pair of objects be symbolized by (a, b), where a is a member of some set A, and b is a member of some set B. Each pair of objects is thought of as **ordered,** meaning that the way the objects are listed in a pair is important. Every day we encounter such ordered pairs of objects, where the two members are distinguished from each other by the role they play: husband-wife pairs, city-state pairs, manufacturer-product pairs, and so on. For an ordered pair (a, b), the order of listing is significant for the role played by each element of the pair, so that the pair (a, b) is not necessarily the same as the pair (b, a).

Now suppose that we have two sets, A and B. *All possible* pairs (a, b) are found, each pair associating a member of A with a member of B. This set of all possible pairings of an $a \in A$ with a $b \in B$ is called the **Cartesian product,** or the **set product,** of A and B:

$$A \times B = \{(a, b) \mid a \in A, b \in B\}.$$

The product $A \times B$ is only a set, but this time the elements are the possible pairs, as symbolized by (a, b). The product $B \times A$ would be a different set, with members (b, a).

This idea of a set of ordered pairs originated with the mathematician and philosopher Descartes (Latin: Cartesius), who used it as the foundation of analytic geometry; hence, the name *Cartesian* product. The "product" part of the name is attributable to the fact that the total number of possible pairs is always the number of members of A *times* the number of members of B.

One of the most important examples of a Cartesian product is known to anyone who has studied elementary algebra. Suppose that we have the set R, which is the set of all possible **real numbers** (that is, the set of all numbers that can be put into exact one-to-one correspondence with points on a straight line). We take the product $R \times R$, which is the set of all pairings of *two* real numbers. Any given pair might be symbolized (x, y), where x is a number called the value on the x-coordinate, or abscissa, and y is a number standing for value on the y-coordinate, or ordinate. This should be a familiar idea, since the product $R \times R$ is used any time we plot points on a graph or a curve. The first set R is the x-axis, and the second set R is the y-axis of the Cartesian coordinates. When a pair such as $(5, 1)$ is used to locate a point on a graph, it means that the number 5 is the

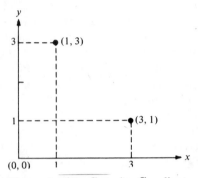

Figure 1.10.1   Cartesian Coordinates

element from the first set of numbers, and 1 is from the second set. The entire set $R \times R$ consists of all points that we *could* locate in the plane where the two axes of the graph lie. Pictorially, the product $R \times R$ and the pairs (3, 1) and (1, 3) might be shown as in Figure 1.10.1. Note that the ordering of a pair is important. The pair (3, 1) is clearly not the same as the pair (1, 3).

The product $R \times R$ involves sets containing only numbers, but the sets in a Cartesian product can contain any types of objects. In dealing with city-state pairs, for example, one set will consist of names of cities and the other set will consist of names of states. Moreover, the sets in a Cartesian product need not contain the same kinds of elements. To illustrate a product of a set of objects with a set of numbers, let

$$A = \{a \mid a \text{ is a man living in the United States}\}$$

and $$G = \{g \mid g \text{ is a weight in pounds}\}.$$

Then

$$A \times G = \{(a, g) \mid a \in A, g \in G\},$$

the set of all possible pairings of a man with a weight.

Given the idea of a product of two sets, it is finally possible to state what we mean by a **mathematical relation:**

A mathematical relation on two sets A and B is a subset of $A \times B$.

In other words, any mathematical relation is a subset of a Cartesian product, *some specific set of pairs out of all possible pairs*. At first glance, this seems to be an extremely trivial idea, but it actually is a remarkably subtle and elegant way to approach a difficult problem.

We often speak loosely of the husband-wife relation, the hand-fits-glove relation, the pitcher-catcher relation, the height-weight relation, the relation of the side of a square to its area, and so on. In each case, the fact of the relation implies that *some pairs out of all possible pairs* make a statement "*a* plays such and such a role relative to *b*" a true one. For any "husband," not all women qualify as "wife"; for any hand, only certain gloves fit; for any side-length of a square, only a certain number can be the area.

On the other hand, a few words of warning are in order before this topic is explored further. It is important not to confuse the idea of a mathematical relation with the idea of a *true relationship* among objects. It is *not necessarily true* that every mathematical relation we might invent must correspond to a real relationship among things. Furthermore, several different relationships, meaning quite different things in our ordinary experience, may show up as the *same* mathematical relation, in that exactly the same pairs qualify for the relation. For example, it *might* happen that for some set of men and some set of women, the relationship "*a* is married to *b*" would involve exactly the same pairs as "*a* files a joint income tax return with *b*." The two *relations* would be identical, but the *relationship* represented means something quite different in each instance.

Being a set, a mathematical relation can be specified in either of the two ways used for any set. In the first place, a relation may be specified by a listing of all pairs that qualify for the relation. For example, if

$$A = \{\text{Chicago, Denver, San Francisco, Seattle, Los Angeles}\}$$

and $B = \{\text{California, Illinois, Colorado, Washington}\},$

then the relation C pairing each city with the appropriate state is specified by the listing

$$C = \{\text{(Chicago, Illinois), (Denver, Colorado), (Seattle, Washington),}$$
$$\text{(San Francisco, California), (Los Angeles, California)}\}.$$

The more usual way of specifying a relation is by a statement of the rule by which a pair qualifies. For example, if A is the set of all women and B is the set of all men, the relation S might be specified by

$$S = \{(a, b) \in (A \times B) \mid a \text{ is married to } b\}.$$

If A is the set of all men in the United States and G is the set of all weights in pounds, a relation T might be specified by

$$T = \{(a, g) \in A \times G \mid \text{the weight of } a \text{ is } g\}.$$

If the Cartesian product is $R \times R$, then the most common way to specify a relation is by a mathematical expression giving the qualifications a number-pair must meet. As examples, consider

$$F = \{(x, y) \mid x + y = 10\},$$

$$V = \{(x, y) \mid x^2 - y^2 = 2\},$$

and $Z = \{(x, y) \mid 3 \leq x \leq 5,\ -2 \leq y \leq -1\},$

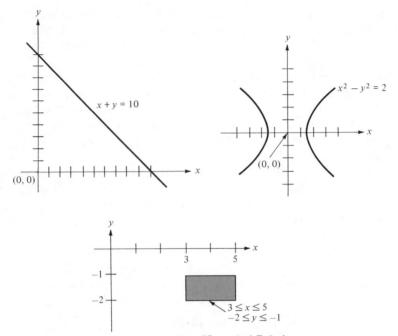

Figure 1.10.2   Three Numerical Relations

which are but a few of the countless relations that can be specified by mathematical rules. When the Cartesian product is $R \times R$, a graph can be used to represent a relation. The graphs corresponding to the numerical relations given by F, V, and Z are shown in Figure 1.10.2. Note that the relation F is represented by a straight line, the relation V is represented by a curve, and the relation Z is represented by a rectangle (including the area within the rectangle).

1.11 THE DOMAIN AND RANGE OF A RELATION

In discussing relations, we need some way to distinguish between the different sets of elements that play different roles. In a relation that is a subset of $A \times B$, it may be true that only *some* elements a enter into one or more (a, b) pairs; what we want is a way to discuss those elements of A that actually *are* paired with b elements in the relation itself. This subset of elements of A actually figuring in the relation is called the **domain,** and it is defined as follows:

Given some relation S, which is a subset of $A \times B$,

   domain of $S = \{a \in A \mid (a, b) \in S$, for at least one $b \in B\}$.

Notice that **the domain of the relation** S **is always a subset of** A. The domain includes each element in A that actually plays a role in the relation, vis-à-vis some b.

For example, let A be the set of all men and B the set of all women. Let the relation $S \subseteq A \times B$ be "a is married to b." Then the domain of S is "all married men," since only the subset of married men in A can figure in at least one pair (a, b) in the relation.

As another example, suppose that the product is $R \times R$, and the relation is given by the rule

$$C = \{(x, y) \mid x^2 + y^2 = 4\}.$$

What is the domain of C? The largest number that x could be is 2 for any (x, y) pair in C, and the smallest number is -2. Thus, x can be any number between -2 and 2 inclusive, but all other numbers in R are excluded. Hence,

$$\text{domain of } C = \{x \in R \mid -2 \le x \le 2\}.$$

The idea of the **range** of a relation is very similar to that of the domain, except that it applies to members of the set B, the second members of pairs (a, b).

The set of all elements b **in** B **paired with at least one** a **in** A **in the relation** S **is called the range of** S**:**

range of $S = \{b \in B \mid (a, b) \in S,$ **for at least one** $a \in A\}$.

In the example of the relation "a is married to b," the range is the set of all married women, a subset of B. In the example of the relation C, for y to be a real number the value of y must lie between -2 and 2, and so the range is

$$\text{range of } C = \{y \in R \mid -2 \le y \le 2\}.$$

It is entirely possible for the domain to be equal to A and for the range to be equal to B in some examples. In others, the range and the domain will be proper subsets of A and B, respectively.

1.12   FUNCTIONAL RELATIONS

We come now to one of the most important mathematical concepts, from the points of view both of mathematics itself and of its applications. This is the idea of a **functional relation,** or **function.** The definition we will use is as follows.

A relation is said to be a *functional relation* **or a** *function* **if each member of the domain is paired with one and only one member of the range.**

That is, in a functional relation each *a* entering into the relation has *exactly one pair-mate b*. Thus, a function is just a special kind of relation. In some mathematical writing this is called a "single-valued function," and a relation is a "multiple-valued function." However, we feel that it is useful to call only the former a "function." This idea will become clearer if we inspect some relations that are functions and some that are not functions.

Given *A* as the set of all men and *B* as the set of all women in some society, the relation

$$\{(a, b) \in A \times B \mid \text{"}a \text{ is married to } b\text{"}\}$$

is a function *if the society is monogamous*, so that each man may have only one wife. Here, each member of the domain, a married man, has one and only one wife, a member of the range. If, on the other hand, the society is polygamous so that a man may have two or more wives, then the relation is not necessarily a function.

Consider the relation defined on pairs of real numbers:

$$\{(x, y) \mid x^2 = y\}.$$

This relation is a function; corresponding to each *x*, there is one and only one *y*, which is the same as the square of *x*. Thus, $x = 2$ can be paired only with $y = 4$, $x = 5$ only with $y = 25$, and so on. Contrast this with the relation

$$\{(x, y) \mid x^2 = y^2\}.$$

In this case, the pair $(2, 2)$ qualifies, but so does the pair $(2, -2)$. Each value of *x* can be associated with *two* values of *y* by this rule, and so the relation is not a function. Figures 1.12.1 and 1.12.2 show how these two relations can be plotted. Notice how in the first example a vertical line

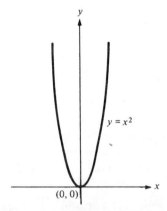

Figure 1.12.1 A Functional Relation

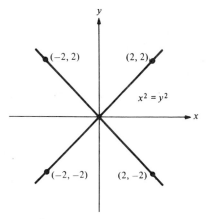

Figure 1.12.2   A Relation That Is Not a Functional Relation

drawn from any point on the x-axis will intercept the curve in at most *one* place, while in the second example *two* interceptions are possible. The same criterion may be applied to the plot of any relation between two numbers in order to determine if it is a function.

There is nothing in the definition of a function that limits the number of times an element b *in the range* may appear in a pair. Thus, in the marriage example, the relation is still a function even though the same woman is married to several men, and in the relation with rule $x^2 = y$, the same y value is associated with two x values (for example, $y = 4$ is paired both with $x = 2$ and $x = -2$). It follows that a given rule relating x and y may specify a function for (x, y) pairs but may specify a nonfunctional relation for (y, x) pairs, or vice versa.

Functions are relations, and hence are **sets of pairs,** and the set itself should be distinguished from the rule specifying the set. The distinction is not always clear in mathematical writing, where commonly both the set and the rule are called functions. However, if you remember that the function itself is the set of pairs, you should have little difficulty in deciding what is meant in a discussion using this word.

The function idea, of course, applies to products of *any* sets. As we have seen, it is quite possible to describe a function where the elements are pairs of people. Other examples can be given where a variety of other things are paired. However, for mathematicians the most interesting functions involve numbers, either as the range of the function or as both range and domain.

An important kind of function is one having numbers as the range and a **set of sets** as the domain. This is called a **set function.** For example, suppose that the name of a state in the United States is given to a set of all persons residing in that state on January 1, 1975. Then the whole collection of the fifty state-sets is a set of sets; let us call this set of sets A. For the moment, let the symbol P stand for the set of all *positive* numbers. Then the product $A \times P$ is the set of all pairings of a *set* with a positive number. Consider the relation

$$\{(a, p) \in A \times P \mid \text{"the total population of } a \text{ was } p \text{ on January 1, 1975"}\}.$$

This relation is a set function, since associated with each and every set of state residents there is one and only one population number p for the date January 1, 1975. One of the most important applications of the idea of a set function is a probability function, to be introduced in the next chapter.

1.13 VARIABLES AND FUNCTIONAL NOTATION

Because of the importance of functional relations both in mathematics and science a special notation has been developed for discussing functions, and we will find this useful in our study of statistics. Before we deal with this notation, the notion of a **variable** must be clarified.

A variable is a symbol that can be replaced by any one of the elements of some specified set. The particular set is called the range of the variable.

Note carefully that the variable is only a stand-in or placeholder, which can *always* be replaced by a particular element from some set of possibilities. Thus, if x is a variable, wherever x appears in a mathematical expression it can be replaced by *one* element from some specified set. This idea should be a familiar one, since algebra deals largely with variables ranging over the real numbers.

Variables may range over any well-defined set. For example, the variable c might symbolize any of a set of colors; the variable g might range over the 365 different dates in the year; the variable y might be any one of the infinite set of possible temperatures in Fahrenheit degrees.

Despite its name, a variable is not something that varies, or wiggles, or scurries around while you work with it. If a variable is used as a symbol in a given mathematical discussion, then replacing the variable by some specific element of its range at one place requires you to replace it with the same element *wherever* the variable appears. The same variable may appear many times in the same discussion, and it retains its identity throughout; if x is replaced by some number in one place, it is replaced by the same number in all other appearances. The name "variable" actually comes from the fact that a symbol represents "various" values. This distinguishes such symbols from mathematical *constants*. Unlike a variable, which can stand for a variety of things, a symbol that can be replaced by *one and only one* number in a given expression is a constant for the expression. For example, consider the well-known mathematical expression

$$c = 2\pi r,$$

where c and r represent the circumference and radius, respectively, of a circle. The symbols c and r here are variables; each stands for any one of an infinite set of *positive numbers*. However, the symbol π is a constant, which can be replaced only with a particular number. Regardless of the values assumed by r and c, π is the same. Note also, in this instance, that c and r are *functionally related* variables; the value that c can assume is dictated by the value of r (and vice versa).

Most of the variables that will concern us here represent sets of numbers, so that given some range of numbers, x stands for any specific member of the range. For instance, the statement "let x be a variable ranging over the real numbers" means that x is a stand-in for any one of the numbers that can be represented as points on a line. The x-axis in a graph is a representation of the range of this variable. When one number, say 10, is selected to replace x, then we say "x assumes the value 10."

Sometimes the range of a numerical variable is specified as other than all real numbers. For example, one might have the variable x, followed by the statement $3 \leq x \leq 5$, meaning that the only values that x symbolizes are those lying between 3 and 5 inclusive (x *ranges* between 3 and 5 inclusive). Or perhaps x is a variable that is integral, meaning that the only numbers that can replace the symbol x are whole numbers.

Variables are the basis for the notation most commonly used for functions. Given are the variables x and y, each with specified range, a relation between the variables is a subset of all possible (x, y) pairs. If this relation is a function, then this fact is symbolized by any of the statements such as

$$y = f(x),$$

$$y = G(x),$$

$$y = \varphi(x),$$

and so on. In words, these expressions all say:

> "There is a rule that pairs each possible value
> of x with at most one value of y."

The different Roman and Greek letters that precede the symbol (x) simply indicate different rules, giving different functions or subsets of all (x, y) pairs. *This notation does nothing more than assert that the function rule exists*. It says "Give me the rule, and the value of x, and I'll give you y."

Notice that a symbol $f(x)$ actually stands for a value of y; this is the value of y that is paired with x by the function rule. The symbol $f(x)$ is *not* the function, but an indicator that the function rule exists, turning a value x into some value $f(x)$, or y.

Sometimes the function rule itself is stated:

$$y = f(x) = x^3 - 2x^2 + 4,$$

$$y = G(x) = k \log (x)x \text{ for } > 0,$$

and
$$y = \varphi(x) = \frac{1}{\sqrt{2\pi}\sigma} \exp\left[-\frac{(x-\mu)^2}{2\sigma^2}\right]$$

are but a few of the countless function rules that might be stated for the variables x and y, both ranging over the real numbers. The terms such as k, e, σ, and μ in these rules are simply constants having the same value in a given function rule regardless of the value that x symbolizes. In each example, the function rule stated permits the association of some y value with a given value of x. *Be sure to notice that these expressions are the rules, not the functions.*

Functional notation has been shown here only for the case where both x and y range over numbers, primarily because this is the most common situation in mathematics. However, the same general form can be used for other functions. Suppose that the variable a stands for any one of a set of men, and the variable y stands for any one of the set of positive numbers that are possible scores on an intelligence test. Granted that each man must have one and only one score, then a function exists relating a and y. This could be symbolized by

$$y = f(a),$$

where a designates a particular man. We would be hard put, however, to state the rule for this function, and so we cannot go beyond the simple assertion that a function exists unless we undertake to list each man paired with his score. In general, such functions are difficult to specify, and this accounts for some of the elaborate attempts made in science to describe relationships as numerical functions which *can* be given mathematical rules.

The idea of a numerical function can be extended without difficulty to the case with three or more variables. Here the domain of the function is a set of *pairs* of numbers. Given variables x, y, and z, each ranging over the real numbers, the statement

$$y = f(x, z)$$

means that a rule exists such that given any pair of numbers x and z in the domain, the value of y is known. For example, it may be that

$$y = f(x, z) = 3x^2 - 2z^2 + 4,$$

so that given this rule, and letting x and z assume some pair of values (x, z), we have completely determined the value of y. Or, it may be that a different rule applies, such as

$$y = g(x, z) = 2x^z,$$

making this second function a very different subset from among all (x, z, y) triples.

1.14   CONTINUOUS VARIABLES AND FUNCTIONS

In statistics it is often necessary to specify that a variable or a function is continuous. Unfortunately, an accurate definition of continuous requires more mathematics than some students can command at this point. However, the following way of thinking about continuous variables and functions, although not really adequate as a definition, will serve our purposes.

Consider a variable x. If there are two numbers u and v such that the range of x is $u \leq x \leq v$, *all real numbers lying between u and v*, then x is said to be continuous in the interval $[u, v]$. If the variable x is continuous over all possible intervals defined by pairs of real numbers u and v, then x is said to be continuous over the real numbers.

The idea of a **continuous function** is best conveyed by a picture of a curve without "gaps" or "breaks" in terms of values of y; that is, if a function $y = f(x)$ is continuous in the interval of numbers $u \leq x \leq v$, then each and every number between $f(u)$ and $f(v)$ also occurs in the range of y values. For each x in the interval, some particular number y can be found by the function rule, if the function is continuous in the interval. A function that is continuous over all possible intervals of real numbers, so that the domain includes all real numbers, is called simply **continuous.** Thus, for example, the function given by $y = 3x^2 - 5$ is continuous over all real numbers, since for every possible x there exists a determinable y, and hence all real values of x are in the domain. On the other hand, a function with the rule $y = 1/(x + 1)$ is **discontinuous,** since for $x = -1$ there is no exactly determinable value of y that can be associated with x, and hence the domain does not include $x = -1$.

The study of discontinuous functions is an important part of mathematics, but most of the functions that concern us here will either have a specified number of possible values that x can assume or will be continuous over the real numbers.

1.15 FUNCTIONS AND PRECISE PREDICTION

In the statistician's use of mathematics, numerical variables symbolize real quantities or magnitudes that can be found (potentially) from measurements of phenomena. Real relationships are summarized as mathematical relations, having some mathematical expressions as the specifying rules. When the statistician deals with a true functional relation between quantities, then this is communicated in its most concise and elegant form by a statement of the function rule.

One of the goals of any science is prediction: given some set of known circumstances, what else can we expect to be true? If the scientific relation is functional in form, then *precise* prediction is possible; given the true value of x in the domain, precisely one value of y will be observed, all other things being constant. However, in this qualification, "all other things being constant," lies one of the central problems of any science.

Each of us, scientist or layman, has learned from infancy that there are relationships in the world about us; things go together, and some things lead to other things. On the other hand, everyone knows that precise prediction of the world is virtually impossible. Even given some information about an object or event, we seldom know *exactly* what its other properties will be. True enough, providing us with information may let us restrict the range of things we *expect* to observe, but the possibility of exact prediction is almost unknown in the everyday world. Nevertheless, the more advanced sciences are very successful in describing relationships *as though* they were mathematical functions. How has this come about? What does the scientist do that makes it possible to make precise predictions from known situations, when the world is so disorderly and unpredictable?

It seems safe to say that the world is full of marvelously complicated relationships and that any event we experience must have its character determined by a vast number of influences. Some of these may exert major forces on what we observe, and others may be quite minor in shaping a given event. Most scientists would subscribe to the idea that, ultimately, if we knew the values of *all* of the relevant variables, *all* of the influences that go into the determination of an event, and the rule that relates these variables to the event itself, then precise prediction would always be possible. It may be that given the complete information about the circumstances, all true relations are functional.

In the physical sciences, it may often be possible to obtain complete information about all of the variables relevant to a particular event, and thus to determine functional relationships between variables. By comparison, the social sciences, the behavioral sciences, and the areas comprising the field of business are not so far along in the use of mathematics to specify real relations. The precision and control in experimentation often possible in the physical sciences are not generally attainable in these areas. In situations in which the full range of relevant factors is quite unknown, statistics becomes a most useful tool. The theory of inferential statistics deals with the problem of "error," the failure of an observation to agree with a true situation. Error is the product of "chance," the influence of the innumerable uncontrolled factors determining a particular event. The concept of "probability" is introduced to evaluate the likelihood that a given observation will disagree to a certain extent with a true value. Instead of being certain or nearly certain about the conclusions reached from observations, as the physical scientist often is, the conclusion drawn by the statistician is much like a bet.

The gambler placing a bet may be faced with a situation in which very little is known. Who knows the real reasons why a coin comes up heads on a given toss or a particular card comes up on the third draw of a poker game? However, given that some things are true, it can be deduced that other things are more or less *likely* to occur, and still other things *unlikely*, regardless of the reasons why. The gambler knows only that there is a certain probability of being right, and also of being wrong, in a given decision, and the wise gambler places his bets accordingly.

In succeeding sections we will continue to compare the task of the statistician to that of the gambler, and to show how statistical theory aids the statistician in making various sorts of "bets." In order to do this, we need some of the ideas of probability theory, which is introduced in the next chapter.

EXERCISES

1. What is required to determine whether a given object is or is not a member of a given set?

2. What is the role of the universal set in set theory? What is the role of the empty set?

3. Specify the sets in (a), (b), and (c) by first listing all their members and then stating a formal rule:
    (a) the set of all positive integers less than or equal to 9,
    (b) the set of all positive even integers less than or equal to 9,
    (c) the set of all positive odd integers less than or equal to 9.
    (d) What is the relation of the sets in parts (b) and (c) to the set in part (a)?

4. Characterize each of the following sets according to whether it is most appropriately considered to be finite, countably infinite, or uncountably infinite:
    (a) the set of all individuals who purchased new automobiles in 1974,
    (b) the possible number of passengers that an airline might carry before having its first fatal accident,

(c)   the true weights, accurate to any number of decimal places, of a specific group of 100 American men,

(d)   the possible true height in inches, accurate to any number of decimal places, of an American woman,

(e)   the possible number of bridge hands that might be dealt before one of the players receives 13 spades,

(f)   the set of distances, measured to the nearest mile, between all possible pairs of points in Florida,

(g)   the set of distances, measured in miles and accurate to any number of decimal places, between all possible pairs of points in California.

5.   Let $A = \{-2, -1, 0, 1, 2\}$, $B = \{x \mid -2 \leq x \leq 2\}$, and $C = \{x \mid x > 0\}$. For each of the following sets, indicate which elements of A are also in the set, and do the same for B and C:

(a) $\{x \mid 3x - 6 = 0\}$,

(b) $\{x \mid x^2 - 1 = 0\}$,

(c) $\{x \mid x^2 - 1 > 0\}$,

(d) $\{x \mid 2x^2 - 3x < 0\}$,

(e) $\{x \mid 5x^3 - 6x^2 + x = 0\}$,

(f) $\{x \mid x/2 \text{ is an integer}\}$.

6.   Let the universal set be the set of real numbers, and let

$$A = \{x \mid x^2 \geq 1\}, B = \{x \mid x^2 + 2x + 1 = 0\}, \text{ and } C = \{x \mid 2x - 3 \leq 7\}.$$

Specify the following sets:

(a) $A \cup B$,

(b) $A \cup B \cup C$,

(c) $C - A$,

(d) $A \cap C$,

(e) $(A \cup \bar{B}) \cap \bar{C}$.

7.   Consider the following sets:

W = set of all employees of a given firm,
A = set of all individuals over 30 years old,
B = set of all individuals with college degrees,
C = set of all individuals working in the firm's Chicago office.

Describe in words the following sets:

(a) $A \cup (B \cup C)$,

(b) $A \cup (\bar{B} \cup \bar{C})$,

(c) $(A \cap \bar{B}) \cup (A \cap B)$,

(d) $A \cap (\bar{B} \cup C)$,

(e) $A \cup (\bar{B} \cap \bar{C})$.

8.   An item produced by a certain machine may have no defects or it may have one or more of three possible types of defects: type A, type B, and type C. In a lot of 900 such items, the numbers of times various combinations of defects are observed are as follows:

A: 39, A and B: 15,
B: 19, A and C: 21,
C: 43, B and C: 12,
             A, B, and C: 8.

Be careful to note than an item may fall into more than one category. For example, the number of items with type A defects, 39, includes those with only type A defects, those with both type A and type B defects, and so on. Translate the following verbal statements into appropriate set symbols and find the number of times the "event" occurs:

(a) type B and type C defects, but no type A defect,
(b) only a type B defect,
(c) no defects,
(d) a type A defect *and* either a type B or a type C defect (but not both B and C),
(e) exactly one defect,
(f) exactly two defects,
(g) at least one defect,
(h) at least two defects,
(i) at most one defect,
(j) at most two defects.

9. In a certain municipal election, three bonding proposals were on the ballot. A voter could vote for any or all of the propositions. Of the total votes cast, 42 percent voted for proposition A, 51 percent for proposition B, and 32 percent for proposition C. In addition, it is known that 13 percent voted for A but not B and not C, 12 percent voted for A and for B but not for C, 26 percent voted for B but not for A or C, and 10 percent voted for C but not for A or B.

(a) What percentage voted for all three propositions?
(b) What percent voted for A and C but not for B?
(c) What percent voted for at least one proposition?

10. In a market research survey, 120 individuals were asked for their preference among three different brands, A, B, and C, of coffee. The respondents were classified according to yearly family income, and the results are as follows (each number represents the number of respondents in that income group who prefer that brand of coffee).

| | | BRAND OF COFFEE | |
	A	B	C
H	18	11	9
M	24	8	10
L	9	19	12

INCOME GROUP (HIGH, MEDIUM, OR LOW)

Counting each respondent as an element of a set, how many respondents are in each of the following sets?

(a) $B \cup C$,
(b) $\bar{A} \cup \bar{B}$,
(c) $(H \cup M) \cap A$,
(d) $(L - A) \cup (H - C)$,
(e) $(H \cup \bar{A}) \cap (H \cup \bar{B})$.

11. Children and teachers in several kindergarten classes were asked to name their favorite color. The numbers of different responses are given below.

		Red (R)	Blue (B)	Yellow (Y)	Other (O)
(M)	Male student	10	5	3	4
(F)	Female student	6	11	2	1
(T)	Female teacher	3	4	1	1

How many members does each of the following sets have?
(a) O,
(b) $R \cup B$,
(c) $M \cup Y$,
(d) $T \cap (B \cup Y)$,
(e) $(\overline{F \cap O})$,
(f) $M \cup (F \cap \bar{B})$.

12. In a study of an industrial organization, a psychologist studied the verbal communication patterns of a group of four workers for a period of one week. In particular, he was interested in the verbal communication between the supervisor and each of the other workers. Let:

A = set of all instances of verbal communication from the supervisor to worker 1 during the period observed and similarly, set B for worker 2 and set C for worker 3.

These three sets and their intersections contained the following number of elements:

A: 30 $A \cap B$: 15
B: 34 $A \cap C$: 12
C: 21 $B \cap C$: 7
                     $A \cap B \cap C$: 5

Find the number of communications from the supervisor
(a) to worker 1, but not to 2 or 3,
(b) to workers 1 and 2, or to workers 1 and 3, but not to all three,
(c) to exactly one of the three workers,
(d) to two or more of the workers.
(e) How many verbal communications in all did the supervisor make to the workers during the week observed?

13. Given that A and B are each subsets of W, which of the following sets of sets must necessarily constitute a partition of the set W?
(a) $\{A, W\}$,
(b) $\{A, \varnothing\}$,
(c) $\{A, \bar{A}\}$,
(d) $\{W, \varnothing\}$,
(e) $\{A - B, B - A\}$,
(f) $\{A - B, A \cap B, B \cap \bar{A},$
     $W - (A \cup B)\}$,
(g) $\{A, B - A, \bar{A} \cap \bar{B}\}$.

14. Consider the following sets:
$A = \{-3, -2, -1, 0, 1, 2, 3\}$,
$B = \{-4, -3, -1, 1, 3, 4\}$,
and $C = \{1, 2, 3, 4, 5\}$.

List the members of the following sets:

(a) $A \cup B \cup C = W$,　　　　(e) $(A \cap B) \cup C$,

(b) $A \cap B$,　　　　　　　　　(f) $(A \cup B) \cap C$,

(c) $A - B$,　　　　　　　　　　(g) $\bar{A} \cap (\bar{B} \cup C)$,

(d) $(B - A) \cup (A - B)$,　　　(h) $(A \cap B) - C$.

15. Using the sets A and C of Exercise 14, let B be a set of real numbers variously defined as follows:

(a) $B = \{x \mid x^2 - 3x + 2 = 0\}$,

(b) $B = \{x \mid x + 1 > 0\}$,

(c) $B = \{x \mid 2 \le 2x^2 - 5 \le 6\}$,

(d) $B = \{x \mid 4x^3 + 10 \ge -7\}$.

For each definition of set B, find $A \cap B$, $B \cap C$, $A \cap B \cap C$, $B \cup (A - C)$, $B \cap (C - A)$.

16. Let the set W consist of all members of the House of Representatives of the 93rd Congress of the United States. Then let:

A = set of all Representatives who were above 60 years of age at the opening of the Congress,

B = set of all women Representatives,

C = set of all Representatives from urban districts, and

D = set of all Democratic Representatives.

Describe the following sets verbally:

(a) $A \cup (B \cap C)$,　　　　　(d) $(C - D) \cap B$,

(b) $(A \cap B) \cup (C \cap D)$,　　(e) $(\bar{A} \cup \bar{C}) \cap (\bar{B} - D)$,

(c) $(\bar{A} \cap B) \cup (\bar{C} \cap \bar{D})$,　　(f) $(B \cap \bar{D}) \cup (B \cap C)$.

17. Draw Venn diagrams to verify that the following theorems on sets hold. Use some systematic shading or coloring to clearly distinguish the relevant sets.

(a) $A \cap (B \cup C) = (A \cap B) \cup (A \cap C)$.

(b) $(A \cap B) \subseteq A \subseteq (A \cup B)$.

(c) $A - B \subseteq A$.

(d) If $A \subset B$ and $C \subset D$, then $(A \cup C) \subset (B \cup D)$.

(e) If $A \supset B$ and $C \supset D$, then $(A \cap C) \supset (B \cap D)$.

(f) $A \cap B = A - (A - B)$.

(g) $A \cup (B - A) = A \cup B$.

(h) $A - (A \cap B) = A - B$.

(i) $A \cap (B - C) = (A \cap B) - C$.

18. For each statement, tell whether it is true (for *any sets* A, B, and C) or false, and if it is false, draw a Venn diagram to demonstrate that it is false.

(a) $(A - C) \cup (B - C) = (A \cup B) - C$.

(b) $A \cap (\overline{A \cap C}) = \emptyset$.

(c) $(\bar{A} \cup \bar{B}) \cap C = (\overline{A \cap B}) \cup C$.

(d) $A \subseteq (A \cap \bar{B}) \cup B$.

(e) $\bar{A} \cup \bar{B} \cup C \subset (\overline{A \cup B \cup C})$.

19. Prove, by set algebra, that

(a) $(\overline{A \cup B}) = (\bar{A} \cap \bar{B})$,　　(b) $(\overline{A \cap B}) = (\bar{A} \cup \bar{B})$.

[*Hint for* (a): Show that $(A \cup B) \cup (\bar{A} \cap \bar{B}) = W$ and that $(A \cup B) \cap (\bar{A} \cap \bar{B}) = \emptyset$, so that the set $(A \cup B)$ must be the complement of $(\bar{A} \cap \bar{B})$.]

20. (a) Given that A, B, and C are *not* mutually exclusive, express their union $A \cup B \cup C$ in terms of the union of three sets that *are* mutually exclusive.

(b) Show (using a Venn diagram if necessary) that even though the three sets A, B, and C do not form a partition of W, subsets exist which, along with $\bar{A} \cap \bar{B} \cap \bar{C}$, do form a partition of W.

21. A salesman's territory consists of Illinois, Indiana, Michigan, and Wisconsin. Let A represent the set of these four states. The salesman is trying to decide on a subset of these states for an upcoming trip. (Of course, one option is not to make the trip.) How many possible subsets are there? List all of the subsets.

22. What distinguishes a Cartesian product from any other set?

23. When is a relation a function?

24. Let $A = \{2, 3, 4, 8, 9\}$. Graph these *mathematical relations*, each of which is a subset of $A \times A$:

(a) $H = \{(2, 9), (4, 9), (8, 9), (9, 9)\}$,

(b) $I = \{(2, 3), (2, 4), (2, 8), (2, 9)\}$,

(c) $J = \{(2, 4), (3, 9)\}$,

(d) $K = \{(2, 3), (3, 4), (4, 8), (8, 9)\}$,

(e) $L = \{(3, 2), (4, 3), (8, 4), (9, 8)\}$,

(f) $H \cap I$,

(g) $J - I$,

(h) $H \cup L$,

(i) $\bar{K}$.

25. Give the domain and range of each of the relations in Exercise 24. Which of these relations are functions?

26. In each of the following, list the elements of the relation and give its domain and range, where $U = \{1, 2, 3, 4, 5, 6, 7, 8, 9, 10\}$ and (x, y) is an element of $U \times U$:

(a) $\{(x, y) \mid x = 4\}$,

(b) $\{(x, y) \mid y = 7\}$,

(c) $\{(x, y) \mid y < x\}$,

(d) $\{(x, y) \mid y = 2x\}$,

(e) $\{(x, y) \mid y = 1/2x\}$,

(f) $\{(x, y) \mid y \neq x + 1\}$.

27. Given the set $A = \{2, 3, 4, 5, 6\}$ and defining $(x, y) \in A \times A$, list the sets of (x, y) pairs that are members of the relations in (a)–(e).

(a) $G = \{(x, y) \mid y/x \text{ is an integer}\}$,

(b) $H = \{(x, y) \mid y \text{ is less than or equal to } x\}$,

(c) $I = \{(x, y) \mid y \text{ is greater than } x\}$,

(d) $J = \{(x, y) \mid y = x + 1\}$,

(e) $K = \{(x, y) \mid y = x - 2\}$.

(f) Give the domain and range of each of the relations in (a)–(e).

(g) Which of the relations in (a)–(e) are functions?

28. Four people are sitting at a table. Let U be the set of people sitting at a table, $U = \{A, B, C, D\}$. The people are seated as in the figure.

Draw the graphs of the following relations that are subsets of $U \times U$ and specify which of the relations are functions:

(a) $\{(x, y) \mid x$ sits next to $y\}$,

(b) $\{(x, y) \mid x$ sits across from $y\}$,

(c) $\{(x, y) \mid x$ sits to the right of $y\}$.

29. Consider the following sets of letters: $\{a, b, c, d, e\} = A$, $\{u, v, w, x, y, z\} = B$, $C = \{a, e, i, o, u\}$. Let $W = A \cup B \cup C$.
List the members of:

(a) $A \cap C$,

(b) $A \cup (B \cap \bar{C})$,

(c) $B \cup \bar{C}$,

(d) $(A \cap B) \cup (\overline{B \cap C})$,

(e) $A - C$,

(f) $A \times C$,

(g) $A \times A$.

30. Given $A = \{x \mid x^2 \leq 16\}$, what is the Cartesian product $A \times A$? Draw the graphs of the following relations in $A \times A$, find the domains and ranges of the relations, and specify which of the relations are functions:

(a) $\{(x, y) \mid x + y = 3\}$,

(b) $\{(x, y) \mid x^2 + y^2 = 4\}$,

(c) $\{(x, y) \mid x^2 = 16 - y\}$,

(d) $\{(x, y) \mid y^3 = 64 - x\}$,

(e) $\{(x, y) \mid y \geq x + 3\}$.

31. A used car dealer has seven cars available for sale. The cars, together with their mileages and prices, are listed in the following table:

Car	Mileage	Price
a: Volkswagen	10,500	$1400
b: Chevrolet	41,000	$ 600
c: Toyota	32,000	$ 850
d: Oldsmobile	8,100	$3200
e: Ford	19,200	$1500
f: Volvo	31,000	$2300
g: Dodge	25,000	$1650

Let A represent the set of cars, M represent the set of mileages, and P represent the set of prices.

(a) What are the elements of $A \times A$ falling into the relation $\{(x, y) \mid x$ has more mileage than $y\}$?

(b) What are the elements of $A \times A$ falling into the relation $\{(x, y) \mid x$ is more expensive than $y\}$?

(c) What elements belong to the intersection of the relations in (a) and (b)?

(d) What are the elements (x, y) of $M \times P$ falling into the relation $\{(x, y) \mid y$ is the price of a car with mileage $x\}$?

(e) If each car has a life expectancy of, say, 60,000 miles before repairs, which elements of $M \times P$ fall into the relation $\{(x, y) \mid$ cost per mile before repairs $\leq$ \$.03$\}$?

32. If each line of the table in Exercise 31 is thought of as an *ordered triple*, how would you symbolize the Cartesian product of which it is an element? How

might one characterize the table in Exercise 31 as a relation on some Cartesian product? [*Hint:* Can a set of *pairs* serve either as the domain or the range of a relation?]

33. A die (one of a pair of dice) has spots on its six sides numbered from 1 to 6. The possible results of tossing a single die can be represented by the set $D = \{1, 2, 3, 4, 5, 6\}$. Find the elements of $D \times D$ that fall into the sets in (a)–(e).
  (a) $\{(x, y) \mid x + y = 7\}$,
  (b) $\{(x, y) \mid x + y = 11\}$,
  (c) $\{(x, y) \mid x + y \leq 4\}$,
  (d) $\{(x, y) \mid x + y > 8\}$,
  (e) $\{(x, y) \mid x + y = 2\} \cup \{(x, y) \mid x + y = 12\}$.
  (f) What fraction of the total number of elements in $D \times D$ does each of these sets of elements represent?
  (g) Can you think of any interpretation that might be given to the fractions, or proportions, in (f)?

2

ELEMENTARY
PROBABILITY
THEORY

Statistical inference and decision involve statements about probability. We all have the words "probably" and "likely" in our vocabularies, and most of us have some notion of the meaning of statements such as "The probability is one-half that the coin will come up heads." However, in order to understand the methods of statistical inference and decision and use them correctly, one must have some grasp of probability theory. This is a mathematical system that is related to the algebra of sets described in the last chapter. Like many mathematical systems, probability theory becomes a useful model when its elements are identified with particular things, in this case the *outcomes of real or conceptual experiments*. Then the theory lets us deduce propositions about the likelihood of various outcomes, if certain conditions are true.

Originally, the theory of probability was developed to serve as a model of games of chance. In this case the experiment in question was rolling a die, spinning a roulette wheel, or dealing a hand from a deck of playing cards. As this theory developed, it became apparent that it could also serve as a model for many other kinds of things having little obvious connection with games, such as results in the sciences. One feature is common to most applications of this theory, however. The observer is uncertain about what the outcome of some observation will be and must eventually infer or guess what will happen. In a sense, then, probability reflects the *uncertainty* about the outcome of an experiment. Indeed, probability can be thought of as the mathematical language of uncertainty.

The uncertainty facing the statistician is like that of the gambler keeping track of the numbers coming up on a roulette wheel. At any given opportunity only the tiniest part of the total set of things of interest can be observed. Because of the observer's human frailty, it is necessary to fall back on logic; *given* that certain things are true, deductions can be made about the truth of other things. The statistician's statements about *all* observations of such and such phenomena are on a par with the gambler's; both are deductions about what should be true, if the initial conditions are met. Furthermore, the gambler, the manager, the engineer, the scientist, and, indeed, the man on the street must make decisions based on incomplete evidence. Each does so in the face of uncertainty about how good those decisions will turn out to be. Probability theory alone does not tell any of these people how they should decide, but it does give ways of evaluating the degree of risk a person takes for some decisions.

The theory of probability is very closely tied to the theory of sets. Indeed, the main undefined terms in the theory, the "events," are simply sets of possibilities. Before we develop this idea further, we need to talk about the ways that events are made to happen: simple experiments.

2.1 SIMPLE EXPERIMENTS

We mean nothing very fancy here by the term **simple experiment,** and there is no implication that a simple experiment need be anything even remotely resembling a laboratory experiment.

A simple experiment is some well-defined act or process that leads to a single well-defined outcome.

Some simple experiments are tossing a coin to see whether heads or tails comes up, cutting a deck of cards and noting the particular card that appears on the bottom of the cut, opening a book at random and noting the first word on the right-hand page, determining the sales of a particular product over a given month, running a rat through a T maze and noting whether the rat turns to the right or to the left, observing the closing price of a common stock on a particular date, lifting a telephone receiver and recording the time until the dial-tone is heard, asking a person about political preferences or brand preferences, giving a person a test and computing the score, counting the number of colonies of bacteria seen through a microscope, and so on, literally without end. The simple experiment may be real (actually carried out) or conceptual (completely idealized), but it must be capable of being described. We also require that each performance of the simple experiment have one and only one outcome, that we know when it occurs, and that the set of all possible outcomes can be specified. Any single performance, or trial, of the experiment must eventuate in one of these possibilities.

Obviously, this concept of a simple experiment is a very broad one, and almost any describable act or process can be called a simple experiment. There is no implication that the act or process even need be initiated by the experimenter, who needs to function only in the role of an observer. On the other hand, it is essential that the outcome, whatever it is, be unambiguous and capable of being categorized among all possibilities.

2.2 EVENTS

The basic elements of probability theory are the possible distinct outcomes of some idealized simple experiment.

The set of all possible distinct outcomes for a simple experiment will be called the *sample space* for that experiment. Any member of the sample space is called a *sample point*, or an *elementary event*.

Every separate and "thinkable" result of an experiment is represented by *one and only one* sample point or elementary event. Any elementary event is *one possible result* of a single trial of the experiment.

For example, we have a standard deck of playing cards that has been shuffled. If the simple experiment is drawing one card from the shuffled deck, then the sample space consists of the 52 separate and distinct cards that we might draw. If the experiment consists of observing the closing price of a common stock on a particular date, then the sample space consists of all the possible closing prices that might be observed. If the experiment consists of stopping people on the street and asking them to name their favorite brand of coffee, then the sample space consists of the possible responses, including the various brands of coffee and responses such as "I have no favorite brand" or "I don't drink coffee."

Often we are interested in the *kind* or *class* of outcome an observation represents. The outcome of an experiment is *measured*, at least by allotting it to some qualitative class. For this reason, the main concern of probability theory is with *sets* of elementary events.

Any set of elementary events is known simply as an event.

Imagine, once again, that the experiment is carried out by drawing playing cards from a standard pack. The 52 sample points (the distinct cards) can be grouped into sets in a variety of ways. The suits of the cards make up 4 sets of 13 cards each. The event "spades" is the set of all card possibilities showing this suit, the event "hearts" is another set, and so on. The event "ace" consists of 4 different elementary events, as does the event "king," and so on

If the experiment consists of observing the closing price of a common stock on a particular date, then events such as "the price is less than 30" or "the price is between 32 and 35" may be of interest. Any set consisting of one or more possible prices of the stock is an event. If the experiment involves stopping people and asking them to name their favorite brand of coffee, events such as "U. S. brand," "foreign brand," or "one of the five top brands in terms of last year's sales" may be of interest.

In short, **events are sets having the elementary events as members.** The elementary events are the raw materials that make up the event. Since *any* subset of the sample space is an event, then some event *must* occur on each and every trial of the experiment.

2.3 EVENTS AS SETS

The symbol S will be used to stand for the **sample space,** the set of all elementary events that are possible outcomes of some simple experiment. Capital letters A, B, and so on will represent events, each of which is some subset of S. Notice how

the sample space S is used like a universal set; for a given simple experiment, every event discussed must be some subset of S.

The set S may contain either a finite or an infinite number of elements.

Since any event A is a subset of S, the operations and postulates of set theory carry over directly to operations on events to define other events. First of all, the set S is an event: S is the "sure event," since it is *bound* to occur (one of its members must turn up on every trial). The event $\emptyset$ is called the "impossible event," since it cannot occur. It is important to remember that the empty or impossible event $\emptyset$ is, nevertheless, an *event* according to our definition, since the empty set is a subset of every set.

By our definition of an event as a subset of S, if A and B are both events, the union $A \cup B$ is also an event. The event $A \cup B$ occurs if we observe a member of A, or a member of B, or a member of both.

In the same way, $A \cap B$ is an event. The event $A \cap B$ requires an outcome that is a member of *both* A and B. Notice that any occurrence of $A \cap B$ is automatically an occurrence of $A \cup B$ but that the reverse is not true.

If A is an event, then so is its complement $\bar{A}$. The events A and $\bar{A}$ cannot occur simultaneously, since $A \cap \bar{A} = \emptyset$. On the other hand, either A or $\bar{A}$ must occur, since $A \cup \bar{A} = S$. The difference between two events, $A - B$, is likewise an event. **In short, each and every subset formed from the elementary events in S is an event.**

Two events A and B, neither equal to $\varnothing$, are said to be mutually exclusive, or disjoint, if $A \cap B$ cannot occur—that is, if $A \cap B = \varnothing$.

In general, K events are mutually exclusive, or disjoint, if *all possible pairs* of events selected from the K events are mutually exclusive. Thus, A, B, and C are mutually exclusive if $A \cap B = \varnothing$, $A \cap C = \varnothing$, and $B \cap C = \varnothing$.

Two or more events are said to be exhaustive if their union must occur.

Thus, A and $\bar{A}$ are exhaustive events, since $A \cup \bar{A} = \text{S}$.

If K events $A_1, A_2, \ldots, A_K$ are both mutually exclusive and exhaustive, we say that they form a K-fold partition of the sample space S.

Note that A and $\bar{A}$ form a two-fold partition of S.

Let us take a concrete example of some events. Suppose that we had a list containing the name of every living person in the United States. We close our eyes, point a finger at some spot on the list, and choose one person to observe. The elementary event is the actual person we see as the result of that choice, and the set S is the total set of possible persons. Suppose that the event A is the set "female" and B is the set "red headed" among this total set of persons. If our chosen person turns out to be female, then event A occurs; if not, event $\bar{A}$. If the person turns out to be red headed, this is an occurrence of event B; if not, event $\bar{B}$ occurs. If the observation shows up as both female and red headed, $A \cap B$ occurs; if either female or red headed, then event $A \cup B$ occurs. If the person is female but not red headed, this is an occurrence of the event $A - B$. Only one thing is sure; we will observe a person living in the United States, so the event S must occur.

Continuing with the above example, suppose that the event C is the set "under 20 years of age," event D is the set "20 to 39 years of age," and event E is the set "at least 40 years of age." The events C, D, and E are mutually exclusive, since no two of them can occur at once. Also, they are exhaustive, since one of them *must* occur. That is, each living person must fall into one and only one of the three classes, or events. Thus, the events C, D, and E form a three-fold partition of S.

In the example just given, the choice of the five events A, B, C, D, and E was completely arbitrary, and the example could have been given just as well with any other scheme for arranging elementary events into event classes. Ordinarily, there is some restricted set of events of interest, but any scheme for arranging elementary events into event classes can be used.

In the next section we are going to define probability in terms of events. However, we want to do this in some way so that probability can be discussed regardless of how the sample space is broken into subsets. This is accomplished by considering all possible subsets of the sample space S.

Given K elementary events in S, there are exactly 2^K events that can be formed, including $\varnothing$ and S.

The set α consisting of all possible subsets of the sample space S will be called the family of events in S.

This set of subsets α will be the basis for the definition of probability to follow.

To illustrate the notion of the family of events in S, suppose that an experiment consists of noting whether the value of U. S. dollars relative to Swiss francs is higher (A), the same (B), or lower (C) than it was exactly one year ago. Hence, $K = 3$, and there are exactly $2^3 = 8$ subsets of the sample space. These 8 subsets are

$$\varnothing, \{A\}, \{B\}, \{C\}, \{A, B\}, \{A, C\}, \{B, C\}, \text{ and } S.$$

Of course, as K increases, 2^K gets large very rapidly, so it is usually not practical to list all of the possible subsets of the sample space as we did in this simple example.

2.4 PROBABILITY FUNCTIONS

We are now ready for a formal definition of what is meant by probability. This definition will be given in terms of a probability function, a pairing of each event with a positive real number (or zero), its probability. The definition is the basis for a *mathematical* theory of probability. The question of how this probability is to be *interpreted* will be discussed a little later.

Definition:

Given the sample space S, and the family α of events in S, a probability function associates with each event A in α a real number $P(A)$, the probability of event A, such that the following axioms are true.

1. **$P(A) \geq 0$ for every A.**
2. **$P(S) = 1$.**
3. **If there exists some countable set of events, $\{A_1, A_2, \ldots, A_K\}$, and if these events are all mutually exclusive, then**

$$P(A_1 \cup A_2 \cup \cdots \cup A_K) = P(A_1) + P(A_2) + \cdots + P(A_K)$$

(the probability of the union of mutually exclusive events is the sum of their separate probabilities).

In essence, this definition states that paired with each event A is a non-negative number, the probability $P(A)$, and that the probability of the sure event S, or $P(S)$, is always 1. Furthermore, if A and B are any two *mutually exclusive* events in the sample space, the probability of their union, $P(A \cup B)$, is simply the sum of their two probabilities, $P(A) + P(B)$.

It is important to remember at this stage that this is a purely formal definition of probability in terms of a function assigning numbers to sets. *Events* have probabilities, and in order to discuss probability we must always discuss the events to which these probabilities belong. When we speak of the probability that a person in the United States has red hair, we are speaking of the probability number assigned to the event "red hair" in the sample space "all persons in the United States." Similarly, when we say that the probability that a coin will come up heads is .50, we are saying that the number .50 is assigned in the probability function to the event "heads" in the sample space "all possible results of tossing a particular coin."

This may seem to be an extremely unmotivated and arbitrary way to discuss probability. We all know that the word "probability" means more than a mere number assigned to a set, and we shall certainly give these probability numbers meaning in the sections to follow. For now, however, let us simply accept this purely formal definition at face value.

Events, like other types of sets, can be illustrated on Venn diagrams. In order to gain a good intuitive grasp of the elementary rules of probability, it is convenient to think of probability in terms of area on a Venn diagram. That is, the area on a Venn diagram that is covered by a set A corresponds to $P(A)$. Since the sample space corresponds to the universal set, which is represented by the entire rectangle in a Venn diagram, Axiom 2 above implies that if area represents probability, then the area of the rectangle must be exactly one. Axiom 1, of course, simply means that there is no such thing as a negative area on a Venn diagram. Figure 2.4.1 illustrates Axiom 3 for $K = 2$. Because A_1 and A_2 are mutually exclusive, the area on the Venn diagram representing their union is equal to the sum of their individual areas. When they are first encountered, equations representing elementary probability theory are easier to understand if they are illustrated on a Venn diagram.

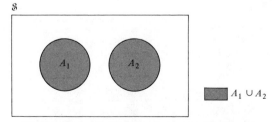

Figure 2.4.1   Probability Axiom 3 with $K = 2$

Given our definition of a probability function, we can begin to deduce other features that probabilities of events must have. Now we shall proceed to derive some consequences of the formal axioms included in our definition above; not only will this demonstrate that deductions do, in fact, follow directly from the axioms of this formal mathematical system, but also we will find the elementary rules of probability extremely useful when we begin to calculate probabilities.

First of all, we will give an informal proof of the following statement:

$$P(\bar{A}) = 1 - P(A). \tag{2.4.1*}$$

For any event A, the probability of the complementary event "not A" is 1 minus the probability of A.

This rule is illustrated in Figure 2.4.2. To derive the rule, we proceed as follows. We know from the algebra of sets that A and $\bar{A}$ are mutually exclusive $(A \cap \bar{A} = \varnothing)$ and that $A \cup \bar{A} = \mathcal{S}$. Then, by Axiom 2 above,

$$P(A \cup \bar{A}) = P(\mathcal{S}) = 1.$$

Furthermore, by Axiom 3, since A and $\bar{A}$ are mutually exclusive,

$$P(A \cup \bar{A}) = P(A) + P(\bar{A}).$$

Then it follows that

$$P(A) + P(\bar{A}) = 1,$$

so that

$$P(\bar{A}) = 1 - P(A).$$

Thus, we have proved an elementary theorem in the formal theory of probability. To illustrate this theorem, if the probability is .25 that a major earthquake will occur in California between January 1, 1980 and January 1, 1990, then the complementary event (no major earthquake during that period) has probability .75.

Next we will show that

$$0 \le P(A) \le 1 \tag{2.4.2*}$$

for any event A in the family of events $\mathcal{S}$. That is, **the probability of any**

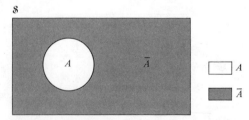

Figure 2.4.2   The Probabilities of an Event and Its Complement

event must lie between zero and one inclusive. Suppose that some event A could be found where $P(A) > 1$; would this lead to a contradiction of one or more of our axioms? If so, then under these axioms no event can have a probability greater than 1. From the theorem just proved,

$$P(A) + P(\bar{A}) = 1.$$

If $P(A)$ should be greater than 1, then it must be true that $P(\bar{A})$ is less than zero; however, Axiom 1 dictates that *any* event must have a probability greater than or equal to zero. Thus, if $P(A)$ is greater than 1, a contradiction is generated, and this means that any event must have a probability lying between 0 and 1 inclusive. In terms of a Venn diagram, it is impossible for the area covered by a set to be negative, and from the definition of a sample space, it is impossible for a set to cover an area larger than that covered by the sample space.

The third theorem we shall prove states that

$$P(\varnothing) = 0. \tag{2.4.3*}$$

The probability of the empty or "impossible" event is zero. To show this we recall that

$$\bar{S} = \varnothing,$$

the set of all elementary events not in S is empty. Then, by Equation (2.4.1),

$$P(\varnothing) = 1 - P(S)$$
$$= 0.$$

Still another important elementary theorem of probability is as follows. If $A \subseteq B$, where A and B are two events in S, then

$$P(A) \leq P(B). \tag{2.4.4*}$$

In other words, **if the occurrence of an event A implies that an event B must occur, so that A is a subset of B, then the probability of A is less than or equal to the probability of B.** The set B can be thought of as the union of *two* mutually exclusive sets:

$$B = (A \cap B) \cup (\bar{A} \cap B).$$

But, since A is a subset of B, $A \cap B = A$, so that

$$B = A \cup (\bar{A} \cap B).$$

Then, by Axiom 3,

$$P(B) = P(A) + P(\bar{A} \cap B),$$

so that

$$P(B) - P(A) = P(\bar{A} \cap B).$$

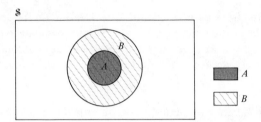

Figure 2.4.3   The Probabilities of an Event and a Subset of that Event

Since $P(\bar{A} \cap B)$ cannot be negative (Axiom 1), $P(B)$ must be greater than or equal to $P(A)$. This rule is illustrated in Figure 2.4.3. For example, suppose that an individual is chosen from a given population and that the individual's income is recorded. If we let A represent "income is greater than \$20,000 per year" and we let B represent "income is at least \$10,000 per year," then $A \subseteq B$. Thus, the probability of observing an income greater than \$20,000 is smaller than the probability of observing an income of at least \$10,000, since the latter event contains all incomes included in the former event *and* some other incomes that are not included in the former event.

Two immediate and useful consequences of the theorem just proved are the following statements.

$$\text{For any pair of events } A \text{ and } B, P(A \cap B) \leq P(A \cup B). \quad (2.4.5^*)$$

$$\text{If } A \subseteq B, \text{ then } P(B - A) = P(B) - P(A). \quad (2.4.6^*)$$

The next theorem is a most important one for all sorts of probability calculations:

given two events A and B, then

$$P(A \cup B) = P(A) + P(B) - P(A \cap B). \quad (2.4.7^*)$$

To prove this, notice that $A \cup B$ can be written in terms of the union of two mutually exclusive sets:

$$A \cup B = A \cup [B - (A \cap B)].$$

By Axiom 3,

$$P(A \cup B) = P(A) + P[B - (A \cap B)].$$

Using Equation (2.4.6),

$$P[B - (A \cap B)] = P(B) - P(A \cap B),$$

and thus $$P(A \cup B) = P(A) + P(B) - P(A \cap B).$$

In other words, **to find the probability that A or B (or both) occurs, we must know the probability of A, the probability of B, and also the probability that A and B both occur.** Be sure to notice, however, that if A and B are *mutually exclusive* events, so that $P(A \cap B) = 0$, then

$$P(A \cup B) = P(A) + P(B),$$

just as provided by Axiom 3.

The rule given by Equation (2.4.7) is illustrated in Figure 2.4.4. Note that $P(A \cap B)$ is subtracted from $P(A) + P(B)$ in order to avoid "double counting." If $P(A \cap B)$ is not subtracted, the area covered by $A \cap B$ is included *twice*. For instance, if $A =$ "person has red hair" and $B =$ "person is male," then in adding $P(A)$ and $P(B)$, we are double counting the probability of a red-haired male. Subtracting $P(A \cap B)$ eliminates this double counting.

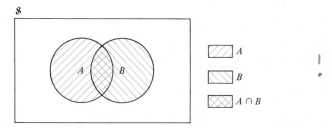

Figure 2.4.4   The Probability of the Union of Two Events

The last theorem to be proved here concerns partitions. If $A_1, A_2, \ldots, A_K$ form a partition of S, then

$$P(A_1) + P(A_2) + \cdots + P(A_K) = 1. \qquad (2.4.8^*)$$

In other words, **if K events form a partition of S, then their probabilities must sum to one.** To prove this theorem, recall that if K events form a partition of S, they must be mutually exclusive and exhaustive. Since $A_1, A_2, \ldots, A_K$ are mutually exclusive,

$$P(A_1 \cup A_2 \cup \cdots \cup A_K) = P(A_1) + P(A_2) + \cdots + P(A_K),$$

by Axiom 3. Furthermore, since they are exhaustive,

$$A_1 \cup A_2 \cup \cdots \cup A_K = S.$$

Then, by Axiom 2,

$$P(A_1 \cup A_2 \cup \cdots \cup A_K) = P(S) = 1,$$

and it follows, on substituting $P(A_1) + P(A_2) + \cdots + P(A_K)$ for

$P(A_1 \cup A_2 \cdots \cup A_K)$, that

$$P(A_1) + P(A_2) + \cdots + P(A_K) = 1.$$

It should be obvious that there is no end to the number of new deductions we could make using the original three axioms and the accumulated theorems such as those just proved. Enormous volumes have been filled with such probability theorems, all deduced in essentially the same way from the same axioms we have been using. Naturally, only some of the most elementary results have been shown here, and these particular results have been chosen because of their simple proofs and because they will be useful in future sections. The idea to be conveyed, however, is that we can deduce all sorts of consequences which *must* be true of these numbers we call probabilities, if these numbers obey the *formal* definition set forth at the beginning of this section. In short, probability theory can perfectly well be studied strictly as an abstract mathematical system without any interpretation at all.

The simple rules we have deduced and the three basic axioms will all be useful to us in learning to calculate probabilities. In the following sections, we will first discuss the *calculation* of probabilities and then turn to the *interpretation* of these numbers called probabilities.

2.5   A SPECIAL CASE: EQUALLY PROBABLE ELEMENTARY EVENTS

Most simple probability calculations actually rest on the "addition" principle embodied in Axiom 3 of Section 2.4. The elementary events making up the sample space S are conceived as the set of all possible *distinctly different* outcomes of the particular simple experiment. Thus, if we somehow knew the probability for each and every elementary event making up the sample space S, then we would also be able to compute the probability of any event in S. **The probability of any event in S is simply the sum of the probabilities of the elementary events qualifying for that event class;** this is an immediate consequence of Axiom 3. Thus, the probability of any event can be found provided that the probability associated with each and every elementary event is known.

However, the practical difficulty in actually computing probabilities is in knowing the probabilities of the elementary events themselves. Fortunately, some of the simple experiments for which we need to calculate probabilities, particularly games of chance, have a feature that does away with this problem. A great many simple experiments are conducted in such a way that it is reasonable to assume that each and every distinct elementary event has the *same* probability. When tickets are drawn in a lottery, for example, great pains are taken to have the tickets well shaken-up in a tumbler before each one is drawn; this mixing operation makes it reasonable to believe that any particular ticket has the same chance of being drawn as any other. In this case, we say that a ticket is **randomly** selected from the tickets in the tumbler. To select a single item *at random* from a set simply means to select an item in such a manner that all items in the set are equally likely to be selected. (In Chapter 5, we will generalize this definition of random selection to more complex experiments.) Cards are thoroughly shuffled and cut, perfectly balanced dice are thrown in a dice-cage, and, in fact, almost all gambling situations have some feature which makes this equal-chances assumption reasonable. As we shall see, even experiments that are not games of chance may be carried out in such a way that each and every elementary event should have the same probability.

When it can be assumed that each and every one of a finite number of possible elementary events has exactly the same probability, then the probability of any event is particularly easy to compute.

If all elementary events in the sample space S have exactly the **same probability, then the probability of any event A is given by**

$$P(A) = \frac{\text{number of elementary events in } A}{\text{total number of elementary events in } S}. \qquad (2.5.1^*)$$

It is easy to see that this rule must hold for the probability of any event provided that the various elementary events are equally probable. Suppose that there are K elements in S. Now consider a subset A consisting of exactly two elementary events, $A = \{a_1, a_2\}$. What is the probability of A? Since all such events are equally likely, $P(a_1) = 1/K$ and $P(a_2) = 1/K$. The set A is the union $a_1 \cup a_2$ of two mutually exclusive sets; thus, by Axiom 3, Section 2.4, $P(A) = (1/K) + (1/K)$, or $2/K$. Proceeding in this way, suppose that the set A contained m events. Here $P(A) = (1/K) + (1/K) + \cdots + (1/K)$, or $1/K$ summed m times, which is simply m/K.

For equally likely elementary events, the probability of an event **A is its relative frequency in the sample space.**

As an illustration, suppose that a rental car agency has 10 cars on hand. Five of these cars are white, 3 are red, and 2 are blue. Each of the 10 cars is reserved for a particular passenger on a flight that is due to arrive shortly. Let an experiment consist of observing the car that is claimed first. We know who has reserved each car, but we have no idea who will arrive at the counter first. Therefore, there is no reason for us to believe that any particular car is more likely to be claimed before any other car. An elementary event here is a particular car being claimed first. Since there are 10 cars, there are 10 elementary events making up the sample space, and each elementary event has probability 1/10.

We are concerned with the 3 events "car claimed first is white," "car claimed first is red," and "car claimed first is blue." What are the probabilities of these events? The event "car claimed first is white" contains 5 elementary events, since 5 of the 10 cars are white. Therefore,

$$P(\text{white}) = \frac{\text{number of white cars}}{\text{total number of cars}} = \frac{5}{10},$$

so that the probability of the event "white" in this experiment is .50. In the same way, we can find

$$P(\text{red}) = .30 \quad \text{and} \quad P(\text{blue}) = .20.$$

Furthermore, using the rules proved in the previous section, it is easy to find

$$P(\text{not white}) = 1 - .50 = .50,$$

and $$P(\text{white or blue}) = .50 + .20 = .70.$$

For another example, consider an experiment consisting of throwing a fair (that is, unloaded) die, marked with spots from 1 to 6 on its sides. The die is constructed and thrown so that each of the sides is equally likely to appear. There are 6 distinct outcomes, so that there are 6 elementary events making up S. The probability of the event "1" is thus 1/6, the event "2" has probability 1/6, and so on. What is the probability of the event "odd number"? An odd number occurs when the die comes up with 1, 3, or 5 spots. That is, the desired probability is

$$P(1 \text{ or } 3 \text{ or } 5 \text{ spots}) = P(\text{odd number}) = 3/6 = 1/2.$$

This section involved the special case in which all of the elementary events are equally likely. In many instances, such an assumption appears realistic, although there are clearly other situations in which the assumption of equally likely elementary events is untenable. If the elementary events are *not* equally likely, then Equation (2.5.1) is not applicable, but it is still possible to compute the probability of any event A by simply adding up the probabilities of the elementary events in A, provided, of course, that these probabilities are somehow known.

2.6 COMPUTING PROBABILITIES

Many simple problems in probability can be reduced to problems in counting. For a sample space containing equally probable elementary events, the computation of probabilities involves two quantities, both of which are *counts of possibilities:* the total number of elementary events and the number qualifying for a particular event class. The key to solving probability problems is to learn to ask: "How many distinctly different ✗ ways can this event happen?" It may be possible simply to list the number of different elementary events that make up the event class in question, but it is often much more convenient to use a rule for finding this number.

A game of chance such as roulette shows how probabilities may be computed simply by listing elementary events. One standard type of roulette wheel has 37 equally spaced slots into which a ball may come to rest after the wheel is spun. These slots are numbered from 0 through 36. Half of the numbers 1 through 36 are red, the others are black, and the zero is generally green.

In playing roulette, a player may bet on any single number, on certain groups of numbers, on colors, and so on. One such bet might be on the odd numbers (excluding zero, of course). The only elementary events in the sample space are the numbers 0 through 36 with their respective colors. Which numbers qualify for the event "odd"? It is easy to count these numbers; they form the set

$$\{1, 3, 5, 7, 9, 11, 13, 15, 17, 19, 21, 23, 25, 27, 29, 31, 33, 35\}.$$

Since there are exactly 18 elementary events qualifying as odd numbers, if each slot on the roulette wheel is equally likely to receive the ball, the probability of "odd" is 18/37, slightly less than 1/2.

As a more complicated example that can be solved by listing, let us find the probability that the winning number on the roulette wheel contains a "3" as one of the digits. The elementary events qualifying are

$$\{3, 13, 23, 30, 31, 32, 33, 34, 35, 36\}.$$

Here the probability is 10/37, if the 37 different elementary events are equally likely.

In many simple problems involving probability, listing can be useful in computing probabilities. If the number of elementary events becomes large, however, then listing is cumbersome, and counting methods such as those to be discussed in the following sections are often more efficient.

2.7   SEQUENCES OF EVENTS

In many problems, an elementary event may be *the set of outcomes of a series or sequence of observations.* Suppose that any trial of some simple experiment must result in one of K mutually exclusive and exhaustive events, $\{A_1, \ldots, A_K\}$. Now the experiment is *repeated* n times. This leads to a sequence of events: the outcome of the first trial, the outcome of the second, and so on in order through the outcome of the nth trial. The outcome of the whole series of trials might be the sequence $(A_3, A_1, A_2, \ldots, A_3)$. This denotes that the event A_3 occurred on the first trial, A_1 on the second trial, A_2 on the third, and so on. The place in order gives the trial on which the event occurred, and the symbol occupying that place shows the event occurring for that trial. For example, the experiment might consist of recording the state (as given by the license plates) in which each car arriving at a particular U. S. National Park is registered. If 6 cars are observed, one possible sequence of outcomes is (Illinois, Colorado, Wisconsin, Illinois, California, Maine).

The idea of a sequence also extends to a series of *different* simple experiments. Here the place in the sequence corresponds to the particular experiment being performed. For example, suppose that as each car arrives at a U.S. National Park, the state of registration, the number of passengers, and the intended length of stay in the park are recorded. Then the sequence of events observed for a single car might be (Michigan, 5, 1 day), for example, where the position describes the particular simple experiment and the symbol in that position represents the observed event.

Remember that each possible sequence is an elementary event if we think of the entire series of trials as the experiment. Each experiment has as its outcome one and only one sequence. Such experiments producing sequences as outcomes are very important to statistics, and so we will begin our study of counting rules by finding how many different sequences a given series of trials could produce. It should be mentioned that the six counting rules to be discussed in the next five sections are merely convenient methods for counting certain possibilities. The rules are not to be interpreted in terms of probabilities, although they will prove useful in the computation of certain probabilities, as we shall see in Section 2.11.

2.8 SEQUENCES: COUNTING RULES 1 AND 2

Suppose that a series of n trials are carried out, and that on each trial any of K events might occur. Then the following rule holds.

COUNTING RULE 1

If any one of K mutually exclusive and exhaustive events can occur on each of n trials, then there are K^n different sequences that may result from a set of trials.

As an example of this rule, consider a coin being tossed. Each toss can result only in an H or a T event ($K = 2$). Now the coin is tossed 5 times, so that $n = 5$. The total number of *possible* results of tossing the coin 5 times is $K^n = 2^5 = 32$ sequences. Exactly the same number is obtained for the possible outcomes of tossing 5 coins simultaneously, if the coins are thought of as numbered and if a sequence describes what happens to coin 1, to coin 2, and so on.

As another example, suppose that each day a security analyst records whether a certain stock increased in price, decreased in price, or did not change in price that day. Each day there are 3 mutually exclusive and exhaustive events that can possibly occur, so that over a period of 1 week, consisting of 5 trading days, there are $3^5 = 243$ possible sequences.

Sometimes the number of possible events in the first trial of a series is different from the number possible in the second, the second different from the third, and so on. It is obvious that if there are different numbers K_1, $K_2, \ldots, K_n$ of events possible on the respective trials, then the total number of sequences will not be given by Rule 1. Instead, the following rule holds.

COUNTING RULE 2

If $K_1, \ldots, K_n$ are the numbers of distinct events that can occur on trials $1, \ldots, n$ in a series, then the number of different sequences of n events that can occur is $(K_1)(K_2) \cdots (K_n)$.

For example, suppose that for the first trial you toss a coin (2 possible outcomes) and for the second you roll a die (6 possible outcomes). Then the total number of different sequences would be $(2)(6) = 12$.

Notice that Counting Rule 1 is actually a special case of Rule 2. If the same number K of events can occur on any trial, then the total number of sequences is K multiplied by itself n times, or K^n.

2.9 PERMUTATIONS: COUNTING RULES 3 AND 4

A rule of extreme importance in probability computations concerns the number of ways that objects may be arranged in order. The rule will be given for arrangements of objects, but it is equally applicable to sequences of events.

COUNTING RULE 3

The number of different ways that n distinct things may be arranged in order is $n! = (1)(2)(3) \cdots (n-1)(n)$ (where $0! = 1$). An arrangement in order is called a permutation, so that the total number of permutations of n objects is $n!$. The symbol $n!$ is called "n factorial."

As an illustration of this rule, suppose that a classroom contained exactly 10 seats for 10 students. How many ways could the student be assigned to the chairs? Any of the students could be put into the first chair, making 10 possibilities for chair 1. But, given the occupancy of chair 1, there are only 9 students for chair 2; the total number of ways chair 1 and chair 2 could be filled is $(10)(9) = 90$ ways. Now consider chair 3. With chairs 1 and 2 occupied, 8 students remain, so that there are $(10)(9)(8) = 720$ ways to fill chairs 1, 2, and 3. Finally, when 9 chairs have been filled there remains only 1 student to fill the remaining place, so that there are

$$(10)(9)(8)(7)(6)(5)(4)(3)(2)(1) = (10)! = 3,628,800$$

ways of arranging the 10 students into the 10 chairs.

For another example, suppose that a teacher has the names of 5 students in a hat. The names are drawn at random one at a time, *without* replacement. Then there are exactly 5! or 120 different sequences of names that might be observed. If all sequences of names are equally likely to be drawn, then the probability for any one sequence is 1/120.

What is the probability that the name of any given child will be drawn first? Given that a certain child is drawn first, the order of the remainder of the sequence is still unspecified. There are $(n - 1)! = 4! = 24$ different orders in which the other children can appear. Thus, the probability of any given child being first in sequences is $24/120 = .20$ or 1/5.

Suppose that exactly *two* of the names in the hat are girls'. What is the probability that the first two names drawn belong to the girls? Given the first two names are girls', there are still $3! = 6$ ways three boys may be arranged. The girls' names may themselves be ordered in two ways. Thus, the probability of the girls' names being drawn first and second is $(2)(6)/120 = 1/10$.

Sometimes it is necessary to count the number of ways that r objects might be selected from among some n objects in all $(r \leq n)$. Furthermore, each different *arrangement* of the r objects is considered separately. Then the following rule applies.

COUNTING RULE 4

The number of ways of selecting and arranging r objects from among n distinct objects is

$$\frac{n!}{(n - r)!}.$$

The reasoning underlying this rule becomes clear if a simple example is taken. Consider a classroom teacher, once again, who has 10 students to be assigned to seats. This time, however, imagine that there are only 5 seats. How many different ways could the teacher select 5 students and arrange them into the available seats? Notice that there are 10 ways that the first seat might be filled, 9 ways for the second, and so on, until seat 5 could be filled in 6 ways. Thus, there are

$$(10)(9)(8)(7)(6)$$

ways to select students to fill the 5 seats. This number is equivalent to

$$\frac{10!}{5!} = \frac{n!}{(n-r)!},$$

the number of ways that 10 students out of 10 may be selected and arranged, but divided by the number of arrangements of the 5 *unselected* students in the 5 *missing* seats.

As an example of the use of this principle in probability calculations, consider an example involving lotteries. In lotteries it is usual for the first person whose name is drawn to receive a large amount, the second some smaller amount, and so on, until some r prizes are awarded. This means that some r names are drawn in all, and the *order* in which those names are drawn determines the size of the prizes awarded to individuals. Suppose that in a rather small lottery 40 tickets had been sold, each to a different person, and only 3 names were to be drawn for first, second, and third prizes. Here $n = 40$ and $r = 3$. How many different assignments of prizes to persons could there be? The answer, by Counting Rule 4, is

$$\frac{40!}{(40-3)!} = \frac{40!}{37!} = (38)(39)(40) = 59{,}280.$$

On how many of these possible sequences of winners would a given person appear as first, second, or third prize winner? If the person were drawn first, the number of possible selections for second and third prize would be $39!/37! = 1482$. Similarly, there would be 1482 sequences in which the person could appear second, and a like number where the person could appear third. Thus, the probability that the person appears in a sequence of three drawings, winning first, second, or third prize is (by Axiom 3, Section 2.4)

$$P \text{ (first, second, or third)} = \frac{3(1482)}{59{,}280} = \frac{3}{40}.$$

2.10 COMBINATIONS: COUNTING RULES 5 AND 6

In a very large class of probability problems, we are not interested in the *order* of events, but only in the number of ways that r things could be selected from among n things, *irrespective of order*. We have just seen that the total number of ways of selecting r things from n and ordering them is $n!/(n - r)!$ by Rule 4. Each set of r objects has $r!$ possible orderings, by Rule 3. A combination of these two facts gives us the following rule.

COUNTING RULE 5

The total number of ways of selecting r distinct combinations of n objects, irrespective of order, is

$$\frac{n!}{r!(n - r)!} = \binom{n}{r}.$$

The symbol $\binom{n}{r}$ is *not* a symbol for a fraction, but instead denotes the number of combinations of n things, taken r at a time. Sometimes the number of combinations is known as a "binomial coefficient," and occasionally $\binom{n}{r}$ is replaced by the symbols ^nC_r or $_nC_r$. However, the name and symbol introduced in Rule 5 will be used here.

It is helpful to note that

$$\binom{n}{r} = \binom{n}{n - r}. \tag{2.10.1}$$

Thus, $\binom{10}{3} = \binom{10}{7}$, $\binom{50}{49} = \binom{50}{1}$, and so on.

As an example of the use of this rule, suppose that a total of 33 men were candiates for the board of supervisors in some community. Three supervisors are to be elected at large; how many ways could 3 men be selected from among these 33 candidates? Here, $r = 3, n = 33$, so that

$$\binom{n}{r} = \frac{(33)!}{3!(30)!} = \frac{(1 \cdot 2 \cdot \ \cdots \ \cdot 32 \cdot 33)}{(1 \cdot 2 \cdot 3)(1 \cdot 2 \cdot \ \cdots \ \cdot 29 \cdot 30)}.$$

Canceling in numerator and denominator, we get

$$\binom{33}{3} = \frac{(31)(32)(33)}{6} = (31)(16)(11) = 5456.$$

If all sets of three men are equally likely to be chosen, the probability for any given set is $1/5456$.

Because of their utility in probability calculations, a table of $\binom{n}{r}$ values for various values ot n and r is included in the Appendix. Although this table shows values of n only up to 20 and values of r up to 10, other values can be found by Equation (2.10.1) and by the relation known as Pascal's rule:

$$\binom{n}{r} = \binom{n - 1}{r - 1} + \binom{n - 1}{r}. \tag{2.10.2}$$

Still other values may be worked out from the table of factorials also included in the Appendix.

Rule 5 can be interpreted as the total number of ways of dividing a set of n objects into two subsets of r and $(n - r)$ objects, since the determination of the first subset of r objects uniquely determines the second subset of $(n - r)$ objects, which consists of the objects remaining after the first subset has been selected. This can be generalized to the situation in which the set of n objects is to be divided into K subsets, where $K \geq 2$.

COUNTING RULE 6

The total number of ways of dividing a set of n **objects into** K **mutually exclusive and exhaustive subsets of** n_1, n_2, $\ldots$, n_{K-1}, **and** n_K **objects, respectively, where**

$$n_1 + n_2 + \ldots + n_K = n,$$

is

$$\frac{n!}{n_1! n_2! \cdots n_K!}.$$

Here the denominator is equal to the product of the $n_i!$'s. It can readily be seen that Rule 5 is a special case of Rule 6, with $K = 2$, $n_1 = r$, and $n_2 = n - r$. We will not present the proof of Rule 6, which just requires successive applications of Rule 5.

As an example of the use of Rule 6, consider an extension of the previous example. Suppose that a total of 33 men were candidates for the board of supervisors. Three supervisors are to be elected at large, with the fourth and fifth highest vote-getters being designated as alternates. How many ways can the 33 candidates be divided into 3 commissioners, 2 alternates, and 28 remaining candidates? Here $n = 33$, $K = 3$, $n_1 = 3$, $n_2 = 2$, and $n_3 = 28$, so that the number of combinations is

$$\frac{n!}{n_1! n_2! n_3!} = \frac{33!}{3! 2! 28!} = \frac{(1 \cdot 2 \cdot \; \cdots \; \cdot 32 \cdot 33)}{(1 \cdot 2 \cdot 3)(1 \cdot 2)(1 \cdot 2 \cdot \; \cdots \; \cdot 27 \cdot 28)}$$

$$= \frac{(29 \cdot 30 \cdot 31 \cdot 32 \cdot 33)}{6 \cdot 2} = (29 \cdot 5 \cdot 31 \cdot 16 \cdot 33) = 2{,}373{,}360.$$

2.11 SOME EXAMPLES: POKER HANDS

The six rules in the preceding four sections provide the basis for many probability calculations when the number of elementary events is finite, especially when sampling is without replacement. One of the easiest examples of their application is the calculation of probabilities for poker hands.

The game of poker as discussed here will be highly simplified and not very exciting to play; the player simply deals 5 cards from a well-shuffled standard deck of 52 cards. Nevertheless, the probabilities of the various hands are interesting and can be computed quite easily. The particular hands we will examine are the following:

1. one pair, with three different remaining cards,
2. full house (three of a kind and one pair),
3. flush (all cards of the same suit).

First of all, we need to know how many different hands may be drawn. A given hand of 5 cards will be thought of as an elementary event; notice that the order in which the cards appear in a hand is immaterial. By Rule 5, since there are 52 different cards in all and only 5 are selected, there are

$$\binom{52}{5} = \frac{52!}{5!47!}$$

different hands that might be drawn. If all hands are equally likely, then the probability of any given one is $1/\binom{52}{5}$.

There are 13 numbers that a pair of cards may show (counting the picture cards), and each member of a pair must have a suit. Thus, by Rule 5 there are $13\binom{4}{2}$ different pairs that might be observed. The remaining cards must show 3 of the 12 remaining numbers and each of the 3 cards may be of any suit. By Rules 1 and 5 there are $\binom{12}{3}(4)(4)(4)$ ways of filling out the hand in this way. Multiplying, we find that there are

$$13\binom{4}{2}\binom{12}{3}4^3$$

different ways for this event to occur. Thus,

$$P(\text{one pair}) = \frac{13\binom{4}{2}\binom{12}{3}4^3}{\binom{52}{5}}.$$

This number can be worked out by writing out the factorials, canceling in numerator and denominator, and dividing. It is approximately equal to .42, so that the chances are slightly better than 4 in 10 of drawing a single pair in 5 cards, if all possible hands are equally likely to be drawn. [Note that $P(\text{one pair})$ is not exactly equal to .42; to two decimal places, it is .42. In this book, fractions are often expressed in decimal form and "rounded off" if necessary, so you should be alert to the approximate nature of certain numerical "equalities."]

This same scheme can be followed to find the probability of a full house. There are 13 numbers that the three of a kind may have and then 12 numbers possible for the pair. The three of a kind must represent three of four suits, and the pair two of four suits. This gives

$$13 \binom{4}{3} 12 \binom{4}{2}$$

different ways to get a full house. The probability is

$$P(\text{full house}) = \frac{13 \binom{4}{3} 12 \binom{4}{2}}{\binom{52}{5}},$$

which is about .0014.

Finally, a flush is a hand of 5 cards all of which are in the same suit. There are exactly four suit possibilities, and a selection of 5 out of 13 numbers that the cards may show. Hence, the number of different flushes is $4\binom{13}{5}$, and

$$P(\text{flush of 5 cards}) = \frac{4 \binom{13}{5}}{\binom{52}{5}},$$

or about .002.

The probabilities of the various other hands can be worked out in a similar way. The point of this illustration is that it typifies the use of these counting rules for actually figuring probabilities of complicated events, such as particular poker hands. Naturally, a great deal of practice is usually necessary before we can visualize and carry out probability calculations from "scratch" with any facility. Nevertheless, many probability calculations depend upon counting how many ways events of a certain kind can occur. Although the example given here utilizes a card game instead of a more important real-world situation, the procedures that are illustrated by the example are of use in a wide variety of situations, as we shall see when we have cause to use counting rules in developing certain probability distributions and statistical techniques.

2.12   THE RELATIVE FREQUENCY INTERPRETATION
OF PROBABILITY

In the previous sections, probability was considered as a mathematical system; given certain probabilities, the theory can be used to determine certain other probabilities. If the probability of obtaining a "head" on a single toss of a coin is 1/2, then the probability of obtaining a "tail" is also 1/2 (assuming that other outcomes, such as the coin landing on edge or the coin disintegrating in mid-air, are not considered as possible events). The mathematical theory can thus be extremely useful in calculating probabilities. However, an important question is still unanswered. What do these probabilities mean? What does it mean to say that the probability of obtaining a "head" on a single toss of a coin is 1/2? If a marble is to be drawn from a box containing 30 red marbles and 70 blue marbles, what does it mean to say that the probability is 3/10 that the marble drawn is red? What does it mean to say that the probability of the event "odd number" in one throw of a die is 1/2? What does it mean to say that the probability of obtaining a "full house" in a 5-card poker hand is .0014? In this section, one possible interpretation of probability is considered: the **relative frequency interpretation.** Later in the chapter, a second possible interpretation is discussed: the **subjective,** or **personal, interpretation.**

In the first of the examples mentioned in the preceding paragraph, if the coin is tossed repeatedly, then how often would we expect to obtain "heads"? Over a long series of tosses, it seems reasonable to expect to observe heads about 50 percent of the time. Similarly, if we keep drawing marbles out of the box with 30 red and 70 blue marbles and if after each draw the marble drawn is replaced in the box before the next drawing occurs, then in the *long run*, after numerous observations, what proportion of the marbles drawn would be expected to be red? It seems sensible to expect that about 3 out of 10 observations will be red. In the same way, in the third example, if we throw the die repeatedly, then it seems reasonable that in the long run, over a large number of such observations, about 1/2, or one in two, of the throws should result in an odd number. Finally, in the poker example, we should expect a full house about 14 times in 10,000 5-card poker hands in the long run. In these examples,

the probability of an event has been interpreted as the relative frequency of occurrence of that event to be expected in the long run.

It is useful to note that relative frequencies satisfy the three axioms of probability. Clearly, a relative frequency cannot be negative, and the relative frequency of the entire sample space is one (if an elementary event not included in the sample space occurs, then the sample space has been incorrectly specified, since it is defined as the set of *all* possible elementary events). Finally, if two events are mutually exclusive, the relative frequency of the union of the two events is equal to the sum of their individual relative frequencies. In a particular series of tosses of a die, if the event "2 or 4 comes up" occurs with relative frequency .30 and the event "6 comes up" occurs with relative frequency .18, then the event "an even number comes up" occurs with relative frequency .48.

Although simple examples of the type given above frequently involve the assumption that all of the elementary events of interest are equally likely, this assumption is not needed in order to talk about a probability as a long-run relative frequency. Indeed, we should carefully distinguish between the long-run relative frequency interpretation of probability and another interpretation, sometimes called the **classical interpretation,** in which the elementary events are assumed equally likely and the probability of an event is identical to the relative frequency of occurrence of the event **in the sample space.** In the marble example, if each marble is thought of as an elementary event, then the probability of drawing a red marble is equal to the number of red marbles, 30, divided by the total number of marbles, 100; this is the relative frequency of the event "red marble" in the sample space. The applicability of the classical interpretation is based on symmetry arguments, which may be reasonable in games of chance and similar situations. In most uncertain situations, however, it is *not* reasonable to assume that all elementary events are equally likely. The long-run relative frequency interpretation of probability does not require this assumption, as we shall see in the next section. In the remainder of this book, the term "relative frequency interpretation" will, unless stated otherwise, refer to relative frequency "in the long run," not relative frequency "in the sample space."

2.13 THE LAW OF LARGE NUMBERS

The connection between probability and long-run relative frequency is a simple and appealing one, and it does form a tie between that basically undefined notion, "the probability of an event," and something we can actually observe, a relative frequency of occurrence. A statement of probability tells us what to *expect* about the relative frequency of occurrence of an event given that enough sample observations are made at random. In the long run, the relative frequency of occurrence of event A should approach the probability of this event, if independent trials are made at random over an indefinitely long sequence. As it stands, this idea is a familiar and reasonable one, but it will be useful to have a more exact statement of this principle for use in future work. The principle was first formulated and proved by James Bernoulli in the early eighteenth century, and it is often called the **law of large numbers.** A more or less precise statement of the law of large numbers is as follows.

If the probability of occurrence of the event A is p, and if n trials are made, independently and under exactly the same conditions, then the probability that the relative frequency of occurrence of A differs from p by any amount, however small, approaches zero as the number of trials grows indefinitely large.

Formally, if r denotes the number of times that A occurs in the n trials, then

$$P\left(\left|\frac{r}{n} - p\right| > \epsilon\right) \to 0 \qquad \text{as } n \to \infty, \qquad (2.13.1)$$

where ϵ is any arbitrarily small positive number. As the number of repetitions, n, increases, the relative frequency of occurrence of the event A, r/n, tends to get closer and closer to the probability, p. For instance, if a symmetric, 6-sided die is tossed repeatedly, the relative frequency of occurrence of the event "6 comes up" is more likely to be close to 1/6 as the number of tosses, n, increases.

The law of large numbers formally expresses in mathematical terms what might be thought of as **statistical regularity.** Such statistical regularity is very important in, for example, the gambling and insurance businesses. Suppose that the death rate for a certain age group in the United States has remained reasonably stable over the past decade at 5 deaths per 100 persons per year. Barring special conditions such as epidemics or new "wonder drugs" (which would violate the requirement of "identical conditions"), an insurance company with a large number of policies among members of the given age group can expect that the relative frequency of death for these policyholders in the coming year will be very close to .05. If only a few policyholders are considered, variations from .05 are not surprising. As more and more policyholders are considered, statistical regularity is observed, as illustrated in Figure 2.13.1; the relative frequency tends to stabilize around .05. If a new drug that cures cancer is introduced, this would tend to reduce the death rate, and the relative frequency might stabilize around, say, .046 instead of .05. The important point is that if repeated groups of, for example, 1000 policyholders are considered, the relative frequencies of death will be very close for the different groups.

The long-run relative frequency interpretation of probability, then, seems intuitively reasonable and empirically sound. If an uncertain situation is repeated a large number of times, the frequency of an event's occur-

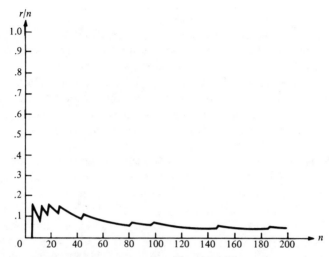

Figure 2.13.1 Statistical Regularity: Relative Frequency of Death as a Function of Sample Size for a Particular Sample

rence should somehow reflect the probability of the event. Most people think of this phenomenon as a "law of averages." Given enough experimental trials or sample observations from the sample space, we expect the relative frequency of occurrence to balance out to approximately equal the probability.

It is important to note, however, that no matter how many repetitions of the uncertain situation are considered, the actual relative frequency of an event need not exactly equal the probability of the event. If a fair coin is tossed 1000 times, the relative frequency of heads is expected to be close to, but not necessarily equal to, one-half. The law of large numbers says that the relative frequency and the probability need be equal only for an infinite number of repetitions, which is a theoretical ideal clearly unattainable in practice.

It is also important not to fall into the error of thinking that an event is ever "due to occur" on any given trial. The independence of trials essentially means that the outcome on any single trial has no bearing on the outcomes on any other trials. In terms of tossing a fair coin, we may say that the fact that it comes up heads on one toss bears no relationship to the outcome of the previous toss, the next toss, or any other toss. Thus, an event cannot be thought of as "due" to occur. But if this is the case, how does Bernoulli's theorem allow for unusual occurrences such as 100 heads in a row? It does so not by compensating for the 100 heads in a row with 100 tails in a row, but rather by a "swamping" procedure, which works as follows. If the first 100 tosses of a sequence are heads, this is an unusual event (provided that the coin is fair). Suppose, however, that in the next 1000 trials, we observe 500 heads and 500 tails. For all 1100 tosses, we have 600 heads and 500 tails, which does not seem as unusual as 100 heads and 0 tails. If we tossed the coin a million more times, with 500,000 heads and 500,000 tails, the total number of heads and tails would be 500,600 and 500,500. This result need not seem too unusual. As the sample size has increased, the original "run" of 100 straight heads has been "swamped" by the subsequent trials, even though the subsequent trials resulted in exactly 50 percent heads and 50 percent tails.

It is interesting to note at this point that in some situations we might have cause to believe the *opposite* of the "event is due to occur" argument. If we met a stranger in a bar who produced a coin and proceeded to toss 100 straight heads, we would *not* think that a tail was "due." We would be much more likely to suspect that the coin was not fair and that the probability of a head on a single toss was greater than $1/2$ (if the coin were two-headed, the probability of a head would be 1). This amounts to questioning the assumption of a fair coin and demonstrates an important use of the law of large numbers. If we have no idea what the probability of a particular event is and we observe a large number of independent trials, we can with some confidence claim that the probability of the event is close to the relative frequency of the event in the sequence of trials. Of course, what constitutes a "large" number of trials depends on the situation and on the desired accuracy. Reasonable accuracy to four decimal places requires a much larger value of n than accuracy to only two decimal places, and thus a large number of trials is needed in the case of a "rare event" with, for instance, probability .0005.

Although it is true that the relative frequency of occurrence of any event must exactly equal the probability only for an infinite number of independent trials, this point must not be overstressed. Even with relatively small numbers of trials we have very good reason to expect the observed relative frequency to be quite close to the probability. The rate of convergence of the relative frequency to the probability is very rapid, even at the lower levels of sample size, although the probability of a small discrepancy between relative frequency and probability is much smaller for extremely large than for extremely small samples. A probability is not a curiosity requiring unattainable conditions, but rather a value that can be estimated with considerable accuracy from a sample. A good bet about the probability of an event is the actual relative frequency we have observed from some n trials, and the larger n is, the better the bet.

In a later section, we will investigate principles such as the law of large numbers in a more formal manner, but this must wait until Chapter 5. As yet, we lack the vocabulary to give a precise meaning to terms, such as "at random" and "independent events," that figure in this theorem (recall that in Section 2.5 we defined "at random" for the case in which there was only a single trial, not for the case with n trials). Nevertheless, accepting this principle at face value for the moment, we can show a simple example of its use.

2.14 AN EXAMPLE OF SIMPLE STATISTICAL INFERENCE

Imagine that we were faced with the following problem. We are given a coin and asked to decide if the coin is fair (that is, if the probability of heads is .50). This is to be decided through the repeated simple experiment of tossing the coin. If the coin actually is fair, then in an infinite number of tosses the relative frequency of the event heads *must* be exactly .50. However, on any given number of tosses, we should only *expect* a relative frequency of .50 to occur. The larger the number of tosses, the more nearly should we expect the relative frequency to approach .50.

Now we toss the coin four times; suppose that four heads occur. This would cast only slight doubt upon the fairness of the coin, since it is relatively likely that the relative frequency of heads will depart widely from .50 on such a small number of trials. However, if after 500 tosses, the number of heads were 500, we would have very serious doubts about the fairness of the coin; although this *could* happen, it is far from what we *expect*. Nevertheless, we still could not be sure. If we tossed the coin 500 million times, still with 100 percent heads, we could be practically certain that the probability of a head for this coin is not .50—*practically* certain, that is, but still not *completely* certain. Only if we tossed the coin an infinite number of times would we be absolutely certain that the coin was not fair, given that the long-run relative frequency differed from .50 in any way.

This hypothetical example exhibits one of the basic features of statistical inference as applied to many real problems. We start out with a theoretical state of affairs that can be represented by a statement of the probability for one or more events. We wish to use empirical evidence to check the truth of that theoretical situation. Therefore, we conduct an experiment in which the event or events in question are possible outcomes, and we take some n observations at random. The theoretical probabilities tell us what to expect with regard to the relative frequencies of the events in question, and we compare the obtained relative frequencies against those expected. To the extent that the obtained relative frequencies of events depart widely from the expected, we have some evidence that the theory is not true. Moreover, given any degree of departure of the observed from the expected relative frequencies, we are more certain that the theory is not true as the sample size increases. However, *we can never be completely sure in our judgment that the theory is false unless we have made an infinite number of observations.*

By the same token, we can never be completely sure that the theory is true. An extremely biased coin (short of a coin which must come up heads with probability 1 or tails with probability 1) *can* give exactly 50 percent heads on any set of n trials. Even though the evidence appears to agree well with the theory, this in no way implies that the theory must be true. Only after an infinite number of trials, when the relative frequency of occurrence of an event matches the theoretical probability exactly, can we assert with complete confidence that the theory is true. Increased numbers of trials lend confidence to our judgments about the true state of affairs, but we can always be wrong, short of an infinite number of observations. A point of interest is that although the "theory" in this example was given as "the probability of heads is .50," it is doubtful that anyone would believe that the probability is *exactly* .50. We are actually interested in whether the probability is *near* .50; by "near" we may mean, for example, that the probability is between .49 and .51, or between .495 and .505. This point is of great importance in the theory of statistical inference and decision, and we will discuss it at length in Chapters 7–9.

The prnciple embodied in the law of large numbers can be used not only to compare empirical results with theoretical probabilities, but also to let us *estimate* true probabilities of events by observing their relative frequencies over some limited number of trials. For example, suppose that we are interested in the political preferences of voters in a particular city with regard to a particular election with just two candidates. We conduct the simple experiment of randomly choosing individuals from the list of voters and asking them whether they prefer candidate X, prefer candidate Y, or have no preference between the two candidates. Before the survey, of course, we do not know exactly how the voters will respond. Now suppose that the first voter that is interviewed states a preference for candidate Y. The observed relative frequencies are .00, 1.00, and .00 for "candidate X," "candidate Y," and "no preference." We then interview a second voter, who prefers candidate X, so the relative frequencies are changed to .50, .50, and .00, respectively. If we keep on randomly selecting voters and interviewing them, we might begin to get a reasonably clear picture of the probabilities for the three possible responses to our question. After 100 trials, for example, the relative frequencies might be .26 for X, .39 for Y, and .35 for no preference. By the time we make 10,000 observations, we can be even more confident that the probabilities are close to the relative frequencies. The larger the number of observations, the less we expect to "miss" in our estimates of the probabilities of the three possible responses.

The statistician making observations in real-world situations is doing something quite similar to drawing marbles blindfolded from a large box containing many marbles of different colors. He cannot see into the box and observe all such phenomena about which he wishes to generalize, but he can observe samples, drawn from among all such elementary events, and generalize from what is actually observed. The generalization is much like a bet, and how good the bet is depends to a great extent on how many sample observations are made. The statistician can never be completely sure that his generalization is the correct one, but the chance of being wrong can be reduced by making a sufficient number of observations.

2.15   THE SUBJECTIVE INTERPRETATION OF PROBABILITY

Relative frequency in the sample space and relative frequency in the long run are but interpretations that can be given to the notion of probability. It is important to remember that these are *interpretations* of the abstract model and that there is no universally agreed upon interpretation. The model per se is a system of relations among numbers that happen to be called probabilities and rules for calculating with these numbers.

The probability concept acquired interpretations in terms of relative frequency because it was originally developed to describe certain games of chance where plays (such as spinning a roulette wheel or tossing dice or dealing cards) are indeed repeated for a large number of trials and where it is reasonable to assume that the elementary events of interest are equally likely. Similarly, there are numerous situations in which statisticians make many observations under essentially the same conditions, and the mathematical theory of probability can be given a relative-frequency interpretation in these situations as well. For instance, a quality-control statistician may observe thousands of items produced by a certain production process, and the weight of each item or simply whether each item is defective or nondefective may be recorded. An actuarial statistician may observe the records of millions of persons, recording for each the number of claims with regard to health insurance or automobile insurance. A medical statistician may observe thousands of persons with a certain disease, recording for each the drug or drugs used and whether or not the person is cured of the disease.

On the other hand, there are many events that can be thought about in a probabilistic sense but that cannot have a probability in terms of the long-run relative-frequency interpretation. Indeed, many everyday uses of the probability concept do not have this relative-frequency connotation at all. You say, "It will probably rain tomorrow," or "The Packers will probably not win the the championship," or "I am unlikely to pass this test." Extending this notion, you might say, "The chances are two in three that it will rain tomorrow," or "The odds are nine to one that the Packers will not win the championship," or "I only have about one chance in 10 of passing this test." These statements appear to be probability statements, and their meanings should be clear to most listeners; but it is very difficult to see how they could describe long-run relative frequencies of outcomes of simple experiments repeated over and over again. The problem is that these events are unique (that is, the situations cannot be duplicated). Although some information may be available regarding past occurrences in *similar* situations, no information in the form of observed frequencies is available regarding repeated trials under *identical* conditions. You may have some information available regarding the past occurrence of rain in a certain location, at a certain time of year, under certain atmospheric conditions (for example, temperature, wind velocity and direction, and humidity), and so on, but it is doubtful that any of this information represents situations exactly identical to the current situation. Similarly, you may have past information about the performance of a football team such as the Packers; but each year new players enter the league, injuries occur, key players retire, and so on. Surely no previous season was played under conditions identical to those of the current season. The same idea holds for the statement involving the passing of a test. Each of the three probability statements given above describes the speaker's degree of belief about a situation that will occur once and once only. In each case it will not be possible to observe repetitive trials of the uncertain situation, so the probability statements cannot be explained in terms of the long-run frequency interpretation of probability.

An interpretation of probability that enables us to explain the probability statements in the preceeding paragraph is the **subjective,** or **personal, interpretation.**

In the subjective interpretation of probability, a probability is interpreted as a measure of degree of belief, or as the quantified judgment of a particular individual.

The probability statements given above thus represent the judgments of the individual making the statement. Since a probability, in this interpretation, is a measure of a degree of belief rather than a long-run relative frequency, it is perfectly reasonable to assign a probability to an event that involves a nonrepetitive situation. As a result, we can think of a probability as representing the individual's judgment concerning what will happen in a single trial of the uncertain situation in question rather than a statement about what will happen in the long run.

It is not necessary, of course, for an experiment to be nonrepetitive for the subjective interpretation of probability to be applicable. Consider once more the four examples that were discussed from a long-run frequency viewpoint in Section 2.12: tossing a coin, drawing marbles from a box, throwing a die, and dealing a poker hand. In each of these examples, the probability of any particular event could be interpreted as a degree of belief, although the experiments are all capable of being repeated. For instance, an individual who is willing to accept certain judgmental assumptions (the die is "fair," the method of tossing the die does not favor any side or sides of the die, and so on) would feel that the probability of the die coming up odd *on a particular throw* is 1/2. Notice that this is a probability statement about a particular throw, not about a long sequence of throws. The point is simply that the subjective interpretation of probability makes sense whether the experiments in question are repetitive or nonrepetitive. Furthermore, the subjective interpretation is operationally more useful than the long-run frequency interpretation because it allows a person to consider individual situations instead of appealing to the "long-run" and to the concept of statistical regularity.

In some respects, subjective probability can be thought of as an *extension* of the relative frequency interpretations of probability. As we have seen, the justification of long-run frequencies is based on certain *assumptions* that are necessary for the proof of the law of large numbers. One important assumption is that the trials comprising the "long run" are independent. As noted in Section 2.13, this essentially means that the outcome on any one trial will in no way affect the outcome on any other trial. Another assumption is that the trials are conducted under identical conditions. Similarly, if probability is interpreted in terms of relative frequencies in the sample space, using Equation (2.5.1), then it must be assumed that the elementary events of interest are equally likely. Although techniques are available for statistically investigating assumptions such as these, it should be pointed out that the decision as to whether the assumptions seem reasonable in any given situation is ultimately a subjective decision. Thus, there is an element of subjectivity in the relative-frequency interpretations of probability. If a person feels that the assumptions are reasonable, it is perfectly acceptable for him to make his subjective probability for a given event equal to the probability determined by the relative-frequency approach, whether this latter probability is based on symmetry arguments or on actual observed relative frequencies. If he does not feel that the assumptions are reasonable or if he has some other information (other than relative frequencies) about the event in question, his subjective probability may differ from the relative-frequency probability. In this respect (and in the respect that the subjective interpretation allows us to make probability statements about nonrepetitive events), subjective probability can be thought of as an *extension* of relative-frequency probability.

For an example involving the assumption of equally likely elementary events, consider a 6-sided die that is not perfectly symmetric. After examining the asymmetries of the die, we might feel that the 6 sides are equally likely to appear face up if the die is tossed, in which case Equation (2.5.1) is applicable for determining probabilities of various events such as "an even number comes up." On the other hand, the asymmetry of the die may lead us to believe that the side marked "1" is considerably more likely to occur than any other side, in which case the elementary events should not be considered to be equally likely and Equation (2.5.1) would not be applicable. In either case, the assignment of probabilities to the events of interest is subjective.

For an example involving observed frequencies, suppose that we are interested in the probability that a given person will be involved in an automobile accident (as the driver of a car, not as a passenger) during the next year. We know from recent data that among drivers in the United States of the same age as the person of interest, the frequency of drivers involved in one or more accidents in a year is 12/100. If we feel that the given person is representative of drivers in the same age group (and this is a subjective judgment on our part), we might be willing to assign the probability .12 to the event that the person will be involved in an accident in the next year. On the other hand, we might feel that the person is a somewhat reckless driver, so that the probability should be, say, .20; or we might want to look up the historical frequencies for drivers living in the same state as this individual (as opposed to all drivers in the United States) or for drivers of the same type of car, and so on. Any or all of these things might be the basis for arriving at the probability in question. The point is that symmetry assumptions or observed relative frequencies may be quite useful in the assessment of probabilities, but the ultimate assessment is subjective; thus, the subjective interpretation of probability can be thought of as an extension of the frequency interpretation. Furthermore, the subjective interpretation is applicable in *any* uncertain situation, whether or not the situation is repeatable. Because many uncertain situations of interest in problems of statistical inference and decision are unique, nonrepetitive situations, the subjective interpretation is extremely useful.

2.16   PROBABILITIES, LOTTERIES, AND BETTING ODDS

Because they represent degrees of belief, subjective probabilities may vary from person to person for the same event. Different individuals may have different degrees of belief, or judgments. Such differences may be due to differences in background, knowledge, and/or information about the event in question. For example, you might state that the probability of rain tomorrow is 1/10, and the meteorologist at the weather bureau might state that the probability of rain tomorrow is 1/2. According to the subjective interpretation of probability, both of these values can properly be considered as probabilities and subjected to the usual mathematical treatment. Such a difference in probabilities is permissible, provided that you and the meteorologist both feel that your respective probabilities accurately reflect your respective judgments concerning the chance of rain tomorrow. In addition, the quantification of these judgments is not as arbitrary as it might first appear. It is true that a person may be somewhat vague when attempting to convert judgments into probabilities, and the subjective interpretation of probability is sometimes criticized on the grounds that the inclusion of subjective judgment results in probabilities that are not "objective," but are in some respects arbitrary. For instance, you may feel that the probability of rain tomorrow is between .10 and .20, but you find it difficult to come up with a single number representing your probability. Even so, there are "objective" procedures, such as lotteries and betting odds, that can be extremely helpful in reducing vagueness and in accurately quantifying judgments.

One possible way of formally defining subjective probability is in terms of **lotteries.** For instance, suppose that you are offered a choice between Lottery I and Lottery II.

> *Lottery I*:   You win $100 with probability 1/2.
> You win $0 with probability 1/2.
>
> *Lottery II*:   You win $100 if it rains tomorrow.
> You win $0 if it does not rain tomorrow.

It is assumed that since the "prize" is the same in both lotteries ($100), you should prefer the lottery that gives you the greater chance of winning the prize. Thus, if you choose Lottery II, then you must feel that the probability of rain tomorrow is greater than 1/2; if you choose Lottery I, then you must feel that this probability is less than 1/2; if you are indifferent between the two lotteries, then you must feel that the probability of rain tomorrow is equal to 1/2. Suppose that you prefer Lottery I. Consider the same lotteries, except that the probabilities in Lottery I are changed to 1/4 and 3/4. If you still prefer Lottery I, implying that you feel that you have a greater chance of winning in Lottery I than in Lottery II, then your subjective probability of rain is less than 1/4. Presumably you could keep changing the probabilities in Lottery I until you are just indifferent between Lotteries I and II; if this happens when the probabilities are .1 and .9, then your subjective probability of rain is .1. In a similar manner, you can assess your probability for any event.

A formal definition of subjective probability can be given in terms of lotteries as follows. Your subjective probability $P(A)$ of an event A is the number $P(A)$ that makes you indifferent between the following two lotteries.

> *Lottery I*:   You obtain X with probability $P(A)$.
> You obtain Y with probability $1 - P(A)$.

> *Lottery II*:   You obtain X if A occurs.
> You obtain Y if A does not occur.

Here X and Y are two "prizes." The only restriction on X and Y is that one must be preferred to the other; if you are indifferent between X and Y, then you are indifferent between the two lotteries regardless of the choice of $P(A)$. In the above example, $X = \$100$, $Y = \$0$, and $A =$ "rain tomorrow."

In order to use the lottery procedure operationally to determine a person's subjective probability for an event A, it is necessary to specify how the outcome in Lottery I is determined. In the first illustration of lotteries given above, suppose that you choose Lottery I; do you win $100 or $0? Probabilities such as those in Lottery I are called **canonical probabilities;** they are probabilities that everyone would agree upon. For instance, if the probabilities are $1/2$ and $1/2$, then the outcome can be decided by the selection of a ball from an urn containing 50 red balls and 50 green balls. If the balls are all of the same size and if they are thoroughly mixed before the drawing, then the chance of drawing a red ball is the same as the chance of drawing a green ball. If the probabilities are $1/10$ and $9/10$ in Lottery I, then an urn with 10 red balls and 90 green balls could be used. To assess your subjective probability of rain tomorrow to the nearest .01, consider the following two lotteries.

> *Lottery I* :   You win $100 if a red ball is drawn from an urn containing R red balls and $100 - R$ green balls.
> You win $0 if a green ball is drawn.

> *Lottery II*:   You win $100 if it rains tomorrow.
> You win $0 if it does not rain tomorrow.

If you can determine a number R that makes you indifferent between Lottery I and Lottery II, then your subjective probability of rain tomorrow is $R/100$. For example, you may feel that if there are fewer than 16 red balls in the urn, you prefer Lottery II; if there are more than 16 red balls, you prefer Lottery I; and if there are exactly 16 red balls (and hence 84 green balls), you are indifferent between the two lotteries. Then your subjective probability of rain tomorrow must be .16.

It is not necessary to use the idea of balls in an urn to determine canonical probabilities. If you feel unsure about the reliability of such a device (perhaps because such devices have been used without proper mixing of balls and have produced decidedly nonrandom results in military draft lotteries, for instance), other devices are available. The most convenient (and perhaps the most reliable) such device is the use of a table of random digits. Such tables are constructed so that at any point in the table, each digit is equally likely to appear (other details, such as independence and equal probabilities for the various possible combinations of digits at successive points in the table, are also required of a table of random digits). For Lottery I, a 2-digit number could be chosen from a table of random digits. If the number is between 00 and 15, you win $100; and if the number is between 16 and 99, you win $0. This is equivalent to the urn with 16 red balls and 84 green balls. The point is that the probabilities in Lottery I must be reasonable to an individual attempting to assess subjective probabilities; thus, canonical probabilities must represent a process about which different persons would agree. Note, by the way, that to assess probabilities to the nearest .001 instead of just to the nearest .01, we could consider an urn with 1000 balls or a 3-digit random number. Of course, it is possible to go only so far in combating vagueness. For some events, it may even be difficult to obtain a subjective probability to the nearest .01. In other words, you may find it difficult to determine a single probability that makes you indifferent between two lotteries. The consideration of lotteries, whether real or hypothetical, should help to reduce vagueness, but it is generally not possible to eliminate it entirely.

Another procedure that is useful in the assessment of subjective probabilities involves **betting odds.** A probability may be thought of in terms of the odds at which a person would be willing to bet. For example, if you say, "The chances are one in 10 that Green Bay will win the championship of the National Football League," this implies that

$$P(\text{Green Bay wins championship}) = 1/10$$

and

$$P(\text{Green Bay does not win championship}) = 9/10.$$

Thus, Green Bay is 9 times as likely to lose the championship as it is to win the championship, in your opinion. This means that you think that 9 to 1 odds against Green Bay winning would be fair odds. The formal relationship between probabilities and odds is as follows.

If the probability of an event is equal to p, then the odds in favor of that event are p to $(1 - p)$.

If the chances are 2 in 3 that it will rain tomorrow, the probability of rain is 2/3 and the odds in favor of rain are 2/3 to 1/3, or 2 to 1. If you have only about 1 chance in 10 of passing a particular test, the probability of your passing the test is 1/10 and the odds in favor of your passing the test are 1/10 to 9/10, or 1 to 9.

It is also possible to convert odds into probabilities.

If the odds in favor of an event are a to b, then the probability of that event is equal to $a/(a + b)$.

If the odds are even that the Dow–Jones Average will rise tomorrow, the odds are a to a, so that the probability that the Dow–Jones Average will rise tomorrow is $a/(a + a)$, or 1/2. If you know the probability of an event, you can calculate the odds in favor of the event, and vice versa. This is important with regard to the assessment of an individual's subjective probabilities, for some persons find it easier and more convenient to think in terms of odds rather than in terms of probability.

It is possible to go one step further and consider actual or hypothetical bets. Suppose that two persons, A and B, make the following bet.

A bets \$3 that the price of IBM common stock goes up tomorrow.

B bets \$2 that the price of IBM common stock does not go up tomorrow.

If you think that this is a "fair" bet (that is, if you are indifferent between A's side of the bet and B's side of the bet), then you apparently feel that 3-to-2 odds in favor of IBM stock going up tomorrow are fair odds. This implies that your probability that IBM stock goes up tomorrow is $3/(3 + 2)$, or .60. Although it is not possible at this point to show formally why this true, a brief intuitive explanation can be given. If you take A's side of the bet, you feel that you have a probability of .60 of winning \$2 (the amount your opponent, B, is wagering), and you have a probability of .40 of losing \$3 (the amount you, as A, are wagering). But $(.60)(\$2)$ is equal to $(.40)(\$3)$, so the amount you "expect" to gain from the bet is \$0. It turns out that if you take B's side of the bet, the amount you "expect" to gain is still \$0. As a result, you are indifferent between the two sides of the bet, and thus you consider the bet to be fair. If your probability that IBM stock goes up tomorrow is greater than .60, the bet is no longer fair in your estimation; you prefer A's side of the bet. On the other hand, if your probability is less then .60, then you prefer B's side of the bet. You only feel that the bet is fair if your probability that IBM stock goes up tomorrow is .60. If subjective probabilities are thought of in terms of betting odds and the corresponding actual or hypothetical bets, then it can be seen that they are not really arbitrary, even though they may differ from person to person.

If the amounts involved in the bet are large or if the odds are very high in favor of one of the events, then a person's attitude toward risk may become a crucial factor in the choice of one side of the bet. This problem will be discussed in Chapter 9 when the notion of utility is considered.

It should be noted that the use of betting situations and lotteries to assess probabilities depends on certain reasonable "axioms of coherence," or "axioms of consistency." It can be shown that if a person behaves in accordance with these axioms, then that person's assessed probabilities must satisfy the mathematical definition of probability given in Section 2.4. For example, it is impossible for a person who obeys the axioms of coherence to assess the probability of rain tomorrow as 1/2 and the probability of no rain tomorrow as 3/4, because these two probabilities must sum to 1. Thus, the axioms of coherence justify the application of the usual mathematical manipulations of probabilities to subjective probabilities. We will not present all of the axioms of coherence, but we will give two examples of such axioms. The axiom of *transitivity* states that preferences are transitive. That is, if you prefer Lottery I to Lottery II and Lottery II to Lottery III, then you must prefer Lottery I to Lottery III. This seems intuitively reasonable. The axiom of *substitutability* states that if you are indifferent between X and Y, then X can be substituted for Y as a prize in a lottery or as a stake in a bet without changing your preferences with regard to the bets or lotteries in question. For example, suppose that you are indifferent between the following two lotteries.

Lottery I: You win $2 for certain.

Lottery II: You win a concert ticket with probability 1/2.
You win $0 with probability 1/2.

Then you should be willing to substitute Lottery II for the $2 stake in the bet considered earlier regarding the price of IBM common stock. Moreover, this substitution should not affect your indifference between the two sides of the bet.

Devices such as lotteries and betting situations have the common property of resulting in statements that can be converted into probabilities. They represent the "behavioral" approach to the quantification of judgment; that is, they force the person who is attempting to determine subjective probabilities to state what *action* would be taken in a particular decision-making situation rather than just to state directly what the probability is. The probabilities can then be inferred from the behavior (that is, from the decisions). In this regard, the subjective interpretation of probability should have great applicability in decision-making situations. The subjective interpretation is also valuable in making decisions because it allows probability statements for nonrepetitive situations as well as for repetitive situations. Important decisions often depend on the outcomes of nonrepetitive experiments. For example, an investment decision may depend on the probability that Congress will raise taxes or the probability that the stock market will go up in the next year. A decision concerning a price increase for a certain product may depend on the probability that other firms producing the product will refuse to go along with the increase. In the next section, some simple examples of the use of probabilities in decision making are presented, examples in which the subjective interpretation of probability is most useful.

2.17   PROBABILITY AND DECISION MAKING

In real-world problems, one is often faced with the task of making inferences or decisions under uncertainty. As you have seen, the uncertainty about any event may be represented quantitatively in the form of a probability, and some procedures for arriving at such probabilities have been discussed. One advantage of expressing uncertainty in terms of probability is that it makes it much easier to communicate the nature of the uncertainty. Instead of presenting long verbal arguments about whether the price of IBM common stock will or will not go up in the coming month, it is simpler to summarize your judgments by saying, "I think that the probability that the price of IBM common stock will go up in the coming month is .65." In business decision making, an important decision may involve uncertainty about a large number of events and may be of concern to quite a few people. For instance, some of the events of interest may involve the stock market, in which case a financial analyst should be consulted; some of the events may involve pending legislation, in which case a person familiar with the political scene should be consulted; some of the events may involve potential sales of a certain product, in which case the sales and advertising managers should be consulted; and so on. The point is that a large-scale problem may involve many uncertainties, and numerous experts may be consulted regarding these uncertainties. The use of probabilities to represent the uncertainties provides a convenient and useful medium for the communication of judgments. Furthermore, in order to apply the formal decision theory that is presented in Chapter 9, it is necessary to express uncertainty in terms of probability. To illustrate this informally, two very simple decision-making problems are presented in this section.

First, consider a decision-making problem involving the drilling of an oil well. The decision maker must decide whether or not to drill for oil at a particular location. To simplify the problem, assume that there are two possible events, "oil" and "no oil" (in an actual drilling situation, one might obtain varying amounts of oil). If the decision is to drill and the event "oil" occurs, the oil obtained will be worth $130,000. The cost of drilling is $30,000, so the net payoff obtained by drilling and striking oil is

$100,000. If the decision is not to drill, the decision maker will exercise an option to sell the drilling rights to an oil wildcatter for a fixed sum of $10,000 plus another $10,000 contingent upon the presence of oil. The consequences of the oil-drilling decision, which depend on whether or not the decision maker drills and whether or not oil is present, can be expressed in a payoff table, as in Table 2.17.1.

Table 2.17.1   Payoff Table for Oil-Drilling Example

| | | EVENT | |
		Oil	*No oil*
	Drill	$100,000	−$30,000
DECISION			
	Do not drill	$20,000	$10,000

The payoffs in Table 2.17.1 are important factors in the decision. Another important factor is the chance of striking oil, so another input to the problem involves the decision maker's uncertainty about the presence of oil. Although the decision maker may have some experience with similar drilling opportunities, this is a unique situation, not a situation involving repetitive trials. To represent the uncertainty, the decision maker should determine his or her subjective probability that there is oil at the particular site in question. Suppose that a geologist is hired to assess the probability of oil. On the basis of similar oil-drilling ventures and geological information regarding the site, the geologist assesses the probability of oil to be 3/10. That is, the odds against oil are 7 to 3.

Formal decision theory prescribes certain criteria for making decisions based on inputs regarding the possible consequences, or payoffs, and inputs regarding the probabilities for the various possible events. These criteria are discussed in Chapter 9. At this point, however, an informal, intuitive discussion of how the decision maker might make a decision in this example can be offered. In drilling, the decision maker will gain $100,000 with probability .30 and lose $30,000 with probability .70. To evaluate this situation, the concept of mathematical expectation, which will be discussed in Chapter 3, is used to compute the "expected payoff" of drilling. Obtaining a payoff of $100,000 with probability .30 and a payoff of −$30,000 with probability .70 corresponds to an "expected payoff" of ($100,000)(.30) + (−$30,000)(.70), or $9,000. For the other possible action, not drilling, obtaining a payoff of $20,000 with probability .30 and a payoff of $10,000 with probability .70 corresponds to an "expected payoff" of $20,000(.30) + $10,000(.70), or $13,000. Since the expected payoff is higher for not drilling than for drilling, the decision should be not to drill.

To demonstrate how the decision depends on the inputs, note that if P(oil) were .40 instead of .30, the best decision would be to go ahead and drill.

For another example, consider the decision by a firm as to whether or not to initiate a special advertising campaign for a certain product. The firm feels that a competitor might introduce a competing product and that if such a new product is introduced, sales of the firm's own product would greatly decrease unless an advertising campaign for the product was in progress. The three actions under consideration are "no advertising," "minor advertising campaign," and "major advertising campaign." Taking into account the cost of advertising and its anticipated effect on sales, the decision maker determines the payoff table given in Table 2.17.2, where the numbers represent net profits, or net payoffs, to the firm. On the basis of currently available information concerning the actions of the competing firm, the decision maker decides that the probability is .60 that the new product will be introduced. The "expected payoffs" for the three actions are $340,000 for "no advertising," $420,000 for "minor advertising campaign," and $440,000 for "major advertising campaign." Thus, the decision should be to launch a major advertising campaign.

Table 2.17.2   Payoff Table for Advertising Example

	EVENT	
	Competitor introduces new product	*Competitor does not introduce new product*
No advertising	100,000	700,000
DECISION *Minor ad campaign*	300,000	600,000
Major ad campaign	400,000	500,000

It should be stressed that in the decision-making examples in this section, the possible effects of attitude toward risk have been ignored. The amounts involved are large enough in the second and third examples to cast serious doubt on the wisdom of ignoring other influences on decision making. The purpose of the examples, however, is to demonstrate the importance of probabilities in decision-making problems, and complicating the examples by bringing in other considerations would not further this purpose. Such considerations are explored in Chapter 9.

The oil-drilling example and the advertising example demonstrate the use of probabilities in decision making. In actual decision-making situations such as these, the situations of interest are often nonrepetitive. For additional examples, consider decisions regarding a bet on a football game, the purchase of some common stock, the creation of a small business firm, or the medical treatment of someone who is ill. In each of these cases, it is not possible to observe repeated trials of the situation at hand, although it may be possible to observe trials involving similar situations, such as the outcomes of previous football games, the past performance of stocks, the success of other business ventures, or the reaction of other patients with similar symptoms to various medical treatments. Because of the nonrepetitive nature of the situations, the probabilities used as inputs to the decision-making procedure must be subjective probabilities.

Of course, not *all* decision-making situations involve nonrepetitive events. Nevertheless, the subjective interpretation of probability can be thought of as an extension of the relative frequency interpretation, as noted in Section 2.3, and any information regarding observed relative frequencies of events can be considered in the determination of subjective probabilities. Often the decision maker has both information regarding observed frequencies *and* other information of a subjective nature. Later in this chapter and in Chapters 8 and 9, the possibility of revising probabilities on the basis of additional information will be discussed. In this way, it is possible to utilize formally all of the available information that is relevant to the decision-making situation.

2.18 CONDITIONAL PROBABILITY

An important concept in the theory of probability is that of **conditional probability.** Often we are interested in the probability that one event will occur, given that a particular second event has occurred or will definitely occur. For instance, you might be interested in the probability that a student will earn at least a "B" average at a particular university, given that the student is a male. You might be interested in the probability that the price of a certain stock will go up, given that taxes remain the same. You might be interested in the probability that sales of a certain firm's product will go down, given that a rival firm introduces a competing product. In each of these cases you are interested in the *conditional probability* of one event, given the occurrence of a second event. The conditional probability of event A, given event B, is denoted by $P(A \mid B)$. For instance, in the case of the firm interested in the effect on sales of the introduction of a competing product, we might have

$$P(\text{sales go down} \mid \text{competing product introduced}) = .40$$

and $P(\text{sales go down} \mid \text{competing product not introduced}) = .10.$

That is, if the competing product is introduced, the probability is .40 that sales will go down, whereas if the competing product is not introduced, the probability is only .10 that sales will go down.

A formal definition of conditional probability is as follows.

Let A and B be events in a sample space made up of a finite number of elementary events. Then the conditional probability of A given B, denoted by $P(A \mid B)$, is

$$P(A \mid B) = \frac{P(A \cap B)}{P(B)}, \qquad (2.18.1^*)$$

provided that $P(B)$ is not zero.

The conditional probability symbol $P(A \mid B)$ is read as "the probability of A *given* B." Note that it is necessary that $P(B) > 0$; it makes no sense to speak of the probability of A given B if B is an impossible event.

From Equation (2.18.1), the conditional probability of A given B is equal to the probability of the intersection of events A and B divided by the probability of the given event B. The probability of an intersection of events is sometimes called a **joint probability,** and a joint probability such as $P(A \cap B)$ is often written in the slightly shorter form $P(A, B)$. Whether $P(A \cap B)$ is called the probability of the intersection of A and B or whether it called the joint probability of A and B, and however it is written, it has the same meaning: the probability that events A and B *both* occur. The conditional probability $P(A \mid B)$, then, is the probability that A and B *both* occur, divided by the probability that B occurs.

As a simple example, suppose that 60 percent of the students at a particular university are male and that 10 percent of the students are male *and* have at least a "B" average. The probability that a student chosen randomly from the student body has a "B" average, given that the student is a male, is therefore .10/.60, or 1/6. For another example, if a fair die is tossed,

$$P(\text{even number comes up} \mid 6 \text{ does not come up}) = \frac{P(\text{even, not } 6)}{P(\text{not } 6)}$$

$$= \frac{P(2 \text{ or } 4)}{P(\text{not } 6)} = \frac{2/6}{5/6} = \frac{2}{5}.$$

Notice that $P(A \mid B)$ and $P(B \mid A)$ represent completely different probabilities:

$$P(A \mid B) = \frac{P(A \cap B)}{P(B)}, \qquad \text{whereas} \qquad P(B \mid A) = \frac{P(A \cap B)}{P(A)}.$$

For instance, in the die example,

$$P(6 \text{ does not come up} \mid \text{even number comes up}) = \frac{P(\text{even, not } 6)}{P(\text{even})}$$

$$= \frac{P(2 \text{ or } 4)}{P(2, 4, \text{ or } 6)}$$

$$= \frac{2/6}{3/6} = \frac{2}{3},$$

which is considerably different from $P(\text{even} \mid \text{not } 6) = 2/5$.

If the only information available concerning the events of interest is in the form of relative frequencies, the concept of conditional probability can be considered in terms of these relative frequencies. Suppose that an experiment is repeated many times under identical conditions, and either (or both) of the events A and B can occur on any single trial. To investigate the probability $P(A \mid B)$, we could consider the relative frequency of occurrence of A, not in the entire sequence of trials, but only in those trials on which B occurs. For instance, suppose that the records of an insurance company indicate that of the one million persons holding health insurance policies with the company, 250,000 were over 50 years old at the beginning of the past year. Of these 250,000 policyholders, 100,000 submitted at least one claim during the year. Thus, among policyholders over 50, the relative frequency of claims was 100,000/250,000, or .4. If you feel that this relative frequency satisfactorily represents all of your information, then for a given period of one year,

$$P(\text{submit at least one claim} \mid \text{over 50 years old}) = .4.$$

Note that the denominator in the relative frequency is 250,000 (the number of policyholders over 50 years old) rather than one million (the total number of policyholders).

In terms of Equation (2.18.1), the conditional relative frequency must equal the relative frequency of the joint event divided by the relative frequency of the given event. In the insurance example presented above, the relative frequency of the joint event "over 50 years old and at least one claim" is 100,000/1,000,000, or .10, and the relative frequency of the given event "over 50 years old" is 250,000/1,000,000, or .25. This yields a conditional relative frequency of .10/.25 = .40, which agrees with the conditional relative frequency determined in the preceding paragraph.

It is also possible to develop the concept of conditional probability by using the subjective interpretation of probability. For example, suppose that you are offered a choice between Lottery I and Lottery II:

Lottery I: If it rains today, you win \$100 with probability 1/2 and you
             win \$0 with probability 1/2.
             If it does not rain today, you win \$25.

Lottery II: If it rains today, you win \$100 if it rains tomorrow and you
             win \$0 if it does not rain tomorrow.
             If it does not rain today, you win \$25.

If it does not rain today, then Lottery I and Lottery II give you the same prize, \$25. If it does rain today, however, then you must choose between winning \$100 with a fixed probability of 1/2 and winning \$100 if it rains tomorrow. Since this portion of each lottery only applies if it rains today, "rain today" is a *given* condition. Thus, if you choose Lottery I, you must feel that $P(\text{rain tomorrow} \mid \text{rain today})$ is greater than 1/2; if you choose Lottery II, then you must feel that this conditional probability is less than 1/2; and if you are indifferent between the two lotteries, then you must feel that the probability is equal to 1/2. By changing the canonical probabilities in Lottery I until you are just indifferent between the two lotteries, you can determine your conditional probability of rain tomorrow, given rain today.

It is also possible to consider conditional probability in terms of conditional bets. Suppose that two persons, A and B, make the following bet:

A bets \$5 that the San Francisco Giants baseball team wins over half of its games in the coming season.

B bets \$2 that the team does not win over half of its games, with the proviso that the bet is called off if any of a stipulated list of key Giant players is injured during the season.

This bet is a conditional bet which is called off if any key players are injured. Thus, if you think that this is a "fair bet," then you apparently feel that 5-to-2 odds in favor of the team winning over half of its games are "fair" odds, given no injuries to key players. This implies that $P(\text{Giants win over half of their games} \mid \text{no injuries to key players})$ is equal to $5/(5 + 2) = 5/7$. As long as the axioms of coherence (mentioned in Section 2.16) are not violated, subjective conditional probabilities must obey Equation (2.18.1).

Another explanation of Equation (2.18.1) does not involve the interpretation of probability at all. The events A and B for any uncertain situation are contained in S, the sample space for that uncertain situation. Their probabilities can be illustrated in terms of a Venn diagram, as discussed in Section 2.4. In Figure 2.18.1, consider what happens if it is known that B has occurred or will occur. Any points of the sample space outside of

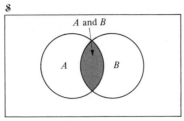

Figure 2.18.1 The Probability of the Intersection of Two Events

B are no longer relevant; in other words, the sample space has been reduced, and B now comprises the entire "new" sample space. The new sample space might be thought of as the sample space conditional upon the occurrence of B. Conditional upon B, then, A can occur only in conjunction with B; this means that A can occur only via the event "A *and* B" (the intersection of A and B). This event is represented in Figure 2.18.1 by the portion of A that intersects B. Thus, the conditional probability of A given B is represented on the Venn diagram by the area of the intersection divided by the entire area of B (this division is necessary because of the reduction of the sample space due to the "given" event B).

In this section we have discussed the concept of the conditional probability of one event given a second event. This notion can be extended to allow more "given" events. For example, we can speak of the probability of event A given events B and C. From the definition of conditional probability, then, since the occurrence of B and C is equivalent to the occurrence of $B \cap C$, we have

$$P(A \mid B, C) = P(A \mid B \cap C) = \frac{P(A \cap B \cap C)}{P(B \cap C)} . \qquad (2.18.2)$$

Here the new sample space is $B \cap C$, since we are given B and C. In general, the new (or reduced) sample space in conditional probability consists of the intersection of the events that are "given."

In the previous sections, probability was discussed with no mention of conditional probability. In a sense, however, *all probabilities are conditional*. That is, they may be conditional upon some assumptions, upon the details of an experiment, upon some action, or upon numerous similar factors. For example, if you say that the probability of rain is 1/3, you should write something like this:

$$P(\text{rain} \mid \text{current atmospheric conditions}) = 1/3.$$

When you say that the probability of getting a full house in a 5-card poker hand is .0014, you should write

$$P(\text{full house} \mid \text{standard, well-shuffled deck of 52 cards}) = .0014.$$

Similarly, considering the probability that a person is involved in an automobile accident in a one-year period, you might be interested in

P(accident | age of the person),

or P(accident | occupation of the person),

or P(accident | number of miles the person drives per year),

and so on.

Of course, as noted above, the notion of conditional probability is not restricted to the case in which there is only one "given" event. For example, consider the conditional probability

P(accident | age of the person, occupation of the person, and number
of miles the person drives per year),

which might be written in the form $P(A \mid B, C, D)$, the probability of event A, given events B, C, and D:

$$P(A \mid B, C, D) = \frac{P(A \cap B \cap C \cap D)}{P(B \cap C \cap D)} \,.$$

Because it is bothersome to write out long expressions such as the one in the preceding paragraph, the "given" conditions often are not written down in probability statements. Often these "given" conditions are fairly obvious, and it is assumed that they are understood by the reader. In some situations, however, particularly when there are competing hypotheses or sets of assumptions, probabilities are written in conditional form. At any rate, you should keep in mind that all probabilities are conditional upon some factors, even if, for the sake of convenience, these factors are not explicitly noted in the probability statement.

2.19 RELATIONSHIPS AMONG CONDITIONAL, JOINT, AND MARGINAL PROBABILITIES

In Equation (2.18.1), $P(A \mid B)$ represents a **conditional probability,** $P(A \cap B)$ represents a **joint probability,** and $P(B)$ represents a **marginal probability.** That is, the probability of one event given another is called a conditional probability, the probability of the intersection of two or more events is called a joint probability, and the probability of a single event such as B is sometimes called a marginal probability. In this section we shall discuss relationships among these types of probabilities and convenient aids, such as tree diagrams and tables, for working with such probabilities.

If we are interested in two events A and B and their complements $\bar{A}$ and

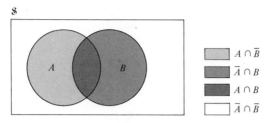

Figure 2.19.1   Joint Probabilities Involving Two Events and Their Complements

$\bar{B}$, there are four possible joint probabilities of interest: $P(A \cap B)$, $P(\bar{A} \cap B)$, $P(A \cap \bar{B})$, and $P(\bar{A} \cap \bar{B})$. Moreover, as can be seen from Figure 2.19.1, the sum of these four probabilities must be one, since they form a partition of the sample space. Other relationships, such as

$$P(A \cap B) + P(A \cap \bar{B}) = P(A), \tag{2.19.1}$$

$$P(\bar{A} \cap B) + P(\bar{A} \cap \bar{B}) = P(\bar{A}), \tag{2.19.2}$$

$$P(A \cap B) + P(\bar{A} \cap B) = P(B), \tag{2.19.3}$$

and
$$P(A \cap \bar{B}) + P(\bar{A} \cap \bar{B}) = P(\bar{B}) \tag{2.19.4}$$

can also be seen from Figure 2.19.1 and are not difficult to prove formally. These four equations express the marginal probabilities of A, $\bar{A}$, B, and $\bar{B}$ in terms of sums of joint probabilities.

For example, suppose that a credit manager randomly selects a small number of accounts from all of the charge accounts of a given department store. For each account that is chosen, the events A = "average outstanding balance during the past year was greater than \$100" and B = "current balance is past due" are of interest. Suppose that $P(A \cap B) = .06$, $P(A \cap \bar{B}) = .24$, $P(\bar{A} \cap B) = .04$, and $P(\bar{A} \cap \bar{B}) = .66$. Since an average outstanding balance greater than \$100 can occur either on an account that is currently past due or on an account that is *not* currently past due, $P(A) = .06 + .24 = .30$. Similarly, $P(B) = .06 + .04 = .10$. Thus, the probability of an account with an average outstanding balance over \$100 is .30, and the probability of an account that is currently past due is .10. These are marginal probabilities.

It is possible to use conditional probabilities in the computation of joint probabilities. Multiplying both sides of Equation (2.18.1) by $P(B)$, we get

$$P(A \cap B) = P(A \mid B)P(B). \tag{2.19.5*}$$

Similarly,

$$P(A \cap B) = P(B \mid A)P(A). \tag{2.19.6*}$$

Extending this to the intersection of three events, we will prove that

$$P(A \cap B \cap C) = P(C \mid A, B)P(B \mid A)P(A). \qquad (2.19.7)$$

To prove this, note from Equation (2.18.2) that

$$P(C \mid A, B) = \frac{P(A \cap B \cap C)}{P(A \cap B)}.$$

Thus,

$$P(C \mid A, B)P(B \mid A)P(A) = \left(\frac{P(A \cap B \cap C)}{P(A \cap B)}\right)\left(\frac{P(A \cap B)}{P(A)}\right)P(A).$$

Canceling terms, we get

$$P(C \mid A, B)P(B \mid A)P(A) = P(A \cap B \cap C),$$

which is what we set out to prove. In a similar manner, we can determine the probability of the intersection of any number of events.

It is sometimes convenient to represent conditional probability in terms of a **tree diagram.** For instance, consider the tree diagram in Figure 2.19.2, which concerns the occurrence or nonoccurrence of rain. Starting at the left side of the diagram, there is a "chance fork," which is a situation in which there are a number of possible outcomes, of which exactly one will occur. The "chance" indicates that there is uncertainty about which outcome will occur. At the first chance fork in this example, either B (rain today) or $\bar{B}$ (no rain today) will occur. In either case, there is a second chance fork at which either A (rain tomorrow) or $\bar{A}$ (no rain tomorrow) will occur. On the right-hand side of the diagram, the four possible events in this example are listed: rain today and tomorrow,

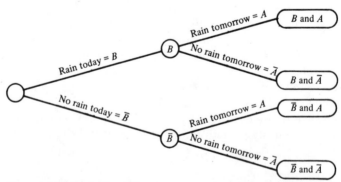

Figure 2.19.2   A Tree Diagram for the Occurrence or Nonoccurrence of Rain

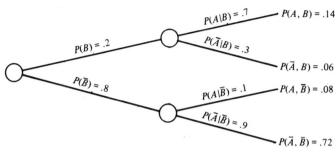

Figure 2.19.3 A Tree Diagram with Probabilities for the Rain Example

rain today but not tomorrow, rain tomorrow but not today, and no rain on either day.

The uncertainty at any chance fork of a tree diagram can be represented in terms of probabilities for the possible outcomes at that fork. At the first fork in the illustration, the probabilities of interest are P(rain today) and P(no rain today). If it rains today, the probabilities at the second fork are P(rain tomorrow | rain today) and P(no rain tomorrow | rain today). If it does not rain today, the probabilities at the second fork are P(rain tomorrow | no rain today) and P(no rain tomorrow | no rain today). Thus, at any given fork in the tree diagram, the probabilities are conditional upon the outcomes at previous forks.

From any of the four endpoints in the tree diagram, it is possible to work backward to see the sequence of occurrences at the chance forks. For instance, "rain today and rain tomorrow" is clearly the result of "rain today" followed by "rain tomorrow given rain today." As a result, the probability of the event "rain today *and* rain tomorrow" (this is a *joint probability*) is the product of P(rain today) and P(rain tomorrow | rain today). This an example of Equation (2.19.5).

For example, suppose that P(rain today) $= .2$, P(rain tomorrow | rain today) $= .7$, and P(rain tomorrow | no rain today) $= .1$. This situation is illustrated in terms of a tree diagram in Figure 2.19.3. An alternative way to present joint probabilities is in terms of a table, as in Table 2.19.1.

Table 2.19.1 Joint Probabilities in Tabular Form

	A	$\bar{A}$	
B	$P(A, B)$	$P(\bar{A}, B)$	$P(B)$
$\bar{B}$	$P(A, \bar{B})$	$P(\bar{A}, \bar{B})$	$P(\bar{B})$
	$P(A)$	$P(\bar{A})$	

Table 2.19.2   Joint Probabilities for Rain Example

	Rain tomorrow	No rain tomorrow	
Rain today	.14	.06	.20
No rain today	.08	.72	.80
	.22	.78	

For the rain example, the probabilities are given in Table 2.19.2. The entries in the body of the table are joint probabilities, whereas the entries on the margins of the table are the marginal probabilities, P(rain today) = .20, P(no rain today) = .80, P(rain tomorrow) = .22, and P(no rain tomorrow) = .78. Each individual probability is obtained by summing the values in the relevant row or column. For instance, since "rain tomorrow" can occur either in conjunction with "rain today" or with "no rain today," Equation (2.19.1) can be used:

P(rain tomorrow)

$= P$(rain today, rain tomorrow) $+ P$(no rain today, rain tomorrow)

$= .14 + .08 = .22.$

For the example presented earlier in this section regarding charge accounts, a tree diagram is presented in Figure 2.19.4. Note that since there is no time element here, as there was in the rain example, the events A and B can be presented in either order on a tree diagram. In Figure 2.19.4, we present B in the left-hand chance fork, then A. This example is also presented in tabular form in Table 2.19.3.

Table 2.19.3   Joint Probabilities for Charge Account Example

	Average balance over $100	Average balance not over $100	
Currently past due	.06	.04	.10
Not currently past due	.24	.66	.90
	.30	.70	

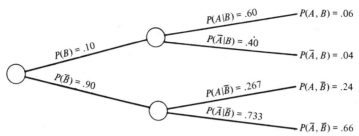

Figure 2.19.4 Tree Diagram for Charge Account Example

Tree diagrams and tables provide a convenient means of representing joint probabilities, conditional probabilities, and marginal probabilities. Moreover, they can be generalized to handle more than three events or events that form a K-fold partition on the sample space, with $K > 2$. In our examples, we have dealt with two events and with 2-fold partitions (for example, A and $\bar{A}$ form a 2-fold partition on the sample space, as do B and $\bar{B}$). If a third event, C, is of interest, a third set of branches can be added on the tree diagram to the right of the first two sets of branches. In the rain example, we might consider rain today, rain tomorrow, and rain two days from now. Or, instead of just A and $\bar{A}$, suppose that we have A_1, A_2, and A_3, where these three sets form a partition on the sample space. Then at the chance fork representing these events, there will be three branches instead of just two. In the rain example, we might consider A_1 = no rain tomorrow, A_2 = measurable rain totaling less than one inch tomorrow, and A_3 = measurable rain totaling at least one inch tomorrow.

2.20 BAYES' THEOREM

Sometimes it is not easy to compute conditional probabilities directly from the conditional probability formula given in Section 2.18. A convenient formula that gives the relationship among various conditional probabilities is **Bayes' theorem,** named after an English clergyman who did early work in probability theory.

The simplest version of Bayes' theorem can be stated as follows: for two events A and B,

$$P(A \mid B) = \frac{P(B \mid A)P(A)}{P(B \mid A)P(A) + P(B \mid \bar{A})P(\bar{A})} . \qquad (2.20.1^*)$$

It is not difficult to derive Bayes' theorem from the conditional probability formulas in the previous sections. From Equation (2.18.1),

$$P(A \mid B) = \frac{P(A, B)}{P(B)}.$$

Similarly,

$$P(B \mid A) = \frac{P(A, B)}{P(A)} \quad \text{and} \quad P(B \mid \bar{A}) = \frac{P(\bar{A}, B)}{P(\bar{A})},$$

so that

$$P(A, B) = P(B \mid A)P(A)$$

and

$$P(\bar{A}, B) = P(B \mid \bar{A})P(\bar{A}).$$

From Equation (2.19.3),

$$P(B) = P(A, B) + P(\bar{A}, B).$$

Thus, using the above formulas for $P(A, B)$ and $P(\bar{A}, B)$, the marginal probability $P(B)$ is

$$P(B) = P(B \mid A)P(A) + P(B \mid \bar{A})P(\bar{A}), \qquad (2.20.2^*)$$

so that

$$P(A \mid B) = \frac{P(A, B)}{P(B)} = \frac{P(B \mid A)P(A)}{P(B \mid A)P(A) + P(B \mid \bar{A})P(\bar{A})}.$$

This completes the derivation of Bayes' theorem.

To illustrate Bayes' theorem, consider a simple example. Suppose that the probability that a student at a particular all-male school passes an examination is .80 if he studies for the examination and .50 if he does not study. Furthermore, suppose that 60 percent of the students in a particular class study for a certain examination. If a student chosen randomly from the class passes the examination, what is the probability that he studied?

This is obviously a problem involving conditional probability. Let B stand for the event "passes the examination" and let A stand for the event "studied for the examination." Of course, $\bar{A}$ then stands for the event "did not study for the examination." The probability that is wanted is the probability that a student studied for the examination, given that he passed. This is $P(A \mid B)$. Since 60 percent of the students in the class studied and the student of interest has been randomly chosen from the class the *unconditional* probability that he studied for the examination, $P(A)$, is .60. Bayes' theorem can be used to determine the *conditional* probability that he studied, given that he passed the examination:

$$P(A \mid B) = \frac{P(B \mid A)P(A)}{P(B \mid A)P(A) + P(B \mid \bar{A})P(\bar{A})} = \frac{.80(.60)}{.80(.60) + .50(.40)}$$

$$= \frac{.48}{.48 + .20} = \frac{.48}{.68}$$

$$= .706.$$

As a second example of the use of Bayes' theorem, consider a medical situation. Suppose that a person chosen at random from the population of a certain city is given a tuberculin skin test and that the reading on the test is positive. Given the results of the skin test, what is the probability that the person has tuberculosis? In order to calculate this probability, you obtain the following information from a medical expert.

1.   The probability that a person with tuberculosis will have a positive reading on a tuberculin skin test is 0.98. That is,

   $P(\text{positive reading} \mid \text{tuberculosis}) = .98.$

2.   The probability that a person without tuberculosis will have a positive reading on a tuberculin skin test is .05. That is,

   $P(\text{positive reading} \mid \text{no tuberculosis}) = .05.$

3.   Of the members of the population in the city of interest, 1 percent have tuberculosis. That is, for a randomly selected person from this population, $P(\text{tuberculosis}) = .01.$

In order to calculate the desired probability, let B represent the event "positive reading on a tuberculin skin test," and let A represent the event "has tuberculosis." The probability $P(A \mid B)$ is then

$$P(A \mid B) = \frac{P(B \mid A)P(A)}{P(B \mid A)P(A) + P(B \mid \bar{A})P(\bar{A})} = \frac{.98(.01)}{.98(.01) + .05(.99)}$$

$$= \frac{.0098}{.0098 + .0495} = \frac{.0098}{.0593} = .165.$$

Before the tuberculin skin test, the probability that the person has tuberculosis is .01, since the only information available is that 1 percent of the members of the population of the city have tuberculosis and the person was chosen at random from this population. After a positive reading on the test, the probability that the person has tuberculosis increases to .165. It follows that after a positive reading, the probability of "no tuberculosis" is $1 - .165$, or .835.

Suppose that the reading on the skin test is negative rather than positive. Since P(positive reading | tuberculosis) = .98, P(negative reading | tuberculosis) = .02. Similarly, P(negative reading | no tuberculosis) = 1 − P(positive reading | no tuberculosis) = 1 − .05 = .95. The new information is now $\bar{B}$ (negative reading) instead of B, and

$$P(A \mid \bar{B}) = \frac{P(\bar{B} \mid A)P(A)}{P(\bar{B} \mid A)P(A) + P(\bar{B} \mid \bar{A})P(\bar{A})} = \frac{.02(.01)}{.02(.01) + .95(.99)}$$

$$= \frac{.0002}{.9407} = .0002.$$

Thus, if the reading on the skin test is negative, the probability of tuberculosis drops from .01 to .0002; if the reading is positive, this probability increases from .01 to .165.

Bayes' theorem can be put into much more general form.

If the J events $A_1, A_2, \ldots, A_J$ form a partition of the sample space S, then

$P(A_i \mid B)$

$$= \frac{P(B \mid A_i)P(A_i)}{P(B \mid A_1)P(A_1) + P(B \mid A_2)P(A_2) + \cdots + P(B \mid A_J)P(A_J)}$$

$$(2.20.3^*)$$

for any event A_i, where i is an integer between 1 and J, inclusive.

To illustrate the general form of Bayes' theorem, consider another medical example. A person chosen at random from the population of a given city is given a chest X-ray, and the results turn out to be "positive" (that is, a "spot" shows up on the X-ray). For the sake of simplicity, assume that the person could have lung cancer, tuberculosis, or neither (it is assumed that no one has *both* cancer and tuberculosis). The following information is available.

1. The probability that a person with lung cancer will have a positive chest X-ray is .90.
2. The probability that a person with tuberculosis will have a positive chest X-ray is .95.
3. The probability that a person without lung cancer or tuberculosis will have a positive chest X-ray is .07.
4. In the city of interest 2 percent of the members of the population have lung cancer and 1 percent have tuberculosis(no one has both diseases).

Using the notation of Equation (2.20.3), let A_1, A_2, and A_3 represent the events "has lung cancer," has tuberculosis," and "has neither lung cancer nor tuberculosis," respectively; and let B represent the event "positive chest X-ray." The probabilities $P(A_1 \mid B)$, $P(A_2 \mid B)$, and $P(A_3 \mid B)$ are

$$P(A_1 \mid B) = \frac{P(B \mid A_1)P(A_1)}{P(B \mid A_1)P(A_1) + P(B \mid A_2)P(A_2) + P(B \mid A_3)P(A_3)}$$

$$= \frac{(.90)(.02)}{(.90)(.02) + (.95)(.01) + (.07)(.97)} = \frac{.0180}{.0954} = .1887,$$

$$P(A_2 \mid B) = \frac{P(B \mid A_2)P(A_2)}{P(B \mid A_1)P(A_1) + P(B \mid A_2)P(A_2) + P(B \mid A_3)P(A_3)}$$

$$= \frac{(.95)(.01)}{(.90)(.02) + (.95)(.01) + (.07)(.97)} = \frac{.0095}{.0954} = .0996,$$

and

$$P(A_3 \mid B) = \frac{P(B \mid A_3)P(A_3)}{P(B \mid A_1)P(A_1) + P(B \mid A_2)P(A_2) + P(B \mid A_3)P(A_3)}$$

$$= \frac{(.07)(.97)}{(.90)(.02) + (.95)(.01) + (.07)(.97)} = \frac{.0679}{.0954} = .7117.$$

Before the chest X-ray, the probabilities for lung cancer, tuberculosis, and neither are .02, .01, and .97, since the person was chosen at random from the population of the city and the only information available is that 2 percent of the members of the population have lung cancer, 1 percent have tuberculosis, and 97 percent have neither disease. After a positive chest X-ray, the probability of cancer increases to .1887, the probability of tuberculosis increases to .0996, and the probability of neither disease decreases to .7117 (the probability of having one disease or the other increases from .03 to .2883).

Bayes' theorem will be utilized in Chapters 8 and 9 to revise prior probabilities on the basis of sample information. The following terminology is used to describe the various probabilities appearing in Bayes' theorem: $P(A_i)$ represents a **prior probability;** $P(B \mid A_i)$ represents a **likelihood,** which involves the additional information B; and $P(A_i \mid B)$ represents a **posterior probability.** A "likelihood" can be interpreted as the likelihood, or probability, of the observed information B, given a particular event A_i. Of course, the likelihood will generally be different for different events. The likelihood of the observed information (a positive reading) given tuberculosis in the first medical example is .98, whereas the likelihood of the observed information given no tuberculosis is .05. The terms "prior" and "posterior" are relative to the observed information.

In terms of the first medical example, the prior probabilities of tuberculosis and no tuberculosis are .01 and .99, respectively. After seeing some additional information (the positive reading on the skin test), the posterior probabilities given this new information are .165 and .835. In terms of the second medical example, the prior probabilities of lung cancer, tuberculosis, and neither are .02, .01, and .97, respectively. After the new information (the positive chest X-ray), the posterior probabilities determined from Bayes' theorem are .1887, .0996, and .7117. The likelihood of the observed information (the positive chest X-ray) given lung cancer is .90, the likelihood given tuberculosis is .95, and the likelihood given neither disease is .07.

Bayes' theorem is a key feature of a modern, decision-theoretic approach to statistics. We will discuss this approach and compare it with the more traditional, or "classical" approach in Chapters 8 and 9.

2.21 INDEPENDENCE

We are now ready to take up the topic of independence. The general idea used heretofore is that independent events are those having nothing to do with each other; the occurrence of one event in no way affects the probability of the other event. But how do we know if two events A and B are independent? If the occurrence of the event A has nothing whatever to do with the occurrence of event B, then we should expect the conditional probability of B given A to be exactly the same as the probability of B, $P(B \mid A) = P(B)$. Likewise, the conditional probability of A given B should be equal to the probability of A, or $P(A \mid B) = P(A)$. The information that one event has occurred does not affect the probability of the other event when the events are independent.

The condition of independence of two events may also be stated in another form. If $P(B \mid A) = P(B)$, then

$$\frac{P(A \cap B)}{P(A)} = P(B), \qquad (2.21.1^*)$$

so that $P(A \cap B) = P(A)P(B)$.

This fact leads to the usual definition of independence.

Events A and B are independent if, and only if, the joint probability $P(A \cap B)$ is equal to the probability of A times the probability of B:

$$P(A \cap B) = P(A)P(B). \qquad (2.21.2^*)$$

For example, suppose that you go into a library and at random select one book (each book having an equal likelihood of being drawn). The sample space consists of all distinct books that a person might select. Suppose that the proportion of books then on the shelves and classified as "fiction" is exactly .15, so that the probability of selecting such a book is also .15. Furthermore, suppose that the proportion of books having red covers is exactly .30. If the event "fiction" is independent of the event "red cover," the probability of the joint event is found very easily:

$$P(\text{fiction and red cover}) = P(\text{fiction})P(\text{red cover}) = (.15)(.30) = .045.$$

Your chances of selecting a red-covered piece of fiction are 45 in 1000.

For another simple example of two independent events, consider two tosses of a fair coin, and let H_1 and H_2 represent "heads comes up on the first toss" and "heads comes up on the second toss," respectively. Clearly, the result of the first toss should have no effect on the second toss, so

$$P(H_2 \mid H_1) = P(H_2) = 1/2.$$

The rain example presented in Section 2.19 is an example involving events that are not independent. From the table of joint and marginal probabilities,

$$P(\text{rain today, rain tomorrow}) = .14,$$

while

$$P(\text{rain roday})P(\text{rain tomorrow}) = (.20)(.22) = .044.$$

In terms of conditional probabilities, which are given in Figure 2.19.3,

$$P(\text{rain tomorrow} \mid \text{rain today}) = .7,$$

whereas $\qquad P(\text{rain tomorrow} \mid \text{no rain today}) = .1.$

According to these probabilities, tomorrow's weather is clearly not independent of today's weather. In terms of a table such as the one presented for the rain example in Section 2.19, independence implies that each joint probability in the body of the table is the product of the corresponding marginal probabilities. As noted above, this is not true for the rain example, since $P(\text{rain today, rain tomorrow}) \neq P(\text{rain today}) \; P(\text{rain tomorrow})$.

This example suggests one of the uses of the concept of independent events. The definition of independence permits us to decide whether or not events are *associated* or *dependent* in some way.

If, for two events A and B, $P(A \cap B)$ is not equal to $P(A)P(B)$, then A and B are said to be associated or dependent.

For example, consider a sample space with elementary events consisting of all the adult male persons in the United States. We wish to answer the question, "Is making over twenty thousand dollars a year independent of having a college education?" Let us call event A "makes over twenty thousand dollars a year," and B "has a college education." Now suppose that the proportion of adult males in the United States who make more than twenty thousand a year is .06; if each adult male is equally likely to be selected, the probability of event A is thus .06. Suppose also that the proportion of adult males with a college education is .30, so that $P(B) = .30$. Finally, the probability of an event $A \cap B$, our observing an adult male who makes over \$20,000 a year and who has a college education is .04. If the events were independent, this probability $P(A \cap B)$ *should* be $(.06)(.30) = .018$; but the actual probability is .04. This leads immediately to the conclusion that these two events are *not* independent (or *are* associated). The probability that event B will occur given event A is much greater than it should be if the events were independent: $P(B \mid A) = .04/.06 = .67$, whereas if the events were independent, it should be true that

$$P(B \mid A) = P(B) = .30.$$

This idea of independence is very important to many of the statistical techniques to be discussed in later chapters. The question of association or dependence of events is central to the scientific question of the relationships among the phenomena we observe (literally, the question of "what goes with what"). The techniques for studying the presence and degree of relatedness among observations will depend directly upon this idea of comparing probabilities for joint events with the probabilities when events are independent.

However, the concept of association of events must not be confused with the idea of causation. Association, as used in statistics, means simply that the events are not independent and that a certain correspondence between their joint and marginal probabilities does not hold. When events A and B are associated, it need not mean that A causes B or that B causes A, but only that the events occur together with probability different from the product of their marginal probabilities. This warning carries force whenever we talk of the association of events; association may be a consequence of causation, but this need not be true.

Sometimes the independence of two events A and B is *conditional* upon a third event (or assumption, or hypothesis) C. That is,

$$P(A, B \mid C) = P(A \mid C)P(B \mid C),$$

but $$P(A, B \mid \bar{C}) \neq P(A \mid \bar{C})P(B \mid \bar{C}).$$

Here we say that A and B are **conditionally independent,** conditional on the event C. Conditional on the event $\bar{C}$, A and B are **conditionally dependent.** For example, suppose that you are a marketing manager and that you are interested in C, the event that a rival firm will not introduce a new product to compete with a certain product your firm produces. You consider the possibility that the rival firm is about to conduct an extensive market survey (event A) and about to hire a large number of new employees (event B). If C is true (the rival firm will *not* introduce a competing product), then you are willing to assume that A and B are independent. If C is not true, then you feel that A and B are *not* independent.

Just as all probabilities are, in a sense, conditional (Section 2.18), all independence or nonindependence can be thought of as conditional upon certain assumptions or hypotheses. Even in a simple experiment such as tossing a coin, the claim that the trials are independent rests upon assumptions such as the assumption that the way the person flips the coin does not depend on the previous outcome. Just as with conditional probabilities, it will be convenient for us not always to write out the given conditions under which events are independent or nonindependent. If it is important in any specific case, we will do so.

The concept of independence can be extended to more than two events. The K events $A_1, A_2, \ldots, A_K$ are said to be **mutually independent** if for all possible combinations of events chosen from the set of K events, the joint probability of the combination is equal to the product of the respective individual, or marginal, probabilities. For example, the three events $A_1, A_2,$ and A_3 are mutually independent if

$$P(A_1, A_2) = P(A_1)P(A_2),$$

$$P(A_1, A_3) = P(A_1)P(A_3),$$

$$P(A_2, A_3) = P(A_2)P(A_3),$$

and $\qquad P(A_1, A_2, A_3) = P(A_1)P(A_2)P(A_3).$

If the first three of these equations are satisfied but the fourth is not, then the three events $A_1, A_2,$ and A_3 are said to be **pairwise independent** but not mutually independent. In this book, the "independence" of a number of events should be interpreted as *mutual* independence unless otherwise stated. Formally, K events $A_1, A_2, \ldots, A_K$ **are mutually independent if for all** $i, j, k,$ **and so on, the following conditions hold (where** $i, j, k,$ **and so on, must all be different):**

$$P(A_i \cap A_j) = P(A_i)P(A_j),$$

$$P(A_i \cap A_j \cap A_k) = P(A_i)P(A_j)P(A_k),$$

$$P(A_i \cap A_j \cap A_k \cap A_l) = P(A_i)P(A_j)P(A_k)P(A_l),$$

$$\begin{matrix} \cdot & & \cdot \\ & & \\ \cdot & & \cdot \\ & & \\ \cdot & & \cdot \end{matrix} \qquad (2.21.3^*)$$

and $\qquad P(A_1 \cap A_2 \cap \cdots \cap A_K) = P(A_1)P(A_2) \cdots P(A_K).$

In other words, the probability of the intersection of any two or more (up to all K) of the K events must equal the product of the respective probabilities of the individual events.

The independence of events may greatly simplify probability calculations in statistical problems, especially when repeated trials of an uncertain situation are being observed. Recall that independence of trials is required in the law of large numbers, which was discussed in Section 2.13. When it is reasonable, this is a most useful assumption, since it implies that the joint probability of the n specific occurrences in a series of n trials is equal to the product of the n individual probabilities. For example, if a fair die is tossed 4 times and the tosses are mutually independent, the probability of 3 sixes followed by a number other than 6 is

$$P(6, 6, 6, \text{non-6}) = P(6 \text{ on toss } 1)P(6 \text{ on toss } 2)$$

$$P(6 \text{ on toss } 3)P(\text{non-6 on toss } 4)$$

$$= \left(\frac{1}{6}\right)\left(\frac{1}{6}\right)\left(\frac{1}{6}\right)\left(\frac{5}{6}\right) = \frac{5}{1296}.$$

If the probability of a hurricane hitting Louisiana during any given year is .20, with this probability being independent of whether or not hurricanes hit Louisiana in previous years, then the probability of 10 successive years without a hurricane in Louisiana is the product of the probabilities for the individual years. Each of these probabilities is .80, so the desired probability is $(.80)(.80) \cdots (.80) = (.80)^{10}$, which is approximately .11.

EXERCISES

1. Explain the statement that probability can be thought of as the mathematical language of uncertainty.

2. Discuss the importance of carefully defining the sample space in any application of probability theory. In the oil-drilling example presented in Section 2.17, the sample space consists of just two events, "oil" and "no oil," in an attempt to simplify the problem. Can you think of some more realistic ways to define the sample space in this example?

3. A coin is tossed n times, and each time either a head or a tail occurs. If the elementary events for this experiment are viewed as the possible *sequences* of results in n tosses, how many different elementary events are there? If $n = 4$, is the event "no heads in 4 tosses" an elementary event? Is the event "2 heads in 4 tosses" an elementary event?

4. A politician is uncertain about the outcome of a particular election. If there are just two candidates in the election, give at least two possible ways of

defining the sample space. If the politician has held public office for some
time and plans to retire from politics if the election is lost, might the sample
space be defined differently than in the case of a younger person who is a candidate for the first time?   { 50% or more, less than 50%) : how much lose by

5. A toll collector on the Pennsylvania Turnpike conducts a simple experiment
    that consists of counting the number of cars passing through a particular
    toll booth up to and including the first car bearing an Illinois license plate.
    For example, if the first 4 cars are not from Illinois but the fifth car is, the
    count stops at 5. What is the sample space for this problem? Is the sample
    space a finite, countably infinite, or uncountably infinite set? If the experiment
    consists of recording the state in which the next car passing through the toll
    booth is licensed, with all unlicensed or foreign-licensed cars classified as "no
    state," what is the sample space? Is this sample space finite, countably in-
    finite, or uncountably infinite?

6. Briefly explain the connection between set theory and probability theory,
    and discuss the usefulness of Venn diagrams to illustrate elementary rules
    of probability.

7. Both a red die and a green die are tossed and r and g represent the numbers
    appearing on the red and green die, respectively. Assume that the dice are
    perfectly symmetric.
    (a) Specify the sample space completely. How many elementary events
         are there?
    (b) Find the probability that $r = g$.
    (c) Find the probability that $r \leq g$.
    (d) Find the probability that $r < 2$ *and* $g \geq 3$.
    (e) Find the probability that $r < 2$ *or* $g \geq 3$ (including the possibility that
         both will occur).
    (f) Find the probability that $r + g = 7$.
    (g) Find the probability that $r + g = 8$ *and* $r - g \geq 0$.

8. A letter is selected at random from the English alphabet (that is, all possible
    letters are equally likely to be chosen). Find the probability that
    (a) the letter is a vowel,
    (b) the letter is a consonant,
    (c) the letter occurs in the last 10 positions of the alphabet (given the con-
         ventional ordering of the alphabet),
    (d) the letter is a consonant falling between the two vowels "a" and "i"
         (not including "a" and "i") in the conventional ordering.

9. Two letters are drawn at random from the alphabet, and the first is replaced
    before the second is drawn. What is the sample space for this experiment?
    Find the probability that the first letter drawn precedes the second letter
    in the conventional ordering of the alphabet. If the first letter is *not* replaced
    before the second is drawn, so that the two letters must be different, find
    the sample space and determine the probability that the first letter precedes
    the second letter in the conventional ordering of the alphabet.

$$\frac{26^2 - 26}{2}$$

Draw 2
ABC···z

A
B
:
z

draw
#1   z

10. Two distinguishable coins are tossed two times each. Find the probability for each of the following events, given that the coins are fair (start by specifying the elementary events making up the sample space):
    (a) at least 2 heads occur,
    (b) at most 2 heads occur,
    (c) the number of heads for coin 1 is less than the number of heads for coin 2.

11. In Exercise 8, Chapter 1, an item is randomly selected from the lot of 900 items; that is, each of the 900 items is equally likely to be the one that is selected. Find the probability that the item selected has
    (a) no defects,
    (b) at least one defect,
    (c) a type A defect (and may or may not have other defects),
    (d) only a type B defect.

12. A young man has a little black book that contains the names and phone numbers of girls who are potential dates. Suppose that he has the names of 3 redheads, 18 brunettes, and 6 blondes in his book, and that the book has a total of 30 pages. He decides to select one page at random from the book in order to determine whom he will call for a date on a particular night, and each name appears on a separate page.
    (a) Specify the sample space. What is the family of events in this sample space?
    (b) What is the probability that he will call a blonde? a redhead? a brunette? (Here, count a blank page as "no call.")
    (c) What is the probability that he will sample a page containing the name either of a redhead or a brunette? What is the probability that the page will *not* contain the name of a blonde?

13. Show by a Venn diagram that
    (a) $P(A \cap \overline{B}) = P(A) - P(A \cap B)$;
    (b) if $B \subseteq A$, then $P(A - B) = P(A) - P(B)$;
    (c) $P(A \cap B) + P(A \cap C) = P[A \cap (B \cup C)] + P(A \cap B \cap C)$.

14. Let A_1, A_2, and A_3 be events defined on some specific sample space S. Find expressions involving only *union, intersection,* and *complementation* for the probability that
    (a) exactly two of A_1, A_2, A_3 occur simultaneously,
    (b) at least two of A_1, A_2, A_3 occur simultaneously,
    (c) at most two of the events A_1, A_2, A_3 occur simultaneously.

15. Compute the probabilities for (a), (b), and (c) in the preceding exercise if:
    (i) $P(A_1) = 1/2, P(A_2) = P(A_3) = 1/4,$
          $P(A_1 \cap A_2) = P(A_1 \cap A_3) = P(A_2 \cap A_3) = 1/8,$
          $P(A_1 \cap A_2 \cap A_3) = 1/32;$
    (ii) $P(A_1) = 1/3, P(A_2) = 2/3,$
          $P(A_1 \cap A_2) = P(A_1 \cap A_3) = P(A_2 \cap A_3) = P(A_1 \cap A_2 \cap A_3) = 0.$

16. A card is drawn from a well-shuffled standard playing deck. An ace is assigned the number one, and face cards count as tens. Find the probability that the card is
    (a) either a spade or the queen of hearts,
    (b) an even numbered card but not a spade or a heart,
    (c) a spade that is not the ace or king, or a heart that is not the queen or jack,
    (d) a numbered card that is a perfect square (consider numbers assigned to the cards without printed numbers).

17. A famous problem in probability theory is the following. If an "ace" is the occurrence of the number "one" on the toss of a die, find the probabilities of
    (a) obtaining at least one ace in four tosses of a fair die,
    (b) obtaining at least one "double ace" in 24 tosses of a pair of fair dice.
    (c) If you were given a chance to gamble on event (a) or event (b), which would you prefer to gamble on? (Assume that you *must* choose one of the two gambles.)

18. Find a general formula, similar to Equation (2.4.7), for the probability of the union of three events, $P(A \cup B \cup C)$.

19. A fair die is to be thrown until a six appears. Find the probability that this will take more than three throws.

√20. Prove that $P(A \cap B) \le P(A) \le P(A \cup B)$ for any events A and B.

21. In one year, the Parliament of a European country included 45 members of the Liberal Party, 38 members of the Conservative Party, and 15 members of the Labor Party. The Prime Minister wished to appoint a commission of 10 persons from Parliament, consisting of 5 Liberals, 3 Conservatives, and 2 Laborites. How many different such committees *might* be appointed? (Leave the answer in symbolic form, and do not bother to work out the exact figure.) If the Prime Minister decided simply to randomly choose a commission of 10 without regard to party, what is the probability of getting 5 Liberals, 3 Conservatives, and 2 Laborites on the commission? What is the probability of getting *no* Conservatives? At least 2 Conservatives? (Again, leave answers in symbolic form.)

√22. If there are 5 people in a room, what is the probability that no two of them have the same birthday? (To simplify the problem, assume that no birthdays fall on February 29.) In general, if there are r people in a room, what is the probability that no two of them have the same birthday? What assumptions did you make in computing these probabilities?

23. If we draw 5 cards from a well-shuffled standard deck of 52 cards, find the probability of obtaining
    (a) 4 of a kind,
    (b) a royal flush (Ace, King, Queen, Jack, and Ten from the same suit).

24. In how many *distinguishable* ways can the letters in the word "sassafras" be ordered?

$$\frac{n!}{n_1! \, n_2! \, n_3! \, n_4!}$$

25. Prove Pascal's rule:

$$\binom{n}{r} = \binom{n-1}{r-1} + \binom{n-1}{r}.$$

26. Prove that

$$n\binom{n}{r-1} = r\binom{n}{r} + (r-1)\binom{n}{r-1}.$$

√27. How many ways can a starting team of 11 players be chosen from a squad of 30 football players, ignoring the position to be played by each player? If the squad consists of 10 backs and 20 linemen, in how many ways can a starting team of 4 backs and 7 linemen be chosen?

28. State how many ways 8 persons can be seated in a row of 8 seats if
    (a) there are no restrictions on the seating arrangement,
    (b) person *A* must sit next to person *B*, but there are no restrictions on the remaining 6 persons,
    (c) there are 4 men and 4 women, and no 2 men (or 2 women) can sit next to each other,
    (d) there are 4 married couples, and each couple must sit together.

29. How many ways can 8 persons be seated around a circular table, taking into account only the *relative* location of the 8 persons (that is, if all 8 persons get up and move one chair to the left, the new arrangement is considered to be indistinguishable from the old arrangement)?

30. Each license plate in a particular state contains two letters followed by a number between 1 and 9999, inclusive.
    (a) What is the maximum number of cars that the state can license under this system?
    (b) If all possible license plates are in use and a car is chosen at random from the cars licensed in the state, what is the probability that the car's license plate contains a W?
    (c) In (b), what is the probability that the number on the car's license plate contains fewer than three digits?

√31. A furniture store sells a particular line of chairs. The chairs are available in two styles (low-back or high-back), three types of upholstery (cloth, vinyl, or leather), and three colors (black, brown, or gold).
    (a) How many distinguishable chairs are available?
    (b) If a customer comes into the store and orders a chair, the probability that it will be a low-back chair is .4, and other probabilities are $P(\text{cloth}) = .2$, $P(\text{vinyl}) = .5$, $P(\text{low-back and vinyl}) = .2$, and $P(\text{low-back and cloth}) = .1$. What is the probability that the chair ordered is a high-back chair with leather upholstery?

32. A salesman's territory consists of Illinois, Indiana, Michigan, and Wisconsin, as in Exercise 21 of Chapter 1. The salesman wants to plan a trip to all four states.

(a) Assuming that each state is visited only once, in how many ways can the trip be made (in the sense of the order in which the states are visited)?

(b) If the salesman discovers that the time available is only sufficient to visit two of the four states, in how many ways can the trip be made? [As in (a), order is still a consideration.]

33. In Section 2.6, the game of roulette was used to illustrate the computation of probabilities in simple situations. In that illustration, there were 37 slots on the roulette wheel. In many casinos, there are 38 slots, with the 38th slot being a "double zero." If a gambler bets on red, find the probability of winning for a 37-slot wheel and for a 38-slot wheel. If the payoff is the same in either case, what are the implications of the difference between the two probabilities?

34. A librarian wants to arrange 3 statistics books, 2 mathematics books, and 4 novels on a bookshelf. Calculate how many arrangements are possible if
(a) the books can be arranged in any manner,
(b) the statistics books must be together, the mathematics books must be together, and the novels must be together,
(c) the novels must be together, but the other books can be arranged in any manner.

35. Why is it that many elementary discussions of, and computations involving, probability rely on the assumption of equally probable elementary events? Is there anything about the theory of probability itself that makes this assumption necessary? What if elementary events cannot be assumed equally probable; does the axiomatic theory of probability still apply?

√ 36. A famous eighteenth-century mathematician, d'Alembert, argued that in two tosses of a coin, heads could appear once, twice, or not at all, and thus each of these three events should be assigned a probability of 1/3. Do you agree with him? Comment on the issues involved and on the implications of the situation with regard to the use of the assumption of equally likely elementary events to determine probabilities.

37. The so-called "gambler's fallacy" goes something like this: in a dice game, for example, a seven has not turned up in quite a few rolls of a pair of honest dice; now a seven is said to be "due" to come up. Why is this a fallacy?

38. The law of large numbers has occasionally been called "the link between the mathematical concept of probability and the real world about us." Discuss this proposition.

39. Take a fair die and throw it 300 times, recording the number which appears at each throw. Find the frequency of each of the six possible events after 10, 20, 50, 100, and 300 throws, and comment on these results in light of the law of large numbers.

40. Comment on the following statement: "The law of large numbers says that if a fair coin is tossed 2 million times, one would be very likely to get exactly 1 million heads."

41. A banker is concerned about a possible increase in interest rates. Find the probability of an increase in interest rates if the odds in favor of such an increase are

   (a) 2 to 1,   (c) 3 to 7,
   (b) 1 to 3,   (d) 99 to 1.

42. An accountant is on trial for embezzlement, and the odds against acquittal are 3 to 2. What is the probability that the accountant will not be convicted?

43. A firm has submitted a bid for construction work on a new rapid transit system. Find the odds in favor of winning the bid if the probability of winning the bid is

   (a) .50,   (c) .20,
   (b) .001,   (d) .875.

44. Explain why the subjective interpretation of probability can be thought of as an extension of the relative frequency interpretation. Are there any restrictions on the types of events for which subjective probabilities can be used?

45. (a) What is your subjective probability that it will rain tomorrow?
   (b) What in your opinion are the odds in favor of rain tomorrow?
   (c) Are your answers to (a) and (b) consistent? If not, why not?
   (d) When "precipitation probabilities" are given out to the public, how do you interpret them? How do you think the average person interprets them?

46. Three football teams are fighting for the league championship. A sportswriter claims that the odds are even, or 1 to 1, that Team A will win the championship, the odds are even that Team B will win the championship, and the odds are even that Team C will win the championship. (Assume that ties for the championship are impossible.) Is this reasonable? Explain your answer.

47. In the preceding problem, the sportswriter states that he is willing to bet for or against Team A at even odds, and likewise for Teams B and C. Even without knowing anything about the teams, would you like to bet with him? Explain.

48. Consider the following bet. One investor bets $10 that the price of XYZ stock will increase in the next year, and a second investor bets $7 that the price of XYZ stock will *not* increase in the next year.

   (a) If an investment analyst prefers the first investor's side of the bet, what can you say about the analyst's probability that the price of XYZ stock will increase in the coming year?
   (b) If the investment analyst thinks that the bet is "fair," what can you say about the analyst's probability that XYZ stock will increase in price?

49. What is your personal probability that the recorded maximum temperature on next January 1 in San Francisco is greater than 50°F? Would you be willing to bet at the odds implied by this probability?

50. How would your answer to Exercise 49 change as a result of obtaining the following additional information?

(a) The average maximum daily temperature during January in San Francisco for the 30-year period from 1931 to 1960 was 56°F.

(b) The information in (a) plus the information that the average minimum daily temperature during January in San Francisco during that 30-year period was 46°F.

51. Discuss the concept of canonical probabilities and their use in the assessment of probabilities via devices such as lotteries. What devices might be used to determine canonical probabilities?

52. Discuss each of the following situations with regard to the different interpretations of probability.

(a) A coin has been tossed 1000 times and exactly 500 heads have been observed. What is the probability of heads on the next toss?

(b) A brand-new coin is obtained at a bank. What is the probability of heads on the next toss?

(c) A coin was tossed at the beginning of the 1970 Stanford–Washington football game to determine choice of kicking versus receiving or choice of goal. What is the probability that the coin came up heads?

(d) A coin is tossed, placed on a table, and covered with a sheet of paper. What is the probability that "heads" is up?

(e) A stranger produces a coin and tosses it 10 times. Each time it comes up heads. What is the probability of heads on the next toss?

(f) You take a brand new coin and toss it 10 times. Each time it comes up heads. What is the probability of heads on the next toss?

53. Discuss each of the following situations with regard to the different interpretations of probability.

(a) In order to be listed on a particular stock exchange, a firm must publicly divulge its total assets and certain other financial information. Thus, the total assets of each firm listed on the exchange is known. If a firm is selected randomly from the firms listed on the exchange, the probability that it has assets exceeding $1 million is .30.

(b) Another group of firms consists of firms that are not listed on the exchange mentioned in (a) and that do not divulge total assets. A firm is to be selected randomly from this group of firms, and an investment analyst feels that the probability that the chosen firm has assets exceeding $1 million is .15.

(c) A weather forecaster studies various weather charts and then states that the probability of rain tomorrow is .20.

(d) The National Weather Service of the United States feeds current weather data into its computer, and on the basis of past situations in which the weather data was very similar to the current data, the computer prints .10 as the probability of rain tomorrow.

54. Discuss the statement "In a sense, all probabilities are conditional."

55. A manager is interested in whether Congress will raise taxes in the next year and in whether the sales of a particular product will be higher in the next year than they were in the past year. Furthermore, the manager's joint probabilities are P (taxes raised, sales higher) = .15, P (taxes raised, sales not higher) = .20, P (taxes not raised, sales higher) = .60, and P (taxes not raised, sales not higher) = .05.
   (a) Find the individual probabilities P (taxes raised), P (taxes not raised), P (sales higher), and P (sales not higher).
   (b) Find the conditional probabilities P (sales higher | taxes raised), P (sales higher | taxes not raised), P (sales not higher | taxes raised), and P (sales not higher | taxes not raised).
   (c) Represent this situation in a tree diagram.
   (d) Represent this situation in a table.
   (e) Are the events "taxes raised" and "sales higher" independent events?

56. Let A be the event "a person is a male," and B the event "a person is blue-eyed" when individuals are chosen at random from some well-defined group of persons. Also let

$$P(A) = .60,$$
$$P(B) = .20,$$
and
$$P(A \mid B) = .10.$$

Find:
   (a) $P(A \cup B)$,   (c) $P(\bar{A} \cup \bar{B})$, $P(\overline{A \cap B})$,
   (b) $P(A \cup \bar{B})$, $P(\bar{A} \cup B)$,   (d) $P\{(A \cap \bar{B}) \cup (\bar{A} \cap B)\}$.

57. Let A be the event that personal income, adjusted for inflation, increases during the coming year, and let B be the event that the unemployment rate increases during the coming year. The probability that personal income increases is .6, the probability that the unemployment rate increases is .15, and the conditional probability that the unemployment rate increases given that personal income does not increase is .25.
   (a) Represent this situation in a table.
   (b) Represent this situation in a tree diagram.
   (c) Find $P(B \mid A)$, $P(A \mid B)$, $P(A \cup B)$, and $P\{(A \cap \bar{B}) \cup (\bar{A} \cap B)\}$.

58. A survey of 200 people in a certain college town has provided the following information about a proposal to allow more liberalized sale of beer:

	Male		Female		Totals
	Student	Nonstudent	Student	Nonstudent	
Favor liberalized sale of beer	70	10	40	0	120
Do not favor liberalized sale	5	30	10	20	65
No opinion	5	0	10	0	15
Totals	80	40	60	20	200

If a person is selected at random from this group, find the probability that the person

(a)   is a male,

(b)   is a female who favors the liberalized sale of beer,

(c)   is a female nonstudent,

(d)   is either a female or a nonstudent,

(e)   is a male who has no opinion, or a female who does not favor the liberalized sale of beer.

59.   If $P(A) = 0$, does it make any sense to consider the conditional probability $P(B \mid A)$? Explain.

60.   Show that if $A \cap B = \varnothing$, then $P(A \mid A \cup B) = \dfrac{P(A)}{P(A) + P(B)}$.

61.   Imagine three identical wallets. One contains two \$5 bills, one contains two \$10 bills, and one contains one \$5 bill and one \$10 bill. You choose a wallet randomly, take a bill from it randomly, and find that it is a \$10 bill. What is the probability that the other bill in the chosen wallet is also a \$10 bill?

62.   Suppose that you are offered the following choice of lotteries.

Lottery I: If the winner of the next presidential election in the United States is a member of the Democratic party, then you win \$10 with probability 1/2 and you lose \$10 with probability 1/2.
       If the winner of the election is not a Democrat, you win \$0.

Lottery II: If the winner of the next presidential election in the United States is a member of the Democratic party, then you win \$10 if the Senate has more Democrats than Republicans at the beginning of the president's term of office and you lose \$10 if the Senate does not have more Democrats than Republicans at that time.
       If the winner of the election is not a Democrat, you win \$0.

Which lottery would you select? What canonical probabilities in Lottery I would make you indifferent between the two lotteries? What subjective probability of yours is being assessed here?

63.   In the medical example in Section 2.20, the new information, which consists of a positive reading on the tuberculin skin test, increases the probability of tuberculosis from .01 to .165. This result is surprising to some people, who claim that they expected a greater increase in the probability of tuberculosis. Explain why the increase is no greater, despite the fact that the tuberculin skin test results in only a .02 chance of a "false negative" reading for a person with tuberculosis and a .05 chance of a "false positive" reading for a person without tuberculosis.

64.   The probability that 1 percent of the items produced by a certain process are defective is .80, the probability that 5 percent of the items are defective is .10, and the probability that 10 percent of the items are defective is .10. An

item is randomly chosen, and it is defective. *Now* what is the probability that 1 percent of the items are defective? That 5 percent are defective? That 10 percent are defective? Suppose that a *second* item is randomly chosen from the output of the process, and it too is defective. Following this second observation, what are the probabilities that 1, 5, and 10 percent, respectively, of the items produced by the process are defective?

65. In a football game, a team with enough time for one more play needs to score a touchdown on that play to win the game. As a fan, you feel that there are only three possible plays that can be used and that the three plays are equally likely to be used. The plays are a long pass, a screen pass, and an end run. Based on past experience in similar situations, you feel that the probabilities of getting a touchdown with these plays are as follows:

Play	$P(\text{touchdown} \mid \text{play})$
Long pass	.50
Screen pass	.30
End run	.10

The team scores a touchdown on the play. What is the probability that they used the long pass? The screen pass? The end run?

66. Urn A is filled with 700 red balls and 300 green balls and Urn B is filled with 700 green balls and 300 red balls. One of the two urns is selected at random (that is, the two urns are equally likely to be selected). The experiment consists of selecting a ball at random (that is, all balls are equally likely) from the chosen urn, recording its color, and then replacing it and thoroughly mixing the balls again. This experiment is conducted three times, and each time the ball chosen is red. What is the probability that the urn that was chosen was Urn A?

67. A firm markets a particular product in a box, but it has been suggested that the packaging be switched from the box to a bag. The bag is cheaper, but it could possibly lead to a decrease in sales if customers prefer the box. The main concern is with profits, and the marketing manager in charge of the product feels that the probability is .8 that the change to a bag will increase profits. The bag is tried in a test-marketing area and leads to a reduction in profits in that area. The probability that this result would occur even if the change would actually increase profits nationally is .4, while if it would not increase profits nationally, the probability is .8. What is the probability of the profitability of the change, given the results from the test-marketing area?

68. A high-ranking U. S. official feels that the probability is .2 that a particular country will devalue its currency in the next six months. A report is then received that says that the devaluation will definitely take place. In the past,

such reports have been received before 60 percent of the actual devaluations and before 20 percent of the rumored devaluations that did not in fact occur. Revise the official's probability that the country will devalue its currency.

69. An investor is interested in the stock of the XYZ corporation. The investor feels that if the stock market goes up in the next year, the probability is .9 that the price of XYZ stock will go up. If the market goes down, the probability is .4 that XYZ will go up. Finally, if the market remains steady, the probability is .7 that XYZ will go up. Furthermore, the investor thinks that the probabilities are .5, .3, and .2 for the market to go up, go down, or remain steady. At the end of the year, the price of XYZ stock has *not* gone up. What is the probability that the stock market as a whole went up?

70. A firm has placed bids on three different contracts. The probabilities of successful bids are .3 for the first contract, .2 for the second contract, and .7 for the third contract. Moreover, there is no reason to assume any relationship between the success or failure of any one bid and the success or failure of any other bid. Find the probabilities of the following events:
    (a) the bids on the first two contracts are successful,
    (b) at least one of the bids on the first two contracts is successful,
    (c) the bid on the third contract is not successful,
    (d) the bid on the first contract fails but the bid on the third contract succeeds,
    (e) exactly one of the bids is successful,
    (f) at least one of the bids is successful,
    (g) exactly two of the bids is successful,
    (h) more than one bid is successful,
    (i) all three bids are successful.

71. Suppose that you have three fair dice, colored red, black, and yellow. Now one of the dice is selected at random and rolled.
    (a) Find the sample space S for this experiment.
    (b) Find the elementary events associated with each of the following events, as well as the probabilities of these events:
        (i)   the die selected is yellow and comes up with a number less than 5,
        (ii) the die selected is red and comes up with an odd number, or is black and comes up with an even number.
    (c) Show that the color of the die and the number appearing on a roll are independent events.

72. The probability that a TV commercial will increase the sales of a particular product by 20 percent is .4 and the probability that it will leave sales unchanged is .6. Also, the probability that a magazine ad will increase sales by 20 percent is .3 and the probability that it will leave sales unchanged is .7. Finally, the probability that a mail-order campaign will increase sales by 10 percent is .6 and the probability that it will leave sales unchanged is .4. Furthermore, the effects on sales of the three procedures are independent.

(a)   If all three procedures are instituted simultaneously, what is the probability that there will be a 30 percent increase in sales? What is the probability that there will be a 40 percent increase in sales? What is the probability that there will be *no* increase in sales?

(b)   If only the TV commercial and the mail-order campaign are used, what is the probability that sales will increase by 30 percent, *given that* the mail-order campaign leads to a 10 percent increase in sales?

73.   The probability that stocks ABC and DEF both increase in price during the next month is $1/6$; the probability that ABC increases in price but DEF does not is $2/9$; and the probability that DEF increases in price but ABC does not is $1/3$.

(a)   What is the probability that neither ABC nor DEF increases in price during the next month?

(b)   Are "ABC increases in price" and "DEF increases in price" independent events? Why or why not?

√ 74.   Explain the difference between mutually exclusive events and independent events. Is it possible for two events A and B to be both mutually exclusive *and* independent? Explain.

75.   Suppose that someone offers to sell you a lottery that will pay you $5 if the price of IBM stock goes up tomorrow and $0 otherwise, with the condition that the lottery is called off and you get your money back if it rains in Seattle tomorrow. How much would you be willing to pay for this lottery? If the lottery is changed so that it is called off only if it does *not* rain in Seattle tomorrow, how much would you pay for it? What do your answers imply about the relationship between price changes of IBM stock and rain in Seattle?

76.   Consider the data of Exercise 10, Chapter 1, and suppose that a person is randomly selected from the 120 survey respondents.

(a)   What is the probability that the person is in the high income group given that Brand A is preferred? Given that Brand B is preferred? Given that Brand C is preferred?

(b)   What is the probability that Brand B is preferred given that the person is in the low income group?

(c)   What is the probability that Brand B is preferred given that the person is *not* in the low income group?

(d)   Are the events "preference for Brand A" and "member of high income group" independent?

(e)   Are the events "medium income group" and "preference for Brand C" mutually exclusive?

77.   In a particular resort town in the summer, the probability that a day is sunny given that the previous day was sunny is .8, and the probability that a day is cloudy given that the previous day was cloudy is .5. For the sake of simplicity, assume that each day is classified either as "sunny" or "cloudy."

(a)   Consider a given weekend. Are the events "sunny on Saturday" and "sunny on Sunday" independent events?

(b) If Friday is sunny, what is the probability that Saturday, Sunday, and Monday will all be sunny?

(c) If Friday is sunny, what is the probability of at most one cloudy day in the next three days?

78. The manager of a store has lost the combination to the store's safe. Since it will take quite a bit of time to get the combination from the main office and it will be quite embarrassing to admit losing the combination, the manager decides to guess at the combination. There are three digits in the combination; let A_i represent the event that the guess is correct for the ith digit, where $i = 1, 2, 3$. The following probabilities are given:

$$P(A_1) = 1/2, \; P(A_2) = P(A_3) = 1/4,$$
$$P(A_1 \cap A_2) = P(A_1 \cap A_3) = P(A_2 \cap A_3) = 1/8,$$

and $P(A_3 \mid A_1 \cap A_2) = 1/4.$

(a) Find the probability that the safe opens when the "guess" is tried.

(b) Find $P(A_2 \mid A_1)$, $P(A_2 \cap A_3 \mid A_1)$, $P(A_2 \cup A_3 \mid A_1)$, and $P(A_3 \mid A_1)$.

(c) Would you say that A_1, A_2, and A_3 are independent events?

(d) Are $\bar{A}_1$, $\bar{A}_2$, and $\bar{A}_3$ independent events?

(e) Suppose that $P(A_1) = 1/3$, $P(A_2) = 2/3$, $P(A_3) = 0$, and $P(A_1 \cap A_2) = P(A_1 \cap A_3) = P(A_2 \cap A_3) = 0$. Are A_1, A_2, and A_3 independent events here?

79. Give an example of three events which are pairwise independent but not mutually independent. [*Hint:* Consider a simple experiment, such as the throwing of two dice.]

80. In Exercise 58, a person is selected at random from the group of 200 respondents to the survey. For this person, let A = "male," B = "student," and C = "favor liberalized sale of beer."

(a) Are A, B, and C mutually independent?

(b) Are A, B, and C pairwise independent?

(c) Are B and C independent?

81. In a summer resort town, the probability of rain on any given day in July is .3. If it rains, the probability of a warm day is .4, but if it does not rain, the probability of a warm day is .8. The sales at a souvenir shop depend upon the weather; the probability of high sales is .6 if it rains and is warm, .6 if it rains and is not warm, .2 if it does not rain and is warm, and .4 if it does not rain and is not warm. Let R denote a rainy day, let W denote a warm day, and let H denote a day with high sales at the souvenir shop.

(a) What is the probability that a day is warm and rainy and the sales are high at the souvenir shop that day?

(b) Find $P(W)$, $P(H)$, and $P(W \cap H)$.

(c) Are W and H independent?

(d) Are W and H conditionally independent, conditional upon rain?

(e) Are W and H conditionally independent, conditional upon no rain?

PROBABILITY
DISTRIBUTIONS

Given the definitions of "event," "sample space," and "probability" in the last chapter, we are now ready to take up a major set of concepts in probability and statistics. First of all, the concept of a random variable will be introduced, and then we will discuss probability distributions of random variables. Finally, measures for summarizing probability distributions will be discussed.

3.1 RANDOM VARIABLES

We defined the **sample space** as **the set of all elementary events that are possible outcomes of some simple experiment.** These outcomes may be expressed in terms of numbers (for instance, the number coming up on one throw of a die; the temperature in degrees centigrade at a given place and time; the price of a certain stock on a given day) or they may be expressed in nonnumerical terms (for instance, the color of a ball drawn from an urn; the sex of a respondent to a questionnaire; the occurrence or nonoccurrence of rain on a particular day).

Imagine that a single number could be assigned to each and every possible elementary event in a sample space S. If we throw a die, the elementary events may correspond to the six possible faces that can come up. If we wished, we could assign to each elementary event the number of spots that appears on the face that comes up; or we could assign the *square* of the number that comes up; or we might even assign the number 1 if a six comes up, 2 if a five comes up, and so on, up to 6 if a one comes up. In all of these cases, each elementary event is associated with exactly one number. The most natural choice in the example, of course, is to assign the number of spots on the face which comes up. The other examples point out the fact that this is not the only choice.

It is important to note that although each elementary event can be associated with only one number, the same number may be associated with more than one elementary event. In the die example, we could assign the number 1 to the events "one comes up," "three comes up," and "five comes up" and assign the number 0 to the remaining elementary events. Thus the number 1 corresponds to the event "odd number comes up" and the number 0 corresponds to the event "even number comes up." This is an example of how more than one elementary event can be associated with a given number. It is also an example of the assignment of numbers to events which are not originally expressed in terms of numbers. The events "odd number comes up" and "even number comes up" are expressed in non-numerical terms. In a similar manner, in the rain example, we could assign the number 1 to the event "rain" and the number 0 to the event "no rain."

The above examples show how a number can be assigned to each elementary event in a sample space. We will use the following terminology.

If the symbol X represents a function which associates a real number with each and every elementary event in a sample space S, then X is called a random variable which is defined on the sample space S. In other words, a random variable X is a real-valued function defined on a sample space.

This notion is illustrated in Figure 3.1.1. The dot in S represents an elementary event, and the function X associates a real number with that elementary event. Recalling the discussion of functions in Chapter 1, we note that a value x may be associated with a number of distinct elementary events, but each elementary event is associated with only one value.

We will use upper case letters, such as X, Y, and Z, to denote random variables, and lower case letters, such as x, y, z, a, b, c, and so on, to denote particular *values* of random variables. The expression $P(X = x)$ symbolizes the probability that the random variable X takes on the particular value x. When there is no chance for confusion, $P(X = x)$ will be abbreviated

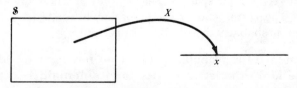

Figure 3.1.1 A Random Variable as a Function from the Sample Space to the Real Line

$P(x)$. A final note with regard to the distinction between a random variable and a particular value of a random variable is that in some contexts there is ambiguity concerning the appropriate notation, although every attempt will be made to be consistent in the distinction among random variables, values of random variables, and other variables (variables about which no uncertainty exists).

The term **random variable** is used here because of its popularity in statistical writing. The terms **uncertain quantity, chance variable,** or **stochastic variable** are sometimes used, and they all mean precisely the same thing: a real-valued function defined on a sample space. We may use the alternative terms on occasion, but for the most part, we will stick with "random variable."

Given a random variable X, events may be defined in terms of X. For example, for any two numbers a and b, "the random variable X takes on a value greater than a but less than b" is an event that has some probability $P(a < X < b)$. Some other events are "$X \leq a$," "$X \geq a$," "$X < a$," "$X > a$," and "$X = a$."

As an example of a random variable, suppose that X symbolizes the height of an American man, measured to the nearest inch. Here, there is some probability that $X \leq 60$, or $P(X \leq 60)$, since there is presumably some set of American men with heights less than or equal to 60 inches. Furthermore, there is some probability $P(70 \leq X \leq 72)$, since there is a set of American men having heights between five feet ten inches and six feet, inclusive. There are also some probabilities $P(X \leq 1)$ and $P(X \leq 120)$. These probabilities are almost surely zero and one, respectively, but the point is that they *do* exist. Here, X symbolizes the numerical value (height in inches) assigned to an American man (an elementary event). This symbol X represents any one of many different such values, and for any arbitrary number a there is some probability that the value of X is less than or equal to a.

3.2 PROBABILITY DISTRIBUTIONS

Consider the set of events $\{A_1, A_2, \ldots, A_K\}$, and suppose that they form a partition of the sample space $\mathcal{S}$. That is, they are mutually exclusive and exhaustive. The corresponding set of probabilities, $\{P(A_1), P(A_2), \ldots, P(A_K)\}$, is a probability distribution. In general,

any statement of a probability function having a set of mutually exclusive and exhaustive events for its domain is a probability distribution.

As a simple example of a probability distribution, imagine a sample space of all employees of a particular firm. An employee is selected at random and classified as "right-handed," "left-handed," or "ambidextrous." The probability distribution might be as given in Table 3.2.1.

Table 3.2.1 Probability Distribution for Hand Preference of Employees

Classification	Probability
Right	.60
Left	.30
Ambidextrous	.10
	1.00

Or, perhaps, the height of an employee drawn at random is measured. Some seven class intervals are used to record the height of employees, and the probability distribution might be as given in Table 3.2.2.

Table 3.2.2 Probability Distribution for Height of Employees

Height in inches	Probability
78–82	.002
73–77	.021
68–72	.136
63–67	.682
58–62	.136
53–57	.021
48–52	.002
	1.000

The set of all possible elementary events forms a partition of S. As a result, the set consisting of the probabilities of all of the elementary events in a sample space is a probability distribution. If we throw a fair die, the distribution would be as given in Table 3.2.3.

Table 3.2.3 Probability Distribution of Number Showing on Die

Face coming up on die	Probability
1	1/6
2	1/6
3	1/6
4	1/6
5	1/6
6	1/6
	———
	1.0

However, suppose that we are interested in the *square* of the number of dots showing on the die. If we define the random variable X to be the square of the number of dots showing, then the probability distribution of X is as given in Table 3.2.4. All values of X other than the values listed in the table have probabilities of zero. Recall that a random variable defined on a sample space associates a real number with each and every elementary event in the sample space. In the example, the probability distribution on the elementary events determines the probability distribution of the random variable X. This exemplifies the fact that, in general,

a set of probabilities, or a probability distribution, for the elementary events in S determines a set of probabilities, or a probability distribution, for any random variable defined on S.

Table 3.2.4 Probability Distribution of Square of Number Showing on Die

x	$P(X = x)$
1	1/6
4	1/6
9	1/6
16	1/6
25	1/6
36	1/6
	———
	1.0

This introduces the important concept of a probability distribution, or simply a distribution, of a random variable.

In the above example, each value of the random variable has the same probability. This need not always be true. Consider the random variable Y, which is defined as the number of heads occurring in two flips of a fair coin. The probability distribution of Y is given in Table 3.2.5. In general, the various values of a random variable are not necessarily equally probable.

Table 3.2.5   Probability Distribution of Number of Heads in Two Tosses of a Fair Coin

y	$P(Y = y)$
0	1/4
1	1/2
2	1/4

In these examples we have specified probability distributions of random variables by simply listing the possible values of the random variables and the corresponding probabilities. If the possible values are few in number, this is an easy way to specify a probability distribution. On the other hand, if there is a large number of possible values, a listing may become cumbersome. In the extreme case, where we have an infinite number of possible values (for example, all real numbers between zero and one), it is clearly impossible to make a listing. Fortunately, there are other ways to specify a probability distribution; in some instances we may be able to specify a mathematical function from which the probability of any value or interval of values can be computed, or we may be able to express the distribution in graphical form. These methods will be discussed in the following sections. First, it is necessary to distinguish between two types of random variables, discrete and continuous.

3.3   DISCRETE RANDOM VARIABLES

There are many situations where the random variable X can assume only a particular *finite* or *countably infinite* set of values; recall from Chapter 1 that this means that the possible values of X are finite in number or they are infinite in number but can be put in a one-to-one correspondence with the positive integers. For example, consider a simple experiment consisting of selecting an individual at random and determining the number of months in the past year during which the individual made at least one trip of 50 miles or more from his or her home. Here, X can assume only the values 0, 1, 2, ... , 12. The numbers 3.68, -17, or π simply cannot be associated appropriately with any elementary event in the sample space. The random variable X can assume only a *finite* set of values. As another example, suppose that the simple experiment consists of tossing a fair coin repeatedly until a head appears. Let X be the number of tosses it takes to get the first head. The first head could appear on the first toss, in which case X would be 1; it might not appear until the one-hundredth toss, in which case X would be 100; in general, it might not appear until the Kth toss, where K is any positive integer. Here we have a *countably infinite* set of values for X.

If the random variable X can assume only a particular finite or countably infinite set of values, it is said to be a discrete random variable.

Not all random variables are discrete. Consider the simple experiment which consists of recording the temperature at a given place and time. Suppose that we can measure the temperature exactly; that is, our measuring device allows us to record the temperature to any number of decimal points. If X is the temperature reading, it is not possible for us to specify a finite or countably infinite set of values as the only possible values for X. If we specify a finite set of values, it is a simple matter to find more values that can be assumed by X, just by taking one of the finite set of values and adding more decimal places. For example, if one of the finite set of values is 75.965, we can determine values 75.9651, 75.9652, and so on, which are also possible values of X. What is being pointed out here is that the possible values of X consist of the set of real numbers, a set which contains an infinite (and uncountable) number of values. In this case, it is said that X is a **continuous random variable.** We will give a more formal definition of a continuous random variable in Section 3.5 after we finish discussing discrete random variables in Sections 3.3 and 3.4.

Given a discrete random variable taking on only the K different values $x_1, x_2, \ldots, x_K$, the following two conditions must be satisfied.

$$1. \quad P(X = x_i) \geq 0, \qquad i = 1, 2, \ldots, K. \qquad (3.3.1^*)$$

$$2. \quad \sum_{i=1}^{K} P(X = x_i) = 1. \qquad (3.3.2^*)$$

That is, the probability of each value that X can assume must be nonnegative and the sum of the probabilities over all of the different values must be 1.00. These two conditions follow directly from the axioms of probability presented in Section 2.4.

For example, let X represent the number of new houses that will be completed by a particular contractor during the coming month. Because of uncertainties concerning the weather and delivery of some appliances, the contractor is unsure as to how many houses will be completed, and the distribution of X is presented in Table 3.3.1. Note that the probabilities in Table 3.3.1 satisfy Equations (3.3.1) and (3.3.2); they are nonnegative, and they sum to one.

Table 3.3.1 Probability Distribution for Number of New Houses

x	$P(X = x)$
0	.02
1	.07
2	.10
3	.14
4	.18
5	.17
6	.10
7	.10
8	.08
9	.04
	1.00

Quite often we are interested in finding the probability that the obtained value of X will fall *between* two particular values, or in some interval. For instance, what is the probability that the contractor will finish at least 3 houses but no more than 5 houses? The various possible values of X are mutually exclusive, so we simply need to add the appropriate probabilities:

$$P(3 \leq X \leq 5) = P(X = 3) + P(X = 4) + P(X = 5)$$
$$= .14 + .18 + .17 = .49.$$

Similarly,

$$P(4 \leq X \leq 8) = P(X = 4) + P(X = 5) + P(X = 6)$$
$$+ P(X = 7) + P(X = 8)$$
$$= .18 + .17 + .10 + .10 + .08 = .63.$$

In general, the probability that X falls in the interval between any two numbers a and b, inclusive, is found by the sum of probabilities for X over *all possible* values between a and b, inclusive:

$$P(a \leq X \leq b) = \text{sum of } P(X = x) \text{ for all } x \text{ such that } a \leq x \leq b$$

$$= \sum_{a \leq x \leq b} P(X = x). \tag{3.3.3*}$$

By the same argument,

$$P(X \geq a) = \text{sum of } P(X = x) \text{ for all } x \text{ such that } x \geq a. \tag{3.3.4}$$

Furthermore,

$$P(X < a) = 1 - P(x \geq a) = \sum_{x < a} P(X = x). \tag{3.3.5}$$

For the contractor example, the probability of completing 5 or more houses is

$$P(X \geq 5) = P(X = 5) + P(X = 6) + \cdots + P(X = 9)$$

$$= .17 + .10 + .10 + .08 + .04 = .49,$$

and the probability of completing less than 5 houses is

$$P(X < 5) = P(X = 0) + P(X = 1) + \cdots + P(X = 4)$$

$$= .02 + .07 + .10 + .14 + .18 = .51.$$

Alternatively, the latter probability could have been computed as follows once the former probability was computed:

$$P(X < 5) = 1 - P(X \geq 5) = 1 - .49 = .51.$$

Be careful to note that in probability calculations such as those given above, it makes a difference whether or not the endpoints of intervals are included. For instance, $P(X < 5)$ was just computed to be .51, but $P(X \leq 5)$ is .68:

$$P(X \leq 5) = P(X = 0) + P(X = 1) + \cdots + P(X = 4) + P(X = 5)$$

$$= P(X < 5) + P(X = 5) = .51 + .17 = .68.$$

Similarly, $P(3 \leq X \leq 5) = .49$, but $P(3 < X < 5) = .18$, $P(3 \leq X < 5) = .32$, and $P(3 < X \leq 5) = .35$. In computing probabilities for intervals of values of a discrete random variable, it is important to note whether neither, one, or both of the endpoints are included in the interval in each instance.

Thus, once the probability distribution of a discrete random variable is specified, probabilities of intervals can be determined. In this section we specified the probability distribution in the example by a listing. This is just one possible way to specify the distribution, and often it is not the most convenient way, as we shall see in the next section.

3.4   PROBABILITY DISTRIBUTIONS OF DISCRETE RANDOM VARIABLES

The probability distribution of a discrete random variable can be specified by a listing of the possible values of the random variable together with the corresponding probabilities. Such a listing is often represented by a graph. Suppose that the set of possible values of a random variable X can be denoted by $\{x_1, x_2, \ldots, x_K\}$. Then the probability distribution can be represented by the set of pairs of values $(x_i, P(X = x_i))$, where $i = 1, 2, \ldots, K$. We can graph such pairs in the Cartesian plane, as shown in Figure 3.4.1. The graph can be thought of as placing a mass of probability $P(X = x_i)$ at the point x_i on the x-axis. As a result, such a graph is often referred to as the graph of a **probability mass function** (PMF), or simply a **probability function.** The advantage of such a graph over a listing is that it is generally easier to read and gives the reader a better notion of the nature of the probability distribution.

For example, consider the PMF of the random variable X in the contractor example in the previous section; this PMF is shown in Figure 3.4.2. As another example, consider the PMF of the random variable Y from Section 3.2: the number of heads in two flips of a fair coin. This PMF is shown in Figure 3.4.3.

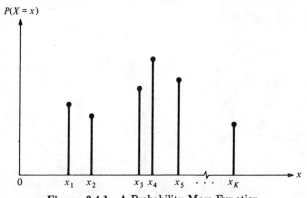

Figure 3.4.1   A Probability Mass Function

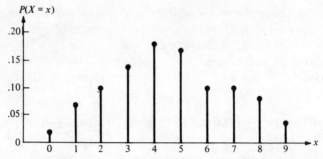

Figure 3.4.2   PMF for Number of New Houses

In some cases it may be possible to express the probability distribution of a discrete random variable in the form of a mathematical function. The simplest example of this procedure concerns the case in which there are exactly K possible values of the random variable X, the integers 1 through K, and each value has probability $1/K$. This distribution can be expressed as follows:

$$P(X = x) = \begin{cases} 1/K & \text{if } x = 1, 2, \ldots, K, \\ 0 & \text{elsewhere.} \end{cases}$$

It is important to specify that the PMF is equal to zero at all values other than $1, 2, \ldots, K$. Since the PMF is a function on the real line, it must be defined at all points on the real line. For an example involving equal probabilities, let X represent the number that comes up on a single toss of a fair die. The probability distribution of X, expressed graphically in Figure 3.4.4, can be expressed in functional form as follows:

$$P(X = x) = \begin{cases} 1/6 & \text{if } x = 1, 2, \ldots, 6, \\ 0 & \text{elsewhere.} \end{cases}$$

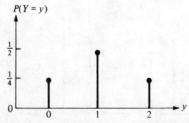

Figure 3.4.3   PMF for Number of Heads in Two Tosses of a Fair Coin

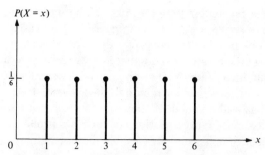

Figure 3.4.4 PMF for Number Showing on a Die

Consider once again the random variable Y, the number of heads in two flips of a fair coin. The PMF of Y can be represented by

$$P(Y = y) = \begin{cases} \dfrac{\dbinom{2}{y}}{4} & \text{if } y = 0, 1, \text{ or } 2, \\ \\ 0 & \text{elsewhere.} \end{cases}$$

There are four possible outcomes (hence the 4 in the denominator), and the number of ways of getting y heads in 2 flips is $\binom{2}{y}$. By calculating the specific values of the PMF from the above formula, you can verify for yourself that the numbers are equal to those in the listing of the distribution of Y given in Section 3.2.

In this section, we have seen how a discrete random variable can be represented in graphical or functional form. Whenever it is possible to use one or both of these two forms, they are usually preferable to a simple listing. In most cases they require less space than a listing and they are easier to understand. When we discuss the most important classes of discrete random variables in Chapter 4, we will rely almost exclusively on functional and graphical representations of probability distributions.

3.5 CONTINUOUS RANDOM VARIABLES

The previous sections dealt with *discrete* random variables, which take on only a finite or countably infinite number of values. There are many variables that, conceptually at least, could take on an uncountably infinite number of values. For example, consider the weight of an object or the temperature at a particular time and place. The underlying sample space in these examples consists not just of integers or decimal expressions given to the nearest one-hundredth; conceptually, variables such as weight and temperature can take on any real number as a value. In practice, limitations of measurement devices imply that the *observed* weight or temperature is discrete. For instance, a balance may only be accurate to the nearest tenth of an ounce. It is still convenient, however, to use *continuous* probability models as approximate representations in such situations.

Formally, a continuous random variable may be defined as follows.

A random variable X is said to be continuous if for every pair of values u and v such that $P(X \leq u) < P(X \leq v)$, it is true that $P(u < X < v) > 0$.

Notice that for a random variable to be continuous in an interval limited by the values u and v, it is necessary that every nonzero interval in this range have a nonzero probability. For this to be true, there must be an infinite number of possible values that the random variable can assume.

As noted above, continuous random variables are idealizations. The fact that not even in the most precise work known can accuracy be obtained to *any* number of decimals limits the actual possibility of encountering a continuous random variable in practice. Why, then, do statisticians so often deal with these idealizations? The answer is that, mathematically, distributions of continuous random variables often are far more tractable than distributions of discrete random variables. The function rules for many continuous variables are relatively easy to state and to study using the full power of mathematical analysis. This is not generally true of discrete distributions. Moreover, continuous distributions are very good approximations to many discrete distributions. This fact makes it possible to organize statistical theory about a few such idealized distributions and to find methods that are good approximations to results for the more complicated discrete situations. Nevertheless, the student should realize that these continuous distributions are mathematical abstractions that happen to be quite useful; they do not necessarily describe real situations exactly.

3.6 PROBABILITY DISTRIBUTIONS OF CONTINUOUS RANDOM VARIABLES

In Section 3.3 it was possible for us to discuss discrete random variables in terms of probabilities such as $P(X = x)$, the probability that the random variable X takes on some value *exactly*. **However, for continuous random variables, the occurrence of any exact value of X may be regarded**

as having zero probability. Instead of considering probabilities for single values, then, it is necessary to consider probabilities for intervals of values, such as $P(a \leq X \leq b)$. Thus, if X represents the maximum official temperature in degrees Fahrenheit at Chicago on December 25, 1990, and if it is assumed that X is continuous, then it is worthwhile to consider probabilities such as $P(20 \leq X \leq 40)$, $P(25 \leq X \leq 35)$, $P(29 \leq X \leq 31)$, $P(29.9 \leq X \leq 30.1)$, $P(29.99 \leq X \leq 30.01)$, and so on, but it is *not* worthwhile to consider $P(X = 30)$. The smaller the interval, the smaller the probability, and the probability that the high temperature is *exactly* 30 is, in effect, zero.

In the continuous case, then, we do not consider the probability that X takes on some value x. Instead, we deal with the so-called **probability density** of X at x, symbolized by

$$f(x) = \text{probability density of } X \text{ at } x.$$

A fairly rough definition of a probability density can be given as follows. For any distribution, it is proper to speak of the probability associated with an interval. Imagine an interval with limits a and b. Then the probability of that interval is

$$P(a \leq X \leq b) = P(X \leq b) - P(X < a).$$

Now suppose that we let the size of the interval be denoted by

$$\Delta x = b - a,$$

so that $\qquad\qquad\qquad\qquad b = a + \Delta x.$

Then the probability of the interval *relative* to Δx is

$$\frac{P(a \leq X \leq b)}{\Delta x} = \frac{P(X \leq a + \Delta x) - P(X < a)}{\Delta x}.$$

Now, suppose that we fix a, but allow Δx to vary; in fact, let Δx approach zero in size. What happens to the probability of this interval relative to the interval size? Both numbers will change, and the ratio will change, but this ratio will approach some **limiting value** as Δx comes close to zero. That is, we can speak of the limit of $P(a \leq X \leq b)/\Delta x$ as Δx approaches zero. *This limit gives the probability density of the variable X at value a.* Loosely speaking, we can say that the probability density at a is the *rate of change* in the probability of an interval with lower limit a, for minute changes in the size of the interval. This rate of change will depend on two things: the function rule assigning probabilities to intervals such as $(X \leq a)$ and the particular "region" of X values we happen to be talking about. Rather than talk about probabilities of X values per se, for continuous random variables it is

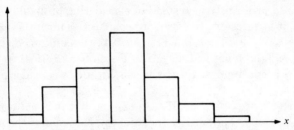

Figure 3.6.1   A Histogram with 7 Intervals

mathematically far more convenient to discuss the probability density and reserve the idea of probability only for the discussion of *intervals* of X values. This need not trouble us very much, as we will generally be interested only in intervals of values in the first place.

The above development of a probability density function (PDF) becomes clearer when it is presented in graphical form. Consider a histogram (bar graph) in which the probability of an interval is represented by the area of the "bar" over the interval (Figure 3.6.1). This area is, of course, given by the product of the height of the bar and the width of the interval; thus, the height of the bar equals the probability of the interval divided by the width of the interval. If each of the intervals is divided into two subintervals, a new histogram could be drawn using the new subintervals, as in Figure 3.6.2. If this process of dividing intervals is repeated over and over, the width of each interval becomes smaller and the histogram begins to look more like a smooth curve. In the limit, it *is* a smooth curve, the probability density function (Figure 3.6.3). Since the height of a bar of the histogram equals the probability of the interval divided by the width of the interval, the limit of this ratio is the height of the density function, as indicated in the previous paragraph.

To illustrate this process, suppose that X represents the weight (in pounds) of a person randomly chosen from the student body of a par-

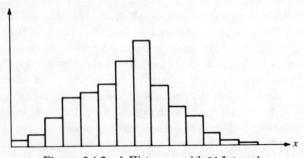

Figure 3.6.2   A Histogram with 14 Intervals

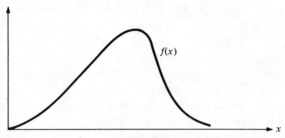

Figure 3.6.3 A Probability Density Function

ticular university. First, the weights might be grouped into 50-pound intervals, yielding a histogram comparable to Figure 3.6.1. Next, to obtain greater precision, the weights might be grouped into 25-pound intervals, yielding a histogram comparable to Figure 3.6.2. This process of reducing the length of each interval can be repeated, yielding histograms with 12.5-pound intervals, 6.25-pound intervals, and so on. As the interval length decreases (and correspondingly, the number of intervals increases), the histogram looks more and more like the smooth curve illustrated in Figure 3.6.3.

Intervals can always be given probabilities, regardless of whether the random variable is discrete or continuous. For continuous random variables the probability of an interval depends on the probability density associated with each value in the interval. The probability of the interval from a to b is

$$P(a \leq X \leq b) = \int_a^b f(x)\ dx. \qquad (3.6.1^*)$$

Since the definite integral of a function over some interval corresponds to the area under the curve of the function in that interval, **we can say that the probability of an interval is the same as the area cut off by that interval under the curve for the probability densities when the random variable is continuous, and the total area is equal to 1.00.** The expression $f(x)\ dx$ can be thought of as the area of a minute interval with midpoint $X = x$, somewhere between a and b. When the number of such intervals approaches an infinite number and their size approaches zero, the sum of all these areas is the *entire* area cut off by the limits a and b. Since there is an infinite number of such intervals, this sum is expressed by the definite integral sign

$$\int_a^b .$$

Readers who are uncomfortable with calculus, then, can think of a definite integral as the limiting form of a sum or as the area under a curve.

Note that this agrees with the definition of the probability of an interval for a discrete variable. In Section 3.3 it was stated that the probability that X lies between two numbers a and b, inclusive, is a sum of probabilities. Since a definite integral is analogous to a sum, the probability of an interval for a continuous random variable is analogous to the probability of an interval for a discrete random variable.

As noted above, when X is continuous, $P(X = a) = 0$ and $P(X = b) = 0$. As a result, the probability of an interval for a continuous random variable is the same whether or nor the endpoints of the interval are included. Obviously, inclusion or exclusion of the endpoints will not affect the area given by

$$\int_a^b f(x) \, dx.$$

In the discrete case, on the other hand, the points *do* make a difference. If X is the number coming up on one throw of a fair die, for instance,

$$P(2 < X < 4) = P(X = 3) = 1/6,$$

while

$$P(2 \leq X \leq 4) = P(X = 2) + P(X = 3) + P(X = 4) = 3/6 = 1/2.$$

It should be strongly emphasized that the height of the density function does not represent probability; indeed, the probability of any single point is 0. **It is the area under the curve that represents probability.** To satisfy the basic axioms of probability theory, a probability density function $f(x)$ must satisfy two properties:

1. $f(x) \geq 0$ for all $x,$ $\qquad\qquad\qquad$ (3.6.2*)

2. $\int_{-\infty}^{\infty} f(x) \, dx = 1.$ $\qquad\qquad\qquad$ (3.6.3*)

That is, $f(x)$ cannot be negative, and the total area under the curve must equal 1. These conditions are analogous to the requirements imposed on a discrete PMF: $P(X = a) \geq 0$ and $\sum_x P(X = x) = 1$. In order to clarify the concept of a continuous random variable and to emphasize the analogy between discrete and continuous random variables, a graphical analysis is quite valuable.

A PDF for some hypothetical distribution is shown in Figure 3.6.4. The two points a and b marked off on the horizontal axis represent limits

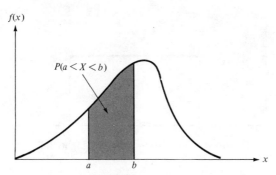

Figure 3.6.4 Probability as Area Under a PDF

of some interval. The shaded portion between a and b shows

$$P(a \leq X \leq b) = \int_a^b f(x) \ dx,$$

the probability that X takes on a value between the limits a and b.

For an example of a specific PDF, consider X, the length of life (in hours) of a light bulb. Suppose that X is uniformly distributed over the interval from 80 to 100. That is, the light bulb is sure to last at least 80 hours, but it will not last longer than 100 hours. Between 80 hours and 100 hours, the probability is uniformly distributed. This means that if we divide the interval from 80 to 100 into subintervals of equal width i, then these subintervals have the same probabilities and we can do this for any value of i. The probability density function of X can be represented in functional form:

$$f(x) = \begin{cases} 1/20 & \text{if } 80 \leq x \leq 100, \\ 0 & \text{elsewhere.} \end{cases}$$

Notice that $f(x)$ satisfies the two requirements of a PDF, since $f(x)$ is nonnegative and

$$\int_{-\infty}^{\infty} f(x) \ dx = \int_{-\infty}^{80} 0 \ dx + \int_{80}^{100} \left(\frac{1}{20}\right) dx + \int_{100}^{\infty} 0 \ dx$$

$$= \int_{80}^{100} \left(\frac{1}{20}\right) dx = \left[\frac{x}{20}\right]\Big|_{80}^{100} = \left(\frac{100}{20}\right) - \left(\frac{80}{20}\right)$$

$$= 1.$$

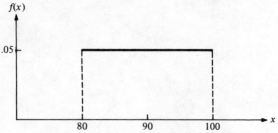

Figure 3.6.5   PDF for Life of a Light Bulb

The graph of $f(x)$ is shown in Figure 3.6.5. Note that the total area under $f(x)$ is simply the area of a rectangle with length 20 and height 1/20. In this example, it is easy to calculate the probability of any interval. Suppose that $80 \leq a < b \leq 100$. Then $P(a \leq X \leq b)$ is the area of a rectangle with length $(b - a)$ and height 1/20; this area is simply $(b - a)/20$. Formally,

$$P(a \leq X \leq b) = \int_a^b \left(\frac{1}{20}\right) dx = \left[\frac{x}{20}\right]\Big|_a^b$$

$$= \frac{(b - a)}{20}.$$

It should be pointed out that it is possible for a random variable to be discrete over part of its range and continuous over another part of its range. Let X represent the amount of rain (in inches) that will fall on a particular summer day in Paris, France. Suppose that $P(X = 0) = .6$ and that the remaining .4 of probability is distributed according to the density function

$$f(x) = \begin{cases} \dfrac{4 - x}{20} & \text{if } 0 < x < 4, \\[2ex] 0 & \text{elsewhere.} \end{cases}$$

The graph of the distribution of X is a combination of a PMF and a PDF, as shown in Figure 3.6.6, and a random variable such as this is often called a **mixed random variable.** Mixed random variables arise in practice because some random variables are continuous except at a few points with positive probability. In the rain example, the quantity of rain is continuous provided that it rains, but there is a positive probability

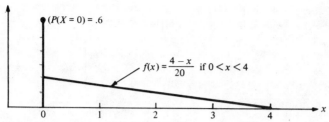

Figure 3.6.6   Probability Distribution of a Mixed Random Variable: The Amount of Rain

(.6) that it will not rain at all. Such random variables can be handled easily with a combination of techniques for working with discrete random variables and techniques for working with continuous random variables.

3.7   CUMULATIVE DISTRIBUTION FUNCTIONS

We have seen that the probability distribution of a random variable X can be represented by a probability mass function (PMF) in the discrete case and by a probability density function (PDF) in the continuous case. There is a useful alternative method for specifying a probability distribution for both discrete and continuous random variables. This involves the concept of cumulative probabilities.

The probability that a random variable X takes on a value at or below a given number x is often written as

$$F(x) = P(X \le x). \tag{3.7.1*}$$

The symbol $F(x)$ represents the function relating the various values of X to the corresponding cumulative probabilities. This function is called the cumulative distribution function (CDF).

A CDF must satisfy certain mathematical properties, the most important of which are:

1. $0 \le F(x) \le 1$;   (3.7.2*)

2. if $a < b$, $F(a) \le F(b)$;   (3.7.3*)

3. $F(\infty) = 1$ and $F(-\infty) = 0$.   (3.7.4*)

The first property reflects the fact that $F(x)$ is a probability, $P(X \le x)$, and properties 2 and 3 follow directly from the definition of a CDF.

Given the PMF of a discrete random variable or the PDF of a continuous random variable, the cumulative distribution function is uniquely determined. If X is discrete,

$$F(x) = P(X \leq x) = \sum_{a \leq x} P(X = a). \qquad (3.7.5^*)$$

If X is continuous,

$$F(x) = P(X \leq x) = \int_{-\infty}^{x} f(x)\ dx. \qquad (3.7.6^*)$$

Consider the example in which X represented the number of heads occurring in two flips of a fair coin. The PMF was

$$P(X = x) = \begin{cases} 1/4 & \text{for } x = 0 \text{ or } x = 2, \\ 1/2 & \text{for } x = 1, \\ 0 & \text{elsewhere.} \end{cases}$$

To compute the CDF, note that $F(x)$ will be 1 for all values of x greater than or equal to 2, since X is certain to be less than or equal to all of these values. Also, $F(x)$ will be zero for all negative values of x, since X cannot possibly be smaller than a negative number (it only takes on the values 0, 1, and 2). Similarly, we can determine the values of the CDF between 0 and 2, with the result shown in Figure 3.7.1.

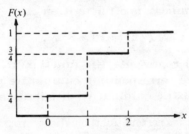

Figure 3.7.1 CDF for Number of Heads in Two Tosses of a Fair Coin

In the discrete case, the CDF is a step function: a step, or discontinuity in the CDF, occurs at any point x such that $P(X = x) > 0$. As a result, there will be K such "steps" in the CDF of X if X can take on any one of K values with positive probability. Furthermore, the amount of the step is equal to $P(X = x)$. To see this graphically, compare the graphs of the PMF and the CDF in the coin example (Figures 3.4.3 and 3.7.1). It can be seen that $F(x)$ has a step of $1 - (3/4) = 1/4$ at $x = 2$, a step of $(3/4) - (1/4) = 1/2$ at $x = 1$, and a step of $(1/4) - 0 = 1/4$ at $x = 0$. Thus, if we know the CDF of a discrete random variable, we can determine the corresponding PMF, and vice versa.

The CDF for the coin example can also be expressed in functional form:

$$F(x) = \begin{cases} 1 & \text{for } x \geq 2, \\ 3/4 & \text{for } 1 \leq x < 2, \\ 1/4 & \text{for } 0 \leq x < 1, \\ 0 & \text{for } x < 0. \end{cases}$$

The CDF, like the PMF, is a function defined on the real line, and in the example it is assigned exactly one value for each real number.

For an example of the CDF of a continuous random variable, recall the situation in which X was uniformly distributed over the interval from 80 to 100, where X represented the life (in hours) of a light bulb. In this case, $F(x)$ represents the probability that the life of the bulb is less than or equal to x hours. Clearly, $F(x)$ is equal to 1 for values of x greater than 100, since the light bulb cannot last longer than 100 hours according to our assumptions in Section 3.6. Similarly, $F(x)$ is equal to 0 for values of x smaller than 80, for we stated that the bulb is sure to last at least 80 hours. For values between 80 and 100,

$$F(x) = \int_{-\infty}^{x} f(x)\, dx = \int_{-\infty}^{80} 0\, dx + \int_{80}^{x} \left(\frac{1}{20}\right) dx$$

$$= \left[\frac{x}{20}\right]\Bigg|_{80}^{x} = \frac{(x - 80)}{20}.$$

Thus, the CDF for this example, which is illustrated in Figure 3.7.2, can

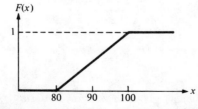

Figure 3.7.2   CDF for Life of a Light Bulb

be stated in functional form as follows:

$$F(x) = \begin{cases} 1 & \text{if } x > 100, \\ (x - 80)/20 & \text{if } 80 \leq x \leq 100, \\ 0 & \text{if } x < 80. \end{cases}$$

Observe that the CDF in this example is a continuous function rather than a step function. This is true for all continuous random variables. In fact, the definition of a continuous random variable can be given in terms of the CDF.

A random variable X is said to be continuous if its cumulative distribution function $F(x)$ is a continuous function. On the other hand, the CDF of a discrete random variable is always a step function rather than a continuous function.

If we know the CDF of a continuous random variable, it is possible to find the density function, $f(x)$. The relationship is given by

$$f(x) = \frac{d}{dx} F(x) = F'(x). \tag{3.7.7*}$$

In general, the derivative of the CDF is equal to the PDF. This follows from a fundamental theorem of calculus which requires that $F(x)$ be a continuous function. The reader not familiar with this theorem can safely take the above statement for granted and not worry about its derivation. A graphical interpretation of Equation (3.7.7) is that the value of the density function at a point x is equal to the slope of the CDF at that point. The derivative of a function at a point x corresponds to the slope of the line tangent to the function at that point.

In the previous example, it follows that

$$f(x) = \frac{d}{dx} F(x) = \begin{cases} \dfrac{d}{dx} (1) = 0 & \text{if } x > 100, \\ \dfrac{d}{dx} \left(\dfrac{x - 80}{20}\right) = \dfrac{1}{20} & \text{if } 80 \leq x \leq 100, \\ \dfrac{d}{dx} (0) = 0 & \text{if } x < 80. \end{cases}$$

This is identical to $f(x)$ as given in Section 3.6. In geometric terms, the value of the density function at any point x is equal to the slope of the line tangent to the CDF at the point x. In the example, $F(x)$ is a straight line for $80 \leq x \leq 100$. At any point x, the tangent line is identical to the line representing $F(x)$. Then, since this line has a given slope which is obviously constant for all x between 80 and 100, the density function must simply be constant for all x between 80 and 100. Compare the graphs of the PDF and CDF (Figures 3.6.5 and 3.7.2).

In general (that is, for distributions other than uniform distributions), density functions are not constant (graphically, they are not horizontal lines). This means that, in general, cumulative distribution functions are not straight lines, as in the above example. Cumulative distribution functions for continuous random variables often have more or less the characteristic S-shape shown in Figure 3.7.3.

The probability that a random variable takes on any value between limits a and b (including b but not a) can be found from

$$P(a < X \leq b) = F(b) - F(a). \qquad (3.7.8^*)$$

This is seen easily if it is recalled that $F(b)$ is the probability that X takes on the value b or below, $F(a)$ is the probability that X takes on the value a or below: their difference must be $P(a < x \leq b)$. For continuous random variables, this probability is unchanged if the $<$ sign is changed to $\leq$ or vice versa, since the probability that X exactly equals any particular number such as a or b is zero. However, for discrete variables, $<$ and $\leq$ signs may lead to different probabilities.

The symbol $F(x)$ can be used to represent the cumulative probability that X is less than or equal to x for either a continuous or a discrete random variable. All random variables have cumulative distribution functions. However, remember that only continuous variables have density functions and that the value of a density function at a particular value of a random variable is *not* a probability. To avoid possible confusion, the density notation $f(x)$ will be reserved for *continuous* variables in all of the following, and $P(x)$ will be used both for discrete variables and for *intervals* of values of continuous variables, as in $P(a \leq X \leq b)$. Occasionally in mathematical statistics the terms "distribution function" or

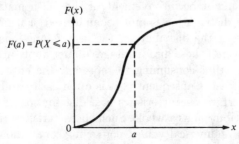

Figure 3.7.3 One Type of CDF for a Continuous Random Variable

"probability function" are used to refer only to the *cumulative* distribution of a random variable, and "density function" is used where we have used *probability* distribution. However, we believe that the terms used here are simpler for the beginning student to learn and use.

Tables of theoretical probability distributions are most often given in terms of the cumulative distribution, or $F(x)$, of a random variable. Given these cumulative probabilities for various values of X, it is easy to find the probability for any interval by subtraction. In particular, when we use the table of the so-called "normal" distribution, we will find that cumulative probabilities are shown.

3.8 SUMMARY MEASURES OF PROBABILITY DISTRIBUTIONS

As we have seen, the probability distribution of a random variable, whether discrete or continuous, can be represented in several alternative ways. If a graphical analysis is desired, a distribution can be represented by a probability mass function (PMF) in the discrete case and by a probability density function (PDF) in the continuous case; or it can be represented by a cumulative distribution function (CDF) in either case. Sometimes it is also possible to represent a probability distribution in functional notation, provided there is some relatively simple mathematical function that can describe the PMF, PDF, or CDF. In the discrete case, a probability distribution can be represented by a simple listing of all possible values of the random variable and the probability corresponding to each value (provided that the random variable takes on only a finite number of values). This last form of representation could also be used as an approximation in the continuous case if the possible values of the random variable are grouped into a finite set of mutually exclusive and exhaustive intervals on the real line.

Thus, there are several ways to represent the probability distribution of a random variable. Each of these methods has some disadvantages. If there are numerous possible values for a discrete random variable, a listing or a graph might be inconvenient; if the mathematical function corresponding to a distribution is quite complicated or if no function can be found to represent the distribution, the use of functional notation might be infeasible. Even if these disadvantages do not apply in a particular case, it may be quite time-consuming to represent the entire probability distribution of interest. Consequently, it is often easier and more efficient to look only at certain characteristics of a distribution than to attempt to specify the distribution as a whole. Such characteristics can be summarized into one or more numerical values, conveying some, though not all, of the information in the entire distribution.

Two such general characteristics of any distribution are its measures of central tendency (or location) and dispersion (or variability). Indices of central tendency are ways of describing the "typical" or the "average" value of the random variable. Indices of dispersion, on the other hand, describe the "spread" or the extent of the difference among the possible values of the random variable.

In the following sections we will discuss several measures of central tendency and dispersion. In many cases, it is possible to adequately describe a probability distribution with a few such measures. It should be remembered, however, that these measures serve only to summarize some important features of the probability distribution; in general, they do not completely describe the entire distribution. In some situations in statistical analysis the entire distribution is of genuine interest, and the summary measures are not sufficient for the problem being attacked. On the other hand, as we will see in Chapter 4, if we know that the distribution is one of a number of frequently encountered types of distributions, it may be possible to get a complete picture of the distribution from a few summary measures.

One of the most common and most useful summary measures of a probability distribution is a measure known as the *expectation* of a random variable. The concept of expectation plays an important role not only as a useful summary measure, but also as a central concept in the theory of probability and statistics.

3.9 THE EXPECTATION OF A RANDOM VARIABLE

A very prominent place in theoretical statistics is occupied by the concept of the mathematical **expectation** of a random variable X.

If the distribution of X is discrete, then the expectation (or expected value) of X is defined to be

$$E(X) = \sum_x xP(X = x) = \sum_x xP(x), \qquad (3.9.1^*)$$

where the sum is taken over all of the different values that the variable X can assume. For a continuous random variable X, the expectation is defined to be

$$E(X) = \int_{-\infty}^{\infty} xf(x)\ dx. \qquad (3.9.2^*)$$

Note the similarity between the discrete and continuous cases. In the discrete case, $E(X)$ is a sum of products of values of X and their probabilities; in the continuous case, $E(X)$ is an integral (which is, after all, a sum in the limiting case) of products of values of X and the corresponding values of the probability density function of X. A mathematical note of interest is that the sum in Equation (3.9.1) or the integral in Equation (3.9.2) may not converge to a finite value, in which case we say that the expectation does not exist. Since the expectations which we will deal with in this book exist in all but a few relatively unimportant cases, we need not concern ourselves with this mathematical problem.

As the discussion in Sections 2.16 and 2.17 suggests, the idea of expectation of a random variable is closely connected with the origin of statistics in games of chance. Gamblers were interested in how much they could "expect" to win in the long run in a game, and in how much they should wager in certain games if the game was to be "fair." Thus, expected value originally meant the expected long-run winnings (or losings) over repeated play; this term has been retained in mathematical statistics to mean the average value for any random variable over an indefinite number of samplings. This holds whether a large number of samplings will actually be conducted or whether the situation is a one-trial affair and we consider hypothetical repetitions of the situation. Over a long series of trials, we can "expect" to observe the expected value. In general, however, on any *single* trial we cannot realistically "expect" the expected value; often the expected value is not even a possible value of the random variable for any single trial, as the following examples illustrate.

The use of expectation in a game of chance is easy to illustrate. For example, suppose that someone is setting up a lottery, selling 1000 tickets at $1 per ticket. He is going to give a prize of $750 to the winner of the first draw. Now suppose that you buy a ticket. How *good* is this ticket in the sense of *how much you should expect to gain*? You can think of your chances of winning and losing as represented by a probability distribution where the outcome of any drawing falls into one of two event categories: This distribution is given in Table 3.9.1, and in Table 3.9.2 the distribu-

Table 3.9.1   Events and Probabilities in a Lottery

Event	Probability
Win	1/1000
Do not win	999/1000

Table 3.9.2   Probability Distribution of Gain from a Lottery

x	$P(x)$
$750	1/1000
−$ 1	999/1000

tion of the random variable X is given. Applying Equation (3.9.1), we find that

$$E(X) = \sum xP(x) = 749 \left(\frac{1}{1000}\right) + (-1) \left(\frac{999}{1000}\right)$$

$$= .749 - .999 = -.25.$$

This amount, a *minus* 25 cents, is the amount which you can expect to gain by buying the ticket, meaning that if you played the game over and over (that is, if you repeatedly bought tickets in lotteries like this one), on the average you would be poorer by about 25 cents per trial.

On the other hand, suppose that the prize offered is $1000, so that the gain in winning is $999. Now the expected value is

$$E(X) = 999 \left(\frac{1}{1000}\right) + (-1) \left(\frac{999}{1000}\right) = 0.$$

Here the game is more worthwhile, since there is at least no amount of money to be lost *or* gained in the long run. Such a game is often called "fair." Obviously, truly fair lotteries and other games of chance are hard to find, since the purpose of most games of chance is to make money for the proprietors, not to break even or to lose money. If the person running the lottery is foolish enough to make the prize $2000, your expected value from buying a ticket is $1, a positive amount.

Instead of a game of chance such as a lottery, consider X, the gain from purchasing 100 shares of a given stock and selling the 100 shares one year from now. The probability distribution of X is given in Table 3.9.3, and the expected value of X is

$$E(X) = (-500)(.03) + (-250)(.07) + (0)(.10) + (250)(.25)$$

$$+ (500)(.35) + (750)(.15) + (1000)(.05) = 367.5.$$

Therefore, the expected gain from purchasing the stock is $367.50.

Table 3.9.3 Probability Distribution of Gain from 100 Shares of a Stock

x	$P(x)$
-500	.03
-250	.07
0	.10
250	.25
500	.35
750	.15
1000	.05

In betting situations and investment situations such as those given in the above examples, the random variable is, of course, gains or losses of amounts of money. Nevertheless, the same general idea applies to any random variable; the expectation is the long-run average value that we should observe. The expectation can be thought of in this way even if the experiment (whether a betting situation or not) is strictly a one-trial affair and will not be repeated. No assumption has been made regarding the *interpretation* of the probabilities used to determine $E(X)$.

3.10   THE ALGEBRA OF EXPECTATIONS

In essence, the expectation of a random variable, whether the random variable is discrete or continuous, is a kind of weighted sum of values, and thus the rules for the algebraic treatment of expectations are basically extensions and applications of the rules of summation. These rules apply either to discrete or to continuous random variables if particular boundary conditions exist; for our purpose these rules can be used without our going further into these special qualifications. Proofs will be presented for the discrete case only, since the proofs for the continuous case are virtually identical, with integrals replacing summation signs. The student is advised to become familiar with these rules. If he does so, he should have little trouble in following simple derivations in statistics such as this book contains.

RULE 1

If $g(X)$ is some function of X, then

$$E(g(X)) = \sum_x g(x)P(x) \qquad \textbf{in the discrete case} \qquad (3.10.1^*)$$

and

$$E(g(X)) = \int_{-\infty}^{\infty} g(x)f(x)\,dx \qquad \textbf{in the continuous case.} \quad (3.10.2^*)$$

The proof of Rule 1 follows from the definition of expectation that was presented in the previous section, but a formal proof utilizes the notion of transformation of variables, which we have not discussed, so no proof will be given here.

To illustrate Rule 1, let $g(X) = 3X^2 + 2$, where X is the number coming up on a single toss of a fair die. The probability distribution of X places probability $1/6$ on each of the values 1, 2, 3, 4, 5, and 6. From Equation

(3.10.1), then,

$$E(g(X)) = [3(1)^2 + 2](\tfrac{1}{6}) + [3(2)^2 + 2](\tfrac{1}{6}) + \cdots + [3(6)^2 + 2](\tfrac{1}{6})$$
$$= 47.5.$$

RULE 2

If a is some constant number, then

$$E(a) = a. \tag{3.10.3*}$$

That is, if the same constant value a were associated with each and every elementary event in some sample space, the expectation or mean of the values would most certainly be a. For a discrete random variable X,

$$E(a) = \sum_x aP(x) = a \sum_x P(x) = a.$$

Rule 2 says that if you will gain \$$a$ no matter what event occurs, your expected gain is obviously \$$a$.

RULE 3

If a is some constant real number and X is a random variable with expectation $E(X)$, then

$$E(aX) = aE(X). \tag{3.10.4*}$$

Suppose that a new random variable is formed by multiplying each value of X by the constant number a. Then the expectation of the new random variable is just a times the expectation of X. This is very simple to show for a discrete random variable X:

$$E(aX) = \sum_x axP(x) = a \sum_x xP(x) = aE(X).$$

In the context of buying shares of a stock, Rule 3 says that your expected gain from buying, say, 100 shares is simply 100 times your expected gain from buying just one share.

RULE 4

If a is a constant real number and X is a random variable, then

$$E(X + a) = E(X) + a. \tag{3.10.5*}$$

This can be shown very simply for a discrete variable. Here,

$$E(X + a) = \sum_x (x + a)P(x)$$
$$= \sum_x xP(x) + a \sum_x P(x)$$
$$= E(X) + a.$$

In the context of buying shares of a stock, Rule 4 says that if an additional commission of, say, $25 is charged for the transaction, then your expected gain is simply reduced by $25.

The expectations of functions of random variables, such as

$$E[(X + 2)^2],$$
$$E(\sqrt{X + b}),$$

and $\qquad\qquad E(b^X),$

to give only a few examples, are subject to the same algebraic rules as summations. That is, the operation indicated within the punctuation is to be carried out *before* the expectation is taken. It is most important that this be kept in mind during any algebraic argument involving summations. In general,

$$E[(X + 2)^2] \neq [E(X) + E(2)]^2,$$
$$E(\sqrt{X}) \neq (\sqrt{E(X)}),$$
$$E(b^X) \neq b^{E(X)},$$

and so on. For the die example used to illustrate Rule 1, $E(3X^2 + 2) = 47.5$. But $E(X) = 1(1/6) + 2(1/6) + \cdots + 6(1/6) = 3.5$, so that $3[E(X)]^2 + 2 = 3(3.5)^2 + 2 = 38.75$. Therefore,

$$E(3X^2 + 2) \neq 3[E(X)]^2 + 2.$$

The next two rules concern two (or more) random variables, symbolized by X and Y or by $X_1, X_2, \ldots, X_K$. Proofs of these rules require the notion of joint probability distributions of two or more random variables. Joint probability distributions will not be discussed until later in the chapter, so the following rules will be presented without proof.

RULE 5

If X is a random variable with expectation $E(X)$ and Y is a random variable with expectation $E(Y)$, then

$$E(X + Y) = E(X) + E(Y). \qquad (3.10.6^*)$$

Verbally, this rule says that the expectation of a sum of two random variables is the sum of their expectations. In an investment situation, if you have an expected gain of $E(X)$ from one investment and an expected gain of $E(Y)$ from a second investment, then the expected gain from making *both* investments is simply the sum of the two individual expected gains.

Rule 5 holds regardless of any relationship (or lack of same) between X and Y. For instance, one of the random variables may even be in a functional relation to the other. If $Y = 3X^2$, then

$$E(X + Y) = E(X + 3X^2)$$
$$= E(X) + E(3X^2)$$
$$= E(X) + 3E(X^2).$$

This principle lets us *distribute* the expectation over an expression which itself has the form of a sum. We will make a great deal of use of this principle, which may be extended to any finite number of random variables.

RULE 6

Given some finite number of random variables, the expectation of the sum of those variables is the sum of their individual expectations. Thus,

$$E(X_1 + X_2 + \cdots + X_K) = E(X_1) + E(X_2) + \cdots + E(X_K). \quad (3.10.7^*)$$

In particular, some of the random variables may be in functional relations to others. Let $Y = 6X^4$, and let $Z = \sqrt{2X}$. Then

$$E(X + 6X^4 + \sqrt{2X}) = E(X) + 6E(X^4) + E(\sqrt{2X}).$$

3.11 MEASURES OF CENTRAL TENDENCY: THE MEAN skip to pg 169.

Imagine a probability distribution of a random variable X. If you were asked to state *one value* that would best communicate the **central tendency,** or **location,** of the distribution as a whole, which value should you choose? Could you determine a rule that would tell you what value to pick for *any* given distribution? You probably could not, since there are several alternatives available, each of which has certain advantages and certain disadvantages. The three most commonly used alternatives are (1) the expectation of X, $E(X)$, which is also called the **mean** of the distribution; (2) the value midway (in terms of probability) between the smallest possible value and the largest possible value, which is called the **median;** and (3) the value of X at which the PMF or PDF reaches its highest point, which is called the **mode.** We will briefly mention other measures of central tendency, but our primary concern will be with the mean, the median, and the mode. In this section we deal with the mean.

In many important instances of probability distributions, the mean or expectation is a parameter. That is to say, the mean enters into the function rule assigning a probability or probability density to each possible value of X. It will sometimes be convenient to use still another symbol when we are dealing with the mean of a random variable, especially in its

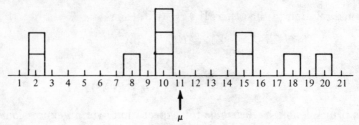

Figure 3.11.1   The Mean as a Center of Gravity

role as a parameter. The small Greek letter mu, μ, will stand for the mean of a probability distribution of a random variable X. That is,

$$E(X) = \mu = \textbf{mean of the distribution of } X. \quad (3.11.1^*)$$

In much of what follows, small Greek letters will be used to indicate parameters of probability distributions, while Roman letters will stand for sample values. The word **parameter** will always indicate a characteristic of a probability distribution and the word **statistic** will denote a summary value calculated from a sample. This distinction will become clearer in Chapter 5, and you need not be too concerned with it at this point.

The mean of a distribution parallels the physical idea of a center of gravity, or balance point, of ideal objects arranged in a straight line. For example, imagine an ideal board having zero weight. Along this board are arranged stacks of objects at various positions. The objects have uniform weight and differ from each other only in their positions on the board. The board is marked off in equal units of some kind, and each object is assigned a number according to its position. A drawing of the board and the objects looks like a PMF (or a histogram). This is shown in Figure 3.11.1. Now given this idealized situation, at what point would a fulcrum placed under the board create a state of balance? That is, what is the point at which the "push" of objects on one side of the board is exactly equal to the push exerted by objects on the other side? This point is the mean of the positions of the various objects:

$$\mu = 2\left(\frac{2}{10}\right) + 8\left(\frac{1}{10}\right) + 10\left(\frac{3}{10}\right) + 15\left(\frac{2}{10}\right) + 18\left(\frac{1}{10}\right) + 20\left(\frac{1}{10}\right) = 11.$$

Here the board would exactly balance if a fulcrum were placed at the position marked 11. Note that since there were piles of uniform objects at various positions on the board, this center of gravity was found in exactly the same way as for the mean of a distribution, since each position (value of x) was, in effect, multiplied by the proportion of objects at that position (the probability of that value), and then these products were summed. In short, the position of any object on the board is analogous to a possible value of X. The mean is then like the center of gravity, or balance point.

This property of the mean is summarized by the statement that **the expectation of the deviation about the mean is zero in any distribution.** The deviation from the mean is simply the signed difference between X and the mean score: $X - \mu$. The expectation of this difference is

$$E(X - \mu) = E(X) - E(\mu)$$
$$= E(X) - \mu,$$

since μ is a constant. But by definition, $\mu = E(X)$, so

$$E(X - \mu) = 0. \qquad (3.11.2^*)$$

Thus, the expectation of the deviation about the mean is zero.

Suppose once again that you were asked to state *one value* to communicate the central tendency of a distribution. If you guessed the mean, your expected error, $E(X - \mu)$, would be zero, though the mean itself might not even be a possible value of X (as in the board example).

Table 3.11.1   Probability Distribution of Income of Employees

x	$P(x)$
6,000	.10
8,000	.40
10,000	.30
15,000	.10
25,000	.05
50,000	.05

For example, let X represent the income (in dollars) of a person chosen at random from the employees of a particular company. The distribution of X is given in Table 3.11.1. The mean of the distribution (that is, the mean income) is

$$E(X) = \sum_x xP(x)$$
$$= 6000(.10) + 8000(.40) + 10,000(.30) + 15,000(.10)$$
$$+ 25,000(.05) + 50,000(.05) = 12,050.$$

The PMF in Figure 3.11.2 illustrates the idea of the mean as the center of gravity in this example.

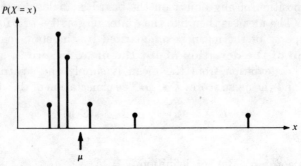

Figure 3.11.2 PMF and Mean of Income of Employees

For an example of the calculation of the mean for a continuous random variable, consider once again the light bulb example from Section 3.6. X represents the length of life (in hours) of a light bulb, and the density function of X is as follows:

$$f(x) = \begin{cases} 1/20 & \text{if } 80 \leq x \leq 100, \\ 0 & \text{elsewhere.} \end{cases}$$

From the definition of the expectation of a continuous random variable,

$$E(X) = \int_{-\infty}^{\infty} xf(x)\,dx = \int_{80}^{100} x\left(\frac{1}{20}\right)dx = \left[\frac{x^2}{40}\right]\Big|_{80}^{100}$$

$$= \frac{(100)^2}{40} - \frac{(80)^2}{40} = \frac{3600}{40} = 90.$$

Thus, the expected life of the light bulb is 90 hours. Since the probability is uniformly distributed over the interval from 80 hours to 100 hours in this example (see Figure 3.6.5), we should not be surprised that the mean is at the midpoint of the interval.

3.12 THE MEDIAN

We defined the median informally as the value of X which is midway (in terms of probability) between the smallest possible value and the largest possible value. In terms of a continuous distribution, this is easy to visualize; the median is the point at which the total area of one under the PDF is divided equally into two segments, each of which has an area of $1/2$. In other words, the probability that X is less than the median is $1/2$, and the probability that X is greater than the median is also $1/2$.

Recall that in the case of a continuous random variable the probability that X is exactly equal to any single value is zero. This is not true for a discrete random variable, so a more formal definition is required for a median. For our purposes, the following definition will suffice.

If

$$P(X \le a) \ge .50 \quad \textbf{and} \quad P(X \ge a) \ge .50, \quad (3.12.1^*)$$

then a is called a median of the distribution of X.

In the continuous case, this amounts to requiring that

$$\int_{-\infty}^{a} f(x) \, dx = \int_{a}^{\infty} f(x) \, dx = .50, \quad (3.12.2)$$

and there is a unique median since $f(x)$ is a continuous function. For example, consider once again the uniform distribution over the interval from 80 to 100. If a is the median, then

$$\int_{-\infty}^{a} f(x) \, dx = \int_{80}^{a} \left(\frac{1}{20}\right) dx = .50.$$

But

$$\int_{80}^{a} \left(\frac{1}{20}\right) dx = \left[\frac{x}{20}\right]\Bigg|_{80}^{a} = \frac{(a - 80)}{20},$$

so that

$$\frac{(a - 80)}{20} = .50,$$

or

$$a - 80 = 10,$$

or

$$a = 90.$$

This is reasonable, since by symmetry, the rectangle representing $f(x)$ in this case is divided into two equal parts by a vertical line at $x = 90$. It is easy to verify that

$$\int_{90}^{\infty} f(x) \, dx = .50.$$

Although we have seen that a continuous distribution has one and only one median, note that in the definition of a median for a discrete random variable, the term "a median" was used instead of "the median." This is because it is possible for a discrete distribution to have more than one point a which satisfies the definition. For example, consider the distribution of incomes presented in Section 3.11. Notice that

$$P(X \leq 8000) = .50, \qquad P(X \geq 8000) = .90,$$
$$P(X \leq 8500) = .50, \qquad P(X \geq 8500) = .50,$$
and $\qquad P(X \leq 9153) = .50, \qquad P(X \geq 9153) = .50.$

In fact, Equation (3.12.1) is satisfied for any value of a such that

$$8000 \leq a \leq 10,000.$$

It is *not* satisfied at the value 10,001, since

$$P(X \geq 10,001) = .20.$$

If you were asked to state *one value* to communicate the central tendency of the distribution of X, and you wanted to use the median, which value would you choose? Any value between 8000 and 10,000 can be thought of as a median; if a single number is desired, a reasonable choice is to take the midpoint of the interval from 8000 to 10,000, which is

$$\frac{8000 + 10,000}{2} = 9000.$$

In general, if the median can assume any value in an interval, we will adopt the convention of taking the midpoint of the interval and calling it the median, even though it is technically only one of numerous possible values which satisfy the definition of a median.

On the other hand, it should be pointed out that in many cases there *is* a unique median in the discrete case as well as in the continuous case. Suppose that the distribution of incomes is slightly different, so that $P(X = 8000) = .39$ instead of .40, and $P(X = 10,000) = .31$ instead of .30. Then the only value satisfying the definition of a median is 10,000.

The median is but one of a class of summary measures known as **fractiles.**

If

$$F(a) = P(X \leq a) \geq f \qquad \textbf{and} \qquad P(X \geq a) \geq 1 - f, \quad (3.12.3^*)$$

then a is called an f fractile of the distribution of X.

For the distribution of incomes, 7000 is a .10 fractile, since

$$P(X \leq 7000) = .10 \qquad \text{and} \qquad P(X \geq 7000) = .90.$$

A median, of course, is simply a .50 fractile. Other fractiles which are encountered frequently are the quartiles (.25 and .75 fractiles), the deciles (.10, .20,..., .90 fractiles), and the percentiles (.01, .02, .03,..., .99

fractiles). The reader can easily verify that for the distribution of incomes, the quartiles are 8000 and 10,000 and the first and ninety-ninth percentiles are 6000 and 50,000. Of course, because this distribution of incomes is discrete, there is not a one-to-one correspondence between values of X and fractiles. For instance, 6000 is not only a first percentile; it is also a second percentile, a .072 fractile, and so on. Also, any value between 6000 and 8000 can be taken as a first decile (.10 fractile); using the rule of thumb given above for determining a median in such a situation, we shall take 7000 and call it the first decile.

For the distribution of the life of a light bulb, a uniform distribution on the interval from 80 to 100, there is a one-to-one relationship between values of X and fractiles. This is because the random variable is continuous. For example, it is easy to verify, using calculus or simply using geometry to find areas of rectangles, that for this distribution, the quartiles are 8500 and 9500 and the first and ninety-ninth percentiles are 8020 and 9980.

Since fractiles are defined in terms of the cumulative distribution function $F(x)$, they can be determined readily from a graph of the CDF. In the discrete case, of course, there may be more than one f fractile for any particular value of f, as we have seen. The median is the most reasonable fractile to use as a measure of central tendency; uses for other fractiles will be discussed in later sections of the book.

3.13 THE MODE

The mode, which is defined as **the value of X at which the PMF or PDF of X reaches its highest point,** is generally the easiest of the three measures of central tendency to compute and interpret. If a graph of the PMF or PDF or a listing of the possible values of X along with their probabilities is available, the determination of the mode is quite simple. If the distribution is expressed in functional form, the determination of the mode may not be quite so easy; in the continuous case, for example, it may require the use of calculus to determine the point at which the PDF attains its maximum value.

From the listing of the distribution of incomes presented in Section 3.11, the mode of the distribution occurs at $X = 8000$. Suppose that the probabilities were changed slightly so that $P(X = 10,000) = .00$ and $P(X = 15,000) = .40$. That is, all employees earning 10,000 are given a raise in pay to 15,000. This change results in two modes, 8000 and 15,000, in the new distribution. By generalizing from this example, you can see that it is possible for a distribution to have several modes, in which case the mode is probably of doubtful value as a measure of central tendency. An extreme example of this is the uniform distribution on the interval from 80 to 100, as encountered in the light bulb example of Section 3.6. The PDF is a horizontal line over the interval from 80 to 100. Technically, we could say that every value of X between 80 and 100 is a mode of the distribution since at each point the definition of a mode is satisfied. In this circumstance, however, it is more common to say that there is *no* mode, since the PDF is no higher at any given point in the interval from 80 to 100 than it is at any other point.

3.14   RELATIONS BETWEEN CENTRAL TENDENCY MEASURES AND THE "SHAPES" OF DISTRIBUTIONS

In discussions of probability distributions, it is often expedient to describe the general "shape" of the density function or mass function. Although the terms used here can be applied to any distribution, it will be convenient to illustrate them by referring to graphs of continuous distributions.

First of all, a distribution may be described by the number of relative maximum points it exhibits, its **modality.** This usually refers to the number of "humps" apparent in the graph of the distribution. Strictly speaking, if the density (or probability or frequency) is greatest at one point, then that value is *the* mode, regardless of whether or not other relative maxima occur in the distribution. Nevertheless, it is common to find a distribution described as bimodal or multimodal whenever there are two or more pronounced humps in the curve, even though there is only one distinct mode. Thus, a distribution may have no modes (Figure 3.14.1), may be unimodal (Figure 3.14.2), or may be multimodal (Figure 3.14.3). Notice that the possibility of multimodal distributions or of distributions in which the mode is at an extreme value, as in Figure 3.14.4, rather than at a "typical" value, reduces the effectiveness of the mode as a description of central tendency.

Another characteristic of a distribution is its symmetry, or conversely, its skewness. A distribution is **symmetric** only if it is possible to divide its graph into two "mirror-image halves," as illustrated in Figure 3.14.5.

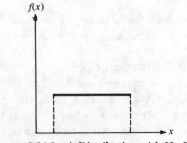

Figure 3.14.1   A Distribution with No Modes

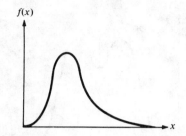

Figure 3.14.2 A Unimodal Distribution

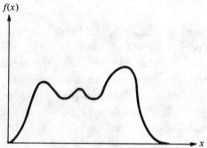

Figure 3.14.3 A Multimodal Distribution

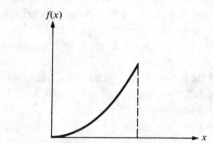

Figure 3.14.4 A Distribution with the Mode at an Extreme Value

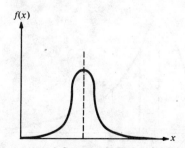

Figure 3.14.5 A Symmetric Unimodal Distribution

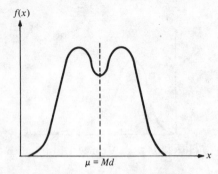

Figure 3.14.6   A Symmetric Bimodal Distribution

Note that in the graph of Figure 3.14.2 there is no point at which the distribution may be divided into two similar parts. When a distribution is symmetric, it will be true that the mean and the median are equal in value. It is not necessarily true, however, that the mode(s) will equal either the mean or the median; witness the example of Figure 3.14.6. On the other hand, a nonsymmetric distribution is sometimes described as **skewed,** which means that the length of one of the **tails** of the distribution, relative to the central section, is disproportionate to the other. For example, the distribution in Figure 3.14.7 is skewed to the right, or **skewed positively.** In a positively skewed distribution, the bulk of the probability falls into the lower part of the range of values of the random variable, with very little probability for intervals containing only extremely high values of the random variable. In general, the mean is greater than the median when the distribution is positively skewed.

On the other hand, it is possible to find distributions skewed to the left, or **negatively skewed.** In such a distribution, the long tail of the distribution occurs among the low values of the variable. That is, the bulk of the distribution shows relatively high scores, although there are a few quite low scores (Figure 3.14.8). Generally, in a negatively skewed distribution, the

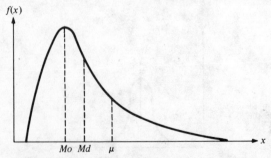

Figure 3.14.7   A Positively Skewed Distribution

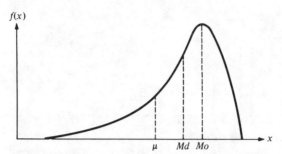

Figure 3.14.8 A Negatively Skewed Distribution

mean is smaller than the median. Thus, a "rough and ready" way to describe the skewness of a distribution is to find the mean and median; if the mean is greater than the median, then the implication is that the distribution is skewed to the right (positively), whereas if the mean is smaller than the median, then the implication is that the distribution is skewed to the left (negatively). If a more accurate determination is needed, other indices reflecting skewness are available; however, we will not discuss such indices. A word of warning: If a distribution is symmetric, then the mean equals the median, but the fact that the mean equals the median does not necessarily imply that the distribution is symmetric.

Describing the skewness of a distribution in terms of measures of central tendency again points up the contrast between the mean and the median as measures of central tendency. The mean is much more affected by the extreme cases in the distribution than is the median. Any alteration of the values of cases at the extreme ends of a distribution will have no effect at all on the median as long as the rank order of the values is roughly preserved; only when values near the center of the distribution are altered is there likelihood of altering the median. This is not true of the mean, which is very sensitive to changes at the extremes of the distribution. The alteration of the probability for a single extreme case in a distribution may have a profound effect on the mean. It is evident that the mean follows the skewed tail in the distribution, while the median does so to a lesser extent. *The occurrence of even a few very high or very low possible values can seriously distort the impression of the distribution given by the mean, provided that one mistakenly interprets the mean as the typical value.* If you are dealing with a nonsymmetric distribution, and you want to communicate a "typical" value, the median would probably be a better measure than the mean.

The distribution of incomes presented in Section 3.11 demonstrates the above point. The graph of the PMF of this distribution is presented in Figure 3.14.9. The distribution is positively skewed, and the mean is larger than the median.

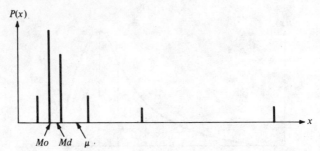

Figure 3.14.9   PMF of Incomes: Positively Skewed

Differences in shape are only rough, qualitative ways of distinguishing among distributions. The only adequate description of a theoretical distribution is its function rule. Distributions that look similar in their graphic form may be very different functions. Conversely, distributions that appear quite different when graphed actually may belong to the same family, as we will see in Chapter 4. A description in terms of modality and skewness is sometimes useful for giving a general impression of what a distribution is like, but it does not communicate fully its essential character.

As the above discussion suggests, there is really no way to say which is the best measure of central tendency for universal use. This depends very much on what we are trying to do and on what we want to communicate in summary form about the distribution. Each of the measures of central tendency is, in its way, a best guess, but the sense of "best" differs with the way error is regarded. If both the size of the errors and their signs are considered, and we want to minimize the average error or the average squared error, then the mean serves as a best guess. If a miss is as good as a mile, and we want to be right as often as possible, then the mode is indicated. If we want to come as close as possible on the average, irrespective of the sign of the error, then the median is a best guess. We will discuss the idea of a summary measure as a "best bet" at greater length in Chapters 6 and 9.

The three measures of central tendency considered here are by no means the only possible measures. Another possible measure is the midrange, which is the midpoint of the possible range of values. If the smallest possible value of X is s and the largest possible value is l, then the mid-range is $(s + l)/2$. A variation of this is the mid-interquartile range, which is the midpoint of the interval from the .25 fractile of X to the .75 fractile of X: (.25 fractile + .75 fractile)/2. The advantage of measures such as the midrange is the ease with which they can be computed. A disadvantage is the fact that the midrange depends solely on the extreme values and not on any probabilities. Another class of measures is the class of trimmed means, which essentially are means computed after some extreme values of the random variable have been eliminated or adjusted toward the center of the distribution. Trimmed means are less sensitive to extreme values than is the untrimmed mean, μ.

There are many potential measures of central tendency, then, each having its advantages and disadvantages. It is impossible to state that one measure is "best" in all situations. *The choice of a measure must depend on the particular distribution and on what the statistician is trying to communicate about the distribution.* The discussion in this section should give the reader some idea of the considerations involved in the choice of a measure in a given situation.

3.15 MEASURES OF DISPERSION: THE VARIANCE

A measure of central tendency, be it mode, median, mean, or some other measure, summarizes only one special aspect of a probability distribution and is not sufficient to completely characterize the distribution. For instance, consider the two distributions shown in Figure 3.15.1. The means, medians, and modes of these distributions are equal, but it is clear that the distributions are not identical. In particular, the two distributions in Figure 3.15.1 differ in that one is more spread out than the other. Probability distributions exhibit **spread** or **dispersion,** the tendency for observations to depart from central tendency. If central tendency measures are thought of as good bets about observations in a distribution, then measures of spread represent the other side of the question; dispersion reflects the "poorness" of central tendency as a description of a randomly selected value, the tendency of values *not* to be like the "average."

The mean is a good bet about the location of a random variable, but *no single value need be exactly like the mean*. A **deviation** from the mean, $x - \mu$, expresses how "off" the mean is as a bet about a particular value, or how

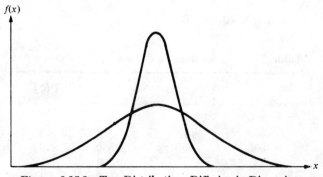

Figure 3.15.1 Two Distributions Differing in Dispersion

much *in error* the mean is as a description of this value. In the same way, we could talk about a deviation from the median, $x - Md$, or perhaps even from the mode. It is quite obvious that the larger such deviations are, on the whole, from a measure of central tendency, the more spread the distribution shows. What we need is an index (or set of indices) to reflect this spread or variability.

First of all, we might simply take the expectation of the deviation about the mean, $E(X - \mu)$, as our measure of variability. This will not work, however, because in Section 3.11 it was shown that in any particular probability distribution, the expected deviation from the mean must be zero.

One device used to get around this difficulty is to take the *square* of each deviation from the mean and then to find the expectation of the squared deviation.

For any distribution, the index $V(X)$, or σ^2, equal to the expectation of the squared deviation from the mean,

$$V(X) = \sigma^2 = E(X - \mu)^2, \tag{3.15.1*}$$

is called the variance of the distribution.

The variance reflects the degree of spread, since σ^2 will be zero if and only if there is only one possible value of X: the mean. The more that the values tend to differ from each other and the mean, the larger the variance will be.

To illustrate the concept of variance, consider X and Y, the profits (in dollars) from purchasing two particular securities. The distributions of X and Y are given in Table 3.15.1. These distributions are graphed in Figures 3.15.2 and 3.15.3. Notice that both distributions are symmetric about 500, which is the mean in each case:

$$E(X) = 0(.10) + 500(.80) + 1000(.10) = 500$$

and $E(Y) = (-500)(.10) + 0(.20) + 500(.40) + 1000(.20)$

$$+ 1500(.10) = 500.$$

Table 3.15.1 Distributions of Profits for Two Securities

x	$P(x)$	y	$P(y)$
0	.1	−500	.1
500	.8	0	.2
1000	.1	500	.4
		1000	.2
		1500	.1

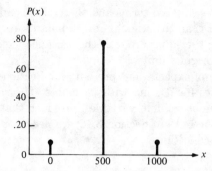

Figure 3.15.2 PMF of Profits from a Stock

Moreover, 500 is also the median and the mode in each case. In terms of measures of central tendency, then, the distributions are identical. However, they are *not* identical in terms of variance:

$$V(X) = (0 - 500)^2(.10) + (500 - 500)^2(.80) + (1000 - 500)^2(.10)$$

$$= (-500)^2(.10) + (0)^2(.80) + (500)^2(.10) = 50,000,$$

but

$$V(Y) = (-500 - 500)^2(.10) + (0 - 500)^2(.20) + (500 - 500)^2(.40)$$

$$+ (1000 - 500)^2(.20) + (1500 - 500)^2(.10)$$

$$= (-1000)^2(.10) + (-500)^2(.20) + (0)^2(.40) + (500)^2(.20)$$

$$+ (1000)^2(.10)$$

$$= 300,000.$$

The variance of Y is six times as large as the variance of X, reflecting the greater dispersion in the distribution of Y. In comparing the two securities, we could say that although the expected gains from the two securities are equal, the security represented by Y is much riskier than the security represented by X. Incidentally, in case you are surprised that the variances seem to be such large numbers in these two cases, it should be noted that

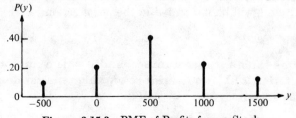

Figure 3.15.3 PMF of Profits from a Stock

the variance involves squared terms and therefore is expressed in terms of squared units rather than the original units (dollars). As we shall see in the next section, taking the square root of the variance yields a measure that is expressed in the original units.

Using the algebra of expectations presented in Section 3.10, it is possible to derive useful rules for dealing with variances. The rules apply to both discrete and continuous random variables, provided that certain boundary conditions are satisfied. As in Section 3.10, we need not concern ourselves with the boundary conditions.

RULE 1

The variance of a random variable is equal to the expectation of the square of the random variable minus the square of the expectation of the random variable:

$$\rightarrow \qquad V(X) = \sigma^2 = E(X^2) - [E(X)]^2. \qquad (3.15.2^*)$$

By definition,

$$V(X) = \sigma^2 = E(X - \mu)^2$$
$$= E(X^2 - 2\mu X + \mu^2)$$
$$= E(X^2) - E(2\mu X) + E(\mu^2).$$

But μ is a constant, so

$$V(X) = E(X^2) - 2\mu E(X) + \mu^2.$$

Then, since $\mu = E(X)$,

$$V(X) = E(X^2) - 2[E(X)]^2 + [E(X)]^2$$
$$= E(X^2) - [E(X)]^2.$$

Rule 1 provides a convenient way to compute the variance of a random variable; it is usually easier to apply Equation (3.15.2) than to apply Equation (3.15.1). This is particularly true if the mean is not a "round" number, since in Equation (3.15.1) it is necessary to subtract the mean from each possible value of the random variable and to square the result. The use of Rule 1 will be illustrated in the next section.

RULE 2

If a is some constant real number, and if X is a random variable with expectation $E(X)$ and variance σ^2, then the random variable $(X + a)$ has variance σ^2:

$$\rightarrow \qquad V(X + a) = V(X) = \sigma^2. \qquad (3.15.3^*)$$

This can be shown as follows:

$$V(X + a) = E[(X + a) - E(X + a)]^2.$$

But

$$E(X + a) = E(X) + a,$$

so that

$$V(X + a) = E[(X + a) - (E(X) + a)]^2$$
$$= E[X - E(X)]^2$$
$$= \sigma^2.$$

RULE 3

If a is some constant real number, and if X is a random variable with variance σ^2, the variance of the random variable aX is

$$V(aX) = a^2 V(X) = a^2 \sigma^2. \tag{3.15.4*}$$

In order to show this, we take

$$V(aX) = E[(aX)^2] - [E(aX)]^2$$
$$= a^2 E(X^2) - a^2 [E(X)]^2$$
$$= a^2 (E(X^2) - [E(X)]^2)$$
$$= a^2 \sigma^2.$$

In short, adding a constant value to each value of a random variable leaves the variance unchanged, but multiplying each value by a constant multiplies the variance by the square of the constant. This is reasonable, since adding a constant merely shifts the distribution along the x-axis. This is quite obvious in a graphical analysis; the dispersion of the distribution remains the same, although the location is shifted (see Figure 3.15.4). On the other hand, multiplying by a constant changes the scale

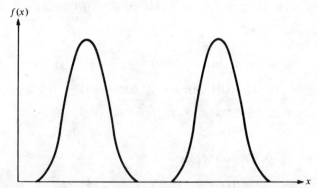

Figure 3.15.4 Two Distributions Differing in Location but Not Dispersion

as well as the location. A value two units from the mean becomes $2a$ units from the mean if each value of the random variable is multiplied by a. The change in scale results in a change in dispersion.

3.16 THE STANDARD DEVIATION

Although the variance is an adequate way of describing the degree of variability in a distribution, it does have one important drawback. The variance is a quantity in squared units of measurement. For instance, if a random variable is expressed in terms of inches, then the mean is some number of inches, and a deviation from the mean is a difference in inches. However, the square of a deviation is in *square-inch units*, and thus the variance, being a mean squared deviation, must also be in square inches. Naturally, this is not an insurmountable problem; taking the positive square root of the variance gives an index of variability in the original units.

The square root of the variance for a distribution is called the standard deviation and is an index of variability in the original measurement units.

The standard deviation of a random variable will be denoted by σ:

$$\sigma = \sqrt{\sigma^2} = \sqrt{E(X - \mu)^2}. \qquad (3.16.1^*)$$

In the example presented in the previous section concerning profits from two securities, the standard deviations are $\sigma_X = \sqrt{50,000} = 223.6$ and $\sigma_Y = \sqrt{300,000} = 547.7$. For another example, we will utilize Rule 1 of the previous section to calculate the variance, and hence the standard deviation, of the distribution of incomes presented in Section 3.11. For this distribution, we found $E(X)$ to be 12,050. Furthermore,

$$E(X^2) = \sum_x x^2 P(x)$$

$$= (6000)^2(.10) + (8000)^2(.40) + (10,000)^2(.30)$$

$$+ (15,000)^2(.10) + (25,000)^2(.05) + (50,000)^2(.05)$$

$$= 237,950,000.$$

Then

$$V(X) = E(X^2) - [E(X)]^2$$

$$= 237,950,000 - (12,050)^2 = 92,747,500,$$

and $\qquad \sigma = \sqrt{92,747,500} = 9631.$

This example illustrates a common computational problem. The numbers involved in these calculations are quite large and thus are cumbersome to work with. However, by using Rule 3 from Section 3.15, we can express the calculations in terms of smaller and more convenient numbers. Suppose that $Y = X/1000$. Then $E(Y) = E(X)/1000 = 12.05$, and

$$E(Y^2) = \sum_y y^2 P(y)$$

$$= 6^2(.10) + 8^2(.40) + 10^2(.30) + 15^2(.10) + 25^2(.05)$$
$$+ 50^2(.05) = 237.95.$$

Then

$$V(Y) = E(Y^2) - [E(Y)]^2$$

$$= 237.95 - (12.05)^2 = 92.7475.$$

But $Y = X/1000$, so $X = 1000Y$, and by Rule 3,

$$V(X) = V(1000Y) = (1000)^2 \ V(Y) = 92,747,500.$$

For an example involving a continuous random variable, consider the distribution of the life of light bulbs presented in Section 3.6. For this distribution, $E(X) = 90$, and

$$E(X^2) = \int_{-\infty}^{\infty} x^2 f(x) \ dx = \int_{80}^{100} x^2 \left(\frac{1}{20}\right) dx = \left[\frac{x^3}{60}\right]\Bigg|_{80}^{100}$$

$$= \frac{(100)^3}{60} - \frac{(80)^3}{60} = 8133.3.$$

Hence,

$$V(X) = E(X^2) - [E(X)]^2 = 8133.3 - (90)^2$$

$$= 8133.3 - 8100 = 33.3,$$

and $\sigma = \sqrt{33.3} = 5.77.$

The reader acquainted with elementary physics may recognize not only that the mean is the center of gravity of a physical distribution of objects, but also that the variance is the **moment of inertia** of a distribution of mass. Furthermore, the standard deviation corresponds to the **radius of gyration** of a mass distribution; the standard deviation is analogous to a resultant force away from the mean. These physical concepts and their associated mathematical formulations have influenced the course of theoretical statistics very strongly and have helped to shape the form of statistical inference as we will encounter it.

3.17   THE MEAN AS THE "ORIGIN" FOR THE VARIANCE

This question may already have occurred to the reader: "Why is the variance, and hence the standard deviation, always calculated in terms of deviations from the *mean?* Why couldn't one of the other measures of central tendency be used as well?" The answer lies in the fact that **the expected squared deviation (that is, the variance) is smallest when calculated from the mean.**

This result may be demonstrated as follows. Suppose that we choose some arbitrary value c and calculate a "pseudo-variance"

$$\sigma_c^2 = E(X - c)^2. \tag{3.17.1}$$

Adding and subtracting μ within the parentheses would not change the value of σ_c^2 at all. However, if we did so, we could expand each squared deviation as follows:

$$(X - c)^2 = (X - \mu + \mu - c)^2 = [(X - \mu) + (\mu - c)]^2$$
$$= (X - \mu)^2 + 2(X - \mu)(\mu - c) + (\mu - c)^2.$$

Substituting into the expression for σ_c^2, we have

$$\sigma_c^2 = E(X - \mu)^2 + 2E(X - \mu)(\mu - c) + E(\mu - c)^2.$$

But $(\mu - c)$ is a constant with regard to the expectations, so that

$$\sigma_c^2 = E(X - \mu)^2 + 2(\mu - c)E(X - \mu) + (\mu - c)^2.$$

The first term on the right above is simply σ^2, the variance about the mean, and the second term is zero, since the expected deviation from μ is zero. On making these substitutions, we find that

$$\sigma_c^2 = \sigma^2 + (\mu - c)^2. \tag{3.17.2}$$

Since $(\mu - c)^2$ is a squared number, it can only be positive or zero, and so σ_c^2 must be greater than or equal to σ^2. The value of σ_c^2 can be equal to σ^2 only when μ and c are the same. In short, we have shown that the variance calculated about the mean will always be smaller than about any other point.

If we are going to use the mean to express the central tendency of a distribution, and if we let the standard deviation indicate the extent of error we stand to make in guessing the mean, then this error-quantity is at its minimum when we guess the mean in place of any other single value. However, it is a curious fact that if we appraise error by taking the absolute difference (disregarding sign) between a value and a measure of central tendency, then **the expected absolute deviation is smallest when the median is used.** This is one of the reasons that squared deviations rather than absolute deviations figure in the indices of variability when the mean is used to express central tendency. When the median is used to express central tendency it is often accompanied by the expected absolute deviation from the median to indicate dispersion, rather than by the standard deviation, which is more appropriate to the use of the mean. The expected absolute deviation is simply

$$\text{A.D.} = E \mid X - Md \mid, \qquad (3.17.3)$$

where the vertical bars indicate a disregarding of sign. Analogically speaking' this measure is to the median as the standard deviation is to the mean·

In addition to the variance, the standard deviation, and the absolute deviation, there are other measures of dispersion which for the most part are not as good as the above measures but are often easier to calculate. The range is defined as the difference between the largest possible value of the random variable and the smallest possible value. This is extremely simple to determine. Unfortunately, the range may not be particularly informative, since it considers only the two extreme values and not the probabilities of the various possible values. Furthermore, without any change in the probabilities, the range can be severely affected by shifting one of the extreme values even further away from the other values. Finally, for many continuous random variables, the range is infinite.

These problems are reduced somewhat by using the interquartile range instead of the range. The interquartile range is the difference between the .75 and .25 fractiles of the distribution; it is not affected by the extreme values as is the range. Another possibility is the interdecile range, the difference between the .90 and .10 fractiles of the distribution. All of these ranges are reasonably easy to calculate in most cases and hence could be useful as "rough and ready" measures of dispersion. In general, however, we will use the variance and standard deviation as measures of dispersion because they are mathematically tractable, whereas the other measures have mathematical properties which make them more difficult to work with.

3.18 THE RELATIVE LOCATION OF A VALUE IN A PROBABILITY DISTRIBUTION: STANDARDIZED RANDOM VARIABLES

A major use of the mean and standard deviation is to transform random variables into standardized random variables, showing the relative status of possible values of a random variable. If we are given the information, "John Doe has a score of 60," we really know very little about what this score means. Is this score high, low, middling, or what? However, if we know something about the distribution of scores, we can judge the location of the score in the distribution. The score of 60 gives quite a different picture when the mean is 30 and the standard deviation is 10 than when the mean is 65 and the standard deviation is 20.

If X is a random variable, then the corresponding standardized random variable is

$$Z = \frac{X - E(X)}{\sigma} = \frac{X - \mu}{\sigma}. \qquad (3.18.1^*)$$

The standardized random variable is a deviation from expectation, relative to the standard deviation. For any value of X, the corresponding value of Z tells us how many standard deviations the value of X is away from the mean, $E(X)$. In general, we will denote *a standardized* random variable by Z.

The two distributions mentioned above give quite different z values to the score of 60:

$$z_1 = \frac{60 - 30}{10} = 3,$$

and

$$z_2 = \frac{60 - 65}{20} = -.25.$$

The conversion of raw values to z values is handy when we wish to emphasize the *location* or *status* of a value in the distribution, and in future sections we will deal with standardized random variables when this aspect of any random variable is to be discussed. The value of Z corresponding to any value of X is found in precisely the same way whether the distribution is discrete or continuous.

Changing each of the values in a distribution to a standardized value creates a distribution having a "standard" mean and "standard" standard deviation. **The mean of a distribution of a standardized random variable is always 0, and the standard deviation is always 1.** This is easily shown as follows:

$$E(Z) = E\left(\frac{X - \mu}{\sigma}\right) = \frac{1}{\sigma} E(X - \mu) = 0, \qquad (3.18.2^*)$$

since μ and σ are constants over the various possible values of X and the expected deviation from the mean is always 0. The standard deviation of a standardized random variable is always the square root of the variance:

$$V(Z) = E[Z - E(Z)]^2 = E(Z^2),$$

by Equation (3.18.2). But then

$$V(Z) = E\left(\frac{X - \mu}{\sigma}\right)^2 = E\left[\frac{(X - \mu)^2}{\sigma^2}\right] = \frac{1}{\sigma^2} E(X - \mu)^2.$$

By definition, $E(X - \mu)^2 = \sigma^2$, so

$$V(Z) = \frac{\sigma^2}{\sigma^2} = 1,$$

and
$$\sigma_Z = \sqrt{V(Z)} = 1. \tag{3.18.3*}$$

The general shape of the probability distribution is not changed by the transformation to standardized values. That is, the standardized distribution corresponding to a positively skewed random variable is also positively skewed, and so on. Often in statistics it is easier to work with a standardized than with a nonstandardized random variable, as we will see in later chapters.

3.19 CHEBYSHEV'S INEQUALITY

There is a very close connection between the size of deviations from the mean and probability, holding for distributions having finite expectation and variance. The following relation is called **the Chebyshev inequality,** after the Russian mathematician who first proved this very general principle:

$$P(|X - \mu| \geq b) \leq \frac{\sigma^2}{b^2}. \tag{3.19.1*}$$

The probability that a random variable X will differ absolutely from its expectation by b or more units $(b > 0)$ is *always* less than or equal to the ratio of σ^2 to b^2. Any deviation from expectation of b or more units can be *no more probable* than σ^2/b^2.

This relation can be clarified somewhat by dealing with the deviation in σ units, making the random variable standardized. If we let $b = k\sigma$, then the following version of the Chebyshev inequality is true:

$$P\left(\frac{|X - \mu|}{\sigma} \geq k\right) \leq \frac{1}{k^2}. \tag{3.19.2*}$$

The probability that a standardized random variable has *absolute* magnitude greater than or equal to some positive number k is *always* less than or equal to $1/k^2$. Thus, given a distribution with some mean and variance, the probability of drawing a case having a standardized value of 2 or more (disregarding sign) must be *at most* 1/4. The probability of a standardized value of 3 or more must be no more than 1/9, the probability of 10 or more can be no more than 1/100, and so on.

This last form of the Chebyshev inequality [(3.19.2)] is quite simple to prove for a discrete random variable. Let

$$Z = \frac{X - \mu}{\sigma}.$$

Then $P(Z = z) = P(X = x)$, $E(Z) = 0$, and $V(Z) = 1$. Now the probability that $|Z|$ is greater than or equal to a positive number k can be written in the form

$$P(|Z| \geq k) = \sum_{(z \geq k)} P(z) + \sum_{(z \leq -k)} P(z).$$

Multiplying both sides of this equation by k^2, we have

$$k^2 P(|Z| \geq k) = \sum_{(z \geq k)} k^2 P(z) + \sum_{(z \leq -k)} k^2 P(z).$$

However, each and every value of z represented in the sum on the right side of the equation above is greater than or equal to k in absolute value. Thus,

$$\sum_{(z \geq k)} k^2 P(z) + \sum_{(z \leq -k)} k^2 P(z) \leq \sum_{(z \geq k)} z^2 P(z) + \sum_{(z \leq -k)} z^2 P(z).$$

Furthermore,

$$\sum_{(z \geq k)} z^2 P(z) + \sum_{(z \leq -k)} z^2 P(z) \leq \sigma_Z^2,$$

since

$$\sigma_Z^2 = \sum_z z^2 P(z),$$

the sum being taken over *all* values of z, not just the values such that $z \geq k$ or $z \leq -k$. Then

$$k^2 P(|Z| \geq k) \leq \sigma_Z^2,$$

or

$$P(|Z| \geq k) \leq \frac{1}{k^2},$$

which is what we set out to prove.

The Chebyshev inequality sets broad limits to the probability associated with extreme deviations of a random variable from its mean. As an example of how this principle might be applied, let X represent the number of days it takes a shipment of goods sent from Los Angeles via surface mail to reach its destination, Zurich, Switzerland. All we know about the distribution of X is the mean, 100, and the standard deviation, 25. It is impossible to make exact probability statements about X, since we do not know the distribution exactly. By using the Chebyshev inequality, however, we can make approximate probability statements in the form of probability inequalities. Consider the value 175; what is the probability that X will be at least this far from the mean (in either direction)? The standardized value relative to 175 for this distribution is

$$z = \frac{x - \mu}{\sigma} = \frac{175 - 100}{25} = 3.$$

Now by setting $k = 3$ in the Chebyshev inequality, we find

$$P(|Z| \geq 3) \leq \frac{1}{9}.$$

The probability of observing a value 3 or more standard deviations from the mean is *no more* than 1/9, regardless of the distribution. In the example, this means that the probability is at most 1/9 that the shipment will arrive in 25 days or less or 175 days or more.

Within how many standard deviations from the mean must *at least one-half* of the probability fall? That is, we want the value k for which

$$P(|Z| \geq k) \leq \frac{1}{2}.$$

By Equation (3.19.2) the value is $\sqrt{2}$, or about 1.4. Regardless of how X is actually distributed, we can state that one half or more of the probability must lie within the approximate limits $100 - (1.4)(25)$ and $100 + (1.4)(25)$, or 65 to 135.

Although this principle is very important theoretically, it is not extremely powerful as a tool in applied problems. The Chebyshev inequality can be strengthened somewhat if we are willing to make assumptions about the general form of the distribution, however. For example, if we assume that the distribution of the random variable is both *symmetric* and *unimodal*, then the relation becomes

$$P(|Z| \geq k) \leq \frac{4}{9} \left(\frac{1}{k^2} \right). \tag{3.19.3*}$$

For such distributions, we can make somewhat "tighter" statements about how large the probability of a given amount of deviation may be. For the example given above, if the distribution were unimodal-symmetric, a value differing by three or more standard deviations from the mean should be observed with probability *no greater than* (4/9)(1/9), or about .05. Similarly, we could find that at least $1 - 4/9$ or $5/9$ of the probability falls within one standard deviation to either side of the mean. Furthermore, $1 - (4/9)(1/4)$ or $8/9$ must fall within two standard deviations, and so on.

3.20 MOMENTS OF A DISTRIBUTION

A truly mathematical treatment of distributions would introduce not only the mean and the variance but also a number of other summary characteristics. These are the so-called **moments** of a distribution, which are simply **the expectations of different powers of the random variable.** Thus, the first moment about the origin of a random variable X is

$$E(X) = \text{the mean.}$$

The second moment about the origin is

$$E(X^2),$$

the third
$$E(X^3),$$

and so on. When the mean is subtracted from X before the power is taken, then the moment is said to be **about the mean;** the variance

$$E[X - E(X)]^2$$

is the second moment about the mean;

$$E[X - E(X)]^3$$

is the third moment about the mean, and so on.

Just as the mean describes the "location" of the distribution on the x-axis, and the variance describes its dispersion, so do the higher moments reflect other features of the distribution. For example, the third moment about the mean is used in certain measures of degree of **skewness;** the third moment will be zero for a symmetric distribution, negative for skewness to the left, positive for skewness to the right. The fourth moment indicates the degree of **peakedness** or **kurtosis** of the distribution, and so on. These higher moments have relatively little use in elementary applications of statistics, but they are important for mathematical statisticians in the study of the properties of distributions and in arriving at theoretical distributions fitting observed data. The entire set of moments for a distribution will ordinarily determine the distribution exactly, and distributions are sometimes specified in this way when their general function rules are unknown or difficult to state.

3.21 JOINT PROBABILITY DISTRIBUTIONS

We are often interested not just in *one* random variable, but in the relationship between *two or more random variables*. In a sample of American men, we might be interested in the relationship between X, the height of a man, and Y, the weight of a man. A medical researcher might be interested in the relationship between X, the number of cigarettes smoked by a person, and Y, the occurrence or nonoccurrence of cancer in that person. A stock market analyst might be interested in the relationship between X, the volume of sales for a given stock on a particular day, and Y, the closing price of that stock on that day. Examples of this nature are numerous. In Chapter 2, joint probabilities of two or more events were discussed; now we will introduce the concept of **joint probability distributions of two or more random variables.**

Suppose that we have two discrete random variables:

$$X, \text{ ranging over the values } x_1, x_2, \ldots, x_J$$

and $\qquad\qquad Y, \text{ ranging over the values } y_1, y_2, \ldots, y_K.$

We can discuss the event that X takes on some value x *and* that Y takes on some value y: this is the event $(X = x, Y = y)$, with probability $P(X = x, Y = y)$. **The set of all such events, together with their probabilities, comprise the joint probability distribution of X and Y.**

Note that in this case there will be JK such events; of course, some may have zero probability, since there may be some combinations $(X = x, Y = y)$ which cannot possibly occur. For example, suppose that an experiment consists of tossing two fair dice. Let X represent the number coming up on the first die, let Y represent the number coming up on the second die, and let W represent the sum of the numbers on the two dice. There are six possible values for X: 1, 2, 3, 4, 5, 6; there are six possible values for Y: 1, 2, 3, 4, 5, 6; and there are eleven possible values for W: 2, 3, 4, 5, 6, 7, 8, 9, 10, 11, 12. There are $6 \cdot 6 = 36$ possible combinations (X, Y), each having probability 1/36. However, there are *not* $11 \cdot 6 = 66$ possible combinations (X, W). The combination $(X = 6, W = 3)$ is impossible, for example; if a 6 shows on the first die, the sum of the numbers on the two dice obviously must be greater than 6, and hence cannot be 3.

As in the case of a single random variable, a joint probability distribution of two discrete random variables can be represented by a listing, in functional form, or by a graph. The listing is similar to the listing of joint probabilities in tabular form, as presented in Chapter 2. For example, suppose that $X = 1$ if a particular stock increases in price and that $X = 0$ otherwise; similarly, $Y = 1$ if a second stock increases in price and $Y = 0$ otherwise. Moreover, the probability that *both* stocks will increase in price

Table 3.21.1   Joint Distribution of X and Y for Stock Example

		Y		
		0	1	
X	0	.35	.05	.40
	1	.10	.50	.60
		.45	.55	

is $P(X = 1, Y = 1) = .50$; the probability that *neither* stock will increase in price is $P(X = 0, Y = 0) = .35$; the probability that the first stock will increase in price but the second will not is $P(X = 1, Y = 0) = .10$; and the probability that the second stock will increase in price but the first will not is $P(X = 0, Y = 1) = .05$. These four (X, Y) combinations, along with their probabilities, make up the joint probability distribution of X and Y. This joint distribution is given in Table 3.21.1. The distribution can also be represented graphically, although the graph here must represent three, rather than two, dimensions (Figure 3.21.1). This is merely an extension of the idea of the graph of a PMF of a single random variable, except that here probabilities are associated with *pairs* of values of two random variables. Since there are two random variables, the total mass, or probability, must be distributed over the possible pairs of values.

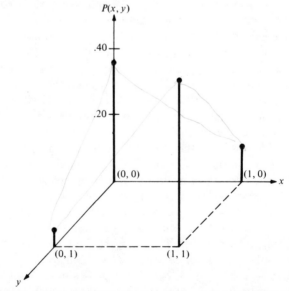

Figure 3.21.1   Joint Distribution of X and Y for Stock Example

In the example, there are just four pairs of values, and the sum of the probabilities over these four pairs is one, as would be expected.

Even though our main interest is in the *relationship between two random variables*, we still might want to know something about each of the two random variables taken individually. Thus, if X is the height of a man and Y is the weight of the same man, we might be interested both in the height-weight relationship and in height and weight as separate random variables. Fortunately, if we know the joint distribution of X and Y, it is possible to find the individual marginal distributions. For example, in the example represented by the above table, the entries in the margins of the table of the joint distribution of X and Y represent the probabilities for X and Y taken separately. For obvious reasons, these are known as **marginal probabilities**, as we pointed out in Section 2.19. It is a simple matter to derive marginal probabilities from joint probabilities:

$$P(X = x_j) = \sum_{k=1}^{K} P(X = x_j, Y = y_k). \qquad (3.21.1)$$

$$P(Y = y_k) = \sum_{j=1}^{J} P(X = x_j, Y = y_k). \qquad (3.21.2)$$

In other words, the marginal probabilities simply represent the sum of the appropriate joint probabilities [that is, the sum of all joint probabilities involving the value of X (or Y) for which the marginal probability is desired].

The set of all values of X together with their marginal probabilities is called the **marginal distribution** of X; the marginal distribution of Y is defined similarly. Notice that this distribution is determined from the joint distribution of X and Y. However, since the effect of the values of Y upon the values of X is summed out by Equation (3.21.1), the marginal distribution is simply the probability distribution of a single random variable, X. If we considered X alone and determined its distribution without looking at any other variables, as in Section 3.3, the result would be identical to the marginal distribution derived from $P(x, y)$.

Using joint probabilities and marginal probabilities, we can determine **conditional probabilities**, just as in Chapter 2.

$$P(X = x \mid Y = y) = \frac{P(X = x, Y = y)}{P(Y = y)}. \qquad (3.21.3)$$

$$P(Y = y \mid X = x) = \frac{P(X = x, Y = y)}{P(X = x)}. \qquad (3.21.4)$$

For instance, in the above example involving the price movements of two stocks,

$$P(X = 1 \mid Y = 1) = \frac{P(X = 1, Y = 1)}{P(Y = 1)} = \frac{.50}{.55} = \frac{10}{11}.$$

The set of values of $P(X = x \mid Y = y)$ for all values of x and a fixed value of y is called the **conditional distribution of** X **given that** $Y = y$. In general, this distribution is different from the marginal distribution of X. Notice that the value y must be specified before the conditional distribution of X given that $Y = y$ can be determined. For the above example, the marginal distribution of X, the conditional distribution of X given that $Y = 0$ and the conditional distribution of X given that $Y = 1$ are given in Table 3.21.2. Similarly, the marginal distribution of Y, the conditional distribution of Y given that $X = 0$ and the conditional distribution of Y given that $X = 1$ are given in Table 3.21.3.

Table 3.21.2 Marginal Distribution of X and Conditional Distributions of X given Different Values of Y

x	$P(x)$	$P(x \mid Y = 0)$	$P(x \mid Y = 1)$
0	.40	7/9	1/11
1	.60	2/9	10/11
	1.00	1	1

Table 3.21.3 Marginal Distribution of Y and Conditional Distributions of Y given Different Values of X

y	$P(y)$	$P(y \mid X = 0)$	$P(y \mid X = 1)$
0	.45	7/8	1/6
1	.55	1/8	5/6
	1.00	1	1

In the discrete case, joint probability distributions are treated just as we treated joint probabilities of events in Chapter 2. The same ideas apply to continuous random variables as well, except that the definitions are stated in terms of probability density functions. Instead of a series of probabilities of the form $P(X = x, Y = y)$, we consider a **joint density**

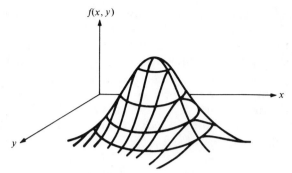

Figure 3.21.2 A Joint Density Function

function $f(x, y)$, such that

$$f(x, y) \geq 0 \quad \text{for all real } x, y \tag{3.21.5}$$

and
$$\int_{-\infty}^{\infty} \int_{-\infty}^{\infty} f(x, y) \; dx \; dy = 1. \tag{3.21.6}$$

The function $f(x, y)$ is a surface in three-dimensional space, with probability represented by *volume* under this surface (between the surface and the xy-plane). A schematic representation of a joint density function as a surface in three dimensions is presented in Figure 3.21.2.

As in the case of a single continuous random variable, the probability of any *single point*, which is a pair of values $(X = x, Y = y)$, is 0; it is necessary to consider joint *intervals* of values, such as $(a \leq X \leq b, c \leq Y \leq d)$. Probabilities of such intervals can be found from the joint density function by double integration:

$$P(a \leq X \leq b, c \leq Y \leq d) = \int_{c}^{d} \int_{a}^{b} f(x, y) \; dx \; dy. \tag{3.21.7}$$

If you are not familiar with double integration, just think of Equation (3.21.7) as a procedure for determining volume under a surface. It is directly analogous to the determination of the area under a curve when only a single continuous random variable is of interest.

Given the joint density function $f(x, y)$ of the continuous random variables X and Y, it is possible to determine the density functions (or **marginal density functions**) of both X and Y, as follows:

$$f(x) = \int_{-\infty}^{\infty} f(x, y) \; dy \tag{3.21.8}$$

and
$$f(y) = \int_{-\infty}^{\infty} f(x, y) \; dx. \tag{3.21.9}$$

Notice the similarity between this operation and the derivation of marginal distributions in the discrete case [Equations (3.21.1) and (3.21.2)]. To obtain the marginal probabilities for the various values of X in the discrete case, we sum the joint probabilities over the possible values of Y. To obtain the marginal density for X in the continuous case, we integrate the joint density over the possible values of Y.

Once we have the joint density and the marginal densities, it is possible to determine the **conditional density functions** in a manner parallel to the determination of conditional probabilities:

$$f(x \mid Y = y) = \frac{f(x, y)}{f(y)} \tag{3.21.10}$$

and
$$f(y \mid X = x) = \frac{f(x, y)}{f(x)}. \tag{3.21.11}$$

Here $f(x \mid Y = y)$ symbolizes the density for the random variable X, given that the random variable Y assumes the value y. In general, the conditional density of one random variable given a particular value of a second random variable is simply the joint density of the two random variables divided by the marginal density of the second random variable, where both densities are evaluated at the given value of the second random variable.

For example, let X represent the proportion of hours during the day that are sunny in New Orleans for the month of May during any given year, and let Y represent the same thing for the month of September during the same year. Suppose that the joint density of X and Y is given by

$$f(x, y) = \begin{cases} 4xy & \text{if } 0 \le x \le 1, 0 \le y \le 1, \\ 0 & \text{elsewhere.} \end{cases}$$

This is a valid joint density function, as it can be shown that Equations (3.21.5) and (3.21.6) are satisfied. The marginal density functions of X and Y are

$$f(x) = \int_{-\infty}^{\infty} f(x, y) \, dy = \int_0^1 4xy \, dy = \left[\frac{4xy^2}{2} \right]_{y=0}^{y=1} = 2x \quad \text{for } 0 \le x \le 1,$$

$$\text{and } f(y) = \int_{-\infty}^{\infty} f(x, y) \, dx = \int_0^1 4xy \, dx = \left[\frac{4x^2y}{2} \right]_{x=0}^{x=1} = 2y \quad \text{for } 0 \le y \le 1.$$

The conditional density functions are

$$f(x \mid y) = \frac{f(x, y)}{f(y)} = \frac{4xy}{2y} = 2x \quad \text{for } 0 \leq x \leq 1, 0 \leq y \leq 1,$$

and $\quad f(y \mid x) = \frac{f(x, y)}{f(x)} = \frac{4xy}{2x} = 2y \quad \text{for } 0 \leq x \leq 1, 0 \leq y \leq 1.$

Note that in this example, $f(x, y) = f(x)f(y)$; the joint density of X and Y is equal to the product of the marginal densities of X and Y. This will not be true in general; when it *is* true, we say that the random variables X and Y are independent, as we shall see in the next section. Because of this independence, it is also true that the marginal density of X is equal to the conditional density of X given *any* value of Y, and similarly for the marginal and conditional densities of Y.

The relationship among *several* random variables is often of some interest. Extending the examples at the beginning of the section, we might be interested in the relationship among three random variables: X, the height of a man; Y, the weight of a man; and W, the age of a man. A medical researcher might be interested in the relationship among four variables: X, the number of cigarettes smoked by a person; Y, the occurrence or nonoccurrence of cancer in that person; W, the age of that person; and U, the occurrence or nonoccurrence of heart disease in that person. A stock market analyst might be interested in the relationship among X, the volume of sales for a given stock on a particular day; Y, the closing price of that stock on that day; W, the volume of sales for the entire stock market on that day; and U, the closing price of some other particular stock on that day. Clearly there are many situations in which the relationship among several random variables is of interest.

The representation of joint probability distributions of three or more random variables is a straightforward extension of the analysis presented for two random variables. In the discrete case, suppose that we have K random variables $X_1, X_2, \ldots, X_K$. The joint distribution of the K random variables can then be represented by

$$P(X_1 = x_1, X_2 = x_2, \ldots, X_K = x_K),$$

where the probability function is defined for all possible combinations of values $(x_1, x_2, \ldots, x_K)$. In accordance with the general axioms of probability, this function must satisfy two requirements:

1. $P(X_1 = x_1, X_2 = x_2, \ldots, X_K = x_K) \geq 0 \quad$ for all $x_1, x_2, \ldots, x_K$

$$(3.21.12)$$

and

2. $\displaystyle\sum_{x_1}\sum_{x_2}\cdots\sum_{x_K} P(X_1 = x_1, X_2 = x_2,\ldots, X_K = x_K) = 1.$ (3.21.13)

To determine the marginal probability function for any single one of the K random variables, we simply sum $P(X_1 = x_1, X_2 = x_2,\ldots, X_K = x_K)$ over the remaining $K - 1$ random variables. For instance, if we wanted to find $P(X_1 = x_1)$,

$$P(X_1 = x_1) = \sum_{x_2}\sum_{x_3}\cdots\sum_{x_K} P(X_1 = x_1, X_2 = x_2,\ldots, X_K = x_K).$$

(3.21.14)

Similarly, to determine the joint probability function for any particular pair of random variables chosen from $X_1, X_2,\ldots, X_K$, just sum $P(X_1 = x_1, X_2 = x_2,\ldots, X_K = x_K)$ over the remaining $K - 2$ random variables.

The continuous case parallels the discrete case, with definite integrals replacing sums. Suppose that $X_1, X_2,\ldots, X_K$ are continuous random variables. Their joint density function can be represented by

$$f(x_1, x_2,\ldots, x_K),$$

where $f(x_1, x_2,\ldots, x_K) \geq 0$ for all real $x_1, x_2,\ldots, x_K$ (3.21.15)

and $\displaystyle\int_{-\infty}^{\infty}\int_{-\infty}^{\infty}\cdots\int_{-\infty}^{\infty} f(x_1, x_2,\ldots, x_K)\, dx_1\, dx_2\ldots dx_K = 1.$ (3.21.16)

[The reader not familiar with the technique of multiple integration need not be overly concerned about the actual process of evaluating definite multiple integrals. It suffices to say that multiple integration is an extension of the integration of functions of one variable to allow us to integrate functions of several variables. Think of this process as being analogous to a multiple summation such as that presented in Equation (3.21.13).]

To determine the marginal density function for any single one of the K random variables, we integrate the joint density over the remaining $K - 1$ random variables. For instance, if we wanted to find the marginal density of X_K, $f(x_K)$,

$$f(x_K) = \int_{-\infty}^{\infty}\int_{-\infty}^{\infty}\cdots\int_{-\infty}^{\infty} f(x_1, x_2, \ldots, x_K)\, dx_1\, dx_2\ldots dx_{K-1}.$$ (3.21.17)

In this section we have introduced the concepts of joint, marginal, and conditional probability distributions. These distributions are important whenever we are interested in the relationship among two or more variables. For instance, suppose that we wanted to predict X, the sales of a particular product during a given time period. The relevant probability distribution would be the distribution of X. However, suppose that we could obtain information regarding the disposable income of consumers during this time period (call this Y). Since the knowledge of the value of Y during this period should help us to predict X, we might now be interested in the joint distribution of X and Y. From this joint distribution, we could compute the conditional distribution of X given that $Y = y$. But we *know* that $Y = y$, so this conditional distribution is the relevant distribution for the prediction of X. Similarly, if we could find out the value of a third related variable, say W, then the distribution of interest would be the conditional distribution of X given that $Y = y$ and $W = w$.

This example illustrates the potential use of distributions involving two or more random variables. Problems such as this will be discussed in Chapter 10, and we will defer until then other specific examples of joint, marginal, and conditional distributions.

3.22   INDEPENDENCE OF RANDOM VARIABLES

We defined the independence of events in Section 2.21, and now we will define independence of random variables in both the discrete and continuous cases. As we shall see, the concept of independence is critical in any attempt to discuss the relationship among random variables.

The definition of independence of discrete random variables is very similar to the definition of independence of events.

Two discrete random variables X and Y are independent if and only if

$$P(X = x, Y = y) = P(X = x)P(Y = y) \qquad (3.22.1^*)$$

for all pairs of values x and y.

Note that for the example involving price changes of stocks that was presented in the preceding section, X and Y are *not* independent. $P(X = 1, Y = 1) = .50$, whereas $P(X = 1)P(Y = 1) = (.60)(.55) = .33$, so that Equation (3.22.1) is not satisfied. It appears that the two stocks have more of a tendency to "move together" (that is, to both go up in price or to both not go up in price) than they would if they were independent.

The same idea applies to continuous random variables as well, except that the definition is stated in terms of probability densities.

A random variable X with density $f(x)$ at value x and a random variable Y with density $f(y)$ at value y are independent if and only if for all (x, y),

$$f(x, y) = f(x)f(y), \qquad (3.22.2^*)$$

where $f(x, y)$ is the joint density function of X and Y.

Alternatively, X and Y are independent if and only if for all (x, y) such that $X = x$ and $Y = y$,

$$f(x \mid y) = f(x), \tag{3.22.3}$$

or, equivalently, $$f(y \mid x) = f(y). \tag{3.22.4}$$

That is, the random variables are independent if and only if the conditional densities are everywhere equal to the marginal densities. All three definitions [(3.22.2), (3.22.3), and (3.22.4)] can be shown to be equivalent by using the definition of a conditional density function. For the example in Section 3.21 involving sunshine in New Orleans, Equations (3.22.2)–(3.22.4) are satisfied, so the proportion of daylight hours with sunshine during May is independent of the proportion of daylight hours with sunshine during September of the same year.

For two random variables, one often has occasion to consider joint intervals such as

$$[(a \leq X \leq b) \quad \text{and} \quad (c \leq Y \leq d)],$$

some value of X lying between a and b and some value of Y between c and d. For independent random variables,

$$P(a \leq X \leq b \text{ and } c \leq Y \leq d) = P(a \leq X \leq b)P(c \leq Y \leq d),$$

$$\tag{3.22.5}$$

just as for any independent events.

The concept of independence can be extended to more than two random variables. The K random variables $X_1, X_2, \ldots, X_K$ are independent if and only if their joint mass (density) function is equal to the product of their marginal mass (density) functions. Formally, if $X_1, X_2, \ldots, X_K$ are discrete random variables, they are independent if and only if

$$P(X_1 = x_1, X_2 = x_2, \ldots, X_K = x_K)$$

$$= P(X_1 = x_1)P(X_2 = x_2) \cdots P(X_K = x_K) \tag{3.22.6}$$

for all sets of values $(x_1, x_2, \ldots, x_K)$. Similarly, if $X_1, X_2, \ldots, X_K$ are continuous random variables with probability density functions $f_1(x_1), f_2(x_2), \ldots, f_K(x_K)$, they are independent if and only if

$$f(x_1, x_2, \ldots, x_K) = f_1(x_1)f_2(x_2) \cdots f_K(x_K) \tag{3.22.7}$$

for all sets of values $(x_1, x_2, \ldots, x_K)$, where $f(x_1, x_2, \ldots, x_K)$ is the joint density for the event $(x_1, x_2, \ldots, x_K)$. It should also be noted that the concepts of pairwise independence and conditional independence, defined in Section 2.21 for events, also apply to random variables by a straightforward extension of the definitions in Section 2.21.

The idea of independence of random variables can be extended to functions of random variables as well. That is, suppose that we have two independent random variables, X and Y. Now imagine some function of X: associated with each possible X value is a new number $U = g(X)$. For example, it might be that

$$U = aX + b,$$

so that U is some linear function of X. Or, perhaps

$$U = 3X^2 + 2X,$$

and so on for any other function rule. Furthermore, let there be some function of Y,

$$W = h(Y),$$

so that a new number W is associated with each possible value of Y. *Then U and W are independent random variables, provided that X and Y are independent.* This principle applies both to continuous and to discrete random variables and to all the ordinary functions studied in elementary mathematical analysis. We will make extensive use of this principle in later sections on sampling random variables.

It is especially interesting to apply this principle to measurements as functions of underlying magnitudes of properties. Imagine that the true magnitudes on two properties for some sample space of objects are represented by the random variables X and Y and that the numerical measurements assigned to these objects are represented by U and W. Furthermore, suppose that U is some function of X and W is some function of Y. Then if the underlying variables X and Y are independent, the measurements U and W must be independent as well.

The idea of independent random variables permeates all of statistical theory. In some instances, X and Y are different kinds of measurements, and we are interested in the association between the attributes they represent. However, equally important is the situation where a series of observations is made, each observation producing one value of the *same* random variable X. The series of outcomes can be regarded as a set of random variables X_1, X_2, X_3, and so on, each having exactly the same distribution. The subscripts refer only to the order of sampling in the sequence, the value observed first, second, and so on. For example, we might be measuring the same individual over time or drawing several cases from the same group of people measured in the same way. If these observations are independent, then for, say, X_1 and X_2,

$$f(x_2 \mid x_1) = f(x_2)$$

and
$$f(x_1 \mid x_2) = f(x_1).$$

The distribution of X is the same regardless of which observation in the sequence or in the total sample we are considering. This amounts to saying that the chances for different intervals of values of X to occur stay the same regardless of where we are in the sampling process. Essentially, this is the concept of independence which was invoked in the discussion of the law of large numbers in Section 2.13.

3.23   MOMENTS OF CONDITIONAL AND JOINT DISTRIBUTIONS

Moments can be defined not only for simple distributions of a single variable, such as $P(x)$ or $f(x)$, but also for conditional distributions such as $P(x \mid y)$ or $f(x \mid y)$. Moments of conditional distributions are based on the concept of **conditional expectation,** which is identical to expectation as we have defined it except for the fact that the expectation is taken with regard to a conditional distribution:

$$E(X \mid y) = \sum_{x} xP(x \mid y) \qquad \text{in the discrete case,} \qquad (3.23.1^*)$$

and $\qquad E(X \mid y) = \int_{-\infty}^{\infty} xf(x \mid y) \, dx \quad \text{in the continuous case.} \quad (3.23.2^*)$

For instance, in the stock-price example of Section 3.21,

$$E(X \mid Y = 0) = 0(7/9) + 1(2/9) = 2/9$$

and $\qquad E(Y \mid X = 1) = 0(1/6) + 1(5/6) = 5/6.$

Recall that if X and Y are independent random variables, then

$$P(x \mid y) = P(x) \quad \text{in the discrete case}$$

and $\qquad f(x \mid y) = f(x) \quad \text{in the continuous case.}$

But if this is true, then from our definition of conditional expectation,

$$E(X \mid y) = E(X). \qquad (3.23.3)$$

Thus, if X and Y are independent, the conditional expectation of X given $Y = y$ is equal to the unconditional expectation of X. By interchanging X and Y, we can see that this also implies that $E(Y \mid x) = E(Y)$ if X and Y are independent. Conditional expectations are extremely useful when we know the value of one variable and would like to predict the value of another variable, as we shall see in Chapter 10.

The concept of moments of a distribution can be extended to joint distributions of two or more random variables. Of particular importance are the expectations of products of powers of the random variables. That is, for a joint distribution of two random variables X and Y, the moments are of the form $E(X^r Y^s)$, where r and s are nonnegative integers. The moment $E(XY)$, where $r = 1$ and $s = 1$, is one moment of special interest in probability theory and statistics.

In order to look at $E(XY)$, we need to define expectation with respect to joint distributions.

If X and Y are random variables with joint probability function $P(x, y)$ in the discrete case or joint density function $f(x, y)$ in the continuous case, and if $g(X, Y)$ is some function of X and Y, then

$$E(g(X, Y)) = \sum_x \sum_y g(x, y) P(x, y) \quad \textbf{in the discrete case} \quad (3.23.4^*)$$

and

$$E(g(X, Y)) = \int_{-\infty}^{\infty} \int_{-\infty}^{\infty} g(x, y) f(x, y) \, dx \, dy \quad \textbf{in the continuous case.}$$

$$(3.23.5^*)$$

Moreover, this definition can be extended to the case of more than two variables in a straightforward manner. From this definition, we can prove that $E(XY)$ can be expressed in a particularly simple form when X and Y are independent.

Given the random variable X with expectation $E(X)$ and the random variable Y with expectation $E(Y)$, then if X and Y are independent,

$$E(XY) = E(X)E(Y). \quad (3.23.6^*)$$

This rule states that if random variables are *statistically independent*, the expectation of the product of these variables is the product of their separate expectations. An important corollary to this principle is as follows.

If $E(XY) \neq E(X)E(Y)$, the variables X and Y are not independent.

The basis for Equation (3.23.6) can be shown fairly simply for discrete variables. Since X and Y are independent, $P(x, y) = P(x)P(y)$. Then

$$E(XY) = \sum_x \sum_y xy P(x) P(y) = \sum_x \sum_y x P(x) y P(y).$$

However, for any fixed x, $yP(y)$ is perfectly free to be any value, so that

$$E(XY) = \sum_x x P(x) \sum_y y P(y) = E(X)E(Y).$$

By extension of Equation (3.23.6) to any finite number of random variables, we have the following result.

Given any finite number of random variables, if all the variables are independent of each other, the expectation of their product is the product of their separate expectations; thus,

$$E(X_1 X_2 \cdots X_K) = E(X_1) E(X_2) \cdots E(X_K). \qquad (3.23.7^*)$$

Very often we are interested in measures of the relationship between two random variables. Such measures are designed to reflect the direction and the strength of the relationship and the extent to which one variable may be used to predict another.

A moment of the joint distribution of X and Y which reflects the direction of their relationship is the covariance, defined as follows:

$$\text{cov } (X, Y) = E[(X - \mu_X) (Y - \mu_Y)]. \qquad (3.23.8^*)$$

This can be put into somewhat more convenient form. Expanding and taking expectations, we find that

$$\begin{aligned}
\text{cov } (X, Y) &= E(XY - \mu_Y X - \mu_X Y + \mu_X \mu_Y) \\
&= E(XY) - \mu_Y E(X) - \mu_X E(Y) + \mu_X \mu_Y \\
&= E(XY) - \mu_X \mu_Y - \mu_X \mu_Y + \mu_X \mu_Y,
\end{aligned}$$

or $\qquad \text{cov } (X, Y) = E(XY) - E(X)E(Y). \qquad (3.23.9^*)$

For the stock-price example of Section 3.21,

$$E(XY) = 0(0)(.35) + 0(1)(.05) + 1(0)(.10) + 1(1)(.50) = .50.$$

But

$$E(X) = .60 \quad \text{and} \quad E(Y) = .55;$$

therefore,

$$\text{cov } (X, Y) = E(XY) - E(X)E(Y) = .50 - (.60)(.55) = .17.$$

If X and Y are independent, $E(XY) = E(X)E(Y)$, so cov $(X, Y) = 0$. That is, the direction of the relationship between X and Y is neither positive nor negative. However, suppose that X and Y are positively related, meaning that a higher value of X *tends* to correspond with a higher value of Y, and a lower value of X *tends* to correspond with a lower value of Y. For example, let X be the income of a given person and let Y represent the number of years of schooling of that person. Then the covariance of X and Y is positive. If X and Y are negatively related (that is, a higher value of X *tends* to correspond with a lower value of Y, and vice versa), then the covariance is negative. For example, let X be the price of steak and let Y be the amount of steak sold by a particular store. The sign of the covariance, then, gives us information as to the direction of the relationship between X and Y.

Unfortunately, the magnitude of the covariance depends very much on the units of measurement for X and Y. Thus, it is very hard to interpret in terms of the *strength* of the relationship between X and Y. For example, X might be in units of inches and Y in units of dollars; the covariance is then in units of inch-dollars. The problem is solved by dividing this covariance by a product involving the same units of measurement as X and Y, which results in an index which is not "contaminated" by the units of measurement.

This measure, which corrects for the scaling of X and Y and thus reflects the strength as well as the direction of the relationship, is the correlation coefficient, which is usually denoted by the Greek letter rho.

$$\rho = \frac{\text{cov } (X, Y)}{\sigma_X \sigma_Y}. \tag{3.23.10*}$$

For the stock-price example of Section 3.21, $V(X) = .24$, $V(Y) = .2475$, and cov $(X, Y) = .17$; therefore,

$$\rho = \frac{.17}{\sqrt{.24} \ \sqrt{.2475}} = .70.$$

It can be shown that

$$-1 \leq \rho \leq +1. \tag{3.23.11*}$$

If the correlation coefficient is equal to $+1$, X and Y are said to be perfectly correlated; if it is -1, they are said to be perfectly negatively correlated; and if it is 0, they are said to be uncorrelated. For values of ρ other than $+1$, -1, and 0, a ρ closer to $+1$ or -1 implies a stronger relationship, and a ρ closer to zero ordinarily implies a weaker relationship. A point of great importance is that ρ measures the strength of the *linear* relationship between two variables. Thus, X and $Y = 2X + 1$ are perfectly correlated because they have a perfect *linear* relationship; on the other hand, X and $U = X^2$ would *not* be perfectly correlated because their relationship is not linear (see Figure 3.23.1). We will pursue the concept of correlation in greater detail in Chapter 10. The purpose of introducing it here is to show that important summary measures exist for joint distributions as well as for distributions of a single random variable.

The covariance of two random variables is useful in the determination of the variance of their sum or difference as well as in the determination of their correlation coefficient. In Section 3.10 we proved that $E(X + Y) = E(X) + E(Y)$. Can we determine a general formula like this for the

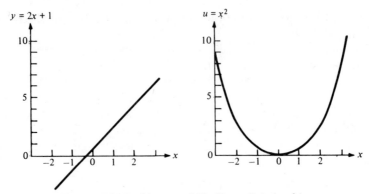

Figure 3.23.1   Linear and Nonlinear Relationships

variance of the sum of two random variables? It turns out that we can, and that, in general, the variance of a sum is not equal to the sum of the variances. From the definition of variance,

$$V(X + Y) = E[(X + Y) - E(X + Y)]^2.$$

But $E(X + Y) = E(X) + E(Y) = \mu_X + \mu_Y,$

so $V(X + Y) = E[(X + Y) - (\mu_X + \mu_Y)]^2$

$$= E[(X - \mu_X) + (Y - \mu_Y)]^2$$

$$= E[(X - \mu_X)^2 + 2(X - \mu_X)(Y - \mu_Y) + (Y - \mu_Y)^2]$$

$$= E(X - \mu_X)^2 + 2E(X - \mu_X)(Y - \mu_Y) + E(Y - \mu_Y)^2.$$

But the first and last terms are simply σ_X^2 and σ_Y^2, and the middle term is twice the covariance, cov (X, Y). Therefore,

$$V(X + Y) = V(X) + V(Y) + 2 \text{ cov } (X, Y). \qquad (3.23.12^*)$$

In a similar manner, it can be shown that

$$V(X - Y) = V(X) + V(Y) - 2 \text{ cov } (X, Y). \qquad (3.23.13^*)$$

If X and Y are independent, their covariance is equal to zero, and we have

$$V(X + Y) = V(X) + V(Y) \qquad (3.23.14^*)$$

and $$V(X - Y) = V(X) + V(Y). \qquad (3.23.15^*)$$

It is interesting to note that unless X and Y are independent, their covariance may have an influence on the variance of their sum and the variance of their difference. If X and Y are positively related (covariance greater than zero), the variance of their sum is *greater* and the variance of their difference is *less* than if X and Y were independent. If they are negatively related, the variance of their sum is less and the variance of their difference is greater than if they were independent.

For example, suppose that

$$V(X) = 10, \qquad V(Y) = 20, \qquad \text{and} \qquad \text{cov } (X, Y) = 5.$$

Then

$$V(X + Y) = 10 + 20 + 2(5) = 40$$

and $\qquad\qquad V(X - Y) = 10 + 20 - 2(5) = 20.$

Since cov (X, Y) is positive, large values of X tend to be associated with large values of Y, and small values of X with small values of Y. Thus, when X and Y are added, some of the resulting values are quite large (large value of X plus large value of Y) and some are quite small (small value of X plus small value of Y), implying a large variance. The variance of $X - Y$ is smaller because the values resulting from (large value of X minus large value of Y) are somewhat similar to the values resulting from (small value of X minus small value of Y). Of course, the situation is reversed when the covariance is negative.

EXERCISES

1. For each of the following situations, indicate whether X is a random variable.
   (a) A drug is administered to a patient at a hospital; $X = 1$ if the patient is still alive 24 hours later and $X = 0$ if the patient is not alive.
   (b) X is the actual maximum temperature, in degrees Fahrenheit to any number of decimal places, in Miami, Florida next June 15.
   (c) The responses to a question concerning a particular product that is being test-marketed are "excellent," "good," "OK," "bad," and "terrible." Let $X = 1$ if the response by a particular consumer is "excellent" or "good," $X = 2$ if the response is "good" or "OK," $X = 3$ if the response is "OK" or "bad," and $X = 4$ if the response is "bad" or "terrible."
   (d) X is the price, in dollars per pound, of sirloin steak at a particular grocery store one year from now.
   (e) X is the minimum temperature in Philadelphia next December 1, in degrees Fahrenheit as reported by the *New York Times*.
   (f) X is the amount of time elapsing, measured in hours and accurate to any number of decimal places, from the beginning of next year until the first snow of the year is recorded by the National Weather Service in Minneapolis, Minnesota.
   (g) X is the price, in dollars per gallon, of milk at a particular grocery store during the next two years.

2. For each random variable in Exercise 1, indicate whether it is discrete or continuous.

3. The random variable X has the following probability distribution:

x	$P(X = x)$
-1	.2
0	.3
3	.2
4	.2
6	.1

(a) Graph the PMF and the CDF of X.
(b) Explain the relationship between the PMF and the CDF.
(c) Find $P(X \geq 2.5)$.
(d) Find $P(-1 < X < 4)$.
(e) Find $P(-1 \leq X \leq 4)$.
(f) Find $P(X < -3)$.
(g) Find $P(X = 1)$.

4. Let X represent the number of years before a particular piece of equipment will need replacement. As an approximation, assume that X can only take on the values 1, 2, 3, and 4 and that the value of the cumulative distribution function of X at those four points is given by the equation $F(x) = x^2/16$.
(a) Find the corresponding probability mass function.
(b) What is the probability that the machine will not need replacement for *at least* three years?

5. The random variable X represents the number of heads occurring in three independent tosses of a fair coin. Represent the distribution of X
(a) by a listing,
(b) by a graph of the PMF,
(c) by a graph of the CDF.
Do the same for Y, the number of heads occurring in *four* independent tosses of a fair coin.

6. Suppose that the face value of a playing card is regarded as a random variable X, with an ace counting as 1 and any face card (Jack, Queen, King) counting as 10. You draw one card at random from a well-shuffled deck. Construct a PMF showing the probability distribution for this random variable and find the probabilities of the following events:
(a) $P(X \leq 6)$,   (d) $P(X \text{ is an even number})$,
(b) $P(X > 4)$,   (e) $P(X \neq 2 \cap X \neq 8)$.
(c) $P[(2 \leq X \leq 7) \cup (X = 10)]$,

7. The random variable X has the probability distribution given by the rule

$$P(X = a) = \begin{cases} 1/k & \text{for } a = 1, 2, 3, 4, 5, \\ 0 & \text{elsewhere.} \end{cases}$$

(a) Is X discrete or continuous?

(b) Find k.

(c) Graph the distribution of X in two ways.

(d) Find the distribution of $Y = (X - 3)^2$ and graph it.

8. A popular brand of toothpaste is available at a reduced price at a certain store, with the condition that each customer can purchase at most two tubes. If a particular customer purchases the toothpaste, let X represent the number of tubes purchased. The cumulative distribution function of X is given by the rule

$$F(x) = \begin{cases} 0 & \text{if } x < 1, \\ \frac{1}{4} & \text{if } 1 \le x < 2, \\ 1 & \text{if } x \ge 2. \end{cases}$$

(a) Find the corresponding PMF.

(b) Find $P(1 < X < 2)$.

9. In Exercise 8, Chapter 1, an item is randomly selected from the lot of 900 items; that is, each of the 900 items is equally likely to be the one that is selected. Let X represent the number of defects in the item that is selected.

(a) Find the PMF of X.

(b) Find the CDF of X.

(c) Find $P(1 \le X < 3)$.

10. In Exercise 32, Chapter 2, suppose that the salesman will either visit all four states in the assigned territory or will not take a trip at all. If the trip is taken, the possible orderings of states in terms of which is visited first, which is visited second, and so on, are equally likely. Furthermore, the odds are 3 to 2 in favor of the trip being taken. Let X represent the number of states visited up to and including Illinois (for instance, if Illinois is the second state visited, then $X = 2$), with $X = 0$ if the trip is not taken.

(a) Find the probability distribution of X and specify it in terms of a listing.

(b) Find the probability that X is at least 3.

(c) Find the probability that X is no greater than 1.

11. A firm requires its employees to have medical checkups at least every three years. An employee is chosen randomly; let $X = 0$ if that employee's last checkup was less than a year ago, $X = 1$ if it was at least a year ago but less than two years ago, and $X = 2$ if it was at least two years ago but less than three years ago. From company records, it is known that the probability distribution of X is

$$P(X = x) = \begin{cases} (x^3 + 3)/18 & \text{if } x = 0, 1, \text{ or } 2, \\ 0 & \text{elsewhere.} \end{cases}$$

(a) Show that this satisfies the requirements for a PMF.

(b) Graph the cumulative distribution function.

(c) Find the probability that the employee's last checkup was less than two years ago.

12. A particular firm is interested in R, the ratio of sales to profits, and the cumulative distribution function for R is

$$F(r) = P(R \leq r) = \begin{cases} 0 & \text{if } r < 1, \\ (r^2 - 1)/8 & \text{if } 1 \leq r \leq 3, \\ 1 & \text{if } r > 3. \end{cases}$$

(a) Is R discrete or continuous?
(b) Find the probability that the ratio is between 1.5 and 2.
(c) Find the probability that the ratio is equal to 1.8.

13. Let X represent the number of hours it takes before a computer program that is submitted to a university's computer center is run and returned to the person who submitted it. The probability distribution of X is

$$P(X = x) = \begin{cases} (6 - x)/15 & \text{if } x = 1, 2, 3, 4, 5, \\ 0 & \text{elsewhere.} \end{cases}$$

(a) Graph this function.
(b) What is the probability that it takes at least 3 hours for the program to be run and returned?
(c) A program is submitted at 7 a.m. and the person submitting the program must go to a class at 9:30 a.m. What is the probability that the program will be run and returned in time to take it to the class?

14. A consumer buys a new color television set, and after some careful analysis of available data regarding breakdowns, the consumer decides that the cumulative distribution function of T, the time in years until major repairs are first needed, is as follows:

$$F(t) = \begin{cases} 0 & \text{if } t < 0, \\ t^3/64 & \text{if } 0 \leq t \leq 4, \\ 1 & \text{if } t > 4. \end{cases}$$

(a) Find and graph the corresponding density function.
(b) If the consumer intends to sell the television set when it is $2\frac{1}{2}$ years old, what is the probability that the set will *not* need any major repairs before it is sold?
(c) Answer (b) if the consumer intends to sell the set when it is 5 years old.

15. Discuss the proposition: "All observed numerical events represent values of discrete variables, and continuous variables are only an idealization."

16. Why is it necessary to deal with probability densities rather than probabilities such as $P(X = x)$ when the variable under consideration is continuous?

17. In Exercise 13, it seems reasonable to assume that X is continuous rather than discrete unless the computer center has a policy of returning programs at regular intervals such as hourly intervals. Assume that X is continuous and that the density function of X is given by

$$f(x) = \begin{cases} 2(5-x)/25 & \text{if } 0 < x < 5, \\ 0 & \text{elsewhere.} \end{cases}$$

(a) Without using integration, show that the area under the curve is equal to one. [*Hint:* Use the fact that the area of a triangle is equal to one-half of the product of the base and the height.]

(b) Find the probability that it takes at least 3 hours for the program to be run and returned.

(c) If the program is submitted at 7 a.m., what is the probability that it will be run and returned in time to take it to a 9:30 class, assuming that the classroom is just across the hall from the computer center?

18. A random variable Y has the density function

$$f(y) = \begin{cases} ky^2 & \text{if } 0 \le y \le 3, \\ 0 & \text{elsewhere.} \end{cases}$$

(a) Find k.

(b) Find the corresponding cumulative distribution function.

19. A grocery chain has just purchased a large shipment of apples, and there is some concern about the rate at which the apples will be sold, since they are less attractive as they lose their freshness. Let X represent the proportion of the shipment that will be sold within the first week, and suppose that the density function of X is given by

$$f(x) = \begin{cases} kx(1-x) & \text{for } 0 < x < 1, \\ 0 & \text{elsewhere.} \end{cases}$$

(a) Find k and graph the density function.

(b) Find $P(1/4 < X < 1/2)$.

(c) Find $P(-1/2 \le X \le 1/4)$.

(d) Find the CDF and graph it.

20. The density function of X is given by the rule

$$f(x) = \begin{cases} x & \text{for } 0 < x \le 1, \\ 2-x & \text{for } 1 < x < 2, \\ 0 & \text{elsewhere.} \end{cases}$$

(a) Find the CDF.

(b) Prove that $f(x)$ satisfies the two requirements for a density function.

(c) Find $P(1/2 < X < 3/2)$.

21. Does the function

$$f(x) = \begin{cases} 2x/3 & \text{for } -1 \le x \le 2, \\ 0 & \text{elsewhere,} \end{cases}$$

satisfy the two requirements for a density function?

22. The maximum temperature (in degrees Fahrenheit) tomorrow in a particular city is denoted by W, and the density function of W is

$$f(w) = \begin{cases} 2(w - 65)/75 & \text{if } 65 \le w \le 70, \\ (80 - w)/75 & \text{if } 70 < w \le 80, \\ 0 & \text{elsewhere.} \end{cases}$$

(a) Graph $f(w)$ and explain its implications for the maximum temperature tomorrow in the city in question.
(b) Find the probability that W is less than 75.
(c) Find the probability that W is between 68 and 72.
(d) Find the CDF of W and graph it.

23. A sales clerk in a store is idle, and the PDF of T, the amount of time until the next customer enters the store, is

$$f(t) = \begin{cases} 10e^{-10t} & \text{if } t \ge 0, \\ 0 & \text{elsewhere,} \end{cases}$$

where T is expressed in hours.

(a) Find the probability that a customer will enter the store within the next 10 minutes.
(b) Graph the density function of T.
(c) Find the cumulative distribution function of T.

24. Let Y denote the number of minutes that a flight from Atlanta to Chicago is late arriving in Chicago (a negative value indicates that the flight arrives early). The CDF of Y is given by

$$F(y) = \begin{cases} 0 & \text{if } y < -10, \\ (y + 10)/60 & \text{if } -10 \le y \le 50, \\ 1 & \text{if } y > 50. \end{cases}$$

(a) Graph $F(y)$.
(b) Find the density function of Y and graph it.
(c) A distribution of this form is often called a "uniform," or "rectangular," distribution. Explain why these adjectives are appropriate and give an interpretation in terms of the situation at hand (the arrival of the flight).
(d) Find the probability that the flight is at least 10 minutes late but no more than 30 minutes late.

(e) A passenger must arrive no more than 20 minutes late in order to catch a connecting flight. What is the probability that the passenger will catch the connecting flight?

25. In Section 3.6 an example of a mixed random variable is given: the amount of rain (in inches) that will fall on a particular summer day in Paris, France. The probability of no rain is .6, and the remaining .4 of probability is distributed according to the density function

$$f(x) = \begin{cases} (4-x)/20 & \text{if } 0 < x < 4, \\ 0 & \text{elsewhere.} \end{cases}$$

(a) Find the probability of more than 2 inches of rain.
(b) Find the cumulative distribution function for the amount of rain.

26. Give some examples of random variables that might be considered to be mixed random variables.

27. Suppose that someone agrees to pay you $10 if you throw at least one six in four tosses of a fair die. How much would you have to pay for this opportunity in order to make it a "fair" gamble? (You pay in advance and then receive either $10 or $0; your payment is not returned.)

28. Explain what is meant by the statement, "on any single trial, we do not expect to observe the expectation exactly."

29. An individual purchases a life insurance policy that will pay $10,000 in case of death in the next year. On the basis of mortality tables for individuals of the same age, sex, occupation, health, and so on, the probability that this particular individual will die in the next year is .01. The insurance policy costs $140.
(a) Let X represent the amount of money paid by the insurance company on this policy. Find $E(X)$ and compare it with the cost of the policy. Given these figures, discuss the individual's decision to purchase the insurance policy.
(b) Given the results in (a), discuss the insurance company's decision to sell the policy for $140.

30. For the distribution of Exercise 3, find the mean, the mode, and the median. Is the distribution symmetric, skewed positively, or skewed negatively?

31. Consider the distribution of X in Exercise 4.
(a) Find the expected number of years before the piece of equipment needs replacement.
(b) An attachment can be purchased to increase the length of life of the piece of equipment by exactly 2 years. Answer (a) if this attachment has been purchased.
(c) A firm owns three pieces of equipment of the type discussed in Exercise 4. The firm plans to use one until it needs replacement, then replace it with the second and use it until it needs replacement, then replace it with the third and use it until it needs replacement, then purchase a new piece of equipment. Find the expected number of years before a new piece of equipment will be purchased.

32. In Exercise 5, find $E(X)$ and $E(Y)$. Do these expectations suggest a general formula for the expected number of heads occurring in n independent tosses of a fair coin? Explain.

33. In Exercise 9, find the expected number of defects, the mode, the median, and the midrange. Compare these values as measures of central tendency, or location, in this situation.

34. For the distribution of Exercise 13, find the mean, the mode, and the median, and comment on the relative value of these measures in this example as measures of the "typical value" of the random variable.

35. Explain why it is possible for a discrete random variable to have more than one median. Is it possible for any random variable to have more than one mode? Is it possible for any random variable to have more than one mean?

36. Can you determine the mode of a distribution by examining the CDF? Explain for both discrete and continuous distributions.

37. What do we mean when we say that the expectation of a random variable can be thought of as a center of gravity?

38. Find the following fractiles of the distribution in Exercise 3:
    (a) .25,   (d) .33,
    (b) .75,   (e) .90.
    (c) .01,

39. For the random variable X in Exercise 8, find $E(X^3)$ and $[E(X)]^3$.

40. For the random variable R in Exercise 12, find the expected ratio of sales to profits, the mode of the distribution, and the median of the distribution. Is the distribution symmetric, positively skewed, or negatively skewed?

41. In Exercise 19, find the mean, the median, the midrange, and the mode of X, the proportion of the shipment of apples that will be sold within the first week after their purchase by the grocery chain. What can you say about the shape of the density function in this example?

42. Find the following fractiles of the distribution in Exercise 17:
    (a) .20,   (c) .75,
    (b) .98,   (d) .40.

43. In Exercise 14, the consumer decides at the time the new color television set is purchased to keep the set long enough so that there is only a probability of .40 that major repairs will be needed before the set is sold. How long will the set be kept? How long will the set be kept if it is decided to keep it long enough so that there is only a .10 probability that major repairs will be needed before it is sold?

44. For the variable in Exercise 25, find the mean, the mode, and the median. Discuss the relative merits of these measures of location in view of the shape of the probability distribution. Also, find the .2, .4, .6, and .8 fractiles of the distribution.

45. Suppose that X represents the daily sales (in terms of number of units sold) of a particular product and that the probability distribution of X is as follows:

x	$P(X = x)$
7,000	.05
7,500	.20
8,000	.35
8,500	.19
9,000	.12
9,500	.08
10,000	.01

Find the expectation and the variance of daily sales. [*Hint:* To find the variance, it is easiest to first find the variance of a different variable, such as $Y = (X - 7000)/1000$.] If the net profit (in dollars) resulting from sales of X items can be given by

$$W = 5X - 38{,}000,$$

find the expectation and the variance of net profit, W.

46. For the distribution of Exercise 3, find the variance, the standard deviation, the expected absolute deviation, and the range. Also, find these same values and find the mean, the mode, and the median under the assumption that the largest value of X is 12 rather than 6. In light of these results and the results of Exercise 30, comment on the relative merit of the different measures of location and dispersion.

47. For the distribution of Exercise 4, find
(a) $E(X)$,   (d) $V(X)$,
(b) $E(X^2)$,   (e) $V(4X + 12)$.
(c) $E(4X + 12)$,

48. Find the variances of the random variables X and Y in Exercise 5.

49. If a random variable has a mean of 10 and a variance of 0, graph its distribution.

50. Prove that in general, $E(X^2)$ is not equal to $[E(X)]^2$. [*Hint:* If they are equal, what can be said about the variance, σ^2?]

51. Criticize the following statement: If the mean of a distribution is 50 and the standard deviation is 10, then the "best bet" about any case drawn at random is 50, and, on the average, we can expect to be in error by 10 points.

52. If the density function of X is given by

$$f(x) = \begin{cases} a + bx^2 & \text{for } 0 \le x \le 1, \\ 0 & \text{elsewhere,} \end{cases}$$

and $E(X) = 2/3$, find a and b.

53. The range is the easiest to compute of all of the measures of dispersion which we discussed. In view of this, why is the range not preferred to the standard deviation, which is much more difficult to compute?

54. For the distribution of Exercise 17, find the variance and the standard deviation.

55. The manager of a motorcycle dealership is considering the possibility of taking a new motorcycle and using it for a year as a rental motorcycle, since several inquiries have been made about rentals. The depreciation, licensing, insurance, and maintenance would amount to a total of $300 for a 365-day year. The motorcycle would rent for $5 per day, and the probability that it will be rented on any particular day is .4 (this probability remains the same for all days in the year, and the rental or nonrental on any given day is independent of the rental or nonrental on any other day).
(a) What is the expected net profit per day?
(b) If Y represents the net profit for the entire year, find $E(Y)$ and $V(Y)$.

56. In Exercise 14, find the expected amount of time until major repairs are needed and find the standard deviation of T.

57. What advantage does the standard deviation have over the variance as a measure of dispersion?

58. A measure not discussed in the text is the coefficient of variation, which is the standard deviation divided by the mean.
(a) Find the coefficient of variation for the variable X in Exercise 3.
(b) If the coefficient of variation of X is known, can the coefficient of variation of cX be found, where c is a constant?
(c) In (a), find the coefficient of variation of $X + 1.8$ and compare it with the coefficient of variation of X.
(d) Compare the standard deviation and the coefficient of variation; in general, what is the effect of dividing the standard deviation by the mean?

59. On a final exam in statistics, the mean was 50 and the standard deviation was 10.
(a) Find the standardized scores of students receiving the grades 50, 25, 0, 100, and 64.
(b) Find the raw grades corresponding to standardized scores of -2, 2, 1.95, -2.58, 1.65, and .33.

60. For the distribution in Exercise 13, find the distribution of the corresponding *standardized* random variable, Z. In what ways are the distributions of X and Z similar and in what ways are they dissimilar?

61. For the distribution of Exercise 3, calculate $P(|X - \mu| \geq 2)$ and compare this with the upper bound for this probability obtained from Chebyshev's inequality.

62. The distribution of X, the percentage increase in gross national product in the United States for a particular year, is as follows:

x	$P(X = x)$
3	1/9
4	2/9
5	3/9
6	2/9
7	1/9

(a) Find $P(|X - \mu| \geq 1.5)$.
(b) Find the upper bound to the probability in (a) obtained from Chebyshev's inequality. p. 170
(c) Find the upper bound to the probability in (a) obtained from the stronger form of Chebyshev's inequality [Equation (3.19.3)].

63. A rough computational check on the accuracy of a standard deviation is that around six times the standard deviation should, in general, include almost the entire range of values for the distribution on which it is based. Do you see any reason why this rule should work? *Must* it be true?

64. For the random variables of Exercises 11 and 13, calculate the third moment about the mean, $E[(X - \mu)^3]$, and discuss the implications of this measure with regard to the skewness of the distributions in these exercises.

65. Give examples of some random variables that might be expected to have roughly symmetric distributions, some that might be expected to have positively skewed distributions, and some that might be expected to have negatively skewed distributions.

66. For the random variable X in Exercise 4, calculate the first three moments about the origin and the first three moments about the mean.

67. Let X represent the sales of a particular product (in units of $10,000) and let Y represent the sales of another product (in units of $10,000). The joint distribution of X and Y is represented by the following table:

			Y		
		1	2	3	4
	1	.12	.18	.24	.06
X	2	.06	.09	.12	.03
	3	.02	.03	.04	.01

(a) Graph the joint PMF of X and Y (you will need a three-dimensional graph).
(b) Are X and Y independent? Explain your answer.
(c) Determine the marginal distributions of X and Y.
(d) Determine the conditional distribution of X given that $Y = 2$. What does this tell you about the conditional distribution of X given *any* particular value of Y?

68. A family is chosen at random from those families in a given area with a particular gasoline credit card. Let X represent the number of cars in the family, and let Y represent the number of licensed drivers in the family. The marginal distributions of X and Y are as follows:

x	$P(X = x)$	y	$P(Y = y)$
1	.60	1	.10
2	.40	2	.50
		3	.30
		4	.10

The conditional probability that there is one car in a family given that there are three licensed drivers is .40; the conditional probability that there are two licensed drivers in a family given that there are two cars is .30; and the joint probability that $X = 1$ and $Y = 4$ is .03.
(a) Find the joint distribution of X and Y.
(b) Find the conditional distribution of Y given that $X = 1$.
(c) Are X and Y independent?

69. In Exercise 13, suppose that we consider another variable in addition to X. Each user of the computer center must specify whether a program will require less than one minute of computer time when submitting the program. Let $Y = 1$ if a particular program is claimed to take less than one minute, and let $Y = 2$ otherwise. To keep the users of the computer honest, the computer center will not run programs with $Y = 1$ for over a minute even if they are not finished in that time. The marginal distribution of X is given in Exercise 13, and the conditional distribution of Y given that $X = x$ (for $x = 1, 2, 3, 4, 5$) is

$$P(Y = y \mid X = x) = \begin{cases} \dfrac{(3 + xy - 2y)}{3x} & \text{if } y = 1, 2, \\ \\ 0 & \text{elsewhere.} \end{cases}$$

(a) Find the joint distribution of X and Y in functional form and in tabular form.
(b) Find the conditional distribution of X given that $Y = 1$ and graph the conditional PMF.

(c) Find the conditional distribution of X given that $Y = 2$ and graph the conditional PMF.

(d) Find the marginal distribution of Y.

(e) On the basis of the results in (a)–(d), what can be said about the relationship of X and Y?

70. Given the marginal distributions in Exercise 69 [that is, the marginal distribution of X from Exercise 13 and the marginal distribution of Y from (d) in Exercise 69], find the joint distribution of X and Y and discuss the relationship of X and Y if the programs are run in the order in which they are submitted, regardless of how long the users claim they will run. [Assume that none of the programs run long enough to affect X via their length, at least for all practical purposes. The main conditions affecting X are the number of programs submitted and the "condition" of the computer (that is, whether the computer is running smoothly or is subject to periodic, short breakdowns).]

71. Let X represent the proportion of a particular market that is obtained by a given brand, and let Y represent the proportion of a market for another product that is obtained by a brand produced by the same firm that produces the brand referred to in the definition of X. The two markets are completely separate, but the firm is interested in the relationship between X and Y because it is well known that the same firm produces both brands. The joint density function of X and Y is

$$f(x, y) = \begin{cases} k(x + y) & \text{for } 0 \le x \le 1, 0 \le y \le 1, \\ 0 & \text{elsewhere.} \end{cases}$$

(a) Find k.

(b) Find the marginal density functions of X and Y.

(c) Find the conditional density function of X, given that $Y = .25$.

(d) Find the conditional density function of X, given that $Y = .50$.

(e) Are X and Y independent?

72. For the joint distribution given in Exercise 67,

(a) find $E(X)$, $E(Y)$, $E(X + Y)$, $E(XY)$, and $E(4X - 2Y)$,

(b) find cov (X, Y), $V(X + Y)$, and $V(X - Y)$,

(c) find $E(X \mid Y = 2)$,

(d) find ρ.

73. For the distribution given in Exercise 68, find cov (X, Y) and ρ.

74. For the distribution given in Exercise 69, find cov (X, Y) and ρ.

75. For the distribution given in Exercise 71, find $E(X)$, $E(Y)$, cov (X, Y), and ρ.

76. From your major field of interest, make up two examples of variables that you might reasonably expect to be independent. Also, make up four examples of variables that should be associated to some degree, two examples involving variables with a positive relationship and two examples involving variables with a negative relationship.

77. There is a real danger in confusing the idea of causation with that of statistical association. Why do you think these two concepts are so often confused? For example, for centuries it was thought that swampy air when breathed caused malaria (hence the name, literally "bad air"). Comment on what this example shows about statistical association and causation.

78. For the distribution of Exercise 4, find the correlation coefficient of X and $Y = (X - 3)^2$. In this example, if we know X, we can easily determine Y with certainty, and vice versa. If this is true, why is the correlation coefficient not equal to $+1$ or -1? What does this example illustrate about correlation and association?

79. If cov $(X, Y) = 1.0$, what can we say about the relationship between X and Y? If cov $(U, W) = 2.0$, does this mean that the relationship between U and W is *stronger* than the relationship between X and Y? Explain.

80. Prove that if X and Y are independent, then their covariance and their correlation coefficient are equal to zero [*Note*: The reverse implication is not true in general.]

81. Prove that cov $(aX, bY) = ab$ cov (X, Y).

82. If $V(X) = 50$, $V(X + Y) = 80$, and $V(X - Y) = 40$, find $V(Y)$ and cov (X, Y).

4

SPECIAL
PROBABILITY
DISTRIBUTIONS

In probability and statistics there are certain classes, or families, of probability distributions that are frequently encountered. These classes of distributions have a wide range of applicability, and as a result it is very useful to study them in their most general form. First, some specific types of discrete distributions will be developed and discussed, and then some continuous distributions will be covered. The first discrete distribution to be introduced is the binomial, and we will dwell somewhat longer on the binomial than on some of the other distributions, both because it is an important distribution and because it is valuable to show in some detail how a theoretical distribution of a discrete random variable can be constructed.

4.1 THE BERNOULLI PROCESS

The simplest probability distribution is one with only two events. For example, a coin is tossed and one of two events, heads or tails, must occur, each with some probability. (This statement presumes, of course, that we are willing to eliminate from consideration such possibilities as the coin landing on edge or the coin disintegrating in mid-air.) At item coming from a particular production process might be inspected, with the two possible events being that the item is defective or that it is not defective. A contractor might bid for a contract, and two events, winning the contract or not winning the contract, are possible. An experiment, or trial, which can eventuate in only one of two outcomes, is called a **Bernoulli trial** (after J. Bernoulli), and an experiment consisting of a series of independent, identical Bernoulli trials is called a **Bernoulli process.**

In general, one of the two events in a Bernoulli process is called a "success" and the other a "failure," or "nonsuccess." These names serve only to tell the events apart; they are not meant to bear any connotation of desirability of the event. In the discussion to follow, the symbol p will stand for the probability of a success, and $q = 1 - p$ will stand for the probability of a failure. In tossing a fair coin, let a head be a success. Then $p = 1/2$ and $q = 1 - p = 1/2$. If the coin is biased, so that heads is twice as likely to come up as tails, then $p = 2/3$ and $q = 1/3$.

Formally, then,

a Bernoulli process consists of a series of independent, dichotomous trials, where the two events at each trial are labeled "success" and "failure," p is the probability of success on a given trial, and p remains unchanged from trial to trial.

Of course, since p remains unchanged from trial to trial, so does q. This feature, requiring that p and q not vary for different trials, is called **stationarity.** Thus, a Bernoulli process may be characterized as being **dichotomous, independent,** and **stationary.**

As can be seen from the above discussion, not every dichotomous process is a Bernoulli process. In a production process, for instance, defective items may tend to occur in "runs," or "batches," in which case the process may not be independent. In a tennis match with two good tennis players, the server tends to have an advantage, so the probability that a particular player wins a game may vary from game to game depending on whether the player in question is serving. Here, the stationarity assumption may be violated. The assumption of independence is also questionable in the tennis example because of the psychological and physiological factors involved in "hot streaks" or "cold streaks" in sporting events. For any given dichotomous process in the real world, of course, it is necessary to investigate carefully the applicability of the Bernoulli assumptions of independence and stationarity before treating the process as though it were a Bernoulli process. Fortunately, although many real-world processes do not obey the assumptions exactly, they often obey the assumptions closely enough to make the notion of a Bernoulli process very useful.

Suppose that we are dealing with a Bernoulli process and that we proceed to make n independent observations. Let n be, say, 5. How many different sequences of 5 outcomes could be observed? The answer, by Rule 1 of Section 2.7, is $2^5 = 32$. *However, here it is not necessarily true that all sequences will be equally probable.* The probability of a given sequence depends upon p and q, the probabilities of the two events. Fortunately, since trials are independent, we can compute the probability of any sequence by the application of Equation (2.21.2).

Suppose that we want to find the probability of the particular sequence of events

$$(S, S, F, F, S),$$

where S stands for a success and F for failure. The probability of first observing an S is p. If the second observation is independent of the first, then by Equation (2.21.2),

$$P(S, S) = p \cdot p = p^2.$$

The probability of an F on the third trial is q, so that the probability of (S, S) followed by F is p^2q. In the same way, the probability of (S, S, F, F) $= p^2q^2$, and that of the entire sequence is $p^2q^2p = p^3q^2$.

The same argument shows that the probability of the sequence (S, F, F, F, F) is pq^4, that of (S, S, S, S, S) is p^5, of (F, S, S, S, F) is p^3q^2, and so on.

Now if we write out all of the possible sequences and their probabilities, a useful fact emerges. **The probability of any given sequence of n independent Bernoulli trials depends only on the number of successes and p, the probability of a success.** That is, regardless of the *order* in which successes and failures occur in a sequence, the probability is

$$p^r q^{n-r}, \tag{4.1.1}$$

where r is the number of successes and $n - r$ is the number of failures.

For example, suppose that 10 percent of the cars produced by a certain manufacturer have defective brakes. If the assumptions of a Bernoulli process are accepted, and if we randomly select cars from this manufacturer, then Equation (4.1.1) can be used to find the probability of a particular outcome. Suppose that we inspect 10 cars and find that the second and the seventh car have defective brakes but that the other cars have good brakes. The probability of this particular sequence is

$$p^2q^8 = (.10)^2(.90)^8.$$

This is also the probability of any other sequence in which precisely 2 of the 10 cars have defective brakes.

For another example, if we toss a fair coin ($p = 1/2$) six times, what is the probability of observing exactly 3 heads followed in order by 3 tails, or (H, H, H, T, T, T)? The answer is

$$p^3q^3 = \left(\frac{1}{2}\right)^3 \left(\frac{1}{2}\right)^3 = \frac{1}{64}.$$

This is also the probability of the sequence (H, T, H, T, H, T), of the sequence (H, T, T, T, H, H), and of any other sequence containing exactly three successes or heads.

An alternative way to consider a Bernoulli process is in terms of **exchangeability** rather than in terms of stationarity and independence. **Exchangeability in the Bernoulli context means that different sequences of Bernoulli trials are considered to be exchangeable, or equally likely, if these sequences consist of the same number of successes and the same number of failures.** This feature of a Bernoulli process may be derived from the assumptions of stationarity and independence or it may be assumed directly by considering exchangeability.

4.2 NUMBER OF SUCCESSES AS A RANDOM VARIABLE: THE BINOMIAL DISTRIBUTION

The probabilities found in the previous section are for *particular sequences*, arrangements of r successes and $n - r$ failures in a certain order. If we want to know the probability of a particular sequence of outcomes of independent Bernoulli trials, that sequence will have the same probability as any other sequence with exactly the same number of successes, given n and p.

In most instances, however, we are not especially interested in particular sequences *in order*. We would like to know probabilities of given numbers of successes *regardless* of the order in which they occur. For example, when a coin is tossed 5 times, there are several sequences of outcomes where exactly 2 heads occur:

$$(H, H, T, T, T),$$
$$(H, T, H, T, T),$$
$$(H, T, T, H, T),$$
$$(H, T, T, T, H),$$
$$(T, H, T, T, H),$$
$$(T, H, H, T, T),$$
$$(T, H, T, H, T),$$
$$(T, T, H, H, T),$$
$$(T, T, H, T, H),$$

and $\qquad (T, T, T, H, H).$

Each and every one of these different sequences must have the same probability, p^2q^3, since each shows exactly two successes and three failures. Notice that there are

$$\binom{n}{r} = \binom{5}{2} = 10$$

different such sequences, exactly as Counting rule 5 from Chapter 2 gives for the number of ways 5 things can be taken 2 at a time.

What we now want is the probability that $r = 2$ successes will occur *regardless* of order. This could be paraphrased as "the probability of the sequence (H, H, T, T, T) *or* the sequence (H, T, H, T, T) *or* any other sequence showing exactly 2 successes in 5 trials." Such "or" statements about mutually exclusive events recall Axiom 3 in Section 2.4. If A and B are mutually exclusive events, then $P(A \cup B) = P(A) + P(B)$. Thus, the probability of 2 successes in any sequence of 5 trials is $P(2$ successes in 5 trials$) = p^2q^3 + p^2q^3 + \cdots + p^2q^3 = \binom{5}{2}p^2q^3$, since each of these sequences has the same probability and there are $\binom{5}{2}$ of them.

When samples of n trials are taken from a Bernoulli process, the number of successes is a discrete random variable. That is, since the various values are counts of successes out of n observations, the random variable can take on only the whole values from 0 through n. We have just seen how the probability for any given number of successes can be found. Now we will discuss the distribution pairing each possible number of successes with its probability. This distribution of the number of successes in n trials is called the **binomial distribution.** Generalizing the result of the preceding example to any r, n, and p, we have the following principle.

In sampling from a Bernoulli process with the probability of a success equal to p, the probability of observing exactly r successes in n independent trials is a binomial distribution, which is of the form

$$P(R = r \mid n, p) = \binom{n}{r} p^r q^{n-r} \qquad \text{for } 0 \le r \le n. \qquad (4.2.1^*)$$

Here R is a random variable representing the number of successes in n trials from a Bernoulli process.

It is easy to show that the binomial distribution, as given in Equation (4.2.1), satisfies the requirements of a probability distribution of a discrete random variable. First, the terms are all nonnegative. Second, the sum of all of the probabilities is one, as we can see by considering the binomial expansion. In algebra you very likely were taught how to expand an expression such as $(a + b)^n$ by the following rule:

$$(a + b)^n = a^n + \frac{n!}{(n - 1)!1!} a^{n-1}b + \frac{n!}{(n - 2)!2!} a^{n-2}b^2 + \cdots$$

$$+ \frac{n!}{1!(n - 1)!} ab^{n-1} + b^n.$$

For example, $(a + b)^3 = a^3 + 3a^2b + 3ab^2 + b^3$ according to this rule. This is the familar "binomial theorem" for expanding a sum of two terms raised to a power.

Notice that the various probabilities in the binomial distribution are simply terms in such a binomial expansion. Thus, if we take $a = p$ and $b = q$, then

$$(p + q)^n = p^n + \binom{n}{n - 1} p^{n-1}q + \binom{n}{n - 2} p^{n-2}q^2 + \cdots + q^n.$$

Since $p + q$ must equal 1.00, then $(p + q)^n = 1.00$, and the sum of all of the probabilities in a binomial distribution is 1.00.

To illustrate the binomial distribution, suppose that in some very large population of animals, 80 percent of the animals have normal coloration and only 20 percent are albino (no skin and hair pigmentation). This may be regarded as a Bernoulli process with "albino" being a success and "normal" a failure. Here the probabilities are $p = .20$ and $q = .80$. A biologist manages to sample this population at random, catching three animals. Suppose that the biologist wants to know the probability distribution of R, the number of albinos in the sample of 3 animals. This distribution is a binomial distribution, and Equation (4.2.1) can be used to calculate the required probabilities:

$$P(R = 0 \mid n = 3, p = .20) = \binom{3}{0}(.20)^0(.80)^3 = (.80)^3 = .512,$$

$$P(R = 1 \mid n = 3, p = .20) = \binom{3}{1}(.20)^1(.80)^2 = 3(.20)(.80)^2 = .384,$$

$$P(R = 2 \mid n = 3, p = .20) = \binom{3}{2}(.20)^2(.80)^1 = 3(.20)^2(.80) = .096,$$

and

$$P(R = 3 \mid n = 3, p = .20) = \binom{3}{3}(.20)^3(.80)^0 = (.20)^3 = .008.$$

If the sampling is random and if the population is so large that sampling without replacement still permits us to regard the results of successive trials as independent, then the biologist has about 51 chances in 100 of observing no albinos, about 38 chances in 100 of observing exactly 1 albino, about 10 chances in 100 of observing exactly 2 albinos, and slightly less than 1 chance in 100 of observing 3 albinos.

For another example, consider an experiment consisting of tossing a fair coin 5 times. Since the coin is fair, $p = 1/2$, and the binomial distribution for R, the number of heads in 5 tosses, is given in Table 4.2.1.

Table 4.2.1 Binomial Distribution of Number of Heads in Five Tosses of a Fair Coin

r	$P(R = r)$
0	$(1/2)^5 = 1/32$
1	$5(1/2)(1/2)^4 = 5/32$
2	$10(1/2)^2(1/2)^3 = 10/32$
3	$10(1/2)^3(1/2)^2 = 10/32$
4	$5(1/2)^4(1/2) = 5/32$
5	$(1/2)^5 = 1/32$
	1

Another example that is formally identical to this last one (with the exception of a different p) but provides different probability values follows. Suppose that among American male college students who are undergraduates, only one in ten is married. A sample of five male students is drawn at random. Let R be the number of married students observed. (We will assume the total set of students to be large enough that sampling can be done without replacement without affecting the probabilities and that observations are independent.) Here, $p = .10$, and the distribution is given in Table 4.2.2.

Table 4.2.2 Binomial Distribution of Number of Married Students in Random Sample of Five Students

r	$P(R = r)$
0	$(9/10)^5 = 59,049/100,000$
1	$5(1/10)(9/10)^4 = 32,805/100,000$
2	$10(1/10)^2(9/10)^3 = 7,290/100,000$
3	$10(1/10)^3(9/10)^2 = 810/100,000$
4	$5(1/10)^4(9/10) = 45/100,000$
5	$(1/10)^5 = 1/100,000$
	1

Contrast the distributions in Tables 4.2.1 and 4.2.2. When p is $1/2$ the distribution shows the greatest probability for $R = 2$ and $R = 3$, with the probabilities diminishing gradually both toward $R = 0$ and $R = 5$. On the other hand, when $p = .10$, the most probable value of R is 0, with a steady decrease in probability for the values 1 through 5. The distribution over such values of R is very different in these two situations even though the probabilities are found by exactly the same *formal* rule. This illustrates that **the binomial is actually a family of theoretical distributions, each following the same mathematical rule for associating probabilities with values of the random variable, but differing in particular probabilities depending upon r and p.**

The unspecified mathematical constants such as n and p that enter into the rules for probability or density functions are **parameters.** Families of distributions share the same mathematical rule for assigning probabilities or probability densities to values of a random variable; in these rules the parameters are simply symbolized as constants. Actually assigning values to the parameters, such as we did for p and n in the distributions above, gives some particular distribution belonging to the family. Thus, *the binomial distribution* usually refers to the family of distributions having the same rule, and *a binomial distribution* is a particular member of this family found by fixing n and p. Many theoretical distributions of interest in statistics can be specified by stating the function rule. The way this simplifies the discussion of distributions will be obvious as we continue; indeed, continuous distributions cannot really be discussed at all except in terms of their function rules.

Binomial probabilities can always be calculated from the function rule given in Equation (4.2.1). Sometimes the calculations become tedious, however, and it is useful to look up binomial probabilities in tables that have been created for that very purpose. For instance, Table V in the Appendix gives values of the binomial probability function for various values of n and p. From Table V, we see that the probability of observing exactly 3 successes in a sample of size $n = 10$ from a Bernoulli process with $p = .30$ is .2668.

Since instructions for using Table V are given in the Appendix, we will not dwell on such details here. It is useful, however, to note that in the table, p only goes as high as .50. How, then, could you find the probability of 6 successes in 10 trials, given that $p = .80$? From Equation (4.2.1), this probability is

$$P(R = 6 \mid n = 10, p = .80) = \binom{10}{6} (.80)^6 (.20)^4.$$

But interchanging the labels "success" and "failure" implies that obtaining 6 successes and 4 failures in 10 trials with $P(\text{success}) = .80$ and $P(\text{failure}) = .20$ is equivalent to obtaining 4 successes and 6 failures in 10 trials with $P(\text{success}) = .20$ and $P(\text{failure}) = .80$. Therefore, 6 defective items in 10 trials when $P(\text{defective}) = .80$ is the same as four nondefective items in ten trials when $P(\text{nondefective}) = .20$. Therefore, the probability of interest can be found as follows:

$$P(R = 6 \mid n = 10, p = .80) = P(R = 4 \mid n = 10, p = .20).$$

From Table V, this probability is .0881. In general, we can write

$$P(R = r \mid n, p) = P(R = n - r \mid n, 1 - p) \qquad (4.2.2^*)$$

for any r, n, and p with $0 \le r \le n$ and $0 \le p \le 1$.

The above discussion involves the calculation of probabilities of the form $P(R = r \mid n, p)$. In Chapter 3 we saw how to find a probability that a random variable lies in an interval by the use of a sum of probabilities, and this idea can be applied to the binomial distribution. For example, we might be interested in the probability of getting at least 2 but no more than 4 heads in 6 tosses of a fair coin. This probability is simply

$$P(2 \le R \le 4 \mid n = 6, p = .5) = P(R = 2 \mid n = 6, p = .5)$$
$$+ P(R = 3 \mid n = 6, p = .5)$$
$$+ P(R = 4 \mid n = 6, p = .5)$$
$$= .2344 + .3125 + .2344 = .7813.$$

That is, the chances are slightly better than 78 in 100 that the number of times heads appears in 6 tosses of a fair coin is between 2 and 4, inclusive. Similarly, suppose that the probability that a car has defective brakes is .10. If a random sample of 20 cars is observed and each car is inspected to see if the brakes are good or defective, then the probability that at most one of the 20 cars has defective brakes is

$$P(R \le 1 \mid n = 20, p = .10) = P(R = 0 \mid n = 20, p = .10)$$
$$+ P(R = 1 \mid n = 20, p = .10)$$
$$= .1216 + .2702 = .3918.$$

Notice that this probability is a *cumulative* probability. If we wanted, we could determine the cumulative distribution function for any given member of the binomial family of distributions. In fact, some tables of the binomial distribution are given in terms of cumulative probabilities, so if you use any tables other than Table V in the Appendix when calculating binomial probabilities, be careful to check whether the table you are working with gives individual probabilities or cumulative probabilities.

It must be reemphasized that the binomial distribution is a *theoretical* distribution. It shows the probabilities for various numbers of successes out of n trials in samples from a Bernoulli process. To the extent that the three basic assumptions of a Bernoulli process (dichotomous process, independent trials, stationary process) are satisfied in any particular real-world application, the binomial distribution is applicable. In many practical situations, the binomial is an appropriate and convenient distribution to assume.

4.3 THE MEAN AND VARIANCE OF THE BINOMIAL DISTRIBUTION

To find the mean and variance of the binomial distribution, first consider the situation in which there is only one Bernoulli trial. Then $P(R = 1) = p$, $P(R = 0) = 1 - p$, and we have

$$E(R) = 1 \cdot p + 0 \cdot (1 - p) = p,$$

$$E(R^2) = 1^2 \cdot p + 0^2 \cdot (1 - p) = p,$$

and $\quad V(R) = E(R^2) - [E(R)]^2 = p - p^2 = p(1 - p).$

Now notice that the binomial distribution is simply the distribution of the number of successes in n independent Bernoulli trials. By the rules of expectation, the mean of the binomial distribution is the sum of the means of the individual Bernoulli trials, or p summed n times:

$$E(R) = np. \tag{4.3.1*}$$

Similarly, because of the independence of the trials, the variance of the binomial distribution is the sum of the variances of the individual Bernoulli trials, or $p(1 - p)$ summed n times:

$$V(R) = np(1 - p) = npq. \tag{4.3.2*}$$

For example, if a fair coin is tossed 10 times, the expected number of heads is $10(1/2) = 5$ and the variance is $10(1/2)(1/2) = 2.5$. If 30 percent of the consumers in a particular large city buy a certain brand of detergent, and a random sample of 100 consumers is taken, then the expected number in the sample that buy the brand in question is $100(.30) = 30$ and the variance is $100(.30)(.70) = 21$. If 1 percent of the items produced by a certain machine are defective and if the assumptions of a Bernoulli process are satisfied, then the expected number of defectives in a sample of 20 items is $20(.01) = .20$ and the variance is $20(.01)(.99) = .198$.

Incidentally, notice that the mean *can* be some value that R cannot take on; the number of successes actually observed may be greater than or less than np, but np serves as a good summary measure of the location, or central tendency, of the distribution of R.

Alternatively, the mean and variance of the binomial distribution can be determined directly from the probability function,

$$P(R = r) = \binom{n}{r} p^r q^{n-r} = \frac{n!}{r!(n-r)!} p^r q^{n-r} \quad \text{for } r = 0, 1, \ldots, n.$$

From the definition of expectation,

$$E(R) = \sum_r rP(r) = \sum_{r=0}^{n} r\left[\frac{n!}{r!(n-r)!} p^r q^{n-r}\right]. \tag{4.3.3}$$

Now, notice that this expression could be factored somewhat, canceling r in the numerator and denominator and bringing an n and a p outside of the brackets:

$$E(R) = \sum_{r=0}^{n} np\left[\frac{(n-1)!}{(r-1)!(n-r)!} p^{r-1} q^{n-r}\right].$$

For $r = 0$, the expression being summed in Equation (4.3.3) is equal to zero, and thus it is possible to change the index of summation r so that it only goes from 1 to n rather than from 0 to n:

$$E(R) = \sum_{r=1}^{n} np\left[\frac{(n-1)!}{(r-1)!(n-r)!} p^{r-1} q^{n-r}\right]$$

$$= np \sum_{r=1}^{n} \frac{(n-1)!}{(r-1)!(n-r)!} p^{r-1} q^{n-r}.$$

But the summation includes all of the terms of a binomial distribution with parameters $n - 1$ and p, and thus the sum must be equal to one (if you do not see this, rewrite the expression in terms of $n' = n - 1$ and $r' = r - 1$). Since the sum is equal to 1,

$$E(R) = np.$$

We can find the variance of the binomial distribution in a similar manner. First, we will determine $E(R^2 - R)$:

$$E(R^2 - R) = \sum_r r^2 P(r) - \sum_r rP(r)$$

$$= \sum_r (r^2 - r)P(r) = \sum_r r(r - 1)P(r).$$

For the binomial distribution, then,

$$E(R^2 - R) = \sum_{r=0}^{n} r(r - 1)\left[\frac{n!}{r!(n-r)!} p^r q^{n-r}\right].$$

The expression being summed is equal to zero if $r = 0$ or $r = 1$, so we can sum from $r = 2$ to $r = n$ instead of from $r = 0$ to $r = n$. Canceling the $r(r - 1)$ and factoring $n(n - 1)p^2$ from the expression within brackets, we obtain

$$E(R^2 - R) = n(n - 1)p^2 \sum_{r=2}^{n} \frac{(n - 2)!}{(r - 2)!(n - r)!} p^{r-2}q^{n-r}.$$

Once again the summation consists of the summation of all of the terms of a binomial distribution, this time a binomial distribution with parameters $n - 2$ and p, so the sum is equal to 1. As a result,

$$E(R^2 - R) = n(n - 1)p^2.$$

But

$$E(R^2 - R) = E(R^2) - E(R),$$

so

$$E(R^2) = E(R^2 - R) + E(R) = n(n - 1)p^2 + np.$$

Therefore,

$$\begin{aligned} V(R) = E(R^2) - [E(R)]^2 &= n(n - 1)p^2 + np - (np)^2 \\ &= n(n - 1)p^2 + np - n^2p^2 \\ &= np^2(n - 1 - n) + np. \\ &= -np^2 + np = np(1 - p) = npq. \end{aligned}$$

Thus, the variance of the binomial distribution is equal to npq, and the standard deviation is equal to $\sqrt{npq}$.

Quite often the variable of interest is not R, the number of successes in a sample of size n, but $Y = R/n$, the *proportion* of successes in a sample of size n. The proportion of successes is a random variable taking on values between 0 and 1. Since $Y = y$ and $R = ny$ are equivalent events, it follows that

$$P(Y = y) = P(R = ny). \tag{4.3.4}$$

For instance, if $n = 8$, $P(Y = .25) = P(R = 2)$. Thus, the distribution of Y can be found from the binomial distribution, since there is a one-to-one correspondence between R and Y. The mean and variance of Y are

$$E(Y) = E\left(\frac{R}{n}\right) = \frac{1}{n} E(R) = \frac{1}{n} (np) = p \tag{4.3.5}$$

and

$$V(Y) = V\left(\frac{R}{n}\right) = \frac{1}{n^2} V(R) = \frac{1}{n^2} (npq) = \frac{pq}{n}. \tag{4.3.6}$$

If $n = 8$ and $p = .30$, $E(Y) = p = .30$ and $V(Y) = pq/n = (.30)(.70)/8 = 21/800$.

4.4 THE FORM OF A BINOMIAL DISTRIBUTION

Although the mathematical rule for a binomial distribution is the same regardless of the particular values of n and p entering into the expression, the "shape" of a histogram or other representation of a binomial distribution depends upon n and p. In general, the probabilities increase for increasing values of R until some maximum point is reached, and then for values beyond this point the probabilities decrease once again. This makes the picture given by the distribution show a single "hump" somewhere between $R = 0$ and $R = n$, with probabilities gradually decreasing on either side of this maximum point. The exact location of this highest probability relative to R depends, of course, on n and p.

One device for locating where this maximum point must occur in any binomial distribution is to find the ratio of probabilities for successive values of R. That is, we take $R = r'$ and $R = r$, where $r' = r - 1$, and put these probabilities in a ratio:

$$\frac{P(R = r \mid n, p)}{P(R = r' \mid n, p)} = \frac{\binom{n}{r} p^r q^{n-r}}{\binom{n}{r'} p^{r'} q^{n-r'}} = 1 + \frac{(n + 1)p - r}{qr}. \quad (4.4.1)$$

When this ratio is greater than 1.00, the probabilities are increasing with increasing r. When the ratio is less than 1.00, the probabilities are decreasing; when the ratio is exactly equal to 1.00, the successive probabilities are exactly the same.

Now consider the situation where R can assume the exact value np (that is, np is an integer). If we take $r = np$, we find that the ratio (4.4.1) is a number greater than 1.00, indicating that the probability for $R = np$ is greater than the probability for $R = np - 1$. On the other hand, if we take $r = np + 1$, we find this ratio is less than 1, meaning that the probability for $np + 1$ is less than that for np. In this way it can be shown that when np is an integer, the probability associated with $R = np$ is greater than that for any other value of R; in this situation, the maximum point in a graph of the binomial distribution occurs at $R = np$. As indicated in Section 4.3, this is also the mean of the distribution. Therefore, we find that in this special situation the value $Y = p$ is the most probable relative frequency of occurrence of an event with probability p.

On the other hand, it may be that R cannot attain the exact value np, since np need not be an integer. However, by exploring the changes in the ratio of probabilities, we would find that in this situation the value $R = r$

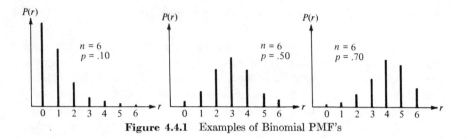

Figure 4.4.1   Examples of Binomial PMF's

with the highest probability is *no more* than p or q away from the value of np; that is,

$$-q \le r - np \le p,$$

where r is the value of R having highest probability in the binomial distribution. Thus, the most probable relative frequency is always a number *close* to the true value of p, even though this exact value may not occur for some values of n.

The PMF of the binomial distribution is symmetric if $p = 1/2$, positively skewed if $p < 1/2$, and negatively skewed if $p > 1/2$. For any value of p other than $1/2$, the skewness is reduced as n becomes larger. For a given value of p, the mean and variance become larger as n increases. For a given value of n, the mean becomes larger as p increases and the variance becomes larger as p gets closer to $1/2$ (in the extreme cases, when $p = 0$ or $p = 1$, the variance is equal to zero, as can be seen by using the formula $V(R) = npq$). Examples of binomial PMF's are presented in Figure 4.4.1.

4.5 THE BINOMIAL AS A SAMPLING DISTRIBUTION

A binomial distribution is one example of a very important kind of theoretical distribution encountered again and again in statistics. These are sampling distributions, where a random variable denotes a possible value of some measure *summarizing* the outcomes of n distinct observations. In the binomial case, this random variable typically is the number of successes in n trials or the proportion of successes in the n trials. In other contexts the random variable may be some other number summarizing a sample result, such as an average or some index of variability. There are many different kinds of sampling distributions, depending on the basic sample space and the way that samples are summarized. Nevertheless, the essential idea is the same: a sample of n observations is drawn, and a number is attached to the sample, in some way summarizing the n outcomes observed. In principle, different values will be found for different samples, and each sample value may be regarded as a value of a random variable. The sampling distribution is a theoretical statement of the probability of observing various intervals of values of such a random variable over all possible samples of the same size.

For example, imagine that all United States adults could be classified into two categories:

Success = prefers foreign movies,

Failure = prefers U. S. movies.

We will assume that each person has one of these two preferences and that there is a very large number of such people. Now we sample United States adults at random, asking each his or her preference. Suppose that our random sample consists of exactly 10 adults. The number of successes observed could be any number from 0 through 10, and there is some probability that can be associated with each possible number, given the true p. The pairing of different sample values we might observe and the probability of each is the sampling distribution. Note, however, that the sampling distribution depends upon how the sampling was done, how we attached the value to a sample, and most of all, what the *true* situation is. If only .30 of the population of adults prefer foreign movies, then the sampling distribution for the number of successes is given by the binomial distribution with $n = 10$ and $p = .30$. This distribution tells us that if we observe 10 successes out of 10 cases, then we have encountered an event that should occur only once in 100,000 samples of 10 people, *if* p is .30. On the other hand, for $p = .30$, we should expect 3 out of 10 to occur relatively often; about 26 in 100 samples should give us this exact value. About 70 in 100 samples should give us 3, 4, or 5 people preferring foreign movies, and so on. In short, a sampling distribution allows us to judge the probability of a particular kind of sample result, given that something is true in general about the population being sampled. Before we can find the sampling distribution, however, we must postulate something true about the general situation; in this case, we postulated that exactly 30 percent of United States adults prefer foreign movies. Given that this is true, the sampling distribution specifies the probabilities of various numbers of adults in the sample who show this preference.

The notion of sampling distributions will be discussed in detail in Chapter 5 and used in problems of inference and decision in the remaining chapters. In the meantime, a little preview of statistical inference and decision in which the binomial rule provides a sampling distribution will be given in the next section. This example cannot be complete at this time simply because all the necessary ingredients have not been introduced. On the other hand, we feel that it is important that the student gain some feel for the use of theoretical distributions as early as possible, to provide a basis for the developments to follow.

4.6 A PREVIEW OF A USE OF THE BINOMIAL DISTRIBUTION

It is now possible to look ahead to an important use of the binomial distribution. This example will deal with a quality control situation in which we must decide, on the basis of sample information, whether or not a process is "out of control" and should be adjusted.

Suppose that the manager of a plant is concerned with a particular manufacturing process within the plant. From time to time the process goes haywire, or out of control, and produces too many defective items. If it is out of control, it can be corrected with an adjustment. However, the manager would have to call in a mechanic to make this adjustment, and this involves an expenditure of money. On the other hand, if the process is in fact out of control, it is expensive to repair the defective items. If the process is not out of control, then the manager *does not* want to pay to have it adjusted; if it is out of control, the manager *does* want to do so.

In order to attack this problem, the manager must be more precise about what is meant by "out of control." Let us assume that the process can be thought of as a Bernoulli process with parameter p, the probability of any single item from the process being defective. Furthermore, assume that the process is in control if p is no greater than .10; otherwise, it is out of control. In other words, the manager has a hypothesis: the process is in control, or in terms of p, p is no greater than .10. If the manager could be sure that the hypothesis was true, one action would be taken (do not adjust the process); and if the manager could be sure that the hypothesis was false, a different action would be taken (call the mechanic to adjust the process). The manager's hypothesis includes all values of p less than or equal to .10; let us consider just the extreme situation in which $p = .10$. This simplifies the problem somewhat and considers the worst possible situation which is still consistent with the hypothesis.

Assume that the manager has absolutely no idea as to whether or not the process is out of control, but that a sample of 10 items from the process can be observed. Assuming the above distribution *and* assuming that the process is independent and stationary (that is, the trials are independent and p remains the same from trial to trial), the manager can calculate the probabilities of the various possible sample results for a sample of $n = 10$, using the binomial distribution. We say that the binomial distribution is being used as a sampling distribution; it is being used to calculate a probability distribution for R, the number of defectives in the

Table. 4.6.1   Distribution of Number of Defectives in Sample
of Size 10 if $p = .10$

r	$P(R = r) = \binom{10}{r}(.10)^r(.90)^{10-r}$
0 defectives	.3487
1	.3874
2	.1937
3	.0574
4	.0112
5	.0015
6	.0001
7 or more	.0000

sample. The resulting distribution is presented in Table 4.6.1. It should be pointed out that the probability of seven or more defective items is not *exactly* zero; to four decimal places, it is zero, however.

Given this theoretical distribution of possible outcomes, we turn to the actual results of the sample. The manager finds that there are five defectives in the sample of ten items. What is the probability that exactly this result should have come up by chance, given that the manager's hypothesis is true? It is the binomial probability for 5 successes in 10 trials, given that p is .10. From the above table, this probability is .0015. This sample result is not a likely occurrence if p is in fact equal to .10.

However, we might be interested in the probability not only of obtaining exactly 5 defective items, but rather in the probability of obtaining *this many or more defective items*. The manager should ask, "What is the probability of a sample result which is *at least* this 'bad,' that is, a sample result with at least 5 defective items?" The answer to this question involves the probability of an interval:

$$P(R \geq 5) = P(R = 5) + P(R = 6)$$
$$+ P(R = 7) + \cdots + P(R = 10) = .0016.$$

Given that $p = .10$, 5 or more defectives should occur only about 16 times in 10,000 independent replications of the experiment. Does this unlikely result cast any doubt on the manager's hypothesis that the process is in control? The answer is yes. For a hypothesis to seem reasonable, it should forecast results that agree fairly well with actual outcomes. If the manager had observed no defectives or one defective in the sample, then there would be little cause to doubt the hypothesis that the process is in control. Even two defectives does not seem to be too unlikely, although it may be enough to cause the manager to call the mechanic, particularly if it is very costly to repair defective items. The occurrence of 5 or more defectives, however, is so unlikely that the manager is almost sure to call the mechanic, unless the mechanic's fee is many times greater than the cost of repairing defective items.

Note that the sample results may cast doubt on the hypothesis; they do *not* disprove it. The results obtained, unlikely as they seem, could occur by chance. Thus, the manager could be making the wrong decision if the mechanic is called. The chances are good that it is the right decision, but we cannot be sure of this. If the mechanic is called, one possible measure of the amount of risk the manager runs of being wrong by abandoning the hypothesis that the process is in control on the basis of the sample evidence is given by the probability of sample results as extreme or more extreme than those actually obtained, if the hypothesis is true. In this case, this probability is .0016. If the probability had been higher, the manager would be less likely to call the mechanic. If one defective item had been observed, for example, the probability is .6513. In this case the results are not so incompatible with the hypothesis, and the manager is less likely to decide to call the mechanic.

The ultimate decision to call or not to call the mechanic depends not only on the sample results, but also on the relevant costs. That is, it depends on the cost of calling the mechanic and the cost of repairing defective items. If the former is much greater than the latter, the manager might decide to stand pat even if the sample result is quite unlikely under the hypothesis. If the latter is much greater than the former, the manager might decide to call the mechanic even if the sample result is not too unlikely under the given hypothesis. We shall discuss this type of question in Chapter 9.

Although this little example should not be taken as a model of sophisticated statistical (and decision-theoretic) practice, it does suggest the use of a theoretical distribution of sample results as an aid in making inferences and decisions on the basis of sample information. The binomial distribution is only one of a number of such sampling distributions we shall employ in this general way.

4.7 NUMBER OF TRIALS AS A RANDOM VARIABLE: THE PASCAL AND GEOMETRIC DISTRIBUTIONS

In the development of the binomial distribution, it was assumed that a fixed number n of Bernoulli trials would be observed, and the random variable of interest was R, the number of successes in n trials. In some situations, the sampling procedure may be slightly different; a sequence of independent Bernoulli trials may be observed, where the observations continue until a fixed number r of successes is observed. Here the number of trials is not fixed; it is a random variable. The distribution of N can be determined by the same line of reasoning used to find the theoretical binomial distribution. For any particular sequence with r successes in n trials, the probability has been found to be

$$p^r q^{n-r}.$$

To determine the binomial distribution, it is necessary to count the number of possible sequences having r successes in n trials: $\binom{n}{r}$. If the sampling is such that the number of successes, r, is predetermined instead of the number of trials, n, then it must be true that the final trial resulted in a success. Otherwise, the trial would not have been observed, since the rth success would have occurred on a previous trial. If the last trial must be a success, then exactly $r - 1$ successes must have occurred in the previous $n - 1$ trials. But the number of possible sequences with $r - 1$ successes in $n - 1$ trials is $\binom{n-1}{r-1}$. Thus, we have the following principle.

In sampling from a Bernoulli process with the probability of a success equal to p, the probability that it will take exactly n trials to observe r successes is

$$P(N = n \mid r, p) = \binom{n-1}{r-1} p^r q^{n-r} \qquad \text{for } 0 < r \le n. \qquad (4.7.1^*)$$

This distribution, the distribution of the number of trials until the rth success, is called the **Pascal distribution,** after the French mathematician Pascal. The mean and variance of the Pascal distribution can be shown to be equal to

$$E(N) = \frac{r}{p} \qquad \text{and} \qquad V(N) = \left(\frac{r}{p}\right)\left(\frac{1}{p} - 1\right) = \frac{r(1 - p)}{p^2}. \qquad (4.7.2^*)$$

The distribution is positively skewed, as can be seen from Figure 4.7.1.

Consider the example in Section 4.2 of the albino and normal animals, where p, the probability of an albino, is equal to .20. If the biologist takes

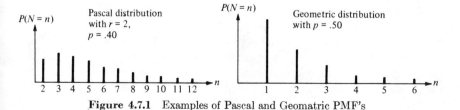

Figure 4.7.1 Examples of Pascal and Geomatric PMF's

a sample of size 3, the probability of observing exactly one albino is a binomial probability:

$$P(1 \text{ albino in 3 animals}) = \binom{3}{1}(.20)^1(.80)^2 = .384.$$

If the biologist decides to sample until a single albino is observed, however, the probability that this will take exactly 3 trials is a Pascal probability:

$$P(3 \text{ animals to get 1 albino}) = \binom{2}{0}(.20)^1(.80)^2 = .128.$$

Even though they describe the same sample (3 trials, 1 albino), the binomial and Pascal probabilities differ because the **sampling plan** is not the same; in one case, the number of trials is fixed before observing the sample, and in the other case, the number of successes is fixed in advance. We say that the **stopping rules** are different. In binomial sampling, we stop sampling after a fixed number of trials (n); in Pascal sampling, we stop sampling after a fixed number of successes (r).

In general, for the Pascal distribution, n and r may be any positive integers with $n \geq r$. The last example demonstrates a special case of the Pascal distribution: the case of $r = 1$, for which sampling continues until the first success. This special case is called the **geometric distribution,** and its probability function is given by

$$P(N = n \mid p) = pq^{n-1}. \tag{4.7.3*}$$

This function is reasonable; the only sequence satisfying the given condition is the sequence consisting of $n - 1$ failures followed by a success, and the probability of this sequence must be pq^{n-1} because of the independence of the trials.

The geometric distribution is applicable in gambling situations in which a gambler is interested in the probability of going, say, 10 trials (a trial might be a spin of a roulette wheel, a deal in a card game, a throw of a pair of dice, and so on) until the first successful gamble. If the gambler knows p, the probability of winning on any given trial, the above probability can be computed by using the geometric distribution. For instance, if $p = 1/2$,

$$P(10 \text{ trials until the first win} \mid p = 1/2) = (1/2)(1/2)^9 = 1/1024.$$

By noting that the mean and variance of the geometric distribution are equivalent to the mean and variance of the Pascal distribution with $r = 1$, since the geometric distribution is a special case of the Pascal distribution, the gambler can compute the expected number of trails until the first win and the variance of the number of trials until the first win:

$$E(N) = \frac{1}{p} = \frac{1}{1/2} = 2 \text{ and } V(N) = \left(\frac{1}{p}\right)\left(\frac{1}{p} - 1\right) = \left(\frac{1}{1/2}\right)\left(\frac{1}{1/2} - 1\right) = 2.$$

It is important to remember that the theoretical distributions we have discussed thus far in this chapter require the assumption that sampling is done from a stationary and independent Bernoulli process. Only if this assumption is satisfied are the three theoretical distributions applicable in the form in which they have been presented. Of course, a mathematical model is seldom a perfect representation of a real-world situation, and yet such models *are* applied to real-world situations. If the assumptions of a model are *approximately* satisfied, then the model still may be useful. Because of the inherent complexity of most real-world problems, it is usually necessary to settle for approximate results. Throughout our study of probability and statistics, we will rely on approximations and on assumptions which are seldom perfectly satisfied. The student should always remember, however, that statistical techniques are based on certain assumptions and that in any application of the techniques the results are only as good as the assumptions.

4.8   THE MULTINOMIAL DISTRIBUTION

The basic rationale underlying the binomial distribution can be generalized to situations with more than two events. This generalization is known as the **multinomial distribution,** having the following rule.

Consider K mutually exclusive and exhaustive events with probabilities p_1, p_2, $\ldots$, p_K. If n observations are made independently and at random, then the probability that exactly n_1 will be of event 1, n_2 of event 2, $\ldots$, and n_K of event K, where $n_1 + n_2 + \cdots + n_K = n$, is given by

$$\frac{n!}{n_1! n_2! \cdots n_K!} (p_1)^{n_1}(p_2)^{n_2} \cdots (p_K)^{n_K}. \qquad (4.8.1^*)$$

Note that the first factor of Equation (4.8.1) corresponds to Counting rule 6 of Section 2.10. Also observe that the binomial distribution corresponds to the multinomial distribution with $K = 2$, $n_1 = r$, and $n_2 = n - r$.

For example, suppose that in a large city, 50 percent of the consumers prefer Brand X coffee, 30 percent prefer Brand Y, and 20 percent prefer Brand Z. In a random sample of 10 consumers, what is the probability that

7 will will prefer Brand X, 2 will prefer Brand Y, and only 1 will prefer Brand Z? From Equation (4.8.1), this probability is

$$\frac{10!}{7!2!1!} \, (.5)^7(.3)^2(.2)^1 = \frac{10 \cdot 9 \cdot 8 \cdot 7 \cdot 6 \cdot 5 \cdot 4 \cdot 3 \cdot 2 \cdot 1}{(7 \cdot 6 \cdot 5 \cdot 4 \cdot 3 \cdot 2 \cdot 1)(2 \cdot 1)(1)} \, (.5)^7(.3)^2(.2)^1$$

$$= .050625.$$

Similarly, the multinomial rule could be used to work out the probability of any other sample distribution of preferences.

For another example, suppose that a defective item from a production process can have 1, 2, 3, or 4 defects. Moreover, suppose that the probabilities for the different numbers of defects (given that an item is defective) are $p_1 = .80$, $p_2 = .18$, $p_3 = .01$, and $p_4 = .01$. In addition, assume that these probabilities are the same for all defective items and that the trials are independent. This can be considered as a multinomial process, and the probability of obtaining exactly 5 items with 1 defect ($n_1 = 5$), 3 items with 2 defects ($n_2 = 3$), 2 items with 3 defects ($n_3 = 2$), and 1 item with 4 defects ($n_4 = 1$) in a sample of 11 defective items is

$$\frac{11!}{5!3!2!1!} \, (.80)^5(.18)^3(.01)^2(.01)^1 = .000053.$$

Although the multinomial rule is relatively easy to state, a tabulation or graph of this distribution is very complicated; here a sample result is not a single number, as for the binomial, but rather a set of $(K - 1)$ numbers (once $n_1, n_2, \ldots, n_{K-2}$, and n_{K-1} are known, n_K is fixed, since the sum of the n_i's must equal n). Thus, the number of possible sample results is very large unless n and K are quite small. In general, this makes it infeasible to consider entire distributions with the multinomial rule, but individual multinomial probabilities can be calculated from Equation (4.8.1).

4.9 THE HYPERGEOMETRIC DISTRIBUTION

Another theoretical probability distribution, the hypergeometric distribution, is similar to the multinomial distribution, but a slight change in the underlying sampling procedure leads to a violation of the assumptions (stationarity, independence) used in the multinomial case. Suppose that we are sampling from some finite set of objects. This set of objects will be referred to as a **population,** so we are sampling from a **finite population.** Examples are drawing cards from a standard 52-card deck of cards, sampling items from a finite set of items that have just been produced by a given machine, sampling people from the residents of a given community, and so on. The multinomial rule (and the binomial rule as well) assumes either that the sampling is done *with replacement* or that the population is infinite, so that the basic probabilities do not change over the trials. However, suppose that we were sampling from a *finite population without replacement*; then the probabilities would change for each observation. By a series of arguments very similar to those used for finding probabilities of poker hands in Section 2.11, we can arrive at a new rule for finding the probabilities of sample results. This rule describes the hypergeometric distribution.

Given a population containing a finite number w of elements, suppose that the elements are divided into K mutually exclusive and exhaustive classes, with w_1 in class 1, w_2 in class 2, ..., w_K in class K. A sample of n observations is drawn at random without replacement and is found to contain n_1 of class 1, n_2 of class 2, ..., n_K of class K. Then the probability of occurrence of such a sample is given by

$$\frac{\binom{w_1}{n_1}\binom{w_2}{n_2}\cdots\binom{w_K}{n_K}}{\binom{w}{n}}, \qquad (4.9.1^*)$$

where $n_1 + n_2 + \cdots + n_K = n$ and $w_1 + w_2 + \cdots + w_K = w$.

To illustrate the hypergeometric distribution, consider the example of the previous section concerning preferences among three brands of coffee. Suppose that we are dealing with a very small population of only 20 consumers and that 10 prefer Brand X, 6 prefer Brand Y, and 4 prefer Brand Z. A sample of 10 consumers is drawn at random *without replacement*. What is the probability that 7 will prefer Brand X, 2 will prefer Brand Y, and only 1 will prefer Brand Z? From Equation (4.9.1), this probability is

$$\frac{\binom{10}{7}\binom{6}{2}\binom{4}{1}}{\binom{20}{10}} = .03897.$$

If the sampling had occurred *with replacement*, the multinomial rule would apply and the probability, as calculated in the previous section, would be .050625. This illustrates that *the sampling scheme adopted can make a real difference in the probability of a given result*.

Despite the above result, the difference between the hypergeometric distribution and the multinomial distribution is a serious practical consideration only when the population contains a small number of members. When the population is large, the selection and nonreplacement of a particular unit for observation has negligible effect on the probabilities of events for successive samplings. For this reason, the hypergeometric probabilities are very closely approximated by binomial or multinomial probabilities when w, the total number of elements in the population, is extremely large. The distinction between these different distributions becomes practically important only when samples are taken from relatively small populations.

In order to compare the binomial and hypergeometric distributions, consider the case with only two classes. It can be shown that the mean and variance of the hypergeometric distribution when $K = 2$ (assuming class 1 represents success, class 2 represents failure, and R is the number of successes in n trials) are

$$E(R) = n\left(\frac{w_1}{w}\right) \quad \text{and} \quad V(R) = \frac{(w-n)}{w-1}\left[n\left(\frac{w_1}{w}\right)\left(1 - \frac{w_1}{w}\right)\right]. \quad (4.9.2^*)$$

Recall that the mean and variance of the binomial distribution are

$$E(R) = np \quad \text{and} \quad V(R) = npq = np(1 - p).$$

Note that the ratio w_1/w represents the proportion of successes in the entire sample space. If we were to take an element at random from the sample space, the probability that it would be a success is simply this ratio, w_1/w. In this sense, w_1/w in the hypergeometric distribution corresponds to p in the binomial distribution, and the similarity in means and variances becomes apparent. If we assume that $p = w_1/w$, the means are equal and the variances differ only by a factor of $(w-n)/(w-1)$, or $1 - [(n-1)/(w-1)]$. But n is the sample size, and w is the population size. Since the two variances differ by a factor of $1 - [(n-1)/(w-1)]$, they become more identical as the ratio $(n-1)/(w-1)$ approaches zero (that is, as the sample size becomes very small in relation to the population size). This agrees with the statement in the preceding paragraph, which essentially says that the distinction between the two distributions becomes practically important only when the population size is small enough (or the sample size is large enough) so that the sample represents a sizable proportion of the population. Sampling from a Bernoulli process can be thought of as sampling from an infinite population, since p remains constant from trial to trial. Appropriately, then, the factor $(w-n)/(w-1)$ is called the **finite population correction,** allowing for the fact that the hypergeometric distribution represents the case in which we are sampling from a finite population. (See the discussion of sampling theory in Chapter 11.)

4.10   THE POISSON PROCESS AND DISTRIBUTION

The binomial distribution is the distribution of the random variable R, the number of successes, or occurrences of a particular event, in a sequence of trials from a stationary and independent Bernoulli process. This distribution has wide applicability, for there are many situations where a sequence of observations can be made and the assumptions underlying the Bernoulli process are satisfied or at least approximately satisfied. Frequently, however, we are not able to observe a finite sequence of trials. Instead, observations take place over a continuum, such as time. Suppose that we are observing the number of cars arriving at a toll booth in a given period of time or the number of deaths occurring from a specific disease in a given period of time. Since the observations take place over time, it is difficult to think of these situations in terms of finite trials, although we are interested in the number of occurrences of a particular event, as in the Bernoulli process.

Suppose that we are interested in the number of occurrences of an event A in a given time period of length t. One way to look at this is to break the time period t into n equal intervals, each of length t/n, and to consider these as n independent, stationary trials from a Bernoulli process. From the binomial distribution, if we know p, the probability of an occurrence of A on any specific trial, we can calculate the probabilities for R, the number of occurrences of A in n trials. Furthermore, the expected number of occurrences is np. Thus, it looks as though the situation of interest (occurrences of an event over a time period of length t) can be treated as a Bernoulli process. However, there is one difficulty that has been ignored. Since the event occurs at various points of time, what would prevent it from occurring twice or more in one of the trials of length t/n? Since the Bernoulli process considers only a dichotomous situation—that is, the occurrence or nonoccurrence of an event, there is no way to allow for the possibility of multiple occurrences within a single trial. This difficulty can be resolved to some extent by making n larger, thus dividing the period of length t into n equal intervals of smaller length. If n is chosen so that the probability of multiple occurrences in any single trial of length t/n is zero for all practical purposes, then the situation resembles that of a Bernoulli process.

As we make n larger and larger, the trials are shorter and shorter in terms of length of time, t/n. As a result, the probability of an occurrence in any single trial must become smaller and smaller. Consider the expected number of occurrences in a period of time t, and call this expectation $E(R)$. From the binomial distribution, $E(R) = np$. The division of the time period into subperiods should not have any effect on the expected number of occurrences in a period of time t, so $E(R)$ should remain constant as n is varied. If this is true, it clear that p must become smaller as n becomes larger, so that np will remain constant. Now, if R is the number of occurrences of A in the given time period, we have, from the binomial distribution,

$$P(R = r \mid n, p) = \binom{n}{r} p^r q^{n-r} \qquad \text{for } r = 0, 1, \ldots, n.$$

It can be shown mathematically that if n is made larger and larger and p smaller and smaller in such a way that np remains constant, then the distribution of R approaches the Poisson distribution, which can be written as follows:

$$P(R = r \mid n, p) = \frac{e^{-np}(np)^r}{r!}. \qquad (4.10.1^*)$$

Since np remains constant, it is customary to set $np = \lambda t$, where λ (Greek λ) is thought of as the intensity of the process (expected rate of occurrence of A) per unit time period and λt is the intensity in a time period of length t. The Poisson distribution can then be written as follows:

$$P(R = r \mid \lambda, t) = \frac{e^{-\lambda t}(\lambda t)^r}{r!} \qquad \text{for } r = 0, 1, 2, \ldots.$$

$$(4.10.2^*)$$

Note that R can now take on any integral value, since there is no upper bound to the number of occurrences of an event in a given time period. The mathematical derivation of the Poisson distribution as a limiting form of the binomial distribution is beyond the scope of this book. However, given the functional form of the distribution as presented above, the mean and variance of the Poisson distribution can be derived in a manner similar to the derivation of the mean and variance of the binomial distribution in Section 4.3.

From the definition of expectation,

$$E(R) = \sum_r r P(r) = \sum_{r=0}^{\infty} r \frac{e^{-\lambda t}(\lambda t)^r}{r!}.$$

Canceling r from the numerator and denominator and factoring out λ, we obtain

$$E(R) = \lambda t \sum_{r=0}^{\infty} \frac{e^{-\lambda t}(\lambda t)^{r-1}}{(r - 1)!}.$$

Since the expression being summed in the first expression for $E(R)$ is zero when $r = 0$, the limits of summation can be changed from 0 and ∞ to 1 and ∞. As a result, the sum in the second expression is simply the summation of all of the terms of a Poisson distribution with parameter λ (to see this, let $r' = r - 1$), and this must be one. But then

$$E(R) = \lambda t. \qquad (4.10.3^*)$$

This is consistent with the argument above, in which it was noted that the expected number of occurrences in a period of time t is equal to np, which is equal to λt.

The variance of the Poisson distribution is equal to the mean:

$$V(R) = \lambda t. \qquad (4.10.4^*)$$

The derivation of the variance of the Poisson distribution parallels the derivation of the variance of the binomial distribution.

We have developed the Poisson distribution by considering a limiting form of the binomial distribution. It is possible to consider the occurrence of an event over a continuum as a process in itself, however, and thus to derive the Poisson distribution without any reference to the binomial. Suppose that we are interested in the number of occurrences of an event in a given period of time of length t. Furthermore, suppose that the occurrence or nonoccurrence of the event in any time interval is independent of its occurrence or nonoccurrence in any other time interval (the independence assumption). Also, suppose that the probability of an occurrence of the event in a time period *of given length* is the same no matter when the period begins and ends (the stationarity assumption). Then we say that **the occurrences of the event are generated by a stationary and independent Poisson process and that the distribution of R, the number of occurrences in a time period of length t, is given by a Poisson distribution, where λ corresponds to the intensity of the process per time period of length one.** The Poisson process has a single parameter, λ, which is called the **intensity of the process** because it represents the expected number of occurrences in a time period of length one. For example, λ might be the number of cars per minute arriving at a toll booth or the number of deaths per year attributed to a specific disease.

When we use the Poisson distribution, it is important to make sure that the units used for λ and t are comparable. For example, if we are concerned with the arrival of telephone calls at a switchboard and t is expressed in minutes, then λ should be expressed in calls per minute. If t is expressed in hours, on the other hand, then λ should be expressed in calls per hour. For instance, suppose that the intensity of the process generating incoming calls is 2 per minute, and we are interested in the probability of receiving exactly 12 calls in a 5-minute period. Here $\lambda = 2$, $t = 5$, and from Equation (4.10.2),

$$P(R = 12 \mid \lambda = 2, t = 5) = \frac{e^{-2(5)}[2(5)]^{12}}{12!} = \frac{e^{-10}10^{12}}{12!}.$$

Alternatively, the same problem could be treated by noting that the intensity per *hour* is 120 and that 5 minutes represent $1/12$ hour. Thus, $\lambda = 120$ and $t = 1/12$, so that

$$P(R = 12 \mid \lambda = 120, t = 1/12) = \frac{e^{-120(1/12)}[120(1/12)]^{12}}{12!} = \frac{e^{-10}10^{12}}{12!}.$$

Therefore, we see that whether the unit of time is taken as a minute or an hour, the results are identical. The important thing to remember is that we must be consistent when considering λ and t; if we express λ in terms of intensity per minute but express t in terms of hours, the results will make no sense.

It is tedious to calculate Poisson probabilities by hand due to such factors as $e^{-\lambda t}$, $(\lambda t)^r$, and $r!$, but tables of Poisson probabilities are available. In these tables, probabilities are given for various values of r and λt. For instance, in the above example, $\lambda t = 10$, and from Table VI in the Appendix,

$$P(R = 12 \mid \lambda t = 10) = .0948.$$

If the process generating incoming calls behaves like a Poisson process with an intensity of 2 calls per minute, then the probability of receiving exactly 12 calls in a particular 5-minute period is .0948.

For any value of λt, we can look at the entire Poisson distribution corresponding to that value. In Table 4.10.1, the Poisson distribution for $\lambda t = 1$ is given (obtained from Table VI in the Appendix). From these probabilities, cumulative probabilities can be determined. If $\lambda t = 1$, for example, $P(R \leq 2) = .3679 + .3679 + .1839 = .9197$. The PMF and CDF of the Poisson distribution with $\lambda t = 1$ are presented in Figure 4.10.1. In general, the Poisson distribution is positively skewed, although it is more nearly symmetric as λ becomes larger. Graphs of Poisson PMF's with $\lambda t = 2$, 5, and 10 are presented in Figure 4.10.2. Note from these graphs that if λt is an integer, the Poisson distribution has two modes, located at $R = \lambda t - 1$ and $R = \lambda t$. It can be shown that if λt is *not* an integer, there is a single mode located at the integer value which is between $\lambda t - 1$ and λt.

Table 4.10.1   Poisson Distribution for $\lambda t = 1$

r	$P(R = r \mid \lambda t = 1)$
0	.3679
1	.3679
2	.1839
3	.0613
4	.0153
5	.0031
6	.0005
7	.0001
> 7	.0000

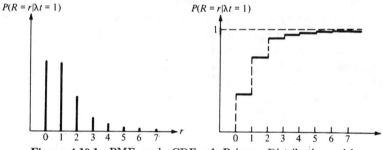

Figure 4.10.1   PMF and CDF of Poisson Distribution with $\lambda t = 1$

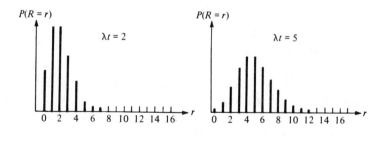

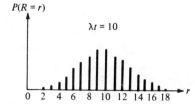

Figure 4.10.2   Examples of Poisson PMF's

4.11 THE POISSON APPROXIMATION TO THE BINOMIAL

The calculation of probabilities is a tedious task for the binomial distribution as well as for the Poisson distribution. Although some special tables are more detailed, the usual tables of the binomial distribution are much more limited than tables of other distributions (that is, only a few values of n and p are considered). Of course, computers can be used to calculate binomial probabilities, but at times it may be inconvenient and time-consuming to use computers for such a task, especially when only one or two particular probabilities are of interest. Because of this, it is sometimes convenient to use other distributions to determine approximations to binomial probabilities. Both the Poisson distribution and the normal distribution provide good approximations to the binomial, although the conditions under which the approximation is good differ in the two cases. The normal distribution is continuous, and we will not discuss its use as an approximation for the binomial until later in this chapter when we cover continuous distributions.

The relationship between the Bernoulli and Poisson processes, and hence between the binomial and Poisson distributions, was discussed in the previous section. We noted that the Poisson distribution could be derived as a limiting form of the binomial if the following three assumptions were simultaneously satisfied:

1. n becomes large (that is, $n \to \infty$),

2. p becomes small (that is, $p \to 0$),

and   3. np remains constant.

Under these conditions, the binomial distribution with parameters n and p can be approximated by the Poisson distribution with parameter $\lambda t = np$. This means that the Poisson distribution provides a good approximation to the binomial distribution if p is close to zero and n is large. Since p and q can be interchanged by simply interchanging the definitions of "success" and "failure," the Poisson is also a good approximation when p is close to one and n is large.

To illustrate the use of the Poisson distribution as an approximation to the binomial distribution, consider the example from Section 4.1 in which 10 percent of the cars produced by a certain manufacturer have defective brakes. Suppose that we randomly sample 10 cars and inspect their brakes. If we are willing to accept the Bernoulli assumptions in this case, the distribution of R, the number of cars in the sample with defective brakes, is a binomial distribution, and the probabilities can be found in Table V in the Appendix. In this case, $n = 10$ and $p = .10$, so if we wish to look at the Poisson approximation, we should let $\lambda t = np = 10(.10) = 1$. The Poisson probabilities can be found in Table VI for $\lambda t = 1$.

For this example, the binomial distribution and the Poisson approximation are given in Table 4.11.1. The two distributions agree reasonably well. If more precision is desired, a possible rule of thumb is that the Poisson is a good approximation to the binomial if $n/p > 500$ (this should give accuracy to at least two decimal places). In the example, $n/p = 100$, and yet the approximation does not seem to be too bad. When the Poisson is not a good approximation, the normal distribution may provide a good approximation, as we will see in Section 4.17.

Table 4.11.1 Binomial Probabilities and Poisson Approximation for Defective Brake Example

r	Binomial $P(R = r \mid n = 10, p = .10)$	Poisson $P(R = r \mid \lambda t = 1)$
0	.3487	.3679
1	.3874	.3679
2	.1937	.1839
3	.0574	.0613
4	.0112	.0153
5	.0015	.0031
6	.0001	.0005
7	.0000	.0001
8	.0000	.0000
9	.0000	.0000
10	.0000	.0000

This section illustrates the fact that it is often possible to use one distribution to approximate another. This is particularly valuable when it is difficult to make probability calculations and when no tables are available for the original distribution. Of course, the use of *any* distribution is usually an approximation of sorts, for the assumptions underlying the distribution are seldom completely satisfied.

4.12 SUMMARY OF SPECIAL DISCRETE DISTRIBUTIONS

In the preceding sections a number of commonly encountered discrete families of distributions have been presented: the binomial, Pascal, geometric, hypergeometric, multinomial, and Poisson distributions. For each of these families, we discussed the conditions under which the distribution might arise, the functional form of the probability function, the parameters of the probability function, the mean and variance of the distribution, and the shape of the distribution for different values of the parameter(s). We have also seen how the distributions are related: the binomial, Pascal, and geometric distributions deal with the Bernoulli process; the hypergeometric distribution deals with samples from a finite population and can be closely approximated by the binomial when the sample size is small relative to the population size; the multinomial distribution represents an extension of the binomial to the case in which there are more than two possible outcomes on each trial; and the Poisson distribution deals with observations made over a continuum, such as time, rather than in a finite sequence of trials—the Poisson can be used to approximate binomial probabilities if the number of trials is large and the parameter p is near zero or one.

In later chapters statistical techniques will be developed for the investigation of samples from these special distributions. These techniques include the estimation of values of the parameters, the testing of hypotheses concerning the parameters or concerning the distribution itself, and the making of decisions when there is uncertainty which can be expressed in probabilistic form with the aid of these distributions. At this point it might be instructive to suggest some practical situations in which these distributions might be applied.

There are numerous situations in which the assumptions of a Bernoulli process seem to be met, and in these situations the binomial, Pascal, and geometric distributions are applicable. One example, as we have seen, is the area of quality control, where we are interested in the items produced by a manufacturing process and, more specifically, in the percentage of defective items. If the probability of an item being defective remains the same over time, and the probability of one item being defective is independent of the past history of defectives and nondefectives, the manufacturing process can be thought of as a Bernoulli process. We might be interested in ascertaining the percentage of defectives; if this is too large the process might be considered "out of control," and adjustments would be necessary.

A gambler might also find the Bernoulli process useful, since the successive trials of many games of chance (for example, roulette) are independent and the chances of winning remain constant from trial to trial. A gambler is usually quite concerned about the possibility of a "losing streak" of such proportions that all of the gambler's capital would be lost. This is the classic problem of "gambler's ruin," which has been studied by many mathematicians. It turns out that the chances of being "wiped out" at the gambling tables depend upon the gambler's beginning capital, the stakes of the game, the probability of winning on any single trial, and the number of trials the gambler plays. If the game is unfair, that is, if the expected gain on any single trial is negative, then in playing for an indefinitely long period of time, the gambler is sure to go broke. Problems of this general nature involve the theory of runs. A run, in terms of a Bernoulli process, is a sequence of consecutive successes or consecutive failures. The gambler may be concerned about long runs of failures. The theory of runs is also applicable in many other situations. It may be used in investigating the assumptions of a Bernoulli process. If a sample of data from a dichotomous process includes quite a few long runs of successes or failures, the assumptions of stationarity and independence may seem doubtful.

There are many other applications involving Bernoulli processes. A medical researcher might be interested in the probability that a certain medical treatment will successfully cure a patient's illness. A professor might be interested in the probability that a student will earn a certain grade on a multiple-choice examination just by pure guesswork. A marketing manager might be interested in the probability that a consumer will buy a particular brand of a product. The examples are endless, and you can no doubt think of some which are directly related to your particular area of interest.

Applications of the multinomial distribution are similar to the above applications except that they deal with situations where there are more than two possible outcomes at each trial. If a manufactured item can possibly have 0, 1, 2, or 3 defects, then a quality-control manager might be interested in the probability of each of these possibilities. If a missile is shot at a target, it might completely destroy, partially destroy, or completely miss the target. The forecasting of election results involves the multinomial distribution when there are more than two candidates for a given elective office. The field of genetics deals with probabilities concerning various characteristics of the offspring of a particular pair of parents. Of course, the multinomial distribution is more difficult to deal with than the binomial distribution, so the set of possible outcomes is often grouped into two classes and the binomial distribution is utilized to make probability statements concerning these two classes. The binomial is also often used to approximate the hypergeometric distribution. In some applications, however, the sample size is reasonably large relative to the population size, and the hypergeometric distribution itself must be used. The hypergeometric distribution is applicable to situations involving sampling without replacement from a finite population. Most sample surveys (opinion polls, questionnaires, and so on) are of this nature. Another example of a hypergeometric application is the estimation of the size of an animal population from recapture data.

The Poisson process is applicable when observations are made over a continuum, such as time, and when the assumptions of stationarity and independence are satisfied. The occurrence of demands for service is often

Poisson-distributed. Examples are the number of cars passing a toll booth in a given time period, the number of customers entering a store in a given time period, and so on. These variables are of interest in determining the number of toll booths to construct, the number of clerks to hire, or the number of items to stock in a store. These topics have been widely studied under the names of *queueing theory* and *inventory theory*. Other random variables that have been shown to obey a Poisson process are the number of misprints in a book and the radioactive disintegration of certain elements.

In any application, it is first necessary to determine which distribution or distributions might be applicable to the situation at hand. In doing this, it is important that the assumptions be carefully investigated, for the probability statements resulting from a given distribution are only as valid as the assumptions underlying the distribution.

4.13 THE NORMAL DISTRIBUTION

Just as there are commonly encountered discrete distributions, there are families of continuous distributions that have wide applicability, especially with regard to the development of statistical techniques. In the following sections a number of such distributions will be discussed, the most important of which is the **normal distribution.** You should keep in mind that a continuous probability distribution represents an idealization, as we pointed out in Section 3.5. In many situations, however, these idealizations are reasonable approximations and are much easier to deal with than the corresponding complicated discrete models.

As noted in the preceding paragraph, an important continuous distribution is the so-called "normal," or "Gaussian," distribution. The normal distribution is completely specified only by its mathematical rule. Quite often, the distribution is symbolized by a graph of the functional relation generated by that rule, and the general picture that a normal distribution presents is the familiar bell-shaped curve of Figure 4.13.1. The horizontal axis represents all the different values of x, and the vertical axis their densities $f(x)$. The normal distribution is continuous for all x between $-\infty$ and ∞, and each conceivable nonzero *interval* of real numbers has a probability other than zero. For this reason, the curve is shown as never quite touching the horizontal axis; the tails of the curve show decreasing probability densities as values grow extreme in either direction from the mode, but any interval representing *any* degree of deviation from central tendency is possible in this theoretical distribution. The normal distribution is symmetric and unimodal, and the mean, median, and mode are all equal. Bear in mind that since the normal distribution is con-

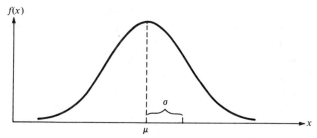

Figure 4.13.1   A PDF for a Normal Distribution

tinuous, the height of the curve shows the probability *density* for each x. However, as for any continuous distribution, the area cut off beneath the curve by any interval is a probability, and the entire area under the curve is one.

The student is advised not to jump to conclusions about the kind of distribution solely on the basis of the graph of a distribution. Although the normal distribution graphs as a bell-shaped curve, every bell-shaped curve is not necessarily a normal distribution. The kind of distribution, the family to which it belongs, depends absolutely on the function rule, and on nothing else. Only if each possible x is paired with a density in the way provided by the normal function rule can one say that the distribution is normal. Other rules may give similar pictures or probabilities that are close to their normal counterparts but, by definition, these distributions are not exactly normal.

The mathematical rule for a normal density function is as follows:

$$f(x) \;=\; \frac{1}{\sqrt{2\pi\sigma^2}}\; e^{-(x-\mu)^2/2\sigma^2}. \qquad (4.13.1^*)$$

This rule looks somewhat forbidding to the mathematically uninitiated, but actually it is not. The π, the e, and the 2, of course, are simply positive numbers acting as mathematical constants. The working part of the rule is the exponent,

$$-\frac{(x-\mu)^2}{2\sigma^2}, \qquad (4.13.2)$$

where the particular value x of the variable X appears along with the two parameters μ and σ^2.

The more that x differs from μ, the larger the quantity in the numerator of this exponent will be. However, this deviation of x from μ enters as a squared quantity, so that *two* different x values showing the same absolute deviation from μ have the same probability density according to this rule. This dictates the symmetry of the normal distribution. Furthermore, the exponent as a whole has a negative sign, meaning that the larger the absolute deviation of x from μ, the smaller the density assigned to x will be by this rule. This dictates that both tails of the distribution show decreasing densities, since the wider the departure of x values from μ, the lower the height of this function's curve will be. However, no real and finite value of X can possibly make the density itself negative or exactly zero; the normal function curve never touches the x-axis, indicating that any interval of numbers will have a nonzero probability. On the other hand, any number, such as e, raised to the zero power is 1; when x exactly equals μ, the density is

$$\frac{1}{\sqrt{2\pi\sigma^2}},$$
(4.13.3)

which is the *largest* density value any x may have. This fact implies that the function curve must be unimodal, with maximum density at the mean, μ. Note that, unlike a probability, a density value such as Equation (4.13.1) can be greater than 1.

It is very important to notice that the precise density value assigned to any x by this rule cannot be found unless the two parameters μ and σ are specified. The parameter μ can be any finite number, and σ can be any finite *positive* number. Thus, like the binomial and the other discrete distributions that we have studied, the normal distribution rule actually specifies a *family* of distributions. Although each distribution in the family has a density value paired with each x by this same general rule, the *particular* density that is paired with a given x value differs with different assignments of μ and σ. Thus, normal distributions may differ in their means (Figure 4.13.2), in their standard deviations (Figure 4.13.3), or in both means and standard deviations (Figure 4.13.4). Nevertheless, given the mean and standard deviation of the distribution, the rule for finding the probability density of any value of the variable is the same.

Since the normal distribution is really a family of distributions, statis-

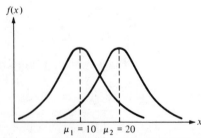

Figure 4.13.2   Two Normal PDF's with Different Means

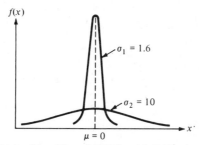

Figure 4.13.3 Two Normal PDF's with Different Variances

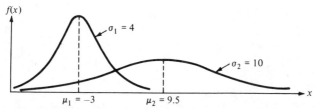

Figure 4.13.4 Two Normal PDF's with Different Means *and* Different Variances

ticians constructing tables or probabilities found by the normal rule find it convenient to think of the variable in terms of a standardized random variable.

The density function of a standardized normal random variable **Z is of the form**

$$f(z) = \frac{1}{\sqrt{2\pi}} e^{-z^2/2}. \tag{4.13.4*}$$

For **standardized normal variables,** the density depends only on the *absolute* value of z; since both z and $-z$ give the same value z^2, they both have the same density. The higher z is in absolute value, the less the associated density is. The standardized form of the distribution makes it possible to use one table of cumulative probabilities for any normal distribution, regardless of its particular parameters, as we shall see in the next section.

4.14 COMPUTING PROBABILITIES FOR THE NORMAL DISTRIBUTION

In Section 3.6 we pointed out that when a distribution is continuous. the probability that X takes on some exact value x is, in effect, zero. For this reason, *all probability statements we will make using a normal distribution will be in terms either of cumulative probabilities or the probabilities of intervals.*

The cumulative probability

$$F(x) = P(X \leq x) = \int_{-\infty}^{x} f(x) \, dx \qquad (4.14.1)$$

is the area under the normal curve in the interval bounded by $-\infty$ and x (Figure 4.14.1). Cumulative probabilities such as this can, of course, be used to find the probability of any interval, since $P(a \leq X \leq b) = F(b) - F(a)$.

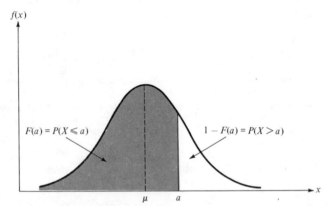

Figure 4.14.1 Probabilities as Areas Under a Normal Density Function

Most tables of the normal distribution give the cumulative probabilities for various *standardized* values. That is, for a given z the table provides the cumulative probability *up to and including that standardized value* in a normal distribution. Table I in the Appendix gives cumulative probabilities for positive values of z. Because of the symmetry of the normal distribution, it is possible to use Table I to determine cumulative probabilities for negative values of z, using the relationship

$$F(-z) = 1 - F(z). \qquad (4.14.2)$$

This relationship is illustrated in Figure 4.14.2. The shaded area on the left is the cumulative probability $P(Z \leq -a)$, where a is positive and $-a$ is negative. Table I gives only the area to the left of a, not the area to the left of $-a$. However, the shaded area on the *right*, which is $1 - F(a)$, is the same as the shaded area on the *left*, $F(-a)$.

The cumulative probabilities for standardized values that are found in Table I can be used to determine cumulative probabilities for any nor-

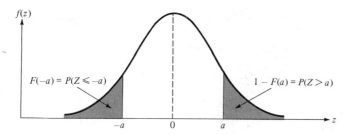

Figure 4.14.2 The Relationship Between $F(a)$ and $F(-a)$ for a Standardized Normal Distribution

mally distributed random variable. The relationship between cumulative probabilities involving Z and cumulative probabilities involving X, where X is normally distributed with mean μ and standard deviation σ, is quite simple:

$$P(X \leq x) = P\left(Z \leq \frac{x - \mu}{\sigma}\right), \qquad (4.14.3^*)$$

or

$$P(Z \leq z) = P(X \leq \mu + z\sigma). \qquad (4.14.4^*)$$

Furthermore, these formulas can be generalized to the case in which *intervals* of values are of interest:

$$P(a \leq X \leq b) = P\left(\frac{a - \mu}{\sigma} \leq Z \leq \frac{b - \mu}{\sigma}\right), \qquad (4.14.5^*)$$

or

$$P(c \leq Z \leq d) = P(\mu + c\sigma \leq X \leq \mu + d\sigma). \qquad (4.14.6^*)$$

Using Table I and the formulas presented in the preceding paragraphs, it is possible to determine probabilities for intervals of values of any normally distributed random variable. For example, suppose that the probability distribution of the price, one year from now, of a particular stock is thought to be a normal distribution with mean $\mu = 50$ and standard deviation $\sigma = 5$. What is the probability that the price will be less than or equal to 57.5? To answer this question, we convert to a standardized value and then use Table I:

$$P(X \leq 57.5) = P\left(Z \leq \frac{57.5 - 50}{5}\right) = P(Z \leq 1.5) = .933.$$

A look at Table I shows that for 1.5 in the first column, the corresponding cumulative probability in the column labeled $F(z)$ is approximately .933. If the standardized value turns out to be negative, an extra step is needed, using Equation (4.14.2). For instance, suppose we are interested in the probability that the price of the stock will be less than or equal to 46:

$$P(X \leq 46) = P\left(Z \leq \frac{46 - 50}{5}\right) = P(Z \leq -.8)$$

$$= 1 - P(Z \leq .8) = 1 - .788 = .212.$$

What if we are interested in the probability that the price will fall between 48 and 60, inclusive? This probability is

$$P(48 \leq X \leq 60) = P\left(\frac{48 - 50}{5} \leq Z \leq \frac{60 - 50}{5}\right) = P(-.4 \leq Z \leq 2.0)$$

$$= F(2.0) - F(-.4)$$

$$= F(2.0) - [1 - F(.4)]$$

$$= .977 - (1 - .655) = .632.$$

By looking at the probabilities of certain intervals of standardized values, we can gain a better impression of the concentration of probability in a normal distribution. For example, what proportion of cases in a normal distribution must lie within one standard deviation of the mean? This is the same as the probability of the interval $(-1 \leq Z \leq 1)$. The table shows that $F(1) = .8413$, and we know that $F(-1)$ must be equal to $1 - .8413$, or $.1587$. Thus,

$$P(-1 \leq Z \leq 1) = F(1) - F(-1) = .8413 - .1587 = .6826.$$

About 68 percent of all cases in a normal distribution must lie within one standard deviation of the mean. Similarly, we can find that **about 95 percent of all cases in a normal distribution must lie within two standard deviations of the mean.**

To find the proportion lying between 1 and 2 standard deviations *above* the mean, we take

$$P(1 \leq Z \leq 2) = F(2) - F(1).$$

From the table, these numbers are .9772 and .8413, so that

$$P(1 \leq Z \leq 2) = .9772 - .8413 = .1359.$$

About 13.6 percent of cases in a normal distribution lie in the interval between 1σ and 2σ above the mean. By the symmetry of the distribution, we immediately know that

$$P(-2 \leq Z \leq -1) = .1359$$

as well.

Beyond what number must only 5 percent of all standardized scores fall?

That is, we want a number b such that the following statement is true for a normal distribution:

$$P(Z > b) = .05.$$

This is equivalent to saying that

$$P(Z \leq b) = .95.$$

A look at the table shows that roughly .95 of all observations must have z scores at or below 1.65. Thus,

$$P(Z > 1.65) = .05.$$

If intervals are mutually exclusive, their probabilities can be added by the "or" rule to find the probability that X falls into either interval. For example, we want to know

$$P(Z < -2.58 \text{ or } Z > 2.58),$$

which is the same as

$$P(|Z| > 2.58).$$

For the first of these intervals, we find

$$F(2.58) = .995,$$

so that

$$F(-2.58) = 1 - .995 = .005.$$

For the other interval, the probability is

$$P(Z > 2.58) = 1 - F(2.58) = .005.$$

The two intervals are mutually exclusive events, and so

$$P(|Z| > 2.58) = .005 + .005 = .01.$$

That is, the probability that Z exceeds 2.58 in absolute value is about one in one hundred. This probability is given by the extreme tails of the distribution in Figure 4.14.3.

It is interesting to compare the probability just found with that given by the Chebyshev inequality for *any* symmetric, unimodal distribution (Section 3.19):

$$P(|Z| \geq 2.58) \leq \frac{4}{9}\left[\frac{1}{(2.58)^2}\right].$$

On working out the number on the right, we get about .067. That is, for *any* unimodal symmetric distribution, the probability of a score deviating 2.58 or more standard deviations from expectation is less than about seven in one hundred. By specifying that the distribution is *normal*, we pin this probability down to the exact value .01. The more we assume or know

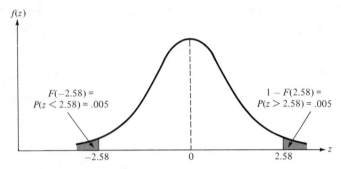

Figure 4.14.3 $P(|Z|>2.58)$ for a Normal Distribution

about the population distribution, the stronger is the statement that can be made about the probability of any degree of deviation from expectation.

4.15   THE IMPORTANCE OF THE NORMAL DISTRIBUTION

The normal distribution is by far the most widely-encountered distribution in inferential statistics. There are at least four very good reasons why this is true.

A population may be assumed to follow a normal law because of what is known or presumed to be true of the measurements themselves. There are two rather broad instances in which the random variable is a measurement of some kind and the distribution is conceived as normal. The first is when we are considering a hypothetical distribution of errors, such as errors in reading a dial, in discriminating between stimuli, or in test performance. Any observation can be assumed to represent a "true" component plus an error. Each error has a magnitude, and this number is thought of as a reflection of pure chance, the result of a vast constellation of circumstances operative at the moment. Any factor influencing performance at the moment contributes a tiny amount to the size of the error and to its direction. Furthermore, such errors are appropriately considered to push the observed measurement up or down with equal likelihood, to be independent of each other over samplings, and to cancel out in the long run. Thus, in theory, it makes sense that errors of measurement or errors of discrimination follow something like the normal rule.

In other instances, the distribution of *true magnitudes* of some trait may be thought of as normal. For example, human heights form approximately a normal distribution, and so, we believe, does human intelligence. Indeed, there are many examples, especially in biology, where the distribution of measurements of a natural trait seems to follow something closely resembling a normal rule. No obtained distribution is ever exactly normal, of course, but some distributions of measurements have relative frequencies very close to normal probabilities. However, the view that the distribution of almost any trait of living things should be normal, although prominent in the nineteenth century, has been discredited. The normal distribution is not, in any sense, "nature's rule," and all sorts of distributions having little resemblance to the normal distribution occur in all fields.

It may be convenient, on mathematical grounds alone, to assume a normal distribution. Mathematical statisticians do not devote so much attention to normal distributions just because they think bell-shaped curves are pretty! The truth of the matter is that the normal function has important mathematical properties shared by no other theoretical distribution. Assuming a normal distribution gives the statistican an extremely rich set of mathematical consequences that can be used in developing methods. Many problems in mathematical statistics either are solved, or can be solved, *only* in terms of a normal distribution. We will find this true especially when we come to methods for making inferences about a variance. The normal distribution is the "parent" of several other theoretical distributions that are important in statistics. In some practical applications the methods developed using normal theory work quite well even when this assumption is not met, despite the fact that the problem can be given a *formal* solution only when a normal distribution is assumed. In other instances, there just is not any simple way to solve the problem when the normal rule does not hold at least approximately.

The normal distribution may serve as a good approximation to a number of other theoretical distributions having probabilities that are either laborious or impossible to work out exactly. For example, in the following section it will be shown that the normal distribution can be used to approximate binomial probabilities under some circumstances, even though the normal distribution is continuous and the binomial is discrete. For very large n, exact binomial probabilities are troublesome to work out; on the other hand, intervals based on the normal function give probabilities that can be used as though they were binomial probabilities.

There is a very intimate connection between the size of a sample, n, and the extent to which a sampling distribution approaches the normal form. Many sampling distributions based on large n can be approximated by the normal distribution even though the population distribution itself is definitely not normal. This is the extremely important principle that we will call the **central limit theorem.** The normal distribution is the *limiting form* for large n for a very large variety of sampling distributions. This is one of the most remarkable and useful principles to come out of theoretical statistics, and we will discuss it in more detail in Chapter 5.

Thus, it is no accident that the normal distribution is the workhorse of inferential statistics. The assumption of normal distributions or the use of the normal distribution as an approximation device is not as arbitrary as it sometimes appears. However, it is important to repeat at this point the caution that although the normal distribution has wide applicability, it is by no means a universal, natural rule. Whenever a normal distribution is used in statistical inference or decision, its applicability should be somehow justified.

4.16   THE NORMAL APPROXIMATION TO THE BINOMIAL

One of the interpretations given to the normal distribution is that it is the limiting form of a binomial distribution as $n \to \infty$ for a fixed p. In sampling from a Bernoulli process, the distribution of R, the number of successes in a fixed number of trials, is a binomial distribution. Suppose that we consider all possible samples of size 10. Next, we consider all possible samples with $n = 10,000$, then all possible samples with $n = 10,000,000$, and so on. For each sample size there will be a different binomial distribution for R, the number of successes.

How does the binomial distribution change with increasing sample size? In the first place, the actual range of the possible number of successes grows larger, since the whole numbers 0 through n form a larger set with each increase in n. Second, the expectation np is larger for each increase in n for a fixed p, and the standard deviation $\sqrt{npq}$ also increases with n.

For any value of n, the number of successes R can be put into standardized form:

$$Z = \frac{R - E(R)}{\sigma_R} = \frac{R - np}{\sqrt{npq}} . \qquad (4.16.1)$$

Then regardless of the size of n, the mean of Z is 0 and the standard deviation is 1. The probability of any value of Z is the same as for the corresponding R. Any interval of values of Z can be given a probability in the particular binomial distribution, and the same interval can also be given a probability for a normal distribution. We can compare these two probabilities for any interval; if the two probabilities are quite close over all intervals, the normal distribution gives a good approximation to the binomial probabilities. On making this comparison of binomial and normal interval probabilities, we should find that the larger the sample size n, the better the fit is between the two kinds of probabilities. **As n grows infinitely large, the normal and binomial probabilities become identical for any interval.**

For example, suppose that a contractor bids on five projects. The probability that the bid will be successful is .5 for each project, and the success or failure of any particular bid is independent of what happens with the other bids. This can be considered as a Bernoulli process with $p = .5$, and the distribution of the number of successful bids is a binomial distribution with $n = 5$ and $p = .5$. This distribution has a mean of $np = 5(.5) = 2.5$ and a standard deviation of $\sqrt{npq} = \sqrt{5(.5)(.5)} = \sqrt{1.25} = 1.12$.

For the moment, let us pretend that this is actually a continuous distribution and that the binomial probability $P(R = r)$ is actually the probability associated with an interval of width 1 centered at r:

$$P(r - \tfrac{1}{2} \leq R \leq r + \tfrac{1}{2}).$$

Now we shall convert R into a standardized random variable Z by taking

$$Z = \frac{R - E(R)}{\sqrt{V(R)}} = \frac{R - 2.5}{1.12}.$$

But then each R interval corresponds to a Z interval, and the probability of the R interval can be stated in terms of the Z interval:

$$P(r - \tfrac{1}{2} \leq R \leq r + \tfrac{1}{2}) = P\left(\frac{r - \tfrac{1}{2} - 2.5}{1.12} \leq \frac{R - 2.5}{1.12} \leq \frac{r + \tfrac{1}{2} - 2.5}{1.12}\right)$$

$$= P\left(\frac{r - 2.5 - \tfrac{1}{2}}{1.12} \leq Z \leq \frac{r - 2.5 + \tfrac{1}{2}}{1.12}\right).$$

$$(4.16.2)$$

Using Equation (4.16.2) and assuming that Z is normally distributed with mean 0 and variance 1, we can determine probabilities for these intervals from tables of the normal distribution and compare them with binomial probabilities from Table V. The binomial probabilities and normal approximations are given for the bidding example in Table 4.16.1.

Table 4.16.1 Binomial Probabilities and Normal Approximation for Bidding Example

r	z interval	Binomial probability	Normal approximation
0	-2.68 to -1.79	.0312	.0367
1	-1.79 to $-.89$	.1563	.1500
2	$-.89$ to 0	.3125	.3132
3	0 to $.89$	.3125	.3132
4	$.89$ to 1.79	.1563	.1500
5	1.79 to 2.68	.0312	.0367

The probabilities for the normal approximation were determined from (4.16.2) with one exception. Since Z ranges from $-\infty$ to $+\infty$, the normal probabilities for the extreme intervals (corresponding to $r = 5$ and $r = 0$) were computed by assuming that the extreme limit was infinite, so that the normal probabilities would sum to one. Thus, the first probability is actually $P(Z \geq 1.79)$ rather than $P(1.79 \leq Z \leq 2.68)$, and the last probability is $P(Z \leq -1.79)$ rather than $P(-2.68 \leq Z \leq -1.79)$.

Observe how closely the normal probabilities approximate their binomial counterparts; each normal probability is within .01 of the corresponding binomial probability. The difference in the probabilities is smallest for the middle intervals and largest for the extremes. Thus, even when n is only 5, the normal distribution gives a respectable approximation to binomial probabilities for $p = .5$.

Now suppose that for this same example, n had been 15. Here the expectation is 7.5 and the standard deviation is $\sqrt{(15)(.25)} = 1.94$. The distribution, with both binomial and normal probabilities, is given in Table 4.16.2. For this distribution, the normal approximation gives an even better fit to the exact binomial probabilities than in the case of $n = 5$. The average absolute difference in probability over the intervals is about .001, whereas when n was only 5, the average absolute difference was about .004. In general, as n is made larger, the fit between normal and

Table 4.16.2 Binomial Probabilities and Normal Approximation for Bidding Example with $n = 15$

r	z interval	Binomial probability	Normal approximation
0	-4.124 to -3.608	.00003	.0002
1	-3.608 to -3.092	.0005	.0012
2	-3.092 to -2.577	.0032	.0036
3	-2.577 to -2.061	.0139	.0147
4	-2.061 to -1.546	.0416	.0409
5	-1.546 to -1.030	.0916	.0909
6	-1.030 to $-.515$	.1527	.1501
7	$-.515$ to 0	.1964	.1984
8	0 to $.515$	.1964	.1984
9	$.515$ to 1.030	.1527	.1501
10	1.030 to 1.546	.0916	.0909
11	1.546 to 2.061	.0416	.0409
12	2.061 to 2.577	.0139	.0147
13	2.577 to 3.092	.0032	.0036
14	3.092 to 3.608	.0005	.0012
15	3.608 to 4.124	.00003	.0002

binomial probabilities improves. In the limit, when n approaches infinite size, the binomial probabilities are exactly the same as the normal probabilities for any interval, and thus we can say that the normal distribution is the limit of the binomial. This is true *regardless of the value of p*.

On the other hand, for any finite n, the more p departs from .5, the less well the normal distribution approximates the binomial. When p is not exactly equal to .5, the binomial distribution is somewhat skewed for any finite sample size n, and for this reason the normal probabilities will tend to fit less well than for p = .5, which always gives a symmetric distribution. For example, notice that for n = 5 and p = .3, illustrated in Table 4.16.3, the correspondence is not as good as before. Here the normal probabilities are still reasonably close to the binomial, but the fit is not as good as for p = .5.

Table 4.16.3 Binomial Probabilities and Normal Approximation for n = 5 and p = .3

r	Binomial probability	Normal approximation
0	.16807	.1635
1	.36015	.3365
2	.30870	.3365
3	.13230	.1379
4	.02835	.0239
5	.00243	.0017

Given a sufficiently large n, normal probabilities may always be used to approximate binomial probabilities irrespective of the value of p, the true probability of a success. The more that p departs in either direction from .5, the less accurate is this approximation for any given n. In practical situations where the sampling distribution is binomial, the normal approximation may ordinarily be used safely if the *smaller* of np, the expected number of successes, or nq, the expected number of failures, is 10 or more. Otherwise, binomial probabilities can either be calculated directly or found from tables that are readily available. Obviously, this rule requires a larger sample size the smaller the value of either p or q, and if we must use the normal approximation to make an inference about the value of p, it is wise to plan on a relatively large sample.

The graphs presented in Figures 4.16.1 and 4.16.2 give some idea of the rate at which the binomial distribution approaches its normal approximation. In Figure 4.16.1, graphs are shown for p = .10 and sample sizes of 10, 20, and 50. Because p = .10, the binomial distribution is positively

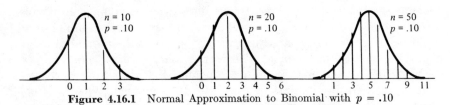

Figure 4.16.1 Normal Approximation to Binomial with $p = .10$

skewed, but the skewness decreases as n becomes larger. This skewness is the reason that for any given n, the normal approximation is less accurate as p approaches 0 or 1. In Figure 4.16.2, graphs are shown for $p = .50$ and sample sizes of 10, 20, and 50. Since $p = .50$, the binomial distribution is symmetric, and thus for a given sample size the normal approximation is better in this case than in Figure 4.16.1. As n increases, the normal approximation becomes more and more accurate.

In a more advanced study of the normal distribution, we would show that as n becomes infinite with fixed p, the mathematical rule for a binomial distribution becomes identical with the rule for a normal density function, with $\mu = np$ and $\sigma = \sqrt{npq}$. Unfortunately, this sort of demonstration is beyond our managing at this point.

In any practical use of the normal approximation to binomial probabilities, it is important to remember that here we regard the binomial distribution as though it were continuous, and we actually find the normal probability associated with an interval with real lower limit of $(r - .5)$ and real upper limit of $(r + .5)$:

$$P\left(\frac{r - np - .5}{\sqrt{npq}} \le Z \le \frac{r - np + .5}{\sqrt{npq}}\right). \qquad (4.16.3)$$

The .5 that is added or subtracted from r is called a **continuity correction** because it allows for the fact that we are using a continuous distribution, the normal distribution, to approximate a discrete distribution, the binomial distribution, that assigns positive probability only to integer values.

As noted above, the normal distribution need not be particularly good as an approximation to the binomial if either p or q is quite small. In these

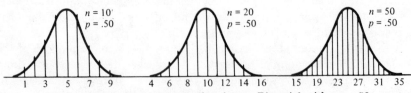

Figure 4.16.2 Normal Approximation to Binomial with $p = .50$

circumstances another theoretical distribution, the Poisson, provides a better approximation to binomial probabilities, as we saw in Section 4.11.

4.17 THE THEORY OF THE NORMAL DISTRIBUTION OF ERROR

The fact that the limiting form of the binomial is the normal distribution actually provides a rationale for thinking of random error as distributed in this normal way. Consider an object measured over and over again independently and in exactly the same way. Imagine that the value X obtained on any occasion is a sum of two independent parts:

$$X = Y + U.$$

That is, the obtained score X is a sum of a constant *true* part Y plus a random and independent *error* component. However, the error portion can also be thought of as a sum:

$$U = g(U_1 + U_2 + U_3 + \cdots + U_n).$$

Here, U_1 is a random variable that can take on only two values:

$U_1 = 1$ when factor 1 is operating,

$U_1 = -1$ when factor 1 is not operating.

The term g is merely a constant, reflecting the "weight" of the error in X. Similarly, the other random errors are attributable to different factors and take on only the values 1 and -1. Now imagine a vast number of influences at work at the moment of any measurement. Each of these factors operates independently of each of the others, and whether any factor exerts an influence at any given moment is purely a chance matter. If you want to be a little anthropomorphic in your thinking about this, visualize old Dame Fortune tossing a vast number of coins on any occasion, and from the result of each deciding on the pattern of factors that will operate. When coin 1 comes up heads, U_1 gets value 1, and if tails, value -1; and the same principle determines the value for every other error portion of the observed score.

Now under this approach, the number of factors operating at the moment we observe X is a number of successes in n independent trials of a Bernoulli experiment. If this number of successes is R, then

$$U = gR - g[n - R] = g[2R - n].$$

The value of U is exactly determined by R, the number of "influences" in operation at the moment, and the probability associated with any value of U must be the same as for the corresponding value of R. **If n is very large,**

then the distribution of U must approach a normal distribution. Furthermore, if any factor is equally likely to operate or not operate at a given moment, so that $p = 1/2$, then

$$E(U) = g[2E(R) - n],$$

so that, since $E(R) = np = n/2$,

$$E(U) = 0.$$

In the long run, over all possible measurement occasions, all of the errors cancel out. Thus,

$$E(X) = E(Y) + E(U) = Y.$$

The long-run expectation of a measurement X is the true value Y, provided that error really behaves in this random way as an additive component of any score.

This is a highly simplified version of the argument for the normal distribution of errors in measurement. Much more sophisticated rationales can be invented, but they all partake of this general idea. Moreover, the same kind of reasoning is sometimes used to explain why distributions of natural traits, such as height, weight, and size of head, follow a more or less normal rule. Here, the mean of some random variable is thought of as the "true" value, or the "norm." However, associated with each individual is some departure from the norm, or error, representing the culmination of all the billions of chance factors that operate on that individual, quite independently of other individuals. Then, by regarding these factors as generating a binomial distribution, we can deduce that the distribution of the random variable should take on a form like the normal distribution. However, this is only a theory about how errors might operate and there is no reason at all why errors must behave in the simple additive way assumed here. If they do not, then the distribution need not be normal in form at all.

4.18   THE EXPONENTIAL DISTRIBUTION

Although the normal distribution is the most important of the continuous distributions, there are nevertheless many instances in which the normal distribution is not applicable. Consider the Poisson process discussed in Section 4.10. The Poisson process is often applicable when we are interested in occurrences of an event over time. In particular, the Poisson distribution is the distribution of the number of occurrences of an event in a given time period of length t, and the single parameter of the

Poisson distribution is λ, the intensity of the process. Suppose that instead of the number of occurrences in a given time period, we are interested in the amount of time until the *first* occurrence. For example, we might be interested in the amount of time until the first incoming telephone call or the amount of time until the first breakdown of a new piece of equipment. Since we assume that the Poisson process is stationary and independent, it is equivalent to say that we are interested in the amount of time *between* occurrences, such as the amount of time between incoming calls or between breakdowns.

Let the random variable T represent the amount of time until the *first* occurrence of a given event A, where the occurrences of A are governed by a stationary and independent Poisson process with intensity λ per unit time period. It is important that T and λ be expressed in comparable units. If T is expressed in hours, then λ must be expressed in occurrences per hour, and so on. Because of the independence and stationarity of the Poisson process, if the intensity per unit of time is λ, then the intensity in T units of time is λT. For example, an intensity of 6 telephone calls per hour is equivalent to an intensity of 18 calls per 3 hours or 3 calls per half-hour or 1 call per 10 minutes.

Recall from Section 4.10 that if occurrences of an event are generated by a stationary and independent Poisson process with intensity λ per unit time period, and we are interested in R, the number of occurrences in a time period of length t, the distribution of R is a Poisson distribution:

$$P(R = r \mid \lambda, t) = \frac{e^{-\lambda t}(\lambda t)^r}{r!} \qquad \text{for } r = 0, 1, 2, \ldots. \qquad (4.18.1)$$

Letting T represent the amount of time until the first occurrence, consider the event $T > t$. If $T > t$, then there must have been no failures in the first t units of time. From the Poisson distribution, the probability of this event is as follows:

$$P(R = 0 \mid \lambda, t) = \frac{e^{-\lambda t}(\lambda t)^0}{0!} = e^{-\lambda t}.$$

Thus,

$$P(T > t) = P(R = 0 \mid \lambda, t) = e^{-\lambda t},$$

and

$$P(T \le t) = 1 - P(T > t) = 1 - e^{-\lambda t},$$

where it is assumed that $t \ge 0$. This is simply the cumulative distribution function of T, which we can rewrite in functional form as

$$F(t) = \begin{cases} 1 - e^{-\lambda t} & \text{if } t \ge 0, \\ 0 & \text{if } t < 0. \end{cases} \qquad (4.18.2)$$

Since $F(t)$ is a continuous function, the density function $f(t)$ is equal to the derivative of $F(t)$:

$$f(t) = \frac{d}{dt} F(t) = \begin{cases} \lambda e^{-\lambda t} & \text{if } t \geq 0, \\ 0 & \text{if } t < 0. \end{cases} \quad (4.18.3^*)$$

This distribution is called the exponential distribution, and it is the distribution of the amount of time until the first occurrence of an event, or, equivalently, the distribution of the amount of time between occurrences of an event, where the occurrences are governed by a Poisson process.

Note that the relationship between the exponential and the Poisson distributions is analogous to the relationship between the geometric and the binomial distributions. The Poisson and binomial distributions concern the number of occurrences of an event. The exponential and geometric distributions concern the amount of time and the number of trials, respectively, until the first occurrence.

It can be shown that the mean and variance of the exponential distribution are

$$E(T) = \frac{1}{\lambda} \quad \text{and} \quad V(T) = \frac{1}{\lambda^2}. \quad (4.18.4)$$

Since the derivation of $E(T)$ and $V(T)$ requires the technique of integration by parts, we will not concern ourselves with the derivation. Note again the parallel between the exponential and the geometric distributions. The mean of the exponential distribution is $1/\lambda$, and the mean of the geometric distribution is $1/p$.

An example of graphs of exponential density functions with $\lambda = 1/2$ and $\lambda = 2$ is presented in Figure 4.18.1. Notice that the exponential

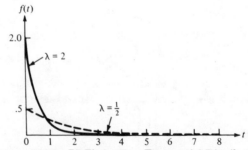

Figure 4.18.1   PDF's for Two Exponential Distributions

distribution is positively skewed, with the mode occurring at the smallest possible value, zero.

To illustrate the exponential distribution, consider the example of Section 4.10 involving the arrival of telephone calls. Suppose that the intensity of the process generating incoming calls is 2 per minute, and we are interested in the distribution of the amount of time (in minutes) until the next call. If the arrival of calls behaves like a Poisson process, this distribution is an exponential distribution with $\lambda = 2$, which is one of the distributions graphed in Figure 4.18.1. What is the probability that the next call will arrive within 5 minutes? From Equation (4.18.2), this probability is

$$P(T \leq 5) = F(5) = 1 - e^{-2(5)} = 1 - .000045 = .999955.$$

Thus, the next call is almost certain to arrive within 5 minutes. Next, suppose that the switchboard operator has to leave the switchboard for 1 minute. What is the probability that no calls arrive during the operator's absence? From the exponential distribution,

$$P(T > 1) = 1 - F(1) = e^{-2(1)} = .1353.$$

Alternatively, we could have used the Poisson distribution (and Table VI):

$$P(R = 0 \mid \lambda = 2, t = 1) = .1353.$$

As noted in Sections 4.10 and 4.12, the Poisson process has many applications. Whenever the variable of interest in these applications is the time until the next occurrence of an event or the time between occurrences of an event, the exponential distribution is applicable. A useful generalization of the exponential distribution is the **gamma distribution,** which can be thought of as the distribution of the amount of time until the rth occurrence of an event in a Poisson process. The exponential distribution is a special case of a gamma distribution with $r = 1$, and the exponential distribution is related to the gamma distribution as the Pascal distribution is related to the geometric distribution (see Section 4.7). However, we will not study the gamma distribution in this book.

4.19 THE UNIFORM DISTRIBUTION

An example of a continuous distribution that was discussed in some detail in Chapter 3 is the distribution of the life (in hours) of a light bulb. It was assumed that this distribution was a uniform distribution over the interval from 80 to 100. Of course, it is possible to consider a more general uniform distribution, one which is defined on an interval from a to b rather than from 80 to 100 (of course, one possibility is that $a = 80$ and $b = 100$), with the obvious restriction that $a < b$. Since the distribution is "uniform," the density function must be constant over the interval from a to b:

$$f(x) = \begin{cases} k & \text{if } a \le x \le b, \\ 0 & \text{elsewhere.} \end{cases} \tag{4.19.1}$$

The only problem, then, is the determination of k. We know that $k \ge 0$, since $f(x)$ cannot be negative, and that the total probability must be equal to one:

$$\int_{-\infty}^{\infty} f(x)\ dx = 1.$$

But
$$\int_{-\infty}^{\infty} f(x)\ dx = \int_{a}^{b} k\ dx = k \int_{a}^{b} dx = k(b - a),$$

so
$$k = \frac{1}{b - a}.$$

The density function of the uniform distribution on the interval from a to b, where $a < b$, is thus

$$f(x) = \begin{cases} \dfrac{1}{b - a} & \text{if } a \le x \le b, \\ \\ 0 & \text{elsewhere.} \end{cases} \tag{4.19.2*}$$

The corresponding cumulative distribution function is equal to

$$F(x) = \int_{-\infty}^{x} f(x)\ dx = \begin{cases} 1 & \text{if } x > b, \\ \displaystyle\int_{a}^{x} \frac{1}{b - a}\ dx = \frac{x - a}{b - a} & \text{if } a \le x \le b, \\ 0 & \text{if } x < a. \end{cases} \tag{4.19.3}$$

Graphs of the PDF and CDF of the uniform distribution are presented in Figure 4.19.1.

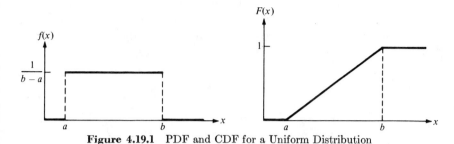

Figure 4.19.1   PDF and CDF for a Uniform Distribution

The mean and variance of the uniform distribution can be calculated easily:

$$E(X) = \int_{-\infty}^{\infty} xf(x)\,dx = \int_{a}^{b} \frac{x}{b-a}\,dx = \frac{1}{b-a}\int_{a}^{b} x\,dx$$

$$= \frac{b^2 - a^2}{2(b-a)} = \frac{(b-a)(b+a)}{2(b-a)}$$

$$= \frac{b+a}{2}, \tag{4.19.4}$$

and

$$E(X^2) = \int_{-\infty}^{\infty} x^2 f(x)\,dx = \int_{a}^{b} \frac{x^2}{b-a}\,dx = \frac{1}{b-a}\int_{a}^{b} x^2\,dx$$

$$= \frac{b^3 - a^3}{3(b-a)} = \frac{(b-a)(b^2 + ab + a^2)}{3(b-a)}$$

$$= \frac{b^2 + ab + a^2}{3}, \tag{4.19.5}$$

so $V(X) = E(X^2) - [E(X)]^2 = \dfrac{b^2 + ab + a^2}{3} - \left(\dfrac{b+a}{2}\right)^2$

$$= \frac{4b^2 + 4ab + 4a^2 - 3b^2 - 6ab - 3a^2}{12}$$

$$= \frac{(b-a)^2}{12}. \tag{4.19.6}$$

Note that the mean, $E(X) = (b + a)/2$, seems intuitively reasonable, for it is the midpoint of the interval from a to b. The distribution is uniform, so we would expect the mean to be located at the midpoint of the interval.

4.20 SUMMARY OF SPECIAL CONTINUOUS DISTRIBUTIONS

Three important families of continuous distributions have been presented in this chapter: the normal, exponential, and uniform distributions. There are numerous other useful families of continuous distributions, some of which will be introduced later (the T, χ^2, F, and beta distributions) and some of which are beyond the scope of this book. It should be emphasized again that continuous distributions are a theoretical idealization and should be recognized as such. In some cases it is possible to relate continuous distributions to real processes, as the exponential is related to the Poisson process and represents the amount of time until the first occurrence of an event. A related distribution that we did not discuss is the gamma distribution, which is the distribution of the amount of time until the rth occurrence of an event in a Poisson process. The normal distribution can be given no single such interpretation; it is strictly a theoretical function which happens to describe adequately the probability distribution of a large number of real-world variables. This necessitates a slightly different approach to the validation of the normal probability model in any application. In previous cases it was important to check the assumptions underlying a particular distribution (for example, stationarity and independence). Since this cannot be done with the normal distribution unless it is being used to approximate some other distribution, such as the binomial, it is necessary to compare sample data with the theoretical model in order to see if the model provides a "good fit" to the data. In the next chapter we will discuss the presentation of sample data in frequency distributions, which are analogous to probability distributions. In later chapters, we will discuss statistical techniques for investigating the "goodness of fit" of a theoretical model to a sample of data.

Because they are usually much easier to work with than discrete distributions, which may become quite complex (in a computational sense), continuous distributions are extremely useful in the development of statistical techniques. This should become clear as the basic statistical theory is presented in the next few chapters. Numerous applications of continuous distributions will be discussed in these chapters, and as a result we will not discuss specific applications of the special continuous families of distributions at this time.

EXERCISES

1. A Bernoulli process is a series of dichotomous trials that are stationary and independent. Explain the terms "stationary" and "independent" with regard to a Bernoulli process. Give some examples of situations in which a series of dichotomous trials is observed but in which the assumptions of stationarity and/or independence seem unrealistic.

2. A multiple-choice examination is constructed so that, in principle, the probability of a correct choice for any item by guessing alone is $1/4$. If the test consists of 10 items, what is the probability that a student will have *exactly* 5 correct answers if the student is just guessing? $p^5 q^5 = \left(\frac{1}{4}\right)^5 \left(\frac{3}{4}\right)^5 =$

3. The probability that an item produced by a certain production process is defective is .05. (a) Assuming stationarity and independence, what is the probability that a lot of 10 items from the process contains no defectives? (b) What is the probability that the first defective item is the fifth item produced?

4. Suppose that the probability that a duck hunter will successfully hit a duck is .25 on any given shot. The hunter decides to hunt until 4 birds have been hit. (a) What is the expected number of shots that will be needed? (b) What is the probability that more than 10 but fewer than 15 shots will be taken? (c) If there is only enough ammunition for 20 shots, what is the probability that the hunter will run out of ammunition before 4 ducks are hit? (c) set-up (not solve)

5. The probability that a particular football team wins a certain game is p, and this probability remains constant from game to game.
   (a) If the outcomes of the games played by the team are independent of each other, for what value of p would the team have at least an even chance of winning all 10 games on its schedule?
   (b) Suppose that p is .70. What is the probability that the team loses no more than 2 of its 10 games?
   (c) If p is .70, what is the probability that the second loss occurs in the 8th game?
   (d) Suppose that p is .70 for home games and .50 for away games. The team plays 5 games at home and 5 games away from home. Now what is the probability that the team will lose exactly 2 games of the 10? What is the expected number of wins?

6. A firm has made a number of bids for contracts. The contracts are awarded completely independently of each other, and p, the probability of a successful bid, is the same for each bid. If the expected number of successful bids is 3 and the variance of the number of successful bids is 2.1, how many bids have been made by the firm and what is p?

7. The probability that a person who is exposed to a certain contagious disease will catch the disease is .20.
   (a) Find the probability that the 20th person exposed to this disease is the 5th one to catch it.

(b) If 20 people are exposed to the disease, what is the expected number of people who will catch the disease? What is the probability that *exactly* 5 catch it? What is the probability that *more than* 5 catch it?

(c) If 1000 people are exposed to the disease, what is the probability that at least 250 catch it? just set-up

(d) What assumptions did you make in answering parts (a)–(c)?

8. Compare the binomial, Pascal, and geometric distributions. Is it possible to determine binomial probabilities by using the Pascal and geometric distributions? Is the event "no successes in 10 trials," which is a binomial event, related to any particular event involving the geometric distribution?

9. A manufacturer of parts for automobiles knows that the proportion of defective parts in the past has been about 10 percent. However, today a random sample of 10 parts has been taken and 2 have been found to be defective. What is the probability of getting 2 or more defective parts by chance in such a sample if the probability of a defective part is only .10? Would you say that the manufacturer's sample suggests strongly that more defective parts than usual are being turned out?

10. A pair of dice is rolled for a total of 4 trials. If the dice are fair, find the probability that the dice total to 7 on exactly 3 of the 4 trials. Find the probability that the dice total to 7 on at most 2 of the 4 trials. What is the expected number of trials (out of 4) that the dice will total to 7?

11. A real-estate speculator invests a large sum of money in a number of parcels of real estate. There are 10 parcels of real estate, and for each one, the probability is .30 that it will turn out to be worthless. The eventual value of any one of the parcels is independent of the eventual value of the others. What is the probability that exactly 3 parcels will turn out to be worthless? What is the probability that at least 5 parcels will turn out to be worthless? If more than 7 of the 10 are worthless, the speculator will be bankrupt, since much of the money used for the investments has been obtained on a short-term loan. What is the probability that the speculator is bankrupt as a result of these investments?

12. The probability that a certain type of surgery is successful is .70.
(a) If 5 patients undergo this type of surgery, what is the probability that 80 percent of them will experience successful surgery?
(b) Answer (a) if the number of patients is 10.
(c) Answer (a) if the number of patients is 20.
(d) Comment on the following statement: "The answers to (a)–(c) should be the same because in each case, p is the same and the desired probability is the probability that 80 percent of the operations are successful."

13. An investment analyst feels that the probability that ABC stock will increase in price during the next 6 months is .4 and the probability that DEF stock will increase in price during the next 6 months is .4. Let $R = 0$ if neither stock increases in price, $R = 1$ if exactly one of the two stocks increases in price, and

$R = 2$ if both stocks increase in price. Do you think that it would be reasonable to use the binomial distribution as the probability distribution of R? Explain your answer.

14. Graph the PMF's of the following binomial distributions and comment on the shapes of these PMF's: (a) $n = 2$, $p = .5$, (b) $n = 10$, $p = .5$, (c) $n = 2$, $p = .1$, (d) $n = 10$, $p = .1$, (d) $n = 2$, $p = .7$, (e) $n = 10$, $p = .7$.

15. A retail establishment that promotes its "customer service" is willing to cash checks with little or no identification from the person cashing the check. As a result, 1 out of 20 checks is returned by the bank and payment is refused.
    (a) On a particular day, 20 checks are cashed. What is the probability that there is at least 1 bad check?
    (b) When a check is returned by the bank, the establishment attempts to contact the individual involved and obtain payment. This procedure results in payment in 80 percent of the cases involving bad checks. On a day when 20 checks are cashed, what is the probability that payment is obtained (either from the bank or later from the individual if the bank refuses to pay) for *all* of the checks?
    (c) In (b), what is the probability that payment is eventually obtained for all but 1 check?

16. A trucker is scheduled to make a 1000-mile trip over roads with very few gas stations. A gas station is located every 200 miles along the route that the trucker will follow, with the first station being 200 miles from the starting point. The truck can go 250 miles on a tank of gas, and it will have a full tank at the beginning of the trip. Because of a gas shortage, the probability is only .8 that any particular gas station is open, but station closings are independent of each other and any station that is open will sell the trucker enough gas to fill the truck's tank.
    (a) What is the probability that the trucker will complete the trip without any delays due to lack of gas?
    (b) If all of the details of the situation are the same except that every 200 miles there are *two* gas stations instead of just one, what is the probability that the trip will be completed without any delays due to lack of gas?

17. In a tennis match, suppose that the outcomes of the games are independent of each other and that for any game, the probability is .7 that the person serving will win. To simplify matters, assume that the players play 9 games, alternating with respect to who serves, and that the player winning the majority of the 9 games wins the match. Player A will serve first, and B is A's opponent.
    (a) Find the distribution of the number of games that A will win.
    (b) Find the distribution of the number of games that B will win.
    (c) What is the probability that A will win the match?
    (d) Is this a Bernoulli process? Explain.
    (e) Do the assumptions given in this problem seem realistic?

18. An MBA candidate schedules job interviews with 10 different firms. For any firm, the probability of receiving a job offer is .25, and independence is assumed.

    (a)   What is the probability that the candidate will receive at least 3 job offers?

    (b)   What is the probability that the first job offer will be received from the fifth firm that interviews the candidate?

    (c)   What is the mean and variance of the distribution of the number of job offers received by the MBA candidate?

19. In Exercise 18, the MBA candidate wants to schedule enough job interviews so that the probability of receiving at least 3 job offers is .9 or greater. What is the smallest number of interviews that will satisfy this condition?

20. Find the mean, the mode, the median, the .25 fractile, and the .75 fractile of the binomial distribution with $n = 15$ and $p = .4$.

21. The probability that an urban resident will vote for a particular candidate for public office is .4, and the probability that a rural resident will vote for this candidate is .7. In each case (urban and rural), assume that voting for or against this candidate follows a Bernoulli process. If 50,000 urban residents vote and 20,000 rural residents vote, what is the expected number of votes that the candidate will obtain?

22. According to a paper company, 5 percent of all trees cut down cannot be used.

    (a)   If 20 trees are cut down in a given period, what is the probability that at least 2 cannot be used?

    (b)   If 10 trees are needed, what is the probability that the company will have to cut down exactly 12 trees in order to get 10 usable trees?

23. For any given match, the probabilities that a soccer team will win, tie, or lose are .3, .4, and .3. If the team plays 10 matches and if the outcomes of the different matches are assumed to be independent, what is the probability that they will win 5, tie 3, and lose 2? What is the probability that they will win exactly 8 matches?

24. What is the probability that in 24 tosses of a fair die, each face will occur exactly 4 times? What is the probability that in 6 tosses, each face will occur exactly once?

25. Just as the binomial distribution is a special case of a much more general principle in algebra (the binomial expansion), the multinomial distribution is a special case of another quite general mathematical principle. Try to state the general rule, and then use this rule to prove that the sum of all probabilities in a multinomial distribution must sum to 1.00. Why are no tables or graphs of multinomial distributions shown in the text?

26. A restaurant offers just 3 main courses: steak, trout, and sweetbreads. The probability that a customer will order steak is .5 and the probability that a customer will order trout is .4.

(a) Assuming independence among customers, even at the same table (is this reasonable?), find the probability that at a table of 4, 2 persons will order steak, 1 will order trout, and 1 will order sweetbreads.

(b) The restaurant serves 200 customers in one evening, and the chef prepares 30 orders of sweetbreads. What is the probability that the restaurant will run out of sweetbreads and will have to give some diners their second choice for a main course?

27. According to genetic theory, each child of a particular pair of parents has a .5 probability of having blue eyes, a .2 probability of having brown eyes, and a .3 probability of having hazel eyes. If these parents have 5 children, what is the probability that all 5 have blue eyes? What is the probability that at least 3 have blue eyes and at least 1 has hazel eyes?

28. A university committee consists of 6 faculty members and 4 students. If a subcommittee of 4 persons is to be chosen randomly from the committee, what is the probability that it will consist of *at least* 2 faculty members? Compute this probability from the hypergeometric distribution *and* from the binomial approximation to the hypergeometric. How good is the approximation?

29. If 12 cards are to be drawn at random without replacement from a standard 52-card deck, what is the probability that there will be exactly 3 cards of each suit? What is this probability if the 12 cards are drawn *with* replacement? (Assume that the deck is thoroughly shuffled before each draw.)

30. One of the key assumptions in the derivation of the binomial distribution is that the various trials are independent. Is this same assumption made in the calculation of probabilities for poker hands or in the derivation of the hypergeometric rule? Explain.

31. An instructor in an introductory statistics class knows that out of 50 students, 12 are freshmen, 25 are sophomores, 11 are juniors, and only 2 are seniors. The students are assigned to 5 extra review sessions randomly, 10 to a session. What is the probability that the first session consists of 1 freshman, 1 sophomore, and 8 juniors? What is the probability that the first session consists only of sophomores *and* that the second session consists of 8 juniors and 2 seniors?

32. Explain the differences among the binomial, multinomial, and hypergeometric distributions, and for each, give examples of realistic situations in which the model might be applicable. When might the binomial distribution serve as a good approximation to the hypergeometric distribution? Explain your answer.

33. In a particular city, juries are supposedly selected randomly from those individuals who are registered to vote. Of the registered voters, half are male and half are female. A list of 20 prospective jurors for a particular month in a particular court is determined, and the list contains the names of 16 males and 4 females. Then, from this list, a jury of 12 individuals is chosen (again,

supposedly randomly) for a certain trial. The jury contains 11 males and 1 female.

(a) What is the probability of obtaining a list with 16 males and 4 females if the list is chosen randomly?

(b) What is the probability of obtaining a jury with 11 males and 1 female if the jury is selected randomly from the list of 16 males and 4 females?

(c) From the list given above, 5 different juries are selected for 5 different trials over a period of 1 month. For each of the trials, the jury consists of 11 males and 1 female. What is the probability of this happening if the jury selection is really random?

34. In a draft lottery, a statistician looked at the month of each birthdate chosen. Of the first 10 birthdates, 4 were in December, 2 were in November, 3 were in October, and 1 was in March. Assuming for convenience that the 12 months of the year have the same number of days, what is the probability of the observed results if the lottery is random?

35. In a lot of 200 new television sets, 20 have some minor defects and the remaining 180 are in perfect condition. A consumer-testing agency randomly chooses 10 sets for inspection. If more than 1 set of the 10 has defects, the agency will give this particular brand of television sets a bad rating. Consider the probability that a bad rating will be given.

(a) Give an exact expression for this probability but do not carry out the calculations.

(b) Find an approximation for this probability.

36. Cars arrive at a certain toll booth according to a Poisson process with intensity 5 per minute. What is the probability that no cars will arrive in a particular minute? What is the probability that more than 8 cars will arrive in a particular minute? What is the probability that exactly 9 cars will arrive in a given 2-minute period, and what is the expected number of arrivals during that period? Comment on the applicability of the assumptions of the Poisson process for this example.

37. Incoming telephone calls at a particular switchboard behave according to a Poisson process with intensity 12 per hour. What is the probability that more than 15 calls will occur in any given 1-hour period? If the person receiving the calls takes a 15-minute coffee break, what is the probability that no calls will come in during the break? (Calculate this probability in two different ways, once using the Poisson distribution and once using the exponential distribution.)

38. Consider the occurrence of misprints in a book, and suppose that they occur at the rate of 2 per page. Assuming a Poisson process, what is the expected number of pages until the first misprint appears? What is the probability that the first misprint will *not* occur in the first page? Comment on the applicability of the Poisson assumptions to the occurrence of misprints or typing errors.

39. Prove that the variance of the Poisson distribution is equal to λt.

40. Explain the similarities and differences between the Bernoulli process and the Poisson process. Give some examples of realistic situations in which the Bernoulli model might be applicable, and do the same for the Poisson model. Explain the terms "stationary" and "independent" with regard to a Poisson process.

41. Graph the PMF's of the following Poisson distributions and comment on the shapes of these PMF's:
    (a) $\lambda t = .5$,   (c) $\lambda t = 3$,
    (b) $\lambda t = 1$,   (d) $\lambda t = 5$.

42. Find the mean, the mode, the median, the .25 fractile, and the .90 fractile of the Poisson distribution with $\lambda t = 5$.

43. An automobile dealership sells an average of 3 cars per day, and the "sales process" is stationary and independent.
    (a) Currently 295 cars have been sold since the beginning of the year. What is the probability that the 300th car of the year will be sold today?
    (b) How many days would it take to make the probability of selling less than 5 cars in that number of days no greater than .10?

44. Accidents occur on a particular stretch of highway at the rate of 3 per week. Compute the following probabilities (i) using the Poisson distribution, (ii) using the exponential distribution:
    (a) the probability that there will be no accidents during a given week,
    (b) the probability that there will be no accidents for two consecutive weeks.

45. In Exercise 44, find the expected number of accidents in the next 4 days and find the expected amount of time until the next accident.

46. Comment on the following statement: "The exponential distribution is to the Poisson distribution as the geometric distribution is to the binomial distribution." p. 222, 255

47. If the median of an exponential distribution is 2, find λ.

48. If the first quartile of an exponential distribution is 2, find λ.

49. Graph the density functions of the exponential distributions with $\lambda = 1$ and $\lambda = 5$ and comment on the shapes of these density functions. For each of the two cases, find the mean, median, mode, and variance.

50. The exponential distribution is "memoryless." For example, in Exercise 44, suppose that the "accident process" is Poisson. Then at any point of time, the distribution of the amount of time until the next accident is exponential with $\lambda = 3$ per week. Whether the last accident just occurred or there has not been an accident for two weeks, the distribution of the amount of time *from now* until the next accident remains the same. Suppose that an item suffers breakdowns according to a Poisson process. What does the "memoryless" property of the exponential distribution imply about the general way in which these breakdowns occur? For instance, does the item wear out with age?

51. A particular type of air pollution is measured in terms of the number of particles per cubic centimeter of air. The occurrence of the particles follows a Poisson process, and the pollution is considered to be severe when the intensity is 10 particles per cubic centimeter. In an investigation of air pollution, 2 cubic centimeters of air were sampled and investigated carefully, and the number of particles in the 2 cm³ was 30. What is the probability of observing 30 particles in 2 cm³ if the intensity is actually 10 particles per cm³?

52. Let T denote the amount of time until the rth occurrence in a Poisson process with intensity λ per unit of time. Then the distribution of T is a gamma distribution, which has a density function of the following form:

$$f(t) = \begin{cases} \dfrac{\lambda\,(\lambda t)^{r-1}e^{-\lambda t}}{(r-1)!} & \text{if } t \geq 0, \\[2mm] 0 & \text{elsewhere.} \end{cases}$$

The mean and variance of this gamma distribution are $E(T) = r/\lambda$ and $V(T) = r/\lambda^2$.

(a) Show that the exponential distribution is a special case of the gamma distribution with $r = 1$.

(b) What distribution related to the Bernoulli process plays a role analogous to the role played by the gamma distribution in the Poisson process?

(c) Graph the density functions of gamma distributions with (i) $r = 2$, $\lambda = 1$, (ii) $r = 5$, $\lambda = 1$, (iii) $r = 2$, $\lambda = 5$, (iv) $r = 5$, $\lambda = 5$.

53. Using the results given in Exercise 52,

(a) find the distribution of the amount of time in minutes until the third telephone call in Exercise 37,

(b) find the expected number of pages until the tenth misprint in the book in Exercise 38,

(c) in Exercise 44, find an expression for the probability that the fourth accident of the year will not occur in the first two weeks of the year (i) using the gamma distribution, (ii) using the Poisson distribution.

54. In a physician's office, the amount of time (in hours) it takes the doctor to examine a patient and to prepare for the next patient is exponentially distributed with $\lambda = 4$. The patients are scheduled at 15-minute intervals, beginning at 9 a.m., and the patients always arrive exactly on time.

(a) When the patient with a 9:15 a.m. appointment on a particular day arrives at the office, what is the probability of having to wait before seeing the doctor?

(b) Consider a clinic where everything is the same as in the physician's office except that there are 5 doctors, 5 patients are scheduled every 15 minutes, and there is no prior assignment of patients to doctors (that is, a patient may be examined by any of the 5 doctors). What is the probability that none of the patients with 9:15 a.m. appointments have to wait before seeing a doctor?

(c) In (b), what is the probability that exactly 2 of the patients with 9:15 appointments have to wait before seeing a doctor?

(d) In (b), consider the probability distribution of the amount of time a patient with an appointment at 11:30 a.m. has to wait before seeing a doctor. What difficulties do you encounter in attempting to specify this distribution? The study of situations such as this is known as queueing theory, or waiting-line theory. ←

55. The instructor in a mathematics course informs the class that two students received grades of 60 and 30 on the mid-term examination and that the standardized values corresponding to these grades are 1 and −1, respectively.

(a) What is the mean and the standard deviation of the grades?

(b) If the grades are approximately normally distributed, what proportion of the grades should lie between 25 and 65?

(c) In (b), between what two grades should the *middle* 50 percent of grades lie?

(d) In (b), if a student receives a grade of 80 on the examination, what proportion of students should receive even higher grades?

56. For the following events in (a)–(c), make a sketch of the regions under the normal curve associated with each event, and use Table I in the Appendix to compute the probability of each such event:

(a) $P(0 < Z \leq 1.96)$, $P(-1.65 \leq Z \leq 0)$, $P(-.63 < Z \leq 1.02)$,

(b) $P(Z \geq 1.3)$, $P(Z \geq .45)$, $P(-.45 < Z \leq 1.30)$,

(c) $P(Z \geq -.32)$, $P(Z \leq .75)$, $P(-.32 < Z \leq .75)$.

(d) Let A be the event $(-.67 \leq Z \leq .67)$ and let B be the event $(.26 \leq Z \leq 1.30)$. Find (to two significant places) the probabilities of the following events:

$$(A \cap B), \quad (A \cap \bar{B}), \quad (\bar{A} \cap B) \cup (A \cap B), \quad (A \cup B).$$

57. The notation $N(\mu, \sigma)$ is often used to denote a normal distribution with mean μ and standard deviation σ, where μ and σ are particular values. If X is distributed as $N(3, 4)$, use the table to find:

(a) $P(1 \leq X \leq 7)$, $P(X \leq 5)$, $P(-2 \leq X \leq 1.5)$,

(b) a number a such that $P[(X - 2) < a] = .95$,

(c) the .05, .25, .67, and .99 fractiles of the distribution.

58. If the .35 fractile of a normal distribution is 105 and the .85 fractile is 120, find the mean and standard deviation of the distribution.

59. Let us assume that at a certain university the actual time that a graduate student spends working toward the Ph.D. degree is approximately normally distributed with a mean of 200 weeks and a standard deviation of 50 weeks. Given 1000 students beginning graduate work toward the Ph.D. at the same time, what number should still be working toward the degree after 300 weeks? Find the number of weeks within which the *middle* 50 percent of students should complete their degrees. (Here we will simply ignore drop-outs and failures.)

60. The mean score of 1000 students on a national aptitude test is 100 and the standard deviation is 15. We will assume that the distribution of such scores is well-approximated by a normal distribution. Find approximately how many students have scores
    (a) between 120 and 155 inclusive,
    (b) greater than 95,
    (c) less than or equal to 175,
    (d) 5 or more points below the score cutting off the lowest 25 percent,
    (e) 10 or more points above the score cutting off the lowest 75 percent,
    (f) between 70 and 120, inclusive.

61. A sample of 4 independent observations is taken from a normal population with mean μ and variance σ^2.
    (a) What is the probability that *all* 4 sample observations will fall between the limits $\mu + \sigma$ and $\mu - \sigma$?
    (b) What is the probability that *all* 4 sample values will fall between the limits $\mu + 1.96\sigma$ and $\mu - 1.96\sigma$?
    (c) What is the probability that *exactly* 1 value will be greater than $\mu + \sigma$?
    (d) What is the probability that 2 sample values will fall below $\mu - 2\sigma$ and the other 2 sample values will fall above $\mu + 2\sigma$?

62. The normal distribution cannot be conveniently explained in terms of assumptions such as stationarity and independence, as can the Bernoulli and Poisson models. How, then, can the use of the normal model be justified in any specific application? Give some examples of realistic situations in which the normal model might be a suitable representation of the real situation.

63. For a particular brand of automobile battery, the distribution of the amount of time from its installation in a car until it fails and must be replaced is a normal distribution with a mean of 3.2 years and a standard deviation of .6 years.
    (a) What is the probability that a battery of this type will last at least 3 years?
    (b) If a firm installs batteries of this type in 10 of its "company cars," what is the probability that none of the 10 fail in less than 2.5 years?
    (c) In (b), what is the probability that exactly 5 of the 10 batteries are still operating after 3.2 years?
    (d) In a very large group of batteries, how long would a battery have to last before failing in order to be in the top 10 percent of all batteries in terms of time until failure?

64. The demand for tickets to a particular concert is normally distributed with mean 5000 and standard deviation 1400.
    (a) If the site of the concert has a maximum capacity of 8000, what is the probability that someone will want to attend the concert but will be unable to purchase a ticket because there are no tickets left?
    (b) The promoters of the concert claim that there is a 30 percent chance that they will lose money on the concert because too few tickets will be sold. How many tickets need to be sold for the promoters to break even?

65. When a copy is made on a particular copying machine, the probability that a piece of paper will be stuck and will cause the machine to stop is .005. A secretary has to make 1 copy of each of 1000 pages. For each of the following probabilities, (i) give an exact expression for the probability (but don't carry out any numerical calculations) and (ii) approximate this probability and give a numerical answer.

(a) The probability that the copying machine will stop at least once because of a piece of paper that is stuck during the time the secretary is using the machine.

(b) The probability that the machine will stop exactly twice during the 1000-page job.

66. An appliance dealer has to place an order for refrigerators. Suppose that 5000 people will enter the appliance dealer's store during the next month, and the probability that any given person who enters the store will purchase a refrigerator is .004. (No customers ever purchase more than one refrigerator in a single visit to the store.) The dealer's current order for refrigerators will be filled immediately, and the next shipment will not arrive until the beginning of the month following the next month. What is the minimum number of refrigerators the store must have in stock if the probability of being out of stock during the next month is to be no more than .05?

67. In Exercise 66, suppose that the product of interest is a popular brand of candy, the store is a candy store, and the probability of purchase is .40 rather than .004. All other details are the same. What is the minimum number of boxes of candy the store must have on hand in order to guarantee that the probability of being out of stock is at most .05?

68. In Exercise 3, give (a) an exact expression (but no numerical calculations) and (b) a numerical result determined from an approximation, for the probability that a lot of 300 items contains more than 10 defective items.

69. In Exercise 12, suppose that 200 patients undergo the type of surgery in question during a period of 1 year. What is the probability that exactly 70 percent of these 200 operations are successful? What is the probability that over 90 percent are successful?

70. Compute the binomial probabilities, the Poisson approximation, and the normal approximation for the following values of n and p:

(a) $n = 10, p = .5$,   (c) $n = 20, p = .5$,

(b) $n = 10, p = .05$,   (d) $n = 20, p = .05$.

Comment on the relative merit of the two approximations in each case.

71. Under what conditions can we approximate a binomial distribution by using a Poisson distribution, and under what conditions can we approximate a binomial distribution by using a normal distribution? Justify your answers with a brief intuitive discussion.

72. An example of a continuous uniform distribution used in the text is that of the life (in hours) of a light bulb. This variable is assumed to be uniform over the interval from 80 to 100.

   (a) What is the probability that the bulb will fail in 86 hours or less?

   (b) What is the probability that the bulb will last *exactly* 93 hours?

   (c) In a lot of 20 bulbs, what is the probability that *exactly* 4 will last more than 95 hours?

   (d) Does this distribution seem realistic for this variable? Discuss the implications of a uniform distribution vis-à-vis the implications of an exponential distribution or a normal distribution for the life of an item.

73. If the random variable X is uniformly distributed on the interval from a to b with mean 6 and variance 3, find a and b.

74. In Exercise 55, assume that the grades are uniformly distributed (ignore the mention of the normal distribution) and answer (a), (b), (c), and (d).

75. A distribution that will be discussed in Chapter 8 is the beta distribution, and the density function of a beta distribution with parameters r and n is

$$f(p) = \begin{cases} \dfrac{(n-1)!}{(r-1)!(n-r-1)!}\, p^{r-1}(1-p)^{n-r-1} & \text{if } 0 \le p \le 1, \\ 0 & \text{elsewhere,} \end{cases}$$

where r and $n - r$ must be positive.

   (a) Graph the density function and find the mean and variance of the beta distribution with $r = 1$ and $n = 2$.

   (b) Graph the density function and find the mean and variance of the beta distribution with $r = 1$ and $n = 3$.

   (c) Graph the density function and find the mean and variance of the beta distribution with $r = 2$ and $n = 4$.

76. The proportion of voters in a certain election who will vote in favor of a particular bond issue has a beta distribution with $r = 3$ and $n = 5$ (see Exercise 75). Find the probability that the bond issue will succeed if it needs favorable votes from at least 50 percent of the voters.

5

FREQUENCY
AND SAMPLING
DISTRIBUTIONS

The preceding three chapters have involved probability theory, and we are now ready to take up the first major set of concepts in statistics, utilizing what we have learned about probability. In developing statistical methods, we will use probability extensively, although the questions of primary interest in statistics are often different from the questions of primary interest in probability theory. Therefore, we will now have to look at matters from a slightly different viewpoint.

In probability theory, certain assumptions are made about the population or process of interest. Recall from Section 4.9 that a population is the set of objects from which the sample is taken; for example, we might sample from a population of students at a given college, from charge account customers for a given firm, or from a particular shipment of cars. Alternatively, in some cases it is convenient to think of the sample as being generated by some process, such as a Bernoulli process or a Poisson process. In any event, given certain assumptions about the population or process of interest, we can then calculate probabilities for various possible sample outcomes. For example, given that the buying behavior of consumers for a particular product can be treated as a Bernoulli process with p, the probability of purchasing the product, being .20 for each consumer, probabilities for sample results can be determined using the distributions developed in Chapter 4. If a random sample of 20 consumers is taken, we can use the binomial distribution to find the probability that, say, exactly 6 of the 20 consumers in the sample actually purchase the product. If we decide to sample until 6 purchases are made, we can use the Pascal distribution to find the probability that the sixth purchase will be made by the twentieth consumer in the sample. We can calculate these probabilities because we have completely specified the details concerning the process; it is assumed to be a Bernoulli process (that is, it is dichotomous, stationary, and independent), *and* we know that $p = .20$.

Typically, the theory of statistics is concerned with a different type of question. In statistics, we generally do *not* know everything about the population or process of interest, but we *do* have *some* relevant information, such as the results of a random sample from the population or process. This information can be investigated from a descriptive standpoint and/or from an inferential viewpoint.

Descriptive statistics consists of a set of procedures for describing and summarizing the information from a sample.

For instance, certain summary measures might be calculated, and the data from the sample could be presented in various sorts of graphical arrays.

Inferential statistics consists of methods for using sample data to make inferences about a population or process.

In the above example concerning consumer purchasing behavior, suppose that we are willing to assume that the Bernoulli process is applicable, but that we do *not* know p, the probability that a particular consumer will purchase the product. Given that we observe 6 purchases in a sample of 20 consumers, we can attempt to make inferences about p. For one thing, we might attempt to make a point estimate, or a "best guess," concerning the value of p. As we shall see in Chapter 6, $6/20 = .30$ turns out to be a reasonable "best guess." Or perhaps someone (the sales manager, for instance) claims that for this product, $p = .20$; the sample information can be used to investigate how reasonable the sales manager's claim, or hypothesis, is, by methods we will discuss in Chapter 7.

In probability theory, then, it is frequently assumed that everything is known about the population or process of interest and that the problem is to calculate probabilities for various possible sample results. In statistics, on the other hand, not everything about the population or process is known, although we may have some sample information, and we may attempt to describe and to use the information to draw inferences about the population. Since it is frequently the case in the real world that we do not know everything of interest about some population or process, the types of questions asked in inferential statistics are very important questions indeed.

In this chapter, we shall begin by discussing the notions of random sampling and measurement scales. Next, some material on descriptive statistics will be presented. The idea of a frequency distribution for sets of observations will be introduced first, together with some of the mechanics for constructing distributions. Next, the idea of a frequency distribution will be shown to be parallel to the idea of a probability distribution; the former concerns observed data, and the latter concerns a theoretical state of affairs (that is, a population or process). Summary measures of frequency distributions will then be discussed and shown to be analogous to the summary measures of probability distributions discussed in Chapter 3. Finally, after the material on descriptive statistics, the very important notion of a sampling distribution will be introduced. This material will be of great importance in the development of the theory of statistical inference and decision in the following chapters.

5.1 RANDOM SAMPLING

Given some simple experiment and the corresponding sample space of elementary events, the set of outcomes of n separate trials is a **sample.** When the sample is drawn *with replacement,* the same member of the population can be observed more than once. When the sample is drawn *without replacement,* the same member of the population can occur no more than once in a given sample. Ordinarily, probability and statistics deal with **random samples.** The word "random" has been used rather loosely in the foregoing discussion, and now the time has come to give it a more restricted meaning in connection with sampling.

It has already been suggested that probability calculations can be made quite simple when the elementary events in a sample space have equal probabilities. The theory of statistics deals with samples from a specified population, and here, too, great simplification is introduced if each distinct sample of a particular size can be assumed to have equal probability of selection. For this reason, the elementary theory of statistics is based on the idea of simple random sampling.

A method of drawing samples such that each and every distinct sample of the same size n has exactly the same probability of being selected is called simple random sampling.

In other words, simple random sampling corresponds to the situation where all possible samples of the same size have exactly the same probability of occurrence. For us, sampling "at random" will always mean simple random sampling, as defined above. This does not mean, however, that the theory of statistics does not apply to situations where samples have unequal probabilities of occurrence; the theory and methods can be extended to any sampling scheme where the probabilities of the various samples are *known,* even though they are unequal. Sampling methods other than simple random sampling can be quite valuable, particularly in areas such as survey sampling. We will discuss some of these alternative sampling schemes in Chapter 11.

Take care to notice that our assuming *samples* of size n to be equally probable does not imply that *events* must be equally probable. Thus, in the voting example in Section 2.14, samples of n voters were assumed equally probable, but this does not mean that the candidates were equally likely to be preferred by the voters. The equal probabilities in the definition of simple random sampling refer to the collection of units being sampled, not to the events that are recorded once the units are sampled.

We shall also have many occasions to require *independent random sampling*. You may recall that independence was formally defined in Section 2.21. Intuitively, we think of the term as follows: trials are independent when the occurrence of an elementary event on a particular trial has *absolutely no connection* with the probability of occurrence of the same or another elementary event on a subsequent trial. A series of tosses of a coin can be thought of as a random and independent sampling of events; what happens on one toss has no conceivable connection with what happens on any other. Each trial of a simple experiment is a sampling of elementary events, and the trials are independent when no connection exists among the particular outcomes of different trials. Recall that independence of trials is an important assumption for the Bernoulli and Poisson processes.

5.2 RANDOM NUMBER TABLES

In practical situations involving a given population, there are a number of schemes for sampling the population randomly and independently. For instance, if the population is not extremely large, we could fill a container with slips of paper or marbles, each one labeled with a particular member of the population. The container could then be shaken to mix the contents thoroughly, and the slips of paper or balls could then be drawn out one at a time, with or without replacement, until exactly n have been drawn. Although a procedure such as this is reasonable in theory, problems such as inadequate mixing of the contents of the container may yield samples that are not, in fact, random samples. For example, procedures of this nature have been used for draft lotteries in the United States, and careful statistical analyses of some of these lotteries suggest that selections have been non-random.

A better procedure for sampling randomly and independently involves the use of a **table of random numbers.** Such a table contains many pages filled with digits from 0 through 9. These tables have been composed in such a way that each digit is approximately equally likely to occur in any spot in the table and there is no systematic connection between the occurrence of any single digit in the sequence and any other. Most books on the design of experiments contain pages of random numbers, and very extensive tables of random numbers may be found in a book prepared by the RAND Corporation; Table VIII in the Appendix was taken from this book.

In order to use these tables for random sampling, we must list the possible members of the population (distinct units that might be observed). Each unit is then given a different number. The table of random numbers is entered by choosing some starting point at random and reading to the right or left, up or down. If the population contains only 10 or fewer individual members, then the digits 0 through 9 themselves are used; if there are 100 or fewer members of the population, digits are grouped by pairs, 00 through 99, and so on. Then the first n nonrepeated random numbers or groups of numbers are chosen, where n is the required sample size. The individual units corresponding to these numbers comprise the sample, which can be regarded as chosen at random.

This method is subject to two drawbacks. In the first place, unless we have an infinitely large table of random numbers, the one that is used must contain either some differences in probability or some dependence among the digits. This makes a sample chosen with any given table only approximately random. Second, and much more important, is the fact that a *listing* of the members of the population (all potential observation units) must be possible. Except in some very restricted situations this is exceedingly difficult, or even impossible, to do. What usually happens is that selection is made from some population much smaller than the one the statistician would really like to use. For example, the statistician might like to make statements about all people of a certain age, but the only group that can really be sampled randomly is college sophomores in a certain locale. Then one of two things must occur: either the statistician assumes that the population actually used is itself a random sample from the larger population, or inferences are restricted to the group that can be sampled randomly. Regardless of which course is taken, in order to make probability statements, the sample must be drawn at random from *some* specific population of potential observations.

Quite apart from their utility for drawing samples, random number tables are interesting as another instance of the notion of "randomness" that is idealized in the theory of probability. A random process is one in which only chance factors determine the particular outcome of any single trial. The possible outcomes are known in advance but not the exact outcome to be realized on any given trial. Nevertheless, built into the process is some regularity, so that each class of outcome (or event) can reasonably be assigned a probability. Perhaps the simplest example of a random phenomenon is the result of tossing a coin. Only chance factors determine whether heads or tails will come up on a particular toss, but a fair coin is so constructed that there is just as much physical reason for heads to appear as for tails, and so we have the justification we need in order to say that, in the long run, heads will appear just as often as tails. This property of the coin is idealized when we say that the probability of heads is .50. A fair die is constructed in such a way that each of the six sides should have the same physical opportunity to appear on a given throw, although unknown factors dictate which side actually does appear on that throw; we idealize this property of the die when we say that the probability of, say, three spots is 1/6. In principle, a random number table can be constructed by using a 10-sided die, each side labeled with a number from 0 through 9. Then any sequence of tosses gives a sequence of random numbers. In practice, of course, random numbers are generated electronically by computer, but the general idea is the same; the physical process is such that no single number is favored for any particular outcome, and chance alone actually dictates the number that does occur in any place in a sequence. When we use random numbers to select samples, we are using this property of randomness, which was inherent in the process by which the numbers themselves were generated. The point is that the ideas of randomness and of probability are not just misty abstractions; we know how to create processes and devise operations that are approximately "random," and we know how to manufacture events with particular probabilities, at least with probabilities that are approximately known.

Random number tables can be used to generate samples with replacement or without replacement. In the former case, if the same number appears two or more times in the chosen set of numbers, then that element is included in the sample two or more times. In the latter case, if a number appears for a second time, it is ignored and we proceed to the next number; each member of the population can occur only once. Unless it is stated otherwise, we will generally assume that sampling is done with replacement. This may seem strange, since in many instances (for example, in business, the social sciences, the behavioral sciences, and related fields) n *different* items are actually sampled and the sampling is obviously done without replacement. In investigating voting preferences, we would not want to include the same voter twice in a sample. However, the number of potential observations (in other words, the size of the population we are sampling from) is often very large, so that the selection and nonreplacement of a unit does virtually nothing to the probabilities of events. For all practical purposes, sampling without replacement can be treated exactly like sampling with replacement in such situations. On the other hand, as we saw in Section 4.9, when the number of units in the population is relatively small, sampling without replacement requires special procedures, since the probabilities of events are altered appreciably with each new observation. This does no real violence to the notion of simple random sampling, however. In the definition of simple random sampling, we do not require that each member of the population be equally likely to be drawn on each trial; instead, we require that *all samples of n members of the population be equally likely to occur.*

5.3 MEASUREMENT SCALES

Prior to any concerns about summarizing data or comparing it with some theoretical state of affairs, the statistician must first measure the phenomena being studied. It therefore seems appropriate to mention the process of measurement at this point and to indicate briefly the role measurement considerations will play in the remainder of this text. We will not even attempt to undertake a thorough discussion of this topic, but we will try to suggest its relevance to statistics.

Whenever the statistician makes observations of any kind, some scheme is employed for classifying and recording what is observed. Any phenomenon or "thing" will have many distinguishable characteristics or attributes, but *the statistician must first single out those properties relevant to the question being studied.* Next, a scheme for measuring these properties must be devised. At its very simplest, *this scheme is a rule for arranging observations into equivalence classes, so that observations falling into the same set are thought of as qualitatively the same and those in different classes as qualitatively different in some respect.* In general, *each observation is placed in one and only one class, making the classes mutually exclusive and exhaustive.*

The process of grouping individual observations into qualitative classes is measurement at its most primitive level. Sometimes this is called **categorical or nominal scaling.** The set of equivalence classes itself is called a **nominal scale.** There are many areas in science where the best we can do is to group observations into classes, giving each a distinguishing symbol or name. For example, taxonomy in biology consists of grouping living things into phyla, genera, species, and so on, which are simply sets or classes. Psychiatric nosology contains classes such as schizophrenic, manic-depressive, paretic, and so on; individual patients are classified on the basis of symptoms and the histories of their diseases. Classification is also prevalent in business and economics: investments are classified in finance (and classes of investments may be further broken down into subclassifications, as with

AA bonds, A bonds, and so on) ; costs may be classified as fixed or variable costs or may be classified in numerous other ways.

Notice that even this kind of measurement can be thought of in terms of a function; the domain of this function is a set of individual observations, and the range is the set of equivalence classes (the nominal scale). The word "measurement" is usually reserved, however, for the situation in which each observation is assigned a *number*; this number reflects a magnitude of some *quantitative* property. There are at least three kinds of numerical measurement that can be distinguished; these are often called ordinal scaling, interval scaling, and ratio scaling. A really accurate description of each kind of measurement is beyond the scope of this book, but we can gain some idea of the distinctions among the different types of measurement.

Imagine a set of objects, and suppose that there is some property that all objects in the set possess, such as value, or weight, or length, or intelligence, or motivation. Furthermore, let us suppose that each object o has a certain amount or degree of that property. In principle we could assign a number, $t(o)$, to any object, standing for the amount that o actually "has" of that characteristic.

Ideally, in order to measure an object o, we would like to determine this number $t(o)$ directly. However, this is not always (or even usually) possible, and so what we do is devise a procedure for pairing each object with another number, $m(o)$, that can be called its **numerical measurement.** The measurement procedure we use constitutes a function rule, telling how to give an object o its $m(o)$ value. But just any measurement procedure will not do; *we want the various $m(o)$ values to reflect the different $t(o)$ values for different degrees of the property.* A measurement rule would be nonsense if it gave numbers having no connection at all with the amounts of some property different objects possess. Even though we may never be able to determine the $t(o)$ values exactly, we would like to find $m(o)$ numbers that will at least be related to them in a systematic way. The measurement numbers must be good reflections of the true quantities, so that information about true magnitudes actually is contained in our numerical measurements.

Measurement operations or procedures differ in the information that the numerical measurements themselves provide about the true magnitudes. Some ways of measuring permit us to make very strong statements about what the differences or ratios among the true magnitudes must be, and thus about the actual differences in, or proportional amounts of, some property that different objects possess. On the other hand, some measurement operations permit only the roughest inferences to be made about true magnitudes from the measurement numbers themselves.

Suppose that we have a measurement procedure or rule for assigning a number $m(o)$ to each object o, and suppose that the following statements are true for any pair of objects o_1 and o_2.

1. $m(o_1) \neq m(o_2)$ only if $t(o_1) \neq t(o_2)$.

2. $m(o_1) > m(o_2)$ only if $t(o_1) > t(o_2)$.

In other words, by this rule we can at least say that if two measurements are unequal, the true magnitudes are unequal, and if one measurement is larger than another, then one magnitude exceeds another. Any measurement procedure for which both statements 1 and 2 are true is an example of **ordinal scaling,** or measurement at the **ordinal level.**

For example, suppose that the objects in question are minerals of various kinds. Each mineral has a certain degree of *hardness*, represented by the quantity $t(o)$. We have no way of knowing these quantities directly, and so we devise the following measurement rule: take each pair of minerals and find out if one *scratches* the other. Presumably, the harder mineral will scratch the softer in each case. When this has been done for each pair of minerals, give the mineral scratched by everything some number, the mineral scratched by all but the first a higher number, and so on, until the mineral scratching all others but scratched by nothing else receives the highest number of all. In each pair the "scratcher" receives a higher number than the "scratchee." (Here we are assuming tacitly that "*a* scratches *b*" and "*b* scratches *c*" implies that "*a* scratches *c*" and that we will get a *simple ordering* of "what scratches what.")

This measurement procedure gives an example of *ordinal* scaling. The possible numerical measurements themselves might be some set of numbers, such as $(1, 2, 3, 4, \ldots)$ or $(10, 17, 24, \ldots)$ or even other symbols having some conventional order, as $(A, B, C, \ldots)$. In any case, the assignment of numbers or symbols to objects is a form of *ranking*, showing which is "more" something. In the example, if one mineral gets a higher number than another, then we can say that the first mineral is harder than the second. Notice, however, that this is really all we can say about the degree or amount of hardness each possesses. If we find two objects where $m(o_1) = 3$ and $m(o_2) = 2$ by this procedure, then it is perfectly true that the difference between the *numbers* is $3 - 2 = 1$, but we have no basis at all for saying that $t(o_1) - t(o_2) = 1$. The difference in hardness between the two minerals could be any positive number. In short, *although the numbers standing for ordinal measurements may be manipulated by arithmetic, the answer cannot neccessarily be interpreted as a statement about the true magnitudes of objects nor about the true amounts of some property.*

However, other measurement procedures give functions pairing objects with numbers where much stronger statements can be made about the true magnitudes from the numerical measurements. Suppose that the following statement, in addition to statements 1 and 2, is true.

3. $t(o) = x$ if and only if $m(o) = ax + b$, where $a > 0$.

That is, the measurement number $m(o)$ is some *linear function* of the true magnitude x (the rule for a linear function is x multiplied by some constant a and added to some constant b). When the statements 1, 2, and 3 are all true, the measurement operation is called **interval scaling,** or measurement at the **interval-scale level.**

Much stronger inferences about magnitudes can be made from interval-scale measurements than from ordinal measurements. In particular, we can say something precise about *differences* in objects in terms of magnitude. For example, finding temperature in Fahrenheit units is measurement on an interval scale. If object o_1 has a reading of 180° and o_2 has 160°, the difference $(180 - 160)$ times a constant actually *is* the difference in "temperature-magnitude" between the objects. It is perfectly meaningful to say that the first object has 20 units more temperature than the second. Moreover, even stronger statements can be made. If $m(o_3) = 140°$, so that

$$m(o_1) - m(o_3) = 40°$$

and
$$m(o_1) - m(o_2) = 20°,$$

then
$$\frac{40°}{20°} = \frac{t(o_1) - t(o_3)}{t(o_1) - t(o_2)} = 2.$$

We can say that in true amount of "temperature," object o_3 is twice as different from o_1 as o_2 is from o_1. Moreover, this statement is not dependent on the particular units (Fahrenheit units). If Centigrade units were used, the readings for o_1, o_2, and o_3 would be 82.22°, 71.11°, and 60°. The difference between $m(o_1)$ and $m(o_3)$ is 22.22°, whereas the difference between $m(o_1)$ and $m(o_2)$ is 11.11°, so object o_3 remains twice as different from o_1 as o_2 is from o_1.

When measurement is at the interval-scale level, any of the ordinary operations of arithmetic may be applied to the differences between numerical measurements, and the results can be interpreted as statements about magnitudes of the underlying property. The important part is the interpretation of a numerical result as a quantitative statement about the property shown by the objects. This is not generally possible for ordinal-scale numbers, but it *can* be done for *differences* between interval-scale numbers. In very simple language, we can do arithmetic to our heart's content on any set of numbers, but the results are not necessarily true statements about amounts of some property objects possess unless interval-scale requirements are met by the procedure for obtaining those numbers.

Interval scaling is about the best we can do in most scientific work, and even this level of measurement is all too rare in business and the social and behavioral sciences. However, especially in the physical sciences, it is sometimes possible to find measurement operations making the following statement true.

4. $t(o) = x$ if and only if $m(o) = ax$, where $a > 0$.

When the measurement operation defines a function such that statements 1 through 4 are all true, then measurement is said to be at the **ratio-scale level.** For such scales, *ratios* of numerical measurements can be interpreted directly as ratios of magnitudes of objects. For example, the usual procedure for finding the length of objects provides a ratio scale. If one object has a measurement value of 10 feet and another has a value of 20 feet, then it is quite legitimate to say that the second object has twice as much length as the first. Notice that this is not a statement one ordinarily makes about the *temperatures* of objects (on an interval scale); if the first object has a temperature reading of 10° and the second 20°, we cannot necessarily say that the second has twice the temperature of the first. Only when scaling is at the ratio level can the full force of ordinary arithmetic be applied directly to the measurements themselves and can the results be reinterpreted as statements about magnitudes of objects.

There are any number of examples that could be adduced to illustrate the differences among these levels of measurement and other possible intermediate levels as well. Our immediate concern, however, is to see the implications of different levels of measurement for probability and statistics.

Most of the statistical techniques introduced in this book are designed for numerical observations, presumably arising from the application of measurement techniques at the interval-scale level. This does not mean that these techniques cannot be applied to numerical data when the interval-scale level of measurement is not met; these techniques can be applied to *any* numerical data where the purely *statistical* assumptions are met. It does mean, however, that caution must be exercised in the interpretation of these statistical results in terms of some property the statistician intended to measure. The statistical conclusion may be quite valid for these data *as numbers*, but the statistician must think quite seriously about the further interpretations that are given to the result.

Of course, it is completely obvious that interval scales simply cannot be attained for some things. It may be that an ordering of objects on some basis is the very best the statistician can do. In this case, ordinal methods like those in Chapter 12 are helpful; many techniques are designed specifically for ordinal data. Even when measurement is only nominal, statistical methods can be applied; some of these are also discussed in Chapter 12. The most common situation is, however, one where numerical measurements are made but where the interval-scale assumptions are hard to justify. In these cases, the user of the statistical techniques should understand the limitations imposed by the measurement device (not by statistics); the road from objects to numbers may be easy, but the return trip from numbers to properties of objects is not.

5.4   FREQUENCY DISTRIBUTIONS

As we have just seen, even in the simplest instances of measurement the statistician makes some observations and classifies them into a set of qualitative measurement classes. These qualitative classes are mutually exclusive and exhaustive, so that each observation falls into one and only one class.

As a summarization of the observations, the statistician often reports the various possible classes, together with the number of observations falling into each. This may be done by a simple listing of classes, each paired with its frequency number, the number of cases observed in that category. The same information may be displayed as a graph, perhaps with the different classes represented by points or segments of a horizontal axis, the frequency shown by a point or a vertical bar above each class. Regardless of how this information is displayed, such a listing of classes and their frequencies is called a **frequency distribution.**

Any representation of the relation between a set of mutually exclusive and exhaustive measurement classes and the frequency of each is a frequency distribution.

Notice that a frequency distribution is a function; each of a set of classes is paired with a number, its frequency. Thus, in principle, a frequency distribution of real or theoretical data can be shown by any of the three ways one specifies any function: an explicit listing of class and frequency pairs, graphically, or by the statement of a rule for pairing a class of observations with its frequency. It should be clear by now that a frequency distribution is analogous to a probability distribution; the former represents the relationship between events (in terms of classes) and their observed frequencies in a given sample, and the latter represents the relationship between events and their probabilities. A frequency distribution, just as a probability distribution, can be specified by a listing, in graphical form, or in functional form.

The set of measurement classes may correspond to a nominal, an ordinal, an interval, or a ratio scale. Although the various possible classes will be qualitative in some instances and quantitative in others, a frequency distribution can always be constructed, provided that each and every observation goes into one and only one class. For example, suppose that some native U. S. citizens are observed, and the state in which each was born is noted. This is like nominal measurement, where the measurement classes consist of the 50 states (and the District of Columbia) into which subjects could be categorized. On the other hand, the subjects might be students in a course and graded according to A, B, C, and so on. Here the frequency distribution would show how many received A, B, and so on. Notice that in this case the measurement actually is ordinal; A is better than B, B better than C. Nevertheless, the frequency distribution is constructed in the same way, except that the various classes are displayed in their proper order.

Even when measurement is at the interval- or ratio-scale level, we actually report frequency distributions in this general way. Suppose that n college students are each weighed, and the weights are noted to the nearest pound. The set of weight classes into which students are placed would perhaps consist of 50 or 60 different numbers; a weight class is the set of students getting the same weight number. There may be only one, or even no students, in a particular weight class. Nevertheless, the frequency distribution shows the pairing of each possible class with some number from 0 to n, its frequency.

The reasons for dealing with frequency distributions rather than raw data are not hard to see. Raw data almost always take the form of some sort of listing of pairs, each consisting of an object and its measurement class or number. Consider a study done on a group of 25 consumers in order to determine their brand preferences. The consumers were classified as preferring brand A, B, C, or D. Table 5.4.1 lists the brand preferences of these 25 consumers. This set of pairs contains all the relevant information, but if there are a great many objects being measured, such a listing is not only laborious but also very confusing to anyone who is trying to get a picture of the set of observations as a whole. If we are interested only in the "pattern" of brand preference in the group, the numbers assigned to the persons are irrelevant, and all that is necessary is the number of individuals stating each brand preference. The data in Table 5.4.1 condense into the frequency distribution in Table 5.4.2, where f represents frequency.

In another study, 2000 families in some city were interviewed about their preferences in art, music, literature, recreation, and so forth. After each interview, the family was characterized as having "highbrow."

Table 5.4.1 Brand Preferences of Consumers

Consumer	Brand preferred
1	A
2	B
3	A
4	A
5	A
6	C
7	D
8	A
9	A
10	A
11	D
12	B
13	D
14	B
15	A
16	B
17	D
18	B
19	D
20	A
21	B
22	B
23	A
24	A
25	D

Table 5.4.2 Frequency Distribution of Brand Preferences

Class	f
A	11
B	7
C	1
D	6
	$\overline{25} = n$

"middlebrow," or "lowbrow" tastes. We can consider this as a case of ordinal-scale measurement, since these three "taste" categories seem to be ordered: "highbrow" and "lowbrow" should be more different from each other than "middlebrow" is from either. A listing of families in terms of these classes would be confusing, if not grounds for libel suits. However, just as before, we can construct a frequency distribution summarizing these data (Table 5.4.3). Notice how much more clearly the characteristics of the group as a whole emerge from a frequency distribution such as this than they possibly could from a listing of 2000 names, each paired with a rating.

Table 5.4.3 Frequency Distribution of "Tastes"

Class	f
Highbrow	50
Middlebrow	990
Lowbrow	960
	2000 $= n$

It is appropriate to point out that a frequency distribution provides clarity at the expense of some information in the data. It is not possible to know from the frequency distribution alone whether a *particular* consumer prefers Brand *B*, or whether a *particular* family is high, middle, or lowbrowed. Such information about *particular* objects is sacrificed in a frequency distribution to gain a picture of the group of measurements *as a whole*. This is true of all descriptive statistics; we want clear pictures of large numbers of measurements, and we can do this only by losing detail about particular objects. The process of weeding out particular qualities of the objects that happen to be irrelevant to our purpose, begun whenever we measure, is continued when we summarize a set of measurements.

5.5 GROUPED DISTRIBUTIONS

In the distributions shown above, there were a few "natural" categories into which all the data fit. It often happens, however, that data are measured in some way that gives a great many categories into which a given observation might fall. Indeed, the number of potential categories of description often vastly exceeds the number of cases observed, so that little or no economy of description would be gained by constructing a distribution showing each *possible* class. The most usual such situation occurs when the data are measured in numerical terms. For instance, in the measurement of height or weight, there are, in principle, an infinite number of numerical classes into which an observation might fall: all the positive real numbers. Even when the measurements result in fewer than an infinite number of possibilities, it is still quite common for the number of available categories to be very large.

For this reason, it is often necessary to form **grouped** frequency distributions. Here the domain of the frequency function is not each possible measurement category, but rather intervals or groupings of categories.

For example, consider a number of people who have been given an intelligence test. The data of interest are the numerical IQ scores. Immediately, we run into a problem of procedure. It could very well turn out that for, say, 150 individuals, the IQ score would be different for each case, and thus a frequency distribution using each different IQ number as a measurement class would not condense our data any more than a simple listing.

The solution to this problem lies in grouping the possible measurement classes into new classes, called **class intervals,** each including a range of score possibilities. Proceeding in this way, the IQs 105, 104, 100, 101, and 102 might turn up in our data. Instead of listing each in a different class, we might put them all into a single class, 100–105, along with any other IQ measurements that fall between these limits. Similarly, we group other sets of numbers into class intervals. Table 5.5.1 shows how the resulting frequency distribution might look for a group of 150 persons. Forming class intervals has enabled us to condense the data so that a simple statement of the frequency distribution can be made in terms of only a few classes. In this distribution, the various class intervals are shown in order in the

Table 5.5.1   Frequency Distribution of IQ Scores

Class interval	Midpoint	f
82–87	84.5	2
88–93	90.5	23
94–99	96.5	22
100–105	102.5	65
106–111	108.5	20
112–117	114.5	10
118–123	120.5	0
124–129	126.5	8
		$150 = n$

first column on the left. The extreme right column lists the frequencies, each paired with the class in that row of the table. The sum of the frequencies must be n, the total number of observations. The middle column contains the midpoint of each class interval; midpoints will be discussed in the next section.

5.6   CLASS INTERVAL SIZE, CLASS LIMITS, AND MIDPOINTS OF CLASS INTERVALS

The first problem in constructing a grouped frequency distribution is deciding upon a class interval size. Just how large shall the range of numbers included in a class be? The class interval size for a grouped distribution is the difference between the largest and the smallest numbers that may go into any class interval. We will let the symbol i denote this interval size.

In the example above, as in most examples to follow, the size i is the same for each class interval. For our example, $i = 6$, meaning that the largest and the smallest score going into a class interval differ by 6 units. Take the class interval labeled 100–105. It may seem that the smallest number going into this interval is 100, and the largest 105; this is not, however, true. Actually, the numbers 99.5, 99.6, 99.7 would go in this interval, should they occur in the data, and the numbers 105.1, 105.2 are also included. On the other hand, 99.2 or 105.8 would be excluded. The interval actually includes any number *greater than or equal to* 99.5 *and less than* 105.5. These limits are called the **real limits** of the interval, in contrast to those actually listed, which are the **apparent limits.** Thus, the real limits of the interval 100–105 are

$$99.5 \text{ to } 105.5,$$

and $$i = 105.5 - 99.5 = 6.$$

In general,

Real lower limit = apparent lower limit minus .5(unit difference),

Real upper limit = apparent upper limit plus .5(unit difference).

The term "unit difference" demands some explanation. In measuring something we usually find some limitation on the accuracy of the measurement, and seldom can we measure with *any* desired degree of accuracy. For this reason, measurement in numerical terms is always rounded, either during the measurement operation itself or after the measurement has taken place. In measuring weight, for example, we obtain accuracy only to the nearest pound, or the nearest one-tenth of a pound, or one-hundredth of a pound, and so on. If we are constructing a frequency distribution where weight has been rounded to the nearest pound, then a unit difference is 1 pound, and the real limits of an interval such as 150–190 would be 149.5–190.5. The i here would be 41 pounds.

On the other hand, suppose that weights were accurate to the nearest one-tenth of a pound; then the unit difference would be .1 pound, and the real class-interval limits would be

$$149.95–190.05.$$

We can use any number for i in setting up a distribution. However, it should be obvious that there is very little point in choosing i smaller than the unit difference (for example, letting $i = .1$ pound when the data are to the nearest whole pounds), or in choosing i so large that all observations fall into the same class interval. Furthermore, i is usually chosen to be a *whole number* of units, whatever the unit may be. Even within these restrictions, there is considerable flexibility of choice. For example, the data of Section 5.5 could have been put into a distribution with $i = 3$, as shown in Table 5.6.1. Notice that this distribution with $i = 3$ is somewhat different from the distribution with $i = 6$, even though they are based on the same set of data. For one thing, here there are 16 class intervals, whereas

Table 5.6.1 Frequency Distribution of IQ Scores

Class interval	Midpoint	f
81–83	82	2
84–86	85	0
87–89	88	9
90–92	91	14
93–95	94	12
96–98	97	10
99–101	100	20
102–104	103	44
105–107	106	13
108–110	109	8
111–113	112	3
114–116	115	7
117–119	118	0
120–122	121	0
123–125	124	3
126–128	127	5
		$150 = n$

there were 8 before. This second distribution also gives somewhat more detail than the first about the original set of data. We now know, for example, that there was no one in the group of persons who had an IQ score between the real limits of 83.5 and 86.5, whereas we could not have known this from the first distribution. We can also tell that more cases showed IQs between 101.5 and 104.5 than between 98.5 and 101.5; conceivably, this could be a fact of some importance. On the other hand, the first distribution gives a simpler and, in a sense, a neater picture of the group than does the second.

The decision facing the maker of frequency distributions is, "Shall I use a small class-interval size and thus get more detail, or shall I use a large class-interval size and get more condensation of the data?" There is no fixed answer to this question; it depends on the kind of data and the uses to which they will be put. However, a convention exists which says that for most purposes *10 to 20 class intervals* give a good balance between condensation and necessary detail, and in practice we usually choose our class-interval size to make about that number of intervals.

There is one more feature of frequency distributions that has not yet been discussed. When we put a set of measurements into a frequency distribution, we lose the power to say exactly what the original numbers were. For the above example, if a person in the group of 150 cases has an IQ of 102, he or she is simply counted as one of the 44 individuals who make up the frequency for the interval 102–104. You cannot tell from the distribution exactly what the IQs of these 44 individuals were, but only that they fell within the limits 102–104. What do we call the IQs of these 44 individuals? We call all of them 103, the midpoint of the interval. **The midpoint of any class interval is that number which falls exactly halfway among the possible numbers included in the interval.**

Quite often the data are such that it is not possible to make a frequency distribution with intervals of a constant size. This most commonly occurs when exact values are not known for some of the observations. For example, suppose that we are measuring temperature, using a thermometer which only goes up to 100 degrees Fahrenheit, and some of the observations are greater than this upper limit. We cannot give exact values for these observations, although we can say that they are greater than 100; that is, they fall in the top class of the frequency distribution, in an interval called "100 or more." This interval is open, since there is no way to determine its upper real limit. Furthermore, there is no way to give such an open interval a midpoint.

It is perfectly correct to show open intervals at either end of a frequency distribution. However, no computations involving the midpoints can be carried out from this distribution unless the cases in the open intervals are dropped from the sample.

In other instances, it may be that there are extreme scores in a distribution that are widely separated from the bulk of the cases. When this is true, a class-interval size that is small enough to show "detail" in the more concentrated part of the distribution will eventuate in many classes with zero frequency before the extreme scores are included. Enlarging the class-interval size will reduce the number of unneccessary intervals but will also sacrifice detail. When this situation arises, it is often wise to have a varying class-interval size, so that the class intervals are narrow where the detail is desired but rather broad toward the extreme or extremes of the distribution. In this case, class-interval limits and midpoints can be found in the usual way, provided that we take care to notice where the interval size changes. Furthermore, midpoints of these intervals can be used for further computations using the methods given in Sections 5.10 and 5.11. For example, suppose that we were measuring daily changes in security prices. In the more concentrated part of the distribution (that is, the part near zero), we might be interested in narrow class intervals. Observations more than, say, 2 units (dollars in this example) from zero might be infrequent enough that we would like wider class intervals for these observations.

5.7 GRAPHS OF DISTRIBUTIONS: HISTOGRAMS

Now we direct our attention for a time to the problem of putting a frequency distribution into graphic form. Often a graph shows that the old bromide, "a picture is worth a thousand words," is really true. If the purpose is to provide an easily grasped summary of data, nothing is as effective as a graph of a distribution.

There are undoubtedly all sorts of ways any frequency distribution might be graphed. A glance at any national news magazine will show many examples of graphs that have been made striking by some ingenious artist. However, we shall deal with three "garden varieties": the histogram, the frequency polygon, and the cumulative frequency polygon.

The **histogram** has already been used in Section 3.6 to represent a probability distribution divided into intervals. Actually, a histogram is just a version of the familiar bar graph. In a histogram each class or class interval is represented by a portion of the horizontal axis. Then over each class interval is erected a bar or rectangle equal in height to the frequency for that particular class or class interval. The histograms representing the frequency distributions in Sections 5.4 and 5.6 are shown in Figures 5.7.1–5.7.3. It is customary to label both the horizontal and the

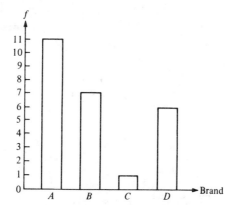

Figure 5.7.1   Histogram of Brand Preferences

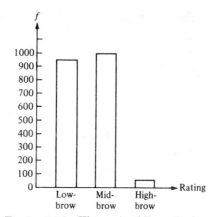

Figure 5.7.2   Histogram of Taste Ratings

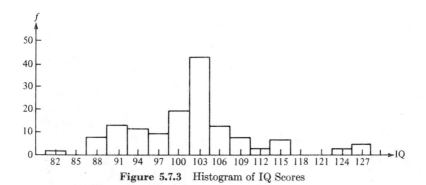

Figure 5.7.3   Histogram of IQ Scores

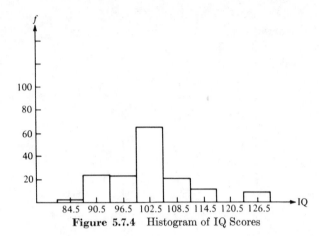

Figure 5.7.4   Histogram of IQ Scores

vertical axes, as shown in these figures, and to give a label to the graph as a whole. In the case where class intervals are employed to group numerical measurements, a saving of time and space may be accomplished by labeling the class intervals by their midpoints (as in Figure 5.7.3) rather than by their apparent or real limits. With categorical measurements, this problem does not arise, of course, since each class can only be given its name or symbol.

If you will look at the two histograms shown in Figures 5.7.4 and 5.7.5, you will get very different first impressions, even though they represent the same frequency distribution. These two figures illustrate the effect that

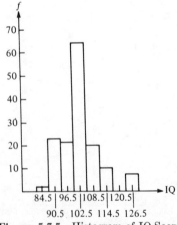

Figure 5.7.5   Histogram of IQ Scores

proportion can have on the viewer's first impressions of graphed distributions. Various conventions about graphs of distributions exist in different fields, of course. In fact, those who make a business of using statistics for persuasion often choose a proportion designed to give a certain impression. However, one convention is to arrange a graph so that the vertical axis is about three-fourths as long as the horizontal axis. This usually will give a clear and esthetically pleasing picture of the distribution.

5.8 FREQUENCY POLYGONS

The histogram is a useful way to picture any sort of frequency distribution, regardless of the scale of measurement of the data. However, a second kind of graph is often used to show frequency distributions, particularly those that are based upon numerical data: this is the **frequency polygon.** In order to construct the frequency polygon from a frequency distribution we proceed exactly as though we were making a histogram. The horizontal axis is marked off into class intervals and the vertical axis into numbers representing frequencies. However, instead of using a bar to show the frequency for each class, a point on the graph corresponding to the midpoint of the interval and the frequency of the interval is found. These points are then joined by straight lines, each being connected to the point immediately preceding and the point immediately following, as in Figure 5.8.1.

The frequency polygon is especially useful when there are a great many potential class intervals, and it thus finds its chief use with numerical data that could, in principle, represent a continuous variable. For a distribution

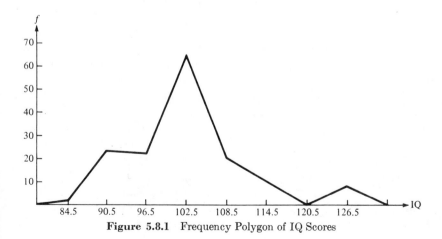

Figure 5.8.1 Frequency Polygon of IQ Scores

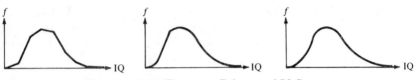

Figure 5.8.2   Frequency Polygons of IQ Scores

based on a very large number of numerical measurements on a potentially continuous scale, if we maintained the same proportions in our graph but employed a much larger number of class intervals, the frequency polygon would provide a picture more like a smooth curve, as shown in Figure 5.8.2, based on several thousand cases. Although the function rule describing the frequency polygon may be extremely complicated to state, the function rule for a smooth curve approximately the same as that of the distribution may be relatively easy to find. For this reason, a frequency polygon based upon interval- or ratio-scale measurements with a relatively large number of classes is sometimes "smoothed" by creating a curve that approximates the shape of the frequency polygon. This new curve may have a rule that is simple to state and that may serve as an approximation of the rule describing the frequency distribution itself. Graphically, the smooth curve may be a good approximation to the frequency polygon, just as a probability density function can sometimes be thought of as a good approximation to the probability mass function of a discrete random variable (Section 3.6).

5.9   CUMULATIVE FREQUENCY DISTRIBUTIONS

For some purposes it is convenient to make a different arrangement of data into a distribution, called a **cumulative frequency distribution,** which is analogous to a cumulative distribution function (Section 3.7). Instead of showing the relation that exists between a class interval and its frequency, **a cumulative frequency distribution shows the relation between a class interval and the frequency at or below its real upper limit.** In other words, the cumulative frequency shows how many cases fall *into this interval or below.* Thus, in the distribution in Section 5.5, the lowest interval shows two cases, and its cumulative frequency is 2. Now the next class interval has a frequency of 23, so that its cumulative frequency is $23 + 2$, or 25. The third interval has a frequency of 22, and its cumulative frequency is $22 + 23 + 2$, or 47, and so on. The cumulative frequency of the very top interval must always be n, since all cases must lie either in the

Table 5.9.1 Cumulative Frequency Distribution of IQ Scores

Class interval	Cumulative frequency
82–87	2
88–93	25
94–99	47
100–105	112
106–111	132
112–117	142
118–123	142
124–129	150

top interval or below. The cumulative frequency distribution for the example of Section 5.5 is given in Table 5.9.1.

Cumulative frequency distributions are often graphed as polygons. The axes of the graph are laid out just as for a frequency polygon, except that the class intervals are labeled by their *real upper limits* rather than by their midpoints. The numbers on the vertical axis are now cumulative frequencies, of course. The graph of the example from Section 5.5 is shown in Figure 5.9.1. Cumulative frequency distributions often have more or less this characteristic **S** shape, although the slope and the size of the "tails" on the **S** will vary greatly from distribution to distribution. Like a frequency polygon, a cumulative frequency polygon is sometimes smoothed into a curve as similar as possible to the polygon. Such a smoothed cumulative distribution curve is sometimes called an **ogive.**

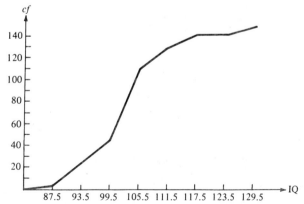

Figure 5.9.1 Cumulative Frequency Distribution of IQ Scores

5.10 MEASURES OF CENTRAL TENDENCY FOR FREQUENCY DISTRIBUTIONS

Any frequency distribution is a summarization of data, but for many purposes it is necessary to summarize still further. Rather than compare entire frequency distributions with each other or with hypothetical distributions, it is generally more efficient to compare only certain characteristics of distributions. As with probability distributions, the two most important characteristics are central tendency and dispersion. In this section we discuss measures of central tendency for frequency distributions, and in the following section we will discuss measures of dispersion.

The measures of central tendency considered for probability distributions have counterparts for frequency distributions. The three most important measures are the sample mean, the sample median, and the sample mode.

The sample mean of a sample of observations is determined by taking the simple arithmetic average of the values; if the sample size is n and the observed values are $x_1, x_2, \ldots, x_n$, then the sample mean m is

$$m = \frac{\sum_{i=1}^{n} x_i}{n}. \tag{5.10.1*}$$

For example, if we randomly select 6 common stocks and record their closing prices at the end of a particular day, and these prices are 42, 57, 14, 24, 60, and 34, then the sample mean is

$$m = \frac{42 + 57 + 14 + 24 + 60 + 34}{6} = \frac{231}{6} = 38.5.$$

Thus, the sample mean is a simple arithmetic average. When we are working with a grouped distribution, the basic idea is the same, but each observation is assumed to equal the midpoint of the interval in which it lies. As a result, each midpoint is included in the calculation of m as many times as the frequency in that interval. If x_j is the midpoint of interval j and f_j is the frequency corresponding to interval j, then the sample mean can be computed for a grouped distribution as follows:

$$m = \frac{\sum_{j} x_j f_j}{n}, \tag{5.10.2}$$

where the summation is taken over all intervals. For the grouped distribution presented in Section 5.5,

$$m = \frac{\begin{array}{c}84.5(2) + 90.5(23) + 96.5(22) + 102.5(65) + 108.5(20) \\ + \ 114.5(10) + 120.5(0) + 126.5(8)\end{array}}{150}$$

$$= \frac{15{,}363}{150} = 102.42.$$

The mean calculated from a distribution with grouped class intervals need not agree exactly with the mean calculated from raw scores. Information is lost and a certain amount of inaccuracy introduced when scores are grouped and treated as though each corresponded to the midpoint of some interval. The coarser the grouping, in general, the more likely is the distribution mean to differ from the raw-score mean. For most practical work the rule of 10 to 20 class intervals gives relatively good agreement. Nevertheless, it is useful to think of the mean calculated from any given distribution as the mean of that *particular* distribution, a particular set of groupings with their associated frequencies.

The median for a set of observed data is ordinarily defined in slightly different ways depending upon whether n is odd or even and upon whether raw data or a grouped frequency distribution is to be described. For a set of raw scores,

when n is odd and the n values are arranged in numerical order, the median corresponds to the $[(n + 1)/2]$th value in order; when n is even and the n values are arranged in numerical order, the median is defined as the number midway between the $(n/2)$th value and the $[(n/2) + 1]$th value.

Then either for odd or even n it will be true that exactly as many values fall above as fall below the median. In the above example concerning the closing prices of common stocks, we must first arrange the prices in order: 14, 24, 34, 42, 57, 60. Now, since n is even, the median is midway between the $(n/2)$th value, or third value, and the $[(n/2) + 1]$th value, or fourth value. Thus, the median is midway between 34 and 42, so it is equal to $(34 + 42)/2$, or 38.

For a grouped distribution, the computation of the median is slightly more involved. First of all, it is necessary to find the interval in which the median is located. This can be done by constructing a cumulative frequency distribution and finding the interval for which the cumulative frequency first equals or exceeds $n/2$. For example, in the cumulative frequency distribution presented in Section 5.9, $n = 150$. For the interval from 94 to 99, the cumulative frequency is 47, which is less than $n/2 = 75$. For the next interval, from 100 to 105, however, the cumulative frequency is 112, which is greater than 75. Therefore, the median lies in the interval from 100 to 105, or using the real limits of the interval, the median is between 99.5 and 105.5.

Since we want a single value for the median, we could simply take the midpoint of the interval, which is 102.5. Recalling that this is a grouped distribution, however, we can probably attain greater accuracy by interpolating within the interval. Since the median should be greater than or equal to 75 cases, it must exceed not only the 47 cases below 99.5, but also 28 of the 65 cases between 99.5 and 105.5. But we do not know the exact values of the 65 cases in question, so a reasonable procedure is to say that the median is 28/65 of the way up the interval from the lower limit. In other words, the median is equal to $99.5 + (28/65)(6)$, or 102.08. This procedure can be summarized by the following formula:

$$\text{Median} = \text{lower real limit} + \left[\frac{(n/2) - cf \text{ below lower limit}}{f \text{ in interval}} \right] i, \quad (5.10.3)$$

where the lower real limit used belongs to the interval containing the median, and the cf refers to the cumulative frequency *up to* the lower limit of the interval. For the example,

$$\text{Median} = 99.5 + \left[\frac{(150/2) - 47}{65} \right] (6) = 102.08.$$

In principle, a median may be found for any distribution in which the variable represents an interval or even an ordinal scale; it is not, in general, applied to a distribution in which the measurement classes are purely categorical, since such classes are unordered.

The **mode** is certainly the most easily computed and the simplest to interpret of all the measures of central tendency. It is merely the midpoint or class name of the most frequent measurement class. If a case were drawn at random from the n observations, then that case is more likely to fall in the **modal class** than any other. So it is that in the graph of any distribution, the modal class shows the highest "peak" or "hump" in the graph. The mode may be used to describe any distribution, regardless of whether the events are categorized or numerical.

Comparisons among the sample mean, the sample median, and the sample mode are analogous to comparisons among the mean, median, and mode of a probability distribution. Such comparisons were discussed in some detail in Sections 3.11–3.14, so we will not repeat them here. A final point of interest is that we have considered the sample mean, median, and mode as descriptive measures of central tendency of sets of data; in Chapter 6 we will consider them as "estimates" of population parameters, and we will see that the sample mean is frequently the most useful measure in this context.

5.11   MEASURES OF DISPERSION FOR FREQUENCY DISTRIBUTIONS

As we pointed out in Section 3.15, an important feature (in addition to central tendency) of a distribution is dispersion, or spread, or variability. For probability distributions, we generally measure this dispersion with the variance or its square root, the standard deviation. Recall that the variance is the expectation of the squared deviation about the mean, $E(X - \mu)^2$.

We define the sample variance to be the average of the squared deviations of the sample values from the sample mean:

$$s^2 = \frac{\sum_{i=1}^{n} (x_i - m)^2}{n}. \tag{5.11.1*}$$

Note that s^2 will be zero if and only if each and every observation has the same value. The more the observations differ from each other and from m, the larger s^2 will be.

Since the sample variance involves squared terms, it has the disadvantage of not being in the same units as the observations and the sample mean. This problem can be resolved by taking the square root of the sample variance, which is the sample standard deviation, s. The sample variance and the sample standard deviation can be computed from a set of data by Equation (5.11.1). This entails finding the mean, subtracting it successively from each score, squaring each result, adding the squared deviations together, and dividing by n to find the variance. Obviously, this procedure is relatively laborious for a sizable number of cases. However, it is possible to simplify the computations by a few algebraic manipulations of the original formulas. For any single deviation, expanding the square gives

$$(x_i - m)^2 = x_i^2 - 2x_i m + m^2,$$

so that, on averaging these squares, we have

$$\sum_i \frac{(x_i - m)^2}{n} = \sum_i \frac{(x_i^2 - 2x_i m + m^2)}{n}$$

$$= \sum_i \frac{x_i^2}{n} - 2 \sum_i \frac{x_i m}{n} + \sum_i \frac{m^2}{n}.$$

However, wherever m appears it is a constant over the sum, and so

$$\sum_i \frac{(x_i - m)^2}{n} = \sum_i \frac{x_i^2}{n} - 2m \sum_i \frac{x_i}{n} + \frac{nm^2}{n},$$

or

$$s^2 = \sum_i \frac{x_i^2}{n} - 2m^2 + m^2 = \frac{\sum_i x_i^2}{n} - m^2. \qquad (5.11.2^*)$$

Finally,

$$s = \sqrt{\frac{\sum_i x_i^2}{n} - m^2}. \qquad (5.11.3^*)$$

These last formulas give a way to calculate the indices s^2 and s with some savings in steps. Another convenient formula for calculating s^2 is

$$s^2 = \frac{n \sum x_i^2 - (\sum x_i)^2}{n^2}. \qquad (5.11.4^*)$$

To illustrate the computation of s^2, consider once again the example from Section 5.10 involving the closing prices of common stocks. Table 5.11.1 is useful in demonstrating the application of Equations (5.11.1), (5.11.2), and (5.11.4) to determine s^2.

For a grouped distribution, each observed value that lies in a particular interval is treated as though it equaled the midpoint of that interval. The sample variance can then be computed as follows:

$$s^2 = \frac{\sum_j (x_j - m)^2 f_j}{n}, \qquad (5.11.5)$$

where x_j is the midpoint of interval j and f_j is the frequency corresponding to interval j. For the grouped distribution presented in Section 5.5, we have already calculated $m = 102.42$ (this calculation is shown in Section 5.10). Now, s^2 can be calculated as in Table 5.11.2. A more convenient

Table 5.11.1 Computation of s^2

x_i	x_i^2	$x_i - m$	$(x_i - m)^2$
42	1764	3.5	12.25
57	3249	18.5	342.25
14	196	-24.5	600.25
24	576	-14.5	210.25
60	3600	21.5	462.25
34	1156	-4.5	20.25
$231 = \sum x_i$	$10{,}541 = \sum x_i^2$	0	$1647.50 = \sum (x_i - m)^2$

$$m = \frac{\sum x_i}{n} = \frac{231}{6} = 38.5$$

$$s^2 = \frac{\sum (x_i - m)^2}{n} = \frac{1647.50}{6} = 274.58$$

$$s^2 = \frac{\sum x_i^2}{n} - m^2 = \frac{10{,}541}{6} - (38.5)^2 = 274.58$$

$$s^2 = \frac{n \sum x_i^2 - (\sum x_i)^2}{n^2} = \frac{6(10{,}541) - (231)^2}{(6)^2} = 274.58$$

Table 5.11.2 Computation of s^2 for Grouped Distribution

x_j	f_j	$x_j - m$	$(x_j - m)^2$	$(x_j - m)^2 f_j$
84.5	2	-17.92	321.1264	642.2528
90.5	23	-11.92	142.0864	3267.9872
96.5	22	-5.92	35.0464	771.0208
102.5	65	.08	.0064	.4160
108.5	20	6.08	36.9664	739.3280
114.5	10	12.08	145.9264	1459.2640
120.5	0	18.08	326.8864	.0006
126.5	8	24.08	579.8464	4638.7712
				11,519.0400

$$s^2 = \frac{\sum (x_j - m)^2 f_j}{n} = \frac{11{,}519.04}{150} = 76.79$$

computational form for calculating s^2 for grouped distributions is

$$s^2 = \frac{\sum x_j^2 f_j}{n} - m^2. \tag{5.11.6}$$

Recall that the variance of a probability distribution is calculated in terms of squared deviations from μ because the expected squared deviation is smallest when the deviations are taken from μ. Similarly, for the sample variance, the average squared deviation is smallest when the deviations are taken from m. However, if we were to consider the average *absolute* deviation rather than the average *squared* deviation, we would find that the average absolute deviation is smallest when the deviations are taken from the sample median:

$$\text{Sample A.D.} = \sum_i \frac{\mid x_i - \text{median} \mid}{n}. \tag{5.11.7}$$

In addition to the sample variance and the sample absolute deviation, there are other measures of dispersion, such as the range (the difference between the highest and lowest observations) and the interquartile range (the difference between the .75 fractile and the .25 fractile of the frequency distribution). In general, however, the sample variance is the most useful of these measures of dispersion. It should be noted at this point that in Chapter 6 we will have cause to use a "modified" sample variance, the modification being a change in the denominator from n to $n - 1$:

$$\hat{s}^2 = \frac{\sum (x_i - m)^2}{n - 1}. \tag{5.11.8*}$$

To differentiate this "modified" sample variance from s^2, we shall denote it by $\hat{s}^2$. Strictly speaking, this is not an "average" squared deviation; we will see in Chapter 6 why it is sometimes considered more useful than s^2. Incidentally, computing forms can be derived for $\hat{s}^2$ as well as for s^2:

$$\hat{s}^2 = \frac{n \sum x_i^2 - (\sum x_i)^2}{n(n - 1)}, \tag{5.11.9}$$

or

$$\hat{s}^2 = \frac{\sum_i x_i^2}{n - 1} - \left(\frac{n}{n - 1}\right) m^2. \tag{5.11.10}$$

5.12 POPULATIONS, PARAMETERS, AND STATISTICS

Recall that a population has been defined as the set of objects from which a sample is taken. Given a population of potential observations, we shall think of the particular numerical score assigned to any particular unit of observation as a value of a random variable; the distribution of this random variable is the **population distribution.** This distribution will have some mathematical form, with a mean μ, a variance σ^2, and all the other characteristic features of a distribution. If you like, you can usually think of the population distribution as a frequency distribution based upon some large but finite number of cases. However, population distributions are almost always discussed as though they were theoretical probability distributions; the process of random sampling of single units with replacement insures that the long-run relative frequency of any value of the random variable is the same as the probability of that value. Later we shall have occasion to idealize the population distribution and treat it as though the random variable were continuous. This is virtually impossible for real-world observations, but we shall assume that it is "true enough" as an approximation to the population state of affairs.

Population values such as μ and σ^2 will be called **parameters of the population.** Strictly speaking, a parameter is a value entering as an arbitrary constant in the particular function rule for a probability distribution, although the term is used more loosely to mean any value summarizing the population distribution. Just as parameters are characteristics of populations, **statistics** are summary measures of samples. The sample mean m, the sample variance s^2, and so on, are examples of statistics.

In Chapters 2–4, we dealt with population distributions and parameters of populations. So far in this chapter, we have discussed frequency distributions and statistics that summarize such distributions. Our concern has been purely descriptive, as we have been using statistics to describe samples. In this book, however, we are more concerned with inferential statistics and decision making than with descriptive statistics. In order to make inferences from a sample to a population, we must somehow relate sample statistics, particularly the sample mean and variance, to their parameter counterparts, the population mean and variance. To do this, we need to introduce the notion of a sampling distribution.

5.13 SAMPLING DISTRIBUTIONS

In discussing statistics such as the sample mean and the sample variance, our primary concern has been with the calculation of values of these statistics once the sample has been observed. Hence, we have used lower case letters to designate the calculated values of the statistics. *Before* the sample is observed, however, a statistic is regarded as a random variable with a particular probability distribution. For instance, the number of successes in a sample of size n from a Bernoulli process is a statistic, and before the sample is taken this statistic is a random variable with a binomial distribution, as we saw in Chapter 4. Similarly, before a sample is taken, the sample mean is a random variable, so we can designate it with a capital M. Just as R has a binomial sampling distribution in the case of a Bernoulli process, M has a probability distribution that depends upon the population or process generating the sample data.

Probability distributions of sample statistics such as R and M must obey the usual laws of probability, like any other probability distributions. In order to distinguish probability distributions of sample statistics from probability distributions of random variables that are not sample statistics, we give the former distributions a special label: **sampling distributions.**

A sampling distribution is a theoretical probability distribution that shows the functional relationship between the possible values of some sample statistic and the probability (density) associated with each value of the sample statistic over all possible samples of a particular size from a particular population.

For a sample of size 1, the value of the single item sampled may be thought of as a sample statistic, in which case the sampling distribution is simply the population distribution. In general, the sampling distribution of a sample statistic will not be the same as the population distribution. We shall see, however, that *the sampling distribution always depends in some specific way upon the population distribution.*

We have already used sampling distributions in the preceding chapters. For example, a binomial distribution is a sampling distribution. Recall that a binomial distribution is based on a two-category population, or a Bernoulli process. A sample of n cases is drawn at random with replacement from such a process, and the number (or proportion) of successes is calculated. The binomial distribution is the sampling distribution showing the relation between each possible sample result and the theoretical probability of occurrence.

Another example of a sampling distribution already introduced is the multinomial distribution. Here, the population has several event categories. A sample of size n is drawn at random with replacement and summarized according to how many of each measurement class occur. The multinomial distribution gives the probability of each possible sample.

Other examples of sampling distributions will be given in this book. A most important distribution we shall employ is the sampling distribution of the sample mean. Here, samples of size n are drawn at random from some population, and each observation is measured numerically. The population distribution has some mean μ, and for each sample, the sample mean m is calculated. **The theoretical distribution that relates the possible**

values of the sample mean to the probability (density) of each over all possible samples of size n is called the sampling distribution of the sample mean. Furthermore, there is a population variance, σ^2. For each sample of size n, the sample variance s^2 is found. The theoretical distribution of sample variances is the sampling distribution of the sample variance. By the same token, *the sampling distribution of any summary characteristic (mode, median, range, and so on) of a sample may be found. This fact is important for the theory of statistics, as it provides ways to make inferences about population parameters on the basis of random samples, in terms of the probability of a sample statistic's value arising by chance from a certain population.* Much of the material in the remainder of this book will deal with various ways of using theoretical sampling distributions to make such inferences.

To clarify further the notion of a sampling distribution, suppose that we are interested in family incomes in a given city. Even if we know precisely the exact population distribution of family incomes, we cannot be certain what the outcome of a sample will be. Suppose that we take a random sample of 100 families and record the income of each family. For this sample, we can calculate the sample mean, sample variance, and any other statistic of interest. Now suppose that we take a second random sample of 100 families and repeat the procedure; this will yield another (most likely different) sample mean, another sample variance, and so on. If we took more and more samples of size 100, we would build up a frequency distribution of the various sample means and another frequency distribution of the various sample variances. As the number of random samples we take gets larger, these frequency distributions will get closer and closer to the corresponding sampling distributions. Of course, this is a very time-consuming way to develop a sampling distribution, so we will develop procedures for determining the sampling distributions of statistics such as the sample mean and sample variance directly from the population distribution for certain situations.

It is a good idea to reiterate at this point that three distinct kinds of distributions have been introduced. The first was the **population distribution,** which is a theoretical distribution describing the probability associated with various values of a random variable for some given population. The second kind of distribution with which we dealt was the **frequency distribution,** or the **sample distribution.** This is a distribution summarizing a set of data by describing the frequencies associated with various classes, based on a randomly chosen subset of a population. Finally, there was the **sampling distribution.** This is a theoretical probability distribution which relates various values or intervals of values of some *sample statistic* to their probabilities of occurrence over all possible samples. In order to know any sampling distribution exactly, we must specify the population distribution. It is very important to bear these distinctions in mind, since all three kinds of distributions will be considered in the following sections.

5.14 THE MEAN AND VARIANCE OF A SAMPLING DISTRIBUTION

Except for a few statistical curiosities that need not concern us here, any random variable has a determinable mean and variance. Since a sample statistic is a random variable, the mean and variance of any sampling distribution are defined in the usual way. That is, let G be any sample statistic; then if the sampling distribution of G is discrete, its expectation or mean is

$$E(G) = \mu_G = \sum gP(g).$$ (5.14.1*)

If the variable G is continuous, then

$$E(G) = \mu_G = \int_{-\infty}^{\infty} gf(g)\, dg,$$ (5.14.2*)

just as for any other continuous random variable.

In the same way, the variance of a sampling distribution for the statistic G can be defined:

$$\sigma_G^2 = E(G - \mu_G)^2,$$ (5.14.3*)

or $$\sigma_G^2 = E(G^2) - [E(G)]^2.$$ (5.14.4*)

The variance of the statistic G gives a measure of the dispersion of particular sample values about the average value of G over all possible samples of size n. The standard deviation σ_G of the sampling distribution reflects the extent to which sample G values tend to be *unlike* the expectation, or are *in error*. To aid in distinguishing the standard deviation of a sampling distribution from the standard deviation of a population distribution, a standard deviation such as σ_G is usually called the **standard error** of the statistic G. When we speak of the standard error of the sample mean, we are referring to the standard deviation of the distribution of possible sample means for all possible samples of size n drawn from a specific population. Similar meanings hold for the standard error of the sample median, the standard error of the sample standard deviation, and so on.

For an example of the determination of the mean and variance of a sampling distribution, consider the sampling distribution of the sample mean, M. Here

$$G = M = \frac{\sum\limits_{i=1}^{n} X_i}{n}.$$

Note that since we are considering the sampling distribution of the sample mean *before* the sample is actually observed, the values in the sample and the sample mean are random variables. Therefore, they are designated by capital letters. Using the algebra of expectations, we find that

$$E(M) = E\left(\frac{\sum\limits_{i=1}^{n} X_i}{n}\right) = \frac{1}{n} E(\sum\limits_{i=1}^{n} X_i) = \frac{1}{n} \sum\limits_{i=1}^{n} E(X_i).$$

But each X_i is simply an observation chosen randomly from the population of interest, so the distribution of each X_i is the population distribution. Thus, if the population mean is designated by μ, as usual, then $E(X_i) = \mu$ for any i, so that

$$E(M) = \frac{1}{n} \sum\limits_{i=1}^{n} \mu = \frac{n\mu}{n} = \mu. \tag{5.14.5*}$$

This statement can be interpreted as follows: **the mean of the sampling distribution of sample means is the same as the population mean.** However, it is not true that the sampling distribution of the mean will be the same as the population distribution; in fact, these distributions will ordinarily be quite different, depending particularly on the sample size.

Intuitively, it seems quite reasonable that the larger the sample size, the more confident we may be that the sample mean is a close estimator of μ. Now we can put that intuition on a firm basis by looking into the effect of sample size on the variance and standard deviation of the distribution of sample means. Using the rules concerning variances presented in Section 3.15,

$$V(M) = V\left[\frac{\sum\limits_{i=1}^{n} X_i}{n}\right] = \frac{1}{n^2} V(\sum\limits_{i=1}^{n} X_i) = \frac{1}{n^2} \sum\limits_{i=1}^{n} V(X_i).$$

But the distribution of each X_i is the population distribution, so if σ^2 stands for the population variance, as usual, then $V(X_i) = \sigma^2$ for each i.

Therefore,

$$V(M) = \frac{1}{n^2} \sum_{i=1}^{n} \sigma^2 = \frac{n\sigma^2}{n^2} = \frac{\sigma^2}{n}.$$

The variance of the sampling distribution of sample means for independent samples of size n is always the population variance divided by the sample size:

$$\sigma_M{}^2 = \frac{\sigma^2}{n}. \qquad (5.14.6^*)$$

It follows directly that the standard error of the sample mean is given by

$$\sigma_M = \sqrt{\sigma_M{}^2} = \frac{\sigma}{\sqrt{n}}. \qquad (5.14.7^*)$$

This is a most important fact, and it gives direct support to our feeling that large samples produce more precise estimators of the population mean than do small samples. When the sample size is only 1, then the variance of the sampling distribution is exactly the same as the population variance. If, however, the sample mean is based on two cases, $n = 2$, then the sampling variance is only $1/2$ as large as σ^2. Ten cases give a sampling distribution with variance only $1/10$ of σ^2, $n = 500$ gives a sampling distribution with variance $1/500$ of σ^2, and so on. If the sample size approaches infinity, then σ^2/n approaches zero. **If the sample is large enough to embrace the entire population, there is no difference between the sample mean and μ.**

In general, the larger the sample size, the more probable it is that the sample mean comes arbitrarily close to the population mean. This fact is closely allied both to the law of large numbers that was discussed in Section 2.13 and the Chebyshev inequality that was discussed in Section 3.19. From the Chebyshev inequality, it is true that for any random variable X,

$$P(\,|\,X - \mu\,| < k) \geq 1 - \frac{\sigma^2}{k^2}, \qquad k > 0. \qquad (5.14.8)$$

Let the random variable be M, and let the variance be $\sigma_M{}^2$. Then

$$P(\,|\,M - \mu\,| < k) \geq 1 - \frac{\sigma_M{}^2}{k^2}. \qquad (5.14.9^*)$$

When n becomes very large, $\sigma_M{}^2$ approaches zero. Thus, regardless of how small k is, the probability approaches 1 that the value of M will be within k units of μ when the sample size grows large.

Often it is useful to transform the sample mean M into standardized form, as follows:

$$Z = \frac{M - \mu_M}{\sigma_M} = \frac{M - \mu}{\sigma/\sqrt{n}}. \qquad (5.14.10^*)$$

For example, suppose that the population distribution of the age of employees of a large firm has a mean of 32 years and a standard deviation of 15 years. For a sample of 25 employees randomly selected from the employees of the firm, $E(M) = 32$ and $V(M) = (15)^2/25 = 9$. If the sample is taken and the sample mean turns out to be 35, the corresponding standardized value is

$$z = \frac{35 - 32}{15/\sqrt{25}} = \frac{3}{3} = 1.$$

If the sample size is 2500 instead of 25, but everything else is unchanged, then

$$z = \frac{35 - 32}{15/\sqrt{2500}} = \frac{3}{.3} = 10.$$

In both cases, the sample mean is exactly 3 units away from the population mean. When the sample size is larger, however, the standardized value is much farther from zero, suggesting that we should be surprised to observe a sample mean 3 units away from μ when the sample size is so large. In fact, for a sample of size 2500 in this situation, the Chebyshev inequality tells us that

$$P(\,|M - \mu| < 3) \geq 1 - \frac{\sigma_M{}^2}{9} = 1 - \frac{(.3)^2}{9} = .99.$$

The probability that the difference between M and μ is less than 3 units is greater than or equal to .99, so a sample mean of 35 in this instance is indeed an unusual event. We will find that reasoning of this nature can be quite valuable in making inferences concerning the population mean, μ.

5.15   STATISTICAL PROPERTIES OF NORMAL POPULATION DISTRIBUTIONS: INDEPENDENCE OF SAMPLE MEAN AND VARIANCE

As suggested earlier, the normal distribution has mathematical properties that are most important for theoretical statistics. For the moment, we shall discuss only two of these general properties, both of which will be useful to know in later sections. The first has to do with the independence of the sample mean and variance, and the second concerns the distribution of combinations of random variables.

Any sample consisting of n independent observations of the same random variable provides both a sample mean and a sample variance. These two values obtained from any sample can be thought of as independent when the population distribution is normal.

> **Given random and independent observations, the sample mean and the sample variance are independent if and only if the population distribution is normal.**

The information contained in the sample mean in no way dictates the value of the sample variance, and vice versa, when a normal population is sampled. Furthermore, **unless the population actually is normal, these two sample statistics are not independent across samples.**

This is a most important principle, since a great many problems concerned with a population mean can be solved only if we know something about the value of the population variance. At least an estimate of the population variance is required before particular sorts of inferences about the value of μ can be made. Unless the estimate of the population variance is statistically independent of the estimate of μ made from the sample, no simple way to make these inferences may exist. This question will be considered in more detail in Chapters 6 and 7. For the moment, suffice it to say that this principle is one of the reasons why statisticians assume normal population distributions.

5.16 DISTRIBUTIONS OF LINEAR COMBINATIONS OF VARIABLES

So far we have emphasized the sample mean as a description of a particular set of data and as an estimator of the corresponding population mean. However, in many situations the sample mean and the mean of the population do not really tell us what we want to find out from the data. It well may be that other ways of weighting and summing the scores or means obtained in one or more samples will answer particular questions about the phenomena under study. Therefore, we need to know something about the sampling distributions of weighted combinations of sample data.

Such weighted sums of sample scores are thought of as values of **linear combinations of random variables.** As a very simple case of a linear combination, imagine two distinct random variables, labeled X_1 and X_2 to show that their values need not be the same. We draw samples of two observations at a time, obtaining one value of X_1 and one of X_2. Then we combine these two numbers into a new value Y in some way such as

$$Y = 3X_1 - 2X_2.$$

We continue to sample (X_1,X_2) pairs of values, and each time we combine the results in the same way, turning each pair of values into a single combined value. This gives a new random variable Y. The range of possible values of Y depends, as you can see, on the ranges of X_1 and X_2. Furthermore, over all possible such samples the probabilities of the various Y values must depend on the *joint* probabilities of (X_1,X_2) pairs. Over all possible samples of (X_1,X_2) pairs drawn at random, there is some probability distribution of Y.

In general, given any n random variables, $(X_1, \ldots, X_n)$, a *linear combination* of values of these variables is a weighted sum,

$$Y = c_1X_1 + c_2X_2 + \cdots + c_nX_n = \sum_{i=1}^{n} c_iX_i, \qquad (5.16.1^*)$$

where $(c_1, \ldots, c_n)$ is any set of n real numbers (not all zero) used as weights. The weights in a linear combination can be any set of n real numbers, provided that at least one weight is not zero. For example, suppose that an investor owns 50 shares of security A and 100 shares of security B. The investor intends to sell both securities exactly one year from now because at that time there will be an opportunity to invest in what appears to be a very profitable real estate venture. Let X_1 represent the change in price of security A during the coming year, and let X_2 represent the change in price of security B during the coming year. The investor's profit during the coming year from the portfolio of two securities is therefore a linear combination. Letting Y represent this profit, we have

$$Y = 50X_1 + 100\,X_2.$$

Next, we consider an important principle involving the normal distribution.

If the n random variables $X_1, \ldots, X_n$ are normally distributed, then any linear combination of these variables is also a normally distributed random variable.

In other words, if we take a sample, obtaining n values, and if the random variable represented by each value has a normal distribution, then any weighted sum of these values is another normally distributed random variable. In the above example concerning a portfolio of securities, if

X_1 and X_2 have normal distributions, then so does Y, the profit from the portfolio. Since normal distributions are often convenient to work with and since linear combinations of variables are encountered quite frequently, the above principle is extremely useful.

One linear combination of particular interest is the sample mean, M. This is a linear combination of n variables, with each weight equal to $1/n$. Hence, an immediate and useful consequence of the above principle is as follows.

Given random samples of n independent observations, each drawn from a normal population, the distribution of the sample means is normal, irrespective of the size of n.

We shall make a great deal of use of this principle. When we can assume that a population with known μ and σ has a normal distribution, a major problem in statistical inference is solved, since we can then give a probability to any interval of values of the sample mean in terms of a normal distribution. This will be illustrated in the next chapter.

While on the subject of linear combinations, we should look into the question of the mean and variance of the sampling distribution of a linear combination of random variables. These two principles do not depend on the normal distribution, however. **Given n random variables and some linear combination,**

$$Y = c_1 X_1 + \cdots + c_n X_n = \sum_{i=1}^{n} c_i X_i,$$

the expected value of Y over all random samples is

$$E(Y) = E(\sum c_i X_i) = \sum E(c_i X_i) = \sum c_i E(X_i)$$
$$= c_1 E(X_1) + c_2 E(X_2) + \cdots + c_n E(X_n). \quad (5\ 16.2^*)$$

The expectation of any linear combination is the same linear combination of the expectations. This follows quite simply from the algebra of expectations, since the expectation of a sum is always a sum of expectations, and the expectation of a constant times a variable is the constant times the expectation of the variable.

In the portfolio example, suppose that the expected change in price for security A is 15 and that the expected change in price for security B is 20. Then the expected profit from the portfolio is

$$E(Y) = 50E(X_1) + 100E(X_2) = 50(15) + 100(20) = 2750.$$

The variance of the sampling distribution of any linear combination can be found by the following rule.

Given n independent random variables with variances $\sigma_1^2, \sigma_2^2, \ldots, \sigma_n^2$, respectively, and the linear combination

$$Y = c_1X_1 + c_2X_2 + \cdots + c_nX_n = \sum_{i=1}^{n} c_iX_i,$$

the distribution of Y has variance given by

$$\sigma_Y^2 = V(\sum c_iX_i) = \sum V(c_iX_i) = \sum c_i^2 V(X_i)$$
$$= c_1^2\sigma_1^2 + c_2^2\sigma_2^2 + \cdots + c_n^2\sigma_n^2. \tag{5.16.3*}$$

The variance of a linear combination of independent random variables is a weighted sum of their separate variances, each weight being the *square* of the original weight given the variable.

In the portfolio example, if $V(X_1) = 100$ and $V(X_2) = 400$, implying that security B is much more volatile than security A, then the variance of the profit from the portfolio is

$$V(Y) = (50)^2 V(X_1) + (100)^2 V(X_2)$$
$$= (2500)(100) + (10,000)(400) = 4,250,000.$$

If this number seems unusually large, remember that it is a variance, which is expressed in squared units. The standard deviation of Y is approximately 2062. From this fact and the information given above, we now know that Y has a mean of 2750 and a standard deviation of 2062. Notice that the calculation of these values required knowledge of the means and variances of X_1 and X_2, but *not* knowledge of the exact form of the distributions of X_1 and X_2. If, in addition, X_1 and X_2 happen to be normally distributed, then the principle given earlier in this section tells us that Y is also normally distributed. In this case, we can use the table of cumulative normal probabilities to calculate probabilities for any intervals of values of Y.

As noted earlier in this section, linear combinations are encountered quite frequently. The sample mean is a primary example, and the portfolio example demonstrates that a variable of interest (here, profit) can often be expressed as a linear combination of some other variables. Another type of linear combination of considerable interest is a simple difference, $Y = X_1 - X_2$. Often we are interested in differences between two populations, such as the difference in the speeds of two machines or the difference in the IQs of two groups of students.

Whatever may motivate our interest in a particular linear combination of random variables, the results of this section provide us with a means of investigating the distribution of the linear combination of interest. We can easily determine the mean and variance, given the means and variances of the variables forming the linear combination. Even more important, if the individual variables are normally distributed, we know that the linear combination is also normally distributed. Of particular interest is that in random sampling from a particular population, if the population distribution is normal, then the sampling distribution of M is also normal.

5.17   THE CENTRAL LIMIT THEOREM

It is quite common for the statistician to be concerned with populations in which the distribution should *definitely not be normal*. We may know this either from empirical evidence about the distribution or because some theoretical issue makes it impossible for the random variable to be normally distributed. An illustration is a distribution of the annual income of individuals, which is generally severely positively skewed. The normal distribution is symmetric, so the assumption of a normal distribution does not make sense in such instances.

Nonetheless, often an inference must be made about the mean of such a population. To do this effectively the statistician needs to know the sampling distribution of the mean, and to know this exactly, it is necessary to specify the particular form of the population distribution. However, if we had enough evidence to permit this, we would be likely to have an extremely good estimate of the population mean in the first place, and we would not need any other statistical methods!

The way out of this apparent impasse is provided by the central limit theorem, which can be given an approximate statement as follows.

If a population has a finite variance σ^2 and mean μ, then the distribution of the sample mean from a sample of n independent observations approaches a normal distribution with variance σ^2/n and mean μ as n increases. When n is very large, the sampling distribution of M is approximately normal.

Absolutely nothing is said in this theorem about the form of the population distribution. Regardless of the population distribution, if n is large enough, the normal distribution is a good approximation to the sampling distribution of the mean. This is the heart of the theorem, since, as we have already seen, the sampling distribution has variance σ^2/n and mean μ for any n.

The sense of the central limit theorem is illustrated by Figures 5.17.1 to 5.17.4. The solid curve in Figure 5.17.1 is a very skewed population distribution in standard form, and the broken curve is a standardized normal

Figure 5.17.1 A Negatively Skewed Population Distribution with a Comparable Normal Distribution

Figure 5.17.2 A Sampling Distribution of M for $n = 2$ with a Comparable Normal Distribution

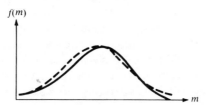

Figure 5.17.3 A Sampling Distribution of M for $n = 4$ with a Comparable Normal Distribution

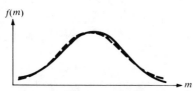

Figure 5.17.4 A Sampling Distribution of M for $n = 10$ with a Comparable Normal Distribution

distribution. The other figures show the standardized form of the sampling distribution of means for samples of size 2, 4, and 10, respectively, from this population, together with the corresponding standardized normal distributions.

Even with these relatively small sample sizes, it is obvious that each increase in sample size yields a sampling distribution more nearly symmetric and tending more toward the normal distribution with the same mean and variance. Any asymmetry decreases with increasing sample size until, in the limit, the normal distribution is reached.

It must be emphasized that in most instances the tendency for the sampling distribution of M to be like the normal distribution is very strong, even for samples of moderate size. Naturally, the more similar to a normal distribution the original population distribution is, the more nearly the sampling distribution of M will be like the normal distribution for any given sample size. However, even extremely skewed or other nonnormal distributions may yield sampling distributions of M that can be approximated quite well by the normal distribution for samples of at least moderate size. In the examples shown in Figures 5.17.1 to 5.17.4, the correspondence between the exact probabilities of intervals in the sampling distribution and intervals in the normal distribution is fairly good even for samples of only 10 observations; for a rough approximation, the normal distribution probabilities might be useful even here in some statistical work. In a great many instances, a sample size of 30 or more is considered large enough to permit a satisfactory use of the normal distribution to approximate the unknown exact probabilities associated with the sampling distribution of M. Thus, even though the central limit theorem is actually a statement about what happens *in the limit* as n approaches an infinite value, the principle at work is so strong that in many instances the theorem is practically useful even for moderately large samples.

The mathematical proof of this theorem is beyond the level of this book, and many eminent mathematicians contributed to its development before it was finally proved in full generality. However, some intuitive feel for why it should be true can be gained from the following example. Here we will actually work out the sampling distribution of the sample mean for a special and very simple population distribution.

Let X represent the number of courses taken by a full-time student at a particular university in the first semester of graduate school, and suppose that the distribution of X is as given in Table 5.17.1. The mean of this little distribution is 4. However, instead of dealing directly with this random variable, let us consider the deviation D of X from μ, or $D = X - \mu$. The distribution of D has a form identical to the distribution of X itself, although the mean D is zero. Since X can assume only three values, each

Table 5.17.1 Probability Distribution of Number of Courses Taken in First Semester of Graduate School

x	$P(x)$
3	1/3
4	1/3
5	1/3

equally probable, there are three possible values of D, also equally probable: -1, 0, and $+1$.

Suppose that sample observations are taken two at a time, independently and at random. The deviation value of the first observation made is D_1 and that of the second is D_2. Corresponding to each (D_1, D_2) pair, there is a mean deviation, $(D_1 + D_2)/2$, which is equal to $M - \mu$. Table 5.17.2 shows the mean deviation value that is produced by each possible joint event. Since X_1 and X_2 are independent, then D_1 and D_2 are also independent, according to Section 3.22. The probability associated with each cell in the table is thus $(1/3)(1/3) = 1/9$.

Now let us find the probabilities of the various values of the *mean* deviation. The value -1 can occur in only one way, and so its probability is $1/9$. On the other hand, $-.5$ can occur in two mutually exclusive ways, giving a probability of $2/9$. In this way we can find the other probabilities and form the distribution of the mean D and the mean X values, as in Table 5.17.3. Here $m = $ mean $d + \mu$. Now notice that whereas in the original distribution there was no distinct mode, in this sampling distribution there *is* a distinct mode at 4, with exact symmetry about this point. What causes the sampling distribution to differ from the population distribution in this way? Look at the table of mean deviations corresponding to joint events once again; there are simply more possible ways for a

Table 5.17.2 Mean Deviations for Sample of Size 2

D_1		-1	0	1
	1	0	.5	1
	0	$-.5$	0	.5
	-1	-1	$-.5$	0

D_2

Table 5.17.3   Distribution of Mean Deviation

m	mean d	$P(m) = P(mean\ d)$
5.0	1.0	1/9
4.5	.5	2/9
4.0	0	3/9
3.5	−.5	2/9
3.0	−1.0	1/9

sample to show a small than an extreme deviation from μ. There are more ways for a sample of two students to occur where the deviations from the average course load tend to "cancel out" than where deviations tend to accumulate in the same direction.

Exactly the same idea could be illustrated for any population distribution, whatever its form. The small and "middling" deviations of M from μ always have a numerical advantage over the more extreme deviations. This superiority in number of possibilities for a small mean deviation increases as n is made larger. Regardless of how skewed or otherwise irregular a population distribution is, by increasing the sample size we can make the advantage given to small deviations so big that it will overcome any initial advantage given to extreme deviations by the original population distribution.

Perhaps this simple example gives you some idea of why it is that the sampling distribution of the mean approaches a unimodal, symmetric form. Of course, this is a far cry from showing that the sampling distribution must approach a *normal* distribution, which is the heart of the central limit theorem. Nevertheless, the basic operation of chance embodied in this theorem is of this general nature; for large n it is much easier to get a small deviation of M from μ than a large one.

As suggested in the preceding section, it is very common for an experimenter to be interested in the means of several samples put into some kind of weighted sum, or linear combination. The central limit theorem applies to such weighted combinations as well.

Given K independent samples, containing $n_1, n_2, \ldots, n_K$ independent observations, respectively, the sampling distribution of any linear combination of means of those samples approaches a normal distribution as the size of each sample grows large.

It is easy to see that the central limit theorem applies to the sampling distribution of each mean separately as the sample size increases. In the limit, each mean has a normal sampling distribution. Then by the principle described in Section 5.16, any linear combination of normal random variables is normal, so that the sampling distribution of the linear combination of means approaches a normal distribution as all the sample sizes grow large.

This is but one of the most elementary extensions of the central limit theorem. It can be shown that a great many of the sampling distributions of all sorts of sample characteristics also approach a normal distribution with increasing sample size. However, in elementary work we will have most occasion to use the principle as it applies to means.

5.18 THE USES OF FREQUENCY AND SAMPLING DISTRIBUTIONS

Frequency distributions are useful primarily in a descriptive sense. That is, they describe the information contained in a particular sample result, in a particular set of observations. When we must report the results of a sample, we generally present a frequency distribution along with some summary measures such as the sample mean and sample variance. There are various ways to specify a frequency distribution and various summary measures of such distributions; these different presentations of the results may give entirely different impressions to the reader. A simple example of this was presented in Section 5.7, in which a slight change in the scales on the axes of a graph led to an entirely different impression. The moral is simple. It is very easy to "lie" with statistics, so it is a good idea to be somewhat wary when presented with the results of a statistical study. Conversely, it is important that the statistician attempt to present the data from such a study as impartially as possible. This means, for example, that when reporting the results of a study of family incomes in a certain community, the statistician should report the median income as well as the mean income, since the sample-mean is greatly affected by a few very high incomes.

In contrast to frequency distributions, sampling distributions are of prime importance in the theory of statistical inference and decision rather than as descriptive methods. When an estimate of a population quantity is desired, we will investigate qualities of competing estimators which are based on their sampling distributions. When the reasonableness of a hypothesis, or a theory, is of interest, inferences about the hypothesis are based directly on values computed from sampling distributions. Finally, in statistical decision theory, sampling distributions provide values that are important inputs to the decision-making process. We will make extensive use of summary measures (means and standard errors) of sampling distributions, probabilities associated with intervals of sample values, and entire sampling distributions, which we will usually attempt to specify in functional form.

It should be pointed out that the discussion of sampling distributions in this chapter is incomplete. For example, we have said nothing about the distribution of the sample mean when we do not know the population variance, partly because a discussion of this distribution requires the introduction of the t distribution, which we have put off until the next chapter. Similarly, we have not discussed the distribution of the sample variance because this involves the χ^2 (chi-square) distribution, which will also be introduced in the next chapter. Other sampling distributions of interest include the distribution of the difference of two sample means (either from independent samples or dependent samples), the distribution of the ratio of two sample variances, the distribution of the median, the distribution of the sample correlation coefficient, and so on. In the following chapters, the most important of these sampling distributions will be presented (insofar as they are within the scope of and can be presented at the mathematical level of this book), along with the implications which they hold for statistical inference and decision.

5.19   TO WHAT POPULATIONS DO OUR INFERENCES REFER?

How do we know the population to which the statistical inferences drawn from a sample apply? If random sampling is to be assumed, *the population is defined by the sample and the manner in which it is drawn. The only population to which the inferences strictly apply is that in which individuals have equal likelihood of appearing in the sample.* It should be obvious that simple random samples from one population may not be random samples from another population. For example, suppose that we wish to sample American college students. We obtain a directory of college students from a Midwestern university and, using a random number table, we take a sample of these students. We are not, however, justified in calling this a random sample of the population of American college students, although we may be justified in calling this a random sample of students at *that* university. *The population is defined not by what we say, but rather by what we do to get the sample.* For any sample, we should always ask the question, "What is the set of potential cases that could have appeared in the sample with equal probability?" If there is some well-defined set of cases which fits this qualification, then inferences may be made to that population. However, if there is some population whose members could not have been represented in the sample with equal probability, then inferences do not *necessarily* apply to that population when methods based on simple random sampling are used. Any generalization beyond the population actually sampled at random must rest on extrastatistical considerations.

From a mathematical statistics point of view the assumption of random sampling makes it possible to determine the sampling distribution of a particular statistic given some particular population distribution. If the various values of the statistic can arise from samples having undetermined or unknown probabilities of occurrence, then the statistician has no way to determine the sampling distribution of that statistic. On the other hand, if each and every distinct sample of n observations has equal probability of occurrence, as in simple random sampling, then the sampling distribution of any given statistic is relatively easy to study, given, of course, the population distribution. Hence, most of the inferential procedures to be discussed in this book utilize the assumption of random sampling. For the next few chapters, this will mean simple random sampling, and in Chapter 11 we will discuss some "fancier" types of random sampling.

EXERCISES

1. Carefully distinguish between descriptive statistics and inferential statistics, and for each of the following situations, indicate whether the approach is descriptive or inferential.
    (a) A statistician reports that for a random sample of families from a community, the average family income is $13,000 per year.
    (b) A sportswriter notes that a particular basketball player took 14 shots in his last game and made 10 of them.
    (c) A marketing manager states that on the basis of a random sample, it appears that Brand X is preferred to Brand Y by the consumers in a given city.
    (d) A firm conducts a survey of 1 percent of the registered voters in a state and predicts that Candidate A will win the senatorial election in that state.

2. Why are random samples of such importance in statistics?

3. A chain of stores has 4 outlets, labeled A, B, C, and D, in a particular area. The president of the chain wants to take a sample of 2 stores from these 4 stores. A fair coin is tossed; if it comes up heads, store A will be included in the sample, and if it comes up tails, store B will be included in the sample. A second fair coin is tossed; if it comes up heads, store C will be in the sample, and if it comes up tails, store D will be in the sample.
    (a) For each store, find the probability of that store being included in the sample.
    (b) Is this procedure simple random sampling? Explain your answer.

4. Distinguish between independent random sampling and nonindependent random sampling. By considering a simple experiment such as the drawing of a series of cards from a well-shuffled deck of cards, give examples of both independent and dependent random sampling. How does this distinction relate to the difference between the binomial and hypergeometric distributions?

5. To generate a table of numbers to be used in sampling, a pair of fair dice is tossed repeatedly, with the last digit of the *sum* of the two numbers appearing on the dice being recorded each time. The resulting table is then used to determine the members of a population that will be included in a sample, just as Table VIII in the Appendix would be used. Is the resulting sample a simple random sample? Explain your answer.

6. With a fair coin, we can generate a table of numbers in the following manner. Let $X_1 = 1$ if the coin comes up heads on the first toss and $X_1 = 0$ if it comes up tails. Similarly, let $X_2 = 1$ if it comes up heads on the second toss and let $X_2 = 0$ if it comes up tails. Define X_3 and X_4 similarly for the third and fourth toss. Now toss the coin 4 times, and find

$$Y = X_1 + 2X_2 + 4X_3 + 8X_4.$$

List the possible values of Y and the probabilities for the different values. Would you call Y a random number? Suppose that you simply ignored values of Y of 10 or greater. Would the resulting numbers be random digits?

7. In many institutions of higher education, a student's record is evaluated by a grade point average. For example, 4 points might be given for each semester hour of A, 3 for each hour of B, 2 for each hour of C, 1 for each hour of D, and 0 for each hour of F. These numerical values are multiplied by the number of hours of each grade earned, summed, and divided by the total number of hours to arrive at a grade point average. Thus, a grade point average of near 4.00 indicates an excellent record and a grade point average below 2.00 indicates a poor record. In light of the discussion of measurement scales given in the text, discuss the merits of this procedure.

8. Indicate the level of measurement you would expect to attain when measuring the following variables:

(a) temperature, (e) brand preference,
(b) income, (f) academic achievement,
(c) sex, (g) color of hair,
(d) height, (h) occupation.

9. A "taste-testing" survey in which respondents are asked to taste two dishes of ice cream and to rate the taste of each on a scale from 1 to 7 is conducted. For each respondent, the rating associated with the dish of ice cream produced according to a new procedure is divided by the rating associated with the dish of ice cream produced according to the old procedure. Next, the average of these ratios is computed. Comment on this procedure.

10. In Exercise 9, it is claimed that if the average ratio is greater than 1, this indicates that on the average, the respondents preferred the ice cream produced according to the new procedure. Does this claim seem reasonable? Why or why not?

11. In a learning experiment, cats were given successive trials at learning a complicated maze. A given cat was trained until it could traverse the maze without error 5 times in a row. If the cat could not go through the maze 5 times successfully after 30 trials, training was stopped. The numbers of trials it took each of 36 cats to learn this maze is given below. Construct a frequency distribution for these data.

Cat	Trials	Cat	Trials	Cat	Trials
1	5	13	30+	25	16
2	12	14	7	26	5
3	7	15	30+	27	30+
4	15	16	9	28	9
5	9	17	30+	29	30+
6	10	18	10	30	19
7	30+	19	13	31	8
8	30+	20	15	32	20
9	8	21	30+	33	30+
10	30+	22	6	34	17
11	6	23	30+	35	30+
12	30+	24	11	36	22

Give a qualitative description of the performance of these animals in the maze on the basis of what this distribution shows.

12. In a pre-election poll, registered voters selected at random were asked whom they would vote for in an election with 5 candidates. Let the candidates be denoted by the letters A, B, C, D, and E; let F represent the response "undecided;" and let G represent the response "don't intend to vote." A sample of 100 voters yielded the following results:

C, A, C, C, E, B, C, A, B, C, G, D, C, C, C, C, A, G, A, B, C, C, E, A, F,
D, C, C, A, F, A, A, A, C, B, C, D, D, A, A, C, A, B, A, C, C, A, E, B, C,
G, G, C, A, B, A, C, C, C, D, C, C, A, E, B, F, C, C, D, C, A, A, C, A, B,
A, A, C, A, B, A, C, C, C, A, A, C, B, D, D, F, A, E, G, C, C, A, B, C, C.

Construct a frequency distribution and display this distribution graphically in terms of a histogram.

13. In one week, 30 patients were admitted to a large state hospital. A list of the patients classified by sex and by tentative diagnosis follows. For each

sex form a frequency distribution by type of diagnosis and display this information via a pair of histograms superimposed on the same graph.

Patient	Sex	Diagnosis	Patient	Sex	Diagnosis
1	M	Senile dementia	16	M	Senile dementia
2	F	Schizophrenia	17	F	Senile dementia
3	M	Schizophrenia	18	M	Manic-depressive
4	M	Manic-depressive	19	F	Schizophrenia
5	F	Senile dementia	20	M	Paretic
6	M	Senile dementia	21	F	Schizophrenia
7	M	Schizophrenia	22	M	Schizophrenia
8	F	Manic-depressive	23	F	Unclassified
9	F	Schizophrenia	24	F	Schizophrenia
10	F	Unclassified	25	F	Schizophrenia
11	F	Schizophrenia	26	F	Unclassified
12	F	Senile dementia	27	M	Senile dementia
13	M	Schizophrenia	28	M	Senile dementia
14	M	Manic-depressive	29	F	Schizophrenia
15	M	Schizophrenia	30	M	Paretic

14. A random sample of 40 long-distance telephone calls was taken from the long-distance calls made by employees of a certain large firm during the past year. For each call, the length of the call was recorded to the nearest minute, with any call of less than a minute being labeled as "1." The results are as follows:

4, 6, 4, 4, 7, 2, 3, 1, 4, 3, 1, 5, 2, 8, 12, 2, 5, 6, 4, 5,
1, 1, 7, 3, 9, 1, 4, 16, 2, 7, 4, 6, 4, 8, 5, 5, 7, 3, 1, 9.

Construct a frequency distribution and discuss the qualitative implications of this frequency distribution.

15. The diameters of parts produced by a certain production process are given below. Starting at 0 and using $i = .0625$, construct a frequency distribution for these data. Repeat the procedure with $i = .1250$ and comment on any differences in the two distributions.

.581	.630	.460	.511	.351	.180	.450	.240
.630	.684	.500	.554	.380	.195	.489	.261
.460	.500	.365	.405	.278	.143	.357	.190
.511	.554	.405	.449	.308	.158	.396	.211
.351	.380	.278	.308	.212	.109	.272	.145
.180	.195	.143	.158	.109	.056	.139	.074
.450	.489	.357	.396	.272	.139	.349	.186
.240	.261	.190	.211	.450	.074	.189	.099

16. The incomes (in dollars) of the families in a particular neighborhood are shown below. Construct a histogram with $i = 1000$, another histogram with $i = 5000$, and another histogram with $i = 333$ to represent these data. Compare the three histograms and discuss.

15,500	12,500	8,600	11,000
21,000	8,600	31,000	16,500
8,700	14,700	10,100	7,300
9,200	9,700	11,300	10,200
16,700	8,800	10,600	14,300
15,400	20,100	13,000	16,700
9,900	9,200	9,400	9,500
13,200	9,500	9,800	10,600

17. In March 1974, a random sample of gasoline stations in the United States was taken, and the price per gallon of premium-grade gasoline was recorded for each station. The data are as follows (each price is given to the nearest cent):

65, 58, 64, 68, 52, 48, 59, 59, 56, 63, 61, 60, 52, 57, 60,
62, 55, 55, 64, 71, 61, 63, 46, 53, 60, 57, 58, 57, 54, 59,
51, 58, 62, 56, 58, 49, 59, 64, 57, 57, 58, 60, 59, 56, 53.

Construct a frequency distribution for the raw data, then construct a frequency distribution for grouped data with $i = 3$ and 60 as a midpoint of an interval and another frequency distribution for grouped data with $i = 5$ and 60 as a midpoint of an interval.

18. Discuss the assumptions about a set of data grouped into a frequency distribution that are implicit when we (a) consider the real, rather than the apparent, limits to be the class boundaries; and (b) use the midpoint of any interval to represent all of the scores grouped into that interval.

19. Comment on the following statement: "The process of data-gathering and reporting often involves the loss or deliberate sacrifice of some potential information."

20. Make up two examples from your area of interest in which a frequency distribution would very likely require the employment of unequal or open class intervals.

21. Suppose that 32 persons participated in a series of gambles and that the net gain for each person is shown below. Construct a histogram, a frequency polygon, and a cumulative frequency polygon for this set of data.

$2.74	−$1.05	$ .25	−$1.39
4.09	− 1.56	.37	− 2.08
.95	− .36	.09	− .48
2.29	− .88	.21	− 1.17
2.27	− .87	.22	− 1.16
.87	.33	− .08	.44
.20	− .08	.02	− .10
1.16	.44	− .10	.59

22.   For the data in Exercise 14, construct a frequency polygon and a cumulative frequency polygon (without grouping the data).

23.   For the data in Exercise 15, construct a frequency polygon and a cumulative frequency polygon corresponding to each of the two frequency distributions determined in that exercise.

24.   For the three frequency distributions obtained in Exercise 16, construct the corresponding frequency polygons and cumulative frequency polygons.

25.   For the data in Exercise 15, compute the sample mean
      (a)   directly from the data,
      (b)   from the grouped frequency distribution with $i = .0625$,
      (c)   from the grouped frequency distribution with $i = .1250$.
      Explain any differences in your three answers.

26.   For the data in Exercise 15, compute the sample median
      (a)   directly from the data,
      (b)   from the grouped frequency distribution with $i = .0625$,
      (c)   from the grouped frequency distribution with $i = .1250$.
      Explain any differences in your three answers. Also, compute the mode according to (b) and (c) and compare the three sample means (from Exercise 25), the three sample medians, and the two sample modes as measures of central tendency.

27.   When we find the median of a grouped distribution by interpolation, what is being assumed about the distribution of values *within* a given class interval? How does this assumption show up in the cumulative frequency polygon?

28.   How does the value of the sample mean computed from a grouped distribution depend on the particular choice of intervals? Does the error introduced by treating each score as equal in value to the midpoint of its interval have any systematic connection with the extent to which the distribution is symmetric?

29.   For the data in Exercise 17, find the sample mean and the sample median (a) for the raw data, (b) for the grouped frequency distribution with $i = 3$, and (c) for the grouped frequency distribution with $i = 5$.

30.   The scores for 12 students on a particular examination are as follows:

18	15	19	27
13	30	24	11
5	16	17	20

Compute the mean, median, standard deviation, and average absolute deviation.

31.   Construct the standardized scores for the 12 students whose raw scores are given in Exercise 30, and then show that the mean of these standardized scores is 0 and the standard deviation of the standardized scores is 1.

32. In a group of 50 boys and 50 girls of high-school age, the number of calories consumed on a particular day by each person can be summarized by the following two frequency distributions.

Class interval	Males	Females
5000–5499	1	1
4500–4999	2	0
4000–4499	4	2
3500–3999	16	5
3000–3499	12	10
2500–2999	7	20
2000–2499	5	8
1500–1999	2	2
1000–1499	1	2
	—	—
	50	50

(a) Compute the sample mean, median, and mode for each of the two frequency distributions.

(b) Compute the sample variance for each of the two frequency distributions.

33. For the data in Exercise 16, find the mean income and the median income. Which do you think is a better measure of central tendency? Why? Also, find the sample variance and standard deviation. [*Hint*: To make this easier, first transform the data so that you will not have to work with such large numbers.]

34. Show that

$$s^2 = \frac{n \sum x_i{}^2 - (\sum x_i)^2}{n^2}.$$

35. Starting with Equation (5.11.8), derive the computational formulas (5.11.9) and (5.11.10).

36. For the data in Exercise 17, find the sample variance (a) for the raw data, (b) for the grouped frequency distribution with $i = 3$, and (c) for the grouped frequency distribution with $i = 5$.

37. A measure of dispersion not discussed in the text is the sample coefficient of variation, which is equal to s/m, the sample standard deviation divided by the sample mean. If the variable of interest is measured in dollars, in what units will the coefficient of variation be expressed? Give an informal interpretation of what the sample coefficient of variation measures.

38. For the data of Exercise 17, calculate the sample coefficient of variation (as defined in Exercise 37) for the raw data.

39. An individual who frequently speculates in real estate purchases some land for $10,000. The land appreciates in value by only 5 percent the first year following the purchase, but the next year it appreciates in value by 75 percent because of a new shopping center nearby. At the end of the second year the speculator sells the land.

   (a) How much money does the speculator receive for the land at the end of the second year?

   (b) For the 2 years in question, let x_1 and x_2 represent the ratio of the value of the land at the end of the year to the value of the land at the beginning of the year. Find the average value of this ratio, $m = (x_1 + x_2)/2$.

   (c) If the ratio in (b) had not fluctuated, but instead had equaled m for both years, how much would have been received for the land at the end of the second year?

   (d) Explain any discrepancy between the answers to (a) and (c).

40. In Exercise 39, find the sample *geometric mean*, which is the square root of the product of the two observations:

$$g = \sqrt{x_1 x_2}.$$

(In general, with n observations $x_1, x_2, \ldots, x_n$, the sample geometric mean is the nth root of the product of the observations: $g = \sqrt[n]{x_1 x_2 \cdots x_n}$.) If the ratio had not fluctuated, but instead had equaled g for both years, how much would have been received for the land at the end of the second year? Compare this value with your answers to (a) and (c) of Exercise 39. What property of the geometric mean is illustrated here? Is that property shared by m, the *arithmetic* mean?

41. A box of loose change is known to contain 100 coins of the following denominations: 35 pennies, 50 nickels, and 15 quarters. You draw a random sample of 10 coins, with replacement. What is the frequency distribution of coins which you should *expect* to get? What is the probability of obtaining this frequency distribution exactly?

42. One thousand college students were classified by sex (700 males, 300 females) and by class in school (400 freshmen, 300 sophomores, 200 juniors, and 100 seniors). If sex is independent of class in school, construct a table showing all possible joint frequencies (such as the frequency of freshman males). Compare this with the frequencies that were actually observed:

	Freshman	Sophomore	Junior	Senior
Male	300	250	100	50
Female	100	50	100	50

43. In your own words, carefully distinguish among a probability distribution, a frequency distribution, and a sampling distribution.

44. In the text, the following distribution of the number of courses taken in the first semester of graduate school was given:

x	$P(X = x)$
3	1/3
4	1/3
5	1/3

A random sample of 2 first-semester graduate students is taken, and the number of courses is recorded for each of the 2 students. Assume that the population from which the sample is drawn is large enough that the sample observations can be treated as independent for all practical purposes. Construct the theoretical sampling distribution of the sample mean based on a random sample of 2 students. [*Hint*: Enumerate the possible combinations of events for samples of size 2, and then translate these into values of the sample mean.]

45. From the population distribution given in Exercise 44, find the sampling distribution of the sample *median* based on 2 independent observations from the population. Compare the standard error of the sample median with the standard error of the sample mean.

46. For the population distribution given in Exercise 44, find the expectation and the variance of the *sample mean* from a sample of size 2
   (a)   directly from the *sampling* distribution of the sample mean,
   (b)   from the mean and variance of the *population* distribution.

47. For the population distribution given in Exercise 44, find the sampling distribution of the *sample variance* for a sample of 2 independent observations by enumerating the possible combinations of events for samples of size 2 and translating these into values of the sample variance.

48. In Exercise 44, assume that the sample size is 3 and find the sampling distributions of the sample mean, sample median, and sample variance.

49. If a customer purchases a book from a display of $3 books at a certain bookstore, the probability that just 1 book is purchased from the display is 1/3, the probability that 2 books are purchased from the display is 1/2, and the probability that 3 books are purchased from the display is 1/6. Let X represent the amount of the sale. If 3 customers are chosen randomly from a very large set of customers purchasing books from this particular display, construct the theoretical sampling distribution of the sample mean for the amount of the sale.

50. In Exercise 49, construct the sampling distribution of the sample median and compute the standard errors of both the sample mean and the sample median.

51. In Exercise 49, find the sampling distribution of the sample standard deviation and find the mean and variance of this sampling distribution.

52. Consider a small population of 4 MBA students who are in the job market. One of the students has no job offers, 1 of the students has 1 job offer, and the remaining 2 students both have 2 job offers. If a random sample of size 2 *without* replacement is to be taken from the population of 4 MBA students, find the sampling distribution of the average number of job offers in the sample.

53. In a very large group of corporate executives, 20 percent have no college education, 10 percent have exactly 2 years of college, 20 percent have exactly 4 years (that is, an undergraduate degree), and 50 percent have 6 years (that is, an undergraduate degree and an MBA). A sample of size 2 (with replacement) is to be taken from this population. Find the sampling distribution of the median number of years in college of the executives in the sample.

54. A municipal building has 10 pay telephones in different locations. The actual usage of the phones on a particular day is as follows.

Number of calls	Number of phones with this number of calls
10	2
12	5
16	3

Two phones are to be picked at random with replacement for a study of phone usage. Find (a) the sampling distribution of the average number of calls per phone in the sample of 2 phones and (b) the variance of this distribution.

55. Answer (a) and (b) in Exercise 54 under the assumption that the sampling is conducted *without* replacement, and compare your answers to those obtained under the assumption of sampling *with* replacement.

56. The number of candy bars sold from a vending machine on any given day is normally distributed with mean 258 and standard deviation 60, and sales on one day are independent of sales on other days. Consider a period of 144 days.
    (a) What is the probability that the sample mean number of candy bars sold for the 144-day period is greater than 263?
    (b) What is the probability that the *total* number of candy bars sold for the 144-day period is greater than 36,000 but less than 38,000?
    (c) If the profit per candy bar sold is 5 cents, what is the probability that the average profit per day over the 144-day period is at least $13?

57. In Exercise 3, Chapter 4, find the sampling distribution of the average number of defectives per lot in a sample of 4 lots of 10 items each.

58. In Exercise 37, Chapter 4, find the sampling distribution of the median number of telephone calls per hour in a sample of 2 hours.

59. The distribution of the number of gallons of milk sold in any given week at a particular grocery store is a normal distribution with mean 4000 and variance 90,000. Furthermore, this distribution is the same at all stores operated by a particular chain, and the milk sales at any one store in the chain are independent of the milk sales at the other members of the chain. The chain operates in cities A, B, and C, with 9 stores in each city. For a given week, find the probability that
   (a) a particular store sells at least 4100 gallons of milk,
   (b) the stores in city A have average sales of at least 4100 gallons of milk,
   (c) the stores in the entire chain have average sales of at least 4100 gallons of milk.

60. In Exercise 59, answer (a), (b), and (c) under the assumption that the 4100 figure represents average sales per week over a 4-week period, where sales for any given week are independent of sales for any other week.

61. In Exercise 53, suppose that we are concerned with the member of the sample of size 2 who has the greater number of years of college experience. Find the sampling distribution of the maximum number of years of college of a member of the sample.

62. In Exercise 59, find the sampling distribution of
   (a) the total sales of milk for a given week for city A,
   (b) the total sales of milk for a given week for the entire chain of stores,
   (c) the difference between the sales of one store, store J, and another store, store K, for a given week,
   (d) the difference between the total sales of the stores in city A and the total sales of the stores in city B for a given week.

63. The type of battery discussed in Exercise 63, Chapter 4, is installed in 9 cars produced by a particular Swedish firm and in 36 cars produced by a particular German firm.
   (a) Find the sampling distribution of the average time until failure for the batteries in the Swedish cars.
   (b) Find the sampling distribution of the average time until failure for the batteries in the German cars.
   (c) What is the probability that the average time until failure is at least .4 years higher for the Swedish cars than for the German cars?
   (d) Find the sampling distribution of the difference between the average time until failure for the Swedish cars and the average time until failure for the German cars.

64. If X is normally distributed with mean 50 and variance 100, Y is normally distributed with mean 80 and variance 25, W is normally distributed with mean 100 and variance 144, and X, Y, and W are independent, find the distribution of:
   (a) $X - Y$,
   (b) $X + Y - W$,
   (c) $6W + 4X - 30Y$,
   (d) $(X + Y + W)/3$.

65. The argument for the normal theory of error presented in Section 4.17 of the text was based on the normal approximation to the binomial, assuming the action of a vast number of "influences" that might affect any given numerical observation. However, can we make a rationale for the normal distribution of error, once again imagining the number of influences to be very large, but without depending directly on the binomial assumption?

66. To demonstrate the central limit theorem, draw 100 samples of size 5 from a random number table and calculate the sample mean for each of the 100 samples. Construct a frequency distribution of sample means. Do the same for 100 samples of size 10 and compare the two frequency distributions. Does the central limit theorem appear to be working?

67. To demonstrate the central limit theorem under a skewed population distribution, follow the same procedure as in Exercise 66 but perform the following transformation before computing sample means:

Random number	x
0, 1, 2, 3	0
4, 5, 6	1
7, 8	2
9	3

68. A particular model of television set is sold at a wide variety of types of stores (appliance stores, discount stores, department stores, and so on), and the price differs considerably from store to store. If a store is chosen at random from all of the stores carrying this particular model, and X represents the price (in dollars) of the television at the chosen store, then X is normally distributed with mean 150 and variance 400. If a sample of 16 stores is selected randomly from the stores carrying this model, let M represent the average price of the television in the sample, and find
    (a) $P(M \geq 150)$,
    (b) $P(145 < M < 149)$,
    (c)   the mean and variance of M.

69. Answer (a)–(c) in Exercise 68 if the sample size is 400, and compare your answers to those obtained when $n = 16$. In each case, how would your answers be affected if the assumption of normality is not applicable?

70. Random samples are taken from two normal populations. The first population has mean 80 and variance 81, and the second population has mean 100 and variance 100. If $n_1 = 9$ is the size of the sample from the first population, $n_2 = 10$ is the size of the sample from the second population, and X_i is the ith member of the sample from the first population, find
    (a)   the distribution of $X_1 + X_2$,
    (b)   the distribution of $X_1 - X_2$,

(c) $P(170 \le X_1 + X_2 \le 185)$,

(d) the distribution of M_1, the sample mean of the sample from the first population,

(e) the distribution of M_2, the sample mean of the sample from the second population,

→ (f) the distribution of $M_1 + M_2$,

→ (g) the distribution of $M_1 - M_2$,

→ (h) $P(M_1 - M_2 > -25)$.

71. If a random sample of size n_1 is taken from a normal population with mean μ_1 and variance σ_1^2, and a random sample of size n_2 is taken from a normal population with mean μ_2 and variance σ_2^2, what is the distribution of the difference between the two sample means, $M_1 - M_2$? (Assume that the two samples are independent of each other.) What if the populations are not normal but the sample sizes are both large?

72. Suppose that the mean and variance of the weight of an item from a certain production process are known but that the exact form of the distribution of weight is not known. Can we use the central limit theorem to approximate probabilities such as the probability that the weight of an item is greater than 55 grams? Explain your answer.

73. A specification calls for a steel rod to be 3 meters long. When a shipment of rods is received, 25 rods are randomly selected from the shipment (shipments usually contain about 1000 rods). If the average length of the 25 rods in the sample is between 2.95 meters and 3.05 meters, the shipment is accepted; otherwise, it is returned to the supplier. The length of rods is normally distributed, and the standard deviation of length is known to be .20 meters. If the actual mean length of rods in a given shipment is 2.90 meters, what is the probability that the shipment will be returned to the supplier?

74. In Exercise 17, the average of the 45 observations is computed, and a newspaper reporter comments that this figure represents the average price paid by gasoline customers at the 45 gasoline stations in the sample. Discuss this statement.

75. Discuss each of the following situations with respect to making inferences about the population of interest on the basis of a sample.

(a) In a survey of voting intentions in a given city, names are chosen randomly from the telephone book for that city.

(b) In a study of the reaction of individuals to certain stimuli, a psychologist takes a random sample of individuals from a basic psychology class.

(c) In a study of the opinions of students at a given university about pass-fail grading, an interviewer stands in the lobby of the student union building and interviews students entering the building.

(d) A newspaper conducts a survey by inserting questions in several editions of the paper and inviting readers to send in their answers.

(e) To study the effectiveness of a new drug, a physician prescribes the drug for several patients and observes their progress.

6

ESTIMATION

In this chapter we discuss the first concepts of statistical inference and decision. In Chapter 5, some basic summary measures of frequency distributions were considered, along with their sampling distributions. Now we will consider the use of such sample statistics to estimate population parameters. Statistical inference, of course, encompasses procedures for generalizing from a sample to a population, and estimation is one such procedure. Two types of estimation will be discussed: (1) point estimation, which consists of the use of a single sample statistic to determine a single value, which is to be used as an estimate of a population parameter, and (2) interval estimation, which involves the determination of an interval of values within which the population parameter must lie with a given "confidence." We will discuss point estimation first and then interval estimation.

6.1 SAMPLE STATISTICS AS ESTIMATORS

If we are interested in a particular population parameter, such as a population mean or a population proportion, the use of a sample statistic to determine a single value, which is to be used as an estimate of the population parameter of interest, is called point estimation.

A single value, or a point in the space of all possible values, is taken as the estimate. A word regarding terminology is in order at this point; a sample statistic being used to estimate a population parameter is called an **estimator** of the parameter, and a specific *value* of the sample statistic, computed from a particular sample, is called an **estimate** of the parameter. Thus, an estimator is a random variable, and we can talk of its probability distribution (which is a sampling distribution, since the estimator is a sample statistic), its expectation, and so on; an estimate is a specific value of this random variable.

How do we use a sample value to make an inference about a population parameter? The fact that the sample represents only a small subset of observations drawn from a much larger set of potential observations makes it risky to say that any estimate is exactly like the population value. As a matter of fact, they very probably will not be the same, as all sorts of different factors of which we are in ignorance may make the sample a poor representation of the population. We lump such factors together under the general heading of *chance* or *random effects*. Nevertheless, samples drawn at random should reflect the population characteristics to some degree, and we need to know how to use the available evidence in the best possible ways to infer the characteristics of the population.

For instance, if we are interested in the mean family income in a large city, it may be too costly to find the exact mean by finding the income of each family in the city. Instead, a random sample of, say, 200 families might be taken, and from this sample a point estimate of μ can be determined. We might choose the sample mean as our point estimate, or we might choose the sample median or yet some other statistic.

In general, there are numerous potential estimators for a population parameter. Certain alternatives are relatively obvious. For instance, each population parameter has its parallel in some sample statistic; the population mean μ has its sample counterpart in the sample mean, the variance σ^2 in the sample variance, the population proportion p in the sample proportion, and so on. The value of a sample statistic contains evidence about the value of the corresponding population value, of course. It should be emphasized, however, that the sample counterpart of a parameter is by no means the only possible estimator of that parameter, and less "obvious" statistics can be useful as estimators in some contexts.

A number of properties have been put forth as desirable for a statistic to have in order to be a good estimator of a population parameter. Four of these properties are **unbiasedness, consistency, efficiency,** and **sufficiency.** Although few statistics satisfy all of these properties, they are considered by most statisticians to be useful properties for estimators. These four criteria for good estimators will be discussed in the next four sections.

6.2 UNBIASEDNESS

Suppose that we are interested in estimating the value of a population parameter θ (Greek theta), and we are considering the use of some sample statistic G as an estimator of the value of θ. Then

G is said to be an unbiased estimator of θ if

$$E\,(G) = \theta. \tag{6.2.1*}$$

Suppose that random samples are taken repeatedly and that the value of G is found for each sample. Then in the long run, if the average value of G is θ, then G is unbiased.

For example, consider the mean of a sample as an estimator of the mean of the population. Here,

$$G = M = \frac{\sum\limits_{i} X_i}{n}$$

and
$$\theta = \mu.$$

Is the sample mean an unbiased estimator of the population mean μ? Yes, since we proved in Section 5.14 that

$$E(M) = \mu. \tag{6.2.2*}$$

The mean of a random sample is an unbiased estimator of the population mean μ.

In exactly the same way, for samples from a Bernoulli process with fixed sample size n, it can be shown that *the sample proportion R/n of cases in a given category is an unbiased estimator of the population proportion p* [this is really just a special case of Equation (6.2.2)]:

$$E\left(\frac{R}{n}\right) = p. \tag{6.2.3*}$$

From the algebra of expectations, $E(R/n) = E(R)/n$. But $E(R)$ is just the mean of a binomial distribution and is equal to np. Thus,

$$E\left(\frac{R}{n}\right) = \frac{np}{n} = p.$$

An example of a **biased** estimator is provided by the sample variance. The sample variance is a biased estimator of the population variance σ^2, since

$$E(S^2) \neq \sigma^2. \tag{6.2.4*}$$

Here S^2 denotes the sample statistic as an estimator (a random variable used to estimate σ^2), while s^2 denotes a specific value of S^2, or a specific estimate of σ^2. This notation is consistent with our use of capital letters to represent random variables and small letters to represent values of random variables.

Equation (6.2.4) can be demonstrated as follows. The expectation of a sample variance is

$$E(S^2) = E\left(\frac{\sum_i X_i^2}{n} - M^2\right) = E\left(\frac{\sum_i X_i^2}{n}\right) - E(M^2).$$

Let us consider the two terms on the extreme right separately. By the algebra of expectations,

$$E\left(\frac{\sum_i X_i^2}{n}\right) = \frac{\sum_i E(X_i^2)}{n}.$$

From the definition of the variance of the population,

$$\sigma^2 = E(X_i^2) - \mu^2,$$

so that

$$E(X_i^2) = \sigma^2 + \mu^2$$

for any observation i. Thus,

$$E\left(\frac{\sum_i X_i^2}{n}\right) = \frac{\sum_i (\sigma^2 + \mu^2)}{n} = \sigma^2 + \mu^2. \tag{6.2.5}$$

Now the variance of the sampling distribution of means is, from Equations (5.14.4) and (6.2.2),

$$\sigma_M^2 = E(M^2) - \mu^2,$$

so that

$$E(M^2) = \sigma_M^2 + \mu^2. \tag{6.2.6}$$

Putting these two results [(6.2.5) and (6.2.6)] together, we have

$$E(S^2) = \sigma^2 - \sigma_M^2. \tag{6.2.7*}$$

But $\sigma_M^2 = \sigma^2/n$, so the expectation of the sample variance can be written in the form

$$E(S^2) = \sigma^2 - \frac{\sigma^2}{n} = \left(\frac{n-1}{n}\right)\sigma^2. \tag{6.2.8*}$$

Thus, on the average the sample variance is *too small* by a factor of $(n-1)/n$.

Since this is true, a way to correct the variance of a sample to make it an unbiased estimator emerges. **An unbiased estimator of the vari-**

ance based on any sample of n independent trials is

$$\hat{S}^2 = \left(\frac{n}{n-1}\right) S^2. \qquad (6.2.9^*)$$

Note that $\hat{S}^2$ is the "modified" sample variance introduced in Chapter 5. The caret or "hat" ($\wedge$) will always be placed over the symbol for a sample variance when it is an unbiased estimator to distinguish it from the ordinary sample variance. It is simple to show that $\hat{S}^2$ is indeed unbiased:

$$E(\hat{S}^2) = \left(\frac{n}{n-1}\right) E(S^2) = \frac{n}{(n-1)} \frac{(n-1)}{n} \sigma^2 = \sigma^2. \qquad (6.2.10^*)$$

Quite often it is convenient to calculate the unbiased variance estimator $\hat{S}^2$ directly, without the intermediate step of calculating S^2. This is done by Equation (5.11.8), (5.11.9), or (5.11.10).

Even though we will be using the square root of $\hat{S}^2$, or $\hat{S}$, to estimate the population σ, it should be noted that $\hat{S}$ is not *itself* an unbiased estimator of σ. That is,

$$E(\hat{S}) \neq \sigma$$

in general. The correction factor used to make $\hat{S}$ an unbiased estimator of σ depends upon the form of the population distribution; for the normal distribution, an approximately unbiased estimator for large n is provided by

$$\text{est. } \sigma = \left[1 + \frac{1}{4(n-1)}\right] \hat{S}. \qquad (6.2.11)$$

Furthermore, special tables exist for correcting the estimate of σ for relatively small samples from such populations. However, the problem of estimating σ from $\hat{S}$ is bypassed, in part, by the methods we will use in making inferences about σ^2, and, if the sample size is reasonably large, the amount of bias in $\hat{S}$ as an estimator of σ ordinarily is rather small. For these reasons we will not bother to correct for the bias in $\hat{S}$ found from the *unbiased* estimator $\hat{S}^2$ of σ^2.

Some modern statistics texts completely abandon the idea of the sample variance S^2 as used here and introduce only the unbiased estimator $\hat{S}^2$ as *the* variance of a sample. However, this is apt to be confusing in some work, and so we will follow the older practice of distinguishing between the sample variance S^2 as a descriptive statistic and as a biased estimator of σ^2 and $\hat{S}^2$ as an unbiased estimator of σ^2.

6.3 CONSISTENCY

An intuitively attractive property for an estimator to possess is that **the sample estimator should have a higher probability of being**

close to the population value θ, the larger the sample size. Statistics that have this property are called consistent estimators. More formally,

the statistic G is a consistent estimator of θ if for any arbitrary number ϵ,

$$P(|\, G - \theta \,| < \epsilon) \to 1 \qquad \text{as } n \to \infty. \tag{6.3.1*}$$

The probability that G is within a certain distance ϵ of the parameter θ approaches 1 as the sample size n approaches infinity, however small the size of the positive number ϵ.

It is not difficult to show that the sample mean M is a consistent estimator of the population mean μ. From Equation (5.14.9),

$$P(|\, M - \mu \,| < k) \geq 1 - \frac{\sigma_M{}^2}{k^2}. \tag{6.3.2}$$

But we know that

$$\sigma_M{}^2 = \frac{\sigma^2}{n},$$

and thus

$$P(|\, M - \mu \,| < k) \geq 1 - \frac{\sigma^2}{nk^2}.$$

When n becomes very large, σ^2/nk^2 approaches zero, and therefore

$$P(|\, M - \mu \,| < k) \to 1 \qquad \text{as } n \to \infty. \tag{6.3.3*}$$

Thus, M *is a consistent estimator of* μ, since this corresponds to Equation (6.3.1) with $G = M$, $\theta = \mu$, and $\epsilon = k$.

The above proof suggests a sufficient (but not necessary) condition for an estimator to be consistent. If G is an unbiased estimator of θ, and if $V(G) \to 0$ as $n \to \infty$, then G is a consistent estimator of θ. To see that this must be true, write the Chebyshev inequality as follows:

$$P(|\, G - E(G) \,| < \epsilon) \geq 1 - \frac{\sigma_G{}^2}{\epsilon^2}. \tag{6.3.4}$$

But G is unbiased, so $E(G) = \theta$. Also, $\sigma_G{}^2 \to 0$ as $n \to \infty$, so as n becomes large, $\sigma_G{}^2/\epsilon^2$ approaches zero, and the value on the right-hand side thus must approach 1, in which case G is a consistent estimator of θ:

$$P(|\, G - \theta \,| < \epsilon) \to 1 \qquad \text{as } n \to \infty. \tag{6.3.5}$$

This shows that the conditions given above are *sufficient* conditions for G to be consistent. To show that they are not *necessary* conditions, we point out that the sample variance S^2 is a consistent estimator of σ^2 even though it is *not* an unbiased estimator of σ^2, thus violating the first condition.

The sample mean, the sample variance, and many other statistics are consistent estimators, as they tend in probability to get closer to the true population value the larger the sample. It is, however, possible to create sample statistics that are not consistent estimators. For example, suppose that we wished to estimate the population mean, and we decided to take the value of the second observation made in any sample as the estimator. Would this be a consistent estimator? No, since the second observation's value would not tend in likelihood to get any closer to the mean value of the population the larger the sample. According to the criterion of consistency, this would not be a good way to estimate the mean. (Would this be an unbiased estimator of the mean, however?)

It should be noted that an important distinction between unbiasedness and consistency is that the former is a fixed-sample property (if an estimator is unbiased, it is unbiased for any fixed sample size), while the latter is an *asymptotic* property (that is, it is concerned only with what happens as the sample size becomes very large). Whenever the sample size is quite small, of course, asymptotic properties such as consistency are not of much interest.

6.4 RELATIVE EFFICIENCY

A third criterion for evaluating a statistic G as an estimator of a parameter θ is that G be efficient relative to other statistics that might be used to estimate θ. You will recall that the standard deviation (or standard error, as it is called when the distribution it describes is a sampling distribution) represents the extent of the difference that chance factors tend to create between a sample estimate and a true parameter value. Good estimators should have sampling distributions with small standard errors, given the n of the sample.

Suppose that two different sample statistics G and H are calculated from the same data and that these two statistics are each unbiased estimators of the same population parameter θ. The theoretical sampling distribution of G determined for samples of size n has a standard error σ_G. There is also a theoretical sampling distribution of H having a standard error σ_H. Each of these two standard errors reflects the tendency of the sample statistic to deviate by chance from the population value. Then, for any n,

the efficiency of the statistic G relative to the statistic H, both as estimators of θ, is given by

$$\frac{\sigma_H{}^2}{\sigma_G{}^2} = \text{efficiency of } G \text{ relative to } H. \qquad (6.4.1^*)$$

The more efficient estimator has the smaller standard error, so that the ratio is greater than 1.00 when the variance of the more efficient estimator appears in the denominator. Other things, such as sample size, being equal, relatively inefficient statistics have relatively larger standard errors than other estimators of the same parameter.

For example, in a unimodal symmetric distribution the population mean and the population median both have the same value. We wish to estimate this value μ by drawing a random sample of size n. We could use either the sample mean (statistic G, let us say) or we could use the sample median (statistic H). In any given sample, these values will very likely not be the same. Which should we use to estimate μ? One possible answer involves efficiency. If the standard error of the mean is represented by σ_M and the standard error of the distribution of sample medians is represented by σ_{Md}, then it is true that for unimodal symmetric population distributions, σ_{Md} is greater than σ_M for any given sample size $n > 2$. For such populations and for a reasonably large n, the magnitude of error in any given estimate of the population mean from the sample mean is likely to be less than in an estimate of μ from the sample median. This implies that for such situations,

$$\frac{\sigma_{Md}^2}{\sigma_M^2} > 1.00; \qquad (6.4.2)$$

the relative efficiency of the mean will be greater than 1.00, indicating that the mean is a more efficient estimator than the median for symmetric unimodal populations. For a specific example, if we assume that the population is normally distributed, then, approximately,

$$\sigma_{Md}^2 = \frac{\pi\sigma^2}{2n}.$$

In this case,

$$\frac{\sigma_{Md}^2}{\sigma_M^2} = \frac{\pi\sigma^2/2n}{\sigma^2/n} = \frac{\pi}{2} = 1.57.$$

This is one reason that so much of statistical inference deals with the mean rather than with the median; the median is less efficient than the mean as an estimator of μ for a symmetric unimodal population distribution, and such distributions are very often assumed in statistical work.

For a normal distribution, then, the sample mean is a more efficient estimator of μ than the sample median. Furthermore, it can be shown that the sample mean is the most efficient estimator of μ; that is, of all unbiased estimators of μ, the sample mean has the smallest standard error. As a result, the sample mean is called a **minimum variance unbiased estimator of μ.**

As presented in this section, the concept of relative efficiency is restricted to unbiased estimators. A more general concept is that of minimum mean-square error. If G is an estimator of θ, the mean-square error of G is

$$\text{MSE } (G) = E(G - \theta)^2. \tag{6.4.3}$$

Observe that if G is unbiased, the mean-square error is identical to the variance of G, since $E(G) = \theta$. If we define the *bias* of an estimator G to be

$$B(G) = E(G) - \theta, \tag{6.4.4}$$

then the mean-square error can be written as the sum of the variance and the square of the bias:

$$\text{MSE } (G) = V(G) + [B(G)]^2. \tag{6.4.5}$$

In some situations, a certain biased estimator may have a smaller MSE than any unbiased estimator, in which case it might be preferred to any unbiased estimator.

6.5   SUFFICIENCY

A concept of great importance in the theory of estimation is that of a sufficient estimator.

A statistic G is said to be a sufficient estimator of the parameter θ if G contains all of the information available in the data about the value of θ.

A sufficient statistic G is a "best" estimator of θ in the sense that G cannot be improved by considering any other aspects of the sample data not already included in the statistic G itself.

A slightly more formal definition of sufficiency may be given as follows. Consider the estimator G of θ once again, and let H be any other sample statistic. We can consider the *conditional* distribution of H given the value of G. Then if the conditional distribution of H given G does not depend in any way on the value of θ, G is a sufficient statistic. In other words, once G is known, knowing any other statistic H gives us no more information about θ, the parameter of interest.

An example of a sufficient estimator is the sample proportion when samples are drawn at random from a Bernoulli process. The sampling distribution of R/n can be found from the binomial rule in this situation, of course. If we are estimating the population value p, then R/n is a sufficient statistic, since there is no information we could add to R/n to make it a better estimator of p. For a normal distribution with known variance σ^2 and unknown mean μ, the sample mean M is a sufficient estimator of μ, the population mean; for this situation the sample mean "wraps up" all of the available information about μ in the sample. This means that in order to estimate μ from a normal distribution with known variance, we need not concern ourselves with the set of individual sample results, $(X_1, X_2, \ldots, X_n)$; all that is needed is to compute the sample mean M and the sample size n. Observe that the entire set of sample results is sufficient but that M and n form a smaller set which is also sufficient. That is, M and n summarize the results of the sample without any loss of information, thereby enabling the statistician to convey all of the available information from the sample in reasonably simple form. Incidentally, if the variance is not known, the sample variance S^2 is needed in addition to M and n for sufficiency.

Sufficient estimators do not always exist, and hypothetical situations can be constructed where no way to find a sufficient estimator of a given parameter can be found. Nevertheless, sufficient estimators, when they do exist, are very important, since if we can find a sufficient statistic, it is ordinarily possible to find an unbiased and efficient estimator based on the sufficient statistic (like sufficient estimators, unbiased and efficient estimators do not always exist). It should be noted that when we consider the decision-theoretic approach to statistics in Chapter 9, a somewhat different view of the estimation problem will be presented; estimation will be thought of as a decision-making problem, and the properties presented in the preceding sections will no longer be of primary interest.

6.6 METHODS FOR DETERMINING GOOD ESTIMATORS

The preceding sections have dealt with some desirable properties for point estimators: unbiasedness, consistency, relative efficiency, and sufficiency. If we have a particular statistic G and would like to use it as an estimator of a parameter θ, we can attempt to find out which, if any, of the four desirable properties G satisfies (or we may even look at other properties such as minimum mean-square error). Unless G is a reasonably simple statistic (that is, unless it can be expressed as a fairly simple mathematical function of the sample values), it may be very difficult to see if it possesses any of the desirable properties. First, the sampling distribution of G may not be known, and the mean and standard error of G may depend on the form of this distribution. Second, even if the sampling distribution is known, it may be difficult to determine the mean and variance or to determine the limiting properties of the sampling distribution as

$n \to \infty$ (which may be necessary to investigate asymptotic properties such as consistency).

Suppose that these problems were not encountered with a particular statistic G, and G was found to be unbiased and consistent. This information is of some value, but it certainly is not enough information to indicate that G is the "best" estimator that we could find in this situation. For example, if we consider a symmetric unimodal population, the sample median is an unbiased and consistent estimator of the population mean, μ. The sample mean, on the other hand, is not only unbiased and consistent, but it is a more efficient estimator of μ than the sample median, as Section 6.4 indicated. How do we know that there is not yet another estimator of μ that is even more efficient, or more desirable in some other way, than M? In other words, if we want to estimate θ, and we have investigated the properties of several alternative estimators of θ, how do we know that there is not yet *another* estimator, unbeknownst to us, which is a "better" estimator than those we have already considered?

The question posed above suggests that some methods for the determination of estimators are needed. If methods for the determination of "good" estimators could be developed, we would be saved some of the trouble of trying to find estimators, determining their properties, and worrying about the possible existence of "better" estimators. Two such methods, the method of maximum likelihood and the method of moments, are discussed in the next two sections. A third method, involving decision theory, will be discussed in Chapter 9. Finally, a fourth method, concerned primarily with a specific class of problems dealing with curve fitting, will be presented in Chapter 10.

Before we discuss these methods, it should be pointed out that according to the criteria we have developed so far (that is, the desirable properties of estimators), it is generally not possible to find a "best" estimator for any particular situation. This is why the terms "good," "better," and "best," when applied to estimators, have been enclosed in quotation marks. Usually there are several alternative estimators possessing some of the desirable properties, but the choice among them is not at all clear-cut. For example, consider two estimators G and H. Suppose that G is unbiased and H is not unbiased, but H has a smaller mean-square error than G. Assume further that it is not possible to modify H to allow for its bias because the bias cannot be expressed in simple form. If you are quite concerned about the amount of bias in H because of the features of the problem at hand, you may prefer G. If, on the other hand, a small mean-square error is important to you, you may prefer H. The former condition might prevail if you have to make an estimate over and over again for different sets of data and you would like to be correct "on the average."

The latter condition might prevail if you have a single set of data and you would like to be as "close" as possible to the true value of the population parameter, even if this means that you will have to accept a biased estimator. If you have a large sample, you might be most concerned about consistency, for this is a large-sample property. At any rate, neither estimator, G nor H, is clearly "better" than the other. The methods to be presented will not necessarily determine "best" estimators; they will determine estimators with certain desirable properties, and it is up to the statistician to decide if these properties are satisfactory for the problem at hand.

6.7 THE PRINCIPLE OF MAXIMUM LIKELIHOOD

Suppose that a random variable X has a distribution that depends upon some population parameter θ. The form of the density function will be assumed known, but not the value of θ. A sample of n independent observations is drawn, producing the set of values $(x_1, x_2, \ldots, x_n)$. Let

$$l(x_1, \ldots, x_n \mid \theta)$$

represent the likelihood or probability (density) of this particular sample result *given* θ. For each possible value of θ, the likelihood of the sample result might be different. Then,

the principle of maximum likelihood requires us to choose as our estimate that possible value of θ making $l(x_1, \ldots, x_n \mid \theta)$ take on its largest value.

In effect, this principle says that when faced with several parameter values, any of which might be the true one for a population, the best "bet" is that parameter value which makes the sample actually obtained have the highest probability. When in doubt, place your bet on that parameter value which makes the obtained result most likely.

This principle may be illustrated very simply for the binomial distribution. Suppose that daily price movements of a security behave according to a Bernoulli process, where the two categories of interest are "price goes up" and "price does not go up." Let p be the probability that the price goes up, and assume that p remains the same from day to day and that successive price changes are independent. Three possible values of p are considered: .4, .5, and .6.

Now suppose that a random sample of 15 days is taken, and the price behavior of the security on these days is recorded. The result is that on 9 of the 15 days, the price went up. What can we say about the three values of p that are being considered? Using the binomial distribution, the probability of the sample result may be calculated for each of the theer hypothetical values of the parameter p. If $p = .5$, the probability is

$$\binom{15}{9} (.5)^9(.5)^6 = .153.$$

For $p = .4$, the probability becomes

$$\binom{15}{9} (.4)^9(.6)^6 = .061.$$

Finally, if p is .6, the probability of 9 successes out of 15 is

$$\binom{15}{9} (.6)^9(.4)^6 = .207.$$

The use of the principle of maximum likelihood to decide among these three possibilities leads to the choice of hypothesis 3, that .6 is the population parameter, since this is the parameter value among the possibilities considered that makes the obtained sample result most likely a priori.

It should be apparent that this approach can be made much more general than in the above example, for we would surely like to be able to consider values of p other than .5, .4, and .6. The parameter p of a Bernoulli process can take on any value between 0 and 1, inclusive. For any value of p in this interval, the likelihood can be written as

$$l(x_1, x_2, \ldots, x_n \mid p) = \binom{n}{r} p^r (1 - p)^{n-r}. \qquad (6.7.1^*)$$

This expression, considered as a function of p, is called the likelihood function for this example. It is important to note carefully the distinction between a **likelihood function** and a **sampling distribution.** A sampling distribution is a distribution over different possible sample results, *given* a value of the parameter of interest. In the above example, we could speak of the sampling distribution of R, the number of days the price goes up in a sample of 15 days, *given* a specific value of p. A likelihood function, on the other hand, is considered as a function of the parameter, and the sample is assumed to have been observed and to be fixed. In the example, the sample result was 9 successes in 15 days; when we calculated likelihoods, we kept the 9 and 15 constant and varied p, using three different values of p. The use of the likelihood function is based on the likelihood principle, which states essentially that the entire evidence of the sample is summarized by the likelihood function. In this context, once the sample is observed, the statistician is not interested in probabilities concerning other sample results that might have occurred but did not in fact occur. The sampling distribution of R for, say, $p = .4$ in the example would involve probabilities for values of R such as 2, 6, 8, and so on. The likelihood function ignores these probabilities, concentrating only on probabilities involving 9 successes.

Once the likelihood function is determined, the next task in the example is to find the value of p for which the likelihood function is maximized. To do this, we shall use the calculus (the reader not interested in the mathematical details can safely skip the derivations). First, since the logarithm of a positive argument is a monotonically increasing function of that argument, maximizing the likelihood function is equivalent to maximizing the logarithm of the likelihood function, which in this case is

$$\log l = \log\binom{n}{r} + r \log p + (n - r) \log (1 - p).$$

To maximize this function with respect to p, we differentiate with respect to p and set the result equal to zero (assuming that the logarithms are natural logarithms; that is, that they are taken to the base e):

$$\frac{d}{dp} \log l = \frac{r}{p} - \frac{n - r}{1 - p} = 0.$$

Thus,

$$\frac{r(1 - p) - (n - r)p}{p(1 - p)} = 0,$$

so

$$r(1 - p) - (n - r)p = 0,$$

or

$$r - rp - np + rp = 0,$$

and hence

$$p = r/n. \tag{6.7.2*}$$

In general, then, the value of p which maximizes the likelihood function is the sample proportion, r/n. We call R/n the **maximum-likelihood estimator** of p. In the example of price changes of a security, the maximum-likelihood estimate based on the sample of size 15 would be 9/15. The likelihood function is graphed in Figure 6.7.1, and you can see from the graph that $9/15 = .6$ is the value of p for which the likelihood function reaches its highest value. Formally, to show that $p = r/n$ is a maximum, note that the second derivative of $\log l$ with respect to p is negative.

For another example, suppose that we are sampling from a normally

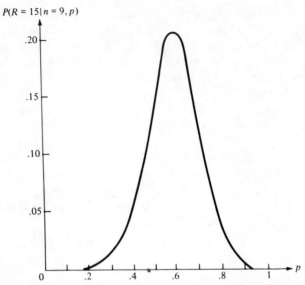

$P(R = 15 | n = 9, p)$

Figure 6.7.1   Likelihood Function for Security Price Example

distributed population with known variance σ^2 and that we wish to find the maximum-likelihood estimator of the population mean, μ, on the basis of a sample of size n from the population: $(x_1, x_2, \ldots, x_n)$. The density function for each observation is, by Equation (4.13.1),

$$f(x_i \mid \mu) = \frac{1}{\sqrt{2\pi\sigma^2}} \exp\left[-\frac{(x_i - \mu)^2}{2\sigma^2} \right]. \tag{6.7.3}$$

Since the trials are assumed to be independent, the likelihood function is simply the product of the n density functions:

$$l(x_1, x_2, \ldots, x_n \mid \mu) = f(x_1 \mid \mu) f(x_2 \mid \mu) \cdots f(x_n \mid \mu)$$

$$= \left(\frac{1}{\sqrt{2\pi\sigma^2}} \exp\left[-\frac{(x_1 - \mu)^2}{2\sigma^2} \right] \right)$$

$$\cdots \left(\frac{1}{\sqrt{2\pi\sigma^2}} \exp\left[-\frac{(x_n - \mu)^2}{2\sigma^2} \right] \right)$$

$$= \left(\frac{1}{\sqrt{2\pi\sigma^2}} \right)^n \exp\left[-\sum_{i=1}^{n} \frac{(x_i - \mu)^2}{2\sigma^2} \right]. \tag{6.7.4*}$$

Now, to maximize $l(x_1, \ldots, x_n \mid \mu)$, we once again take the logarithm and use the calculus to maximize log l:

$$\log l = n \log \left(\frac{1}{\sqrt{2\pi\sigma^2}} \right) - \frac{\sum_{i=1}^{n} (x_i - \mu)^2}{2\sigma^2}.$$

$$\frac{d}{d\mu} \log l = - \frac{1}{2\sigma^2} \sum_{i=1}^{n} 2(x_i - \mu)(-1) = \frac{\sum (x_i - \mu)}{\sigma^2}.$$

Setting this equal to zero, we find that

$$\sum_{i=1}^{n} (x_i - \mu) = 0,$$

or

$$\sum x_i - n\mu = 0,$$

which implies that

$$\mu = \frac{\sum x_i}{n}. \tag{6.7.5*}$$

Thus, for populations having a normal distribution, the sample mean is a maximum-likelihood estimator of μ. The reader can check the second-order conditions to verify that this does indeed produce the maximum value for the likelihood function.

The principle of maximum likelihood is quite useful in theoretical statistics, since general methods exist for finding the value of θ that maximizes the likelihood of a particular sample result. The principle gains its greatest importance from the fact that if a *sufficient* estimator of a parameter exists, the maximum-likelihood estimator is based on this sufficient statistic. In addition, maximum-likelihood estimators possess certain desirable large-sample properties.

Finally, maximum-likelihood estimators possess the following useful *invariance property*. If G is a maximum-likelihood estimator of θ and the function $h(\theta)$ is a function which has an inverse [that is, there is a one-to-one relationship between values of θ and the corresponding values of $h(\theta)$], then $h(G)$ is a maximum-likelihood estimator of $h(\theta)$. For example, if a sample of size n is taken from a normally distributed population, it can be shown that the sample variance S^2 is a maximum-likelihood estimator of the population variance σ^2. The property of invariance implies that S, the sample standard deviation, is also a maximum-likelihood estimator of σ, the population standard deviation. Note that this example shows that maximum-likelihood estimators are not always unbiased, since S^2 is not unbiased (Section 6.2).

The principle of maximum likelihood is introduced here not only because of its importance in estimation but also because of the general point of view it represents about inference. This point of view is that *true population situations should be those making our empirical results likely; if a theoretical situation makes our obtained data have very low likelihood of occurrence, then doubt is cast on the truth of the theoretical situation.* Theoretical propositions are believable to the extent that they accord with actual theoretical state of affairs, and we obtain this result nevertheless, then we are led back to examine our theory. Good theoretical statements accord with observation by giving predictions having high probabilities of being observed.

Naturally, the results of a single experiment, or even of any number of experiments, cannot prove or disprove a theory. Replications, variants, different ways of measuring the phenomena, and so on, must all be brought into play. Even then, proof or disproof is never absolute. Nevertheless, the principle of maximum likelihood is in the spirit of empirical science, and it runs throughout the methods of statistical inference.

6.8 THE METHOD OF MOMENTS

A second method for determining point estimators is the **method of moments,** which essentially amounts to equating sample moments and population moments and solving the resulting equations for the parameter(s) of interest. Formally, the kth population moment (about the origin) of the random variable X is defined as $E(X^k)$, which can be determined by using the usual rules of expectation:

$$E(X^k) = \begin{cases} \sum x^k P(X = x) & \text{if } X \text{ is discrete,} \\[2ex] \displaystyle\int_{-\infty}^{\infty} x^k f(x)\, dx & \text{if } X \text{ is continuous.} \end{cases} \qquad (6.8.1^*)$$

The kth population moment is sometimes denoted by μ_k. If a sample of size n is taken from the population and the values $x_1, x_2, \ldots, x_n$ are observed, the kth sample moment is defined as m_k, where

$$m_k = \frac{\displaystyle\sum_{i=1}^{n} x_i{}^k}{n}. \qquad (6.8.2^*)$$

Given these definitions of m_k and μ_k, then, the method of moments consists of equating m_1 and μ_1, m_2 and μ_2, and so on, until enough equations are obtained to permit a solution in terms of θ, the parameter being estimated. Since the number of equations required varies with the form of the distribution of X and with the specific parameter(s) of interest, it is best to demonstrate the method of moments with some examples.

The simplest example of the application of the method is in the case in which we wish to estimate a population moment, μ_k. It is obvious that the estimate will simply be the corresponding sample moment, m_k. In this way, for example, the sample mean is determined to be the estimate of the population mean according to the method of moments. Similarly, m_2 is the estimate of μ_2. Given these two estimates, we can determine an estimate of the population variance, σ^2. Specifically, we will show that the method of moments determines the sample variance s^2 as the estimate of σ^2. First, note that

$$\sigma^2 = E(X^2) - [E(X)]^2$$

$$= \mu_2 - (\mu_1)^2.$$

Using the method of moments, we set $\mu_1 = m_1$ and $\mu_2 = m_2$:

$$\text{est. } \sigma^2 = m_2 - m_1{}^2.$$

From our definition of sample moment,

$$m_2 = \frac{\sum x_i{}^2}{n},$$

so

$$\text{est. } \sigma^2 = \frac{\sum x_i{}^2}{n} - m^2.$$

But, as we have seen, the right-hand side of this equation is equal to s^2, and thus s^2 is the estimate of σ^2 determined by the method of moments. It should be pointed out that if we extended the notion of the method of moments to allow consideration of moments about the mean as well as moments about the origin, the derivation just presented would be unnecessary. Defining the kth population moment about the mean as $E(X - \mu)^k$ and the kth sample moment about the mean as

$$\sum_{i=1}^{n} \frac{(x_i - m)^k}{n},$$

we find that if $k = 2$, we get σ^2 and s^2, respectively. Notice that the population moments are taken about the population mean and the sample moments are taken about the sample mean.

As another example of the application of the method of moments, consider the binomial distribution. To estimate the parameter p, note that

$E(R) = np$ and $m = r$. Equating these and solving for p, we derive

$$\text{est. } p = \frac{r}{n}.$$

For most of the distributions we have studied, it is necessary to look only at the first one or two moments in order to determine an estimator. An advantage of the method of moments, then, is that it is not too difficult in most cases to find an estimator. If we desire to estimate several parameters simultaneously or if the formulas for the population moments are complicated or not known, the method of moments may not be easy to apply. In these situations, however, it is generally at least as difficult, if not more so, to find a maximum-likelihood estimator (for example, it may be quite difficult, or even impossible, to differentiate l or $\log l$). In addition, while the method of moments seems intuitively satisfying (we would hope that a sample moment would provide a reasonably "good" estimator of the corresponding population moment), it does not provide as much of a "guarantee" of desirable properties as does the principle of maximum likelihood. As a result, it is not usually used if it is feasible to use the principle of maximum likelihood instead. In reasonably uncomplicated situations, though, the two methods often result in identical or virtually identical estimators. In estimating the parameter p of a Bernoulli process, this is the case, as we have seen, and it is also true in estimating μ or σ^2 of a normal distribution or λ of a Poisson process, to mention but a few examples.

6.9 PARAMETER ESTIMATES BASED ON POOLED SAMPLES

Sometimes we have several independent samples and each provides an estimate of the same parameter or set of parameters. When this happens, there is a real advantage in pooling the sample values to obtain an estimate. These pooled estimates are actually weighted averages of the estimates from the different samples. The big advantage lies in the fact that the sampling error will tend to be smaller for the pooled estimate than for any single sample's value taken alone.

Suppose that there are two independent samples, based on n_1 and n_2 observations, respectively. From each sample a mean m is calculated to estimate μ. Then the **pooled estimate** of the mean is

$$\text{Pooled } m = \text{est. } \mu = \frac{n_1 m_1 + n_2 m_2}{n_1 + n_2}. \tag{6.9.1}$$

For J independent samples, where any given sample j gives an estimate

m_j of μ, the pooled estimate is

$$\text{est. } \mu = \frac{\sum_j n_j m_j}{\sum_j n_j} .$$

(6.9.2)

The fact that the pooled estimator is likely to be better than the single sample estimators taken alone is shown by the standard error of the pooled mean. For J independent samples, with n_1, n_2, ..., n_J as the respective sample sizes, the standard error of the pooled estimator is

$$\frac{\sigma}{\sqrt{\sum_j n_j}} ,$$

(6.9.3*)

which is obviously smaller than any σ/n_j, the standard error of a single one of the J estimators. Naturally, these standard errors refer to the situation where the respective samples all come from the *same* population with mean μ and variance σ^2.

In the same general way, it is possible to find pooled estimators of σ^2. For two independent samples, a pooled estimate is

$$\text{est. } \sigma^2 = \frac{n_1 s_1^2 + n_2 s_2^2}{n_1 + n_2 - 2} ,$$

(6.9.4)

which is an unbiased estimate of σ^2. Pooled estimators can be found for other parameters as well, and we will make use of the idea of pooling in Chapter 7.

6.10   SAMPLING FROM FINITE POPULATIONS

It has been mentioned repeatedly that most sampling problems deal with populations so large that the fact that samples are taken without replacement of single cases can safely be ignored. However, it may happen that the population under study is not only finite, but relatively small, so that the process of sampling without replacement has a real effect on the sampling distribution.

Even in this situation the sample mean is still an unbiased estimator regardless of the size of the population sampled. Hence, no change in procedure for estimation of the mean is needed.

However, for finite populations the sample variance is biased as an estimator of σ^2 in a way somewhat different from the former, infinite-population, situation. When samples are drawn *without replacement* of individuals, the unbiased estimate of σ^2 is

$$\text{est. } \sigma^2 = \frac{n(w-1)}{(n-1)w} s^2, \tag{6.10.1}$$

where w is the *total number* of elements in the population.

Another difference from the infinite-population situation involves the variance of the sampling distribution of means. For a population of size w, from which samples of size n are drawn, the variance of the sample mean is

$$\sigma_M{}^2 = \left(\frac{w-n}{w-1}\right)\frac{\sigma^2}{n} . \tag{6.10.2}$$

The variance of the mean tends to be *somewhat smaller* for a fixed value of n when sampling is from a finite population than when it is from an infinite population. Note that here the size of $\sigma_M{}^2$ depends both on w, the total number in the population, and upon n, the sample size. Note that $\sigma_M{}^2$ is simply σ^2/n multiplied by the "finite population correction" defined in Section 4.9.

For example, suppose that a production manager is interested in the mean diameter of a population of 100 steel rods. It is too troublesome and costly to measure the precise diameter of each rod in the population in order to calculate μ exactly. Thus, the manager decides to take a random sample of 10 rods and to use the sample mean as an estimate of μ. If the sample is taken *without* replacement, then the variance of M is given by Equation (6.10.2), with $w = 100$ and $n = 10$:

$$\sigma_M{}^2 = \left(\frac{100-10}{100-1}\right)\frac{\sigma^2}{10} = \left(\frac{90}{99}\right)\frac{\sigma^2}{10} .$$

On the other hand, if the sample is taken *with* replacement, so that effectively it can be thought of as coming from an infinite population, the variance of M is simply $\sigma^2/n = \sigma^2/10$. In the two different sampling situations, the variances of M differ by a factor of 90/99, which is about .909. Notice that if $n = 50$, the two variances would differ by a factor of $(100 - 50)/(100 - 1) = 50/99$, which is slightly greater than one-half. As noted in Section 4.9, the finite population correction becomes more important as the proportion of the population that is sampled increases. When 50 percent of the population is sampled, the finite population correction has a much greater effect on $\sigma_M{}^2$ than when 10 percent is sampled.

6.11 INTERVAL ESTIMATION: CONFIDENCE INTERVALS

In estimating the mean μ of a normal population, the sample mean M has been shown to be a maximum-likelihood estimator and a method-of-moments estimator. M also satisfies the four desirable properties of point estimators listed in Sections 6.2–6.5. Thus, the sample mean is a good estimator of μ. Nevertheless, it is clear that the sample mean ordinarily will not be exactly equal to the population mean because of sampling error. Similarly, a sample variance will seldom exactly equal the corresponding population variance, a sample proportion will seldom exactly equal the corresponding p from the population, and so on. The point estimate alone does not give us any idea of the magnitude of the possible sampling error. In sampling the weights of packages in a manufacturing situation, can we be reasonably confident that the sample mean is no more than 1 ounce from the population mean? Or, will we be lucky if the sample mean is within 5 pounds of the population mean?

One way to investigate the accuracy of a given estimator is to consider the entire sampling distribution of the estimator. For instance, the amount of dispersion in the sampling distribution gives an indication of the sampling error. At the other extreme, as noted above, just presenting the value of the sample statistic of interest tells the reader nothing about the accuracy of estimation. An intermediate procedure that provides a convenient way of indicating the general magnitude of the sampling error in any given situation is interval estimation.

An interval estimate, or a confidence interval, is a random interval of values with a given probability of covering the true population parameter.

When there is a large degree of sampling error, the confidence interval estimated from any sample will be large; the range of values likely to cover the population parameter is wide. On the other hand, if sampling error is small the parameter is likely to be covered by a small estimated range of values; in this case we can feel confident that the true parameter has been "trapped" within a small range of estimation for any given sample.

To understand the notion of a confidence interval, consider this simple situation. Imagine a normally distributed population, and suppose that we are interested in μ, the population mean. For convenience, assume that the population variance σ^2 is known. For a sample of only one item, the single value of X that is observed serves as an estimator of μ. Moreover, since this is a single value that has been randomly drawn from a normally distributed population with mean μ and variance σ^2, the distribution of X before the sample is taken is identical to the population distribution. Thus, from the tables of the normal distribution, we can determine probabilities for various intervals of values of X. For example, from Table I,

$$P(-1.96 \leq Z \leq +1.96) = .95.$$

That is, about 95 percent of the area under the standard normal density function is between -1.96 and $+1.96$. But $Z = (X - \mu)/\sigma$, so that

$$P\left(-1.96 \leq \frac{X - \mu}{\sigma} \leq +1.96\right) = .95. \qquad (6.11.1)$$

Algebraic manipulations within the parentheses yield

$$P(-1.96\sigma \leq X - \mu \leq +1.96\sigma) = .95,$$

or $\qquad P(-X - 1.96\sigma \leq -\mu \leq -X + 1.96\sigma) = .95,$

or $\qquad P(X + 1.96\sigma \geq \mu \geq X - 1.96\sigma) = .95$

(to change the sign of each term within the parentheses, we must reverse the inequalities). The last expression can be written as

$$P(X - 1.96\sigma \leq \mu \leq X + 1.96\sigma) = .95. \qquad (6.11.2^*)$$

That is, over all possible samples, the probability is about .95 that the interval from $X - 1.96\sigma$ and $X + 1.96\sigma$ will include μ. Therefore, this interval is called a **95 percent confidence interval for μ.** The two boundaries of the interval, $X - 1.96\sigma$ and $X + 1.96 \sigma$, are called **95 percent confidence limits.**

How can we interpret the probability in Equation (6.11.2) and similar statements involving confidence limits? Under the long-run frequency interpretation of probability, μ is a fixed value rather than a random variable. Thus, the probability statement is really not about μ, but about *samples*. The population either does or does not have a μ equal to a given number, and according to the long-run frequency interpretation of probability, it does not make sense to say "the probability is such and such that the true mean takes on the value a." Before the sample is observed, however, X *is* a random variable. In theory, we could consider all possible samples, find X for all samples, and determine a 95 percent confidence interval for each and every sample. The confidence limits for any particular sample depend on the value of X in that sample. In short, over all possible samples there will be many possible 95 percent confidence intervals. Some of these confidence intervals will represent the event "covers the true mean" and others will not. If one such confidence interval is sampled at random, then the probability is .95 that it covers the true mean. The long-run frequency interpretation of a confidence interval, then, is as follows. If samples were drawn repeatedly under identical conditions, and if a 95 percent confidence interval were computed for each of these samples, then in the long run 95 percent of these confidence intervals would "cover," or include, the true value of μ.

It should be pointed out that according to the subjective interpretation of probability, it is possible to consider μ as a random variable and to interpret Equation (6.11.2) as a probability statement about μ. Prior to observing the sample, both X and μ are random variables; after observing X, μ is still a random variable and it is possible to make probability statements about μ. This interpretation will be discussed in more detail in Chapters 8 and 9.

In the above discussion, we have used the terminology "a confidence interval" rather than "the confidence interval." This is because confidence intervals are not unique. For instance, a 95 percent confidence interval for μ in the above situation might be defined in other ways. You can verify for yourself from Table I in the Appendix that the probabilities for other intervals, such as $-1.75 \leq Z \leq 2.33$ and $-2.06 \leq Z \leq 1.88$, are also approximately .95. Using the same procedure as in the case of the interval from -1.96 to 1.96, we find that

$$P(X - 1.75\sigma \leq \mu \leq X + 2.33\sigma) = .95$$

and $\qquad P(X - 2.06\sigma \leq \mu \leq X + 1.88\sigma) = .95.$

Thus, the intervals from $X - 1.75\sigma$ to $X + 2.33\sigma$ and from $X - 2.06\sigma$ to $X + 1.88\sigma$ are also 95 percent confidence intervals for μ. In principle, 95 percent confidence limits for μ might be found in any number of ways, since many different ways exist to find areas equal to .95 under a normal distribution. If so many different ways of defining a 95 percent confidence interval are possible, what is the advantage of choosing the interval from $X - 1.96\sigma$ to $X + 1.96\sigma$? The answer is that, in this case, this particular interval is the *shortest* possible interval yielding "95 percent confidence." Naturally, there is an advantage in pinning the population parameter within the narrowest possible range with a given probability. Moreover, this interval is centered about X, which is a "good" point estimator of μ in this particular situation.

In determining confidence intervals, there is no reason to limit ourselves to 95 percent confidence intervals. For our sample of size 1, for instance, a 99 percent confidence interval for μ is given by $(X - 2.58\sigma, X + 2.58\sigma)$ and a 60 percent confidence interval for μ is given by $(X - .84\sigma, X + .84\sigma)$. Notice that the 99 percent confidence interval is larger than the 95 percent confidence interval but the 60 percent confidence interval is smaller than the 95 percent confidence interval. In order to have greater confidence (that is, larger probability for the interval to include μ), it is necessary to include more possible values.

In general, if

$$P(-a \leq Z \leq a) = 1 - \alpha, \qquad (6.11.3)$$

then the interval

$$X \pm a\sigma. \qquad (6.11.4^*)$$

is a 100 $(1 - \alpha)$ percent confidence interval for μ.

The $(1 - \alpha)$ terminology is related to hypothesis testing and will become clear in the next chapter. The **confidence coefficient** of the interval is $1 - \alpha$.

If we want a confidence interval to have a confidence coefficient of $(1 - \alpha)$, then for the interval given by Equation (6.11.4), a must be the $1 - (\alpha/2)$ **fractile of the standard normal distribution.** For a 95 percent confidence interval, $100(1 - \alpha) = 95$, implying that $1 - \alpha = .95$, or $\alpha = .05$. But if $\alpha = .05$, then $\alpha/2 = .025$, so a should be the $1 - .025 = .975$ fractile of the standard normal distribution. From Table I, $F(1.96)$ is approximately .975, which is why the 95 percent confidence interval symmetric about X has limits of the form $X \pm 1.96\sigma$.

The example given in this section, that of estimating a parameter with a sample of size 1, is highly artificial, so no practical examples will be presented at this point. The general ideas presented here, however, are of broad applicability, and we will use these ideas in the following sections to determine confidence intervals for parameters in various types of situations.

6.12 CONFIDENCE INTERVALS FOR THE MEAN WHEN σ^2 IS KNOWN

The simplest generalization of the example in the previous section is to the situation where a random sample of n independent trials is taken. We will still assume that the population is normally distributed and that the population variance is known. For a sample of size n, the sample mean M is a "good" point estimator of μ. Moreover, we know from Chapter 5 that the sampling distribution of M in sampling from a normal population with known variance is a normal distribution with mean μ and variance σ^2/n. Thus, by an argument identical to that used in the previous section, a 95 percent confidence interval for μ has limits

$$M \pm 1.96 \frac{\sigma}{\sqrt{n}} . \tag{6.12.1*}$$

This can easily be seen by comparing Equation (6.12.1) with Equation (6.11.2) and noticing that both X and M have mean μ, but whereas the standard error of X is just σ, the standard error of M is $\sigma/\sqrt{n}$.

Of course, we are not always interested in a 95 percent confidence interval. For any level of confidence, the limits of a confidence interval for μ can be determined from the formula

$$M \pm a \frac{\sigma}{\sqrt{n}} , \tag{6.12.2*}$$

where a is the $1 - (\alpha/2)$ fractile of the standard normal distribution and the interval estimate is a $100(1 - \alpha)$ percent confidence interval.

For example, suppose that we are interested in the weights of items produced by a certain machine. We are willing to assume that the population of weights is normally distributed, and from past experience concerning the variability in the weight of the output of the machine, we will assume that σ is known to be 8 grams. A random sample of 16 items produced by the machine is observed. When the weight of each item in grams is determined by using an extremely accurate balance, the sample mean is found to be 282 grams. Therefore, a 95 percent confidence interval for μ, the population mean, has limits of

$$282 \pm 1.96 \left(\frac{8}{\sqrt{16}} \right) = 282 \pm 3.92.$$

Thus, 278.08 and 285.92 are the limits of a 95 percent confidence interval. For an 80 percent confidence interval, $1 - \alpha = .80$, so that $\alpha = .20$. For this value of α, a is the .90 fractile of the standard normal distribution. From Table I, then, $a = 1.28$, and an 80 percent confidence interval for μ has limits of

$$282 \pm 1.28 \left(\frac{8}{\sqrt{16}} \right) = 282 \pm 2.56.$$

The interval is $(279.44, 284.56)$.

The width of any confidence interval for μ depends upon the standard error of M, and anything that makes this standard error smaller reduces the width of the interval. Thus, any increase in sample size, operating to reduce $\sigma_M = \sigma/n$, makes the confidence interval shorter. In the above example, when $n = 16$, a 95 percent confidence interval has limits

$$M \pm 1.96(2).$$

However, if $n = 64$, the limits are

$$M \pm 1.96(1),$$

and if $n = 1600$, the limits are

$$M \pm 1.96(.2).$$

Continuing in this way, if the sample size becomes extremely large, the width of the confidence interval approaches zero.

A practical result of this relation between the standard error of the mean and sample size is that *the population mean may be estimated within any desired degree of precision, given a large enough sample size*. This principle has already been introduced in terms of the Chebyshev inequality (Section 3.19), but it can be stated even more strongly when the sampling distribution has some known form, such as the normal distribution.

For instance, how many cases should we sample if the probability is to be .99 that the sample mean lies within $.1\sigma$ of the true mean? That is, the experimenter wants

$$P(\mid M - \mu \mid \leq .1\sigma) = .99.$$

Assuming that the sampling distribution is nearly normal, which it should be if the sample size is large enough, this condition is equivalent to requiring that the 99 percent confidence interval have limits

$$M - .1\sigma$$

and $$M + .1\sigma.$$

But a symmetric 99 percent confidence interval for μ is of the form

$$M \pm 2.58 \frac{\sigma}{\sqrt{n}},$$

so that $$.1\sigma = 2.58 \frac{\sigma}{\sqrt{n}}.$$

Solving for n, we find

$$\sqrt{n} = 25.8,$$

or $$n = 665.64.$$

In short, if the experimenter makes 666 independent observations, the probability of the estimate being wrong by more than $.1\sigma$ is only 1 in 100. Notice that we do not have to say exactly what σ is in order to specify the desired accuracy in σ units and to find the required sample size.

If the experimenter is willing to take a somewhat larger chance of an absolute error exceeding $.1\sigma$, then the required n is smaller. If the probability of an error exceeding $.1\sigma$ in absolute magnitude is to be .05, then

$$.1\sigma = 1.96 \frac{\sigma}{\sqrt{n}},$$

so that $n = 384$.

This idea may be turned around to find the accuracy *almost* insured by any given sample size. Given that n is, say, 30, within how many σ units of the true mean will the sample estimate fall with probability .95? Here, letting k stand for the required number of σ units,

$$k\sigma = 1.96 \frac{\sigma}{\sqrt{n}},$$

or
$$k = \frac{1.96}{\sqrt{30}} = \frac{1.96}{5.48} = .358.$$

For $n = 30$, the chances are about 95 in 100 that our estimate of μ falls within about .36 population standard deviations of the true value, provided that the sampling distribution is approximately normal.

It is obvious that the accuracy of estimation in an experiment can be judged in this way. If the sample size is known, then we can judge how "off" the estimate of the mean is likely to be in terms of the population standard deviation. Furthermore, if we have some idea of the desirable accuracy of estimation, then the sample size can be set so as to attain that degree of accuracy with high probability. This is a very sensible and easy way to determine necessary sample size. Unfortunately, it requires the statistician to think to some extent about the population and what is going to be done with the results. For the problem at hand, just how much error in estimation can be tolerated? We will be better able to answer this question in Chapter 9, when decision theory is presented. One of the important decisions facing the statistician is the determination of sample size, and the decision depends on the trade-off between costs, or losses, due to imprecise estimates and on the cost of sampling. For any contemplated sample, if the anticipated increase in precision is "worth more" to the statistician than the cost of taking the sample, then the sample should be taken. This idea will be elaborated upon in Chapter 9.

In the situation where σ is known, an analogy to the idea of finding confidence intervals for the mean is tossing rings of a certain size at a post. The size of the confidence interval is like the diameter of a ring, and random sampling is like random tosses at the post. The predetermined size of the ring determines the chances that the ring will cover the post on a given try, just as the size of the confidence interval governs the probability that the true value μ will be covered by the range of values in the estimated interval. Reducing the size of σ_M by increasing n is like improving our aim, since we are more likely to cover the true value with an interval of a given size in terms of σ_M. Later, when we deal with situations where the population σ is unknown, the analogy becomes that of tossing rings with a certain *distribution* of sizes at a post; here, the probability that the ring covers the post on a given try depends on the distribution of ring-sizes used, just as the probability that the confidence interval covers the true mean depends on the distribution of estimated confidence intervals over random samples. This idea will be elaborated upon later in this chapter.

Throughout this section, we have assumed that the population of interest is normally distributed. Thus, the sampling distribution of M is a normal distribution, and values of a in Equation (6.12.2) can be found from tables of the normal distribution. If the population is *not* normally distributed, then we cannot say for sure whether M is normally distributed. The basic ideas involving confidence intervals are still valid, but it may be less convenient to calculate confidence limits if normality cannot be assumed. Fortunately, because of the central limit theorem, even if the population is not normal, the distribution of M may be approximately normal if the sample size is not too small. Therefore, except for small samples, the normality assumption does not seem crucial. If the sample is not large enough to make the distribution of M normal or approximately so, then other distributions may be appropriate for determining confidence intervals. A caution is that if the distribution of M is not symmetric, then the smallest intervals giving a particular confidence coefficient may have lower and upper limits that are not equidistant from the sample estimate. In determining confidence intervals for the variance of a normal population later in this chapter, we will encounter a skewed sampling distribution, the chi-square distribution. Here, the upper and lower limits of the confidence interval are not equidistant from the sample estimate.

6.13   THE PROBLEM OF UNKNOWN σ^2: THE T DISTRIBUTION

If we are sampling from a normally distributed population and the mean μ and the variance σ^2 of the population are known, then the standardized random variable

$$Z = \frac{M - \mu}{\sigma/\sqrt{n}} \qquad (6.13.1^*)$$

has a normal distribution. As a result, we can use the facts we know about normal distributions in making inferences based on this sampling distribution. In many situations, however, the variance of the population is *not known exactly*, in which case the standard error of the sample mean cannot be determined. One possible way to allow for this uncertainty about σ^2 is to use an *estimator* of σ^2, namely the sample variance S^2, which we have defined as follows:

$$S^2 = \frac{\sum (X_i - M)^2}{n}.$$

If this is done, the corresponding random variable becomes

$$T = \frac{M - \mu}{S/\sqrt{n-1}}. \tag{6.13.2*}$$

If we use $\hat{S}^2$ instead of S^2 as an estimator of σ^2, then we can write

$$T = \frac{M - \mu}{\hat{S}/\sqrt{n}}. \tag{6.13.3*}$$

There is an important difference between Z and T. For Z, the numerator $M - \mu$ is a random variable, the value of which depends upon the particular sample drawn; the denominator, $\sigma/\sqrt{n}$, is a constant, since it is the same regardless of the particular sample of size n that is observed. For T, the numerator is a random variable identical to the numerator of Z, but the denominator is also a random variable. The particular value of S or $\hat{S}$ is computed from the values observed in the sample. Over different samples, the same value of M will yield precisely the same value of Z but may yield different values of T because different samples with the same M may have different sample variances.

The solution to the problem of the nonequivalence of T and Z rests on the study of T itself as a random variable. Recall that we are assuming that the population is normally distributed. As a result, from Section 5.15, we know that M and S are independent random variables, so that the numerator and denominator of T are independent. Without such independence, which requires a normal population, the sampling distribution of T is difficult to specify exactly. However,

under normality, the distribution of T is quite well known. The density function for T is given by the rule

$$f(t) = G(\nu) \left[1 + \frac{t^2}{\nu} \right]^{-(\nu+1)/2}. \tag{6.13.4}$$

Here, $G(\nu)$ stands for a constant number which depends *only* on the parameter ν (Greek nu); how this number is found need not really concern us. Let us focus our attention on the "working part" of the rule, which involves only ν and T. This looks very different from the normal distribution function rule in Section 4.13. As with the normal function rule, however, a quick look at this mathematical expression tells us much about the distribution of T.

First of all, notice that T enters this rule only as a squared quantity, showing that the distribution of T must be symmetric, since a positive and a negative value having the same absolute size must be assigned the same density by this rule. Second, since all the constants in the function

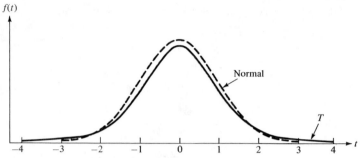

Figure 6.13.1   Distribution of T with $\nu = 4$ and Standardized
Normal Distribution

rule are positive numbers, and the entire term involving T is raised to a
negative power, the largest possible density value is assigned to $T = 0$.
Thus, $T = 0$ is the mode. If we inferred from the symmetry and uni-
modality of this distribution that the mean of T is also 0, we should be
quite correct, provided that $\nu > 1$. In short, the T distribution is a unimodal, symmetric, bell-shaped distribution having a graphic form much
like a normal distribution, even though the two function rules are quite
dissimilar. Loosely speaking, the curve for a T distribution differs from
the standardized normal in being "plumper" in extreme regions and
"flatter" in the central region, as Figure 6.13.1 shows.

Notice that the only unspecified constants in (6.13.4) are ν and $G(\nu)$,
which depends only on ν. This is a **one-parameter distribution**; the
single parameter is ν, called the **degrees of freedom.** *Ordinarily, in most
applications of the T distribution to problems involving a single sample, ν is
equal to $n - 1$, one less than the number of independent observations in the
sample.* In principle, the value of ν can be any positive number, and it just
happens that $\nu = n - 1$ is the value for the degrees of freedom for the
particular T distributions we will use first. Later we will encounter problems calling for T distributions with other numbers of degrees of freedom.
Like most theoretical distributions, the T distribution is actually a family
of distributions, with general form determined by the function rule but
with particular probabilities dictated by the parameter ν. For $\nu > 1$, the
mean of the distribution of T is 0. For $\nu > 2$, the variance of the T distribution is $\nu/(\nu - 2)$, so that the smaller the value of ν, the larger the
variance. As ν becomes large, the variance of the T distribution approaches
1.00, which is the variance of the standardized normal distribution.

Incidentally, the random variable T is often called "Student's T," and
the distribution of T, "Student's distribution." This name comes from the
statistician W. S. Gossett, who was the first to use this distribution in an
important problem and who first published his results in 1908 under the

pen-name "Student." Usually the random variable is denoted by a lower-case t. In order to attempt to maintain our distinction between random variables and their values, we use a capital T for the random variable and a lower-case t for the values.

Unlike the table of the standardized normal function, which suffices for all possible normal distributions, tables of the T distribution must actually include many distributions, each depending on the value of ν, the degrees of freedom. Consequently, tables of T are usually given only in abbreviated form; otherwise, a whole volume would be required to show all the different T distributions we might need.

Table II in the Appendix shows selected percentage points of the distribution of T in terms of the value of ν. Different ν values appear along the left-hand margin of the table. The top margin gives values of the cumulative distribution function,

$$F(t) = P(T \leq t).$$

To illustrate the use of Table II, suppose that we are working with a T distribution with 9 degrees of freedom. Therefore, we must refer to the row corresponding to $\nu = 9$. To find the .90 fractile of this T distribution, look in the column headed .90 and find the value 1.383. This means that for a T distribution with 9 degrees of freedom, the .90 fractile is 1.383, so that 90 percent of the area under the density function is to the left of $t = 1.383$ and 10 percent of the area is to the right of $t = 1.383$ (see Figure 6.13.2).

Notice that in Table II, only values of $F(t)$ greater than .50 are given. By symmetry, however, fractiles in the lower half of the distribution can be found. For instance, the .10 fractile of the T distribution with 9 degrees of freedom would simply be -1.383. The T distribution, like the normal distribution, is symmetric about zero, so probabilities for intervals of values of T can be computed in a similar fashion to computations for intervals of Z, as discussed in Section 4.14.

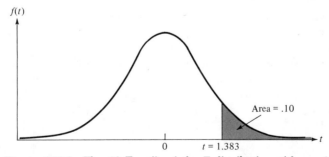

Figure 6.13.2   The .90 Fractile of the T distribution with $\nu = 9$

Sometimes we are given a value t, and we want to find the probability that T is less than or equal to t. For instance, suppose that $\nu = 12$ and that $t = 2.681$. From Table II, we find that $P(T \leq 2.681) = .99$ when $\nu = 12$. Next, for $\nu = 12$ again, suppose we are interested in $P(T \leq 1.05)$. Looking along the row corresponding to $\nu = 12$, we cannot find the value 1.05. The closest values to 1.05 are 0.695 and 1.356, and the corresponding cumulative probabilities are

$$P(T \leq 0.695) = .75 \qquad \text{and} \qquad P(T \leq 1.356) = .90.$$

Since 1.05 is between 0.695 and 1.356, we can infer that $P(T \leq 1.05)$ must be between .75 and .90. However, from Table II, this is all the information that is available about $P(T \leq 1.05)$. As noted above, it is necessary to give tables of the T distribution in abbreviated form, so that it is not always possible to determine probabilities exactly for all intervals. Of course, if we needed more accuracy, we could find more extensive tables of the T distribution. For the purposes of this book, Table II will suffice.

From Table II, you can see how the fractiles of the T distribution change as ν increases. As ν increases, the variance of the T distribution [which is equal to $\nu/(\nu - 2)$ for $\nu > 2$] decreases, approaching a limiting value of one. As the variance decreases, the distribution becomes more concentrated about the mean, zero, so that the various fractiles move closer to zero. For instance, the .95 fractile is 2.015 when $\nu = 5$, 1.812 when $\nu = 10$, 1.725 when $\nu = 20$, and 1.697 when $\nu = 30$. In the limit, as ν approaches infinity, the .95 fractile approaches 1.645. The limiting fractiles of the T distribution as $\nu \to \infty$ are presented in the bottom row of Table II. Moreover, these limiting fractiles are equal to the corresponding fractiles of the standard normal distribution because of the following fact: **as ν grows large, the T distribution approaches the standard normal distribution.**

The practical result of this convergence of the T and the normal distributions is that T *can* be treated as Z provided that the sample size is substantial. The normal probabilities are quite close to, although not identical with, the exact T probabilities for large ν. On the other hand, when the sample size is small the normal probabilities cannot safely be used, and instead we use Table II.

How large is "large enough" to permit use of the normal tables? If the population distribution is truly normal, a sample size of 30 or 40 permits a quite accurate use of the normal tables. Beyond this sample size, the normal probabilities are extremely close to the exact T probabilities. Of course, if a great deal of accuracy is desired, the T tables should be used even for larger sample sizes. For most purposes, a ν of 30 or 40 should be large enough to permit the use of the normal tables.

Recall that the stipulation is made that the population distribution be *normal* when the T distribution is used, even when the normal approximations are substituted for the exact T-distribution probabilities. This requirement that the population be normal limits the usefulness of the T distribution, since this is an assumption that we can seldom really justify in practical situations. Fortunately, when the sample size is fairly large and when that the parent distribution is roughly unimodal and symmetric, the T distribution apparently still gives an adequate approximation to the exact (and often unknown) probabilities of intervals for T under these circumstances. However, you should insist on a relatively *larger* sample size the *less* confident you are that the normal rule holds for the population, if you plan to use the T distribution. In effect, if the sample size is large enough so that the normal probabilities are good approximations to the T probabilities anyway, then the form of the parent distribution is more or less irrelevant. However, often the sample size is so small that the T distribution must be used, and here it *is* somewhat risky to make inferences on the basis of the T distribution unless the population is more or less normally distributed.

6.14 THE CONCEPT OF DEGREES OF FREEDOM

Before we proceed to the uses of the T distribution, we should examine the notion of **degrees of freedom**. The degrees of freedom parameter reflects the fact that a T statistic involves a sample standard deviation as the basis for estimating σ. Recall the basic definition of the sample standard deviation:

$$ s = \sqrt{\frac{\sum (x_i - m)^2}{n}} . $$

The sample standard deviation is based upon a sum of squared deviations from the sample mean. If, however, we take a sum of *unsquared* deviations from the sample mean, we obtain the following result:

$$ \sum (x_i - m) = \sum x_i - \sum m = \sum x_i - nm = \sum x_i - \sum x_i = 0 $$

(the sum of deviations about the mean must be zero).

These two facts have an important consequence. Suppose that you are told that $n = 4$ in some sample and that you are to guess the four deviations from the mean m. For the first deviation you can guess any number, and suppose you say

$$ d_1 = 6. $$

Similarly, quite at will, you could assign values to two more deviations, say

$$d_2 = -9$$

and
$$d_3 = -7.$$

However, when you come to the fourth deviation value, you are *no longer free* to guess any number you please. The value of d_4 *must* be

$$d_4 = 0 - d_1 - d_2 - d_3,$$

or
$$d_4 = 0 - 6 + 9 + 7 = 10.$$

In short, given the values of any $n - 1$ deviations from the mean, which could be any set of numbers, the value of the last deviation is completely determined. Thus, we say that there are $n - 1$ degrees of freedom for a sample variance, reflecting the fact that only $n - 1$ deviations are "free" to be any number and that given these free values, the last deviation is completely determined. It is not the sample size per se that dictates the distribution of T, but rather the number of degrees of freedom in the estimate of σ. We will encounter the concept of degrees of freedom again later in this chapter in connection with the χ^2 and F distributions.

6.15   CONFIDENCE INTERVALS FOR THE MEAN WHEN σ^2 IS UNKNOWN

When we are sampling from a normal population and we do not know the variance of the population, we cannot use the normal distribution and Equation (6.12.2) even though the population itself is normally distributed. Instead, the T distribution is used to establish confidence limits for the mean. For some specified value of ν, then, $100(1 - \alpha)$ percent confidence limits for μ are found from

$$M \pm a(\text{est. } \sigma_M), \tag{6.15.1*}$$

where

$$\text{est. } \sigma_M = \frac{S}{\sqrt{n - 1}} = \frac{\hat{S}}{\sqrt{n}}$$

and a is the $1 - (\alpha/2)$ fractile of the T distribution with ν degrees of freedom. Note the similarity between (6.15.1) and (6.12.2). The only differences are that in (6.15.1), est. σ_M replaces σ_M and a is a fractile of the T distribution instead of the standard normal distribution. If n is large, of course, a can be looked up in the table of the normal distribution because normal probabilities provide very close approximations for T probabilities.

For example, consider once again the example of Section 6.12 involving the weights of items produced by a certain machine. In that example we assumed that from past experience concerning the variability in the weight of the output of the machine, the population standard deviation was known to be 8 grams. In some instances it may be realistic to assume that we know σ but not μ; however, in general it seems more realistic that *both* parameters would not be known. In the example, a random sample of 16 items is observed and the weight of each item in grams is determined. From the 16 observed weights, the sample mean is found to be 282 grams and the sample standard deviation is found to be 9.4 grams.

Since $n = 16$, we are working with a T distribution with $n - 1 = 15$ degrees of freedom. For a 95 percent confidence interval, $1 - \alpha = .95$, so that $\alpha = .05$, and a is the $1 - .025 = .975$ fractile of the T distribution with 15 degrees of freedom. From Table II, then, $a = 2.131$. Thus, the limits of a 95 percent confidence interval for μ are, from Equation (6.15.1),

$$282 \pm 2.131 \left(\frac{9.4}{\sqrt{15}} \right) = 282 \pm 5.17,$$

and the interval is $(276.83, 287.17)$. For an 80 percent confidence interval, $\alpha = .20$, so that a is the $1 - .10 = .90$ fractile of the T distribution with $\nu = 15$. This fractile is, from Table II, 1.341, and the limits of the interval are

$$282 \pm 1.341 \left(\frac{9.4}{\sqrt{15}} \right) = 282 \pm 3.25.$$

An 80 percent confidence interval for μ, then, is $(278.75, 285.25)$. In a similar fashion, intervals could be determined for other levels of confidence.

6.16   CONFIDENCE INTERVALS FOR DIFFERENCES BETWEEN MEANS

Often in statistical work, we are interested not just in a single mean, but in the difference between two means. For example, we may be interested in the difference in the production rates of two machines, the difference in the IQ's of children from two different schools, or the difference in performance of two different types of securities, and so on. The comparison of two populations is of interest more often than the investigation of a single population, and the comparison of primary interest concerns the means of the populations. Suppose that two populations, 1 and 2, are of interest. We draw a sample of size n_1 from population 1 and an *independent* sample of size n_2 from population 2, and we consider the difference between their sample means, $M_1 - M_2$. Now suppose that we continue to draw pairs of independent samples of these sizes from these populations. For each pair of samples drawn, the difference $M_1 - M_2$ is recorded. What is the distribution of such sample *differences* that we should expect in the long run? In other words, what is the sampling distribution of the difference between two means?

You may already have anticipated the form of the sampling distribution of the difference between two means, since all of the groundwork for this distribution has been laid in Section 5.16. The difference between sample means drawn from independent samples is actually a linear combination:

$$(1)M_1 + (-1)M_2.$$

Let us apply the results of Section 5.16 to this problem. In the first place,

$$E(M_1 - M_2) = E(M_1) - E(M_2) = \mu_1 - \mu_2, \qquad (6.16.1^*)$$

which accords with Equation (5.16.2) for any linear combination. Second, from Equation (5.16.3),

$$V(M_1 - M_2) = (1)^2\sigma^2{}_{M_1} + (-1)^2\sigma^2{}_{M_2}$$

$$= \sigma^2{}_{M_1} + \sigma^2{}_{M_2}. \qquad (6.16.2^*)$$

Hence, the standard error of the difference, $\sigma_{\text{diff.}}$, is

$$\sigma_{\text{diff.}} = \sqrt{\sigma^2{}_{M_1} + \sigma^2{}_{M_2}} = \sqrt{\frac{\sigma_1{}^2}{n_1} + \frac{\sigma_2{}^2}{n_2}}, \qquad (6.16.3^*)$$

provided that samples 1 and 2 are completely independent. Notice that there is no requirement at all that the samples be of equal size.

These statements about the mean and the standard error of a difference between means are true regardless of the form of the parent distributions. However, the form of the sampling distribution can also be specified under the following condition.

If the distribution for each of two populations is normal, then the distribution of differences between sample means is normal.

This result follows quite simply from the principle in Section 5.16 for linear combinations. When we can assume both populations to be normal, the form of the sampling distribution is known to be *exactly* normal.

On the other hand, one or both of the original distributions may not be normal; in this case the central limit theorem comes to our aid.

As both n_1 and n_2 grow infinitely large, the sampling distribution of the difference between means approaches a normal distribution regardless of the form of the original distributions.

In short, when we are dealing with two very large samples, the question of the form of the original distributions becomes irrelevant, and we can approximate the sampling distribution of the difference between means by a normal distribution.

In the above discussion, it was assumed not only that the two populations were normal but also that their respective variances were known. If the first assumption holds but the second assumption does not hold, then Equation (6.16.3) cannot be used because the variances are not known. The mean of the sampling distribution of $M_1 - M_2$ is still $\mu_1 - \mu_2$, but since the variances are not known, an estimate of the standard error is needed. To simplify the determination of this estimate, one further assumption is helpful: the assumption that the variances of the two populations are equal.

When $\sigma_1 = \sigma_2 = \sigma$,

$$\sigma_{\text{diff.}} = \sqrt{\frac{\sigma^2}{n_1} + \frac{\sigma^2}{n_2}} = \sqrt{\sigma^2 \left(\frac{1}{n_1} + \frac{1}{n_2} \right)}. \tag{6.16.4*}$$

Now, as we showed in Section 6.9, when we have two or more estimates of the same parameter σ^2, the *pooled* estimate is actually better than either one taken separately. From Equation (6.9.4), it follows that

$$\text{est. } \sigma^2 = \frac{n_1 s_1^2 + n_2 s_2^2}{n_1 + n_2 - 2} \tag{6.16.5}$$

is a "good" estimate of σ^2 based on the two samples. Hence,

$$\text{est. } \sigma_{\text{diff.}} = \sqrt{\text{est. } \sigma^2 \left(\frac{1}{n_1} + \frac{1}{n_2} \right)}$$

$$= \sqrt{\left(\frac{n_1 s_1^2 + n_2 s_2^2}{n_1 + n_2 - 2} \right) \left(\frac{n_1 + n_2}{n_1 n_2} \right)}. \tag{6.16.6}$$

With this estimate of $\sigma_{\text{diff.}}$, the sampling distribution of interest is a T distribution with

$$\nu = (n_1 - 1) + (n_2 - 1) = n_1 + n_2 - 2$$

degrees of freedom.

In order to justify the use of the T distribution in problems involving a difference between means, we must make two assumptions: the populations sampled are normal, and the population variances are homogeneous, σ^2 having the same value for each population. The first assumption, that of a normal distribution in the populations, is apparently the less important of the two. As long as the sample size is even moderate for each group, quite severe departures from normality seem to make little practical difference in the conclusions reached. Naturally, the results are more accurate the more nearly unimodal and symmetric the population distributions are. Thus, if we suspect radical departures from a generally normal form, then we should plan on larger samples.

On the other hand, the assumption of homogeneity of variances is more important. For samples of equal size, relatively big differences in the population variances seem to have relatively small consequences for the conclusions derived from a T statistic. On the other hand, when the variances are quite unequal, the use of different sample sizes can have serious effects on the conclusions. The moral should be plain; given the usual freedom about sample size in experimental work, *when in doubt use samples of the same size.*

However, sometimes it is not possible to obtain an equal number in each group. Then one way to solve this problem is by the use of a correction in the value for degrees of freedom. This is useful when we cannot assume equal population variances and when the samples are of different sizes. In this situation, however, the separate standard errors are computed from each sample and the pooled estimate is not made. Then

$$\text{est. } \sigma_{\text{diff.}} = \sqrt{\text{est. } \sigma^2{}_{M_1} + \text{est. } \sigma^2{}_{M_2}},$$

and the corrected number of degrees of freedom is found from

$$\nu = \frac{(\text{est. } \sigma^2{}_{M_1} + \text{est. } \sigma^2{}_{M_2})^2}{(\text{est. } \sigma^2{}_{M_1})^2/(n_1 + 1) + (\text{est. } \sigma^2{}_{M_2})^2/(n_2 + 1)} - 2. \quad (6.16.7)$$

This need not result in a whole value for ν, in which case the use of the nearest whole value for ν is sufficiently accurate for most purposes. Of course, if the samples are large enough so that the normal approximation is applicable, then the normal tables can be used and it is not necessary to compute ν.

From this discussion of the sampling distribution of $M_1 - M_2$, we can see that confidence intervals for differences in means can be determined either from the normal tables or from the T tables, depending on whether or not the sample size is large and on whether or not the variances of the two populations are known. In either case, of course, it is assumed that the two populations are normally distributed.

If $\sigma_1{}^2$ and $\sigma_2{}^2$ are known, then a $100(1 - \alpha)$ percent confidence interval for $\mu_1 - \mu_2$ is given by

$$(M_1 - M_2) \pm a \sqrt{\frac{\sigma_1{}^2}{n_1} + \frac{\sigma_2{}^2}{n_2}}, \quad (6.16.8^*)$$

where a is determined from the normal tables from our choice of $(1 - \alpha)$.

As in Section 6.12, a is the $1 - (\alpha/2)$ fractile of the standard normal distribution.

For example, suppose that we wish to compare the mean gasoline mileage (in miles per gallon) of one particular car (that is, one particular make and model) with that of another. Samples of 5 cars of the first kind and 7 cars of the second kind are selected at random from large inventories of the cars that are being held by the manufacturers and awaiting shipment to dealers. The 5 cars in the first sample and the 7 cars in the second sample are driven under carefully controlled conditions, and the mileage per gallon is determined for each car. From past data concerning similar cars, the assumption of normality of the populations seems reasonable, and we have considerable information about the variability in mileage per gallon. We are willing to assume that the population variances are known, with $\sigma_1^2 = 4.5$ and $\sigma_2^2 = 4.9$. The sample means turn out to be 18 and 20, respectively, so a 90 percent confidence interval for $\mu_1 - \mu_2$ is given by

$$(18 - 20) \pm 1.645 \sqrt{\frac{4.5}{5} + \frac{4.9}{7}},$$

which simplifies to $(-4.08, .08)$. This interval is 4.16 units wide. If we wanted a more accurate (that is, a narrower) 90 percent confidence interval, we could increase one or both of the sample sizes.

If σ_1^2 and σ_2^2 are *not* known but we can assume that they are equal, then we must use an estimate of the standard error of the difference between means, and a $100(1 - \alpha)$ percent confidence interval for $\mu_1 - \mu_2$ is given by

$$(M_1 - M_2) \pm a \sqrt{\left(\frac{n_1 S_1^2 + n_2 S_2^2}{n_1 + n_2 - 2}\right)\left(\frac{n_1 + n_2}{n_1 n_2}\right)}, \qquad (6.16.9^*) \leftarrow$$

where the degrees of freedom, ν, equals $n_1 + n_2 - 2$. Here a is determined from the T tables; it is the $1 - (\alpha/2)$ fractile of the T distribution with ν degrees of freedom.

In the above example, suppose that the population variances are not known but are assumed to be equal and that the two sample variances are 4.8 and 5.0, respectively. The estimated standard error of the difference between sample means is found by the pooling procedure of Equation (6.16.6):

$$\text{est. } \sigma_{\text{diff.}} = \sqrt{\frac{5(4.8) + 7(5.0)}{10}\left(\frac{5 + 7}{5(7)}\right)} = \sqrt{2.02} = 1.42.$$

We are dealing with a T distribution with $n_1 + n_2 - 2$ degrees of freedom, or 10 degrees of freedom; thus, if we want a 90 percent confidence interval for $\mu_1 - \mu_2$, the limits of the interval are

$$(18 - 20) \pm a(1.42).$$

But $a = 1.812$, from Table II, so the limits are -2 ± 2.57, and the interval is $(-4.57, .57)$.

6.17 APPROXIMATE CONFIDENCE INTERVALS FOR PROPORTIONS

Forming an exact confidence interval for a proportion is a considerably more complicated business than it is for a single mean or for a difference between two means. For small samples, this is because the distribution we are dealing with is a discrete distribution rather than a continuous distribution. For large samples (large enough so that the normal approximation is applicable), it is because the standard error of a proportion R/n depends on the true population parameter p (recall that the standard error of the mean does not depend on μ). Thus, since a sample only provides us with an *estimate* of p, the true standard error is unknown unless p is known. Obviously, this could not be the case, for if p were known, we would certainly not be attempting to determine a confidence interval for p!

When n is small, confidence intervals for p can be determined directly from tables of the binomial distribution. An understandable discussion of this procedure would require several examples, so we will confine ourselves to the large-sample case. When n is large, so that the sampling distribution of the proportion is approximately normal, $100(1 - \alpha)$ percent confidence limits for a proportion are given approximately by

$$\frac{n}{n + a^2}\left[\frac{R}{n} + \frac{a^2}{2n} \pm a\sqrt{\frac{R(n - R)}{n^3} + \frac{a^2}{4n^2}}\right], \qquad (6.17.1)$$

where a is the $1 - (\alpha/2)$ fractile of the standard normal distribution.

This rather mysterious-looking form for finding the confidence limits for a proportion can be justified as follows. Since the sampling distribution of R/n for large samples can be assumed approximately normal, the appropriate confidence limits should be approximately

$$\frac{R}{n} \pm a\sqrt{\frac{pq}{n}}.$$

The possible values of p corresponding to the limits thus satisfy the relation

$$\left(p - \frac{R}{n}\right)^2 = a^2\left(\frac{pq}{n}\right).$$

We can substitute $1 - p$ for q, so that the equation is given in terms of p, R, n, and a. Notice that the equation is a quadratic equation in terms of p, since it can be expressed in the form

$$(n + a^2)p^2 - (2R + a^2)p + \left(\frac{R^2}{n}\right) = 0.$$

Solving for p by the usual rule for solving a quadratic equation, we get

$$\frac{(2R + a^2) \pm \sqrt{(2R + a^2)^2 - 4(R^2/n)(n + a^2)}}{2(n + a^2)},$$

which simplifies to Equation (6.17.1).

If n is very large, Equation (6.17.1) may be replaced by the simpler approximation

$$\frac{R}{n} \pm a \sqrt{\frac{R(n - R)}{n^3}}. \tag{6.17.2}$$

For example, suppose that we are interested in the proportion of consumers who will purchase a new product, and we are willing to assume that the purchase behavior can be treated as a Bernoulli process. If a sample of 100 consumers is taken, and exactly 40 of the consumers in the sample purchase the product, then the sample proportion, .40, can be used to estimate the population p. If a 95 percent confidence interval for p is wanted, we can use Equation (6.17.1):

$$\frac{100}{100 + (1.96)^2}\left[\frac{40}{100} + \frac{(1.96)^2}{200} \pm 1.96 \sqrt{\frac{40(60)}{(100)^3} + \frac{(1.96)^2}{4(100)^2}}\right],$$

which simplifies to (.309, .498). Using the approximation in Equation (6.17.2), we get

$$\frac{40}{100} \pm 1.96 \sqrt{\frac{40(60)}{(100)^3}},$$

which simplifies to (.304, .496). Thus, in this instance, the two formulas result in limits that differ only at the third decimal place.

6.18   THE CHI-SQUARE DISTRIBUTION

Thus far we have been concerned primarily with the sampling distribution of the sample mean M and with differences in sample means. In some instances, the sampling distribution of the sample variance is of interest. If the population from which we are sampling is normally distributed, this sampling distribution can be expressed in terms of a χ^2 (chi-square) distribution. We will first discuss the theoretical development of the chi-square distribution and then show its relationship to the sampling distribution of the sample variance.

Suppose that the random variable X is normally distributed with mean $E(X) = \mu$ and variance $V(X) = E(X - \mu)^2 = \sigma^2$. A sample of size 1 is to be taken, and we consider the **squared standardized value**

$$Z^2 = \frac{(X - \mu)^2}{\sigma^2}. \tag{6.18.1}$$

Let us call this squared standardized value $\chi^2_{(1)}$, so that

$$\chi^2_{(1)} = Z^2. \tag{6.18.2}$$

Now we will look into the sampling distribution of this variable $\chi^2_{(1)}$.

First of all, what is the range of values that χ^2 might take on? The original normal variable X ranges over all real numbers, and this is also the range of the standardized variable Z. However, $\chi^2_{(1)}$ is always a squared quantity, and so its range must be all the *nonnegative* real numbers. We can also infer something about the form of the distribution of $\chi^2_{(1)}$. The bulk of the cases (about 68 percent) in a normal distribution of standardized scores must lie between -1 and 1. Given a z between -1 and 1, the corresponding $\chi^2_{(1)}$ value lies between 0 and 1, so that the bulk of this sampling distribution will fall in the interval between 0 and 1. This implies that the form of the distribution of $\chi^2_{(1)}$ will be skewed, with a high probability for a value in the interval from 0 to 1 and relatively low probability in the interval from 1 to ∞. The graph of the distribution of $\chi^2_{(1)}$ is represented in Figure 6.18.1.

Figure 6.18.1 pictures **the chi–square distribution with 1 degree**

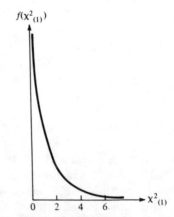

Figure 6.18.1   The Distribution of χ^2 for $\nu = 1$

of freedom. The distribution of the random variable $\chi^2_{(1)}$, where

$$\chi^2_{(1)} = \frac{(X - \mu)^2}{\sigma^2} \tag{6.18.3}$$

and X is normally distributed with mean μ and variance σ^2, is a chi-square distribution with 1 degree of freedom.

Now let us go a little further. Suppose that samples of *two* cases are drawn independently and at random from a normal distribution. We find the squared standardized value corresponding to each observation:

$$Z_1{}^2 = \frac{(X_1 - \mu)^2}{\sigma^2},$$

$$Z_2{}^2 = \frac{(X_2 - \mu)^2}{\sigma^2}.$$

The *sum* of these two squared standardized values is found and designated as the random variable $\chi^2_{(2)}$:

$$\chi^2_{(2)} = \frac{(X_1 - \mu)^2}{\sigma^2} + \frac{(X_2 - \mu)^2}{\sigma^2} = Z_1{}^2 + Z_2{}^2. \tag{6.18.4}$$

If we look into the distribution of the random variable $\chi^2_{(2)}$, we find that the range of possible values extends over all nonnegative real numbers. However, since the random variable is based on *two* independent observations, the distribution is somewhat less skewed than for $\chi^2_{(1)}$; the probability is not so high that the sum of two squared standardized values should fall between 0 and 1. This is illustrated in Figure 6.18.2. This illustrates a **chi-square distribution with 2 degrees of freedom.**

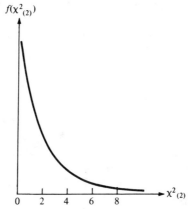

Figure 6.18.2 The Distribution of χ^2 for $\nu = 2$

Finally, suppose that we take n independent observations at random from a normal distribution with mean μ and variance σ^2 and define

$$\chi^2_{(n)} = \frac{\sum_{i=1}^{n}(\dot{X}_i - \mu)^2}{\sigma^2} = \sum_i Z_i^2. \qquad (6.18.5^*)$$

The distribution of this random variable has a form that depends upon the number of independent observations.

In general, for n independent observations from a normal population, the sum of the squared standardized values for the observations has a chi–square distribution with n degrees of freedom.

Notice that the standardized values must be relative to the *population mean* and to the *population standard deviation*. Chi-square distributions for larger numbers of degrees of freedom have the form illustrated in Figure 6.18.3.

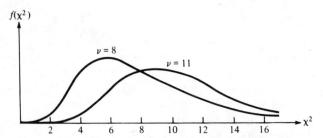

Figure 6.18.3   The General Form of the Distribution of χ^2 for Larger Numbers of Degrees of Freedom

The function rule assigning a probability density to each possible value of χ^2 is given by

$$f(\chi^2) = h(\nu) \exp\left(-\frac{\chi^2}{2}\right) (\chi^2)^{(\nu/2)-1} \quad \text{for } \chi^2 \geq 0, \quad \nu > 0. \qquad (6.18.6^*)$$

As shown in Figures 6.18.1 through 6.18.3, the plot of this density function presents the picture of a very positively skewed distribution, at least for relatively small values of ν, with a distinct mode at the point $\nu - 2$ for $\nu > 2$. However, as ν is increased, the form of the distribution appears to be less skewed to the right. In this function rule, the value $h(\nu)$ is a constant depending only on the parameter ν. Unlike the normal and the T distributions, it is not as easy to infer the characteristics of the distribution of χ^2 from this function rule. However, the thing to notice is that there is only one value other than χ^2 that must be specified in order to find the density: the parameter ν. Like the T distribution, the distribution of χ^2 depends only on the degrees of freedom, the parameter ν. The members of the family of chi-square distributions all follow the same general rule, but the exact form of the distribution depends on the number of degrees of freedom, ν.

From the original definition of χ^2, it is quite easy to infer what the mean of a chi-square distribution must be:

$$E[\chi^2{}_{(n)}] = E(\sum_1^n Z_i{}^2) = \sum_1^n E(Z_i{}^2). \tag{6.18.7}$$

However, for any i, $E(Z_i{}^2) = 1$, by the definition of a standardized random variable. Thus,

$$E[\chi^2{}_{(n)}] = n. \tag{6.18.8*} \leftarrow$$

If a chi-square variable has a distribution with ν degrees of freedom, then the expectation of the distribution is simply ν.

The variance of a chi-square distribution with ν degrees of freedom is always

$$V[\chi^2{}_{(\nu)}] = 2\nu. \tag{6.18.9*} \leftarrow$$

The degrees of freedom ν provides all the information we need to specify the particular chi-square distribution completely.

Like the T distribution, the particular distribution of χ^2 depends on the parameter ν, and it is difficult to give tables of the distribution for all values of ν that we might need. Thus, Table III in the Appendix, like Table II, is a condensed table, showing certain fractiles of the χ^2 distribution for various values of $F(\chi^2)$. The rows of Table III list various degrees of freedom ν and the column headings are cumulative probabilities $F(\chi^2)$, just as in Table II. Because the χ^2 distribution is not symmetric, it is necessary to list cumulative probabilities less than .50 as well as cumulative probabilities greater than .50.

For example, suppose that we are dealing with a chi-square distribution with 5 degrees of freedom, and we want to find the .75 fractile of this distribution. Looking in Table III under the row for $\nu = 5$ and the column for $F(\chi^2) = .75$, the .75 fractile is 6.62568. As another example, look at the row labeled $\nu = 2$ and the column headed .90. The entry in this combination of row and column indicates that in a distribution with 2 degrees of freedom, chi-square values of 4.605 or less should occur with probability of about 9 in 10 under random sampling. Finally, look at the row for $\nu = 24$ and the column for .005; the cell entry indicates that chi-square values of 9.886 or less occur with probability of only about .005 in a distribution with 24 degrees of freedom.

When several independent random variables each have a chi-square distribution, then the distribution of the *sum* of the variables is also known. This is a most important and useful property of variables of this kind, and it can be stated more precisely as follows.

If a random variable $\chi^2_{(\nu_1)}$ has a chi–square distribution with ν_1 degrees of freedom and an independent random variable $\chi^2_{(\nu_2)}$ has a chi–square distribution with ν_2 degrees of freedom, then the new random variable formed from the sum of these variables,

$$\chi^2_{(\nu_1+\nu_2)} = \chi^2_{(\nu_1)} + \chi^2_{(\nu_2)}, \tag{6.18.10*}$$

has a chi-square distribution with $\nu_1 + \nu_2$ degrees of freedom.

In short, the new random variable formed by taking the sum of two independent chi-square variables is itself distributed as χ^2, with degrees of freedom equal to the sum of those for the original distributions. A little thought should convince you that this must be so; if $\chi^2_{(\nu_1)}$ is the sum of squares of ν_1 independent variables, each normally distributed with mean 0 and variance 1, and $\chi^2_{(\nu_2)}$ is based on the sum of ν_2 independent such squares, then $\chi^2_{(\nu_1+\nu_2)}$ must be a sum of $\nu_1 + \nu_2$ squared values of independent normal variables, each with mean 0 and variance 1. Therefore, $\chi^2_{(\nu_1+\nu_2)}$ qualifies as a chi-square variable by definition.

As the number of degrees of freedom grows infinitely large, the distribution of χ^2 approaches the normal distribution. You should hardly find this principle surprising by now! Actually, the same mechanism is at work here as in the central limit theorem. For any ν greater than 1, a chi-square variable is equivalent to a *sum* of ν independent random variables. This fact is of more than theoretical interest when very large samples are used. For very large ν, the probability for any interval of values of $\chi^2_{(\nu)}$ can be found from normal standardized values of the form

$$z = \frac{\chi^2_{(\nu)} - \nu}{\sqrt{2\nu}}, \tag{6.18.11*}$$

since the mean and variance of a chi-square distribution are ν and 2ν, respectively. However, this approximation is not good unless ν is extremely large. A somewhat better approximation procedure is to find a fractile of χ^2 by using the formula

$$\chi^2 = \tfrac{1}{2}\{z + \sqrt{2\nu - 1}\}^2, \tag{6.18.12}$$

where z is the corresponding fractile of the standard normal distribution. The necessary values of z are given at the bottom of Table III and can also be found from Table I.

6.19   THE DISTRIBUTION OF THE SAMPLE VARIANCE FROM A NORMAL POPULATION

In this section, the sampling distribution of the sample variance will be studied. It will be assumed that the population actually sampled is normal, and the results in this section apply, strictly speaking, only when this assumption is true.

Recall that the sample variance was defined by

$$S^2 = \frac{\sum\limits_{i=1}^{n} (X_i - M)^2}{n}$$

Suppose, however, that we actually know the value of μ. For any single deviation from μ, $(X_i - \mu)$, it is true that

$$(X_i - \mu) = (X_i - M) + (M - \mu), \qquad (6.19.1)$$

since adding and subtracting the same number to any given number does not change the value. Now, for any squared deviation from μ,

$$(X_i - \mu)^2 = [(X_i - M) + (M - \mu)]^2$$
$$= (X_i - M)^2 + (M - \mu)^2 + 2(X_i - M)(M - \mu). \quad (6.19.2)$$

Summing over all of the squared deviations from μ in the sample, we have

$$\sum_i (X_i - \mu)^2 = \sum_i (X_i - M)^2 + \sum_i (M - \mu)^2$$
$$+ 2 \sum_i (X_i - M)(M - \mu). \qquad (6.19.3)$$

The value of $(M - \mu)^2$ is constant over all i, so

$$\sum_i (M - \mu)^2 = n(M - \mu)^2.$$

Furthermore, since $(M - \mu)$ is the same for all i,

$$\sum_i (X_i - M)(M - \mu) = (M - \mu) \sum_i (X_i - M) = 0$$

because the sum of the deviations about a sample mean must be 0. Hence,

$$\sum_i (X_i - \mu)^2 = \sum_i (X_i - M)^2 + n(M - \mu)^2. \qquad (6.19.4^*)$$

If we divide the entire expression by σ^2, we have

$$\frac{\sum_i (X_i - \mu)^2}{\sigma^2} = \frac{\sum_i (X_i - M)^2}{\sigma^2} + \frac{n(M - \mu)^2}{\sigma^2}. \qquad (6.19.5)$$

Now look at the term on the left in this expression. Given n independent trials from a normal population distribution,

$$\frac{\sum_i (X_i - \mu)^2}{\sigma^2} = \chi^2_{(n)}, \qquad (6.19.6^*)$$

so that this sum of squared deviations from μ, divided by σ^2, qualifies as a chi-square variable with n degrees of freedom.

Furthermore, if the population distribution is normal, the sampling distribution of M must be normal with mean μ and variance σ^2/n. Thus,

$$\frac{n(M - \mu)^2}{\sigma^2} = \frac{(M - \mu)^2}{(\sigma^2/n)} = \chi^2_{(1)}. \qquad (6.19.7^*)$$

Using Equations (6.19.6) and (6.19.7) in Equation (6.19.5), we have

$$\chi^2_{(n)} = \frac{\sum_i (X_i - M)^2}{\sigma^2} + \chi^2_{(1)}, \qquad (6.19.8)$$

or

$$\chi^2_{(n)} = \frac{nS^2}{\sigma^2} + \chi^2_{(1)}. \qquad (6.19.9^*)$$

The term on the left is a chi-square variable with n degrees of freedom, and the term on the extreme right is a chi-square variable with 1 degree of freedom. Finally, it can be shown, by use of the addition property of chi-square variables and the fact that M and S^2 are independent in samples from a normally distributed population, that **the ratio nS^2/σ^2 is a chi-square variable with $n - 1$ degrees of freedom:**

$$\frac{nS^2}{\sigma^2} = \chi^2_{(n-1)}. \qquad (6.19.10^*)$$

If, instead of S^2, we use the unbiased estimator $\hat{S}^2$, we find that

$$\frac{(n - 1)\hat{S}^2}{\sigma^2} = \chi^2_{(n-1)}. \qquad (6.19.11^*)$$

Moreover, this same idea applies to the sampling distribution of any estimate of a variance, provided that the basic population sampled is normal. Thus, it is also true that when a pooled estimate of σ^2 is made from two independent samples of n_1 and n_2 cases,

$$\frac{(n_1 + n_2 - 2) \text{ est. } \sigma^2}{\sigma^2} = \chi^2_{(n_1+n_2-2)}. \qquad (6.19.12^*)$$

The pooled sample estimate of σ^2, multiplied by the degrees of freedom and divided by the true value of σ^2, is a chi-square variable with $\nu = n_1 + n_2 - 2$.

The sampling distribution of the unbiased estimator $\hat{S}^2$ is actually the distribution of the variable

$$\hat{S}^2 = \frac{\chi^2_{(\nu)}\sigma^2}{\nu}, \qquad (6.19.13^*) \quad \leftarrow$$

a chi-square variable multiplied by σ^2 and divided by degrees of freedom. However, inferences are usually made in terms of the distribution of $(n - 1)\hat{S}^2/\sigma^2$, since this is just a chi-square variable; thus, we can get by with only one chi-square table for this as well as for other uses of this distribution.

6.20   CONFIDENCE INTERVALS FOR THE VARIANCE AND STANDARD DEVIATION

Finding confidence intervals for the variance is quite simple, provided that the normal distribution rule holds for the population. Suppose that we have a sample of n independent observations, and we want 95 percent confidence limits for σ^2. For samples from a normal distribution, it must be true that

$$P\left[a \leq \frac{(n - 1)\hat{S}^2}{\sigma^2} \leq b\right] = .95, \qquad (6.20.1^*) \quad \leftarrow$$

where a and b are the .025 fractile and the .975 fractile, respectively, of the chi-square distribution with $n - 1$ degrees of freedom. But the terms within the square brackets can be manipulated algebraically as follows:

$$P\left[a \leq \frac{(n - 1)\hat{S}^2}{\sigma^2} \leq b\right] = P\left[\frac{1}{a} \geq \frac{\sigma^2}{(n - 1)\hat{S}^2} \geq \frac{1}{b}\right]$$

$$= P\left[\frac{1}{b} \leq \frac{\sigma^2}{(n - 1)\hat{S}^2} \leq \frac{1}{a}\right]$$

$$= P\left[\frac{(n - 1)\hat{S}^2}{b} \leq \sigma^2 \leq \frac{(n - 1)\hat{S}^2}{a}\right].$$

That is, the probability is .95 that the true value of σ^2 will be covered by an interval with limits

$$\frac{(n-1)\hat{S}^2}{b} \quad \text{and} \quad \frac{(n-1)\hat{S}^2}{a}.$$

In general, a $100(1-\alpha)$ **confidence interval for** σ^2 **is of this same form:**

$$\left[\frac{(n-1)\hat{S}^2}{b}, \frac{(n-1)\hat{S}^2}{a}\right], \tag{6.20.2*}$$

where a **and** b **are the** $\alpha/2$ **fractile and the** $1-(\alpha/2)$ **fractile of the chi–square distribution with** $n-1$ **degrees of freedom.** In terms of S^2 rather than $\hat{S}^2$, the interval is

$$\left[\frac{nS^2}{b}, \frac{nS^2}{a}\right]. \tag{6.20.3*}$$

These limits can always be turned into confidence limits for σ by replacing each limiting value by its square root.

For example, consider the example of Section 6.15 involving the weights of items produced by a certain machine. It was assumed in that example that the population was normally distributed, and a sample of size 16 was considered, yielding a sample standard deviation of 9.4 grams. Suppose that we want a 95 percent confidence interval for σ^2. Thus, we need to find the .025 and .975 fractiles of the chi-square distribution with $n-1=15$ degrees of freedom. From Table III, these values are $a = 6.26214$ and $b = 27.4884$, and, from Equation (6.20.3), a 95 percent confidence interval for σ^2 is

$$\left[\frac{16(9.4)^2}{27.4884}, \frac{16(9.4)^2}{6.26214}\right] = (51.43, 225.76).$$

There is an important difference between the confidence intervals given by Equations (6.20.2) and (6.20.3) and the confidence intervals determined in previous sections by the use of the normal and T distributions. Since the normal and T distributions are symmetric, the confidence interval symmetric about a point estimator such as M or $M_1 - M_2$ is also the shortest possible interval with the given degree of confidence. In the case of confidence intervals for σ^2, on the other hand, the chi-square distribution is used, and this distribution is asymmetric. Notice that the procedure outlined in this section gives us intervals that are *symmetric in terms of probability*; that is, a and b in Equations (6.20.2) and (6.20.3) are determined so as to "cut off" an equal amount of probability, $\alpha/2$, in each tail of the chi-square distribution. Because of the asymmetry of the distribution, however, the limits of the interval are not equidistant from the sample variance. In the above example, the sample variance is $(9.4)^2$, or 88.36. The limits of the 95 percent confidence interval calculated from Equation (6.20.3) are 51.43 and 225.76; these values clearly are not equidistant from 88.36. The upper limit is much further from 88.36 than is the lower limit, reflecting the positive skewness of the distribution.

Regardless of whether confidence intervals are found by the method given here or by an alternative method giving shortest intervals, it is true that the larger the sample size, the narrower is the interval with a given probability of covering the true value of σ^2. For this reason, when we are estimating σ^2 for any purpose, the larger the sample size, the more confident we can be of coming close to the value of σ^2 with our sample estimate $\hat{s}^2$ or s^2.

In Section 6.13 it was pointed out that although the rationale for the T distribution demands the assumption of a normal population distribution, in practice the T distribution may be applied when the parent distribution is not normal, provided that n is at least moderately large. *This is not the case for inferences about the variance, however.* We run a considerable risk of error in using the chi-square distribution either to test a hypothesis about a variance or to find confidence limits unless the population distribution is normal or approximately so. The effect of the violation of the normality assumption is usually minor for large n but can be quite serious when inferences are made about variances for moderate n. This is an important principle, and it will be emphasized several times in succeeding sections.

6.21 THE F DISTRIBUTION

Some problems center on the value of a single population variance, but often a comparison of the variances of *two* populations is of interest. The F distribution is used to make inferences of this sort.

Imagine two distinct populations, each having a normal distribution. We draw two independent random samples; the first sample, from population 1, contains n_1 cases, and that from population 2 consists of n_2 cases. From each sample, an unbiased estimate of the population variance is computed, and from the results of Section 6.19, we know that

$$\frac{(n-1)\hat{S}_1^2}{\sigma_1^2} = \chi^2_{(n-1)},$$

so that

$$\hat{S}_1^2 = \frac{\sigma_1^2 \chi^2_{(\nu_1)}}{\nu_1}.$$

Similarly,

$$\hat{S}_2^2 = \frac{\sigma_2^2 \chi^2_{(\nu_2)}}{\nu_2}.$$

Now for each possible pair of samples, one from population 1 and one from population 2, we take the *ratio* of $\hat{S}_1^2$ to $\hat{S}_2^2$ and call this ratio the random variable F:

$$F = \frac{\hat{S}_1^2}{\hat{S}_2^2} = \frac{\text{est. } \sigma_1^2}{\text{est. } \sigma_2^2}. \qquad (6.21.1^*)$$

What sort of distribution should the random variable F actually have when the hypothesis that $\sigma_1^2 = \sigma_2^2$ is true? First of all, notice that when we put the two variance estimates in a ratio, this is actually the ratio of two independent chi-square variables, each divided by its degrees of freedom:

$$F = \frac{[\chi^2_{(\nu_1)}/\nu_1]}{[\chi^2_{(\nu_2)}/\nu_2]}, \qquad (6.21.2^*)$$

provided that $\sigma_1^2 = \sigma_2^2$.

Showing F as a ratio of two chi-square variables, each divided by its ν, is actually a way of defining the F variable.

A random variable formed from the ratio of two independent chi-square variables, each divided by its degrees of freedom, is said to be an F ratio and to follow the rule for the F distribution.

When this definition is satisfied, meaning that both parent populations are normal, have the same variance, and that the samples drawn are independent, then the theoretical distribution of F values can be found.

The density function for F is much too complicated to allow us to gain much idea of the form of the distribution by looking at its mathematical expression. Suffice it to say that the density function for F depends only upon two parameters, ν_1 and ν_2, which can be thought of as *the degrees of freedom associated with the numerator and the denominator of the F ratio*. The range for F is the set of *nonnegative real numbers*. The expectation of F is $\nu_2/(\nu_2 - 2)$, provided that $\nu_2 > 2$. In general form, the distribution for any fixed ν_1 and ν_2 is asymmetric, although the particular "shape" of the density function varies considerably with changes in ν_1 and ν_2.

Since the distribution of F depends upon two parameters ν_1 and ν_2, it is even more difficult to present tables of F distributions than those of χ^2 or T. Tables of F are usually encountered only in drastically condensed form, including very few fractiles of the distribution. For instance, Table IV in the back of this book presents only three fractiles: the .95 fractile, the .975

fractile, and the .99 fractile. These fractiles are given for various combinations of ν_1 and ν_2, with each fractile given in a separate table. The columns of each table give values of ν_1, the degrees of freedom for the numerator of F, and the rows give values of ν_2, the degrees of freedom for the denominator.

The use of Table IV can be illustrated in the following way. Suppose that two independent samples are drawn, containing $n_1 = 10$ and $n_2 = 6$ cases, respectively. The degrees of freedom associated with the two variances are $\nu_1 = 10 - 1 = 9$ and $\nu_2 = 6 - 1 = 5$. For the ratio

$$F = \frac{\hat{S}_1{}^2}{\hat{S}_2{}^2},$$

the degrees of freedom must be 9 for the numerator and 5 for the denominator. Now suppose that $F = 7.00$. Does this fall into the upper .05 of values in an F distribution with 9 and 5 degrees of freedom? We turn to Table IV and find the page for the .95 fractile. Then we look at the column for $\nu_1 = 9$ and the row for $\nu_2 = 5$. The tabled value is 4.77. Our obtained F value of 7.00 exceeds 4.77, and thus our sample result falls among the upper 5 percent in an F distribution.

Suppose that we are interested in the *lower* rather than the upper tail of an F distribution. Let $F_{(\nu_1;\nu_2)}$ stand for an F ratio with ν_1 and ν_2 degrees of freedom, and let $F_{(\nu_2;\nu_1)}$ be an F ratio with ν_2 degrees of freedom in the *numerator* and ν_1 degrees of freedom in the *denominator*. The numbers of degrees of freedom in the numerator and the denominator are simply reversed for these two F ratios. Then it is true that, for any positive number c,

$$P(F_{(\nu_1;\nu_2)} \geq c) = P\left[F_{(\nu_2;\nu_1)} \leq \frac{1}{c}\right]; \qquad (6.21.3^*)$$

the probability that $F_{(\nu_1;\nu_2)}$ is greater than or equal to c is the same as the probability that the reciprocal of $F_{(\nu_1;\nu_2)}$ is less than or equal to the *reciprocal* of c. Practically, this means that the *value required for F in the lower tail of some particular distribution can always be found by finding the reciprocal of the corresponding value required in the upper tail of a distribution with numerator and denominator degrees of freedom reversed.*

For instance, suppose that we want the F value that cuts off the *bottom* .05 of sample values in a distribution with 7 and 10 degrees of freedom. We find this by first locating the value which cuts off the *top* .05 in a distribution with 10 and 7 degrees of freedom: this is 3.64, from Table IV. Then the F value we are looking for in the *lower* tail when 7 and 10 are the degrees of freedom is the reciprocal of 3.64, $1/3.64 = .27$. Thus, .27 is the .05 fractile of the F distribution with 7 and 10 degrees of freedom.

6.22 CONFIDENCE INTERVALS FOR RATIOS OF VARIANCES

Finding confidence intervals for ratios of variances requires a direct application of the F distribution. If two populations are normally distributed and we take independent samples from the two populations of size n_1 and n_2, obtaining variance estimators $\hat{S}_1^2$ and $\hat{S}_2^2$, then the ratio

$$F = \frac{\hat{S}_1^2/\sigma_1^2}{\hat{S}_2^2/\sigma_2^2}$$

has an F distribution with $n_1 - 1$ and $n_2 - 1$ degrees of freedom, from Section 6.21. But this implies that

$$P\left(a \leq \frac{\hat{S}_1^2/\sigma_1^2}{\hat{S}_2^2/\sigma_2^2} \leq b\right) = 1 - \alpha, \qquad (6.22.1^*)$$

where a and b are the $\alpha/2$ and the $1 - (\alpha/2)$ fractiles of the F distribution with $n_1 - 1$ and $n_2 - 1$ degrees of freedom. Equation (6.22.1) is equivalent to the following expression:

$$P\left(\frac{a\hat{S}_2^2}{\hat{S}_1^2} \leq \frac{\sigma_2^2}{\sigma_1^2} \leq \frac{b\hat{S}_2^2}{\hat{S}_1^2}\right) = 1 - \alpha. \qquad (6.22.2^*)$$

Thus, a $100(1 - \alpha)$ percent confidence interval for σ_2^2/σ_1^2 has limits

$$\frac{a\hat{S}_2^2}{\hat{S}_1^2} \quad \text{and} \quad \frac{b\hat{S}_2^2}{\hat{S}_1^2}. \qquad (6.22.3^*)$$

For example, consider the example from Section 6.16 concerning gasoline mileages of two types of cars. It is assumed that the populations are normally distributed, and samples of $n_1 = 5$ cars of the first type and $n_2 = 7$ cars of the second type are selected at random from large inventories of the cars that are being held by the manufacturers and awaiting shipment to dealers. The sample variances are $S_1^2 = 4.8$ and $S_2^2 = 5.0$. Since the limits in Equation (6.22.3) require the unbiased sample variances, it is necessary to convert as follows:

$$\hat{S}_1^2 = \left(\frac{n_1}{n_1 - 1}\right) S_1^2 = \left(\frac{5}{4}\right)(4.8) = 6.0$$

and $\qquad \hat{S}_2^2 = \left(\dfrac{n_2}{n_2 - 1}\right) S_2^2 = \left(\dfrac{7}{6}\right)(5.0) = 5.83.$

In this situation, we are working with the F distribution with 4 and 6 degrees of freedom. If we want a 95 percent confidence interval, we need to find the .025 and .975 fractiles of this distribution. From Table IV, the .975 fractile is $b = 6.23$. The .025 fractile can be found from Equation (6.21.3) by taking the reciprocal of the .975 fractile of the F distribution with 6 and 4 degrees of freedom. From Table IV, this latter fractile is 9.20, so $a = 1/9.20$. From Equation (6.22.3), then, a 95 percent confidence interval for $\sigma_2{}^2/\sigma_1{}^2$ has limits

$$\left(\frac{1}{9.20}\right)\left(\frac{5.83}{6.0}\right) \quad \text{and} \quad 6.23\left(\frac{5.83}{6.0}\right).$$

The interval is (.106, 6.05). This is a very wide interval, suggesting that the small sample sizes do not provide us with a very accurate estimate of the ratio of variances.

Despite the risk of needless repetition, it is well to emphasize the importance of the normal-distribution assumption in inferences about population variances. Neither the chi-square nor the F distributions can safely be used for *variance* hypotheses unless the population distribution is normal or the sample sizes are quite large. This warning does not necessarily extend to all uses of these two distributions, but it is valid when the primary question to be answered is about one or more variances.

6.23   RELATIONSHIPS AMONG THE THEORETICAL SAMPLING DISTRIBUTIONS

Now that the major sampling distributions have been introduced, the connections among some of these theoretical distributions can be examined in more detail. Over and over again, the normal, T, chi-square, and F distributions have proved their utility in the solution of problems in statistical inference. Remember, however, that none of these distributions is empirical in the sense that someone has taken a vast number of samples and found that the sample values actually do occur with exactly the relative frequencies given by the function rule. Rather, it follows mathematically (read "logically") that if we are drawing random samples from certain kinds of populations, various sample statistics *must* have distributions given by these function rules. Like any other theory, the theory of ·sampling distributions deals with "if-then" statements. This is why the assumptions we have introduced are important; if we wish to apply the theory of statistics to making inferences from samples, then we cannot expect the theory necessarily to provide us with correct results unless the conditions specified in the theory hold true. As we have seen, from a practical standpoint these assumptions may be violated to some extent in our use of these theoretical distributions as approximations, especially for large samples.

Apart from the general requirement of random sampling of independent observations, the most usual assumption made in deriving sampling distributions is that the population distribution is normal. The chi-square, the T, and the F distributions all rest on this assumption. The normal distribution is, in a real sense, the "parent" distribution to these others. This is one of the main reasons for the importance of the normal distribution; the normal function rule not only provides probabilities that are often excellent approximations to other probability (density) functions, but it also has highly convenient mathematical properties for deriving other distribution functions based on normal populations.

The chi-square distribution rests directly upon the assumption that the population is normal. As you will recall from Section 6.18, the chi-square variable is basically a sum of squares of independent *normal* variables, each with mean 0 and with variance 1. At the elementary level, the problem of the distribution of the sample variance can be solved explicitly only for normal populations; this sampling distribution depends on the distribution of χ^2, which in turn rests on the assumption of a normal distribution of single observations. Furthermore, in the limit, the distribution of χ^2 approaches a normal form.

There are close connections in theory between the F distribution and both the normal and chi-square distributions. Basically, the F variable is a ratio of two independent chi-square variables, each divided by its degrees of freedom. Since a chi-square variable is itself defined in terms of the normal distribution, the F distribution also rests on the assumption of two (or more) normal populations.

The T distribution has links with the F, chi-square, and normal distributions. The T statistic for a single mean can be written as

$$T = \frac{(M - \mu)/\sigma}{\sqrt{(\widehat{S}^2/n\sigma^2)}} = \frac{(M - \mu)/\sigma_M}{\sqrt{(\widehat{S}^2/\sigma^2)}} . \qquad (6.23.1^*)$$

The numerator of T is obviously merely Z, a standardized score in the normal sampling distribution of means. However, consider the term in the denominator. From Section 6.19, we know that

$$\frac{(n - 1)\widehat{S}^2}{\sigma^2} = \chi^2_{(n-1)},$$

so that

$$\frac{\widehat{S}^2}{\sigma^2} = \frac{\chi^2_{(n-1)}}{n - 1} .$$

Thus,

$$T = \frac{Z}{\sqrt{\chi^2_{(n-1)}/(n - 1)}} . \qquad (6.23.2^*)$$

In general, a T variable is a standard normal variable Z in ratio to the square root of a chi-square variable divided by ν. Let us look at T^2 for a single mean in the light of this definition:

$$T^2 = \frac{Z^2}{\chi^2/(n-1)}. \qquad (6.23.3)$$

The numerator of T^2 is, by definition, a chi-square variable with 1 degree of freedom, and the denominator is a chi-square variable divided by its degrees of freedom, $\nu = n - 1$. Furthermore, these two chi-square variables are independent, by the principle stating that for a normal population, $\hat{S}^2$ is independent of M (Section 5.15). Thus, for a single mean, T^2 qualifies as an F ratio with 1 and $n - 1$ degrees of freedom. In general,

$$T^2_{(\nu)} = F_{(1,\nu)}; \qquad (6.23.4^*)$$

the square of T with ν degrees of freedom is an F variable with 1 and ν degrees of freedom.

The important relationships among these theoretical distributions are summarized in Table 6.23.1, showing how the distribution represented in the column depends for its derivation upon the distribution represented in the row.

Table 6.23.1   Relationships Among the Theoretical Sampling Distributions

Distribution	*Chi-square*	*F*	*T*
Normal	Parent, and limiting form as $\nu \to \infty$. Defined as sum of normal and independent Z^2 values.	Parent, making values in numerator and denominator independent χ^2/ν values.	Parent, and limiting form as $\nu \to \infty$. Numerator is normal Z.
Chi-square		Variables in numerator and denominator are independent χ^2/ν.	Denominator is $\sqrt{\chi^2/\nu}$.
F			$T^2_{(\nu)} = F_{(1/\nu)}$

EXERCISES

1. Distinguish between the terms "estimator" and "estimate," and complete the following sentence: "In estimating the mean μ of a normally distributed population, the sampling distribution of the sample mean, an (estimate, estimator) of μ, is normal, and the value of the sample mean calculated from a particular sample turns out to be 124, an (estimate, estimator) of μ.

2. A production manager is willing to assume that the weight of an item is normally distributed with known variance, but the mean μ is not known. A random sample of four independent observations, X_1, X_2, X_3, and X_4, is taken. Consider the following two sample statistics:

$$G = \frac{(X_1 + X_2 + X_3 + X_4)}{4}$$

and

$$H = \frac{(4X_1 + 3X_2 + 2X_3 + X_4)}{10}.$$

(a) Is G an unbiased estimator of μ? Prove your answer.
(b) Is H an unbiased estimator of μ? Prove your answer.
(c) Find the variance of G and the variance of H.
(d) Which of the two estimators is more efficient as an estimator of μ? What is the relative efficiency?

3. It is assumed that X, the amount of time (in hours) required to complete a certain task, is normally distributed. A random sample of n individuals is chosen, and X_i represents the time required for the ith individual to complete the task, where $i = 1, 2, \ldots, n$. Let G simply equal X_1, the time required by the first individual to complete the task; let H be the average time required by the first two individuals; and let M be the sample mean, the average time required by the n individuals in the sample. The sample statistics G, H, and M are potential estimators of μ, the population mean. For each of the three estimators, answer each of the following questions.
(a) Is the estimator sufficient?
(b) Is the estimator unbiased?
(c) Is the estimator consistent?
(d) What is the variance of the estimator?

4. Do the estimators G and H in Exercise 3 and the estimator H in Exercise 2 seem to be intuitively "good" estimators? Compare these estimators with the sample mean.

5. In Exercise 37, Chapter 4, suppose that the intensity of the process is not known. A sample of 1-hour periods is taken, and for each 1-hour period the number of incoming telephone calls is recorded. The sample consists of n 1-hour periods, and the number of incoming calls in the ith period is denoted by R_i. If you wish to estimate λ, the intensity per hour of the process, what estimator or estimators might you consider using?

6. In Exercise 5, show that the sample mean, $\sum R_i/n$, is an unbiased estimator of λ. Find the variance of this estimator. Is it a consistent estimator of λ?

7. What is meant by the statement that "consistency, unlike the other three properties of estimators considered in the text, is a large-sample property"?

8. Using Equations (6.4.3) and (6.4.4), derive Equation (6.4.5).

9. Using the definition of bias given by Equation (6.4.4), find the bias of S^2 as an estimator of σ^2.

10. If G is an unbiased estimator of θ, under what conditions would it be possible for G^2 to be an unbiased estimator of θ^2? Explain your answer.

11. We have a sample of n independent observations from a certain population, and we want to estimate the population mean μ. Under what conditions is the linear combination

$$G = \sum_{i=1}^{n} a_i X_i$$

an unbiased estimator of μ? [*Hint*: Find the conditions which the a_i must satisfy for $E(G) = \mu$.]

12. In Exercise 11, find the variance of G. Assuming that the a_i are such that G is unbiased, how can we choose the a_i in such a manner as to make the variance of G as small as possible?

13. The density function of the random variable X is given by

$$f(x) = \begin{cases} \dfrac{2b(c - bx)}{c^2} & \text{if } 0 \le x \le \dfrac{c}{b}, \\ 0 & \text{elsewhere.} \end{cases}$$

It turns out that this distribution, with two positive parameters b and c, yields

$$E(X) = \frac{c}{3b}$$

and

$$\sigma_X{}^2 = \frac{c^2}{18b^2}.$$

(a) Given a sample of size n, consisting of independent observations, would you say that M is a *sufficient* estimator of the parameter c?

(b) Suppose that it is known that $c = 3$. Is the sample mean for n observations then an unbiased estimator for the value of b? Why or why not?

(c) When $c = 3$, it is true that $b = E(X)/(2\sigma_X{}^2)$. Would you say then that $M/(2\hat{S}^2)$ calculated from a sample of size n is an unbiased estimator of b? Why or why not?

(d) When $b = 3$, is M a consistent estimator of c?

14. In Exercise 44, Chapter 5, the following distribution of the number of courses taken in the first semester of graduate school was given:

x	$P(X = x)$
3	1/3
4	1/3
5	1/3

A random sample of two first-semester graduate students is taken, and the number of courses is recorded for each of these graduate students. The population of first-semester graduate students is large enough that the sample observations can be treated as independent for all practical purposes.

(a) Are the sample mean, the sample median, and the sample midrange (the average of the largest and smallest values in the sample) all unbiased estimators of the population mean?

(b) Find the variance of the sample mean, the sample median, and the sample midrange.

(c) Find the relative efficiency of the sample mean as compared with the sample median and the relative efficiency of the sample mean as compared with the sample midrange.

15. Consider a sample of two independent observations from the distribution given in Exercise 14.

(a) Is the sample range (the difference between the largest and smallest values in the sample) unbiased as an estimator of the population variance?

(b) Compute the mean-square errors of the three sample statistics S^2, $\hat{S}^2$, and the sample range as estimators of σ^2. Which of the three statistics has the smallest mean-square error?

16. Do Exercise 14 for a sample of size 3 from the population $(X,$ amount of sale$)$ given in Exercise 49, Chapter 5.

17. Do Exercise 15 for a sample of size 3 from the population $(X,$ amount of sale$)$ given in Exercise 49, Chapter 5.

18. Prove that the kth sample moment is an unbiased estimator of the kth population moment of a random variable (provided, of course, that the kth population moment exists).

19. In a sample of size n from a Bernoulli process, show that the sample proportion R/n is a consistent estimator of p.

20. In Exercise 7, Chapter 4, suppose that p is not known but it is known that 500 people have been exposed to the disease and 82 of the 500 people have caught the disease. Determine an estimate of p. What can you say about the accuracy of this estimate?

21. For sampling from a Bernoulli process with fixed r, the distribution of N is a Pascal distribution, as discussed in Section 4.7. In this situation, $(r-1)/(N-1)$ is an unbiased estimator of p. Explain why this might be considered intuitively to be a "reasonable" estimator of p.

22. Give intuitive interpretations of the four criteria given in Sections 6.2–6.5 as "good properties" for point estimators. For each criterion, give a realistic situation for which that criterion might be considered to be particularly important.

23. In sampling from a Bernoulli process with fixed sample size n, consider the probability $P(R = r \mid n, p)$. In sampling from a normally distributed popula-

tion with known variance, consider the density $f(m \mid \mu)$. Probabilities such as $P(R = r \mid n, p)$ and densities such as $f(m \mid \mu)$ can be interpreted in terms of a sampling distribution or in terms of a likelihood function. Explain the difference between the two interpretations.

24. For a sample of size n from an exponential distribution with parameter λ, find the maximum-likelihood estimator of λ. Suppose that λ is the intensity per hour of the arrival of cars at a toll booth. If a sample of size 10 is taken, with the following amounts of time between arrivals, compute the maximum-likelihood estimate of λ.

$$4 \quad 3 \quad 6 \quad 1 \quad 1 \quad 4 \quad 2 \quad 6 \quad 1 \quad 3$$

If the cost per hour of maintaining the toll booth can be expressed as $\lambda^2 + 2\lambda + 5$, what is the maximum-likelihood estimate of the cost per hour of maintaining the toll booth?

25. On the basis of a sample of size n of unit time intervals from a Poisson process, as in Exercise 5, determine an estimator of the Poisson parameter λ by using (a) the principle of maximum likelihood, (b) the method of moments.

26. In the security price example in Section 6.7, assume that the sample results are as given in the text but that the sampling was done with r fixed rather than with n fixed. For both the discrete and continuous cases considered for the security price example in Section 6.7, use the principle of maximum likelihood to determine an estimate. Compare the resulting estimates with the corresponding estimates determined under the assumption of sampling with fixed n.

27. In Exercise 38, Chapter 4, suppose that λ is not known. However, it is known that the publisher of the book deals with only two printing firms, one of which has a misprint rate of 1.5 per page and one of which has a misprint rate of 2 per page. A random sample of 8 pages from the book yields a total of 13 misprints upon close checking and rechecking. Use the principle of maximum likelihood to determine an estimate of λ.

28. The gamma distribution is discussed in Exercise 52, Chapter 4. For a random sample of size 1 from a gamma distribution, find the maximum-likelihood estimator of λ and the method of moments estimator of λ.

29. The life of a light bulb, as measured by the number of working hours from the time it is first used, is assumed to be uniformly distributed with parameters a and b:

$$f(x) = \begin{cases} 1/(b - a) & \text{if } a \le x \le b, \\ 0 & \text{elsewhere.} \end{cases}$$

A random sample of n light bulbs is chosen from a very large shipment of light bulbs, and the life of each of the bulbs is determined by using it until it fails. The sample results are denoted by $x_1, x_2, \ldots, x_n$. Use the method of moments to determine estimates for a and b.

30. In the text, Equation (6.9.1) provides a way to form a pooled estimate of a mean based on two nonoverlapping samples. Try to arrive at a similar expression that applies when the first sample, of size n_1, and the second sample, of size n_2, have exactly n_3 cases in common. [*Hint*: A Venn diagram may be helpful.]

31. In Exercise 12, Chapter 4, suppose that p is not known. Two random samples of patients undergoing the type of surgery in question are observed, with n_1 patients in the first sample and n_2 patients in the second sample. The number of successful operations in the first sample is R_1, while the number of successes in the second sample is R_2. Using the information from these two samples, determine a pooled estimator of p that is unbiased.

32. In sampling from a finite population without replacement, the variance of the sample mean is somewhat smaller than it is in the infinite-population situation with the same sample size. Give an intuitive explanation for this result.

33. Do Exercise 14 under the assumption that the population consists of only 6 first-year graduate students, 2 of whom take 3 courses, 2 of whom take 4 courses, and 2 of whom take 5 courses. Discuss any differences between the answers to Exercise 14 and the answers to this exercise.

34. Let X represent the age (in years) of an individual randomly selected from a large group of executives, and assume that X is normally distributed and that the variance of X is 225. A random sample of 9 executives is observed, with ages 42, 56, 68, 56, 48, 36, 45, 71, and 64.
    (a) Compute the sample mean and sample variance.
    (b) Name three intuitively reasonable point estimators for μ, the mean age in the population, and determine the corresponding estimates from the above data.
    (c) Which of the three estimators given in (b) would you prefer? Why?
    (d) Find an 80 percent confidence interval for μ.
    (e) Find a 95 percent confidence interval for μ.

35. In a certain population of men, the mean weight (in pounds) is μ and the variance is 400. If it is assumed that weight is normally distributed in this population and if a random sample (with replacement) of 25 persons from the population yields a sample mean weight of 168 and a sample variance of 441, find 50 percent and 98 percent confidence intervals for the population mean weight μ.

36. Based on historical data, the variability of the diameter of a part produced by a particular machine can be represented by a standard deviation of .1 inch. However, nothing is known about the population mean. A sample of size 100 is taken, and the sample mean diameter is 8.4 inches. Find 95 percent and 75 percent confidence intervals for the population mean, assuming that the diameters are normally distributed and that the sample observations are independent.

37. An educator is interested in the mean IQ in a given population (say, the freshman class at a certain university). It is too expensive to test the entire population, but the assumption is made that IQ is normally distributed with variance 121 in the population of interest. The educator wants to take a random sample of members of the population in order to determine a 90 percent confidence interval for mean IQ. How large a sample is needed if the width of the confidence interval is to be no greater than 5? If the width is to be no greater than 1?

38. Suppose that you find a 90 percent confidence interval for the parameter μ, based on a particular sample, and that the confidence limits are 50 and 60. In terms of the long-run frequency interpretation of probability, how should this confidence statement be interpreted? Might a different interpretation be given by an advocate of subjective, or personal, probability? Explain.

39. A random sample of 100 families was selected from a large city, and the amount of money spent for food during a particular week was determined for each of the families. On the basis of this information, $120 and $137 are the limits of a 95 percent confidence interval for the mean amount of money spent for food by families in the given city during that week. For each of the following statements, determine whether the statement is justified by the information just given.
    (a) If a family is selected at random from the city, the odds are 19 to 1 that the family spent between $120 and $137 for food that week.
    (b) If random samples of size 100 are drawn repeatedly, 95 percent of the sample means will fall between $120 and $137.
    (c) Approximately 95 percent of the families in the city spent between $120 and $137 on food during the week in question.
    (d) If random samples of size 100 are drawn repeatedly and 95 percent confidence intervals are computed for all such samples, approximately 95 percent of these intervals will include the true mean amount of money spent for food by families in the given city during the week in question.
    (e) The odds are approximately 19 to 1 that the average amount spent for food by the 100 families in the sample during the following week was between $120 and $137.

40. In the process of estimating the mean of a population, a statistician wants the probability to be .95 that the sample mean will come within $.2\sigma$ of the true mean. What sample size should be used? If the sample size used is 100, within what distance of the true mean (in σ units) can the statistician be 95 percent sure that the sample mean falls?

41. A professional organization is interested in μ, the mean age of its membership. The relevant data are not available or easily obtainable for the entire membership, so a sample must be taken. It is assumed that the standard deviation of age is 10 years.
    (a) How large must the sample be in order for a 68 percent confidence interval for μ to be no more than 2 years wide?
    (b) If a 95 percent confidence interval for μ calculated from a particular sample is (42.1, 42.7), find the sample size and the sample mean.

42. Given a sample mean of 45, a population variance of 25, and a sample of size 30 drawn from a normal population, establish a 99 percent confidence interval for the population mean based on this sample. Can you find *another* pair of limits for which it will also be true that the probability of these limits covering the true mean is .99? If so, are these limits closer together or farther apart than the limits originally found? Does this suggest a general principle?

43. In determining a $100(1 - \alpha)$ percent confidence interval for the mean of a normal population, assuming that the variance is known, how large a sample is needed to make the confidence interval 1/3 as large as it is when the sample size is n? What are the implications of this result?

44. Discuss the following statement: "The determination of sample size in an experiment involves a balancing of the cost of taking the sample and the desired precision of the results of the sample."

45. Why is it that the probability that $(M - \mu)/\sigma_M$ is greater than 1.65 is only .05 for samples of size 5 from a normal distribution, whereas the probability that $(M - \mu)/$est. σ_M is greater than 1.65 is approximately .09 for samples of the same size from the same population?

46. Does it seem reasonable that the T distribution should have "fatter tails" than the normal distribution? Why?

47. In Exercise 34, assume that the variance of X is unknown and find 80 percent and 95 percent confidence intervals for μ. Compare these with the intervals computed in parts (d) and (e) of Exercise 34.

48. In Exercise 35, suppose that the population variance is not known. Find 50 percent and 98 percent confidence intervals for μ and compare them with the intervals found in Exercise 35.

49. In most realistic situations, do you think it is reasonable to assume that the mean of a random variable is unknown but that the variance is known? Can you give some circumstances in which this state of affairs might be somewhat plausible?

50. For the T distribution with 12 degrees of freedom, find
    (a)   the .10, .60, and .995 fractiles,
    (b)   the mean and variance,
    (c) $P(T < -.695)$,
    (d) $P(-2.179 \leq T \leq 1.356)$.

51. For the T distribution with 6 degrees of freedom, can you find $P(T \geq 1)$ from Table II in the Appendix? Can you find the .20 fractile of the distribution? Explain.

52. Discuss the concept of degrees of freedom in the context of the determination of an interval estimate for the mean of a normally distributed population when the variance of the population is not known. In a sample of size n, why aren't there n degrees of freedom instead of $n - 1$?

53. Let Y represent the age (in years) of an individual randomly selected from a large group of union officials, and assume that the variance of Y is 100. A

random sample of 4 union officials is observed, with ages 55, 63, 76, and 68. From this information and the information presented in Exercise 34,

(a) give a point estimate for the difference between the mean age of executives and the mean age of union officials,

(b) find 80 percent and 99 percent confidence intervals for this difference in means, assuming normality.

54. Do Exercise 53(b) under the assumption that the variances are unknown in both populations but are assumed to be equal. Does equality of variances appear to be a reasonable assumption in this instance?

55. Do Exercise 53(b) under the assumption that the variances are unknown and are *not* assumed to be equal.

56. In Exercise 63, Chapter 5, assume that the parameters of the distribution are not known. The sample of 9 Swedish cars yields a sample mean battery life of 3.6 years and a sample standard deviation of .8 years, whereas the sample of 36 German cars yields a sample mean of 3.0 years and a sample standard deviation of .6 years. Find a 90 percent confidence interval for the difference between mean battery life in cars produced by the Swedish firm and mean battery life in cars produced by the German firm.

57. Suppose that daily changes in the Dow-Jones Average of Industrial Stocks are normally distributed and that the change on any given day is independent of the change on any other day. A random sample of 81 daily changes is obtained, with sample mean .20 and sample variance 1.50. Find a 90 percent confidence interval for μ, the population mean. Suppose that a second random sample is obtained, with a sample size of 25, a sample mean of .15, and a sample variance of 1.20. Using the information from both samples, find a 90 percent confidence interval for μ. How reasonable are the assumptions in this problem?

58. An IQ test is given to a randomly selected group of 10 freshmen at a given university and also to a randomly selected group of 5 seniors at the same university. For the freshmen, the sample mean is 120 and the (unbiased) "modified" sample variance is 196. For the seniors, the sample mean is 128 and the (unbiased) "modified" sample variance is 121. Determine a 95 percent confidence interval for $\mu_S - \mu_F$, where μ_F is the population mean IQ for the freshmen and μ_S is the population mean IQ for the seniors. (Assume that the population variances are equal for the two populations.)

59. Independent random samples are taken from the output of two machines on a production line. The weight of each item is of interest. From the first machine, a sample of size 36 is taken, with a sample mean weight of 120 grams and a sample variance of 4. From the second machine, a sample of size 64 is taken, with a sample mean weight of 130 grams and a sample variance of 5. It is assumed that the weights of items from the first machine are normally distributed with mean μ_1 and variance σ^2 and that the weights of items from the second machine are normally distributed with mean μ_2 and variance σ^2 (that is, the variances are assumed to be equal). Find a 99 percent confidence interval for the difference in population means, $\mu_1 - \mu_2$.

60. Assuming that the variances of two normally distributed populations are equal, develop a formula for finding a confidence interval for the *sum* of the means. [*Hint*: Consider the sampling distribution of $\mu_1 + \mu_2$.] Use this to determine a 99 percent confidence interval for the sum of the means in Exercise 59.

61. In a sample of 100 items from a production process, 4 are found to be defective. Assuming that the production process behaves like a Bernoulli process, determine 95 percent and 99.7 percent confidence intervals for p, the probability of obtaining a defective item on any single trial.

use approximate $z = 3.0$

62. In Exercise 20, find approximate 60 percent and 90 percent confidence intervals for p.

63. What difficulties are encountered in attempting to determine confidence intervals for p, the parameter of a Bernoulli process, on the basis of a small sample of, say, only 5 observations?

64. Show that the fact that $[n/(n-1)]S^2$ is an unbiased estimator of σ^2 and that nS^2/σ^2 is a chi-square variable implies that the expectation of a chi-square random variable with $n-1$ degrees of freedom is simply $n-1$.

65. For samples from normal populations, the variance of the sampling distribution of $\hat{S}^2$, the unbiased estimator of σ^2, is $2\sigma^4/(n-1)$. Prove this, using the fact that the mean of a chi-square random variable is equal to the number of degrees of freedom. [*Hint*: Remember that $V[\chi_{(\nu)}^2] = 2\nu$.]

66. A sample of size 8 is chosen randomly from a normally distributed population with mean 24 and variance 9. The standardized value corresponding to each observed value is computed by subtracting 24 and dividing the result by 3. Let W denote the sum of the squares of the 8 standardized values, and find
    (a) $P(W \geq 20.09)$,
    (b) $P(2.73 < W < 5.07)$,
    (c) $P(W > 2.18)$.

67. In Exercise 35, if S^2 is the sample variance of the sample of size 25, find
    (a) $P(S^2 \leq 198.40)$,
    (b) $P(304.64 \leq S^2 \leq 687.68)$,
    (c) $P(S^2 > 531.20)$,
    (d) $E(S^2)$.

68. In Exercise 47, find 80 percent and 95 percent confidence intervals for σ^2, the variance of age in the population of executives. In light of these intervals, discuss the variance assumption that was made in Exercise 34.

69. In Exercise 48, find a 90 percent confidence interval for σ, the standard deviation of weight in the population of interest.

70. Two independent samples from the same normal population are pooled to form the following estimator of the population variance:

$$G = \frac{n_1 S_1^2 + n_2 S_2^2}{n_1 + n_2 - 2}.$$

Discuss the sampling distribution of this pooled estimator of σ^2.

71. Would the sampling distribution in Exercise 70 change if the samples were drawn from normal populations with different means but with the same variance? Explain.

72. Using the data in Exercise 57, find a 90 percent confidence interval for σ^2 based on
    (a)   the first sample only,
    (b)   the second sample only,
    (c)   both samples.
    Repeat the process under the assumption that the sample sizes are 10 and 5, respectively.

73. In Exercise 17, Chapter 5, determine 90 percent confidence intervals for the mean and variance of the price per gallon of premium-grade gasoline, assuming that the population is normally distributed.

74. In Exercise 16, Chapter 5, determine 95 percent confidence intervals for the mean and variance of income. Do you think that the assumption of normality is reasonable in this case on the basis of the given data? Why or why not?

75. For the F distribution with 5 degrees of freedom in the numerator and 8 degrees of freedom in the denominator, find
    (a)   the .025 and .99 fractiles of the distribution,
    (b)   the mean of the distribution,
    (c) $P(F \leq 3.69)$.

76. In Exercise 58, find a 95 percent confidence interval for the ratio of the variance of the freshman population to the variance of the senior population $(\sigma_F{}^2/\sigma_S{}^2)$.

77. Given the fact that the expected value of an F variable with ν_1 and ν_2 degrees of freedom is equal to $\nu_2/(\nu_2 - 2)$ for $\nu_2 > 2$, prove that the variance of a T variable with ν degrees of freedom is $\nu/(\nu - 2)$.

78. If X and Y are independent and normally distributed and if we let $U = cX + d$ and $W = bY + g$, discuss the relationship between the F ratio for $\sigma_X{}^2$ and $\sigma_Y{}^2$ and the F ratio for $\sigma_U{}^2$ and $\sigma_W{}^2$. Consider the following two cases:
    (a) $c = b$ and $d = g$,
    (b) $c \neq b$ and $d \neq g$.
    How would the T distribution for the difference between means be changed by the transformation from X and Y to U and W?

79. For the data in Exercise 59, find 90 percent and 98 percent confidence intervals for the ratio of the variances of weights from the two machines.

7

HYPOTHESIS
TESTING

The general motivation for inferential statistics should be clear by now: How can we say something about the population, given only the sample evidence? We have seen that certain sample statistics yield good estimates of particular population parameters, and these point estimates are often the first inferences made from a set of data. Of course, even "good" point estimates tend to differ from the parameters being estimated because of sampling fluctuations, and confidence intervals give us some indication of the precision of a particular point estimate. Point estimates and interval estimates, then, provide us with methods for saying something about the population on the basis of some sample evidence.

Often a statistician is not interested solely in estimation, but in deciding if some hypothetical population situation, or hypothesis, seems reasonable in light of the sample evidence. Sometimes the problem is to judge which of several possible hypotheses is best supported by the evidence at hand. Statistical procedures for attacking a problem of this nature fall under the heading of hypothesis testing. We will discuss hypothesis testing in this chapter, concentrating on situations in which there are two competing hypotheses concerning the population parameter of interest.

7.1 STATISTICAL HYPOTHESES

A **statistical hypothesis** is usually a statement about one or more population distributions, specifically about one or more parameters of such population distributions. *It is always a statement about the population, not about the sample.* The statement is called a hypothesis because it refers to a situation that *might* be true. Statistical hypotheses are almost never *exactly* equivalent to real-world hypotheses, which are usually statements about phenomena or their underlying bases. Quite commonly, however, statistical hypotheses grow out of or are implied by real-world hypotheses.

We shall use the following scheme to indicate a statistical hypothesis: a letter H followed by a statement about parameters, the form of the distribution, or both, for one or more specific populations. For example, one hypothesis about a population could be written

H: population in question is normally distributed with $\mu = 48$ and $\sigma = 13$,

and another could be written

H: population in question is represented by a Bernoulli process with $p = .5$.

Hypotheses that *completely* specify a population distribution are known as **simple hypotheses.** In general, the sampling distribution of any statistic is completely specified given a simple hypothesis and n, the sample size.

We will also encounter hypotheses such as

H: population is normal with $\mu = 48$.

Here, the exact population distribution is not specified, since no requirement is put on σ, the population standard deviation. When the population distribution is not determined completely, the hypothesis is known as **composite.** Composite hypotheses are more difficult to deal with than simple hypotheses, but in many situations simple hypotheses are quite unrealistic.

Hypotheses may also be classified by whether they specify *exact* parameter values or merely a *range* or *interval* of such values. For example, the hypothesis that $\mu = 100$ would be an **exact hypothesis,** although the hypothesis that $\mu \geq 100$ would be **inexact.** We will study both exact hypotheses and inexact hypotheses in this chapter; exact hypotheses will be considered first and will be used to introduce many of the basic notions involved in hypothesis testing.

7.2   CHOOSING A WAY TO DECIDE BETWEEN TWO EXACT HYPOTHESES

The shift from estimation to hypothesis testing requires quite a bit of new terminology. Instead of plunging right into this terminology, we will illustrate some of the ideas of hypothesis testing with a simple example that involves two exact hypotheses. Some of the numbers chosen are a bit extreme in order to illustrate the notions involved as clearly as possible. In the next section we will look at the general hypothesis testing situation and generalize some of the notions that are being illustrated in the example presented in this section.

An economist has two alternative theories about some economic behavior. According to Theory I, 80 percent of all consumers should reduce their savings in response to a tax increase, whereas according to Theory II, only 40 percent should exhibit this behavior. In reality there may be other theories, and it is possible that neither Theory I nor Theory II is correct. For the purposes of the example, however, imagine that one or the other of the hypotheses must be true.

The two theories suggest the two competing hypotheses

$$H_0: p = .80$$

and $$H_1: p = .40,$$

where p is the proportion of consumers in the population that will reduce their savings in response to a tax increase. The subscripts 0 and 1 have no particular meaning here; they are merely indices to let us tell the hypotheses apart.

A tax increase occurs, and in order to investigate the two theories and to test one against the other, the statistician takes a random sample of consumers. The reaction of each consumer in the sample to the tax increase is to be ascertained. Given each of the hypotheses, the random sampling procedure yields a Bernoulli process, so that for a fixed sample size n, the distribution of R, the number of consumers in the sample reducing their savings in response to the tax increase, is a binomial distribution. Of course, the exact distributions differ under the two hypotheses, since $p = .8$ for the first hypothesis and $p = .4$ for the second hypothesis.

For the sake of simplicity, assume there are to be only 10 consumers in the sample. Moreover, to make the sample statistic directly comparable with p, consider the sample *proportion*, R/n. The sampling distribution of R/n under each of the two hypotheses can be found from tables of the binomial distribution, and the resulting distributions are presented in Table 7.2.1.

How shall the statistician choose between the two hypotheses, or two alternative theories, upon seeing the sample evidence? A decision rule that tells the statistician when to choose H_0 and when to choose H_1 is needed. There are many such decision rules that could be formulated, but

Table 7.2.1 Binomial Distributions with $n = 10$ for $p = .8$ and $p = .4$

r	r/n	$F(r/n \mid p = .8)$	$P(r/n \mid p = .4)$
0	.0	.0000 (+)	.006
1	.1	.0000 (+)	.040
2	.2	.000 (+)	.121
3	.3	.001	.215
4	.4	.006	.251
5	.5	.026	.200
6	.6	.088	.111
7	.7	.201	.042
8	.8	.302	.011
9	.9	.268	.002
10	1.0	.107	.0001 (+)

for the moment we will consider only the following four possibilities.

DECISION RULE 1

If R/n is less than .8, choose H_1; if R/n is greater than or equal to .8, choose H_0.

DECISION RULE 2

If R/n is less than .6, choose H_1; if R/n is greater than or equal to .6, choose H_0.

DECISION RULE 3

If R/n is less than .4, choose H_1; if R/n is greater than or equal to .4, choose H_0.

DECISION RULE 4

If R/n is less than .2 or greater than .8, choose H_1; if R/n is between .2 and .8, inclusive, choose H_0.

Notice that each decision rule *completely* specifies what choice is to be made for any possible sample outcome. In contrast, for example, a rule that says to choose H_1 if the sample proportion is less than .2 and to choose H_0 if the sample proportion is greater than .7 is incomplete in the sense that it provides no guidance when the sample proportion is between .2 and .7, inclusive.

Notice also that these are just four of the very large number of possible decision rules and that there is no particular reason that these particular four rules were chosen as examples. As the chapter progresses, we will develop the notion of how we can arrive at a "good" decision rule; in fact, investigating these four decision rules in this example will provide some initial guidance in this direction.

Regardless of which decision rule is selected, there are two ways that the statistician *could* be right and two ways of making an *error*. This is diagrammed in Table 7.2.2. If H_0 is chosen and H_1 is actually true, an error is made. Furthermore, if H_1 is chosen and H_0 is true, exactly the opposite error is made. The other two possibilities lead, of course, to correct decisions.

Table 7.2.2   Errors in Hypothesis-Testing Situation

TRUE SITUATION

		H_0	H_1
	H_0	correct	error
DECISION			
	H_1	error	correct

Given the sampling distributions under the two hypotheses, and given any decision rule, we can find the probabilities of the two kinds of error. For example, consider the first decision rule, which requires that H_1 be chosen if the sample proportion is less than .8 and H_0 be chosen otherwise. If H_1 is really true, what is the probability that H_0 will be chosen and an error will be made? Decision rule 1 tells the statistician to choose H_0 if R/n is greater than or equal to .8. The probability of erroneously choosing H_0, then, is the probability that R/n is greater than or equal to .8, *given* that H_1 is true (that is, *given* that $p = .4$). From the sampling distribution of R/n given that $p = .4$, we see that the probability of R/n being greater than or equal to .8 is about .013. This is determined as follows:

$$P(R/n \geq .8 \mid p = .4) = P(R/n = .8 \mid p = .4)$$
$$+ P(R/n = .9 \mid p = .4)$$
$$+ P(R/n = 1 \mid p = .4)$$
$$= .011 + .002 + .000 = .013,$$

where all of the values are rounded to three decimal places. Therefore, if H_1 is true, the probability is .013 of *erroneously* deciding that H_0 is true.

In the same way, we can find the probability of the error made in choosing H_1 when H_0 is true. By Decision rule 1, H_1 is chosen whenever R/n

Table 7.2.3 Probabilities of Erroneous and Correct Decisions Under Decision Rule 1

TRUE SITUATION

		H_0	H_1
DECISION	H_0	.677	*.013*
	H_1	*.323*	.987

is less than or equal to .7; in the distribution under H_0, this interval of values has a probability of about .323. Thus, the probabilities of erroneous and correct decisions under Decision rule 1 are as given in Table 7.2.3. The probability of a correct decision is $1 - P(\text{error})$ for either of the possible true situations (the two error probabilities appear in italics). Notice that by using this rule the statistician is far more likely to make an erroneous judgment when H_0 is true than when H_1 is true.

Exactly the same procedure gives the erroneous and correct decision probabilities under Decision rule 2; see Table 7.2.4. By this second rule, the probability of error is relatively smaller when H_0 is true than it is for Rule 1. However, look what happens to the probability of the other error! This illustrates a general principle in the choice of decision rules. **Any change in a decision rule that makes the probability of one kind of error smaller will ordinarily make the other error probability larger** (other things, such as sample size, being equal).

Table 7.2.4 Probabilities of Erroneous and Correct Decisions Under Decision Rule 2

TRUE SITUATION

		H_0	H_1
DECISION	H_0	.966	*.166*
	H_1	*.034*	.834

Before trying to choose between Rules 1 and 2, let us write down the probabilities for Decision rules 3 and 4. First, Decision rule 3 is represented in Table 7.2.5. Here the probability of erroneously choosing H_1 when H_0 is true becomes even smaller, but the probability of the other type of error, erroneously choosing H_0, is extremely large.

Table 7.2.5 Probabilities of Erroneous and Correct Decisions Under Decision Rule 3

TRUE SITUATION

		H_0	H_1
DECISION	H_0	.999	.618
	H_1	.001	.382

Next, consider Decision rule 4 (Table 7.2.6). Even on the face of it this decision rule does not look sensible. The probability of error is large when H_0 is true, and the statistician is almost *sure* to make an error if H_1 is the true situation! This illustrates that not all decision rules are reasonable. Here, regardless of what is true the statistician has a larger chance of making an error using this rule than in using Rules 1, 2, or 3. Rules such as Rule 4 are called **inadmissible** by decision theorists. We need confine our attention only to the relatively "good" Rules 1, 2, and 3.

Table 7.2.6 Probabilities of Erroneous and Correct Decisions Under Decision Rule 4

TRUE SITUATION

		H_0	H_1
DECISION	H_0	.625	.952
	H_1	.375	.048

By taking a careful look at Decision rule 4, we could have predicted that it would not be a good rule. Since H_0 specifies that $p = .8$ and H_1 specifies that $p = .4$, small values of R/n tend to support H_1 and large values of R/n tend to support H_0. Yet, under Decision rule 4, the two largest possible values of R/n, .9 and 1.0, lead to the choice of H_1. In other words, the two values of R/n that should provide the most evidence in favor of H_0 as opposed to H_1 actually cause the statistician to choose H_1! Intuitively, this seems quite unreasonable, and intuition is supported by the high error probabilities of Rule 4 as compared with the first three rules.

Notice that for each of the first three rules, H_1 is chosen for "small" values of R/n, and H_0 is chosen for "large" values of R/n. The only difference is that the dividing line between "small" and "large" changes from rule to rule; for Rule 1, .8 is the dividing line, for Rule 2 it is .6, and for Rule 3 it is .4. In this example, any "good" decision rule must be of the general form of the first three rules considered here, so that the only feature distinguishing "good" rules from each other is the dividing line. Technically, the dividing line is called the **critical value** of R/n. In order to make a decision, the statistician need only compare the observed sample proportion with the critical value.

Now that we have determined what a "good" decision rule looks like in this example, we can immediately eliminate from consideration all other decision rules. That is, we need not consider a decision rule that instructs us to, say, choose H_1 when R/n is greater than .6 and to choose H_0 otherwise, because we know that large values of R/n should favor H_0 and small values should favor H_1, not vice versa. Nevertheless, the statistician's problem of how to choose a decision rule is still not solved, for there are still many "good" decision rules even after the others have been discarded. Since in this example the distinguishing feature among different "good" decision rules is the dividing line, or the critical value of the sample proportion, there are as many "good" rules as there are critical values of R/n. Decision rules 1, 2, and 3 provide three different critical values, and there are yet others. For the moment, however, we will confine our discussion to Rules 1, 2, and 3.

There is a real problem in deciding among Rules 1, 2, and 3. Rule 1 is good for making the error probability small when H_1 is true, but it is somewhat risky when H_0 represents the true state of affairs. Rule 3, on the other hand, yields a very small error probability when H_0 is true, but it also provides a very high risk of error when H_1 is true. Rule 2 is "between" Rules 1 and 3, as might be expected from the fact that the critical value for Rule 2, .6, is between the critical values for Rules 1 and 3, .8 and .4. In general, we see that as the critical value decreases, the probability of erroneously choosing H_0 increases and the probability of erroneously choosing H_1 decreases. Obviously, any choice among "good" decision rules should have something to do with the relative seriousness of the two types of errors. It might be that making an error is a minor matter when H_0 is true, but very serious given H_1. In this case Rule 1 might be preferable to Rules 2 and 3. If, on the other hand, erroneously choosing H_1 when H_0 is true is much more serious than the other type of error, then Rule 1 might be disastrous and Rule 3 might be the best of the three rules. In short, any rational way of choosing among "good" decision rules should involve some notion of the loss involved in making an error.

The example presented in this section should give you some idea of what hypothesis testing is all about. In the following section, some additional terminology will be introduced and several of the notions encountered in this example will be discussed in more general terms.

7.3   TYPE I AND TYPE II ERRORS

The process of comparing two hypotheses in the light of sample evidence is traditionally called a **statistical test** or a **significance test**. Moreover, it is common say that we are testing the hypothesis H_0 *against* the alternative H_1. Sometimes a single hypothesis, H_0, is of primary interest, and H_1 is thought of as an alternative hypothesis against which to test H_0. Even though we will sometimes speak of testing a single hypothesis, however, in the hypothesis testing framework the statistician acts as though the decision is between two hypotheses. Either H_0 can be chosen or H_1 can be chosen, and in practice this choice is usually stated in a slightly different form. **Either H_0 is accepted by the statistician or it is rejected in favor of H_1.**

A decision rule, then, tells the statistician when to accept H_0 and when to reject H_0 in favor of H_1. The decision rule involves a sample statistic, and it tells which values of the statistic lead to acceptance of H_0 and which values lead to rejection of H_0. The set of values leading to rejection is known as the **region of rejection.** Specifying the region of rejection for a decision rule completely specifies the decision rule, as it is assumed that any value of the statistic not included in the rejection region will lead to acceptance. Thus, it is common to speak of decision rules in terms of their regions of rejection. In the example of the previous section, the rejection regions of Rules 1–4 are, respectively, "R/n less than .8," "R/n less than .6," "R/n less than .4," and "R/n less than .2 or greater than .8."

In hypothesis testing, the statistician can make two kinds of errors: rejecting the hypothesis H_0 when it is in fact true, or accepting H_0 when it is in fact false. We shall call the former a Type I error and the latter a Type II error.

A Type I error is committed when H_0 (the tested hypothesis) is falsely rejected, and a Type II error is committed when H_0 is falsely accepted.

For any rejection region, we can determine the probability of each of the two kinds of error. We shall use the following notation:

$$\alpha = P(\text{Type I error}) = P(\text{rejecting } H_0 \mid H_0 \text{ is true}),$$

$$\beta = P(\text{Type II error}) = P(\text{accepting } H_0 \mid H_0 \text{ is false}).$$

Thus, the probabilities of errors and correct decisions can be represented in tabular form as in Table 7.3.1. Since the error probabilities are α and β, the probabilities of correct decisions are $1 - \alpha$ and $1 - \beta$. (One important warning; the sum $\alpha + \beta$ has no meaning, and it is incorrect to assume that this sum equals 1.) In the example of the previous section, Decision

Table 7.3.1   Probabilities of Erroneous and Correct Decisions

TRUE SITUATION

		H_0	H_1
	Accept H_0	$1 - \alpha$	β
DECISION			
	Reject H_0	α	$1 - \beta$

rule 1 yields an α of .323 and a β of .013; for Rule 2, $\alpha = .034$ and $\beta = .166$; for Rule 3, $\alpha = .001$ and $\beta = .618$; and for Rule 4, $\alpha = .375$ and $\beta = .952$.

It is possible, then, to summarize the "performance" of a decision rule by specifyng the values of α and β for the rule. Ideally, we would like to have $\alpha = 0$ and $\beta = 0$. It is generally impossible to get perfect precision and to eliminate all sampling error, so we must settle for making α and β as small as possible. First of all, we can eliminate all decision rules that are **dominated** by other decision rules. A rule is said to be dominated if there is another decision rule that has both a lower value of α *and* a lower value of β. In the example of the previous section, Rule 4 is dominated by Rule 1, for instance. (It is also dominated by Rules 2 and 3.) It is important to note that in order to be eliminated from consideration, a rule must be dominated by a *single* other rule. In the example, Rule 1 gives a smaller β than does Rule 2, and Rule 3 gives a smaller α than does Rule 2. However, this does *not* mean that Rule 2 is dominated. It is not dominated by either Rule 1 or Rule 3 alone, since it has a lower α than Rule 1 and a lower β than Rule 3.

Technically, it should be noted that it is not necessary for both error probabilities to be strictly larger for one rule than for another in order to have dominance. If one error probability is larger for the first rule than for the second rule, but the second error probability is the same for both rules, then the first rule is dominated by the second rule. In the example, if we could find a decision rule with $\alpha = .323$ and $\beta = .100$, for instance, it would be dominated by Rule 1.

Any decision rule that is dominated by another is called **inadmissible,** as noted in the previous section. Once the inadmissible rules are eliminated from consideration, the remaining rules (that is, the admissible rules) are all "good" rules in the sense that they are not dominated by any other rules. This can be represented on a graph, as in Figure 7.3.1. The horizontal axis gives values of α, and the vertical axis gives values of β. The shaded region contains all possible (α,β) pairs that can be attained for a particular sampling scheme. Notice that any point that is *inside* the shaded area but not on the solid curve that represents the lower left-hand boundary of the

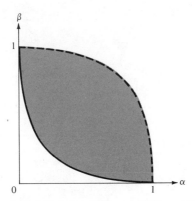

Figure 7.3.1   Probabilities of Error for Different Decision Rules

shaded area is dominated by some point on the curve. To see this, take any point inside the shaded area and draw a line from that point to the origin. The line *must* cross the solid curve, and the point at which the line crosses the curve has both a smaller α *and* a smaller β than the point originally chosen as the end-point of the line. This is illustrated in Figure 7.3.2. The set of "good" rules, or admissible rules, is thus represented on the graph by the solid curve that is the lower left-hand boundary of the shaded region.

Before the question of choosing among the admissible decision rules is discussed, it should be noted that the statistician has some control over the curve in Figures 7.3.1 and 7.3.2. Once all the details of the sampling procedure are specified, the curve is fixed. However, by altering these details, the statistician has some control over the curve. For instance, increasing the sample size leads to greater precision, as we have seen in Chapters 5 and 6, so that increasing the sample size will generally shift the curve representing the lower left-hand boundary of the shaded region closer to the

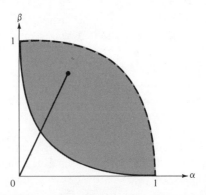

Figure 7.3.2   Point Inside Shaded Area Dominated by Point on
Admissible Set

origin, thus giving the statistician more attractive (α,β) pairs to choose from. Any other means of increasing the precision of the sample information, such as using a more accurate balance when weights are of interest, will also shift the curve toward the origin. In general, the statistician must weigh the advantage of shifting the curve toward the origin against the disadvantage of the higher cost that is generally required to attain increased precision. This question will be discussed further later in this chapter and in Chapter 9, when a procedure for formally determining an "optimal" sample size will be discussed.

Assuming that the sampling procedure has been specified and that the curve representing admissible decision rules is fixed, how can the statistician choose a *single* decision rule to use? The choice among the admissible rules involves a trade-off between α and β; in order to decrease one of the error probabilities, it is necessary to accept an increase in the other. As suggested at the end of the previous section, a choice of a single rule should depend in some manner on the relative seriousness of the two types of errors. If a Type I error is judged to be much more serious than a Type II error, for instance, then we will probably be willing to make α small, even at the expense of increases in β. In Chapter 9, losses will be formally introduced, and the choice of a decision rule will be directly related to such losses. This will provide a formal decision-theoretic approach to hypothesis testing. In this chapter the emphasis will be on how to set up "good" decision rules in various situations, and convention rather than decision theory will be used in the choice of a particular rule in any given situation. The notion of conventional procedures for determining a decision rule will now be discussed.

7.4 CONVENTIONAL DECISION RULES

In "classical" statistics, given some hypothesis H_0 to be tested, the region of rejection is usually found as follows.

Convention

Set α, the probability of falsely rejecting H_0, equal to some small value. Then, in accordance with the alternative H_1, choose a region of rejection such that the probability of observing a sample value in that region is equal to α when H_0 is true. The obtained result is said to be significant beyond the α level if the sample statistic falls within that region.

Ordinarily, the values of α used are .05 or .01, although, on occasion, larger or smaller values are employed. Even though only one hypothesis,

H_0, may be exactly specified, and this determines the sampling distribution employed in the test, in choosing the region of rejection we act as though there were two hypotheses, H_0 and H_1. The alternate hypothesis H_1 dictates which portion or tail of the sampling distribution contains the rejection region for H_0. In some problems, the region of rejection is contained in only one tail of the distribution, so that only extreme deviations in a given direction from expectation lead to rejection of H_0. In the example of Section 7.2, for example, only "small" values of R/n should lead to the rejection of H_0. In other problems, big deviations of either sign are candidates for the rejection region, so that the region of rejection lies in both tails of the sampling distribution.

The conventions about the permissible size of the α probability of Type I error actually grew out of a particular sort of experimental setting. Here, it is known in advance that one kind of error is extremely important and is to be avoided. In this kind of experiment, these conventional procedures make sense when viewed from the decision-making point of view. Furthermore, designation of the hypothesis H_0 as the "null" hypothesis and the arbitrary setting of the level of α can best be understood within this context. An example frequently used to demonstrate this approach concerns the trial of an accused criminal. The two errors that can be made are: (a) set a guilty person free, and (b) convict an innocent person. According to the concepts of justice which have evolved in the United States, (b) is a much more serious error than (a). As a result, the hypothesis that the accused person is innocent is taken as the "null" hypothesis. (Historically, the hypothesis to be tested has been called the null hypothesis, primarily because it is a hypothesis that the statistician often sets out to "disprove"; H_0 is called the **null hypothesis** and H_1 is called the **alternative hypothesis.** This terminology serves no useful purpose, especially if hypothesis testing is viewed as a decision-making procedure, so for the most part we will avoid it.) Then α is the probability of convicting an innocent person, and the philosophy of the system of courts in the United States is to make α a very small, even if this results in a fairly high β. This is clear from the many safeguards in the system which are designed to protect the accused, and from the claim that a person is presumed innocent until proven guilty (the word "proven" is used rather loosely here, implying that P(guilty) is quite high, but not necessarily equal to 1). An interesting point with respect to this example is that courts in some countries appear to take the opposite approach, considering error (a) to be more serious and presuming a person guilty unless proven innocent. The choice of a "null" hypothesis, then, is very much dependent upon a set of values, which can be thought of as implying a set of losses.

As another example of an experimental setting where a Type I error is clearly to be avoided, imagine that we are testing a new medicine, with the goal of deciding if the medicine is safe for the normal adult population. By "safe" we mean that the medicine fails to produce a particular set of undesirable reactions on all but a very few normal adults. Now in this instance, deciding that the medicine is safe when actually it tends to produce reactions in a relatively large proportion of adults is certainly an error to be avoided. Such an error might be called "abhorrent" to the statistician and to the general public. Therefore, the hypothesis "medicine unsafe" or its statistical equivalent is cast in the role of the null hypothesis, H_0, and the value of α is chosen to be extremely small, so that the abhorrent Type I error is very unlikely to be committed. A great deal of evidence against the null hypothesis is required before H_0 is to be rejected. The statistician has complete control over Type I error, and regardless of any other feature the study of the medicine may have, the statistician can be confident of taking very little risk of asserting that H_1, or "medicine safe," is true when actually H_0, or "medicine unsafe," is true.

In other words, the conventional practice of arbitrarily setting α at some very small level is based on the notion that one kind of error is extremely important and must be avoided if possible. This is quite reasonable in some contexts, such as the study of the safety of a new medicine or the guilt of an accused man. Of course, in other contexts it is not so reasonable. In general, if we want to consider losses formally, we have to go beyond the conventional procedure of this section and consider decision-theoretic procedures that will be presented in Chapter 9. The decision-theoretic approach eliminates the need for conventions, since it provides a way to select the optimal action in any situation. It thus eliminates the need to consider such designations as "null" hypotheses. However, it does require the determination of losses, and there are many hypothesis-testing problems where the primary objective is inference rather than decision making and where any losses are vague or hard to determine. Thus, we defer consideration of the complete decision-theoretic approach until Chapter 9. In the remainder of this chapter, we will discuss testing under the assumption that α has been determined (although not necessarily small, such as in the convention). You should keep in mind that the determination of α (and β) depends on the relative seriousness of the two types of errors.

7.5 DECIDING BETWEEN TWO HYPOTHESES ABOUT A MEAN

Before we leave the problem of deciding between two exact hypotheses, one more example will be given, this time involving the mean of a population. Suppose that the workers in a particular plant each turn out finished products at the rate of X per hour, where X is normally distributed with mean 138 and standard deviation 20. A considerable amount of past data is available to support the normality assumption and the particular mean and variance. An engineer has come up with a slightly different working procedure that is claimed to increase the output by 4 per hour, to a mean of 142 per hour, while leaving the standard deviation and the shape of the distribution unchanged. A statistician is called in to investigate the engineer's claim.

Given the above information, the statistician sets up the following two hypotheses:

$$H_0: \mu = 138$$

and $$H_1: \mu = 142.$$

A sample of 100 man-hours is taken, and α is set at .05. How can the rejection region be determined?

First, since the hypotheses involve a mean, and we know that M, the sample mean, is a good estimator of μ, the sample statistic of interest here is M. Moreover, since the population is assumed to be normally distributed with known standard deviation 20, the sampling distribution of M for a sample of $n = 100$ is a normal distribution with mean μ and standard error $\sigma_M = 20/\sqrt{100} = 2$. Note that the sample size is large enough so that even if the assumption of normality of the population is violated somewhat, the central limit theorem implies that the sampling distribution of M can be assumed to be normal.

Under H_0, then, M is normally distributed with mean 138 and standard error 2, whereas under H_1, M is normally distributed with mean 142 and standard error 2. These two potential sampling distributions of M are presented in Figure 7.5.1. From looking at the hypotheses and at these sampling distributions, it is clear that low values of M favor H_0 and high values of M favor H_1. Therefore, the rejection region for a "good" decision rule should be of the following form: reject H_0 if $M \geq c$. Thus, the next step is the determination of c, the critical value of M.

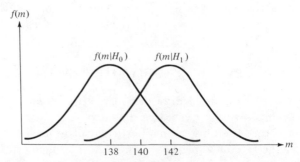

Figure 7.5.1 Sampling Distributions of M Under H_2 and H_1

The requirement that $\alpha = .05$ implies that

$$\alpha = P(\text{reject } H_0 \mid H_0 \text{ true}) = P(M \geq c \mid \mu = 138) = .05.$$

But $\qquad P(M \geq c \mid \mu = 138) = P\left(Z \geq \dfrac{c - 138}{2}\right),$

so we want

$$P\left(Z \geq \frac{c - 138}{2}\right) = .05,$$

where Z has a standard normal distribution since M is normally distributed. From Table I in the Appendix, it is approximately true that

$$P(Z \geq 1.65) = .05,$$

so we have $\qquad \dfrac{c - 138}{2} = 1.65,$

or $\qquad\qquad\qquad c = 138 + 1.65(2) = 141.30.$

Given a critical value of 141.30 for M, β can be calculated as follows:

$$\beta = P(\text{accept } H_0 \mid H_1 \text{ true}) = P(M < 141.30 \mid \mu = 142)$$

$$= P\left(Z < \frac{141.30 - 142}{2}\right) = P(Z < -.35) = .36.$$

The two error probabilities, $\alpha = .05$ and $\beta = .36$, are illustrated in Figure 7.5.2. It turns out that the probability of a Type II error is much greater than the probability of a Type I error. In this example, it may be that

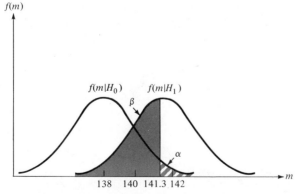

Figure 7.5.2 Probabilities of Error as Areas Under Sampling Distributions for $c = 141.3$, $n = 100$

acceptance of H_0 means that current working procedures will be retained, whereas rejection of H_0 means that the proposed new working procedure will be used. The desire to keep the probability of a Type I error low reflects an unwillingness to shift to a new procedure without being reasonably confident that it is in fact better than the current procedure. A Type I error here means that the workers will have to be trained in the new procedure and the procedure will lead to no improvements in output. Thus, the expenses of training the workers, both in terms of direct monetary outlay and possible worker dissatisfaction, will be suffered for naught. On the other hand, a Type II error means that the current procedure will be retained even though the new procedure is in fact better, implying that possible gains due to increases in productivity are being foregone.

By looking at Figure 7.5.2, you can see how changing the critical value of M will alter the error probabilities. As the critical value is moved to the left, the shaded area under the left-hand curve, which represents α, will increase. At the same time, the cross-hatched area under the right-hand curve, which represents β, will decrease. For instance, if the critical value is reduced from 141.30 to 140, the error probabilities will be

$$\alpha = P(M \geq 140 \mid \mu = 138)$$

$$= P\left(Z \geq \frac{140 - 138}{2}\right) = P(Z \geq 1) = .16$$

and $\qquad \beta = P(M < 140 \mid \mu = 142)$

$$= P\left(Z < \frac{140 - 142}{2}\right) = P(Z < -1) = .16.$$

This situation is illustrated in Figure 7.5.3.

After considering the potential seriousness of the errors, the statistician decides that α should be kept at .05 but that $\beta = .36$ is much too large. As noted in the previous section, one way to reduce one error without

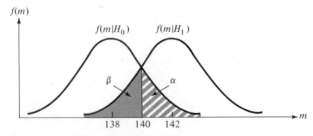

Figure 7.5.3 Probabilities of Error as Areas Under Sampling Distributions for $c = 140$, $n = 100$

increasing the other is to take a larger sample and hence gain greater precision. Suppose the statistician contemplates changing the sample size from 100 to 400. In that case the standard error of M is reduced to $20/\sqrt{400} = 1$, and the critical value of M is determined from

$$\frac{c - 138}{1} = 1.65,$$

so that
$$c = 138 + 1.65(1) = 139.65.$$

The probability of a Type II error is now

$$\beta = P(\text{accept } H_0 \mid H_1 \text{ true}) = P(M < 139.65 \mid \mu = 142)$$

$$= P\left(Z < \frac{139.65 - 142}{1}\right) = P(Z < -2.35) = .01.$$

Increasing the sample size from 100 to 400 while holding α constant reduces β from .36 to .01. The error probabilities in this situation are presented graphically in Figure 7.5.4.

This example illustrates the steps followed in a simple hypothesis-testing situation. First, the real-world situation is considered and translated into a statistical hypothesis H_0 and an alternative hypothesis H_1. Where appropriate, certain assumptions (such as normality or a known population variance) are made. Next, the sampling procedure is specified, and the sample statistic most relevant to the test is determined. The assumptions should make it possible to specify the sampling distribution of the sample

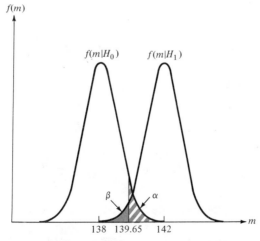

Figure 7.5.4 Probabilities of Error as Areas Under Sampling Distributions for $c = 139.65$, $n = 400$

statistic of interest given H_0 and the sampling distribution given H_1. The next step is to determine a region of rejection, and this can be broken down into two steps. In the first step, the general form of "good" rejection regions, or admissible rejection regions, is determined. In the second step, a specific region is chosen from among this admissible set. In the example, admissible rejection regions are of the form $M \geq c$. If α is then fixed, c must be the $1 - \alpha$ fractile of the sampling distribution of M given that H_0 is true. Finally, the sample itself is obtained, the statistic in question is computed, and the decision is made. A sample result falling into the region of rejection is said to be a *significant* result under the hypothesis H_0.

7.6 LIKELIHOOD RATIOS AND HYPOTHESIS TESTING

In Section 6.7, the notion of a likelihood function was introduced in the discussion of the determination of maximum-likelihood estimates. In this section, we shall see how the notion of a likelihood ratio is useful in statistical hypothesis testing. It is most convenient to introduce this concept in terms of examples that have been presented in the earlier sections of this chapter.

In the example of Section 7.2, an economist is interested in the hypotheses

$$H_0: p = .80$$

and

$$H_1: p = .40,$$

where p represents the proportion of consumers in a certain population who will reduce their savings in response to a tax increase. A random sample of 10 consumers is taken from the population, and the sampling distributions of R/n given H_0 and given H_1 are simply binomial distributions. For any given value of R, the probabilities

$$P(r/n \mid H_0) = P(r/n \mid p = .8) = \binom{n}{r} .8^r .2^{n-r}$$

and

$$P(r/n \mid H_1) = P(r/n \mid p = .4) = \binom{n}{r} .4^r .6^{n-r}$$

can be found from tables of the binomial distribution.

The probability $P(r/n \mid H_0)$ can be thought of as the likelihood of the particular sample result r if H_0 is true, and $P(r/n \mid H_1)$ can be thought of as the likelihood of the particular sample result r if H_1 is true. For a test of H_0 versus H_1, then, the likelihood ratio is simply the ratio of these two individual likelihoods. Letting Ω denote the likelihood ratio, we have

$$\Omega = \frac{P(r/n \mid H_0)}{P(r/n \mid H_1)} = \frac{\binom{n}{r}(.8)^r(.2)^{n-r}}{\binom{n}{r}(.4)^r(.6)^{n-r}},$$

which can be simplified to

$$\Omega = \left(\frac{.8}{.4}\right)^r \left(\frac{.2}{.6}\right)^{n-r} = \frac{2^r}{3^{n-r}}.$$

For $n = 10$, the values of Ω corresponding to the possible values of R are given in Table 7.6.1. Note that as R increases, Ω also increases. High values of R are more likely under H_0 than H_1, whereas low values of R are more likely under H_1 than H_0. Thus, the likelihood in the numerator increases relative to the likelihood in the denominator as R increases.

Table 7.6.1 Likelihood Ratio as a Function of r in Consumer Savings Example

r	Ω
0	.00002
1	.0001
2	.0006
3	.0037
4	.0219
5	.1317
6	.7901
7	4.7407
8	28.4444
9	170.6667
10	1024.0000

In terms of a statistical test, we discovered in Section 7.2 that for this example, H_0 should be rejected in favor of H_1 for "small" values of R, where the exact notion of "small" depends on the relative seriousness of the two types of error. But as R gets smaller, so does Ω. Therefore, H_0 should be rejected for small values of Ω. This result is a very general, important result that is not limited to this particular example.

In a statistical test of H_0 versus H_1, one should reject H_0 for "small" values of the likelihood ratio and accept H_0 for "large" values of the likelihood ratio, where the critical value dividing "small" and "large" values should depend on the relative seriousness of the two types of error. That is, the rejection region should be of the form "$\Omega \leq k$," where k is a critical value. Tests with rejection regions of this form are called likelihood-ratio tests, and likelihood-ratio tests possess very desirable properties. For instance, in testing an exact hypothesis versus an exact alternative, a test is inadmissible if it is not a likelihood-ratio test.

This important result is known as the **Neyman-Pearson lemma** after J. Neyman and E. S. Pearson, who played major roles in the development of the notion of likelihood-ratio tests.

We will not present a proof of the Neyman-Pearson lemma here, but we note that it holds in the above example. Admissible decision rules in the example are of the form "reject H_0 if $R \leq c$," where c is some critical value. But as R gets smaller, so does Ω, so $R \leq c$ is equivalent to $\Omega \leq k$, where k is a critical value in terms of the likelihood ratio.

For another illustration of the use of likelihood ratios in hypothesis testing, consider the example of Section 7.5. Here the hypotheses of interest are

$$H_0: \mu = 138$$

and $$H_1: \mu = 142,$$

where μ is the mean of a normally distributed population with standard deviation 20. A sample of size 100 is taken, and the sample statistic of interest is M. The sampling distribution of M under H_0 is a normal distribution with mean 138 and standard deviation $20/\sqrt{100} = 2$, and the sampling distribution of M under H_1 is a normal distribution with mean 142 and standard deviation 2. Therefore, for any particular value of M, the likelihood ratio is

$$\Omega = \frac{f(m \mid H_0)}{f(m \mid H_1)} = \frac{\dfrac{1}{2\sqrt{2\pi}} e^{-(m-138)^2/2(4)}}{\dfrac{1}{2\sqrt{2\pi}} e^{-(m-142)^2/2(4)}}.$$

Canceling terms and simplifying, we get

$$\Omega = \frac{e^{-(m-138)^2/8}}{e^{-(m-142)^2/8}} = e^{-[(m-138)^2-(m-142)^2]/8}$$

$$= e^{-[(m^2-276m+19{,}044)-(m^2-284m+20{,}164)]/8}$$

$$= e^{-(8m-1120)/8} = e^{-(m-140)} = e^{140-m}.$$

From Section 7.5, we know that "good" (that is, admissible) decision rules for this example are of the form "reject H_0 if $M \geq c$." But as the value of M increases, the value of

$$e^{140-m}$$

decreases, so "reject H_0 for large values of M" is equivalent to "reject H_0 for small values of Ω." As in the preceding example, the Neyman-Pearson lemma applies and the rejection region for an admissible decision rule must be of the form

$$\Omega \leq k, \tag{7.6.1*}$$

where k is some critical value. In other words, in order for M to be larger than some critical value c, Ω must be smaller than some other critical value k.

Graphically, we are dealing with the ratio of the heights of two normal density functions in this example. For instance, if $\alpha = .05$, it was shown in Section 7.5 that c, the critical value of M, is 141.30. At this critical value, the two likelihoods are

$$f(m = 141.3 \mid H_0) = \frac{1}{2\sqrt{2\pi}} e^{-(141.3-138)^2/2(4)} = .051$$

and $\qquad f(m = 141.3 \mid H_1) = \frac{1}{2\sqrt{2\pi}} e^{-(141.3-142)^2/2(4)} = .188,$

as shown in Figure 7.6.1. The likelihood ratio at this value of M is

$$\Omega = \frac{.051}{.188} = .27,$$

and since 141.30 is the critical value of M, .27 is equal to k, the critical value of the likelihood ratio. That is, if $\alpha = .05$ in this example, H_0 should be rejected if Ω is less than or equal to .27. Here α is given by

$$\alpha = P(M \geq 141.30 \mid \mu = 138) = P(\Omega \leq .27 \mid \mu = 138).$$

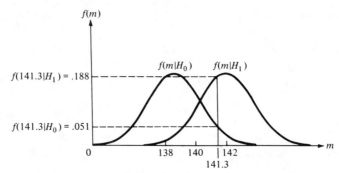

Figure 7.6.1   Likelihood Ratio as a Ratio of Densities

In general, the relationship between α, the probability of a Type I error, and k, the critical value of Ω, is

$$\alpha = P(\Omega \le k \mid H_0 \text{ is true}). \tag{7.6.2}$$

Note that for any value of M, Ω is just a ratio of two normal densities. At no point in this section has it been necessary to compute *areas* under a density function; only the *heights* are of interest when computing a likelihood ratio. In Figure 7.6.1, we could find likelihood ratios for values of M other than 141.30 by simply shifting the vertical line to the right or to the left. For instance, if the value of M decreases from 141.30, the likelihood under H_0 increases and the likelihood under H_1 decreases, so the likelihood ratio increases. At 140, the two normal density functions intersect, and the likelihood ratio is equal to 1. Below 140, Ω becomes even larger. On the other hand, if the value of M *increases*, Ω decreases.

Admittedly, we have glossed over some details here. In the second example, the formal development of the likelihood ratio should begin with two likelihoods of the form given by Equation (6.7.4). These likelihoods can be simplified to yield the same likelihood ratio obtained here. At this point, however, the details of the determination of a likelihood ratio are not as important as the notion of a likelihood ratio and the result that "good" tests are of the form "$\Omega \le k$."

The consideration of inexact hypotheses complicates matters somewhat because the probabilities or densities in the numerator and denominator of the likelihood ratio are no longer uniquely defined. For example, instead of $f(m \mid \mu = 138)$, we may have an expression of the form $f(m \mid \mu \le 138)$, which is not uniquely defined. One way of circumventing this difficulty is to use a procedure that is very closely allied to the maximum-likelihood principle discussed in Section 6.7. The likelihood ratio of interest is a ratio of maximum likelihoods, with the maximum likelihood of the particular sample result *given* the hypothesis H_0 in the numerator and the maximum likelihood of the sample result over all possible values of the relevant parameters in the denominator.

In our example involving the mean output of workers, consider the hypotheses

$$H_0: \mu \leq 138$$

and $$H_1: \mu > 138.$$

If the sample mean is 141.3, what is the likelihood ratio when it is considered as a ratio of maximum likelihoods? In Figure 7.6.1, the likelihood associated with $\mu = 138$ is shown as the height of the density function on the left at $\mu = 138$. But for any value of μ less than 138, the density function of M will be shifted even further to the left, and the height of the density function at the value 141.3 will be even smaller. Therefore, the maximum likelihood under H_0 is .051, the likelihood associated with $\mu = 138$.

Next, we must find the maximum likelihood over all values of μ. From Chapter 6, we know that the maximum likelihood estimator of μ, the mean of a normal distribution with known variance, is simply the sample mean. Therefore, the maximum likelihood over all values of μ is simply the value of the likelihood at the actual sample mean. In the example, this is

$$f(m = 141.3 \mid \mu = 141.3) = \frac{1}{2\sqrt{2\pi}} \, e^{-(141.3-141.3)^2/2(4)} = .199.$$

Defining Ω as a ratio of maximum likelihoods, we have

$$\Omega = \frac{.051}{.199} = .26.$$

As in the case of exact hypotheses, H_0 is rejected for small values of Ω. However, the critical value of Ω is different when the hypotheses are inexact. In this situation it turns out that the critical value is .26 for the inexact hypotheses, and this is very close to the critical value we arrived at for the exact hypotheses (.27). This closeness is coincidental, however; if the exact H_1 stated that $\mu = 150$ instead of $\mu = 142$, the critical value of Ω for the exact hypotheses would be quite different from .27.

Although the consideration of inexact hypotheses introduces some complications in terms of likelihood ratios, such complications need not concern us here. The key point to remember is that the tests of hypotheses discussed in this chapter and later chapters are equivalent to likelihood-ratio tests. These are the "best" available tests for the hypotheses considered, meaning that when the assumptions underlying these tests are true, no other available procedure would answer our question about H_0 and H_1 better.

7.7 THE POWER OF A STATISTICAL TEST

Suppose that we have a hypothesis, H_0, which we are interested in testing, and that it is an exact hypothesis concerning a particular parameter, which we shall call θ:

$$H_0: \theta = \theta_0.$$

For any given rejection region and any particular value of θ, say θ_1, we can compute

$$P(\text{reject } H_0 \mid \theta = \theta_1).$$

If $\theta_1 = \theta_0$, this probability is simply $P(\text{reject } H_0 \mid H_0 \text{ is true})$, or α. On the other hand, if $\theta_1 \neq \theta_0$, we can think of a test of H_0 versus the alternative $H_1: \theta = \theta_1$, in which case the probability is

$$P(\text{reject } H_0 \mid H_1 \text{ is true}) = 1 - P(\text{accept } H_0 \mid H_1 \text{ is true})$$

$$= 1 - \beta.$$

For any value of θ_1, then, the probability $P(\text{reject } H_0 \mid \theta = \theta_1)$ can be computed, and this probability is called the power of the test of H_0 against the alternative H_1.

In the example in Section 7.5, the hypotheses were $H_0: \mu = 138$ and $H_1: \mu = 142$. When $\alpha = .05$, H_0 should be rejected when the sample mean is greater than or equal to 141.3. Under this decision rule, $\beta = .36$, so the power of the test against the specific alternative $\mu = 142$ is .64.

The power of a test of H_0 is not unlike the power of a microscope. It reflects the ability of a decision rule to detect from evidence that the true situation differs from a hypothetical one. Just as a high-powered microscope lets us distinguish gaps in an apparently solid material that we would miss with low power or the naked eye, so does a high-powered test of H_0 almost insure us of detecting when H_0 is false. Pursuing the analogy further, any microscope will reveal "gaps" with more clarity the larger these gaps are; the larger the departure of H_0 from the true situation H_1, the more powerful is the test of H_0, other things being equal.

For example, suppose that we are interested in the hypothesis

$$H_0: \mu = \mu_0,$$

and we consider the alternative

$$H_1: \mu = \mu_1,$$

where for convenience we will assume that $\mu_1 > \mu_0$. Under either hypothesis, the sampling distribution of M is assumed to be normal with a known standard error σ_M. Since we know that large values of M favor H_1 and small values favor H_0, the rejection region will be of the form $M \geq c$. If $\alpha = .05$, the critical value is $\mu_0 + 1.65\sigma_M$, and the power of the test can be computed for any value of μ_1.

First of all, suppose that $\mu_1 = \mu_0 + \sigma_M$. The power is then

$$P(\text{reject } H_0 \mid \mu = \mu_0 + \sigma_M) = P(M \geq \mu_0 + 1.65\sigma_M \mid \mu = \mu_0 + \sigma_M)$$

$$= P\left(Z \geq \frac{(\mu_0 + 1.65\sigma_M) - (\mu_0 + \sigma_M)}{\sigma_M}\right)$$

$$= P(Z \geq .65) = .26.$$

This is illustrated in Figure 7.7.1; the shaded area under the right-hand curve is β, and the unshaded area under this curve is $1 - \beta$, the power.

Against a different alternative hypothesis, the power would be different. For example, for $\mu_1 = \mu_0 + 3\sigma_M$,

$$P(\text{reject } H_0 \mid \mu = \mu_0 + 3\sigma_M) = P(Z \geq -1.35) = .91.$$

The value $\mu_0 + 3\sigma_M$ is sufficiently greater than μ_0 so that if the true value of μ is actually $\mu_0 + 3\sigma_M$, the test has a high probability of detecting this fact.

The power of the test of any true value of μ_1 can be found in the same way. Often, to show the relation of power to the true value of μ_1, so-called **power functions** or **power curves** are plotted. One such curve is given in Figure 7.7.2, where the horizontal axis gives the possible values of μ in terms of μ_0 and σ_M, and the vertical axis indicates the power for that alternative. Notice that for this particular decision rule, the power curve rises for increasing values of μ_1 and approaches 1.00 for very large values. In any

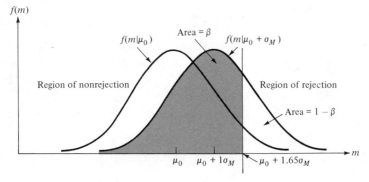

Figure 7.7.1 Power as Area Under a Sampling Distribution

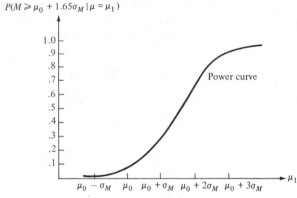

Figure 7.7.2   A Power Curve

statistical test where the region of rejection is in the direction of the true value covered by H_1, the greater the discrepancy between the tested hypothesis and the true situation, the greater the power.

In mathematical statistics, a general theory of hypothesis testing exists that extends the elementary ideas discussed here to much more complex situations. Among other things, this theory specifies desirable characteristics that a test statistic might possess and indicates methods for finding such test statistics, all within the general framework outlined in this chapter. For example, one desirable property of a test is that it be **most powerful for the particular true alternative** to the hypothesis H_0. That is, among all of the different ways that we could devise to test some particular H_0 against some particular true alternative value covered by H_1, the most powerful test for any fixed value of α affords the smallest probability of Type II error, other things such as sample size being equal. **Uniformly most powerful tests** for any fixed value of α are those which give a smaller probability of Type II error than any other test regardless of the value which happens to be true among those covered by H_1, other things being equal.

Two other properties of interest are **unbiasedness** and **consistency**. A test statistic (and decision rule) is said to afford an unbiased test of H_0 if the use of this statistic (and rule) makes the probability of rejecting H_0 be at its smallest when H_0 is actually true. That is, an unbiased test will have **minimum** power when H_0 is true. In somewhat imprecise terms, a consistent test is one for which the probability of rejecting a *false* H_0 approaches 1.00 as the sample size approaches ∞. That is, a consistent test always gains power against any true alternative as the sample size is increased.

7.8   THE EFFECT OF α AND SAMPLE SIZE ON POWER

Since β will ordinarily be small for large α, as we have seen in the discussion earlier in this chapter, it follows that setting α larger makes for relatively more powerful tests of H_0. For example, the two power curves given in Figure 7.8.1 for the same situation show that if α is set at .10 rather than at .05, the test with $\alpha = .10$ will be more powerful than that for $\alpha = .05$ over all possible true values under alternative H_1. Making the probability of error in rejecting H_0 larger has the effect of making the test more powerful, other things being equal.

In principle, if it is very costly to make the mistake of overlooking a true departure from H_0 but not very costly to reject H_0 falsely, we could (and perhaps should) make the test more powerful by setting the value of α at .10, .20, or more. This is not ordinarily done in scientific research (particularly in the physical sciences), however. There are at least two reasons why α is seldom taken to be greater than .05. In the first place, the problem of relative losses incurred by making errors is seldom faced in scientific research; hence, conventions about the size of α are adopted. The other important reason is that given some fixed α, the power of the test can be increased either by increasing sample size or by reducing the standard error of the test statistic in some other way. Of course, in many problems in business, the social sciences, and the behavioral sciences, it is feasible to assess losses due to erroneous decisions. Nevertheless, in these problems, the statistician may still be interested in increasing the power of a test by increasing sample size rather than by increasing α.

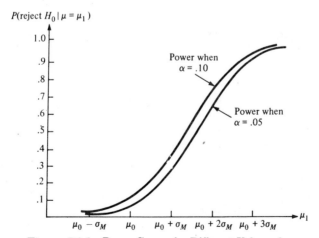

Figure 7.8.1   Power Curves for Different Values of α

The example of Section 7.5 demonstrates how increasing the sample size can increase the power of a test. In that example, the sample size was increased from 100 to 400 while α was held constant. As a result, β was reduced from .36 to .01, implying that the power of the test against the alternative $\mu = 142$ was increased from .64 to .99. This is due to the reduction in the standard error of M that is a direct result of the increase in sample size (recall that $\sigma_M = \sigma/\sqrt{n}$).

The disadvantages of an arbitrary setting of α can thus be offset, in part, by the choice of a large sample size. Other things being equal and regardless of the size chosen for α, the test may be made powerful against any given alternative H_1 in the direction of the rejection region, provided that n can be made very large (see Figure 7.8.2).

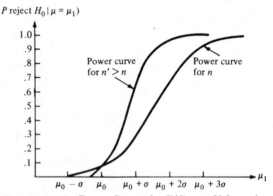

Figure 7.8.2   Power Curves for Different Values of n

Once again, however, it is not always feasible to obtain very large samples. Samples of substantial size are often costly for the statistician, if not in money, then in effort. Our ability to attain power through large samples only partly offsets the failure to choose a decision rule according to costs, or losses, in such research; large samples may not really be necessary in some research using them, especially if the error thereby made improbable is actually not very important.

Even given a fixed sample size, the statistician has another device for attaining power in tests of hypotheses. *Anything that makes σ, the population standard deviation, small will increase power, other things being equal.* This is one of the reasons for the careful control of conditions in good experimentation. By making conditions constant, the statistician rules out many of the factors that contribute to variation in the observations. Statistically, this amounts to a relative reduction in the size of σ for some experimental population. Ruling out some of the error variance from the observations decreases the standard error of M and thus *increases* the power of the test against alternative hypotheses. Experiments in which the variability attributable to experimental or sampling error is small are said to be precise; the result of such precision is that the statistician is quite likely to be able to detect when something of interest is happening. The application of experimental controls is like restricting inferences to populations with smaller values of σ^2 than otherwise, and thus control over error variation through careful experimentation implies powerful statistical tests. It follows that controlled experiments in which there is little "natural" variation in the materials observed can attain statistical power with relatively few observations, while those involving extremely variable material may require many observations to attain the same degree of power.

7.9   TESTING INEXACT HYPOTHESES

The primary notions of errors in inference and the power of a statistical test have just been illustrated for an extremely artificial situation, in which a decision must be made between two exact alternatives. Such situations are almost nonexistent in the real world. Instead, the statistician is far more likely to be called on to evaluate inexact hypotheses, each of which encompasses a whole range of possible true values. What relevance, then, does the discussion of the exact two-alternative case have to what statisticians actually do? The answer is that the statistician makes inferences as though deciding between two exact alternatives, even though the primary interest lies in judging between inexact hypotheses. Thus, the mechanism we have been using for decisions between exact hypotheses is exactly the same as for any other set of alternatives.

An example should clarify this point. As a fairly plausible situation, imagine a production manager who must decide whether or not to replace a machine with a new machine. From past experience, the manager knows that the old machine produces at the mean rate of 100 units per minute, and the standard deviation of the rate per minute is 15. Furthermore, the distribution of values is approximately normal.

The new machine which is being considered is a newer model, and the manager has reason to suspect that it may produce at a higher rate than the old machine. In order to investigate this supposition, the manager obtains permission from the manufacturer of the new machine to use the machine on a trial basis. In doing so, sample information will be obtained which will be useful in decision making. Basically, the question to be answered is, "Does the new machine produce at a higher mean rate than the old machine?" We will assume that there is no reason to think that the standard deviation might be different for the new machine, nor the general form (approximately normal) of the distribution.

The answer to this question is tantamount to a decision between two *inexact* hypotheses:

$$H_0: \mu \leq 100$$

$$H_1: \mu > 100.$$

That is, the new machine has either a mean rate less than or equal to that of the old machine or a mean rate greater than the old machine.

In choosing a decision rule, the manager decides to set α equal to .01. A result greater than 100 tends to favor H_1, and so the rejection region should correspond to the upper 1 percent of a sampling distribution of M. In terms of standardized values, this corresponds to values of Z greater than or equal to 2.33.

What, however, is the hypothesis being tested? As written, H_0 is inexact, since it states a whole region of possible values for μ. As a first step, let us simplify H_0 to an exact hypothesis, the hypothesis that $\mu = 100$. The sampling distribution of M under the exact hypothesis is a normal distribution with mean 100 and standard error $15/\sqrt{n}$, and the rejection region in terms of standardized values is simply $Z \geq 2.33$. Note that once α is specified, the rejection region can be specified without being given an exact alternative hypothesis. This is because α only relates to the sampling distribution under H_0, not the sampling distribution under H_1. In effect, the situation is as in Table 7.9.1.

Table 7.9.1   Error Probabilities in Machine Example

		TRUE SITUATION	
		$\mu = 100$	$\mu > 100$
DECIDE	$\mu = 100$	.99	(β)?
	$\mu > 100$	.01	$(1 - \beta)$?

Although there is a unique value of α in this situation, there is *not* a unique value of β. The alternative hypothesis is inexact, so in order to obtain a unique value of β, we would need to find a probability of the form $P(\text{accept } H_0 \mid \mu > 100)$. Because we are conditioning on an interval of values of μ rather than a single value, this probability cannot be determined. However, the probabilities $P(\text{accept } H_0 \mid \mu = \mu_1)$ and $P(\text{reject } H_0 \mid \mu = \mu_1)$ *can* be determined for all possible values of μ_1 greater than 100, and a graph of the latter probability is a power function. Therefore, although there is no unique β, we can look at the power of the test against each possible value in the alternative set.

Next, we need to generalize the above result from the exact hypothesis $\mu = 100$ to the inexact hypothesis $\mu \leq 100$. The statistician has no real interest in the hypothesis that $\mu = 100$. Nevertheless, this is a useful dummy hypothesis in the sense that *if* $\mu = 100$ can be rejected with $\alpha = .01$, then *any other hypothesis that $\mu < 100$ can be rejected with $\alpha < .01$.* In other words, if the test leads to rejection of $\mu = 100$, then we can be even more confident in rejecting any value of μ less than 100.

But what does this do to β? Given some μ_1 covered by H_1, *the power of the test of $\mu = 100$ is less than the power for any other hypothesis covered by H_0 with fixed α.* If we test any other exact hypothesis embodied in H_0, such as $\mu = 90$, then neither α nor β would exceed those for $\mu = 100$. Testing the exact hypothesis with given α and β probabilities can be regarded as testing *all* hypotheses covered by H_0, with *at most* α and β probabilities of error (the β depending on the *true* mean, of course). This is illustrated in Figure 7.9.1.

Suppose now that a sample of 200 trials yields a mean of 103.5. The standard error of the mean is $\sigma_M = 15/\sqrt{200} = 1.06$. Thus, the standardized value corresponding to the sample mean is

$$z = \frac{103.5 - 100}{1.06} = 3.30.$$

Since this value exceeds the critical value of 2.33, the hypothesis that $\mu = 100$ can be rejected with $\alpha < .01$, and any other hypothesis covered by H_0 can also be rejected with α less than .01. If the statistician decides that the mean is greater than 100, this could be an error, but the probability of such an error is less than .01.

On the other hand, suppose that the sample mean had been only 102. Here Z would have been only 1.89, which is not large enough to place the sample in the region of rejection for H_0 if the probability α is to be .01.

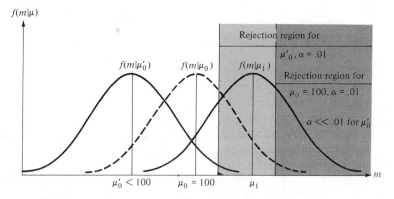

Figure 7.9.1 Rejection Regions for Different Means Under H_0

Under the decision rule using $\alpha = .05$, this would lead to a rejection of the hypothesis, but not by the rule originally decided upon.

The same problem can be put on a much more realistic basis, particularly with regard to choosing a sample size, if we have an idea of some minimal value of μ we would like to be very sure to detect as a significant result if true. That is, suppose the statistician in this problem knows that it is absolutely essential that the new machine be purchased if its mean rate of production is at least 103. This is important enough to require the power of the test to be at least .95 when μ is at least 103. Does the choice of a sample size and of a rejection region meet this qualification?

For $n = 200$, the standard error of the mean is 1.06. Consequently, the critical value of the mean is $c = 100 + 2.33(1.06) = 102.47$. If the value of μ were actually 103, then in this sampling distribution under the true value of μ the critical value would correspond to

$$z = \frac{102.47 - 103}{1.06} = -.5.$$

In a normal distribution, 69 percent of all values must fall at or above $-.5$ standard errors from the mean. Hence, the power of this test against the true alternative $\mu = 103$ is only .69. If the statistician wants this power to be .95 for $\mu = 103$, then a larger sample is needed. In fact, the sample size required can be found from

$$-1.65 = \frac{c - 103}{\sigma/\sqrt{n}}$$

and

$$2.33 = \frac{c - 100}{\sigma/\sqrt{n}}.$$

Solving for n, we find that the required sample size is 396. Then for this sample size we can find the following probabilities of errors and correct decisions; these probabilities are given in Table 7.9.2. In circumstances

Table 7.9.2   Probabilities of Erroneous and Correct Decisions

TRUE SITUATION

		$\mu \leq 100$	$100 < \mu < 103$	$\mu \geq 103$
DECIDE	H_0	$1 - \alpha \geq .99$	$.05 < \beta < .99$	$\beta \leq .05$
	H_1	$\alpha \leq .01$	$.01 < 1 - \beta < .95$	$1 - \beta \geq .95$

where there *is* such an important range of possible values of μ, representing a situation we want to be very sure to detect if true, then we can adjust the sample size or α value or both so as to to make the power against such alternatives as great as necessary.

7.10 ONE-TAILED REJECTION REGIONS

In the example of Section 7.9 two inexact hypotheses were compared, having the forms

$$H_0: \mu \leq \mu_0$$

and

$$H_1: \mu > \mu_0.$$

Here the entire range of possible values for the parameter under study (in this case, the population mean) was divided into two parts, that above and that below (or equal to) an exact value μ_0.

In this instance, the appropriate region of rejection for H_0 consists of values of M relatively much larger than μ_0. Such values have a rather small probability of representing means covered by the hypothesis H_0 but are more likely to represent means in the range of H_1. Such a rejection region consisting of sample values in a particular direction from the expectation given by the values included in H_0 is called a **directional** or **one-tailed** rejection region. For the particular hypotheses compared in Section 7.9, the rejection region was one-tailed, since only the right or "high-value" tail of the sampling distribution under H_0 was considered in deciding between the hypotheses.

For some questions, the two inexact hypotheses are of the form

$$H_0: \mu \geq \mu_0$$

and

$$H_1: \mu < \mu_0.$$

Once again the region of rejection is one-tailed, but this time the lower or left tail of the sampling distribution contains the region of rejection for H_0. The choice of the particular rejection region thus depends both on α and on the alternative hypothesis H_1.

Tests of hypotheses using one-tailed rejection regions are also called directional. The direction of the value of the statistic from μ_0 is important in directional tests since the sample result must show not only departure from expectation under H_0 but also a departure in the right direction to be considered strong evidence against H_0 and for H_1.

Directional hypotheses are implied when the basic question involves terms like "more than," "better than," "increased," or "declined." The essential question to be answered by the data has a clear implication of a difference or change in a specific direction. For example, in the problem of Section 7.9, the production manager wanted specifically to know if the new machine produced at a higher rate than the old machine, indicating a directional hypothesis.

Many times, however, the statistician goes into a problem without a clearly defined notion of the direction of difference to expect if H_0 is false. Questions such as "Did something happen?" "Is there a difference?" or "Was there a change?" are asked without any specification of expected direction. Next we will examine techniques for nondirectional hypothesis testing.

7.11 TWO-TAILED TESTS OF HYPOTHESES

Imagine a study carried out on the "optical dominance" of human subjects. There is interest in whether or not the dominant eye and the dominant hand of a subject tend to be on the same or different sides. Subjects are to be tested for both kinds of dominance and then classified as "same side" or "different side" in this respect. We will use the letters "S" and "D" to denote these two classes of subjects.

From knowledge about the relative frequency of each kind of dominance in the population of interest, the statistician reasons that if there actually is no tendency for eye dominance to be associated with hand dominance, then we should expect 58 percent S and 42 percent D. However, little is known about what to expect if there is some connection between the two kinds of dominance. To try to investigate this question, our statistician draws a random sample of 100 subjects, each with a full set of eyes and hands, and classifies them.

The question at hand may be put into the form of two hypotheses, one exact and one inexact:

$$H_0: p = .58,$$

$$H_1: p \neq .58.$$

The first hypothesis represents the possibility of no connection between the two kinds of dominance, and the inexact alternative is simply a statement that H_0 is not true, since H_1 does not specify an exact value of p.

The α chosen is .05. The statistician then is faced with the choice of a rejection region for the hypothesis H_0. Either a very high percentage or a very low percentage of S subjects in the sample would tend to discount

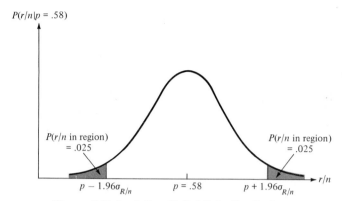

Figure 7.11.1 A Two-Tailed Rejection Region

the credibility of the hypothesis that $p = .58$ and would thus lend support to H_1. The rejection region should be arranged so that H_0 will be rejected when extreme departures from expectation of *either* sign occur. This calls for a rejection region in *both* tails of the sampling distribution of R/n, the sample proportion, when $p = .58$. Thus, the rejection region is called **two-tailed.** Since the sample size is relatively large, the binomial distribution of the sample proportion may be approximated by a normal distribution, and the rejection region may be diagrammed as shown in Figure 7.11.1.

Since the total probability of Type I error has been set at .05, the rejection region in the upper tail of the distribution will contain *the highest 2.5 percent* and the lower tail *the lowest 2.5 percent of sample proportions,* given that $p = .05$. In short, each region contains exactly the proportion $(1/2)\alpha$, or $\alpha/2$, of all samples under H_0. Consequently, either a very large or a very small sample proportion will lead to a rejection of H_0, and the total probability of Type I error is .05. The standardized value cutting off the upper rejection region is 1.96, and that for the lower is -1.96. Any sample giving a standardized value beyond these two limits will lead to a rejection of H_0.

Since the basic sampling distribution is binomial, the standard error of R/n given that $p = .58$ is

$$\sigma_{R/n} = \sqrt{\frac{pq}{n}} = \sqrt{\frac{(.58)(.42)}{100}} = .049$$

(Section 4.3). Using the normal approximation to the binomial, we find

$$Z = \frac{(R/n) - p}{\sigma_{R/n}}$$

Now suppose that the sample proportion of S subjects is .69. Then

$$z = \frac{.69 - .58}{.049} = 2.24.$$

This value exceeds 1.96, and so the result is said to be significant beyond the 5 percent level. If the statistician concludes that eye and hand dominance do tend to be related, then the risk of being wrong is less than .05.

Suppose, however, that the sample result had come out to be only .53. Then

$$z = \frac{.53 - .58}{.049} = -1.02,$$

making the result nonsignificant.

Although the example just concluded dealt with a hypothesis about a proportion, exactly the same procedure applies to two-tailed tests of means. An exact and an inexact hypothesis are posed:

$$H_0: \mu = \mu_0,$$

$$H_1: \mu \neq \mu_0.$$

The exact hypothesis is tested by forming a region of rejection consisting of two portions, each having probability $\alpha/2$ when H_0 is true. A sample result falling into either portion of the rejection region is said to be significant at the α level.

Now let us return to the relative merits of one- and two-tailed tests. In many circumstances calling for one-tailed tests, the form of the question or of the sampling distribution makes it clear that the only alternative of logical or practical consequence must lie in a certain direction. For example, when we turn to the problem of testing many means for equality simultaneously, it will turn out that the only rejection region making sense lies in one tail of the particular sampling distribution employed. In other circumstances the question involves considerations of "which is better" or "which is more," where the discovery of a difference from H_0 in one direction may have real consequences for practical action, although a difference in another direction from H_0 may indicate nothing. For example, consider a treatment for some disease. The cure rate for the disease is known, and we want to see if the treatment *improves* the cure rate. We have not the slightest practical interest in a possible decrease in cures; this, like no change, leads to nonadoption of the treatment. The thing we want to be sure to detect is whether or not the new treatment is really better than what we have. In many such problems where different practical actions depend on the sign of the deviation from expectation, a one-tailed test is clearly called for. Otherwise, the two-tailed test is often safest for the statistician asking only "Is there a difference?"

7.12 ONE- AND TWO-TAILED TESTS: POWER, OPERATING CHARACTERISTIC, AND ERROR CURVES

In deciding whether a hypothesis should be tested with a one- or two-tailed rejection region, the primary concern of the statistician must be the original question. By and large, many tests done in scientific research are non-directional, simply because research questions tend to be framed this way. However, there are also many situations where one-tailed tests are clearly indicated by the question posed. In a decision-theoretic context, one-tailed tests such as the test in the example of Section 7.9 are more commonly encountered than two-tailed tests.

The powers of one- and two-tailed tests of the same hypothesis will be different, given the same α level and the same true alternative. If a one-tailed test is used, and the true alternative is in the direction of the rejection region, then the one-tailed test is more powerful than the two-tailed test over all such possible values of μ. In a way, we get a little statistical credit in the one-tailed test for asking a more searching question.

Power curves for one- and two-tailed tests for means are compared in Figure 7.12.1. Sometimes, curves called **operating characteristic** (OC) **curves** or **error curves** are used by statisticians instead of power curves. Any one of the three types of curves can be derived from either of the other two, given a specific hypothesis-testing situation and a choice of a rejection region. Recall that the power function is the probability of rejecting H_0 as

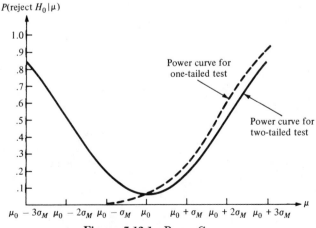

Figure 7.12.1 Power Curves

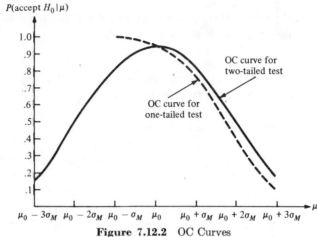

Figure 7.12.2   OC Curves

a function of the parameter of interest. In the situation depicted in Figure 7.12.1, $\alpha = .05$, and the power curve is

$$P(\text{reject } H_0 \mid \mu).$$

Similarly, the operating characteristic curve is

$$P(\text{accept } H_0 \mid \mu) = 1 - P(\text{reject } H_0 \mid \mu) = 1 - \text{power curve},$$

and the error curve is

$$P(\text{committing an error} \mid \mu).$$

Operating characteristic curves and error curves corresponding to the power curves in Figure 7.12.1 (that is, corresponding to a situation with $\alpha = .05$) are presented in Figures 7.12.2 and 7.12.3. The relationships

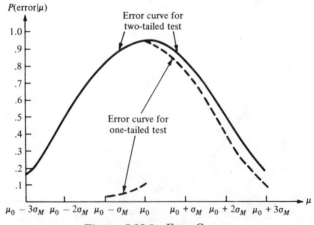

Figure 7.12.3   Error Curves

Table 7.12.1 Relationships Between Probabilities of Error and Power Curve, OC Curve, and Error Curve

	Region included in H_0	Region included in H_1
Power curve	α	$1 - \beta$
Operating characteristic curve	$1 - \alpha$	β
Error curve	α	β

between the curves and the probabilities of error, α and β, are summarized in Table 7.12.1 Note that for a two-tailed test, the operating characteristic and error curves differ only at one point: the point corresponding to H_0. The operating characteristic curve is a smooth curve in this case, whereas the error curve has a "gap" at μ_0; at this point, the error curve takes on the value α, while the operating characteristic curve takes on the value $1 - \alpha$. From the graphs and the relationships between the curves and the probabilities of errors, it is clear that a low error curve is desirable. As for power curves, we would like the power curve to be high in the region covered by H_1 and "low" in the region covered by H_0. Of course, since the OC curve is one minus the power curve, we would like the OC curve to be high in the region covered by H_0 and low in the region covered by H_1.

7.13 REPORTING THE RESULTS OF TESTS

The general problem of interest in hypothesis testing should be clear to you by now. The statistician wants to know if the sample results deviate from the results expected under some hypothesis H_0 to such a degree that H_0 should be rejected in favor of some alternative hypothesis H_1. First, it is necessary to decide upon some sample statistic which will be the basis for the test and to determine the sampling distribution of this statistic under the assumption that H_0 is true. Next, α is determined, either by considering the relative seriousness of the losses associated with the two types of errors or by choosing to follow the "classical" statistician's convention. From the sampling distribution and the value of α, a rejection region can be found. Finally, the sample results are observed and the value of the sample statistic of interest is calculated. If the sample statistic is in the rejection region, H_0 is rejected; if not, H_0 is accepted (or, of course, the statistician could choose a third option: the option of suspending judgment until more information is obtained).

The end result of this testing procedure is expressed by the decision: accept, reject, or suspend judgment. If the statistician reports this decision and nothing else, very little is being told about the sample. Suppose that H_0 is rejected at the .05 level. What does this tell us? It tells us that the sample result fell in a category of results that would be expected to occur by chance only 5 percent of the time, given the truth of the hypothesis. Was the result "just barely" in the rejection region? Was it such an extreme result with relation to H_0 that it would have been in a rejection region of size .01? A region of size .001? We cannot say.

Consider a concrete example, the machine example of Section 7.9. In terms of the standardized normal random variable, the rejection region for $\alpha = .01$ consisted of all standardized values greater than or equal to 2.33. The z value computed from the sample turned out to be 3.30, so that the hypothesis was rejected at the .01 level. Consider this question: "What is the smallest value of α for which the hypothesis would have been rejected?" This is obviously just the probability of a standardized value greater than or equal to 3.30; this probability is .0005. **We shall call this probability the p-value associated with this specific hypothesis and sample result.**

First, what does the p-value tell us? It tells us how "unusual" the sample result is as compared with the sampling distribution under the assumption that H_0 is true. It answers our question: "How unlikely is this sample result given the situation initially postulated?"

Second, of what importance is the p-value for reporting purposes? It provides more information than the simple reporting of acceptance or rejection at some level of significance. Furthermore, it enables readers to choose their own levels of significance (perhaps on the basis of losses which may be quite different from those of the person reporting the results) and to determine whether H_0 would have been rejected at *those* levels of significance.

For any value of α,

> **if $\alpha \geq p$-value, the hypothesis should be rejected,**

and   if $\alpha < p$-value, the hypothesis should be accepted.

Finding a p-value from a given sample result is just the opposite of the procedure for finding a rejection region from a given value of α. It can be found by asking this question: "If the result obtained fell exactly on the borderline between the region of rejection and the region of acceptance, what would the level of significance be?" The resulting number is equal to the p-value. You should be careful to note that the calculation of a p-value depends on whether the test is one-tailed or two-tailed. In the above example, suppose that the production manager wanted to make a two-tailed test of the hypothesis that the machine would produce at a rate of 100 units per minute. In this case, a rejection region would include portions of both tails of the sampling distribution, and the p-value corresponding to a z value of 3.30 would be twice as large as for the one-tailed test. The p-value would be $2(.0005) = .001$. For both the one- and two-tailed tests, the sample results obtained would be extremely unlikely if H_0 were true. A low p-value, then, casts doubt upon the hypothesis H_0; conversely, a high p-value tends to support the hypothesis.

Consider one more example, concerning the hypotheses

$$H_0: p \geq .5$$

and $$H_1: p < .5,$$

where p represents the proportion of voters in a particular city who currently intend to vote for a particular bond issue. A random sample of 10 voters is taken, resulting in 4 individuals who claim they will vote for the bond issue. Since we have a directional test, with the lower tail being the tail of interest, and since the sampling distribution of the number of "successes" is a binomial distribution,

$$p\text{-value} = P(R \leq 4 \mid n, p) = \sum_{r=0}^{4} \binom{10}{r} (.5)^r (.5)^{10-r}.$$

From tables of the binomial distribution, this cumulative probability is found to be .3770. Now, for any α,

$$\text{if} \quad \alpha \geq .3770, \quad \text{reject } H_0,$$

and $$\text{if} \quad \alpha < .3770, \quad \text{accept } H_0.$$

It is interesting to note that if the test had been directional but in the opposite direction, then the p-value would have been

$$p\text{-value} = P(R \geq 4 \mid n, p) = \sum_{r=4}^{10} \binom{10}{r} (.5)^r (.5)^{10-r} = .8281.$$

Finally, had the test been two-tailed, the p-value would have been

$$p\text{-value} = 2P(R \leq 4 \mid n, p) = .7540.$$

In summary, then, the p-value concept is quite valuable in hypothesis testing. If the statistician does not have to make a decision, but merely wants some idea of how likely or unlikely the observed sample result is if some hypothesis H_0 is true, then the p-value may be suitable. It provides a measure of how "unusual" the observed result seems if H_0 is true.

A particular alternative need not be specified, although we do have to specify whether the alternative is directional or nondirectional, and if directional, in which direction.

Suppose that the statistician *does* have to make a decision and that after evaluating the losses associated with the two types of errors, a value of α is determined. By computing the p-value from the observed sample statistic and comparing this with α, we arrive at the same decision that would be arrived at by determining a rejection region and checking to see if the observed value falls in this region. Furthermore, it is just as easy to find a p-value from the tables as it is to find a rejection region. At any rate, in most cases the two approaches are equally time-consuming. Then what is the advantage of the p-value approach? If, in addition to making a decision, the statistician intends to report findings, then the p-value may be useful as a reporting device. It allows others to make their own decisions, based on their own choices of α. Furthermore, it gives the readers of the report an idea of how "unusual" the observed result was in light of the tested hypothesis.

7.14 SAMPLE SIZE AND p- VALUES

Statistically, a p-value always means the same thing with regard to H_0. For instance, a p-value of .01 indicates that the sample results are unlikely given H_0. However, the *practical* interpretation of a p-value depends crucially on the sample size. For a test of

$$H_0: \mu = 100$$

against
$$H_1: \mu \neq 100,$$

for example, a p-value of .01 associated with a sample of size 9 would imply quite different things than a p-value of .01 associated with a sample of size 900. A small sample leading to a p-value of .01 might imply not only that $\mu = 100$ was unlikely, but also that other values around 100 (for example, 101) were also unlikely. On the other hand, a p-value of .01 associated with a large sample would imply that μ was not exactly equal to 100 but would not be as likely to rule out values of μ close to but not equal to 100.

Suppose that the population in question is normally distributed with a standard deviation of 10. For a sample of size 9, a sample mean of 108.6 would imply a p-value of .01. Notice that this sample result tends to make $\mu = 101$ seem unlikely, although not quite as unlikely as $\mu = 100$. If 108.6 and 91.4 were taken as the critical values of M, the power of the test against the alternative that $\mu = 101$ would be .0133. Thus, although

α is very small (.01), the β corresponding to a true mean of 101 is very large (.9867).

On the other hand, if $n = 900$, a sample mean of 100.86 would imply a p-value of .01. Notice that for $\alpha = .01$, the critical values of M are much closer to 100 in this case than in the case with $n = 9$. Although H_0 appears just as unlikely per se in each case, some alternative hypotheses close to 100 seem more likely with the larger sample size. With $n = 900$ and $\alpha = .01$, the power of the test against the alternative that $\mu = 101$ is .6628, and the corresponding value of β is .3372.

By observing that the critical values of M will get closer and closer to 100 as n is increased and that the power against any specified alternative will be greater as n is increased, we note the following fact; if we want to "disprove" H_0, it is just necessary to take a large sample to make it highly likely that H_0 will be rejected. A large sample will be quite sensitive to slight deviations from H_0, whereas a smaller sample will be sensitive only to larger deviations from H_0. If the true mean is 101, for example, it will generally take a larger sample before H_0 is rejected than it would if the true mean were, say, 110.

This result makes intuitive sense; it takes a larger sample to detect small differences than to detect large differences. In a way, this is unfortunate, since it means that we can be almost sure to reject an exact hypothesis such as $\mu = 100$ simply by taking a large enough sample. If this is the case, then it may not be a good idea to take *too* large a sample, particularly if we are not concerned with small deviations from the hypothesized value. Admittedly, this statement sounds almost like heresy! Statisticians are often trained to think that large samples are desirable, and many statistical procedures are valid only for large samples. Furthermore, it is true that *as long as the statistician's primary interest is in precise estimation, then the larger the sample the better.* When we want to come as close as we possibly can to the true parameter values, we can always do better by increasing sample size.

On the other hand, many statistical studies are exploratory in nature. The statistician wants to investigate associations, to map out the main relationships in some area, not to waste time and effort by concluding that an association exists when the degree of prediction actually afforded by that association is negligible. In short, the statistician would like a "significant" result to represent not only a nonzero association (a deviation from the "null" hypothesis), but an association of considerable size. *When the sample size is very large, there is a real danger of detecting trivial associations as significant results.* If the statistician wants significance to be very likely to reflect a sizable association in the data and also wants to be sure not to be led by a significant result into some blind alley, then it is necessary to pay attention to *both* aspects of sample size. In general, the statistician would like the sample size to be *large* enough to give confidence that the associations of interest will indeed show up, while being *small* enough so that trivial associations will be excluded from significance.

Stripped of the language of decision theory, all that a "significant" result or a small *p*-value implies is that we have observed something relatively unlikely given the hypothetical situation. Statistical significance is a statement about conditional probability, nothing else. It does not guarantee that something important, or even meaningful, has been found. Conventions about significant results should not be turned into canons of good statistical practice. It is interesting to speculate how many of the early discoveries in science would have been statistically significant in the experiments in which they were first observed. Even in the crude and poorly controlled experiment, some departures from expectation stand out simply because they are interesting and suggest things that the experimenter might not be able to explain. These are matters that warrant looking into further regardless of what the conventional rule says to decide.

Must the statistician actually make a decision about what is true from the sample data? Naturally, choosing to suspend judgment and wait for more evidence is a decision to adopt a course of action. However, why cannot the statistician adopt this time-honored strategy of the scientist more often than is apparently done? In many applied situations the opportunity to suspend judgment is just not available to the statistician. A decision must be made here and now on the basis of what little evidence is available. Fortunately, however, statisticians often do have the privilege of waiting for more evidence.

The occurrence of a "significant" result in a significance test is not always a command to decide something. All that the significance test per se can give is a probability statement about obtaining a particular result if the given hypothesis is true. If we are using the rule represented by the rejection region, then this probability statement is also a statement of the probability of one kind of error. But here the direct contribution of the procedures discussed in this chapter stops; the actual decision should depend on other factors, such as potential losses and personal probabilities, that are not a part of the formal mechanism of the "classical" statistical test.

Why, then, do statisticians bother with significance tests at all? *Regardless of what we are going to do with the information—change our opinion, adopt a course of action, or whatever—we may want to know relatively how probable is a result like that obtained, given a hypothetical true situation.* Basically, a "classical" significance test gives this information, and that is all. The conventions about significance level and regions of rejection can be regarded as ways of defining "improbable." The occurrence of a significant result in terms of these conventions is really a signal. Until we discuss decision theory in more detail, then, the student is advised to think of "classical" hypothesis testing in the way that it has traditionally been used by most statisticians: a conventional signaling device saying, for a significant result, "Here is something relatively unlikely given the situation initially postulated but relatively much more likely under an alternative situation."

Now that the general theory and philosophy of hypothesis testing have been presented, the rest of the chapter will deal with specific applications of the theory: tests of particular types of hypotheses, such as hypotheses concerning means.

7.15 TESTS CONCERNING MEANS

Some of the examples presented earlier in the chapter involved means of normal distributions. In these examples, we "fudged" a bit on the usual situation; we assumed that σ^2 was somehow known, so that the standard error of the mean was also known exactly. Now we must face the cold facts of the matter; for inferences about the population mean, σ^2 is seldom known. Instead, we must use the only substitute available for σ^2, which is an estimate, such as our unbiased estimate $\hat{s}^2$, calculated from the sample.

Recalling that if the population is normally distributed, then the random variable

$$T = \frac{M - \mu}{\text{est. } \sigma_M} = \frac{M - \mu}{\hat{S}/\sqrt{n}} \qquad (7.15.1^*)$$

has a T distribution with $n - 1$ degrees of freedom, we see that the only change in the testing procedure will be to calculate T instead of Z and to look in the tables of the T distribution rather than the normal tables. Otherwise, the rejection regions (or p-values) are determined in exactly the same manner as when the population variance was assumed to be known. Of course, if the sample size is large enough for the T distribution to be well-approximated by the normal distribution, we can calculate T but look in the normal tables.

For example, suppose that a manufacturer of refrigerators claims that at a particular setting, the mean temperature in the freezer compartment of a certain model of refrigerator is 10 degrees Fahrenheit. Thus, we decide to test

$$H_0: \mu \leq 10$$

versus $$H_1: \mu > 10.$$

We take a random sample of 16 refrigerators from a given lot of refrigerators and record the temperature in the freezer compartment at the particular setting of interest. The sample mean is 10.24 and the unbiased sample variance is .36. We assume that the distribution of temperatures is normally distributed. The value of T is then

$$t = \frac{10.24 - 10}{.6/4} = 1.6.$$

(Notice that the "10" in this calculation represents the "dividing line" between H_0 and H_1.) Looking in the T tables under 15 degrees of freedom, we see that for a one-tailed test to the right, this value corresponds to a p-value between .05 and .10. For $t = 1.341$, the p-value is .10, and for $t = 1.753$, the p-value is .05. Therefore, given the limits of the table, we can only say in this instance that the p-value is between .05 and .10, unless we try to interpolate. Thus, if $\alpha \geq .10$, H_0 should be rejected; if $\alpha \leq .05$, H_0 should be accepted; and if $.05 < \alpha < .10$, we cannot say for sure what the decision should be.

When the T distribution is used to test a hypothesis, the assumptions should be kept in mind, primarily the assumption that the population is normally distributed. If large samples are available, this should be no problem, but for small samples the statistician should exercise caution in this regard (see Section 6.13).

7.16 TESTS CONCERNING DIFFERENCES BETWEEN MEANS

As we pointed out in Section 6.16, statisticians are often interested not just in a single mean, but in the difference between means. For example, suppose that the president of a chain of retail stores is interested in the difference in mean sales at two of the stores in the chain. He formulates the following hypotheses:

$$H_0: \mu_1 - \mu_2 = d_0,$$
$$H_1: \mu_1 - \mu_2 \neq d_0.$$

For instance, he might let $d_0 = 0$, so H_0 corresponds to the hypothesis that there is no difference between the means. Although this is the most common choice of d_0, any other value could be chosen. Furthermore, the president is willing to assume that the distributions of sales at the two stores are both normal, with respective known variances $\sigma_1^2 = 10,000$ and $\sigma_2^2 = 7200$. He then takes independent random samples of size $n_1 = 10$ and $n_2 = 12$ from the sales records of the two stores, observing the following sample results:

$$m_1 = 8000, \qquad m_2 = 8050,$$
$$s_1^2 = 9000, \qquad s_2^2 = 8000.$$

Since the variances are known, the test statistic is simply

$$z = \frac{(m_1 - m_2) - d_0}{\sigma_{\text{diff.}}} = \frac{(m_1 - m_2) - d_0}{\sqrt{\sigma_1^2/n_1 + \sigma_2^2/n_2}}. \qquad (7.16.1^*)$$

If $d_0 = 0$,

$$z = \frac{8000 - 8050}{\sqrt{10,000/10 + 7200/12}} = \frac{-50}{40} = -1.25.$$

The p-value corresponding to this z value for a two-tailed test is

$$P(|Z| \geq 1.25) = .212.$$

Thus, the president will reject H_0 if $\alpha \geq .212$ and will accept H_0 otherwise.

Suppose, however, that the president had not known σ_1^2 and σ_2^2, but that he felt that they could be assumed to be equal. This is much more realistic, and it requires the use of a T statistic with $n_1 + n_2 - 2$ degrees of freedom (see Section 6.16):

$$T = \frac{(M_1 - M_2) - d_0}{\text{est. } \sigma_{\text{diff.}}} = \frac{(M_1 - M_2) - d_0}{\sqrt{\left(\dfrac{n_1 S_1^2 + n_2 S_2^2}{n_1 + n_2 - 2}\right)\left(\dfrac{n_1 + n_2}{n_1 n_2}\right)}}. \qquad (7.16.2^*)$$

For this example, $n_1 + n_2 - 2 = 20$, and

$$t = \frac{8000 - 8050}{\sqrt{1705}} = \frac{-50}{41.3} = -1.21.$$

Looking up the corresponding p-value in the T tables, we see that for a two-tailed test, the p-value is between .20 and .50, with the indication being that it is near the lower end-point, .20. If the alternative hypothesis H_1 had been $H_1: \mu_1 - \mu_2 < 0$, the p-value would be between .10 and .25. On the other hand, if H_1 was of the form $H_1: \mu_1 - \mu_2 > 0$, then the p-value would be between .75 and .90.

As noted above, it is not necessary for d_0 to be equal to zero. If the manager of store 2 claimed that sales were at least 200 units higher than the sales at store 1, the president might set up the hypotheses as follows:

$$H_0: \mu_1 - \mu_2 \leq -200,$$

$$H_1: \mu_1 - \mu_2 > -200.$$

Now, given the sample results, t is calculated:

$$t = \frac{(8000 - 8050) - (-200)}{\text{est. } \sigma_{\text{diff.}}} = \frac{150}{41.3} = 3.63.$$

For this example, the p-value is less than .001, casting much doubt on the claim of the manager of store 2.

Note that the T test in this section required three major assumptions: (a) normality of the population distributions, (b) independence of the two samples as well as independence within each sample, and (c) equality of the population variances. Assumptions (a) and (c) are more important for small samples than for large samples. If the sample size is small or if assumption (c) seems quite unreasonable, then as noted in Section 6.16, we can consider the sample variances separately rather than pooling them and we can use a slightly different formula for the degrees of freedom in the T statistic. The estimate of $\sigma_{\text{diff.}}$ is

$$\text{est. } \sigma_{\text{diff.}} = \sqrt{\frac{S_1^2}{n_1 - 1} + \frac{S_2^2}{n_2 - 1}} = \sqrt{\frac{\hat{S}_1^2}{n_1} + \frac{\hat{S}_2^2}{n_2}}, \qquad (7.16.3)$$

and the degrees of freedom are given by Equation (6.16.7). As for assumption (b), the assumption of independence, in some situations the statistician intentionally takes samples that are related rather than independent. This case is considered in the following section.

7.17   PAIRED OBSERVATIONS

Sometimes it happens that samples from two populations are "paired" samples. By this, we mean that each observation in the first sample is related in some way to exactly one observation in the second sample, so the samples are not independent. For example, suppose that the two samples are two examinations given to the same class, so that we can pair the score of any single member of the class on the first examination with his or her score on the second examination. Or, consider a questionnaire administered to a group of married couples. One sample could be the wives and one the husbands, in which case a wife could be thought of as "paired" with her husband if husband-wife differences were of interest.

Given two samples matched in this pairwise way, either by the statistician or otherwise, it is still true that the difference between the means is an unbiased estimate of the population difference (in two matched populations):

$$E(M_1 - M_2) = \mu_1 - \mu_2.$$

However, the matching and the consequent *dependence* within the pairs change the standard error of the difference. This can be shown quite simply. By definition, the variance of the difference between two sample means is

$$\sigma_{\text{diff.}}^2 = E(M_1 - M_2 - \mu_1 + \mu_2)^2,$$

which is the same as

$$E[(M_1 - \mu_1) - (M_2 - \mu_2)]^2.$$

Expanding the square, we have

$$E(M_1 - \mu_1)^2 + E(M_2 - \mu_2)^2 - 2E(M_1 - \mu_1)(M_2 - \mu_2).$$

The first of these terms is just $\sigma_{M_1}{}^2$, and the second is $\sigma_{M_2}{}^2$. However, what of the third term? This term is cov (M_1, M_2), the covariance of the means (Section 3.23). Then, for matched samples,

$$\sigma_{\text{diff.}}{}^2 = \sigma_{M_1}{}^2 + \sigma_{M_2}{}^2 - 2 \text{ cov } (M_1, M_2).$$

This unknown value of cov (M_1, M_2) could be something of a problem, but actually it is quite easy to bypass this difficulty altogether. Instead of regarding this as two samples, we simply think of the data as coming from one sample of *pairs*. Associated with each pair i is a difference

$$D_i = (Y_{i1} - Y_{i2}), \tag{7.17.1}$$

where Y_{i1} is the observed value for the member of pair i who is in group 1, and Y_{i2} is the observed value for the member of pair i who is in group 2. Then an ordinary T test for a *single* mean is carried out using the values of D_i. That is,

$$M_D = \frac{\sum_i D_i}{n}$$

and

$$S_D{}^2 = \frac{\sum_i D_i{}^2}{n} - M_D{}^2.$$

Then

$$\text{est. } \sigma_{M_D} = \frac{S_D}{\sqrt{n-1}} = \frac{\hat{S}_D}{\sqrt{n}},$$

and T is found from

$$T = \frac{M_D - E(M_D)}{\text{est. } \sigma_{M_D}} \tag{7.17.2}$$

with $n - 1$ degrees of freedom. *Be sure to notice that here n stands for the number of differences, which is the number of pairs.*

Naturally, the hypothesis is about the true value of $E(M_D)$, which is always $\mu_1 - \mu_2$. Thus, any hypothesis about a difference can be tested in this way, provided that the samples used are matched *pairwise*. Similarly, confidence limits are found just as for a single mean, using M_D and σ_{M_D} in place of M and σ_M.

For example, suppose that a company president is concerned about the value of a training program that the company runs for new employees. He decides to administer an examination to the 25 members of a particular class in the training program. Furthermore, he decides to administer the same test twice: once at the beginning of the training program, and once at the end. It is thought that the training program should result in an average score increase of more than 30 points; if it does not, the president may eliminate the program. The hypotheses the president is interested in are as follows:

$$H_0: \mu_2 - \mu_1 \leq 30,$$

$$H_1: \mu_2 - \mu_1 > 30.$$

The subscripts 1 and 2 refer to the two times the examination is given. The class of 25 takes the examination twice, the difference in score is computed for each member, and the results are summarized as follows:

$$m_D = 36.16,$$

$$\text{est. } \sigma_{M_D} = 4.24.$$

From these results, t can be calculated:

$$t = \frac{36.16 - 30}{4.24} = 1.45.$$

Looking in the T tables under $n - 1 = 24$ degrees of freedom, we find that the p-value is between .05 and .10. The evidence appears to favor H_1 somewhat. Of course, in order to make a decision one way or the other about the hypotheses, the president would have to consider his losses and determine α. If $\alpha \leq .05$, he should accept H_0 and seriously consider eliminating the training program; if $\alpha \geq .10$, he should reject H_0 and continue the program; if $.05 < \alpha < .10$, we need more precise tables to tell us what he should do.

7.18   THE POWER OF T TESTS

The idea of the power of a statistical test was discussed in detail in preceding sections only in terms of the normal distribution. Nevertheless, the same general considerations apply to the power of tests based on the T distribution. Thus, the power of a T test increases with sample size, increases with the discrepancy between the null hypothesis value and the true value of a mean or a difference, increases with any reduction in the true value of σ, and increases with any increase in the size of α, given a true value covered by H_1.

Unfortunately, the actual determination of the power for a T test against any given true alternative is more complicated than for the normal distribution. The reason is that when the null hypothesis is false, each value of T computed involves $E(M)$ or $E(M_1 - M_2)$, which is the exact value given by the null (and false) hypothesis. If the true value of the expectation could be calculated into each T, then the distribution would follow the T function tabled in the Appendix. However, when H_0 is false, each value of T involves a different expectation; this results in a somewhat different distribution, called the **noncentral T distribution.** The probabilities of the various values of T cannot be known unless one more parameter, δ, is specified besides ν. This is the so-called noncentrality parameter

$$\delta = \frac{\mu - \mu_0}{\sigma_M}, \tag{7.18.1}$$

which expresses the difference between the true expectation μ and that given by the null hypothesis, or μ_0, in terms of σ_M. For a hypothesis about a difference and for samples of equal size,

$$\delta = \frac{(\mu_1 - \mu_2) - (\mu_{01} - \mu_{02})}{\sigma_{\text{diff.}}}. \tag{7.18.2}$$

This makes the determination of the power of a T test considerably more troublesome than for tests using the normal distribution. In particular, information about σ for each population is required. However, for most uses that are likely to be made of the power concept in applications, especially in determining sample size, the normal approximation may still suffice. For more accurate work, most advanced texts in statistics give tables of the power function of T in terms of both ν and δ.

7.19   TESTING HYPOTHESES ABOUT A SINGLE VARIANCE

Just as for the mean, it is possible to test hypotheses about a single population variance (and, of course, a standard deviation). The exact hypothesis tested is

$$H_0: \sigma^2 = \sigma_0^2,$$

where σ_0^2 is some specific positive number. The alternative hypothesis may be either directional or nondirectional, depending, as always, on the original question.

As usual, some value of α is decided upon, and a region of rejection is adopted depending both on α and the alternative H_1. The test statistic itself is

$$\chi^2_{(n-1)} = \frac{(n-1)\hat{S}^2}{\sigma_0^2}. \qquad (7.19.1)$$

This test statistic has a χ^2 distribution with $n - 1$ degrees of freedom (see Section 6.19).

For example, there is some evidence that women tend to be a less variable, more homogenous group than men. We might ask this question about height: "It is well known that men and women in the United States differ in terms of their mean height; is it true, however, that women show less variability in height than do men?" Now let us assume that from the records of the Selective Service System we actually *know* the mean and standard deviation of height for American men between the ages of 20 and 25 years. However, such complete evidence is lacking for women. Assume that the standard deviation of height for the population of men 20 to 25 years is 2.5 inches. For this same age range, we want to ask if the population of women shows this same standard deviation, or if women are *less* variable, with their distribution having a smaller σ. The null and alternative hypotheses can be framed as

$$H_0: \sigma^2 \geq 6.25 \qquad (\text{or } [2.5]^2)$$

and $H_1: \sigma^2 < 6.25.$

Imagine that we plan to draw a sample of 30 women at random, each between the ages of 20 and 25 years, and measure the height of each. The test statistic will be

$$\chi^2_{(29)} = \frac{(29)\hat{S}^2}{6.25}.$$

What is the region of rejection? Here, *small* values of χ^2 tend to favor H_1, that women actually are less variable than men. Hence, we want to use a region of rejection on the left (or small-value) tail of the chi-square distribution with 29 degrees of freedom. If $\alpha = .01$, then from Table III in the Appendix, the critical value of χ^2 leading to rejection of H_0 should be 14.257, so we should reject H_0 if $\chi^2 \leq 14.257$.

Now the actual value of $\hat{S}^2$ obtained turns out to be 4.55, so that

$$\chi^2_{(29)} = \frac{(29)(4.55)}{6.25} = 21.11.$$

This value is larger than the critical value decided upon, and we cannot reject H_0 if α is to be .01. In fact, the p-value in this case is found from the tables to be between .10 and .25.

This example illustrates that, as with the T and the normal distribution, either or both tails of the chi-square distribution can be used in testing a hypothesis about a variance. Had the alternative hypothesis in this problem been

$$H_1: \sigma^2 \neq 6.25,$$

then the rejection region would lie in both tails of the chi-square distribution. The rejection region on the lower tail of the distribution would be bounded by a chi-square value corresponding to $\nu = 29$ and $F(\chi^2) = .005$, which is 13.321; the rejection region on the upper tail would be bounded by the value for $\nu = 29$ and $F(\chi^2) = .995$, which is 52.336. Any obtained χ^2 value falling *below* 13.121 or *above* 52.336 would let one reject H_0 beyond the .01 level. For this two-tailed test, the p-value would be twice as large as for the above one-tailed test; that is, it would be between .20 and .50.

A final note concerning the use of the χ^2 statistic to test hypotheses concerning a single variance is that the population must be normally distributed. If the population is not normally distributed, then the χ^2 statistic might not be applicable unless the sample size is very large (see Section 6.20).

7.20 TESTING HYPOTHESES ABOUT TWO VARIANCES

In many situations the statistician is concerned with hypotheses involving two variances. In the example of the preceding section, if the standard deviation of height for men was not known, then the hypothesis of interest would be the hypothesis that the variance of height for women is greater than or equal to the variance of height for men. To test this hypothesis, we would take independent samples from the population of men and the population of women.

Suppose that independently of the sample results for women presented in Section 7.19, a sample of 21 men was taken, resulting in a value of $\widehat{S}^2$ equal to 9.10. If the subscripts 1 and 2 refer to men and women, respectively, then the hypotheses of interest are

$$H_0: \sigma_2{}^2 \geq \sigma_1{}^2$$

and

$$H_1: \sigma_2{}^2 < \sigma_1{}^2.$$

If it is assumed that the two populations are normally distributed, then the ratio

$$F = \frac{\widehat{S}_1{}^2/\sigma_1{}^2}{\widehat{S}_2{}^2/\sigma_2{}^2} \qquad (7.20.1^*)$$

has an F distribution with $n_1 - 1$ and $n_2 - 1$ degrees of freedom (Section 6.21). But the dividing line between the two hypotheses is at $\sigma_1^2 = \sigma_2^2$, so the value of F in the example is simply

$$\frac{\hat{s}_1^2}{\hat{s}_2^2} = \frac{9.10}{4.55} = 2.00,$$

with 20 and 29 degrees of freedom. From the F tables, we find that this is slightly greater than the value required for one-tailed significance at the .05 level. Thus, the p-value is slightly less than .05.

Had the alternative hypothesis been one-tailed in the opposite direction, then we would have had to consider the required value of F on the lower tail of the distribution, using the procedure given in Section 6.21. Since the choice of the sample variance to put in the numerator is arbitrary (we could just as easily have chosen $F = \hat{S}_2^2/\hat{S}_1^2$ with $n_2 - 1$ and $n_1 - 1$ degrees of freedom), it is convenient for one-tailed hypotheses to put the larger sample variance in the numerator, so that it is only necessary to consider the upper tail of the F distribution. In doing this, of course, it is imperative that you make sure that the larger sample variance is the one which would be *expected to be larger in light of the alternative hypothesis, H_1.*

Once again, you should note the caution that the use of the F distribution in inferences about two variances presupposes that the two populations are normally distributed (refer to Section 6.22). In fact, all of the specific testing procedures discussed in Sections 7.15–7.20 require the assumption that the populations of interest are normally distributed. For small samples, the statistician should investigate the applicability of this assumption in any specific situation. If the assumption seems unreasonable, then we may not be justified in using the Z, T, χ^2, and F tests. For large samples, the central limit theorem comes to the rescue; by now you should see what we meant when we said that the central limit theorem has important implications for statistical inference.

EXERCISES

1. Why might there be differences between statistical hypotheses and real-world hypotheses? How does this relate to the general problem of building a statistical model of a real-world situation?

2. In the example of Section 7.5, X is the rate per hour at which finished products are turned out by workers in a particular plant under a new working procedure. It is assumed that X is normally distributed and that the standard deviation of X is 20. The following hypotheses are of interest:

$$H_0: \mu = 138$$

and

$$H_1: \mu = 142.$$

For a sample of 100 man-hours, find the probabilities of the two possible types of error for each of the following decision rules.

(a) Choose H_1 if M, the sample mean output rate, is greater than 140; choose H_0 otherwise.
(b) Choose H_1 if $M > 141$; choose H_0 otherwise.
(c) Choose H_1 if $M > 143$; choose H_0 otherwise.
(d) Choose H_1 if $M < 140$; choose H_0 otherwise.
(e) Choose H_1 if $136 < M < 145$; choose H_0 otherwise.

3. For each of the decision rules in Exercise 2,
    (a) determine whether it is dominated by any of the other four rules;
    (b) if it is not dominated by any of the other four rules, see if you can find some other rule that dominates it.

4. In Exercise 27, Chapter 6, consider the following decision rules for a random sample of 8 pages, and compute the error probabilities for each rule.
    (a) Decide that $\lambda = 1.5$ if R, the total number of misprints in the 8 pages, is less than 16; decide that $\lambda = 2$ otherwise.
    (b) Decide that $\lambda = 1.5$ if $R < 13$; decide that $\lambda = 2$ otherwise.
    (c) Decide that $\lambda = 1.5$ if $R > 18$; decide that $\lambda = 2$ otherwise.
    (d) Decide that $\lambda = 1.5$ if $R < 14$; decide that $\lambda = 2$ otherwise.

5. Are any of the decision rules in Exercise 4 inadmissible? Explain.

6. Find the general form an admissible decision rule must have in Exercise 4, and find the error probabilities for each possible admissible rule. How might we choose among these rules?

7. The proponent of a new drug claims that if a patient is given this drug, the probability that the patient will be cured is .9. Others, however, claim that the probability of being cured is only .7. Denote the proponent's claim by H_0 and the other claim by H_1. Determine α and β for each of the following decision rules.
    (a) Reject H_0 if R, the number of patients cured in a random sample of 10 patients who are given the drug, is less than 8; accept H_0 otherwise.
    (b) Reject H_0 if R is no greater than 7; accept H_0 otherwise.
    (c) Accept H_0 if R is at least 9; reject H_0 otherwise.

8. In Exercise 2, find the error probabilities for each of the decision rules when the sample size is 400 instead of 100. Discuss the differences in error probabilities for the two sample sizes.

9. In Exercise 7, the physician taking the random sample of 10 patients wants the probability of claiming that the proponent's claim is wrong when in fact it is actually correct to be no greater than .10.
    (a) Find a decision rule that will make the probability in question as large as possible given the constraint that it cannot be greater than .10.
    (b) Does the decision rule in (a) attain the limiting value of .10? If not, why can't you find a rule for which the probability in question is *exactly* 10?

10. In Exercise 2, (a) find a decision rule for which $\alpha = .05$ and (b) find a decision rule for which $\beta = .05$.

11. Distinguish between the two types of errors which are possible in hypothesis testing. In general, is it reasonable to assume that one type of error is always more serious than the other? If there are two types of error, why does the "classical" convention presented in Section 7.4 only involve one type of error? After α is fixed, does the statistician have any control over β? Explain.

12. Some statisticians claim that we must *never*, under any circumstances, *accept* the "null" hypothesis as true. We can reject the null hypothesis with sufficient evidence, they agree, but apparently we are in some sort of limbo when a result fails to be significant. Comment on this stand. Does it ever seem reasonable to believe that a hypothesis such as H_0: $\mu = 0$ is *exactly* true? If not, how could H_0 be modified to make it seem reasonable that it *could* be exactly true?

13. Discuss the following proposition: "Much of the controversy surrounding the use of conventions such as that given in Section 7.4 can be explained in terms of whether hypothesis testing is viewed as a "signaling device," as discussed in Section 7.14, or as a formal decision-making procedure."

14. A bookbag is filled with 100 poker chips. You know that either 70 of the chips are red and the remainder blue, or 70 are blue and the remainder red. Ten chips are to be drawn successively with replacement. You must decide whether the composition of the bookbag is 70R–30B or 70B–30R. Find a decision rule that will make the probabilities of the possible errors in inference small and equal.

15. Comment on the following statement: "If everything else is held constant, and only admissible decision rules are considered, making α smaller leads to a larger β, and vice versa, because the sum of the two error probabilities must remain the same."

16. In what ways can the statistician make both α and β smaller simultaneously?

17. Suppose that you know that exactly one of the following two hypotheses must be true:

$$H_0: \mu = 100, \quad \sigma = 15,$$

or $$H_1: \mu = 105, \quad \sigma = 15.$$

The population sampled has a normal distribution, and the sample size is to be 25. If the probability of a Type I error is to be .10, what is the critical value of M? If M turns out to be 103, what is the p-value?

18. Let X denote the weight of a carton that is randomly selected from the cartons hauled during a certain month by a trucking firm. Consider the hypotheses

$$H_0: \mu = 50$$

and $$H_1: \mu = 60,$$

where weight is expressed in pounds, X is assumed to be normally distributed, and the variance of X is assumed to be 400. A random sample of 4 cartons is taken, and M represents the sample mean weight of the 4 cartons.

(a) If $\alpha = .10$, find the critical value of M and compute β.

(b) If $\alpha = .10$ and H_1 states that $\mu = 80$ instead of $\mu = 60$, find the critical value of M and compute β.

(c) In view of the results in (a) and (b), comment on the "classical" convention and on how the test is affected by changes in H_1.

19. Comment on the statement that "A low value of α, such as .01, guarantees that a statistician will not make many errors."

20. In Exercise 4, find the value of the likelihood ratio corresponding to each value of R, where the hypotheses are H_0: $\lambda = 1.5$ and H_1: $\lambda = 2$. If H_0 is rejected whenever the likelihood ratio is less than or equal to .5 and accepted otherwise, find α and β and specify the rejection region in terms of R.

21. In Exercise 7, find the likelihood ratio if the number of patients cured in a random sample of 10 patients turn out to be 6. What is the likelihood ratio if all 10 patients are cured?

22. In Exercise 14, 1 chip is drawn at random from the bookbag, and the chip is red. Let H_0 represent the hypothesis that the composition of the bookbag is 70R–30B, and let H_1 represent the hypothesis that the composition of the bookbag is 70B–30R.

(a) Find the likelihood ratio for the sample of 1 chip.

(b) Find the likelihood ratio if the sample of size 1 yields a blue chip.

(c) Find the likelihood ratio if a sample of size 2 with replacement yields 2 red chips.

(d) Find the likelihood ratio if a sample of size 3 with replacement yields 2 red chips and 1 blue chip.

(e) Is there any relationship between the likelihood ratio found in (a) and the likelihood ratio found in (c)? Explain.

(f) Is there any relationship between the likelihood ratio found in (d) and those found in (b) and (c)? Explain.

23. In Exercise 18, find the critical value of the likelihood ratio if the probability of deciding that $\mu = 60$ when in fact $\mu = 50$ is to be .01.

24. A consumer research firm claims that the average life of a particular type of flashlight battery is 60 hours. The manufacturer of the battery, on the other hand, claims that the average life is 80 hours. The consumer research firm and the manufacturer agree that the standard deviation is 40 hours. In order to investigate the two claims, the consumer research firm takes a random sample of 100 batteries and determines the life of each battery.

(a) If the probability of deciding erroneously that the consumer research firm is right when the manufacturer is actually right is to be .10, find the rejection region.

(b) How large a sample would be needed to make $\alpha = \beta = .05$, and what would the rejection region be?

25. In Exercise 24, find the rejection region if the probability of deciding erroneously that the manufacturer is right when the consumer research firm is actually right is to be .10.

26. Why should there be any systematic relationship between the likelihood ratio and a "good" rejection region in a hypothesis-testing situation? Does the Neyman-Pearson lemma presented in Section 7.6 seem intuitively reasonable? Why are rejection regions such as $\Omega \geq .8$ or $1.1 < \Omega < 1.4$ inadmissible?

27. In Exercise 14, 10 chips are to be drawn randomly and with replacement from the bookbag. Give a general formula for the likelihood ratio in this situation, where the hypotheses are H_0: 70R–30B and H_1: 70B–30R. From this formula, show that a rejection region of the form $\Omega \leq k$ is equivalent to a rejection region of the form $R \leq c$, where R is the number of red chips in the sample of size 10. [*Hint*: Consider the expression $\Omega \leq k$ and attempt to manipulate it algebraically to arrive at the result $R \leq c$.]

28. Let X represent the amount of time required for a customer to transact his or her business with a teller at a particular bank. Assume that X is exponentially distributed, and consider the hypotheses H_0: $\lambda = .5$ and H_1: $\lambda = .3$, where X is measured in minutes.
    (a) For a sample of 1 customer, give a general expression for the likelihood ratio in this example.
    (b) If the customer in (a) transacts his or her business in 4 minutes, find the likelihood ratio.

29. In Exercise 28, find the region of rejection if $\alpha = .10$. What is the probability of making a Type II error? [*Hint*: Be careful; as the observed value of X gets larger and larger, what is implied about the value of the parameter λ?]

30. Let X represent height, in inches, where the population of interest is a particular population of male basketball players. In this population, height is normally distributed with variance 9, and the hypotheses of interest are H_0: $\mu \leq 75$ and H_1: $\mu > 75$. A sample of size 100 is to be drawn randomly from the population.
    (a) If $\alpha = .05$, what is the critical value of M?
    (b) What is the power of this test if the true value of μ is equal to 75.5?
    (c) What is the power if the true value of μ is 76?
    (d) Draw the power curve for this test.

31. In Exercise 30, find the critical value of M and draw the power curve if $\alpha = .20$. Compare this power curve and the one obtained in Exercise 30.

32. In Exercise 30, find the critical value of M and draw the power curve if $\alpha = .05$ and you are testing

$$H_0: \mu \geq 75$$

against

$$H_1: \mu < 75.$$

Also, find the region of rejection and draw the power function if $\alpha = .05$ and you are testing

$$H_0: \mu = 75$$

against

$$H_1: \mu \neq 75.$$

33. In Exercise 7, consider the hypotheses H_0: $p = .9$ and H_1: $p < .9$. That is, the alternative to the proponent's claim is simply the claim that p is less than .9. It is decided to reject H_0 if the number of patients cured in a random sample of 10 patients who are given the drug turns out to be less than 7.
   (a) Find the power of this test against the alternative that $p = .8$.
   (b) Find the power curve for this test.
   (c) Find α.

34. The sample is taken in Exercise 33, and 7 out of the 10 patients are cured.
   (a) What is the probability of this sample result if H_0 is true?
   (b) Give a general expression for the probability of this sample result given p.
   (c) Find the value of p that maximizes the probability in (b) and calculate the maximum value of the probability.
   (d) Compute the ratio of the probabilities determined in (a) and (c), and go through the same procedure for a sample in which only 5 of the 10 patients are cured. Compare the two ratios in light of the discussion in Section 7.6 of likelihood ratios for inexact hypotheses.

35. Draw the "ideal" power curve for the test of

$$H_0: \mu \le \mu_0$$

against $$H_1: \mu > \mu_0.$$

(That is, consider the best possible situation as far as power is concerned; for example, what would you like the power to be against any $\mu > \mu_0$?) Also, draw the "ideal" operating characteristic curve and the "ideal" error curve.

36. In the example in Section 7.5, suppose that you want to know if the new working procedure changes the mean rate per hour at which finished products are turned out. Thus, consider the hypotheses H_0: $\mu = 138$ and H_1: $\mu \ne 138$.
   (a) If $\alpha = .10$ and $n = 100$, find the rejection region for a test of H_0 versus H_1.
   (b) If n is 100 and the sample mean is 140.2, find the p-value.

37. In Exercise 36, find the power of the test in (a) against the specific alternative $\mu = 139$. For this test, find the power curve, the operating characteristic curve, and the error curve.

38. A safety expert wants to investigate the effect of a lower speed limit on the accident rate. Under the old speed limit, accidents along a particular stretch of road averaged 5 per week. A new speed limit is being used on a trial basis, and there have been 11 accidents in 3 weeks. The "accident process" is assumed to be a Poisson process.
   (a) The safety expert wants the probability of claiming that the accident rate has been reduced when in fact the mean accident rate has not changed to be no greater than .05. For what values of R (the number of accidents in the 3-week period under the new speed limit) should it be claimed that the accident rate has been reduced?
   (b) For the sample that was actually observed (11 accidents in 3 weeks), what is the p-value?

39. Suppose that you want to test

$$H_0: \mu = 300$$

against
$$H_1: \mu \neq 300,$$

where the population in question is normal with variance 10,000. You plan to take a sample of size 4. Consider the following two rejection regions:

region 1: $M \leq 200$ or $M \geq 400,$

region 2: $297 \leq M \leq 303.$

For each of these two rejection regions, draw the power curve. Comment on the difference between the two curves.

40. A statistician wants to test the hypothesis

$$H_0: \mu = 3$$

against
$$H_1: \mu \neq 3$$

for a normally distributed population with variance 1.44. Given $\alpha = .10$ and $n = 36$, how powerful would this test be if $\mu = 4$? Given $\alpha = .10$ and $n = 16$, how powerful would this test be if $\mu = 4$? If the statistician wants the test to have power of .80 if $\mu = 3.5$, what sample size should be used, given that $\alpha = .10$?

41. In Exercise 73, Chapter 5, a specification calls for a steel rod to be 3 meters long. The length is assumed to be normally distributed, the standard deviation of length is known to be .20 meters, and rods that are too short or too long are of no use to the firm purchasing the rods. To investigate the mean length of rods in a shipment, 25 rods are chosen randomly and measured If the sample mean is too large or too small, the shipment is returned to the. supplier. For what values of the sample mean should the shipment be returned to the supplier if the probability of returning a shipment for which the mean length is exactly 3 meters is to be only .01?

42. In Exercise 41, what is the power of the test against a true mean of 2.8? Against what other possible value of μ does the test have exactly the same power?

43. In Exercise 41, draw the power curve, the operating characteristic curve, and the error curve for the following three rules:

rule 1:   accept the shipment if $2.95 \leq M \leq 3.05,$
rule 2:   accept the shipment if $2.90 \leq M \leq 3.10,$
rule 3:   accept the shipment if $2.90 \leq M \leq 3.05.$

Can you think of a situation in which a rule such as the third rule might be preferable to rules such as the first two rules? Explain.

44. In Exercise 41, suppose that it is easy to shorten rods that are too long, but rods that are too short are of no use to the firm purchasing the rods. Does this change your answer to Exercise 41? If not, why not, and if so, how is your answer changed?

45. Explain the relationship among power curves, operating characteristic curves, and error curves.

46. A person who has just been exposed to the notion of hypothesis testing for the first time in a statistics course decides to investigate the gasoline mileage obtained by a particular car. The price of gasoline has just increased considerably, and the individual is concerned about the cost per mile of running the car. Let X denote the mileage per gallon for a tank of gasoline, where the tank holds 15 gallons. The individual records the mileage per gallon for 4 tanks of gasoline, assumes that X is normally distributed and that $V(X) = 16$, considers the hypotheses H_0: $\mu = 20$ and H_1: $\mu \neq 20$ (the manufacturer claims the car should get 20 miles per gallon), and sets $\alpha = .01$.
    (a) Find the rejection region for this test.
    (b) If $M = 26$, what should the person decide?
    (c) If $M = 14$, what should the person decide?
    (d) Discuss the entire procedure chosen by the novice statistician in this exercise and indicate whether you agree or disagree with the various steps in the procedure. How might you change the procedure?

47. Explain the relationship between confidence intervals and two-tailed tests of hypotheses. Could you develop "one-sided" confidence intervals to relate to one-tailed tests? Explain.

48. An experimenter found that in a random sample of 300 women students in college, 35 percent smoked cigarettes. Can the experimenter reject the hypothesis H_0: $p = .40$ in favor of the alternative hypothesis H_1: $p \neq .40$ at the .05 significance level?

49. In a random sample of 400 consumers in a given city, 250 preferred Brand A cars to Brand B cars. Test

$$H_0: p = .50$$

against
$$H_1: p \neq .50$$

with $\alpha = .20$.

50. In Exercise 49, the hypotheses H_0: $p \leq .50$ and H_1: $p > .50$ appear to be of greater interest than the hypotheses considered in that exercise. If $\alpha = .20$, find a rejection region for the one-tailed situation.

51. Discuss the relationship between predetermined significance levels (values of α) and the use of p-values in hypothesis testing. What are the advantages of the use of p-values?

52. Discuss the statement: "Ideally, the choice of α should be based on the relative seriousness of the losses associated with the two possible errors. The larger the loss associated with a Type I error is, or the smaller the loss associated with a Type II error is, the smaller α should be, all other things being equal."

53. In Exercise 46, suppose that $M = 21$. What is the p-value associated with this sample mean? If you decided that $\alpha = .01$, would you reject H_0 in favor of H_1? If someone else decided that $\alpha = .35$, would he or she reject H_0 in favor of H_1 on the basis of the same data?

54. In Exercise 30, find the p-value if the sample mean turns out to be
    (a) 75.3,   (c) 74.5,
    (b) 76.0,   (d) 75.0.

55. In Exercise 40 with $n = 16$, find the p-value if the sample mean turns out to be
    (a) 2.9,   (c) 3.5,
    (b) 3.1,   (d) 2.7.

56. In Exercise 41, find the values of the sample mean for which the shipment will *not* be returned if (a) $n = 4$, (b) $n = 100$, (c) $n = 400$, (d) $n = 1600$. Given that everything else is held constant, discuss the implications of changes in the number of rods sampled before deciding whether to accept a shipment or to send it back to the supplier.

57. Discuss the following statement: "When the sample size is very large, there is a real danger of detecting trivial associations as significant results."

58. A psychologist in a public school system is asked by the superintendent to provide evidence to the school board that the students will profit from more intensive training. There are 4000 students in the city school system and 794 in this particular high school. The psychologist decides to test the hypothesis that the mean IQ of the students is 110. There is very good reason to believe, from the standardization of the test used, that whatever the mean IQ value is, the variance of IQ scores for these students is 116. Furthermore, in consultation with the superintendent, the psychologist decides that the probability of Type I error should be .05. Since the only definite course of action to be taken is to recommend special training if there is evidence that the mean IQ is greater than 110, and no particular action will be taken if the mean IQ is less than or equal to 110, the psychologist decides to do a one-tailed test. Some 30 of the high school students are selected completely at random, given the test, and found to have a mean IQ of 121. The psychologist reports to the superintendent that the students have an IQ significantly greater than 110 and recommends that the board be requested to proceed with the plans for accelerated programs.

    Answer the following questions briefly, with particular attention to the italicized words:
    (a) Is the psychologist's conclusion at all warranted?
    (b) What *sampling distribution* is relevant to the statistical test, and what *assumptions* are necessary?
    (c) What, precisely, is the *null* hypothesis? the alternative hypothesis?
    (d) What is the relevant *statistic*?
    (e) What is the *test statistic*?
    (f) What is the relevant *parameter*?
    (g) What is the *population* under consideration?
    (h) What is the p-value of the result? What is the *probability of Type I error* that is being tolerated when the psychologist rejects the null hypothesis?
    (i) What is the *probability of Type II error* in this test if the true mean IQ is 113?

(j)   What is a 99 *percent confidence interval for the mean* IQ as calculated from these data?

(k)   Could the psychologist reject the hypothesis that the mean IQ is actually 108? If so, at what level?

(l)   What does it mean, in this situation, to say that the statistical test basically gives a *conditional probability of the form* $P(M \geq 121 \mid \mu = 110, \sigma^2 = 116)$?

(m)   What is the *region of rejection* and what is the *critical value* for this test?

59. A drug manufacturer is concerned over the possibility that a new medication might have the undesirable side-effect of elevating a person's body temperature. The medication will be marketed if the manufacturer could be quite sure that the mean temperature of healthy individuals after they take the medication would be 98.6 degrees Fahrenheit or less; otherwise, it will be withheld. The drug is administered to a random sample of 60 healthy individuals, and the sample mean temperature and sample variance are 98.4 and .38, respectively.

(a)   Assuming that the distribution of temperatures is normal, what is the *p*-value?

(b)   The drug manufacturer wants the probability of marketing the drug to be only .10 when the mean temperature of healthy individuals after taking the drug is really 98.6. What decision should be made in light of the sample information?

60. How would your answers to Exercise 59 change if $n = 20$ but everything else was the same?

61. In Exercise 59, if $\alpha = .10$, find the region of rejection. Can you find the exact power curve for this test? Can you approximate the power curve? Explain.

62. Suppose that you open 16 soft-drink bottles and find that the contents average 11.6 ounces, with a sample variance of 1.0. What can you say about the statement "contents 12 oz" which is on each bottle? State the relevant hypotheses and determine the *p*-value. What assumptions did you make in computing this *p*-value? Set up a test for which the probability of unjustly accusing the bottling firm of failing to put an average of 12 oz in its bottles is only .05.

63. In Exercise 36, assume that the population variance is unknown. A sample of size 100 yields a sample mean of 140.8 and a sample variance of 321. Find the *p*-value.

64. The average diameter in a large shipment of circular parts is of interest, and the hypotheses being considered are $H_0: \mu \leq .25$ and $H_1: \mu > .25$. A sample is taken, and the results are given in Exercise 15, Chapter 5. Find the *p*-value.

65. Do Exercise 41 under the assumption that the population standard deviation is not known but that the sample standard deviation is .20. Explain any difference between the result obtained in this exercise and the result obtained in Exercise 41.

66. A random sample of ten independent observations from a normal distribution produced the following values:

$$51, 50, 49, 43, 56, 46, 45, 30, 55, 52.$$

If $\alpha = .20$, test the hypothesis that the mean is 50 against the alternative that it is not 50. Also, determine an 80 percent confidence interval for the mean. What is the p-value for the test of H_0: $\mu = 50$ against H_1: $\mu \neq 50$?

67. List the assumptions underlying the small sample test of a hypothesis about a single mean and the small sample test of a difference between two means. Evaluate the apparent importance of these assumptions.

68. Why is it that a T test cannot be applied to test a hypothesis about a proportion?

69. The chamber of commerce in a given town, A, claims that the mean income in that town is at least \$2000 higher than in a neighboring town, B. In response, the chamber of commerce in town B takes a random sample of size 9 in town A and a random sample of size 12 in town B, with the following results:

Town A		Town B	
12,000	14,500	13,100	11,900
17,200	13,100	9,400	14,700
20,000	9,700	17,300	12,000
14,500	16,100	14,100	10,600
10,200		12,500	8,900
		10,300	18,300

Assume that the variance of income is the same in both towns. If the chamber of commerce in town B wants the probability of refuting town A's claim when in fact it is true to be only .05, should they publicly refute the claim or should they simply ignore it? What is the p-value in this case?

70. In Exercise 69, suppose it is known that the standard deviation of income is 2200 in each town. What should the chamber of commerce of town B do? What should the chamber of commerce do if it is known that the standard deviation of income is 2400 in town A and 1600 in town B? Under what conditions might it be reasonable to assume that the standard deviations are known?

71. The output of each of two machines is normally distributed. If the new machine has a higher mean output than the old machine, you will purchase the new machine. Thus, you wish to test H_0: $\mu_N \leq \mu_0$ against H_1: $\mu_N > \mu_0$. You are willing to assume that the variability of output is the same for both machines. You observe the following sample results:

Old machine: 27, 44, 35, 37, 56, 19, 32, 45

New machine: 50, 33, 43, 61, 25, 51, 38, 41, 35, 27.

What is the p-value? What other facts might you want to know before you make the final decision concerning the purchase of the new machine?

72. In Exercise 71, if you decide that the mean output of the new machine would have to be at least 5 units higher than the mean output of the old machine in order to justify considering its purchase, what are the hypotheses of interest and what is the p-value?

73. A detergent manufacturer (the Clean Soap Co.) is concerned about claims made by a rival firm (Bright Detergent, Inc.) that *Bright* detergent results in cleaner clothes than Clean detergent. Twelve dirty sheets are randomly assigned to the two detergents and washed (in identical washing machines). After washing, the cleanliness of each shirt is measured on a "Bright-O-Meter," with the following results (a cleaner sheet will have a higher rating):

Bright detergent	Clean detergent
8	7
7	5
9	8
8	10
6	6
10	6

Assuming normality and equality of variances, state the relevant hypotheses and determine the p-value. On the basis of this, what can the Clean Soap Co. infer? Should they publicly claim that Bright's claim is false?

74. The same test is given to two classes of students in statistics. The first class (41 students) has a mean score of 75 with a sample variance of 100; the second class (21 students) has a mean score of 80 with a sample variance of 121. Did one class perform substantially better than the other? State the relevant hypotheses and the assumptions which you are making, and compute the p-value. Also, compute a 90 percent confidence interval for the difference in the means.

75. An educator has developed a new IQ test and would like to compare it to a certain widely used IQ test. Both tests are given to a randomly chosen group of 5 students, with the following results:

Child	Old test	New test
1	128	123
2	103	113
3	110	115
4	120	130
5	115	135

The new test is much shorter than the old test, and it would be easier to use. Of course, this makes no difference unless the results of the two tests are consistent. State the relevant hypotheses and assumptions and determine the p-value.

76. The government is interested in changes in research and development expenditures in the past decade. Eight large companies are randomly selected and the following data are collected on expenditures on research and development (in units of $100,000):

Company	1965	1975
A	12	14
B	8	9
C	14	17
D	9	8
E	17	20
F	20	28
G	7	11
H	11	14

Test the hypothesis that mean R & D expenditures have increased by less than $200,000 against the alternative that the mean has increased by more than that. What assumptions have you made, and what is the p-value? Determine a 90 percent confidence interval for the difference in means.

77. In Exercise 56, Chapter 6, test the hypothesis (at the $\alpha = .01$ level) that the average battery life in cars made by the Swedish firm is not as long as the average battery life in cars made by the German firm against the alternative hypothesis that the average battery life is as long or longer in the Swedish cars as it is in the German cars. Also, find the p-value.

78. A statistician is contemplating the use of two independent groups of equal size in a study. The two groups are from populations with the same standard deviation, denoted by σ. The test to be used is to have power of .90 of detecting a difference of $\pm \sigma / 20$ between the two means, and α is .01. How large should each sample (that is, each group) be?

79. A large car rental company wants to compare two brands of gasoline. Two random samples of 15 cars, each to be driven 2000 miles by customers, are chosen independently. The cars in one sample use Brand A, the others use Brand B, and the number of miles per gallon (mpg) is computed for each car. For Brand A, the average mpg for the cars in the sample is 15.0 and the sample standard deviation of mpg is 1.2; for Brand B, these figures are 13.5 and .9, respectively.

The rental company has a deal whereby it can get Brand B slightly cheaper than Brand A so that it will pay them to switch to Brand A only if Brand A

averages at least 1 mpg more than Brand B in the population. On the basis of the *sample*, if the company decides to switch to Brand A, what is the probability that an error is being made?

80. For many years a program of psychotherapy has been conducted at a state hospital. It has been found that among patients treated successfully in the past, the average time under treatment was 220 hours, with a standard deviation of 40 hours. A new treatment has just been developed, and for a sample of size 25, the mean was approximately the same although the sample standard deviation was only 25 hours. Assuming normality, can we say that the standard deviation for the new treatment is significantly less than 40 hours if $\alpha = .05$? What is the p-value?

81. A manufacturer makes parts under a government contract which specifies a mean weight of 100 grams and a tolerance of 1 gram. That is, the government will reject parts unless the weight is between 99 and 101 grams. To meet these tolerances, the manufacturer wants the variance of weight to be no more than .5 gram. A sample of 5 parts yields weights of 99.6, 101.2, 100.4, 98.8, and 100.0 grams. It is assumed that the weights are normally distributed. Find the p-value for the test of

$$H_0: \sigma^2 \leq .5$$

against $\qquad\qquad H_1: \sigma^2 > .5.$

82. In Exercise 81, the manufacturer decides that it is too costly to reduce the variance below .5 gram, so that the hypotheses of interest are

$$H_0: \sigma^2 = .5$$

and $\qquad\qquad H_1: \sigma^2 \neq .5.$

Find the p-value. If $\alpha = .20$, would you accept or reject H_0? If $\alpha = .01$?

83. In Exercise 82, are the lower and upper limits of the acceptance region equidistant from .5? If not, would it be possible to determine an acceptance region such that these limits *would* be equidistant from .5? Do you think they should be equidistant from .5? Comment on the issues involved here.

84. A randomly selected group of 30 boys of junior high-school age and an independent group of 30 senior high-school boys were each given the same test of mechanical ability. The sample mean and sample variance in the junior high group were 168 and 1521, respectively, and the sample mean and sample variance in the senior high group were 173 and 1089, respectively. At the .01 level of significance, can we say that the two groups are significantly different in variability?

85. National norms for the test mentioned in Exercise 84 exist, showing that the adult male population has a standard deviation of 28 on this test. Can we say, with $\alpha = .05$, that junior high-school boys are more variable in their test performance than the population of adult males? Are the senior high-school boys more variable than the adult males? Are the combined junior and senior high-school groups more variable than the adult males?

86. A random sample of 25 U. S. men born in 1935 showed a distribution of height with a sample standard deviation of 3.1 inches. On the other hand, a random sample of 43 U. S. men born in 1945 showed a sample standard deviation of 2.5 inches. Is this evidence (at the .05 level) that these two generations of men have different degrees of variability in height?

87. The daily high temperatures in a midwestern city were found for a sample of 30 days in the summers of 1930–1945. These temperatures had a sample standard deviation of 4.8 degrees Fahrenheit. Another random sample of 20 daily high temperatures taken in the same city during the summers of 1946–1961 had a sample standard deviation of 8.6 degrees Fahrenheit. Can we say that the daily high temperature in the summers of the second 16-year period was significantly more variable than in the first 16-year period? Use $\alpha = .01$.

88. In Exercise 69, are the variances significantly different at the .05 level?

89. In Exercise 71, are the variances significantly different at the .01 level?

90. Suppose that two independent random samples were to be compared in terms of their variability. If the values in one of these samples were multiplied by the constant k, what would be the effect on the F-ratio based on the variances of the two samples? State whether this suggests a way to test a hypothesis of the form

$$H_0: \sigma_1 = k\sigma_2.$$

How would you proceed?

91. Discuss the importance of the normality assumption in each of the tests discussed in this chapter which involve such an assumption. Is the assumption more crucial for some tests than for others? What other assumptions are important?

92. Suppose that $\alpha = .05$ and that K separate tests of hypotheses are made, quite independently of each other Show that when all of the null hypotheses are true, the probability of one or more significant results *by chance alone* (that is, even if all of the hypotheses being tested are exactly true) is equal to

$$1 - (.95)^K.$$

Also, show that in general, for any value of α, the corresponding probability is

$$1 - (1 - \alpha)^K.$$

8

BAYESIAN
INFERENCE

The inferential procedures presented in Chapters 6 and 7 are based on sample information. In general, certain assumptions are made about the population or process being sampled from, and these assumptions determine sampling distributions for statistics such as the sample mean, sample variance, or sample proportion. On the basis of these sampling distributions, properties of estimators and tests of hypotheses can be investigated, and "good" estimators and tests can be found. This approach is empirical, since it uses only empirical evidence: the evidence contained in samples from the population or process of interest. It is called the **sampling-theory** or **classical** approach to statistical inference. The term "sampling theory" describes the methods better than "classical," which is not very meaningful in this context. The latter term, however, is used more frequently in the statistical literature, so we will use it in this book. In its broadest sense, classical statistics consists of inferential and decision-making procedures that formally are based solely on sample evidence, although other information may be used in an informal manner (in determining hypotheses of interest, in deciding to use a specific model such as a Bernoulli model or a normal model, and so on).

Another approach to statistical inference and decision exists, and it is different from the classical approach primarily in that information other than sample information is formally utilized. In this approach, which has gained many new followers in recent years, sample information is combined with other available information, and the resulting combination of information is the basis for inferential and decision-making procedures. The mechanism used to combine information is Bayes' theorem, which was encountered in Chapter 2. As a result, the term **Bayesian** is used to describe

471

this general approach to statistics. In this chapter we will be concerned primarily with the combination of sample information with other available information and with inferential procedures based on the combined information. Insofar as possible, we will attempt to avoid decision-theoretic considerations in discussing these procedures and comparing them with the corresponding classical procedures. In Chapter 9, decision theory will be discussed, and we will reexamine Bayesian procedures from the standpoint of decision theory.

8.1 BAYES' THEOREM FOR DISCRETE RANDOM VARIABLES

In Section 2.20, **Bayes' theorem** was discussed in terms of events. It is also possible to interpret Bayes' theorem in terms of the conditional distribution of a random variable, discrete or continuous. In this section we present Bayes' theorem in terms of a discrete random variable.

Suppose that we are interested in making inferences about a parameter θ, and we are willing to assume that θ can only take on J possible values $\theta_1, \ldots, \theta_J$. In the terminology of Chapter 7, these J values might be thought of as J competing hypotheses concerning θ, with H_i represented by $\theta = \theta_i$ for $i = 1, \ldots, J$. Furthermore, assume that the information concerning θ can be summarized by a probability distribution consisting of a set of probabilities $P(\theta = \theta_i)$. This probability distribution is called the **prior distribution of** θ. We then observe a sample, and the outcome of the sample can be summarized by y, the observed value of a sample statistic Y. The **likelihood function** is then of the form $P(y \mid \theta_i)$ if Y is discrete or $f(y \mid \theta_i)$ if Y is continuous. We will use the $P(y \mid \theta_i)$ notation in this section, but you should keep in mind that Y could be continuous. On the other hand, we have assumed that θ is discrete.

Given the information in the preceding paragraph, we would like to combine the prior distribution and the likelihood function to arrive at a **posterior distribution** consisting of a set of probabilities $P(\theta_i \mid y)$ representing a combination of the prior information and the sample information.

The mechanism used to combine information is Bayes' theorem, which can be expressed as follows for a discrete random variable θ:

$$P(\theta_i \mid y) = \frac{P(y \mid \theta_i) P(\theta_i)}{\sum_{j=1}^{J} P(y \mid \theta_j) P(\theta_j)} . \qquad (8.1.1^*)$$

If the set of values of θ is countably infinite, then the summation index in the denominator of Equation (8.1.1) goes from 1 to ∞, of course.

Equation (8.1.1) enables us to revise probabilities concerning a parameter θ on the basis of new information, which may be summarized by the sample statistic Y. Note that Equation (8.1.1) is identical to Equation (2.20.3) with the event A_i represented by $\theta = \theta_i$ and the event B represented by $Y = y$. An example of Bayes' theorem with $J = 3$ was given in Section 2.20, and a detailed example will be presented in Section 8.2. Before proceeding to this latter example, we will elaborate a bit upon the inputs to Bayes' theorem (the prior distribution and the likelihood function) and upon some differences between the classical approach to statistics and the Bayesian approach to statistics.

In classical statistics, all inferences are based on the sample information, and some classical techniques are based directly upon what we have called the *likelihood function*. Recall that in estimation, the likelihood function is used to find maximum-likelihood estimators (Section 6.7); in hypothesis testing, the likelihood function is used to develop likelihood ratio tests (Section 7.6). The likelihood function is also used as an input to Bayes' theorem, and it is the input representing the sample information. For any θ_i, $P(y \mid \theta_i)$ can be thought of as the likelihood of the sample result y, given θ_i.

In the Bayesian approach to statistics, the inclusion of the prior distribution enables us to base inferences and decisions on all of the information that is available, whether or not the information is in the form of "sample information." *The motivation for Bayesian methods is essentially the desire to base inferences and decisions on any and all available information, whether it is sample information or information of some other nature.* This motivation is particularly strong in decision theory. Erroneous decisions may be quite costly, and it may not be reasonable to ignore pertinent information just because it is not "objective" sample information. As this last statement implies, the classical approach is associated with the frequency interpretation of probability. Because it is impossible to talk about information in terms of frequency probabilities unless the information arises from a random sample from a well-defined population or process, classical statisticians formally admit only sample information in inferential and decision-making procedures. Information other than sample information is rejected as "nonobjective."

Nevertheless, the statistician often has some information about a parameter θ prior to taking a sample. This information may be of a subjective nature, in which case the prior distribution will consist of a set of subjective probabilities. In the classical approach to statistics, all probabilities should be based on the long-run frequency interpretation of probability, and hence subjective probabilities are not admissible. Most statisticians advocating the Bayesian viewpoint contend that all available information about a parameter, whether it be of an objective or subjective nature, should be utilized in making inferences or decisions. Decisions often must be made in situations where there is little or no sample information but a great deal of information of a more subjective nature.

Since followers of the frequency interpretation of probability do not admit subjective probabilities, they do not admit probability statements about a population parameter θ. Frequentists contend that θ has a certain value, which may be unknown, and that it is senseless to talk of the probability that θ equals some number; either it does or it does not. This was pointed out in Section 6.11 when we discussed the interpretation of confidence interval statements. The subjectivist, on the other hand, *does* think of θ as a random variable and thus allows probability statements concerning θ. A confidence interval for θ can be interpreted as a probability statement concerning θ; for example, we might say that the probability is .90 that a particular population mean μ is greater than 123 but less than 141. Incidentally, in order to maintain consistency with the preceding chapters, we will continue to denote parameters by symbols such as μ, λ, p, and so on, even when the parameters are regarded as random variables. This means that when parameters are considered as random variables, we will generally not follow our convention of using capital letters to represent random variables.

In order to be able to express all information in probabilistic terms, most (but not all) Bayesians follow the subjective interpretation of probability, as presented in Sections 2.15–2.16. Often, much of the available information consists of the judgments of one or more persons. If probabilities are interpreted as degrees of belief of individuals, this sort of information can be formally included in the analysis. In our discussion of Bayesian methods, we will utilize the subjective interpretation of probability. It should be remembered, however, that the terms "Bayesian" and "subjective" are not synonymous; it is possible to develop the Bayesian approach without using subjective probability.

The subjective interpretation of probability, of course, does not prevent the utilization of the sampling distributions which have been discussed in previous chapters. If a statistician subjectively accepts the assumptions (such as normality, known variance, and so forth) underlying a particular sampling distribution, then that distribution can be given a subjective interpretation. Since the Bayesian *will* utilize the sampling distributions most commonly encountered in classical statistics and, in addition, will use other information, the Bayesian approach can be thought of as an *extension* of the classical approach. In this and the following chapter, comparisons of Bayesian and classical methods will be made so that the student can readily understand the similarities and differences between the methods. We will see that under certain conditions, Bayesian and classical methods produce similar or even identical results. Of course, there is a difference in interpretation even though the numbers may be the same. As you may have guessed, the "certain conditions" amount essentially to stating that there is no information available other than sample information, or any such information is of little importance relative to the sample information.

8.2 AN EXAMPLE OF BAYESIAN INFERENCE AND DECISION

The example considered in this section comes from the area of quality control. Suppose that a production manager is concerned about the items produced by a certain manufacturing process. More specifically, he is concerned about the proportion of these items that are defective. From past experience with the process, he feels that p, the proportion of defectives, can take on only four possible values: .01, .05, .10, and .25. These four values of the parameter p might be explained in terms of malfunctions in the production process. Either (1) there are no malfunctions, in which case $p = .01$ and the process is "in control"; (2) there is a "type x" malfunction, in which case $p = .05$; (3) there are *two* malfunctions, "type x" and "type y," in which case $p = .10$; or (4) there are *three* malfunctions, "type x," "type y," and "type z," in which case $p = .25$. This is obviously a simplification of the real situation. It is highly unlikely that p would take on only the four given values and no intermediate values. It might be more realistic to think of p as a continuous random variable rather than as a discrete random variable (and this is done later in this chapter), but for the purposes of this example p is considered to be discrete.

Although the production manager has never formally determined the proportion of defectives in any "batch" from the manufacturing process, he has observed the process and he has some information concerning p. Suppose that he feels that this information can be summarized in terms of the following degree-of-belief, or subjective, probabilities for the four possible values of p:

$$P(p = .01) = .60,$$

$$P(p = .05) = .30,$$

$$P(p = .10) = .08,$$

and
$$P(p = .25) = .02.$$

In words, the probability is .60 that 1 percent of the items are defective, the probability is .30 that 5 percent of the items are defective, and so on. For instance, he might arrive at these probabilities by considering various lotteries or bets, as discussed in Section 2.16. These four probabilities constitute the production manager's prior distribution of p.

In addition to this prior information, the production manager decides to obtain some sample information from the process. In doing so, he assumes that the process can be thought of as a Bernoulli process, with the assumptions of stationarity and independence appearing reasonable. That is, the probability that any one item is defective remains constant for all items produced and is independent of the past history of defectives from the process.

A sample of 5 items is taken from the production process, and 1 of the 5 is found to be defective. How can this information be combined with the prior information? Under the assumptions that the production manager has made, the sampling distribution of the number of defectives in 5 trials, given any particular value of p, is a binomial distribution. The likelihoods are thus

$$P(R = 1 \mid n = 5, p = .01) = \binom{5}{1} (.01)(.99)^4 = .0480,$$

$$P(R = 1 \mid n = 5, p = .05) = \binom{5}{1} (.05)(.95)^4 = .2036,$$

$$P(R = 1 \mid n = 5, p = .10) = \binom{5}{1} (.10)(.90)^4 = .3280,$$

and $$P(R = 1 \mid n = 5, p = .25) = \binom{5}{1} (.25)(.75)^4 = .3955.$$

In this example, Bayes' theorem can be written in the form

$$P(p_i \mid y) = \frac{P(y \mid p_i)P(p_i)}{\sum_j P(y \mid p_j)P(p_j)},$$

where y represents the sample result, 1 defective in 5 trials. [Here, the role of the parameter value θ_i in Equation (8.1.1) is played by p_i.] Notice that the numerator is the product of a likelihood and a prior probability and that the denominator is just the sum of all of the possible numerators.

To determine the posterior distribution, it is convenient to set up the following table and to apply Bayes' theorem in tabular form. The first column in Table 8.2.1 lists the possible values of p (that is, the values of p having nonzero prior probabilities). The second and third columns give the prior probabilities and the likelihoods, and the fourth column is the product of these two columns. Finally, the fifth column is obtained from the fourth column by dividing each individual entry by the sum of the four entries.

Table 8.2.1 Bayes' Theorem in Tabular Form for Quality-Control Example

(1) p	(2) Prior probability	(3) Likelihood	(4) (Prior probability) $\times$ (likelihood)	(5) Posterior probability
.01	.60	.0480	.02880	$.02880/.12403 =$ .232
.05	.30	.2036	.06108	$.06108/.12403 =$ .492
.10	.08	.3280	.02624	$.02624/.12403 =$ .212
.25	.02	.3955	.00791	$.00791/.12403 =$ .064
	1.00		.12403	1.000

This determines the posterior probabilities, since Bayes' theorem has the form

$$\text{posterior probability} = \frac{(\text{prior probability})(\text{likelihood})}{\sum (\text{prior probability})(\text{likelihood})}.$$

This example illustrates some features of Bayes' theorem. For one thing, the prior probabilities must sum to 1 (since they form the prior distribution, which must be a proper probability distribution), whereas there is no such restriction on the sum of the likelihoods. The likelihoods used in the example come from different binomial distributions (one from the binomial distribution with $n = 5$ and $p = .01$, one from the binomial distribution with $n = 5$ and $p = .05$, and so on), so there is no restriction on their sum. The posterior probabilities, however, must sum to 1 (since they form the posterior distribution, which must be a proper probability distribution), and this is why it is necessary to divide by $\sum$ (prior probability)(likelihood) in Bayes' theorem. The sum of the entries in column 4 of the above table is not equal to 1, but when each entry is divided by this sum, the resulting values (column 5) do add to 1. This process of making the probabilities sum to 1 is sometimes referred to as *normalizing* the probabilities.

Another feature of Bayes' theorem that is illustrated here is that if the prior probability of any value of p is 0, the posterior probability of that value will also be 0, since the posterior probability is a multiple of the prior probability (the multiplier being the likelihood divided by the normalizing sum). Since the prior probability of $p = .20$ is 0, for example, the posterior probability of $p = .20$ is 0. The moral is obvious; when determining prior probabilities for discrete probability models, make sure that any value that you consider to be even remotely possible has a nonzero prior probability. As noted above, this production example is highly simplified and unrealistic because values such as .02, .09, .20, and so on, are assigned prior probabilities of 0 (and thus designated as essentially "impossible" events).

Returning to the production manager, suppose that he decides to take yet another sample in order to obtain more information about the production process. In particular, he takes another sample of size 5 and obtains 2 defectives. The posterior probabilities calculated after the first sample are now the prior probabilities, and the likelihoods are once again determined via the binomial distribution. The posterior probabilities are given in Table 8.2.2.

Table 8.2.2   Bayes' Theorem for Second Sample in Quality-Control Example

p	Prior probability	Likelihood	(Prior probability) × (likelihood)	Posterior probability
.01	.232	.0010	.00023	.005
.05	.492	.0214	.01053	.244
.10	.212	.0729	.01545	.359
.25	.064	.2637	.01688	.392
	1.000		.04309	1.000

It is interesting to observe the changes in the probabilities as new sample information is obtained. The original prior distribution, which is presented in graphical form in Figure 8.2.1, indicates that the production manager thought that the low values of p(.01 and .05) were more likely than the higher values. The mean, $E(p)$, was .034. The first sample of size 5, with 1 (that is, 20 percent) defective, shifts the probabilities so that .05 appears to be the most likely value of p (.01 is the most likely value according to the production manager's original prior distribution). After the first sample,

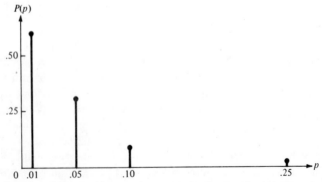

Figure 8.2.1   Prior Distribution of p in Quality-Control Example

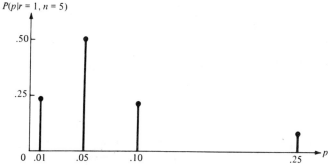

Figure 8.2.2 Posterior Distribution of p after First Sample in Quality-Control Example

the mean is .064, as compared with the prior mean of .034 (the distribution is illustrated in Figure 8.2.2). The second sample of size 5, with 2 (that is, 40 percent) defectives, shifts the probabilities so that the highest possible value, .25, becomes the most likely, while the lowest value now has a probability of only .005, compared with the original prior probability of .60 assessed by the production manager. After the second sample, the mean of the posterior distribution is .146 (this posterior distribution is illustrated in Figure 8.2.3). In other words, the sample results (a total of 3 defectives in 10 items) change the probabilities considerably, making the high value of p, .25, much more likely than it seems under the original prior distribution. This does not mean that the sample will always change the probabilities so drastically. If the production manager had observed 0 defectives or 1 defective instead of 3 defectives in the sample of 10 items, the posterior probabilities would be much closer to the prior probabilities. You can verify this by performing the calculations.

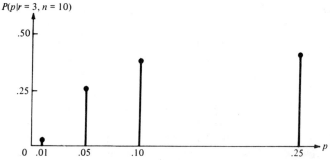

Figure 8.2.3 Posterior Distribution of p after Second Sample in Quality-Control Example

Table 8.2.3   Bayes' Theorem for First and Second Samples Combined in Quality-Control Example

p	Prior probability	Likelihood	(Prior probability) $\times$ (likelihood)	Posterior probability
.01	.60	.0001	.00006	.005
.05	.30	.0105	.00315	.245
.10	.08	.0574	.00459	.359
.25	.02	.2503	.00501	.391
	1.00		.01281	1.000

In this example, the production manager took 2 samples and revised his probabilities after each sample. A useful feature of Bayes' theorem is that he could have waited and revised his probabilities just once, using the combined results of the 2 samples (that is, 3 defectives in 10 trials). When applying Bayes' theorem for independent trials, the same result is obtained whether the calculations are made after each trial or just once at the end of all of the trials.

Starting with the original prior probabilities and using all 10 trials to determine the likelihoods, the results are given in Table 8.2.3. These posterior probabilities are identical to the probabilities obtained from revising twice, once after each sample. [The only differences (at the third decimal place) in the posterior probabilities actually presented in the two tables are due to rounding errors in the calculations.] This feature of Bayes' theorem is very important, since it reduces the number of applications of the theorem required to revise probabilities on the basis of several samples. Of course, the production manager may want to revise after each sample, primarily because the resulting probabilities may help him to decide whether or not to take another sample, as we shall see in the next chapter.

Now that the posterior probabilities have been calculated, let us consider the following question. Why might the production manager have been so interested in p in the first place? There are a number of explanations, but they all relate to the fact that a high value of p could be quite costly to the firm. It could be quite costly to repair defective items, perhaps so costly that they are scrapped entirely. In addition, if any defective items are shipped to the firm's customers, the customers might be displeased and might take their business elsewhere.

Often the quality-control statistician or the production manager simply wants an estimate of p. Such an estimate can be determined from the posterior distribution by taking some summary measure, such as the mean, median, or mode, of the posterior distribution. The comments in Chapter 3 and Chapter 6 concerning the relative merit of various possible measures (or estimators) apply here. In contrast to the classical statistician (who would base an estimate solely upon the sample information, as reflected by the likelihood function), our Bayesian statistician bases an estimate on all available information, including that reflected by the prior distribution. The posterior distribution is based on both prior and sample information.

It may be that the production manager must make a decision concerning the process instead of just estimating p. Suppose that he has three possible actions: (1) he can leave the process as is; (2) he can have a mechanic make a minor adjustment in the process that will guarantee that p will be no higher than .05 (specifically, if p is .10 or .25, the adjustment reduces it to .05; if p is .05 or .01, the adjustment has no effect); or (3) he can have the mechanic make a major adjustment in the process that will guarantee that p will be .01. In the first paragraph of this section, the four possible values of p were related to malfunctions in the production process. In terms of that explanation of the values of p, a major adjustment corrects all malfunctions while a minor adjustment corrects only "type y" and "type z" malfunctions.

Suppose that the production manager has just received an order for 1000 items and that he must choose one of the above three actions before the production run is started. There are certain costs that are relevant to his decision. First, the profit to the firm is $.50 per item sold, and the cost of replacing defectives is a fixed amount, $2 per item. The other relevant costs are the costs of making adjustments in the process: a minor adjustment costs $25, while a major adjustment costs $100. From these figures, it is clear that the profit to be realized from the run of 1000 items is 1000 ($.50) = $500. From this amount the cost of replacing defectives and the cost of any adjustments must be subtracted. If p is .01, the expected number of defectives is .01(1000) = 10, and the cost of replacing these is $2 per item, or a total of $20. Similarly, for the other three possible values of p, the expected costs of replacing defectives are $100, $200, and $500, respectively. From this information, the production manager can determine a "payoff table," or "reward table," which shows the net profit to the firm for each combination of an action and a possible event. For example, if $p = .10$ and a minor adjustment is made, the adjustment changes p to .05, so the expected cost of replacing defectives is $100. The cost of the minor adjustment is $25, so the payoff is $500 − $100 − $25 = $375. The entire payoff table is given in Table 8.2.4.

In Chapter 9, various criteria for making decisions will be discussed. Suffice it to say at this point that a reasonable criterion is to choose the act for which the expected payoff or reward (expected net profit to the

Table 8.2.4   Payoff Table for Quality-Control Example

EVENT (VALUE OF p)

		.01	.05	.10	.25
	Major adjustment	$380	$380	$380	$380
ACTION	*Minor adjustment*	$455	$375	$375	$375
	No adjustment	$480	$400	$300	$0

firm in this example) is largest. The expectations should be taken with respect to the production manager's posterior distribution of p, since this represents all of the information available to him concerning p, the state of nature. The posterior probabilities of the four states of nature are .005, .245, .359, and .391. The expected rewards (denoted by ER) are then

$$ER(\text{major adjustment}) = \$380,$$

$$ER(\text{minor adjustment}) = .005(\$455) + .995(\$375) = \$375.40,$$

and $ER(\text{no adjustment}) = .005(\$480) + .245(\$400) + .359(\$300)$

$$= \$208.10.$$

This indicates that the production manager should have the mechanic make a major adjustment. This is reasonable, since the probability of a high proportion of defectives (.10 or .25) is quite high. It is interesting to note that if the production manager had obtained no sample information, he would base his decision on the prior distribution, in which case no adjustment would be made. You can verify this by calculating the expected rewards with respect to the prior distribution of p.

This quality-control example illustrates the use of Bayes' theorem to revise probabilities on the basis of sample information and the use of the resulting posterior probabilities as inputs in a decision-making situation. We emphasize again that the example was purposely simplified and that we will discuss a potentially more realistic approach to the problem later in the chapter.

8.3   THE ASSESSMENT OF PRIOR PROBABILITIES

In the discrete case, the prior distribution consists of a set of probabilities. The only restrictions are that the probabilities must be nonnegative and must sum to 1. Under these restrictions, how can the statistician determine a prior distribution?

The prior probabilities should reflect the prior information about the parameter in question. If this information is primarily in the form of sample results, then the prior probabilities should be close to the observed relative frequencies. In the example of the previous section, suppose that the production manager has been away from the plant and is not aware of the recent performance of the manufacturing process. He does know, however, that in the past, the process has been in control 60 percent of the time, there has been a type x malfunction 30 percent of the time, there have been type x *and* type y malfunctions 8 percent of the time, and there have been type x, type y, and type z malfunctions 2 percent of the time.

If there is little or no sample information, then the probabilities should be based on whatever other relevant information is available. In the quality-control example, the process may have been in control 95 percent of the time in the past, but current information causes the production manager to assess a probability of only .60 that it is in control now. The ultimate choice of a prior distribution by a statistician is a subjective choice, no matter what form the prior information takes, so it will be assumed that the prior probabilities are subjective probabilities. As noted in Chapter 2, a subjective probability reflects the degree of belief of a given person (the decision maker) about a given proposition (say, for example, the proposition that $p = .10$). But if probabilities are based on the judgments of a person, what is needed is a way for the person to quantify these judgments (that is, to express them in probabilistic terms).

Numerous techniques are available for the quantification of judgment; some of these were described in Sections 2.16 and 2.18. A person who understands the concept of probability might simply be able to assign probabilities directly to the various possible values of the uncertain quantity of interest. In the example in Section 8.2, the production manager might say, "I think the probability is .60 that the proportion of defectives produced by this production process is .01."

Of course, there is an element of vagueness in the determination of subjective prior probabilities. You may feel certain that the probability of rain tomorrow is between .10 and .30, for instance, but you find it difficult to pick a single number to represent your probability. In situations in which vagueness is present, devices such as those discussed above may prove useful. Would you bet for or against rain tomorrow at odds of 4 to 1 in favor of "no rain"? If you would bet for rain at these odds, then your probability of rain must be greater than .20. If you would prefer a lottery ticket paying $5 if it rains tomorrow to a second lottery ticket paying $5 if no heads are obtained in two tosses of a fair coin, then your probability of rain must be greater than .25, since the probability of obtaining no heads in two tosses of a fair coin is .25. Considering betting situations and lotteries of this nature helps to combat the problem of vagueness. In the quality-control example, the production manager might say, "I think the odds are 3 to 2 in favor of the proportion defective being .01." From Section 2.16, the probability corresponding to such odds is .60. In terms of lotteries, the production manager should be indifferent between (1) a lottery ticket that will pay $100 if a ball drawn at random from an urn with 60 red balls and 40 blue balls turns out to be red and (2) a lottery ticket that will pay $100 if the process turns out to be in control. If the problem at hand is quite important, then it is well worth it for the decision maker to take special care in an attempt to express prior judgments in terms of probabilities.

It should be emphasized that whatever the technique or techniques used to assess the prior probabilities, the probabilities must be nonnegative and must sum to 1. If these requirements are not satisfied, then the probabilities are said to be *inconsistent*. An assessor who obeys certain reasonable *axioms of coherent assessment* (such as, if A is considered to be more likely than B, and B more likely than C, then A should be considered to be more likely than C—this is an axiom of transitivity) will always be consistent in a decision-making sense. Such an assessor (sometimes referred to as the "rational man") will have no inconsistencies in probability assessments. Some experiments have indicated that individuals do not always assess probabilities in a rational manner. For example, they frequently assess probabilities that do not sum to 1, perhaps because of carelessness, vagueness, or a misunderstanding of the assessment procedures. At any rate, it should not present a major problem, since inconsistencies can be removed if the assessor is made aware of their existence. If someone pointed out an arithmetic error to you, you would no doubt correct the error. Similarly, if someone pointed out that your prior probabilities summed to, say, 1.20, you could change the probabilities so that they would sum to 1. In this regard, it is sometimes useful to attempt to assess the same set of probabilities in more than one way in order to check your assessments. For instance, a probability distribution can be assessed once by using lotteries and a second time by considering odds. If the resulting two distributions differ in any respect, you should reconcile the differences. Once again, this is analogous to performing checks on arithmetic calculations, although the arithmetic calculations do have a single correct answer, which is not the case with subjective probability assessments.

The question of whose prior probabilities are being assessed has not been raised in this section. Generally, the person faced with the inferential or decision-making problem of interest (that is, the statistician or the decision maker) is the person assessing the prior probabilities. In many situations, however, particularly for important problems, it may be worthwhile to hire an expert to assess prior probabilities. For instance, you might hire a meteorologist to assess the probability of rain, a stock-market analyst to assess probabilities for future stock prices, a political analyst to assess probabilities involving potential legislation by Congress, and so on. This raises the question of the evaluation of probability assessments. If experts frequently assess probabilities for events within their spheres of expertise, it is possible to evaluate these probabilities in light of the events that actually occur. Carrying this one step further, the statistician or decision maker might be able to calibrate the experts by investigating their past performance at assessing probabilities. For instance, if the decision maker feels that a particular meteorologist consistently underestimates the probability of rain by approximately .1, then the probabilities received from the meteorologist might be increased by .1. This is a crude form of calibration, of course, and more sophisticated techniques are available. The point is that although an expert may be brought in to make probability assessments, the ultimate choice of prior probabilities rests with the statistician or decision maker.

In summary, then, anyone who understands the basic concept of probability should be able to assess a prior distribution for a discrete random variable. When we are dealing with continuous random variables, the assessment of a prior distribution may be considerably more difficult, as we will see later in this chapter. The main thing that you should keep in mind is that the prior distribution should reflect all of the relevant information available to the statistician prior to observing a particular sample.

8.4 THE ASSESSMENT OF LIKELIHOODS

The assessment of prior probabilities is an important problem in Bayesian inference and decision. The determination of another input to the Bayesian model, the likelihoods, is also quite important. In many situations it is possible to represent the data-generating process that generates new sample information in terms of a well-known statistical model. For example, in Section 8.2 the Bernoulli model was utilized. The Bernoulli process requires assumptions of stationarity and independence, as indicated in Section 4.1. Of course, real-world data-generating processes are unlikely to obey *exactly* the assumptions of any statistical model. Thus, it must be recognized that a statistical model, like any type of mathematical model, is an idealization that we cannot expect to realize exactly in practice. However, in a great many situations, such a model may be a close approximation to reality and its use may be justified. Certainly the use of the Bernoulli model in Section 8.2 simplified the determination of likelihoods, since the likelihoods could be expressed in terms of the binomial distribution.

For an example involving a process other than the Bernoulli process, consider a problem involving the owner of an automobile dealership. The owner is concerned about the performance of his salesmen; in particular, he is concerned with their success or lack of success in selling cars. From past experience with the dealership, he feels that salesmen can be divided roughly into three categories, which he calls "great," "good," and "poor." A "great" salesman sells cars at a rate of one every other day. It should be noted that this does not mean that such a salesman will sell one car exactly every other day; a "great" salesman may have periods of several days with no sales or he may have several sales in a single day. On the average, however, he sells about one car every two days. A "good" salesman sells cars at a rate of one every fourth day, and a "poor" salesman sells cars at a rate of one every eighth day. The owner is willing to assume that the car-selling process behaves like a Poisson process with intensity λ per day, where there are three possible values of λ: $\frac{1}{2}$ (the "great" salesman), $\frac{1}{4}$ (the "good" salesman), and $\frac{1}{8}$ (the "poor" salesman).

The owner of the dealership has just hired a new salesman. Based on past experience with salesmen and on a personal interview with the new employee, the owner judges that the odds are 4 to 1 against his being a great salesman but that the odds are even that he could be a good salesman. The owner feels that such odds would be fair odds for bets concerning the performance of the salesman. This implies that the probability that he is a great salesman is $1/(1 + 4) = .2$ and the probability that he is a good salesman is $1/(1 + 1) = .5$. It follows that since the only other alternative is being a poor salesman, the probability of this alternative is .3. The owner's prior distribution for λ, then, may be represented by the following probabilities:

$$P(\lambda = \tfrac{1}{2}) = .2,$$

$$P(\lambda = \tfrac{1}{4}) = .5,$$

and
$$P(\lambda = \tfrac{1}{8}) = .3.$$

Suppose that the owner has hired the new salesman for a trial period of 24 working days. During those 24 days, the salesman sells 10 cars. In applying Bayes' theorem to revise the distribution of λ, the owner can use the Poisson distribution to determine the likelihoods, since he has assumed that the process is Poisson. Thus, from Table VI, the likelihoods are

$$P(R = 10 \mid t = 24, \lambda = \tfrac{1}{2}) = \frac{e^{-12}(12)^{10}}{10!} = .1048,$$

$$P(R = 10 \mid t = 24, \lambda = \tfrac{1}{4}) = \frac{e^{-6}(6)^{10}}{10!} = .0413,$$

and $\quad P(R = 10 \mid t = 24, \lambda = \tfrac{1}{8}) = \frac{e^{-3}(3)^{10}}{10!} = .0008.$

Bayes' theorem can be written in the form

$$P(\lambda_i \mid y) = \frac{P(y \mid \lambda_i)P(\lambda_i)}{\sum\limits_{j} P(y \mid \lambda_j)P(\lambda_j)},$$

where y represents the sample result, 10 sales in 24 days. The posterior probabilities are determined in Table 8.4.1.

Table 8.4.1   Bayes' Theorem for Automobile-Salesman Example

λ	Prior probability	Likelihood	(Prior probability) $\times$ (likelihood)	Posterior probability
$\tfrac{1}{2}$	.2	.1048	.02096	.501
$\tfrac{1}{4}$	.5	.0413	.02065	.493
$\tfrac{1}{8}$	.3	.0008	.00024	.006
	1.0		.04185	1.000

The prior probabilities for "great salesman," "good salesman," and "poor salesman" are .2, .5, and .3. After seeing the sample information, the revised, or posterior, probabilities for these three situations are .501, .493, and .006, respectively. Informally, this means that the new salesman sold enough cars in the 24-day period to virtually convince the owner that he is not a poor salesman and to greatly increase the odds, in the owner's estimation, in favor of his being a great salesman. Using the prior distribution, the odds are 4 to 1 *against* his being a great salesman; using the posterior distribution, the odds are approximately even. The prior and posterior expectations of λ are .26 and .37. Prior to seeing the sample information, the owner would guess that, on the average, the salesman would sell about .26 cars per day, or about 1 car every 3.8 days. After the salesman sells 10 cars in 24 days, the owner feels that, on the average,

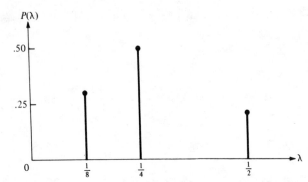

Figure 8.4.1 Prior Distribution of λ for Automobile-Salesman Example

the salesman will sell about .37 cars per day, or about 1 car every 2.7 days. The prior and the posterior distributions of λ are illustrated in Figures 8.4.1 and 8.4.2, respectively.

In order to illustrate the use of discrete prior distributions with a continuous data-generating process, suppose that a retailer is interested in the distribution of weekly sales at one of his stores. In particular, he is willing to assume that the random variable X, weekly sales (expressed in terms of dollars), is normally distributed with unknown mean μ and known variance $\sigma^2 = 90,000$. It should be stressed that this is a subjective assumption. Perhaps the retailer feels that the normality assumption is justified because he has found that for other stores that he has investigated, X has tended to be approximately normally distributed. Furthermore, at various other stores that are quite similar to the store in question, mean weekly sales have differed from store to store, but the variance has

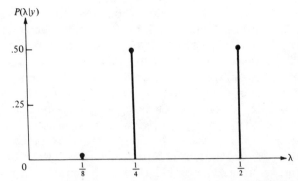

Figure 8.4.2 Posterior Distribution of λ for Automobile-Salesman Example

been reasonably stable at about 90,000. Hence, the retailer is willing to make the simplifying assumption that σ^2 is known. Finally, in assuming an independent normal data-generating model, he is implicitly assuming that there is no trend or seasonal effect to worry about. That is, he feels that for the type of store in question, sales generally do not change greatly over time and are not affected by such things as the weather, holidays, and so on. These are the usual types of assumptions that must be considered in the selection of a statistical model.

Suppose that the retailer decides to consider only five potential values of μ: 1100, 1150, 1200, 1250, and 1300. Based on experience with similar stores, he assesses the following prior probabilities:

$$P(\mu = 1100) = .15,$$

$$P(\mu = 1150) = .20,$$

$$P(\mu = 1200) = .30,$$

$$P(\mu = 1250) = .20,$$

and

$$P(\mu = 1300) = .15.$$

The retailer decides that he would like to obtain more information about the sales of the store, so he takes a sample from the past sales records of the store, assuming that weekly sales from different weeks can be regarded as independent. For a sample of 60 weeks, he finds the average weekly sales (the sample mean) to be $M = 1240$. Prior to seeing this sample, the conditional distribution of M, given μ, was a normal distribution with mean μ and variance $\sigma^2/n = 90,000/60 = 1500$. Thus, the likelihood function is of the form

$$f(1240 \mid \mu, \sigma/\sqrt{n} = 38.73) = \frac{1}{38.73\sqrt{2\pi}} e^{-(1240-\mu)^2/2(1500)}.$$

But if we let $z = (1240 - \mu)/38.73$, the likelihood function can be written as follows:

$$f(1240 \mid \mu, \sigma/\sqrt{n} = 38.73) = \frac{1}{38.73}\left(\frac{1}{\sqrt{2\pi}} e^{-z^2/2}\right).$$

In other words, the height of a normal density function at a particular value of the random variable of interest is equal to the height of the standard normal density function at the corresponding standardized value, divided by the standard deviation:

$$f(x \mid \mu, \sigma^2) = f\left(\frac{x - \mu}{\sigma} \,\middle|\, 0, 1\right) \Big/ \sigma. \qquad (8.4.1^*)$$

In the example, the standard deviation of the distribution of M is $\sigma/\sqrt{n} = 38.73$. Using values of the standard normal density function from Table VII, the likelihoods for the example are

$$f(1240 \mid \mu = 1100, \sigma/\sqrt{n} = 38.73) = f\left(\frac{1240 - 1100}{38.73}\,\middle|\,0, 1\right)\middle/ 38.73$$

$$= \frac{f(3.61 \mid 0, 1)}{38.73} = \frac{.0006}{38.73},$$

$$f(1240 \mid \mu = 1150, \sigma/\sqrt{n} = 38.73) = f\left(\frac{1240 - 1150}{38.73}\,\middle|\,0, 1\right)\middle/ 38.73$$

$$= \frac{f(2.32 \mid 0, 1)}{38.73} = \frac{.0270}{38.73},$$

$$f(1240 \mid \mu = 1200, \sigma/\sqrt{n} = 38.73) = f\left(\frac{1240 - 1200}{38.73}\,\middle|\,0, 1\right)\middle/ 38.73$$

$$= \frac{f(1.03 \mid 0, 1)}{38.73} = \frac{.2347}{38.73},$$

$$f(1240 \mid \mu = 1250, \sigma/\sqrt{n} = 38.73) = f\left(\frac{1240 - 1250}{38.73}\,\middle|\,0, 1\right)\middle/ 38.73$$

$$= \frac{f(-.26 \mid 0, 1)}{38.73} = \frac{.3857}{38.73},$$

and $f(1240 \mid \mu = 1300, \sigma/\sqrt{n} = 38.73) = f\left(\frac{1240 - 1300}{38.73}\,\middle|\,0, 1\right)\middle/ 38.73$

$$= \frac{f(-1.55 \mid 0, 1)}{38.73} = \frac{.1200}{38.73}.$$

The application of Bayes' theorem can be represented in tabular form when the prior distribution is discrete. In Table 8.4.2, note that the 38.73 in the denominator of each likelihood cancels out in the application of Bayes' theorem. In general, multiplying or dividing all of the likelihoods by the same constant does not affect the posterior probabilities, since only the *relative* values of the likelihoods are needed for Bayes' theorem. The prior and posterior distributions of μ are graphed in Figures 8.4.3 and 8.4.4.

Table 8.4.2 Bayes' Theorem for Sales Example

μ	Prior probability	Likelihood	(Prior probability) $\times$ (likelihood)	Posterior probability
1100	.15	.0006/38.73	.00009/38.73	.001
1150	.20	.0270/38.73	.00540/38.73	.032
1200	.30	.2347/38.73	.07041/38.73	.412
1250	.20	.3857/38.73	.07714/38.73	.450
1300	.15	.1200/38.73	.01800/38.73	.105
	1.00		.17104/38.73	1.000

Statistical models such as the Bernoulli, Poisson, and normal models play an important role in problems of inference and decision. They greatly simplify the determination of likelihoods, and, to the extent that different persons would agree about the use of a particular model to represent a particular real-world situation, they result in likelihoods about which many people might be in agreement. That is, given that different persons are willing to accept the assumptions of a certain model, they should agree about the likelihoods determined from that model. As a result, the determination of likelihoods is considered by many to be less controversial than the determination of prior probabilities. However, the willingness to accept the assumptions of a model and to consider that model as a satisfactory representation of reality is a subjective choice. Whether judgments are used to select a model or whether they are used to assess probabilities directly without using a model, the element of subjectivity is still present. If a statistical model seems to be realistic in any given situation, it may simplify the assessment of likelihoods, but it should be emphasized that the decision to use such a model remains subjective.

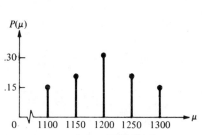

Figure 8.4.3 Prior Distribution of μ for Sales Example

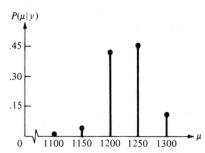

Figure 8.4.4 Posterior Distribution of μ for Sales Example

In some situations, it may not be possible to represent the data-generating process by some well-known statistical model, in which case the likelihoods must be assessed directly, much as prior probabilities are assessed. For a simple example involving the assessment of likelihoods, suppose that the president of a firm is concerned that a competing firm might be planning to introduce a new product that could seriously affect the sales of the first firm's products. The president feels that the chances are only 1 in 5 that the competitor is actually planning to introduce a new product, so that his prior probabilities are $P(\text{new product}) = .2$ and $P(\text{no new product}) = .8$. After assessing these probabilities, he receives the additional information that the competing firm is building a new plant. In order to use Bayes' theorem to revise his probabilities, the president must assess the likelihoods associated with this new information. One of these likelihoods is the probability that the competing firm would build a new plant, given that it plans to introduce a new product; the other likelihood is the probability that the new plant would be built, given no plans regarding a new product. After careful consideration of the competitor's situation, the president decides that if a new product is to be introduced, the odds would be 3 to 2 in favor of the building of a new plant; on the other hand, if a new product is not to be introduced, the odds would be 4 to 1 against the building of a new plant. Therefore, the president's subjective likelihoods are

$$P(\text{new plant} \mid \text{new product}) = \frac{3}{(3 + 2)} = .6$$

and

$$P(\text{new plant} \mid \text{no new product}) = \frac{1}{(1 + 4)} = .2.$$

Using the notation of Section 8.1, let $\theta = 1$ if the new product is to be introduced and let $\theta = 0$ if the new product is not to be introduced; the sample information y consists of the knowledge that the new plant is being built. Bayes' theorem can then be applied, and the details are presented in Table 8.4.3.

Table 8.4.3   Bayes' Theorem for New-Product Example

θ	Prior probability	Likelihood	(Prior probability) × (likelihood)	Posterior probability
1	.2	.6	.12	.43
0	.8	.2	.16	.57
	1.0		.28	1.00

The new information concerning the building of the plant increases the probability that $\theta = 1$ (that is, the probability that the competitor is planning to introduce a new product) from .2 to .43. Note that since likelihoods can be taken to be any positive multiple of the relevant conditional probabilities, it is only necessary to give the *relative* likelihoods. In other words, instead of assessing $P(\text{new plant} \mid \text{new product}) = .6$ and $P(\text{new plant} \mid \text{no new product}) = .2$, it is sufficient to say that the new plant is 3 times as likely given the new product as it is given no new product. In general, it should be easier to assess relative likelihoods than to assess all of the conditional probabilities of interest.

The discussion in this section concerning the assessment of likelihoods applies whether the prior and posterior distributions are discrete or continuous. The examples all involved discrete prior distributions because we have not yet discussed continuous prior distributions. In the following section we introduce Bayes' theorem for continuous random variables.

8.5   BAYES' THEOREM FOR CONTINUOUS RANDOM VARIABLES

In many problems involving statistical inference and decision, it is more realistic and more convenient to assume that the random variable of interest is continuous. As we have pointed out, measurement procedures are such that actual observed random variables can never be truly continuous. In many situations, however, the measurement may be precise enough so that for all practical purposes we can assume that a random variable is continuous. Indeed, in many cases it seems quite realistic to assume that a random variable is continuous. For example, in the quality-control illustration in Section 8.2, it seems much more realistic to assume that p, the proportion defective, is a continuous random variable than to assume that it just takes on the four values .01, .05, .10, and .25.

If θ, the parameter of interest, is continuous, the prior and posterior distributions of θ can be represented by density functions. Furthermore, if sample information involving θ can be summarized by the sample statistic Y, then the posterior density is simply the conditional density of θ given the observed value y of Y. From Equation (3.21.10), this can be written as

$$f(\theta \mid Y = y) = \frac{f(\theta, y)}{f(y)} . \tag{8.5.1}$$

The conditional density of one random variable given a value of a second random variable is simply the joint density of the two random variables divided by the marginal density of the second random variable.

Notice that the joint density, $f(\theta, y)$, and the marginal density, $f(y)$, are not known; we know only the prior distribution, $f(\theta)$, and the likelihood function, $f(y \mid \theta)$. Fortunately, the required joint and marginal densities can be written in terms of the prior distribution and the likelihood function (see Section 3.21):

$$f(\theta, y) = f(\theta)f(y \mid \theta) \qquad (8.5.2)$$

and

$$f(y) = \int_{-\infty}^{\infty} f(\theta, y) \, d\theta = \int_{-\infty}^{\infty} f(\theta)f(y \mid \theta) \, d\theta. \qquad (8.5.3)$$

The usual form of Bayes' theorem for continuous random variables is obtained by replacing the numerator and denominator of Equation (8.5.1) by the expressions in Equations (8.5.2) and (8.5.3), respectively.

Bayes' theorem for continuous random variables, then, is as follows:

$$f(\theta \mid y) = \frac{f(\theta)f(y \mid \theta)}{\displaystyle\int f(\theta)f(y \mid \theta) \, d\theta}. \qquad (8.5.4^*)$$

The densities $f(\theta \mid y)$ and $f(\theta)$ represent the posterior distribution and the prior distribution, respectively, and $f(y \mid \theta)$ represents the likelihood function.

These terms have the same interpretation for continuous random variables as they have for discrete random variables, although here the prior and posterior distributions must be proper density functions. That is, they must be nonnegative, and the total area under the curve, determined by integrating the density function over its entire domain, must be equal to 1. The integral in the denominator of Equation (8.5.4) serves the same purpose as the sum in the denominator of Equation (8.1.1); it makes the posterior distribution a proper probability distribution. As in the discrete case, the likelihood function is a function of θ, with y fixed (equal to the observed value of y). It can be seen that Bayes' theorem for continuous probability models is analogous to Bayes' theorem for discrete probability models.

In the discrete case, Bayes' theorem can be expressed in words as

$$\text{posterior probability} = \frac{(\text{prior probability})\,(\text{likelihood})}{\sum (\text{prior probability})\,(\text{likelihood})},$$

and the corresponding statement for continuous random variables is

$$\text{posterior density} = \frac{(\text{prior density})\,(\text{likelihood})}{\int (\text{prior density})\,(\text{likelihood})}.$$

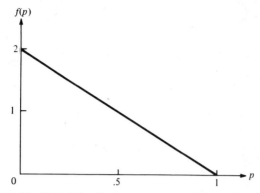

Figure 8.5.1   Prior Distribution of p for Market-Share Example

To illustrate Equation (8.5.4), suppose that $\theta = p$ represents the market share (expressed as a decimal, not as a percentage) of a new brand of a certain product. The new brand is considerably different from the other brands of the product, so the share of the market that it will attract is by no means certain. The marketing manager handling the new brand thinks that it might, if it "catches on," attract virtually the entire market for the product (that is, p might be close to 1). On the other hand, it might not be successful at all (that is, p might be close to 0), or it might be only moderately successful. The manager assumes that p is continuous and assesses a prior distribution for p with a triangular-shaped density function (Figure 8.5.1):

$$f(p) = \begin{cases} 2(1 - p) & \text{if } 0 \leq p \leq 1, \\ 0 & \text{elsewhere.} \end{cases}$$

In order to obtain some empirical information about p, a sample of 5 consumers of the product in question is taken; 1 purchases the new brand and the other 4 purchase other brands. The marketing manager is willing to assume that the sample is random and that the process of purchasing the product can be satisfactorily represented by a Bernoulli process. Thus, the likelihood function can be represented in terms of the binomial distribution with $R = 1$ fixed and p varying:

$$f(y \mid p) = P(R = 1 \mid n = 5, p) = \binom{5}{1} p^1 (1 - p)^4 = 5p(1 - p)^4.$$

This likelihood function is graphed in Figure 8.5.2.

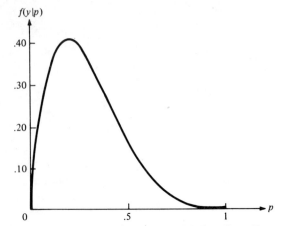

Figure 8.5.2   Likelihood Function for Market-Share Example

Applying Equation (8.5.4),

$$f(p \mid y) = \frac{f(p)f(y \mid p)}{\int f(p)f(y \mid p) \, dp} = \frac{[2(1-p)][5p(1-p)^4]}{\int_0^1 [2(1-p)][5p(1-p)^4] \, dp}$$

$$= \frac{10p(1-p)^5}{10 \int_0^1 p(1-p)^5 \, dp} = \frac{p(1-p)^5}{\int_0^1 p(1-p)^5 \, dp}$$

if $0 \le p \le 1$, and $f(p \mid y) = 0$ elsewhere. The integral in the denominator is equal to $(1!)(5!)/(7!) = 1/42$; thus, the posterior density function of p is

$$f(p \mid y) = \begin{cases} 42p(1-p)^5 & \text{if } 0 \le p \le 1, \\ 0 & \text{elsewhere.} \end{cases}$$

This posterior distribution is illustrated in Figure 8.5.3. A comparison of Figures 8.5.1 and 8.5.3 shows how the new information changes the marketing manager's distribution.

Conceptually, Equation (8.5.4) provides a convenient way to revise density functions in the light of sample information. In practice, however, it may prove quite difficult to apply this formula. If $f(\theta)$ and $f(y \mid \theta)$ are not fairly simple mathematical functions, it may be a hard task to carry out the integration in the denominator of Equation (8.5.4). In some situations, it may be necessary to resort to advanced mathematical techniques to determine the exact posterior distribution. One way to avoid this diffi-

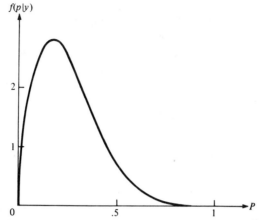

Figure 8.5.3 Posterior Distribution of p for Market-Share Example

culty is to use discrete approximations of continuous probability models; this approach, which falls under the general heading of numerical methods, is considered briefly later in this chapter. A second approach, involving the restriction of the prior distribution to a certain family of distributions that depends on the form of the likelihood function, is discussed in the remainder of this section and in the following sections.

Because of potential difficulties in the application of Equation (8.5.4), Bayesian statisticians have developed the concept of **conjugate prior distributions,** which essentially are families of distributions that ease the computational burden when used as prior distributions. Of course, the posterior distribution depends on the likelihood function as well as on the prior distribution. Usually, once certain assumptions are made about the population or process that is being sampled (for example, the assumption that the population is normally distributed or the assumption that the process is a stationary and independent Bernoulli process), the likelihood function is uniquely determined according to the chosen statistical model. In other words, if a particular data-generating model is specified, the likelihood function is known. For any particular model and likelihood function, the Bayesian statistician attempts to determine a conjugate family of prior distributions. This is a set of distributions, each of which can be combined with the given likelihood function so as to yield a posterior distribution that is of the same "form" as the prior distribution in the sense that it also is a member of the conjugate family.

It is important to remember that a conjugate prior distribution is "conjugate" only with respect to a given likelihood function (that is, a given statistical model). Conjugate families of distributions corresponding to numerous likelihood functions have been developed. Only two situations are discussed in detail here. The first situation involves sampling from a stationary and independent Bernoulli process, in which case the conjugate family is the family of beta distributions. The second situation involves sampling from a normally distributed population with known variance, in which case the conjugate family is the family of normal distributions. These two cases are discussed in the following sections.

8.6   CONJUGATE PRIOR DISTRIBUTIONS FOR THE BERNOULLI PROCESS

In previous chapters we have discussed stationary and independent Bernoulli processes, including estimation and hypothesis-testing procedures involving the Bernoulli parameter p, the probability of success on a Bernoulli trial. These procedures utilized only sample information from the Bernoulli process. In Section 8.2 we presented an example of Bayesian techniques involving a Bernoulli process under the assumption that the prior distribution was discrete. It is often unrealistic, however, to limit p to a finite number of values. Theoretically, p can assume any real value from 0 to 1, so that the prior distribution should be continuous rather than discrete. Using Equation (8.5.4), the statistician can determine a continuous prior distribution and then find the posterior distribution following an observed sample. This could be a difficult task unless the prior distribution is a member of the family of distributions which is conjugate relative to the Bernoulli process. This conjugate family is the family of **beta distributions.**

The density function of the beta distribution of p with parameters r and n, where $n > r > 0$, is

$$f(p) = \begin{cases} \dfrac{(n-1)!}{(r-1)!(n-r-1)!}\, p^{r-1}(1-p)^{n-r-1} & \text{if } 0 \le p \le 1, \\[2ex] 0 & \text{elsewhere.} \end{cases}$$

$$(8.6.1^*)$$

Note that the beta distribution has two parameters, r and n. It is necessary that $n > r > 0$, but n and r need not be integers. Incidentally, if n and r are not integers, the factorial terms in Equation (8.6.1) must be replaced by mathematical functions known as gamma functions, but we need not be concerned with this detail, as the general form of the beta density function remains the same.

The mean and the variance of a beta distribution with parameters r and n are

$$E(p \mid r, n) = \frac{r}{n} \qquad (8.6.2^*)$$

and

$$V(p \mid r, n) = \frac{r(n - r)}{n^2(n + 1)}. \qquad (8.6.3^*)$$

The shape of the beta distribution depends on r and n. If $r = n/2$, the distribution is symmetric. If $r < n/2$, the distribution is positively skewed (the "long tail" of the density function is to the right), while if $r > n/2$, the distribution is negatively skewed (the "long tail" is to the left). If $r > 1$ and $n - r > 1$, the distribution is unimodal (that is, it has a single mode) with the mode [the value p for which $f(p)$ is largest] equal to $(r - 1)/(n - 2)$. If $r \leq 1$ or $n - r \leq 1$, the distribution is either unimodal with mode at 0 or 1, U-shaped with modes at 0 *and* 1, or simply the uniform distribution on the unit interval [that is, $f(p)$ is constant; this occurs when $r = 1$ and $n = 2$]. Thus, the family of beta distributions can take on a wide variety of shapes as r and n vary. Some examples of beta distributions with expectation r/n equal to 1/2 and 1/20 are presented in Figures 8.6.1 and 8.6.2.

A useful fact is that cumulative beta probabilities and values of the beta density function can be determined from tables of the binomial distribution provided that r and n are integers.

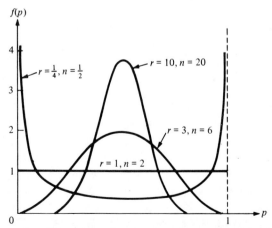

Figure 8.6.1   Examples of Beta Density Functions with $r/n = .5$

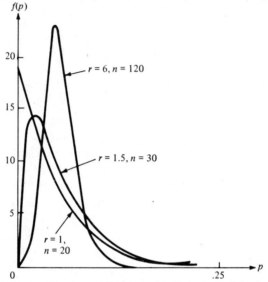

Figure 8.6.2   Examples of Beta Density Functions with $r/n = .05$

$$F_\beta(p \mid r, n) = \begin{cases} P_{\text{bin}}(R \geq r \mid n - 1, p) & \text{if } p \leq 1/2, \\ P_{\text{bin}}(R < n - r \mid n - 1, 1 - p) & \text{if } p \geq 1/2, \end{cases} \quad (8.6.4^*)$$

and $f_\beta(p \mid r, n) = (n - 1)P_{\text{bin}}(R = r - 1 \mid n - 2, p)$. $\qquad (8.6.5^*)$

For example, if $r = 2$ and $n = 5$, the values of the beta CDF and the beta density function at $p = .3$ are

$$F_\beta(.3 \mid r = 2, n = 5) = P_{\text{bin}}(R \geq 2 \mid 4, .3) = .3483$$

and $f_\beta(.3 \mid r = 2, n = 5) = 4P_{\text{bin}}(R = 1 \mid 3, .3) = 4(.4410) = 1.764.$

Now that we have discussed the beta distribution, we are prepared to return to the main topic of this section, which is the use of beta prior distributions for p, the parameter of a Bernoulli process. First, a word on notation is in order. We often will denote prior distributions and parameters of prior distributions (such as the mean of the prior distribution) with single primes. Posterior distributions and parameters of posterior distributions shall be denoted with double primes. For example, $f'(\theta)$ and $f''(\theta \mid y)$ represent the prior and posterior distributions of θ; μ_θ' and μ_θ'' [or $E'(\theta)$ and $E''(\theta)$] represent the means of the prior and posterior distributions, respectively. Of course, you should remember that the terms "prior" and "posterior" are relative terms, relating to a particular sample result.

The main result of this section is as follows.

Suppose that a statistician is sampling from a stationary and independent Bernoulli process. Then if the prior distribution of p is a beta distribution, the posterior distribution will also be a beta distribution. In particular, if the prior distribution is a beta distribution with parameters r' and n', and if the sample consists of r successes in n trials, then the posterior distribution is a beta distribution with parameters r'' and n'', where

$$r'' = r' + r \tag{8.6.6*}$$

and
$$n'' = n' + n. \tag{8.6.7*}$$

This result can be proved by applying Equation (8.5.4):

$$f''(p \mid y) = \frac{f'(p)f(y \mid p)}{\int_0^1 f'(p)f(y \mid p) \, dp} = \frac{f'(p)P(r \mid n, p)}{\int_0^1 f'(p)P(r \mid n, p) \, dp}$$

$$= \frac{\dfrac{(n'-1)!}{(r'-1)!(n'-r'-1)!} p^{r'-1}(1-p)^{n'-r'-1} \binom{n}{r} p^r (1-p)^{n-r}}{\int_0^1 \dfrac{(n'-1)!}{(r'-1)!(n'-r'-1)!} p^{r'-1}(1-p)^{n'-r'-1} \binom{n}{r} p^r (1-p)^{n-r} \, dp}.$$

$$\tag{8.6.8}$$

The terms not involving p can be moved outside of the integral in the denominator and canceled with the like terms in the numerator, leaving

$$f''(p \mid y) = \frac{p^{r'-1}(1-p)^{n'-r'-1} p^r (1-p)^{n-r}}{\int_0^1 p^{r'-1}(1-p)^{n'-r'-1} p^r (1-p)^{n-r} \, dp}. \tag{8.6.9}$$

Defining r'' and n'' as in Equations (8.6.6) and (8.6.7), we have

$$f''(p \mid y) = \frac{p^{(r'+r)-1}(1-p)^{(n'+n)-(r'+r)-1}}{\int_0^1 p^{(r'+r)-1}(1-p)^{(n'+n)-(r'+r)-1} \, dp}$$

$$= \frac{p^{r''-1}(1-p)^{n''-r''-1}}{\int_0^1 p^{r''-1}(1-p)^{n''-r''-1} \, dp}. \tag{8.6.10}$$

The integral in the denominator of Equation (8.6.10) is a special function known in mathematics as a beta function and is equal to $(r'' - 1)!(n'' - r'' - 1)!/(n'' - 1)!$, so that

$$f''(p \mid y) = \frac{(n'' - 1)!}{(r'' - 1)!(n'' - r'' - 1)!} p^{r''-1}(1 - p)^{n''-r''-1}. \quad (8.6.11)$$

Thus, the posterior distribution of p is a beta distribution with parameters r'' and n''.

In the market-share example of Section 8.5, the prior distribution is a beta distribution with $r' = 1$ and $n' = 3$, and the sample results are $r = 1$ and $n = 5$. Therefore, the posterior distribution is a beta distribution with $r'' = 2$ and $n'' = 8$. Given Equations (8.6.6) and (8.6.7), it is not necessary to go to the bother of using Equation (8.5.4) when the prior distribution is a beta distribution and the sample comes from a Bernoulli process.

How does the sample information change the marketing manager's distribution in the market-share example? The prior and posterior distributions are illustrated in Figures 8.5.1 and 8.5.3. Since the mean of a beta distribution with parameters r and n is simply r/n, the mean of the prior distribution is $r'/n' = 1/3 - .33$ and the mean of the posterior distribution is $r''/n'' = 2/8 = .25$. The sample results shift the mean downward, making lower values of p appear more likely. Note that the sample mean is $r/n = 1/5 = .20$. The posterior mean always lies between the prior mean and the sample mean when the prior distribution is a beta distribution and the sample is taken from a Bernoulli process.

The variance of the beta distribution with parameters r and n is given by Equation (8.6.3). In the example, the prior variance is .056 and the posterior variance is .021. In most, but not all, cases, the posterior variance will be smaller than the prior variance. This is intuitively reasonable, since the sample provides new information, and increased information should reduce the uncertainty, which might be measured by the variance, about the quantity of interest. Thus, it might be expected that additional information would reduce the variance, just as the variance of a sample mean is smaller as the sample size becomes larger. In some situations, notably when the sample information greatly shifts the mean of the posterior distribution toward $\frac{1}{2}$, the variance of the posterior beta distribution may be larger than the variance of the prior beta distribution.

In this chapter, we have tacitly assumed that the samples generated from a Bernoulli process are samples with a fixed number of trials, n, so that the sample statistic of interest is R and the distribution of R is a binomial distribution. From Section 4.7, we know that another way of sampling from a Bernoulli process is to fix r and to let N, the number of trials, be the random variable, in which case the distribution of N is a Pascal distribution.

In comparing Equations (4.2.1) and (4.7.1), we see that the only difference between the mathematical expression for the binomial distribution and that for the Pascal distribution is that the combinatorial term $\binom{n}{r}$ in the binomial distribution is changed to $\binom{n-1}{r-1}$ in the Pascal distribution. In comparing Equation (8.6.8) with Equation (8.6.9), we see that terms such as this, since they do not involve p, cancel out in the application of Bayes' theorem. Therefore, the results of this section hold whether the sampling plan is binomial or Pascal. The only important part of the likelihood function for Bayesian purposes is the part involving p, and the likelihood function in the Bernoulli case can be taken as simply $p^r(1 - p)^{n-r}$. The procedure used to tell us when to stop sampling is called a **stopping rule,** and if the stopping rule has no effect on the posterior distribution, then it is said to be **noninformative.** The two stopping rules we have discussed (sample until you have n trials, and sample until you have r successes) are both noninformative.

8.7   THE USE OF BETA PRIOR DISTRIBUTIONS: AN EXAMPLE

We are now prepared to consider a somewhat more realistic modification of the quality-control example presented in Section 8.2. Suppose that the production manager feels that the prior information concerning p, the proportion of defective items produced by the process, can be well-represented by a beta distribution with parameters $r' = 1$ and $n' = 20$ (procedures for assessing such distributions are discussed in Section 8.11). This distribution has a mean of $r'/n' = .05$ and a variance of $r'(n' - r')/n'^2(n' + 1) = .0023$ (the standard deviation is thus .048). The graph of the density function is presented in Figure 8.7.1. This continuous prior distribution seems more realistic than the discrete distribution used in Section 8.2, in which only four possible values of p were considered. Notice that the density function reaches its highest value at $p = 0$ and decreases as p increases.

Following the assessment of the prior distribution, the production manager takes a sample of 5 items from the production process, observing 1 defective item. Under the assumption that the production process behaves as a stationary and independent Bernoulli process, with p representing the probability of a defective item on any trial, the posterior distribution is a beta distribution with parameters

$$r'' = r' + r = 1 + 1 = 2$$

and

$$n'' = n' + n = 20 + 5 = 25.$$

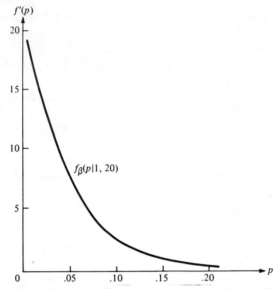

Figure 8.7.1 Prior Distribution of p for Quality-Control Example

This distribution is shown in Figure 8.7.2. The mean of the distribution is $2/25 = .08$ and the variance is $2(23)/25^2(26) = .0028$. Note that the variance of the posterior distribution is slightly larger than the variance of the prior distribution; this happens because of the shift in the mean. The posterior mean is .03 units to the right of the prior mean. Notice also that the mode of the posterior distribution is no longer 0; instead, it is equal to 1/24, or about .042. In general, if $r > 1$ and $n - r > 1$, the mode of the beta distribution with parameters r and n is $(r - 1)/(n - 2)$.

Suppose that the production manager decides to take a second sample

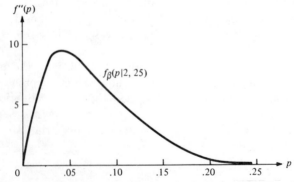

Figure 8.7.2 Posterior Distribution of p Following First Sample in Quality-Control Example

of size 5 and that 2 defective items are observed in this second sample. The posterior distribution calculated after the first sample is now the prior distribution with respect to the new sample, so $r' = 2$ and $n' = 25$. After the second sample, the posterior distribution is a beta distribution with parameters

$$r'' = r' + r = 2 + 2 = 4$$

and

$$n'' = n' + n = 25 + 5 = 30.$$

This distribution is shown in Figure 8.7.3. The mean of the distribution is $4/30 = .133$, and the variance is $4(26)/(30)^2(31) = .0037$. Once again the mean is shifted to the right and the variance increases. The mode is now $3/29 = .103$.

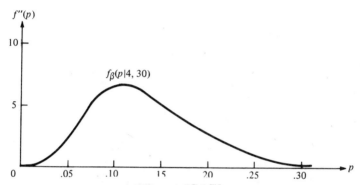

Figure 8.7.3   Posterior Distribution of p Following Second Sample in Quality-Control Example

Just as in the discrete case, the production manager could wait and revise his original prior distribution once after observing both samples. If he does this, the samples can be summarized by $r = 3$ and $n = 10$, and the parameters of the posterior beta distribution are

$$r'' = r' + r = 1 + 3 = 4$$

and

$$n'' = n' + n = 20 + 10 = 30.$$

This is identical to the posterior distribution determined from two successive applications of Bayes' theorem.

If the production manager wants an estimate of p, he should base his estimate on the posterior distribution. For example, he might want to use the mean of the posterior distribution, which in this case is .133. Suppose that instead of just estimating p, he must make a decision concerning the process, a decision that depends on the value of p. It is convenient to modify the example from Section 8.2 so that the costs are the same but the actions available to the manager are slightly different. The two actions are as follows: (1) he can leave the process as is, or (2) he can have a mechanic make an adjustment in the process that will halve the proportion of defectives, p—this adjustment costs $100.

The profit from a run of 1000 items is 1000($.50) = $500, less the costs of replacing defectives and adjusting the process (if the second action is chosen). The expected number of defectives in a run of 1000 is $1000p$, and the cost of replacing a single defective is $2, so the total expected cost of replacing defectives is $2000p$ if the process is not adjusted. If the adjustment is made, the proportion of defectives is cut in half, so the total expected cost of replacing defectives is $2000p/2 = $1000p$. Thus, the profits to the firm associated with the two actions are

$$\text{Profit(no adjustment)} = \$500 - \$2000p$$

and

$$\text{Profit(adjustment)} = \$500 - \$1000p - \$100.$$

The corresponding expected profits, or expected rewards, for the two acts are

$$ER\text{(no adjustment)} = \$500 - \$2000E(p)$$

and

$$ER\text{(adjustment)} = \$400 - \$1000E(p).$$

From the posterior distribution, $E(p) = .133$, so the respective expected payoffs are $234 and $267. In order to maximize expected profit, the production manager should have the process adjusted.

If the production manager had not obtained any sample information, the decision would have been based on the prior distribution. The prior mean is $E(p) = .05$, and the expected payoffs are $400 and $350. Solely on the basis of the prior information, then, the production manager should leave the process as is. As in the decision-making example of Section 8.2, the sample information revises the prior distribution in such a way as to make large values of p more likely and small values less likely. Since large values of p are more costly due to the costs of replacing defectives, the expected payoffs calculated from the posterior distribution are considerably smaller (for both actions) than the expected payoffs calculated from the prior distribution.

This example illustrates the use of Bayes' theorem to revise a continuous probability distribution and the application of the resulting posterior distribution in a decision-making problem. We will consider decision theory in more detail later, but first we will discuss Bayesian techniques for a normal population or process.

8.8 CONJUGATE PRIOR DISTRIBUTIONS FOR THE NORMAL PROCESS

Although it may often be useful, Bayesian inference for a Bernoulli process is limited in application to those situations in which the assumptions of a Bernoulli process are satisfied. That is, there must be two possible outcomes at each trial (or if there are more than two, the outcomes must be divided into two sets), the probabilities of these two outcomes must remain constant from trial to trial, and the trials must be independent. In many situatious the data-generating process is such that there are more than two possible outcomes at each trial; in particular, any real number within some specified interval (which may be the entire real line) may be a potential outcome. An example is sampling from a normal population, or normal process. In previous chapters the wide applicability of the normal distribution has been discussed. In this section we will see how prior information can be easily combined with sample information from a normal process if the prior information can also be expressed in the form of a normal distribution.

Suppose that a statistician is sampling from a normal process or population. The statistician is willing to assume that σ^2, the variance of the process, is known, and although μ, the mean of the process, is not known, some prior information is available concerning μ.

If the prior distribution of μ is a normal distribution, the posterior distribution will also be a normal distribution. In particular, if the prior distribution is a normal distribution with mean m' and variance σ'^2, and if the sample size and sample mean from a random sample from the normal process are n and m, then the posterior distribution is a normal distribution with mean m'' and variance σ''^2, where

$$m'' = \frac{(1/\sigma'^2)\ m' + (n/\sigma^2)m}{(1/\sigma'^2) + (n/\sigma^2)} \qquad (8.8.1^*)$$

and

$$\sigma''^2 = \frac{1}{(1/\sigma'^2) + (n/\sigma^2)} \cdot \qquad (8.8.2^*)$$

The posterior mean is a weighted average of the prior mean and the sample mean, the weights being reciprocals of the prior variance and of σ^2/n, the variance of the sampling distribution of M, respectively. That is, the reciprocal of the posterior variance is equal to the sum of the reciprocals of the prior variance and of the variance of M.

Be sure not to confuse σ^2 and σ'^2; the former is the variance of the data-generating process or the population, whereas the latter is the prior variance of μ, the *mean* of the process or the population. To clarify this difference, consider a transformation to a new parameter n':

$$n' = \frac{\sigma^2}{\sigma'^2}. \tag{8.8.3*}$$

Therefore,

$$\sigma'^2 = \frac{\sigma^2}{n'}, \tag{8.8.4}$$

so that the prior variance can be written in terms of n' and the process variance σ^2. The prior distribution is thus a normal distribution with mean m' and variance σ^2/n'. Now, if n'' is defined as

$$n'' = \frac{\sigma^2}{\sigma''^2}, \tag{8.8.5*}$$

then Equation (8.8.1) can be written in the form

$$m'' = \frac{(n'/\sigma^2)m' + (n/\sigma^2)m}{(n'/\sigma^2) + (n/\sigma^2)},$$

or

$$m'' = \frac{n'm' + nm}{n' + n}. \tag{8.8.6*}$$

Furthermore, Equation (8.8.2) becomes

$$\frac{\sigma^2}{n''} = \frac{1}{(n'/\sigma^2) + (n/\sigma^2)} = \frac{1}{(n' + n)/\sigma^2} = \frac{\sigma^2}{n' + n},$$

implying that

$$n'' = n' + n. \tag{8.8.7*}$$

Comparing Equations (8.8.6) and (8.8.7) with Equations (8.8.1) and (8.8.2), we see that the prior distribution is expressed in terms of m' and n' rather than m' and σ'^2, and the posterior distribution is expressed in terms of m'' and n'' rather than m'' and σ''^2. How can n' (and n'') be interpreted? From Equation (8.8.4), n' appears to be the sample size required to produce a variance of σ'^2 for a sample mean, because the variance of the sample mean from a sample of size n' is equal to σ^2/n'. This suggests an interpretation for the prior distribution. The prior distribution is roughly equivalent to the information contained in a sample of size n' with a sample mean of m'. Under this interpretation, Equations (8.8.6) and (8.8.7) can be thought of as formulas for pooling the information from two samples. The pooled (posterior) sample size is equal to the sum of the two individual sample sizes (one from the prior distribution, one from the sample), and the pooled (posterior) sample mean is equal to a weighted average of the two individual sample means (one from the prior, one from the sample). This does not mean that the prior information must consist solely of a prior sample from the process; it merely indicates that the prior information can be thought of as being roughly equivalent to such sample information. It should be noted that n' need not be an integer (recall that for a beta prior distribution, r' and n' do not have to be integers). For instance, if the variance of the data-generating process is $\sigma^2 = 110$ and the variance of the prior distribution is $\sigma'^2 = 4$, then $n' = \sigma^2/\sigma'^2 = 27.5$.

Using the new parameter introduced in Equation (8.8.3), the relative weights of the prior information and the sample information can be investigated. If you want to estimate μ, for example, a reasonable estimate is the posterior mean m'', which is always between the prior mean m' and the sample mean m. Under what conditions is m'' closer to m' than to m, and under what conditions is m'' closer to m? From Equation (8.8.6), m'' is a weighted average of m' and m, with the weights being equal to $n'/(n' + n)$ and $n/(n' + n)$, respectively. Therefore, if $n' > n$, the prior mean is given more weight, and the posterior mean m'' is closer to m' than to m. If $n' < n$, the sample mean is given more weight, and m'' is closer to m than to m'. If $n' = n$, m'' is exactly midway between m' and m.

In terms of the above interpretation of the prior distribution, these results become obvious; when two samples are pooled, the one with the larger sample size automatically receives more weight in the determination of the pooled mean. In determining the posterior distribution, the Bayesian statistician is pooling information. If there is more prior information than sample information (where information in this context is inversely related to variance), then the posterior distribution is affected more by the prior information (as expressed in terms of a prior distribution) than by the sample information (as expressed in terms of a likelihood function). Also, note that since $n'' = n' + n$ must be larger than n', the posterior variance σ^2/n'' must be smaller than the prior variance σ^2/n'. If information is thought of as inversely related to variance, this result is intuitively appealing because it implies that we are *gaining* information by sampling.

8.9   THE USE OF NORMAL PRIOR DISTRIBUTIONS: AN EXAMPLE

To illustrate the procedures presented in Section 8.8, we will consider once again the example of Section 8.4 involving a retailer who is interested in the distribution of weekly sales at one of his stores. In Section 8.4, the prior distribution of μ was assumed to be discrete, though it is far more realistic to assume that μ is continuous. From informal conversations with the manager of the store and from knowledge about sales at similar stores, the retailer decides that the mean of the prior distribution, m', is 1200 and that the standard deviation, σ', is 50. Furthermore, he feels that his prior distribution is symmetric about the mean and is shaped roughly like a normal density function, so he assumes that the distribution is normal. Be sure to distinguish between $\sigma^2 = 90{,}000$, which represents the variance of weekly sales, and $\sigma'^2 = 2500$, which represents the variance of the retailer's prior distribution of μ, *mean* weekly sales.

The retailer may be interested in the distribution of sales for several reasons. Perhaps he must decide whether to keep the store or to sell it, whether to open more stores of a similar nature or in a similar neighborhood, or whether to bring in a new manager for the store. Naturally, such decisions necessitate the consideration of a loss function and of relevant variables other than sales. To keep this example from becoming too complicated, assume that the retailer does not have to make a decision at this particular moment. Instead, he just wants some idea about the mean of the distribution of weekly sales, and he decides to look at an interval of values of μ.

Given the prior distribution, an interval that is centered at m' with a probability of .95 can be determined. This interval is simply

$$(m' - 1.96\sigma', \ m' + 1.96\sigma'),$$

or

$$(1200 - 1.96(50), \ 1200 + 1.96(50)),$$

or

$$(1102, 1298).$$

This is called a **credible interval.** Based on the prior distribution, then, the interval from 1102 to 1298 is a 95 percent credible interval for μ.

On the basis of the sample of 60 weeks with sample mean $m = 1240$, the retailer can revise his prior distribution. Following this sample, the posterior distribution for μ is a normal distribution with mean m'' and variance σ''^2, where

$$m'' = \frac{(1/\sigma'^2)m' + (n/\sigma^2)m}{(1/\sigma'^2) + (n/\sigma^2)} = \frac{(1/2500)(1200) + (60/90{,}000)(1240)}{(1/2500) + (60/90{,}000)}$$

$$= 1225$$

and $$\sigma''^2 = \frac{1}{(1/\sigma'^2) + (n/\sigma^2)} = \frac{1}{(1/2500) + (60/90{,}000)}$$

$$= 937.5.$$

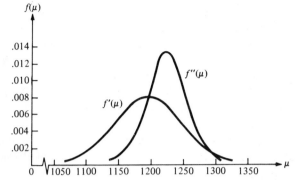

Figure 8.9.1 Prior and Posterior Distributions of μ in Sales Example

The prior and posterior distributions of μ are graphed in Figure 8.9.1. The limits for a 95 percent credible interval based on the posterior distribution are

$$m'' - 1.96\sigma'' \quad \text{and} \quad m'' + 1.96\sigma'',$$

or

$$1225 - 1.96(30.6) \quad \text{and} \quad 1225 + 1.96(30.6),$$

or

$$1165 \quad \text{and} \quad 1285.$$

Notice that the width of the 95 percent credible interval based on the prior distribution is $1298 - 1102 = 196$, whereas the width of the interval based on the posterior distribution is $1285 - 1165 = 120$. This reflects the fact that the introduction of the sample information reduces the standard deviation from 50 to 30.6. Incidentally, the classical 95 percent *confidence* interval, based solely on the sample information, has limits of approximately 1164 and 1316, as determined from Equation (6.12.1). The width of this confidence interval is $1316 - 1164 = 152$, which is considerably greater than that of the credible interval just found.

For the moment, consider n' and n'' instead of σ'^2 and σ''^2. We have

$$n' = \frac{\sigma^2}{\sigma'^2} = \frac{90,000}{2500} = 36,$$

so that

$$n'' = n' + n = 36 + 60 = 96.$$

Thus, the "equivalent prior sample size" is 36, while the size of the actual sample is 60. In the determination of the posterior mean, the prior mean receives a weight of $36/96 = .375$, whereas the sample mean receives a weight of $60/96 = .625$. This explains why the mean of the posterior distribution is closer to m than to m'.

One aspect of this example may seem particularly unrealistic—the assumption that the variance of the process is known. The retailer may have *some* information about the variance, but it is highly unlikely that it is known for certain. Unfortunately, a formal Bayesian approach to inference for a normal process is somewhat more complex if the variance is not assumed to be known. If the variance is not known, the retailer would have to assess a *joint* prior distribution for the variance and the mean and modify this distribution on the basis of the sample information. In this case, the sample variance would be of interest as well as the sample mean, since it provides information concerning the process variance. At any rate, assessing and revising a *joint* prior distribution is more complicated than assessing and revising a univariate prior distribution, and we will not take up this topic.

Does this mean that we will be unable to carry out a Bayesian analysis for the normal process with unknown variance? In some cases, yes. In other cases, however, the central limit theorem comes to our aid once again. In Equations (8.8.1) and (8.8.2), σ^2 only appears in connection with the sample information, not the prior information. If we replace σ^2 by the unbiased sample variance $\hat{S}^2$, which is an estimator of σ^2, the sampling distribution is a T distribution rather than a normal distribution. If the sample size n is large enough, though, this T distribution is virtually identical to the standard normal distribution. If n is large enough, then, we can substitute $\hat{S}^2$ for σ^2 in Equations (8.8.1) and (8.8.2) and proceed as though σ^2 were known. For example, if the retailer did not know σ^2 but observed $\hat{S}^2 = 96,000$, you can verify that the posterior mean and variance would be 1224 and 975.6, respectively.

8.10   CONJUGATE PRIOR DISTRIBUTIONS FOR OTHER PROCESSES

In the past four sections, we discussed the use of conjugate prior distributions for two particular statistical models, the Bernoulli model and the normal model with known variance. Similar procedures have been developed for numerous other models. For instance, consider the Poisson model. The random variable of interest is λ, the intensity of the process, and the conjugate family for the Poisson process is the family of **gamma distributions,** which have density functions of the form

$$f(\lambda) = \begin{cases} \dfrac{e^{-\lambda t'}\,(\lambda t')^{r'-1}t'}{(r'-1)!} & \text{if } \lambda \geq 0, \\ 0 & \text{elsewhere,} \end{cases} \qquad (8.10.1)$$

where $r' > 0$ and $t' > 0$. Just as in the beta distribution, the factorial term $(r - 1)!$ technically should be a gamma function, to allow for noninteger values of r, but you need not be concerned with this technicality. The mean and the variance of the gamma distribution are $E(\lambda \mid r', t') = r'/t'$ and $V(\lambda \mid r', t') = r'/t'^2$.

For instance, in the automobile-salesman example presented in Section 8.4, suppose that the owner of the dealership feels that his prior information about the average number of sales per day for the salesman in question can be represented by a gamma distribution with $r' = 4$ and $t' = 16$. The mean and the variance of the distribution are $E(\lambda \mid r', t') = .25$ and $V(\lambda \mid r', t') = .0156$, and the distribution may be interpreted in terms of equivalent prior information. The owner's prior information is roughly equivalent to a sample of 16 days with 4 sales. On the basis of this prior distribution and the sample of 10 sales in 24 days, the owner's posterior distribution for λ is a gamma distribution with $r'' = r' + r = 4 + 10 = 14$ and $t'' = t' + t = 16 + 24 = 40$. Thus, the posterior mean and variance are $r''/t'' = .35$ and $r''/t''^2 = .00875$.

This Poisson example demonstrates the use of conjugate prior distributions for a model other than the Bernoulli model or the normal model with known variance. For any data-generating process that can be described by a general statistical model, the application of Bayes' theorem is quite simple if the prior distribution is a member of the appropriate conjugate family of distributions. However, it should be stressed that the applicability of conjugate prior distributions in any given situation depends in part on the applicability of a particular statistical model, such as the Bernoulli model or the normal model. This is because the conjugate family of distributions depends on the form of the likelihood function, which in turn depends on assumptions concerning a statistical model. In some situations, no standard statistical model is applicable, and it is necessary to assess likelihoods directly (see Section 8.4), in which case no conjugate family can be found. Furthermore, even if a certain model is applicable to the data-generating process and even if the corresponding conjugate family is known, it may be that no member of the family adequately represents the assessor's prior judgments. These factors operate to reduce the applicability of conjugate prior distributions, of course. Methods for the assessment of prior distributions are discussed in the following section.

8.11 THE ASSESSMENT OF PRIOR DISTRIBUTIONS

In all of the examples, it has been assumed that the prior distribution is given; that is, the statistician has already determined the prior distribution. Suppose that this is not true. Instead, suppose that the statistician has certain prior information and wants to express this information in terms of a prior distribution. How can the statistician's judgments be expressed as a probability distribution? In Section 8.3 we discussed the assessment of individual probabilities. If the random variable in question is discrete, the statistician's distribution will consist of a number of such probabilities. In the continuous case, the distribution is represented by a density function or a cumulative distribution function.

The most obvious way to attempt to assess a continuous probability distribution is simply to specify the density function or the cumulative distribution function. One way to do this is to specify the functional form of the density function. The relationship between subjective judgments and a mathematical function, if such a relationship does exist, is usually not at all obvious. Directly attempting to specify the functional form of the distribution, then, might not be such a good approach. A second approach involves the graph of the density function and/or the cumulative distribution function. The assessor may have some idea of the general shape of the distribution. Perhaps a rough graph of the PDF or CDF can be drawn. Of these two, the PDF is probably more meaningful to most assessors; that is, they can translate their prior information into a PDF more easily than into a CDF. One way to determine a PDF is to assess probabilities for certain intervals, draw a histogram corresponding to these probabilities, and attempt to fit a smooth density function to the histogram. The relationship between a histogram and a density function was pointed out in Section 3.6. Figure 8.11.1 illustrates the assessment of a PDF by this *grouping and smoothing* technique. For instance, if you want to assess a probability distribution for θ, the mean sales (in dollars) of a particular item, you might assess $P(\theta < 10{,}000)$, $P(10{,}000 \leq \theta \leq 12{,}000)$, $P(12{,}000 \leq \theta \leq 14{,}000)$, and so on, construct a histogram, and then try to find a continuous distribution that is a good approximation to the histogram. In some situations, much of the prior information may consist of historical data that can be represented in terms of a histogram. As indicated in Chapter 2, historical frequencies may be helpful in the assessment of probabilities, although other prior information may cause the prior distribution to differ somewhat from the historical data.

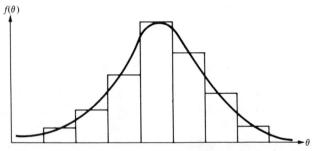

Figure 8.11.1   Assessment by "Grouping and Smoothing"

In determining a PDF or CDF, it might be helpful to consider certain summary measures of the distribution. In particular, some measure of location, such as the mean, median, or mode of the prior distribution, would be useful. This would give the assessor some idea of where the "center" of the distribution is. Next, a measure of dispersion might be considered. Measures such as the variance may be difficult to think about in this context, but it may be easy to think in terms of credible intervals. The assessor might ask, "Can I find an interval centered at the mean that includes 50 percent, or 75 percent, or 95 percent, and so forth, of the prior probability?"

The use of credible intervals suggests the assessment of fractiles of the prior distribution. Remember that an f fractile of the distribution of the continuous variable X is a point a such that $P(X \leq a) = f$. The .50 fractile, for example, is the median of the distribution, which for a continuous random variable divides the area under the density function into two equal parts, each with area one-half. Intuitively, the median of the assessor's prior distribution is the value that the assessor feels is equally likely to be exceeded or not exceeded. Once the median is assessed, the .25 and .75 fractiles are the values that divide the two halves of the distribution in half again, and so on. The advantage of this technique is that it merely requires the assessor to determine a series of fractiles, and the assessment of any single fractile is similar to the assessment of an individual probability. After a number of fractiles are assessed and plotted on a CDF graph, the assessor can then draw a rough curve through them and thus determine the entire distribution.

For example, suppose that a statistician wants to assess a prior distribution for the proportion of defectives produced by a production process. First, the statistician feels that the proportion p is just as likely to be above .05 as below .05. This implies indifference between receiving a lottery ticket that pays \$100 if p is less than .05 and receiving a lottery ticket that pays \$100 if p is greater than .05. In other words, the odds are even. This implies that .05 is the median of the distribution. Next, looking just at the interval below .05 (that is, the interval from 0 to .05), the statistician decides that .035 divides *this* interval into two equally likely subintervals, so that .035 is the .25 fractile of the distribution. Similarly, the .75, .125, .375, .625, .875, and so on, fractiles are assessed. These points are then plotted on a graph and an attempt is made to draw a rough CDF through them. To check the assessments, the statistician might consider various points on the curve to see if they reflect the prior information reasonably well. As noted in Section 8.3, there may be some vagueness on the part of the assessor, but devices such as hypothetical lotteries or bets should help to reduce this vagueness.

At this point, the discussion of the assessment of prior distributions is interrupted to ask a question. Why do we want a prior distribution in the first place? Presumably, we want to represent our prior uncertainty in probabilistic terms in order to make an inferential statement or to make a decision, and we want to base the inferential statement or decision on all of the information that happens to be available. We may intend to take a sample and use Bayes' theorem to combine the prior distribution and the likelihood function, forming a posterior distribution. But if that is the case, then it is convenient if the prior distribution is a member of a conjugate family of distributions. If we are sampling from a Bernoulli process, for example, and if we are interested in p, the probability of a success on any single trial, the analysis is greatly simplified if the prior distribution happens to be a beta distribution. For sampling from a normal process with known variance, the conjugate family is the normal family of distributions.

A subjectively assessed prior distribution is not necessarily a member of a conjugate family of distributions. In general, it does not follow any mathematical function exactly. However, it may be possible to find a member of the conjugate family that is a "good fit" to the assessed distribution. Thus, the concept of conjugate distributions provides a convenient model that may be realistic in many situations. If the family of conjugate distributions is fairly "rich," including a variety of distributions, then there is a good chance that the prior distribution can be approximated reasonably well by a member of the conjugate family.

It is possible for the statistician to consider the conjugate family before assessing a distribution. If it is felt that the prior information is roughly equivalent to sample information from the process or population of interest, then it may be possible to determine the prior parameters directly. If a production manager interested in the proportion of defective items produced by a certain process feels that the prior information is roughly equivalent to a sample of 40 items with 3 defectives, then a beta distribution with $r' = 3$ and $n' = 40$ can be used as the prior distribution.

If it is not feasible to think in terms of "equivalent sample information," another alternative is to consider the mean and variance of the distribution. If the production manager thinks that the prior distribution can be represented by a beta distribution and that the mean and variance are .08 and .0016, respectively, then Equations (8.6.2) and (8.6.3) may be used:

$$\frac{r'}{n'} = .08$$

and
$$\frac{r'(n' - r')}{n'^2(n' + 1)} = (.04)^2 = .0016.$$

Solving these two equations simultaneously, we find that $r' = 3.6$ and $n' = 45$. Notice that this technique is analogous to the method of moments, which we used to determine estimators.

In the normal case, of course, the mean and variance are themselves the prior parameters, so there are no equations to solve. However, the assessor might wish to assess a pair of fractiles and use the tables of the normal distribution to find m' and σ'. For a very simple example, suppose that you are interested in θ, the maximum temperature next July 4 in Chicago. You feel that the .50 fractile, or the median, is 86 and that the .75 fractile is 92. Furthermore, you decide that your distribution of θ is approximately normal. Since the normal distribution is symmetric, the mean and the median are equal, so that the mean is 86. From the table of the cumulative standard normal distribution, the .75 fractile of the standard normal distribution is approximately .67. This implies that the .75 fractile of any normal distribution is .67 standard deviations, or $.67\sigma'$, to the right of the mean. But 92 is 6 units to the right of the mean, 86, so

$$6 = .67\sigma'.$$

Therefore,

$$\sigma' = 8.96.$$

Your distribution for θ is thus a normal distribution with mean 86 and standard deviation 8.96.

There are no ironclad rules to follow in assessing prior distributions, and the procedures discussed in this section provide different ways to approach the assessment problem. The technique used in any specific situation should depend on the form of the prior information. In addition, it should be noted that assessing a prior distribution, like assessing a likelihood, is part of the general model-building process, and the desired "accuracy" depends on the situation at hand. In some cases, the ultimate inferences or decisions do not change much even for large changes in the prior distribution. It is then said that the inferential or decision-making procedure is *insensitive* to variations in the prior distribution. In other instances, of course, slight changes in the prior distribution may cause changes in the inferences or decisions, and the procedure is quite *sensitive* to such changes. A **sensitivity analysis** is often quite useful in an inferential or decision-making problem, for it indicates aspects of the problem requiring great care and other aspects for which a great deal of "accuracy" is not needed. Some research concerning sensitivity analysis has indicated that in a wide variety of situations, inferences and decisions are reasonably insensitive to moderate variations in the prior distribution. This implies that a conjugate prior distribution, while it may not be a perfect fit to a subjectively assessed distribution, will often be a "satisfactory" fit.

8.12   DISCRETE APPROXIMATIONS OF CONTINUOUS PROBABILITY MODELS

As noted in Section 8.5, Equation (8.5.4), Bayes' theorem for continuous probability models, may often be difficult to apply because of potential problems in evaluating the integral in the denominator of the equation. If the prior distribution is a conjugate distribution, this difficulty does not arise. Otherwise, it may be necessary to evaluate the integral numerically, which essentially amounts to approximating the prior density function by a discrete mass function and applying Bayes' theorem for discrete probability models. The approximation is accomplished by dividing the set of possible values of the parameter of interest into a number of intervals and determining the probability of each interval, which is then assumed to be a mass at the midpoint of the interval.

For an illustration, consider a modification of the example presented in Section 8.9 in which a retailer assessed a normal prior distribution for the uncertain quantity μ, mean weekly sales at a particular store. Instead of a normal prior distribution for μ, suppose that the retailer's prior distribution is represented by the bimodal density function shown in Figure 8.12.1. A distribution such as this might be reasonable if some extraneous consideration is expected to affect mean sales. For instance, the retailer's best guess about mean sales might be 1400 provided that consumers expect that Congress will reduce personal income taxes, and his best guess might be 1000 provided that consumers do not expect that taxes will be reduced. The bimodal distribution results because he is uncertain as to whether or not consumers expect that Congress will reduce taxes; the two "halves" of the distribution are equal (the distribution is symmetric), so apparently he feels that the probability that consumers expect a tax reduction is about 1/2.

For a very rough discrete approximation to the prior distribution, consider intervals of width 100 centered at 800, 900, and so on, up to 1600. The prior probabilities of these intervals can be determined by integrating

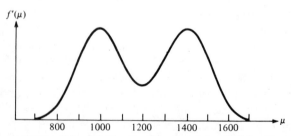

Figure 8.12.1   Bimodal Density Function for μ in Sales Example

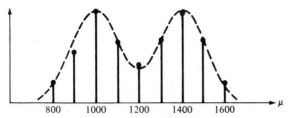

Figure 8.12.2   Discrete   Approximation   for   Bimodal   Density
Function in Sales Example

the prior density function over each interval. The functional form of the
prior density function has not been given, however. Very often the prior
distribution is presented only in graphical form, and the functional form
of the density function is not known. If this is the case, the probabilities of
the chosen intervals can be determined approximately by drawing the
prior density function on graph paper and by counting the number of
"squares" under the curve in each interval. The resulting discrete approximation for the example is given in Table 8.12.1 (the mass function
is graphed in Figure 8.12.2 together with the prior density function that
is being approximated, with the vertical scales adjusted to facilitate comparisons of the two functions).

Table 8.12.1   Discrete   Approximation   for   Bimodal   Density
Function for μ in Sales Example

Interval	Midpoint	Prior probability
750–850	800	.03
850–950	900	.12
950–1050	1000	.19
1050–1150	1100	.12
1150–1250	1200	.08
1250–1350	1300	.12
1350–1450	1400	.19
1450–1550	1500	.12
1550–1650	1600	.03

In order to apply Bayes' theorem to this discrete approximation, the
likelihoods must be determined. But as in the normal example in Section
8.4, the likelihoods are simply heights of normal density functions of M,
given different values of μ, evaluated at the observed sample mean. Suppose that in this case the sample mean for a sample of size 27 turns out to
be 1100. Using Equation (8.4.1) and Table VII in the Appendix, the likeli-

hood for $\mu = 1000$ is

$$f(1100 \mid 1000, 3333) = f\left(\frac{1100 - 1000}{\sqrt{3333}}\middle| 0, 1\right)\middle/ \sqrt{3333}$$

$$= \frac{f(1.73 \mid 0, 1)}{57.7} = \frac{.0893}{57.7}$$

$$= .001548,$$

where 3333 is the variance of M, $\sigma^2/n = 90,000/27$. The entire set of likelihoods for the example is as follows:

$$f(1100 \mid 800, 3333) = f(5.20 \mid 0, 1)/57.7 = 0,$$

$$f(1100 \mid 900, 3333) = f(3.47 \mid 0, 1)/57.7 = .000017,$$

$$f(1100 \mid 1000, 3333) = f(1.73 \mid 0, 1)/57.7 = .001548,$$

$$f(1100 \mid 1100, 3333) = f(0 \mid 0, 1)/57.7 = .006913,$$

$$f(1100 \mid 1200, 3333) = f(-1.73 \mid 0, 1)/57.7 = .001548,$$

$$f(1100 \mid 1300, 3333) = f(-3.47 \mid 0, 1)/57.7 = .000017,$$

$$f(1100 \mid 1400, 3333) = f(-5.20 \mid 0, 1)/57.7 = 0,$$

$$f(1100 \mid 1500, 3333) = f(-6.93 \mid 0, 1)/57\ 7 = 0,$$

and $$f(1100 \mid 1600, 3333) = f(-8.67 \mid 0, 1)/57.7 = 0.$$

The values given as 0 are not *exactly* 0, but they are 0 for all practical purposes. Bayes' theorem for discrete probability models can now be applied; see Table 8.12.2.

Table 8.12.2   Bayes' Theorem for Discrete Approximation in Sales Example

μ	Prior probability	Likelihood	(Prior probability) × (likelihood)	Posterior probability
800	.03	0	0	0
900	.12	.000017	.00000204	.002
1000	.19	.001548	.00029412	.235
1100	.12	.006912	.00082944	.662
1200	.08	.001548	.00012384	.099
1300	.12	.000017	.00000204	.002
1400	.19	0	0	0
1500	.12	0	0	0
1600	.03	0	0	0
	1.00		.00125148	1.000

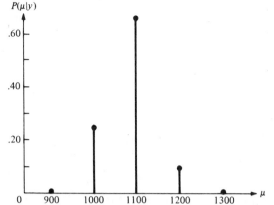

Figure 8.12.3   Posterior Distribution for μ from Discrete Approximation in Sales Example

The posterior distribution is graphed in Figure 8.12.3. Of course, the accuracy of the discrete approximation can be increased by taking a larger number of intervals so that the intervals are narrower. This example illustrates the use of discrete approximations when the prior distribution is not a member of the relevant conjugate family of distributions. Discrete approximations may also be quite useful when the data-generating process cannot be satisfactorily represented by a well-known statistical model, in which case a conjugate distribution cannot be found. Moreover, with the aid of a computer it is convenient to use discrete approximations and to investigate the sensitivity of the results to such things as changes in the choice of intervals.

8.13   REPRESENTING A DIFFUSE PRIOR STATE

Suppose that a decision maker wants to assess a prior distribution in a situation in which there is very little or no prior information. More specifically, the prior information is such that it is "overwhelmed" by the sample information. Then it is said that the decision maker has a **diffuse** state of prior information. The situation described is not necessarily an informationless state in the usual meaning of the word; it is informationless in a relative sense. When we say that someone's prior distribution is diffuse, we mean only that it is *diffuse relative to the sample information.*

A good example of diffuseness concerns the determination of the weight of a potato. If you were given a potato and asked to assess a distribution of its weight, you would clearly have some information about the weight. On the other hand, your information would probably be of a rather vague nature. You could probably specify some limits within which you were

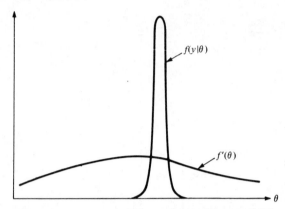

Figure 8.13.1   A Prior Distribution that is Diffuse Relative to the Sample Information

sure the weight would lie, and your distribution might have a peak, or mode, somewhere within these limits. It is doubtful, however, that your distribution would have a sharp "spike" anywhere.

If the potato were weighed on a balance of known precision (say, a standard deviation, or standard error of measurement, of $\frac{1}{2}$ gram), how would your posterior distribution look following a single weighing? Since your distribution is probably quite "spread out," or diffuse relative to the likelihood function, the sample information receives much more weight than the prior distribution. In fact, *the posterior distribution depends almost solely on the sample information.* Notice that the sample size here is only 1, but the precision of the balance is so high that the likelihood function is much more precise than the prior distribution. It should be noted that increased precision in a sample can result from a larger sample size or from more precise measurement.

Graphically, the situation is represented in Figure 8.13.1. Let θ represent the weight of the potato and let y represent the sample result, the weight obtained from one reading of the balance. Relative to the likelihood function, the prior distribution $f'(\theta)$ is "flat." It is not strictly a uniform distribution, but it *is* virtually uniform *relative* to $f(y \mid \theta)$. From Section 8.5, the posterior density function is proportional to the product of the prior density and the likelihood function:

$$f''(\theta \mid y) \; \propto \; f'(\theta)f(y \mid \theta), \qquad\qquad (8.13.1)$$

where the $\propto$ sign is read "is proportional to." The constant of proportionality is the reciprocal of the normalizing integral,

$$\int_{-\infty}^{\infty} f'(\theta)f(y \mid \theta) \; d\theta.$$

But the prior distribution is essentially "flat," so it can be approximated by a constant function, $f'(\theta) = k$. The posterior density is then proportional to the likelihood function:

$$f''(\theta \mid y) \propto f(y \mid \theta). \qquad (8.13.2)$$

The proportionality in Equation (8.13.2) is not a *strict* proportionality because $f'(\theta)$ is not *strictly* uniform. Unless the decision-making problem at hand is quite sensitive to slight variations in the prior distribution, though, this approximation can safely be used.

If the decision maker's prior distribution is diffuse relative to the sample information, it should not make too much difference exactly how the mathematical form of the density function is specified, provided of course that the function chosen is relatively "flat." In order to simplify the process of revising the distribution on the basis of sample information, the decision maker would prefer the distribution to be a member of the conjugate family of distributions. The *exact* form of the prior distribution is not too important because of the diffuseness, so the statistician might as well ease the computational burden by selecting a diffuse conjugate distribution.

If we are sampling from a Bernoulli process, the beta family of distributions is conjugate, and we will take the improper beta distribution with $r' = 0$ and $n' = 0$ to represent diffuseness concerning p, the parameter of the Bernoulli process. This distribution, which is of the form $f'(p) = p^{-1}(1 - p)^{-1}$, is improper because the requirement that $n' > r' > 0$ is not satisfied and the area under the curve is not equal to 1 (the integral representing this area does not converge). Nevertheless, unless the sample consists of all successes or all failures, the posterior distribution with parameters $r'' = r$ and $n'' = n$ *will* be proper. Incidentally, an alternative choice for a diffuse beta prior is a beta distribution with parameters $r' = 1$ and $n' = 2$; this distribution is strictly uniform on the unit interval.

If we are sampling from a normal process with known variance, the normal family of distributions is conjugate, and we will take the improper normal distribution with $n' = 0$ (implying infinite prior variance) to represent diffuseness concerning μ. Again, although this distribution is improper, the posterior distribution determined through Equations (8.8.6) and (8.8.7) will be a proper normal distribution, with $m'' = m$, $n'' = n$, and $\sigma''^2 = \sigma^2/n$.

8.14   DIFFUSE PRIOR DISTRIBUTIONS AND SCIENTIFIC REPORTING

When a diffuse prior distribution is used in a Bayesian analysis, the posterior distribution is approximately proportional to the likelihood function, as pointed out in Section 8.13. Thus, any inferences and/or decisions based on the posterior distribution will in reality depend almost solely on the sample information, as summarized by the likelihood function. But a classical statistician bases inferences and decisions solely on the sample information. Under a diffuse prior state, then, Bayesian and classical statistical procedures are essentially based on the same set of information.

As you shall see, under a diffuse prior distribution Bayesian techniques often result in *numerical* results identical to classical results. However, there is still an important difference in *interpretation*. The Bayesian thinks of θ as a random variable and is willing to make probability statements concerning θ. The prior and posterior distributions are probability distributions of θ. The classical statistician, on the other hand, claims that θ is a fixed parameter, not a random variable, and that it makes no sense to talk of the probability of values of θ occurring. We shall explore the resulting differences in interpretations of inferential statements in the next few sections.

Diffuse distributions are valuable to the Bayesian for several reasons. Their first and most obvious use occurs when there is very little prior information relative to the information contained in the sample. A second, equally important use of diffuse prior distributions relates to scientific reporting. When reporting the results of a statistical analysis, the statistician is faced with this question: What should be reported? One possible answer is to report the posterior distribution and any resulting decisions. The entire posterior distribution itself is much more informative than any summarizations in the form of point or interval estimates. Furthermore, the posterior distribution is informative in the sense that it reflects not only the sample results, but also the prior judgments of the statistician. However, someone else might want to investigate the results of the statistical analysis, to assess a prior distribution (which may be quite different from the statistician's distribution), and to use Bayes' theorem to revise the distribution. Unless the sample results are reported separately (that is, other than as a part of the posterior distribution), others will not be able to conduct their own analyses.

One possible solution to the problem is for the statistician to carry out a separate analysis using a diffuse prior distribution. By removing the effect of the statistician's own prior judgments, this analysis will enable others to include their own prior judgments and to proceed accordingly. Also, it will show approximately what results would have been obtained using classical techniques. Alternatively, the statistician could report the likelihood function, since all of the evidence of the sample is included in the likelihood function. Bear in mind that the statistician's inferences and decisions should be based on the statistician's *own* posterior distribution.

For the benefit of others, however, it is useful to present the results that follow from the use of a diffuse prior distribution. Such results may be of some interest even to the statistician, who can then investigate the influence of the prior distribution on the results.

This discussion of diffuse prior distributions with respect to scientific reporting holds only if there is substantial agreement on the form of the likelihood function. That is, there must be substantial agreement on the use of a particular statistical model to represent the data-generating process (in some situations, of course, no well-known statistical model is appropriate). The choice of such a model is ultimately subjective, as emphasized in Section 8.4, so the likelihoods determined from the model are ultimately subjective in nature. To the extent that different people agree about the choice of a statistical model, then, they also agree about the likelihoods, whereas they are less likely to agree about prior probabilities. Thus, the reporting of results based on a diffuse prior distribution may be of some general interest.

8.15 THE POSTERIOR DISTRIBUTION AND ESTIMATION

With respect to inference, the Bayesian statistician considers the entire posterior distribution (or any probability determined from this distribution) as an inferential statement, and Bayes' theorem is the basic inferential mechanism. *If a point estimate is desired, it will of course be based on the posterior distribution rather than just on the sample information.* Since properties such as unbiasedness and efficiency involve only the sample information, they are not of much interest to the Bayesian, although in a rough sense, they do have Bayesian "counterparts." For instance, a Bayesian counterpart of efficiency might be the desire to have a small posterior variance. Some potential estimators based on the posterior distribution are the posterior mean, the posterior median, the posterior mode, and so on. If the posterior distribution of μ, the mean of a normal process, is a normal distribution, then the posterior mean m'' is the obvious choice as an estimate of μ. If the posterior distribution of p, the parameter of a Bernoulli process, is a beta distribution, then the posterior mean r''/n'' is an estimate of p, as is the posterior mode, which is equal to $(r'' - 1)/(n'' - 2)$ provided that $r'' > 1$ and $n'' - r'' > 1$.

If the prior distribution is diffuse, estimates based on the posterior distribution may be related to classical estimates. In sampling from a normal process with a diffuse normal prior distribution with $n' = 0$, m'' is equal to m, the usual classical estimate of μ. In sampling from a Bernoulli process with fixed sample size n, the usual classical estimate of p is r/n, which is equal to the posterior mean r''/n'' under a diffuse beta prior distribution with $r' = n' = 0$. On the other hand, if the sampling is done with a fixed r rather than a fixed n, a frequently-used classical estimate is $(r - 1)/(n - 1)$, whereas under the Bayesian approach the posterior distribution is identical under binomial sampling and Pascal sampling, as noted in Section 8.6.

One important method for determining "good" classical estimators is the method of maximum likelihood, discussed in Chapter 6. This method finds the value of θ that maximizes the likelihood function. But under a diffuse prior distribution, the posterior distribution is approximately proportional to the likelihood function, so the maximum-likelihood estimate is approximately equal to the posterior mode. The interpretation differs in the two approaches, however. The mode of the posterior distribution is "the most likely value of θ," whereas the maximum-likelihood estimate is "the value of θ that makes the *observed sample results* appear most likely." The Bayesian makes a probability statement about θ, while the classical statistician makes a probability statement concerning the sample results.

How about interval estimation? From the posterior distribution, the probability of any interval of values of θ can be determined. To find an interval containing 90 percent of the probability, we could consider the interval from the .05 fractile to the .95 fractile, or the interval from the .01 fractile to the .91 fractile, and so on. The conventional choice of an interval containing $100(1 - \alpha)$ percent of the probability is the interval from the $\alpha/2$ fractile to the $1 - (\alpha/2)$ fractile. Under a diffuse distribution, this corresponds to the classical notion of a $100(1 - \alpha)$ percent confidence interval.

For a normal process with unknown mean μ and known variance σ^2, the usual classical 95 percent confidence limits for μ are

$$m - 1.96\sigma/\sqrt{n} \quad \text{and} \quad m + 1.96\sigma/\sqrt{n}.$$

Under a normal posterior distribution for μ with mean m'' and variance $\sigma''^2 = \sigma^2/n''$, the limits of the corresponding Bayesian interval are

$$m'' - 1.96\sigma/\sqrt{n''} \quad \text{and} \quad m'' + 1.96\sigma/\sqrt{n''}.$$

Under a diffuse normal prior distribution with $n' = 0$, we have $m'' = m$ and $n'' = n$, and thus the Bayesian limits are identical to the classical limits. Once again, the interpretation is different. The Bayesian would say that "the probability is .95 that μ lies within the specified interval," and the classical statistician would say that "in the long run, 95 percent of all such intervals will contain the true value of μ."

The classical statement is based on long–run frequency considerations, and the Bayesian statement is a statement which concerns not the long run, but the specific case at hand. To differentiate between the two, we call the classical interval a confidence interval and the Bayesian interval a credible interval.

(Incidentally, some followers of the frequency school of thought do interpret some interval estimation statements as probability statements about parameters, calling such probabilities "fiducial" probabilities. We will not concern ourselves with the notion of "fiducial" probabilities, however.)

If the prior distribution is not diffuse, point estimates and interval estimates based on the posterior distribution may differ from similar estimates based on the likelihood function. In the production process example of Section 8.2, the maximum-likelihood estimate of p is $r/n = 3/10 = .30$, and the posterior mean turns out to be .146. In the modification of the same example, presented in Section 8.7, the maximum-likelihood estimate is .30 and the posterior mean is $r''/n'' = 4/30 = .133$. In the example concerning the weekly sales of a store (Section 8.9), the maximum-likelihood estimate, the sample mean, equals 1240. The mean of the posterior distribution is 1225. A 95 percent *credible* interval based on the posterior distribution is

$$(m'' - 1.96\sigma'', m'' + 1.96\sigma''),$$

or $$(1165, 1285).$$

A 95 percent *confidence* interval based on the sample information alone is

$$\left(m - \frac{1.96\sigma}{\sqrt{n}}, m + \frac{1.96\sigma}{\sqrt{n}}\right),$$

or $$(1164, 1316).$$

These examples demonstrate that the inclusion of prior information can have quite an effect on point and interval estimates. In the next chapter we will look at estimation in yet another light: estimation as a decision-making procedure. To do this it will be necessary to include not only prior information, but also information concerning the potential "losses" due to errors in estimation. Before discussing decision theory, let us consider other inferential procedures based on the posterior distribution.

8.16   PRIOR AND POSTERIOR ODDS RATIOS

Instead of estimating a certain parameter, a statistician may wish to test a hypothesis concerning that parameter. In Chapter 7, we presented the classical theory of hypothesis testing. If hypothesis testing is thought of as a choice between two actions, accepting or rejecting a given hypothesis, then it is a decision-making procedure. As we pointed out in Chapter 7, a formal decision-theoretic approach to hypothesis testing necessitates the consideration of losses due to erroneous decisions. The decision to accept or reject a hypothesis should depend not only on the sample information, but also on the losses. The formal decision-theoretic approach will be presented in the next chapter. In this and the following sections, we will treat hypothesis testing on a more informal level.

Suppose that we are interested in the proportion of votes that a politician will receive in a forthcoming election. Let us call this proportion p. The politician's advisors think that there are two possible values for p: .40 and .60. Furthermore, the prior probabilities of these two values are

$$P'(.40) = .25$$

and
$$P'(.60) = .75.$$

A sample of 40 voters is taken, and 18 of them claim that they will vote for this politician. Table 8.16.1 shows the computation of the posterior probabilities (the likelihoods are based on the assumption that the process is a Bernoulli process).

Table 8.16.1 Bayes' Theorem for Voting Example

Value of p	Prior probability	Likelihood	(Prior probabilty) $\times$ (likelihood)	Posterior probability
.40	.25	.10255	.02564	.628
.60	.75	.02026	.01519	.372
	1.00		.04083	1.000

Let the two possible values of p correspond to two hypotheses,

$$H_0: p = .40$$

and
$$H_1: p = .60.$$

Then the **prior odds ratio** of H_0 to H_1 is defined as the ratio of the prior probabilities,

$$\Omega' = \frac{P'(H_0)}{P'(H_1)} = \frac{.25}{.75} = \frac{1}{3}. \qquad (8.16.1^*)$$

If this ratio is greater than 1, then H_0 is more likely than H_1; if it is less than

1, H_1 is more likely than H_0; and if it is equal to 1, H_0 and H_1 are equally likely. The same holds for the **posterior odds ratio** of H_0 to H_1,

$$\Omega'' = \frac{P''(H_0 \mid y)}{P''(H_1 \mid y)} = \frac{.628}{.372} = 1.69. \qquad (8.16.2^*)$$

Also, from Section 7.6, we know that the **likelihood ratio** is of the form

$$\Omega = \frac{P(y \mid H_0)}{P(y \mid H_1)} = \frac{.10255}{.02026} = 5.06. \qquad (8.16.3^*)$$

But, from Bayes' theorem, the posterior probability of a hypothesis is proportional to the product of the prior probability of that hypothesis and the likelihood associated with that hypothesis. Therefore, the ratio of two such posterior probabilities must equal the ratio of the prior probabilities times the ratio of the likelihoods. Thus, we can now write Bayes' theorem in terms of odds:

$$\Omega'' = \Omega'\Omega. \qquad (8.16.4^*)$$

In words, the posterior odds ratio is equal to the product of the prior odds ratio and the likelihood ratio.

You can verify for yourself that this is true for the example.

For another example, suppose that the occurrence of accidents along a particular stretch of highway is assumed to follow roughly a Poisson process, and you feel that the intensity of accidents, λ, is either 2 per week or 3 per week. Your prior judgments are such that you feel that the odds in favor of $\lambda = 3$ are 2 to 1. Thus, for $H_0: \lambda = 2$ and $H_1: \lambda = 3$, the prior odds ratio is

$$\Omega' = \frac{P'(\lambda = 2)}{P'(\lambda = 3)} = \frac{1}{2}.$$

You then obtain sample information in the form of the accident records for the highway for a 6-week period. In this 6-week period, 17 accidents occurred, so the likelihood ratio is

$$\Omega = \frac{P[R = 17 \mid \lambda t = 2(6) = 12]}{P[R = 17 \mid \lambda t = 3(6) = 18]} = \frac{.0383}{.0936} = .4092.$$

Thus, the posterior odds ratio is

$$\Omega'' = \Omega'\Omega = (.5)(.4092) = .2046.$$

This implies that the odds in favor of $\lambda = 2$ are .2046 to 1, so the odds in favor of $\lambda = 3$ are 1 to .2046, or 4.9 to 1. The sample information increased the odds in favor of $\lambda = 3$ from 2 to 1 to 4.9 to 1.

For a third example, consider the weight of items produced by a certain manufacturing process. Suppose that this weight is normally distributed with mean μ and variance $\sigma^2 = 4$; furthermore, it is assumed that the mean weight μ is either 9 or 10 (perhaps it is 9 if the process is "in control," but it is 10 if the process is "out of control"). The prior probabilities are $P(\mu = 9) = .85$ and $P(\mu = 10) = .15$, for the process tends to be out of control about 15 percent of the time. Thus, for $H_0: \mu = 9$ and $H_1: \mu = 10$, the prior odds ratio is

$$\Omega' = \frac{P'(\mu = 9)}{P'(\mu = 10)} = \frac{.85}{.15} = 5.67.$$

A sample of size 25 is taken from the process, and the mean weight of the 25 items is 9.2. The likelihood ratio is a ratio of two normal density values:

$$\Omega = \frac{f(9.2 \mid \mu = 9, \sigma^2/n = .16)}{f(9.2 \mid \mu = 10, \sigma^2/n = .16)}.$$

Using Equation (8.4.1) and Table VII in the Appendix,

$$\Omega = \frac{f\left(\dfrac{9.2 - 9}{.4} \middle| 0, 1\right) \middle/ .4}{f\left(\dfrac{9.2 - 10}{.4} \middle| 0, 1\right) \middle/ .4} = \frac{f(.5 \mid 0, 1)}{f(-2.0 \mid 0, 1)}$$

$$= \frac{.3521}{.0540} = 6.52.$$

From Equation (8.16.4),

$$\Omega'' = \Omega'\Omega = 5.67(6.52) = 36.97.$$

These examples illustrate the use of Equation (8.16.4) to determine a posterior odds ratio. Of course, the examples are somewhat artificial in the sense that the hypotheses are exact. For inexact hypotheses, the direct application of Equation (8.16.4) poses some problems in terms of the likelihood ratio. When the hypotheses are *exact*, the likelihoods are simply probabilities or densities, such as $P(R = 17 \mid \lambda t = 12)$ or $f(9.2 \mid \mu = 9, \sigma^2/n = .16)$. On the other hand, when the hypotheses are *inexact*, the likelihoods of the form $P(y \mid H_0)$ or $f(y \mid H_0)$ are conditioned on *intervals* of values rather than single values, leading to expressions such as $P(R = 17 \mid \lambda t > 12)$ or $f(9.2 \mid \mu \leq 9.4, \sigma^2/n = .16)$. As a result, for inexact hypotheses we will not determine the posterior odds ratio via Equation (8.16.4). Instead, we will first compute the posterior distribution of the parameter and *then* calculate the posterior probabilities of the two hypotheses in order to form Ω''.

In the weight example presented earlier in this section, for instance, μ could be assumed to be continuous. Suppose that a normal prior distribution for μ is assessed, Equations (8.8.1) and (8.8.2) are used to revise the distribution, and the resulting posterior distribution of μ turns out to be a normal distribution with mean $m'' = 9.2$ and variance $\sigma''^2 = .0225$. If the hypotheses

$$H_0: \mu \geq 9.4$$

and
$$H_1: \mu < 9.4$$

are of interest, then

$$\Omega'' = \frac{P''(\mu \geq 9.4)}{P''(\mu < 9.4)} = \frac{P\left(Z \geq \dfrac{9.4 - 9.2}{.15}\right)}{P\left(Z < \dfrac{9.4 - 9.2}{.15}\right)} = \frac{P(Z \geq 1.33)}{P(Z < 1.33)} = \frac{.092}{.908} = .101.$$

Similarly, in the election example presented earlier in this section, it would be more realistic for the politician to assume that p is continuous. The hypotheses of interest might be

$$H_0: p \leq .5$$

and
$$H_1: p > .5.$$

Suppose that the prior distribution for p is a beta distribution with $r' = 3$ and $n' = 6$ and that the sample information consists of 10 voters, 3 of whom claim that they will vote for the politician in question. Thus, the posterior distribution is a beta distribution with $r'' = r' + r = 3 + 3 = 6$ and $n'' = n' + n = 6 + 10 = 16$. The posterior probability of H_0 can be calculated by using Equation (8.6.4) and Table V:

$$P''(H_0) = P''(p \leq .5 \mid r'' = 6, n'' = 16) = P_{\text{bin}}(R \geq 6 \mid n = 15, p = .5)$$

$$= \sum_{r=6}^{15} P(R = r \mid n = 15, p = .5) = .8491.$$

$P''(H_1)$ is simply $1 - P''(H_0)$, so the posterior odds ratio is

$$\Omega'' = \frac{P''(H_0)}{P''(H_1)} = \frac{.8491}{.1509} = 5.63.$$

In this section we have discussed the determination of posterior odds ratios. Next, we will relate these ratios to hypothesis testing and compare briefly the classical and Bayesian approaches to hypothesis testing.

8.17 THE POSTERIOR DISTRIBUTION AND HYPOTHESIS TESTING

As we saw in Section 7.6, the likelihood ratio plays an important role in the development of classical hypothesis-testing procedures. The logical extension of this approach in Bayesian terms is to multiply the likelihood ratio by a prior odds ratio to arrive at a posterior odds ratio. Alternatively, of course, the Bayesian can simply compute a posterior distribution and calculate a posterior odds ratio from that distribution, as we did for inexact hypotheses in Section 8.16.

The classical statistician rejects H_0 for small values of the likelihood ratio [Equation (7.6.1)]. It is clear that small values of Ω'' tend to favor H_1 and large values of Ω'' tend to favor H_0.

In general, the Bayesian uses rejection regions of the form

$$\Omega'' \leq k, \tag{8.17.1}$$

where k is some critical value.

Of course, if the prior odds ratio happens to be equal to 1, then $\Omega'' = \Omega$, and Equation (8.17.1) is equivalent to Equation (7.6.1). A prior odds ratio of 1 might be thought of as a diffuse prior state of information, since neither hypothesis is favored. If the prior odds ratio differs from 1, the posterior odds ratio differs from the likelihood ratio, and hence Bayesian and classical inferences will not necessarily be the same numerically (in terms of interpretation, they will never be the same).

The choice of the critical value k depends on the losses associated with the two possible errors. Such losses will be introduced in Chapter 9, so we will postpone further discussion of k until then. In the absence of any information about the relative seriousness of the two types of error, we might simply let $k = 1$, in which case H_0 will be accepted if it is more likely than H_1 and rejected otherwise.

In order to compare the classical and Bayesian approaches in further detail, we shall consider only one-tailed tests (two-tailed tests will be discussed in the next section). In the example considered in Section 8.9, a retailer is interested in weekly sales at a particular store, where $\sigma^2 = 90,000$. The prior distribution of μ, mean weekly sales, is a normal distribution with $m' = 1200$ and $\sigma'^2 = 2500$. A sample of 60 weeks with sample mean $m = 1240$ is observed, and the posterior distribution is a normal distribution with mean $m'' = 1225$ and a variance of $\sigma''^2 = 937.5$. For the details of the determination of the posterior distribution, see Section 8.9.

Suppose that the retailer is interested in the hypotheses

$$H_0: \mu \geq 1250$$

and

$$H_1: \mu < 1250.$$

From the posterior distribution,

$$P''(H_0) = P''(\mu \geq 1250) = P\left(Z \geq \frac{1250 - 1225}{\sqrt{937.5}}\right)$$

$$= P(Z \geq .82) = .21$$

and $$P''(H_1) = P''(\mu < 1250) = 1 - P''(\mu \geq 1250)$$

$$= 1 - .21 = .79.$$

Next, we will consider the retailer's problem from a classical standpoint. From the sample information, the p-value is

$$P\left(M \leq 1240 \mid \mu = 1250, \frac{\sigma^2}{n} = 1500\right) = P\left(Z \leq \frac{1240 - 1250}{\sqrt{1500}}\right)$$

$$= P(Z \leq -.26) = .40.$$

This value, which is computed from the sampling distribution of M, indicates how "unusual" the sample results are if H_0 is true. The p-value bears no relation to $P''(H_0)$ or $P''(H_1)$, but of course the prior distribution is not diffuse. If we take a diffuse prior distribution (an improper normal distribution with $n' = 0$), the posterior distribution of μ is a normal distribution with $m'' = m = 1240$ and variance $\sigma''^2 = \sigma^2/n = 1500$. Using this posterior distribution,

$$P''(H_0) = P''(\mu \geq 1250) = P\left(Z \geq \frac{1250 - 1240}{\sqrt{1500}}\right)$$

$$= P(Z \geq .26) = .40.$$

For this example, then, the p-value is equal to $P''(H_0)$ under a diffuse prior distribution. However, although the *results* are identical numerically,

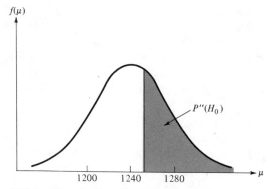

Figure 8.17.1 Posterior Distribution Following Diffuse Prior Distribution in Sales Example

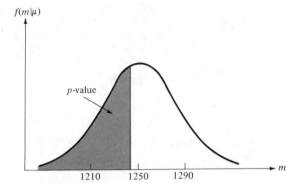

Figure 8.17.2   Sampling Distribution of M given that $\mu = 1250$ in Sales Example

the *interpretation* is not. The difference is demonstrated in Figures 8.17.1 and 8.17.2. The curve centered about $\mu = 1240$ (Figure 8.17.1) is the posterior distribution following a diffuse prior distribution, and the shaded area represents $P''(H_0)$. The density function centered about $M = 1250$ (Figure 8.17.2) is the sampling distribution of M given that $\mu = 1250$, and the shaded area represents the p-value. After the sample is observed, the classical statistician is looking at the observed value of M and comparing it with other values that might have occurred but did not occur to get some idea of how "unusual" the result seems. The probability statement is in terms of the sample mean, not in terms of μ. The Bayesian, on the other hand, is concerned only with the observed value of M and not with other values that did not occur, and the probability statement is in terms of μ, the parameter of interest.

If the prior distribution is diffuse, the results obtained from the classical approach in a one-tailed test are similar to the results obtained using the posterior distribution. In some instances the results are identical numerically, as in the above example, although the interpretations differ. In the next section we will consider problems that arise for two-tailed tests.

8.18   THE POSTERIOR DISTRIBUTION AND TWO-TAILED TESTS

The preceding section pertained to one-tailed tests, and we will now consider the Bayesian approach to two-tailed tests. In the example involving the retailer, suppose that the hypotheses of interest are

$$H_0: \mu = 1250$$

and

$$H_1: \mu \neq 1250.$$

From the posterior distribution,

$$P''(H_0) = P''(\mu = 1250) = P(Z = .82) = 0$$

and $\qquad P''(H_1) = P''(\mu \neq 1250) = P(Z \neq .82) = 1.$

At first glance, it appears that H_0 is essentially impossible. But even if $m'' = 1250$ (certainly no other value of m'' would make H_0 appear more likely), it turns out that $P''(H_0) = P(Z = 0) = 0$. Since the hypothesis H_0 consists of just a single point and the posterior distribution of μ is continuous, the probability of H_0 must be zero.

The above result reflects the absurdity of the hypothesis H_0. Under a two-tailed testing situation there usually is a difference between the *statistical* hypotheses and the corresponding *real-world* hypotheses. What the retailer probably means by H_0 is "μ is close to the value 1250." It is doubtful that the intended meaning is "μ is *exactly* equal to the value 1250." Under the classical approach to hypothesis testing, it is convenient to let H_0 be an exact hypothesis, even though such a hypothesis, taken literally, is quite unrealistic.

Expressing H_0 as an exact hypothesis in a two-tailed test leads to some difficulties with respect to results obtained via classical procedures. For instance, for a test of H_0: $\mu = 1250$ versus H_1: $\mu \neq 1250$ in the example involving the retailer, if α is set at .05 the rejection region is of the following form: reject H_0 if M is *not* in the interval from

$$1250 - 1.96 \frac{\sigma}{\sqrt{n}} \qquad \text{to} \qquad 1250 + 1.96 \frac{\sigma}{\sqrt{n}}.$$

But as the sample size increases, this interval becomes narrower and narrower, and the test becomes more sensitive to very small deviations of μ from 1250. This problem, which is caused by the fact that H_0 is unrealistic and the convention of holding α constant, means that the statistician is virtually certain of rejecting H_0 if the sample size is large enough (see Section 7.14).

How would a Bayesian approach a two-tailed testing situation? One alternative would be to place a mass of probability at the exact value specified by H_0. The distribution of μ would then be a "mixture" of distributions, a discrete distribution consisting of a mass at a single point and a continuous distribution representing the remaining probability. Unless the statistician feels that such a distribution accurately reflects the available information, it is better to modify H_0 and H_1 so that they correspond with the real-world hypotheses of interest. For example, the retailer might be interested in

$$H_0\text{: } 1240 \leq \mu \leq 1260$$

and $\qquad H_1\text{: } \mu < 1240 \text{ or } \mu > 1260.$

From the posterior distribution,

$$P''(H_0) = P''(1240 \leq \mu \leq 1260)$$

$$= P\left(\frac{1240 - 1225}{30.6} \leq Z \leq \frac{1260 - 1225}{30.6}\right)$$

$$= P(.49 \leq Z \leq 1.14) = .185,$$

and $$P''(H_1) = 1 - P''(H_0) = .815,$$

so that $$\Omega'' = \frac{.185}{.815} = .227.$$

Hypotheses such as these can be handled within the classical framework, but the tests are not always easy to develop.

By modifying the hypotheses and making them realistic, then, the Bayesian can make inferences from the posterior distribution in a two-tailed testing situation. In fact, the hypothesis H_0 need not be restricted to a single interval. In general, if the entire set of possible values of a parameter θ is partitioned into two mutually exclusive and exhaustive sets A and B, the hypotheses can be written in the form

$$H_0: \theta \in A$$

and $$H_1: \theta \in B.$$

Then the posterior probabilities of interest are

$$P''(H_0) = P''(\theta \in A)$$

and $$P''(H_1) = P''(\theta \in B) = 1 - P''(H_0).$$

These equations provide a general framework for hypothesis testing in terms of the Bayesian approach; it is necessary only to calculate certain posterior probabilities once the posterior distribution has been determined.

EXERCISES

1. What distinguishes the Bayesian approach to statistical inference from the classical approach, and what role is played by Bayes' theorem in Bayesian inference?

2. With respect to Bayes' theorem, what do the prior distribution, the likelihoods, and the posterior distribution represent? Explain the statement, "The terms *prior* and *posterior*, when applied to probabilities, are relative terms, relative to a given sample." Is it possible for a set of probabilities to be both prior *and* posterior probabilities? Explain.

3. It is believed that p, the proportion of consumers (in a large population) who will purchase a certain product, is either .2, .3, .4, or .5. Furthermore, the prior probabilities for these four values are $P(p = .2) = .2$, $P(p = .3) = .3$, $P(p = .4) = .3$, and $P(p = .5) = .2$. A random sample of size 10 is taken from the population, and of the 10 consumers, 3 state that they will buy the product. What are the posterior probabilities? In calculating the posterior probabilities, what assumptions did you need to make?

4. You feel that p, the probability of heads on a toss of a particular coin, is either .4, .5, or .6. Your prior probabilities are $P(.4) = .1$, $P(.5) = .7$, and $P(.6) = .2$. You toss the coin three times and obtain heads once and tails twice. What are the posterior probabilities? If you then toss the coin three *more* times and once again obtain heads once and tails twice, what are the posterior probabilities? Also, compute the posterior probabilities by pooling the two samples and revising the original probabilities just once; compare with your previous answer.

5. In Exercise 4, suppose that you feel that in comparing $p = .6$ and $p = .4$, the odds are 3 to 2 in favor of $p = .6$. Furthermore, in comparing $p = .4$ and $p = .5$, the odds are 5 to 1 in favor of $p = .5$. On the basis of these odds, find your prior distribution for p.

6. A medical researcher feels that p, the probability that a person who is exposed to a certain contagious disease will catch it, is either .10, .12, .14, .16, .18, or .20. The researcher's prior probabilities are $P(p = .10) = .05$, $P(p = .12) = .08$, $P(p = .14) = .13$, $P(p = .16) = .30$, $P(p = .18) = .34$, and $P(p = .20) = .10$. From this prior distribution, find $E(p)$ and $V(p)$. An experiment is then conducted in which 20 persons are exposed to the disease and 2 catch it. On the basis of this new information, find the researcher's posterior distribution for p and find the mean and the variance of this posterior distribution.

7. A veteran public official is up for reelection and is concerned about p, the proportion of the votes cast that will be for him. In the past, he has run for office 20 times; he received 48 percent of the vote 4 times, 50 percent of the vote 4 times, 52 percent of the vote 5 times, 54 percent of the vote 4 times, 56 percent of the vote twice, and 58 percent of the vote once. In the upcoming election, however, his opponent is young, vigorous, and has a large campaign budget. As a result, the veteran official feels that the past frequencies slightly overstate his chances. He feels that this opponent will obtain about 2 percent more of the votes than his past opponents, so that his prior distribution consists of the probabilities $P(p = .46) = 4/20$, $P(p = .48) = 4/20$, $P(p = .50) = 5/20$, and so on. He then conducts a small survey of a randomly selected group of 100 voters, 49 of whom state that they will vote for him and 51 of whom state that they will vote for his young opponent. Find the posterior distribution of p and use this distribution to calculate the expected proportion of votes the veteran public official will obtain.

8. A security analyst feels that daily price changes of a particular security are independent and that the probability that the security's price increases on any given day is either .4, .5, or .6. If it is .4, the security is classified as a

poor investment; if it is .5, the security is classified as an average investment; and if it is .6, the security is classified as a good investment. The security analyst feels that the security is equally likely to be an average investment or a good investment and that it is twice as likely to be a good investment as it is to be a bad investment. New information then becomes available; out of 15 days, the security's price increases 9 times. Find the posterior probability that the security is a good investment, that it is an average investment, and that it is a poor investment.

9. An executive of a paper company is unsure about the proportion of trees in a large forest that are useless to the company. On the basis of previous experience with different forests, it is felt that the probability that 5 percent of the trees are useless is .2, the probability that 10 percent of the trees are useless is .7, and the probability that 15 percent of the trees are useless is .1. It is assumed that the useless trees are more or less randomly distributed throughout the forest. A sample of 20 trees is taken, and when these trees are cut down, it is found that exactly 3 of them are useless to the paper company. What is the executive's posterior distribution for the proportion of trees that are useless?

10. In Exercise 9, suppose that the sample size of 20 was not determined in advance. Instead, trees were cut down until the third useless tree was found, and the third useless tree turned out to be the twentieth tree that was cut down. Find the posterior distribution and compare it with the posterior distribution found in Exercise 9. What general principle is being illustrated here?

11. Explain the statement, "Likelihoods are only relative (to each other), so they can be taken to be the appropriate conditional probabilities $P(y \mid \theta)$ for fixed y and different values of θ or they can be taken to be any positive multiple of these conditional probabilities."

12. In Exercise 6, find the researcher's posterior distribution for p if the experiment includes 1000 people and 118 of them catch the disease. [Hint: Use an approximation to determine the likelihoods.]

13. In Exercise 9, suppose that the executive feels a priori that the three possible proportions that are being considered are equally likely. A sample of 20 trees yields 3 useless trees, and a second sample of 20 trees yields no useless trees.
    (a) Find the posterior distribution following the first sample.
    (b) Revise the distribution found in (a) on the basis of the second sample.
    (c) Find the posterior distribution following the combined sample of 40 trees without using the distribution found in (a).

14. Accidents along a particular stretch of highway occur roughly according to a Poisson process, and the intensity of the process is either 2, 3, or 4 accidents per week. A state police official's prior probabilities for these three possible intensities are .25, .45, and .30, respectively. If the official observes the highway for a period of 3 weeks and if 10 accidents occur, what are the posterior probabilities?

15. In Exercise 37, Chapter 4, suppose that the switchboard operator is unsure whether the intensity of incoming calls is 10, 11, 12, or 13 per hour. The operator feels that 12 is the most likely value and that 12 is twice as likely as 13, twice as likely as 11, and 5 times as likely as 10. Find the prior distribution for λ. On the basis of a sample of 30 minutes with 7 calls, revise this distribution of λ by using Bayes' theorem.

16. The owner of a service station is concerned about the rate at which the mechanics at the station provide service for the customers. The owner feels that all of the mechanics work at about the same speed, which is either 2, 2.5, or 3 cars per hour. Furthermore, 2.5 cars per hour is twice as likely as each of the other two values, which are assumed to be equally likely. In order to obtain more information, the owner observes two mechanics for a 3-hour period, noting that the first mechanic services 7 cars during that period and the second mechanic services 6 cars. Assuming that the servicing process follows the Poisson model, find the posterior distribution of the rate at which the mechanics provide service.

17. Given the limitations of the Poisson table in the Appendix, how could you use this table to determine the posterior distribution in Exercise 16 if the sample consisted of a 40-hour period during which the first mechanic serviced 82 cars and the second mechanic serviced 87 cars?

18. The owner of the automobile dealership discussed in Section 8.4 has just hired another new salesman. Little is known about the new salesman, so it is felt that the past data (10 percent "great," 40 percent "good," and 50 percent "poor") accurately reflect the owner's judgments about him. In the first 12 days on the job, the new salesman sells 5 cars. What is the posterior distribution? The owner decides to keep the new salesman on the payroll if the expected number of cars sold per day (according to the posterior distribution) is at least .32. Should the new salesman be kept on the payroll?

19. In Exercise 36, Chapter 4, suppose that after seeing a sample in which 8 cars arrive at the toll booth in a 2-minute period, the posterior distribution for λ, intensity per minute, is $P(\lambda = 4) = .40$, $P(\lambda = 5) = .50$, and $P(\lambda = 6) = .10$. What was the prior distribution for λ before seeing this sample?

20. Let p represent the proportion (rounded to the nearest tenth) of students at Stanford University who consumed at least one glass of beer during the past week. Thus, the possible values of p are taken to be 0, .1, .2, .3, .4, .5, .6, .7, .8, .9, and 1.0. Assess your subjective probability distribution for p. If you take a random sample of 10 Stanford students and find that 3 of them consumed at least one glass of beer during the past week, use this sample to revise your distribution.

21. In Exercise 14, choose a stretch of highway with which you are familiar and assess your subjective probability distribution for λ, the intensity of accidents per week, assuming that λ takes on only integer values.

22. In Exercise 7, consider an upcoming election about which you have some knowledge and assess your subjective probability distribution for p, the proportion of votes cast that a particular candidate will obtain. Assume that the possible values of p are given to two decimals and that the last digit must be even.

23. Let λ represent the number of planes per minute landing at O'Hare Field in Chicago during the hours of 5 P.M. to 7 P.M. on a Friday night next June. Assuming for simplicity that λ can only take on integer values, assess your prior distribution for λ. If you are then told that in a randomly chosen 30-second period during the above-stated hours, exactly 4 planes landed, revise your distribution for λ intuitively without using any formulas. Then, assuming a Poisson process, revise your distribution formally according to Bayes' theorem. Explain any differences in the two revised distributions. What are the possible implications of these differences with respect to this example and with respect to Bayesian inference in general?

24. Why are models such as the Bernoulli and the Poisson models useful with regard to the assessment of likelihoods? What are the advantages and disadvantages of using such models?

25. A production manager who is interested in the mean weight of items turned out by a particular process feels that the weight of items from the process is normally distributed with mean μ and that μ is either 109.4, 109.7, 110.0, 110.3, or 110.6. The production manager assesses prior probabilities of $P(\mu = 109.4) = .05$, $P(\mu = 109.7) = .20$, $P(\mu = 110.0) = .50$, $P(\mu = 110.3) = .20$, and $P(\mu = 110.6) = .05$. From past experience, the process variance is assumed to be $\sigma^2 = 4$. He randomly selects 5 items from the process and weighs them, with the following results: 108, 109, 107.4, 109.6, and 112. Find the production manager's posterior distribution and compute the means and the variances of the prior and posterior distributions.

26. A judge is presiding over an antitrust case involving the effect of a particular merger on competition in the brewing industry. Let $\theta = 2$ if the merger has a severe effect on competition, $\theta = 1$ if it has a minor effect, and $\theta = 0$ if it has no effect. The judge feels that $\theta = 0$ and $\theta = 1$ are equally likely and that each is 3 times as likely as $\theta = 2$. An economist is called in as a witness to provide additional information, and the gist of the economist's testimony is that the merger is not harmful to competition in the brewing industry. The judge considers this to be important evidence, but it is not accepted completely at face value because the economist is being retained by the main firm involved in the merger. In particular, the judge feels that the testimony given by the economist is most likely if, in fact, $\theta = 0$ and least likely if, in fact, $\theta = 2$. Furthermore, the new information is viewed as 4 times as likely if $\theta = 0$ than if $\theta = 2$ and twice as likely if $\theta = 0$ than if $\theta = 1$. After the economist testifies, what is the judge's probability distribution for θ?

27. In Exercise 26, a second economist retained by the main firm involved in the merger testifies that the merger is not harmful to the brewing industry. The judge feels that this new testimony is twice as likely if $\theta = 0$ than if $\theta = 2$

and 1.5 times as likely if $\theta = 0$ than if $\theta = 1$. After the second economist testifies, what is the judge's probability distribution for θ? Compare the likelihoods in Exercise 26 with the likelihoods in this exercise, and comment on any differences. Might the likelihoods assessed by the judge for the second economist's testimony be different if the second economist were not being retained by any of the firms involved in the merger or if the *first* economist had stated that the merger is harmful to the brewing industry? Explain your answer.

28. In Bayes' theorem, why is it necessary to divide by $\sum$ (prior probability) $\times$ (likelihood) in the discrete case and by $\int$ (prior density) $\times$ (likelihood) in the continuous case?

29. In Exercise 4, assume that the coin is tossed just once. Apply Bayes' theorem given that the toss results in heads, and apply Bayes' theorem given that the toss results in tails. In each case, consider the denominator of Bayes' theorem, the sum of the products of the prior probabilities and the respective likelihoods. What interpretation can be given to these sums? Explain.

30. In Exercise 18, suppose that you are just given the prior probabilities and that the salesman's actual performance is not known yet. What is the owner's probability distribution for the number of cars the salesman will sell in the first 2 days on the job? [*Hint*: Since λ is not known, this distribution cannot be conditional on a particular value of λ; it is a *marginal* distribution.]

31. In the market-share example in Section 8.5, find the posterior distribution if the sample consists of one consumer who does not purchase the brand.

32. You are interested in p, the proportion of station wagons among the registered vehicles in a particular state. Your prior distribution for p is a normal distribution with mean .05 and variance .0004. To obtain more information, a random sample of 50 registered vehicles is taken, and 3 are station wagons.
    (a) In using Equation (8.5.4) to revise your distribution of p, what difficulties are encountered?
    (b) How might you avoid such difficulties in this situation? Can they always be avoided?

33. Discuss the importance of conjugate families of distributions in Bayesian statistics.

34. Find the mean and the variance of the beta distribution with parameters $r = 2$ and $n = 6$, and graph the density function. Do the same for the following beta distributions:

$$r = 4, \; n = 6; \qquad r = 4, \; n = 12; \qquad r = 8, \; n = 12.$$

Explain how the different values of r and n affect the shape and the location of these four distributions.

35. If the mean and the variance of a beta distribution with parameters r and n are $2/3$ and $1/72$, respectively, find r and n.

36. Let p represent the probability that a person's body temperature will increase after a certain drug is taken. If the distribution of p is a beta distribution with $r = 12$ and $n = 20$, find
    (a) $P(p \geq .6)$,   (c) the value of the density function at
    (b) $P(.5 < p < .7)$, $p = .7$.

37. In Exercise 3, suppose that the prior distribution could be represented by a beta distribution with $r' = 4$ and $n' = 10$. Find the posterior distribution. Also, find the posterior distribution corresponding to the following beta prior distributions:

$$r' = 2, \; n' = 5; \qquad r' = 8, \; n' = 20; \qquad r' = 6, \; n' = 15.$$

In each of the four prior distributions considered in this exercise, the mean of the prior distribution is .40. How, then, do you explain the differences in the means of the posterior distributions?

38. In Exercise 4, suppose you feel that the mean of your prior distribution is $\frac{1}{2}$ and that the variance of the distribution is $\frac{1}{20}$. If your prior distribution is a member of the beta family, find r' and n' and determine the posterior distribution following the sample of size 6. Graph the density functions and find the mean and the variance of the posterior distribution.

39. In sampling from a Bernoulli process, the posterior distribution is the same whether we sample with n fixed (binomial sampling) or with r fixed (Pascal sampling). Explain why this is true. Suppose that a statistician merely samples until he or she is tired and decides to go home. Would the posterior distribution still be the same (that is, is the stopping rule noninformative)?

40. In Exercise 6, the medical researcher decides to treat p as a continuous random variable. The prior distribution of p is a beta distribution with mean .167 and $n' = 6$. Find the posterior distribution of p following the sample of 20 persons and calculate the mean and variance of this posterior distribution.

41. In Exercise 20, let p be continuous and assess your subjective probability distribution for p. Revise this distribution on the basis of the sample of 10 students, 3 of whom consumed at least one glass of beer during the past week.

42. In Exercise 22, assume that p is continuous and assess your subjective probability distribution for p. Can you find a beta distribution that expresses your judgments reasonably well?

43. In Exercise 25, if μ is assumed to be continuous and if the prior distribution for μ is a normal distribution with mean 110 and variance .4, find the posterior distribution.

44. You are attempting to assess a prior distribution for the mean of a process, and you decide that the .25 fractile of your distribution is 160 and the .60 fractile is 180. If your prior distribution is normal, determine the mean and the variance.

45. In reporting the results of a statistical investigation, a statistician reports that the posterior distribution for μ is a normal distribution with mean 52 and variance 10 and that a sample of size 4 with sample mean 55 was taken from a normal population with variance 100. On the basis of this information, determine the statistician's *prior* distribution.

46. Explain the parametrization of a normal prior distribution in terms of n' and m', as given in Section 8.8. How does this parametrization help us to see the relative weights of the prior and the sample information in computing the mean of the posterior distribution? For the prior and posterior distributions in Exercise 43, express the distributions in terms of n' and m'. How could you interpret this prior distribution in terms of an equivalent sample?

47. In assessing a distribution for the mean height of a certain population of college students, a physical-education instructor decides that the distribution is normal, the median is 70 inches, the .20 fractile is 67 inches, and the .80 fractile is 72 inches. Can you find a normal distribution with these fractiles? Comment on the ways in which the instructor could make the assessments more consistent.

48. A data-generating process is a normal process with unknown mean μ and with known variance $\sigma^2 = 225$. A sample of size $n = 9$ is taken from this process, with the sample results 42, 56, 68, 56, 48, 36, 45, 71, and 64. If your prior judgments about μ can be represented by a normal distribution with mean 50 and variance 14, what is your posterior distribution for μ? From this distribution, find $P(\mu \geq 50)$ and $P(\mu \geq 55)$.

49. The number of customers entering a certain store on a given day is assumed to be normally distributed with unknown mean μ and unknown variance σ^2. As the store manager, you feel that your prior distribution for μ is a normal distribution with mean 1000 and that $P(900 \leq \mu \leq 1100) = .95$. You then take a random sample of 10 days, observing a sample mean of 900 customers and a sample variance of 50,000. Find your approximate posterior distribution for μ. Aside from the usual argument concerning the applicability of the normal model, why is your posterior distribution only approximate?

50. Using Equation (8.5.4), prove that if the data-generating process of interest is a normal process with unknown mean μ and known variance σ^2, if the prior distribution of μ is a normal distribution with mean m' and variance σ^2/n', and if the sample information consists of a sample of size n from the process with sample mean m, then the posterior distribution of μ is a normal distribution with mean m'' and variance σ^2/n'', where n'' and m'' are given by Equations (8.8.6) and (8.8.7).

51. A Florida orange grower is concerned with the diameter of oranges (in inches). The diameter X of any orange can be looked upon as a normally distributed random variable with mean μ and variance 1.44. The orange grower feels that there is a .40 chance that μ is less than 2.65 and a .80 chance that μ is less than 3.30. Moreover, the prior distribution for μ is normal. Sixteen oranges are selected at random and their diameters are measured. The sample mean diameter is 2.2 inches. Find the orange grower's posterior distribution for μ.

52. A statistician is interested in the mean μ of a normal population, and the prior distribution is normal with mean 800 and variance 12. The variance of the population is known to be 72. How large a sample would be needed to guarantee that the variance of the posterior distribution is no larger than 1? How large a sample would be needed to guarantee that the variance of the posterior distribution is no larger than .1?

53. An auditor believes the average credit card balance outstanding is probably over $10. More precisely, the auditor's judgments about the average balance can be represented by a normal distribution with a mean of 12 and a standard deviation of 2. A sample of 25 accounts is drawn at random, and the average balance on these accounts is $9. It is known from historical data that the standard deviation of credit card balances is $5, and the distribution of outstanding balances is assumed to be a normal distribution. What is the auditor's posterior distribution for the average credit card balance?

54. The president of a firm owning a chain of grocery stores feels that X, the sales of a particular item in a given store on a particular day, is normally distributed with mean μ. The prior distribution of μ is a normal distribution, and a sample yields a sample mean of 180. The president's posterior distribution is a normal distribution with mean 200. What can be said about the mean of the prior distribution?

55. A statistician decides that the prior distribution for a certain parameter θ is an exponential distribution and that the .67 fractile of this distribution is 2. Find the exact form of the distribution.

56. In Exercise 14, suppose that the official's prior distribution of λ, the intensity of accidents per week, is a gamma distribution with mean 3.5 and variance .5. Find the posterior distribution of λ and determine the mean and the variance of this posterior distribution.

57. Try to assess a probability distribution for θ, the maximum temperature tomorrow in the city where you live. In assessing the distribution, use two or three of the methods proposed in the text and compare the results. Save the distribution and look at it again after you find out the true value of θ. Do this for 3 or 4 consecutive days and comment on any difficulties that you encounter in attempting to express your subjective judgments in probabilistic form.

58. In assessing a prior distribution for p, the proportion of votes that will be cast for a particular candidate in a statewide election, a political analyst feels that the mean of his prior distribution is .45. Furthermore, the analyst feels that if a random sample of 2000 voters yields 960 who state that they will definitely vote for the candidate and 1040 who state that they will *not* vote for the candidate (the sample includes no undecided voters), the mean of the posterior distribution would be .47. Assuming that the process behaves approximately as a Bernoulli process, that stated voting intentions are representative of actual voting behavior, and that the political analyst's prior distribution is a member of the beta family of distributions, find the exact form of the prior distribution.

59. Comment on the following statement: "One cannot speak of sensitivity except in connection with a particular decision-making situation." Can you think of an example in which the decision-making procedure would be quite insensitive to changes in the prior distribution? Can you think of an example in which it would be quite sensitive?

60. Discuss the importance of discrete approximations to continuous prior distributions in Bayesian analysis. Since the concept of conjugate prior distributions greatly simplifies the analysis, why is it ever necessary to use discrete approximations?

61. In Exercise 6, suppose that the medical researcher assesses a normal prior distribution for p with mean .16 and variance .0009. Using a discrete approximation to this prior distribution with intervals of width .01, find the approximate posterior distribution of p following the sample of size 20 with $r = 2$. Repeat the procedure with intervals of width .05 and comment on the differences in the approximate posterior distributions.

62. In Exercise 32,
   (a) try to find a beta distribution to approximate the normal prior distribution (for example, you might find the beta distribution with the same mean and variance as this normal distribution), discuss the "closeness" of the approximation, and find the posterior distribution;
   (b) use a discrete approximation to the normal prior distribution with intervals of width .01, discuss the "closeness" of the approximation, and find the posterior distribution.

63. Explain the notion of diffuseness. Give a few examples of situations in which your prior distribution would effectively be diffuse and a few examples in which it would definitely *not* be diffuse relative to a given sample.

64. The beta distribution with $r' = n' = 0$ and the beta distribution with $r' = 1$ and $n' = 2$ have both been used by statisticians as "diffuse" beta distributions. Discuss the advantages and disadvantages of each.

65. Discuss the relationship between classical statistical procedures and Bayesian procedures under a diffuse prior distribution, both with regard to numerical results and with regard to the interpretation of results.

66. Discuss the propositions, "Personal probabilities of the statistician have no place in rigorous scientific investigation," and "All probabilities are interpretable as personal probabilities."

67. Discuss: "The Bayesian approach to statistics can be thought of as an *extension* of the classical approach."

68. In Exercise 3, find a point estimate for p, the proportion of consumers who will purchase the product, based on
   (a) the prior distribution alone,
   (b) the sample information alone,
   (c) the posterior distribution.

69. Do Exercise 68 under the assumption that the prior distribution of p is a beta distribution with $r' = 4$ and $n' = 10$.

70. In Exercise 14, find a point estimate for λ, the intensity of occurrence of accidents along a particular stretch of highway, based on
    (a) the prior distribution alone,
    (b) the sample information alone,
    (c) the posterior distribution.

71. In Exercise 48, find
    (a) a point estimate for μ based on the prior distribution alone,
    (b) a point estimate for μ based on the sample information alone,
    (c) a point estimate for μ based on the posterior distribution,
    (d) a 90 percent credible interval for μ based on the prior distribution alone,
    (e) a 90 percent confidence interval for μ based on the sample information alone,
    (f) a 90 percent credible interval for μ based on the posterior distribution.
    Comment on any differences in your answers to (a) through (c) and to (d) through (f).

72. In Exercise 51, find
    (a) an 86.6 percent confidence interval for μ,
    (b) a point estimate for μ based on the posterior distribution,
    (c) a 38.3 percent credible interval for μ.

73. *Carefully* distinguish between a classical confidence interval and a Bayesian credible interval and also between the classical and Bayesian approaches to point estimation.

74. A marketing manager who is interested in p, the proportion of consumers who will buy a particular new product, considers the following two hypotheses:

$$H_0: p = .10$$

and
$$H_1: p = .20.$$

The prior probabilities are $P(p = .10) = .85$ and $P(p = .20) = .15$, and a random sample of 8 consumers results in 3 consumers who state that they will buy the product if it is marketed.
    (a) What is the prior odds ratio?
    (b) What is the likelihood ratio?
    (c) What is the posterior odds ratio?
    (d) The manager decides that the posterior probability of H_0 must be no larger than .40 to make it worthwhile to market the product. Should the product be marketed?
    (e) On the basis of the sample information alone, should the product be marketed? If the decision is made on this basis, what is implied about the prior distribution?

75. Suppose that you are uncertain about which of two authors, A or B, wrote a particular essay, and you feel that you would be indifferent between the following lotteries.

> *Lottery I:*   You win $10 if A wrote the essay, $0 otherwise.
> *Lottery II:*   You win $10 with probability .3, $0 with probability .7.

You know from past analyses of the writings of A and B that A uses a certain key word with an average frequency of 4 times in 100 words, and B uses this word with an average frequency of 2 times in 100 words (assume that the process is stationary and independent). In the particular essay of interest, there are 500 words and the key word appears 8 times. You are interested in finding out which author wrote the essay.
   (a) What statistical model would you use to represent the data-generating process?
   (b) Express the "real-world" hypotheses in terms of the statistical model chosen in (a).
   (c) Find the prior odds ratio.
   (d) Find the likelihood ratio.
   (e) Find the posterior odds ratio.
   (f) What would you conclude about the authorship of the essay?

76. Prove Equation (8.16.4). [*Hint:* Use Equation (8.1.1).]

77. In Exercise 66, Chapter 2, Urn A contains 700 red balls and 300 green balls, while Urn B contains 700 green balls and 300 red balls. One of the two urns is selected at random, and 10 balls are drawn at random with replacement from the urn that has been selected. Seven of the 10 balls are red and 3 are green.
   (a) Without performing any calculations, subjectively assess the posterior odds ratio of H_0 (the selected urn contains 700 red balls and 300 green balls) to H_1 (the selected urn contains 700 green balls and 300 red balls).
   (b) Calculate the posterior odds ratio in (a) formally by using Bayes' theorem.
   (c) Compare your answers to (a) and (b) and comment on any difference between the answers.

78. In Exercise 25, assume that the prior distribution is discrete with $P(\mu = 109) = .2$ and $P(\mu = 110) = .8$. For H_0: $\mu = 109$ and H_1: $\mu = 110$, find
   (a) the prior odds ratio,
   (b) the likelihood ratio,
   (c) the posterior odds ratio.

79. In Exercise 74, it would be more realistic to consider hypotheses such as

$$H_0: p \leq .15$$

and $$H_1: p > .15.$$

If the prior distribution is a beta distribution with $r' = 2$ and $n' = 12$, find the posterior distribution and use this posterior distribution to find the posterior odds ratio of H_0 to H_1.

80. In Exercise 43, suppose that the production manager is interested in the hypotheses $H_0: \mu \leq 110$ and $H_1: \mu > 110$. From the prior distribution, find the prior odds ratio; also, from the posterior distribution, find the posterior odds ratio. If the losses involved in a decision-making problem involving the production process are such that H_0 should be accepted only if the posterior odds ratio is greater than 3, what decision should be made?

81. In testing a hypothesis concerning the mean of a normal process with known variance against a one-tailed alternative, discuss the relationship between the classical p-value and the posterior probability $P''(H_0)$, (a) if the prior distribution is diffuse, (b) if the prior distribution is not diffuse.

82. Comment on the following statement: "A hypothesis such as $\mu = 100$ is not realistic and should be modified somewhat; otherwise, the Bayesian approach to hypothesis testing may not be applicable."

83. An automobile manufacturer claims that the average mileage per gallon of gas for a particular model is normally distributed with $m' = 20$ and $\sigma' = 4$, provided that the car is driven on a level road at a constant speed of 30 miles per hour. A rival manufacturer decides to use this as a prior distribution and to obtain additional information by conducting an experiment. In the experiment, it is assumed that the variance of mileage is $\sigma^2 = 96$ and the mileage is normally distributed. The experiment is conducted on 10 randomly chosen cars, with the sample mean (the average mileage in the sample) equaling 18. Find the posterior distribution for μ, the average mileage per gallon of gas. On the basis of this posterior distribution, what is the probability that μ is greater than or equal to 20? What is the probability that μ is between 19 and 21?

84. In Exercise 43, suppose that the manager is interested in the hypotheses $H_0: \mu = 110$ and $H_1: \mu \neq 110$. From the posterior distribution, what is the posterior odds ratio of H_0 to H_1? What is the posterior odds ratio if the hypotheses are $H_0: 109 < \mu < 111$ and $H_1: \mu \leq 109$ or $\mu \geq 111$?

85. A statistician is interested in $H_0: \mu = 50$ and $H_1: \mu \neq 50$. The prior distribution consists of a mass of probability of .25 at $\mu = 50$, with the remaining .75 of probability distributed uniformly over the interval from $\mu = 40$ to $\mu = 60$. Find the prior probability that:
    (a) $45 < \mu < 55$,
    (b) $47 \leq \mu \leq 50$,
    (c) H_0 is exactly true,
    (d)   the hypothesis $49 < \mu < 51$ is true.

86. In Exercise 51, find the prior and posterior odds ratios for $H_0: 2.8 \leq \mu \leq 3.2$ versus $H_1: \mu < 2.8$ or $\mu > 3.2$.

87. In Exercise 53, consider the following three hypotheses:

$$H_1: \mu \leq 10,$$

$$H_2: 10 < \mu < 12,$$

and $$H_3: \mu \geq 12.$$

(a) Find the posterior probabilities of these three hypotheses.

(b) If a decision problem facing the auditor involves three actions (corresponding to the three hypotheses), and the hypothesis to be selected is the one which is most probable, which hypothesis would be selected under the posterior distribution?

(c) In part (b), which hypothesis would be selected on the basis of the *prior* distribution?

88. A statistician is interested in the difference in the means of two normal populations, μ_1 and μ_2. Each population has variance 100, and the two populations are independent. A sample of size 25 is taken from the first population, with sample mean $m_1 = 80$, and a sample of size 25 is taken from the second population, with sample mean $m_2 = 60$.

(a) Using the classical approach to hypothesis testing presented in Chapter 7, test H_0: $\mu_1 - \mu_2 \leq 0$ against the alternative H_1: $\mu_1 - \mu_2 > 0$, with $\alpha = .10$.

(b) Let $\mu = \mu_1 - \mu_2$, and suppose that the prior distribution of μ is normally distributed with mean $m' = 10$ and variance $\sigma'^2 = 50$. Find $P'(H_0)$ and $P'(H_1)$.

(c) Find the posterior distribution of $\mu = \mu_1 - \mu_2$.

(d) From the posterior distribution, find the posterior odds ratio of H_0 to H_1.

89. Exercise 88 suggests a Bayesian approach to inferences regarding the difference between two means, under the conditions that the two populations of interest are normally distributed and the variances are known. Using the notation of Exercise 88, determine a general formula for finding the posterior distribution of the difference between two means if the prior distribution is normally distributed and has mean m' and variance σ'^2.

9

DECISION
THEORY

In Chapters 6 and 7 we discussed classical inference, which consists of procedures that are formally based solely on sample information. In Chapter 8, we saw that the Bayesian approach to inferential problems utilizes prior information as well as sample information and bases inferences on the posterior distribution. At various points in these chapters, we hinted at the decision-theoretic approach and the use of likelihood functions and posterior distributions in making decisions. A grasp of the concepts of inferential statistics is a useful prerequisite to a full understanding of the many statistical implications of decision theory. The preceding chapters should have provided that grasp, so we are finally ready to take the plunge.

Why are we interested in **decision theory?** Life is a constant sequence of decision-making situations. Every action you take, with the exception of a few involuntary physiological actions, such as breathing, can be thought of as a decision. Of course, most of these decisions are quite minor decisions, because the consequences involved are not very important. For example, consider the decision whether to walk up a flight of stairs or to take the elevator. Unless you have a heart condition or some other physical reason for taking the elevator, this is not an important decision. The relevant factors entering into the decision include the exertion required to walk up the stairs, the availability of the elevator, whether you are in a hurry, and so on. Minor decisions such as this are usually made intuitively without conscious thought. Other examples are the decision to smoke a cigarette, drive home via a certain route, have a particular beverage with your evening meal, and so on. For some individuals, these decisions may require some thought. A person trying to break the smoking habit may find the decision as to whether or not to smoke a cigarette to be a difficult decision. For most people, though, such decisions are usually made out of habit. They do not consciously think about the various possible actions because they have faced the situation (or similar situations) many times in the past.

Other decisions require some thought but can still be made intuitively. Examples are a choice of brands of a product, a choice of movies on a Saturday night, and so on. Eventually, you will face a decision-making problem that requires some serious thought. Consider a major purchasing decision, such as the decision to buy a new car. In such a situation, you evaluate such factors as the condition of your present car, the cost of a new car, the enjoyment of owning a new car, and so on. You probably would make this decision subjectively, but it is the type of decision for which the formal decision theory to be discussed in this chapter might prove useful. Similar problems are investment decisions (either by individuals or by corporations), medical decisions (whether or not to undergo surgery), and job decisions (for the individual, whether or not to accept a job offer; for the corporation, whether or not to make such an offer). It may not always be easy to specify the problem formally and apply the formal decision-making procedures. Whether to proceed at an informal level or at a formal level is a decision in itself. Whenever the consequences of making a wrong decision could be quite serious, however, decision theory should prove worthwhile.

9.1 CERTAINTY VERSUS UNCERTAINTY

This chapter deals with decision making under the condition of **uncertainty.** This condition refers to uncertainty about the true value of a variable related to the decision, or, in simpler terms, uncertainty about the actual state of the world, or state of nature.

Formally, a consequence of a decision, which may be expressed in terms of a payoff or a loss to the decision maker, is the result of the interaction of two factors: (1) the decision, or the action, selected by the decision maker; and (2) the event, or state of the world, that actually occurs.

In general, the decision-making problem is easier to solve if the decision maker knows which event will occur. For a trivial example, suppose that you have to decide whether or not to carry an umbrella when you leave home in the morning. There are two possible actions: carry the umbrella or leave it at hone. Also, there are two relevant events, or states of the world: rain or no rain. If you know for certain that it will rain, you will definitely carry your umbrella. If you know for certain that it will not rain, you will leave the umbrella at home. In this situation, you are making a decision under the condition of **certainty.**

For another example, suppose that you are faced with a decision concerning the purchase of common stock. For the sake of simplicity, assume that you intend to invest exactly $1000 in a single common stock and to hold the stock for exactly 1 year, at which time you will sell it at the market price. Furthermore, assume that you are considering only 3 stocks, A, B, and C, each of which currently sells for $50 a share. Thus, you intend to buy 20 shares of 1 of the 3 stocks. Your selection of a single stock from the three constitutes your *decision*, or *action*. The prices of the 3 stocks 1 year from now constitute the *state of the world*. The combination of your action and the state of the world determines your payoff. Suppose that at the end of 1 year, the prices of stocks A, B, and C are $60, $40, and $50. The payoffs for the three possible actions are then $+$200, $-$200, and $0, respectively. For this state of the world, the best decision is to buy stock A, for this results in the highest payoff. If you know the state of the world with certainty, the decision can be made in this manner, and this is once again decision making *under certainty*. In this example, if you know what the prices of the stocks will be in 1 year, then you will simply buy the stock that will give you the maximum payoff—that is, the stock that will increase in value the most during the coming year.

It should be stressed that **decision making under certainty** is not always as trivial as it seems to be in these examples. For instance, suppose that a manufacturer must ship a certain product from a number of factories to a number of warehouses. Each factory produces a certain number of units of the product, and each warehouse requires a certain number of units. Furthermore, the cost of shipping from any given factory to any given warehouse depends on the amount shipped, the particular factory, and the particular warehouse. The decision-making problem is this: What shipping pattern minimizes the total transportation costs? That is, what is the least expensive way to transport the product from the factories to the warehouses? In this problem there is no uncertainty; the amounts produced at the various factories, the amounts needed at the various warehouses, and the costs of shipping are all known. Even under certainty, this decision-making problem is clearly not trivial to solve (although it becomes much *more* complex if the amounts produced, the amounts needed, or the costs are uncertain). It may be solved by a technique known as *linear programming*, a special (and very important) case of *mathematical programming*, which is a general approach to certain types of decision-making problems. Other decision-making problems under certainty require the use of different types of mathematical optimization procedures. Since we are concerned with the case of uncertainty rather than certainty, such procedures are not discussed here, although you should realize that decision making under certainty includes many important and by no means mathematically trivial problems.

In the problems that concern us here, the actual state of the world is *not* known with certainty. In the investment example, unless you are fortunate enough to receive copies of the *Wall Street Journal* 1 year in advance or unless you possess extrasensory perception, you could not know for certain what the prices of the 3 stocks would be 1 year hence. Thus, treating the purchase of common stock as a decision under certainty is not at all realistic. You no doubt have ideas regarding the prices, ideas that may be based on your impressions of the economy in general, of particular industries, and of specific firms within the industries. Your knowledge may stem from years and years of careful study of the stock market, or it may consist of a hot tip from a "friend." At any rate, you have some knowledge, but not perfect knowledge. Hence, you are faced with a problem of **decision making under uncertainty.**

There are numerous examples of decision making under uncertainty. In Section 2.17, examples were presented concerning an oil-drilling decision under uncertainty about the presence of oil and concerning an advertising decision under uncertainty about the actions of a competing firm. Most investment decisions are made under uncertainty about the eventual return on the investment. Production decisions are made under uncertainty about the demand for a product. Numerous medical decisions are made under uncertainty (for instance, uncertainty about the outcome of surgery or the effect of a drug). The area of decision making under uncertainty clearly includes many important decision-making problems.

9.2 PAYOFFS AND LOSSES

We have noted that the consequences of a decision can be expressed in terms of either payoffs or losses. It is important to define explicitly what is meant by these two terms. A **payoff,** or **reward,** can be interpreted in the usual manner; if it is expressed in monetary units, it represents the net change in your total wealth as a result of your decision and the actual state of the world. This can be either positive or negative. It is important to remember that payoffs are always expressed in net terms rather than gross terms, so that all relevant factors can be taken into consideration. For instance, when making a decision about the price of a particular product and the quantity to be produced, a firm should take into account all relevant costs as well as the gross profits attributable to the item of interest.

In the oil-drilling example presented in Section 2.17, there are two possible actions (drill and do not drill) and two events (oil and no oil). If the decision maker drills and strikes oil, the oil will be worth $130,000 and the cost of drilling is $30,000, so the net payoff is $100,000. If the decision is not to drill, the drilling rights will be sold for a fixed sum of $10,000 plus another $10,000 contingent upon the presence of oil. The payoff table, as presented in Section 2.17, is given in Table 9.2.1.

Table 9.2.1   Payoff Table for Oil-Drilling Example

		EVENT	
		Oil	*No oil*
DECISION	*Drill*	$100,000	−$30,000
	Do not drill	$20,000	$10,000

A payoff table consists of the set of payoffs for all possible combinations of actions and states of the world.

If there are m actions and n states of the world, the payoff table will have mn entries. In the oil-drilling example, there are 4 entries. We will denote payoffs by R (for "reward") because P is used to represent probability. For instance,

$$R(\text{drill, no oil}) = -\$30,000.$$

How can the above payoff table be converted to an equivalent **loss table?** "Loss" in this context refers to **opportunity loss.** For any combination of an action and a state of the world, the question is: Could you have obtained a higher payoff, given that particular state of the world? If the answer is no, then your loss is zero. If the answer is yes, then **your loss is the positive difference between the given payoff and the highest possible payoff under that state of the world.** Keep in mind that in this book, the term "loss" is always to be interpreted in terms of opportunity loss. Negative values in a payoff table are to be regarded as negative payoffs, negative rewards, or costs; in terms of opportunity loss, they may correspond to losses of zero.

Consider the first column of the above payoff table. Given that oil is present, the higher of the two payoffs is $100,000, corresponding to the action "drill." This is the best the decision maker can do if there is oil, so the associated loss is zero. If the decision is "do not drill," the payoff is $80,000 *lower*, so the loss corresponding to the lower left-hand cell of the payoff table is $80,000. Similarly, in the second column, the higher payoff is $10,000 and the other payoff is −$30,000, so the losses are $0 and $40,000 for these

Table 9.2.2   Loss Table for Oil-Drilling Example

EVENT

		Oil	No oil
	Drill	$0	$40,000
DECISION			
	Do not drill	$80,000	$0

two payoffs, respectively. The **loss table** is given in Table 9.2.2. A loss table is sometimes called a **regret table,** since the entries reflect the decision maker's regret at not having made the decision that turns out to be optimal under the actual state of the world.

For another example, consider the stock-purchasing decision from the previous section. Furthermore, for the sake of simplicity, assume that there are only four possible states of the world. In the first (state I), the prices of the stocks A, B, and C at the end of 1 year are, respectively, $60, $50, and $40. For state II, the prices are $80, $60, and $40. For state III, the prices are $20, $40, and $10. Finally, for state IV, the prices are $60, $60, and $60. This results in the payoff table in Table 9.2.3., where the payoffs are expressed in dollars. For example, if stock A is bought and if state II occurs, the price goes up from $50 to $80, or $30 per share. Since 20 shares will be bought, the payoff is $600. This payoff table corresponds to the loss table in Table 9.2.4.

The procedure for converting a payoff table to a loss table should be clear by now. First, work with each column of the table separately. In each column, find the highest payoff (which may be negative, zero, or positive). The opportunity loss corresponding to this payoff is zero, and the loss corresponding to any other payoff in the same column is obtained by subtracting that payoff from the highest payoff in the column. Formally, if R(action i, state j) and L(action i, state j) represent the payoff

Table 9.2.3   Payoff Table for Stock Example

STATE OF THE WORLD

		I	II	III	IV
	Buy A	200	600	−600	200
ACTION	Buy B	0	200	−200	200
	Buy C	−200	−200	−800	200

Table 9.2.4 Loss Table for Stock Example

STATE OF THE WORLD

		I	II	III	IV
	Buy A	0	0	400	0
ACTION	Buy B	200	400	0	0
	Buy C	400	800	600	0

and the loss, respectively, corresponding to the combination of the ith row (ith action) and the jth column (jth state of the world) in the table, then

L(action i, state j)

$$= \max_k \; R(\text{action } k, \text{ state } j) - R(\text{action } i, \text{ state } j). \quad (9.2.1)$$

Here, $\max_k R$(action k, state j) is the largest payoff in the jth column of the payoff table.

The loss tables presented above demonstrate some interesting characteristics of losses. First, losses cannot be negative. This can be seen from the way in which they are defined; each loss is either zero or the (positive) difference between two payoffs. Payoffs can be negative, zero, or positive; losses, on the other hand, can only be zero or positive (there is no such thing as a negative regret!). In addition, if all the payoffs in a column are equal, the corresponding losses will all be zero. This is because no payoff is higher than another, so if that particular state of nature occurs, it makes no difference which action is chosen.

An alternative way to present the payoffs or losses in a decision-making problem is in terms of a tree diagram. For example, tree diagrams for the oil-drilling and stock-purchasing examples are presented in Figures 9.2.1 and 9.2.2. In each case, the first fork of the tree corresponds to the action chosen by the decision maker and the second fork corresponds to the state of the world. Thus, each terminal branch at the right-hand side of the tree diagram corresponds to a combination of a particular action and a particular state of the world. The numbers at the end of these terminal branches are the corresponding payoffs (alternatively, losses could be used in place of payoffs). Tree diagrams are particularly useful in representing complex decision-making problems with sequences of actions and events over time, as we shall see when the concept of the value of information is discussed later in this chapter.

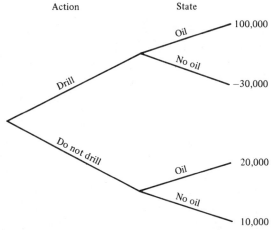

Figure 9.2.1 Tree Diagram for Oil-Drilling Example

Given a payoff table, a loss table, or a tree diagram, we can attempt to find out if any of the actions are **inadmissible.** Recall that the concept of inadmissibility was discussed in Sections 7.2 and 7.3 in the context of hypothesis testing. In general, an action is said to **dominate** another if for *each possible state of the world*, the first action leads to at least as high a

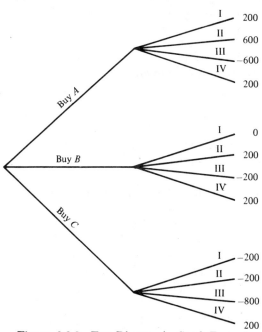

Figure 9.2.2 Tree Diagram for Stock Example

payoff (or at least as small a loss) as the second action, and if for at least one state of the world, the first action leads to a higher payoff (or smaller loss) than the second action. If one action dominates another, then it would never be reasonable to choose the second action (you could always do at least as well with the first action, and for at least one state of the world, you could do better with the first action). Therefore, the second action is said to be inadmissible. Notice that the last requirement in the definition of dominance (that the first action leads to a *higher* payoff for at least one state of the world) is necessary to prevent actions with identical payoffs from being classified as "dominated by each other."

In the stock-purchasing example, the third action (buy C) is dominated by both of the other two actions, as you can see by looking at either the payoff table or the loss table. The first action is better than the third action for states I, II, and III, and it is just as good as the third action for state IV. The same is true of the second action as compared with the third action. Hence, the third action is *inadmissible*; there is no need to consider it further. Be careful to note that in order to be inadmissible, an action must be dominated by another *single* action, not just by a combination of other actions. In the advertising example in Section 2.17, the minor advertising campaign has a smaller payoff than the major advertising campaign if the competing firm introduces a new product, and it has a smaller payoff than no advertising if the competing firm does not introduce a new product. However, "minor ad campaign" is not dominated by "major ad campaign" alone or by "no advertising" alone. Thus, it is *not* inadmissible.

In making a decision, it is necessary to consider only admissible acts, acts that are not dominated by any other act.

The practical importance of this is that the elimination of inadmissible acts may greatly reduce the "dimension" of the decision-making problem by enabling a decision maker to confine attention to the admissible acts. For instance, in setting up the problem, a decision maker may consider, say, 15 actions. When a payoff table or loss table is considered, some actions (perhaps 10 of the actions) turn out to be inadmissible. Thus, these 10 actions can be eliminated from further consideration, leaving a much simpler problem (at least computationally) involving a choice among only 5 actions.

9.3   NONPROBABILISTIC CRITERIA FOR DECISION MAKING UNDER UNCERTAINTY

If a table or a tree diagram of payoffs or losses is known in a decision problem and if the actual state of the world is known for certain, the decision is obvious; choose the action giving the highest payoff (smallest loss) for that particular state of the world. In terms of the table, choose the action corresponding to the largest entry in the relevant column of the payoff table or the smallest entry in the relevant column of the loss table. Because of the way in which losses were defined in Section 9.2, the same action is selected in both cases.

Under uncertainty, however, we cannot confine our attention to a single column of the table, and several decision-making criteria for the uncertainty case have been developed. In this section we consider criteria that do not depend on the use of probabilities, and in the next section probabilistic decision-making criteria will be discussed.

One decision rule, called the maximin rule, says: for each action, find the smallest possible payoff, and then choose the action for which this smallest payoff is largest.

The payoff table for the quality-control example of Section 8.2 is presented in Table 9.3.1, and the smallest payoffs for the three actions are 380, 375, and 0. Thus, the maximin rule advises the decision maker to make a major adjustment in the production process. This is a conservative decision: "Let us make the major adjustment so we will not have to worry about what could happen if we did not make the adjustment and p happened to be .10 or .25."

In effect, the maximin rule tells the decision maker to assume that for each action, the worst possible state of the world will occur. This is an extremely conservative approach, since it only looks at the lowest payoff in each row and ignores the magnitudes of the other payoffs. It is easy to construct payoff tables for which the maximin rule gives intuitively unreasonable results. In Table 9.3.2, for example, the smallest payoffs for the two actions are −1 and 0, so the second action is selected by the maximin criterion. The possible gain of $100,000 is completely ignored. These two actions could be thought of as "do something" and "stay as is."

Table 9.3.1 Payoff Table for Quality-Control Example

| | | STATE OF THE WORLD (VALUE OF p) | | | |
		.01	.05	.10	.25
	Major adjustment	380	380	380	380
ACTION	*Minor adjustment*	455	375	375	375
	No adjustment	480	400	300	0

Table 9.3.2   Payoff Table for which Maximin Rule Seems Unreasonable

STATE OF THE WORLD

	100,000	−1
ACTION	0	0

According to the maximin rule, we should never take an action if there is a possibility of a negative payoff, since we could always do better in a maximin sense by doing nothing and obtaining a payoff of zero. This is most unrealistic!

A second decision rule, called the **maximax** rule, assumes that the best will happen, not the worst.

According to the maximax rule, the decision maker should find the largest possible payoff for each action and then choose the action for which this largest payoff is largest.

This rule maximizes the maximum payoff. In the quality-control example, the maximum payoffs are 380, 455, and 480, and the maximax rule thus chooses the third action, "no adjustment." This is a risky decision: "Let us go for the highest payoff, $480, and hope that p is not .10 or .25."

Notice that the maximax rule always chooses the action corresponding to the highest payoff in the entire table. Just as the maximin rule ignores the high payoffs, the maximax rule ignores the low payoffs, assuming that for each action, the best possible state of the world will occur. In the payoff table in Table 9.3.3, the highest payoffs for the two actions are 100,000 and 100,001; the maximax criterion chooses the second action. Thus, in order to gain one extra dollar (from $100,000 to $100,001), the decision maker is taking the chance of ending up with nothing, whereas a payoff of at least $99,999 is assured if the first action is taken. In this example, then, the maximax rule appears to be unreasonable.

The maximin and maximax rules deal with the payoff table.

Table 9.3.3   Payoff Table for which Maximax Rule Seems Unreasonable

STATE OF THE WORLD

	100,000	99,999
ACTION	100,001	0

Table 9.3.4 Loss Table for Quality-Control Example

STATE OF THE WORLD

		.01	.05	.10	.25
	Major adjustment	100	20	0	0
ACTION	*Minor adjustment*	25	25	5	5
	No adjustment	0	0	80	380

A third rule, called the minimax loss, or minimax regret criterion, says: for each action, find the largest possible loss, and then choose the action for which this largest loss is smallest.

The loss table in the quality-control example is given in Table 9.3.4. The maximum losses for the three actions are 100, 25, and 380. According to the minimax loss rule, a minor adjustment should be made in the production process. In this example, the three decision criteria considered in this section result in three different decisions.

As the name implies, the minimax rule minimizes the maximum possible loss. The minimax loss rule is neither as conservative as the maximin rule nor as risky as the maximax rule. To illustrate this, look at the loss tables in Tables 9.3.5 and 9.3.6, which correspond to the payoff tables in Tables 9.3.2 and 9.3.3. In each case the minimax loss rule chooses the first action.

Table 9.3.5 Loss Table Corresponding to Table 9.3.2

STATE OF THE WORLD

	0	1
ACTION		
	100,000	0

Table 9.3.6 Loss Table Corresponding to Table 9.3.3

STATE OF THE WORLD

	1	0
ACTION		
	0	99,999

Other nonprobabilistic decision-making criteria, some of which may be better than the above rules, have been proposed. For instance, one rule is a combination of the maximin and maximax rules, using an "optimism-pessimism" index. Of course, all of these rules ignore probabilities concerning the various possible states. In the quality-control example, for instance, if $P(p = .01)$ is very close to 1 (that is, the production manager is almost sure that the process is in control), then "no adjustment" appears to be the best course of action. On the other hand, if $P(p = .25)$ is very high (the chances are very good that the process has gone completely "haywire"), then "no adjustment" is a very unattractive choice. The rules in the next section take into account probabilities of the states of the world.

9.4 PROBABILISTIC CRITERIA FOR DECISION MAKING UNDER UNCERTAINTY

If some information, but not perfect information, is available regarding the states of the world, then the decision maker is operating under uncertainty. If the uncertainty is expressed in terms of probabilities, then these probabilities can be used as inputs to the decision-making process (recall that probability was called the mathematical language of uncertainty in Chapter 2). In particular, they can be used to calculate expected payoffs or expected losses of the potential actions.

The expected payoff, or expected reward (ER) criterion says to choose the act with the highest expected payoff.

This criterion has been used in decision-making examples in previous chapters.

The expected loss (EL) criterion says to choose the act with the smallest expected loss.

It is not necessary to debate the relative merits of the ER and EL criteria, since they are equivalent.

For any decision–making problem, the ER and EL criteria always yield identical decisions.

As we shall see when utility theory is discussed, the primary justification for the use of ER (or EL) as a decision-making criterion is the fact that under certain assumptions about an individual's utility function, maximizing ER or minimizing EL is equivalent to maximizing expected utility.

For the oil-drilling example presented in Section 2.17, suppose that the probability of oil is .30. The expected payoffs are then

$$ER(\text{drill}) = .30(\$100,000) + .70(-\$30,000) = \$9000$$

and $ER(\text{do not drill}) = .30(\$20,000) + .70(\$10,000) = \$13,000.$

The loss table for this example is presented in Section 9.2, and the expected losses are

$$EL(\text{drill}) = .30(\$0) + .70(\$40,000) = \$28,000$$

and $EL(\text{do not drill}) = .30(\$80,000) + .70(\$0) = \$24,000.$

Not drilling has the larger expected payoff and smaller expected loss. Furthermore, its ER is \$4000 higher than the ER of drilling, and its EL is \$4000 lower than the EL of drilling. This demonstrates a relationship between expected payoffs and expected losses; if the ER of one action is c units *higher* than the ER of a second action, then the EL of the first action will be exactly c units *lower* than the EL of the second action. In other words, there is a direct relationship between differences in expected payoffs and the differences in the corresponding expected losses. We will explore this relationship in greater detail later in this chapter when the value of perfect information is discussed. At this point, the main thing to keep in mind is that the ER and EL criteria lead to identical actions. This is useful, since in some problems it is convenient to think in terms of payoffs, whereas in other problems it is more convenient to think in terms of losses.

Consider once again the example involving the proportion of defective items, p, produced by a certain process. From Section 8.2, the prior probabilities are

$$P(p = .01) = .60,$$

$$P(p = .05) = .30,$$

$$P(p = .10) = .08,$$

and $\qquad P(p = .25) = .02.$

Using these probabilities and the payoff table given in Sections 8.2 and 9.3, the expected payoffs are

$$ER(\text{major adjustment}) = 380,$$

$$ER(\text{minor adjustment}) = .60(455) + .40(375) = 423,$$

and $\quad ER(\text{no adjustment}) = .60(480) + .30(400) + .08(300)$
$$+ .02(0) = 432.$$

The loss table is given in Section 9.3, and the expected losses are

$$EL(\text{major adjustment}) = .60(100) + .30(20) + .08(0)$$
$$+.02(0) = 66,$$

$$EL(\text{minor adjustment}) = .60(25) + .30(25) + .08(5) -$$
$$+ .02(5) = 23,$$

and $\quad EL(\text{no adjustment}) = .60(0) + .30(0) + .08(80)$
$$+.02(380) = 14.$$

If additional information is obtained, such information should be used to revise the probabilities. In Section 8.2, the following posterior probabilities were calculated after a sample of size 10 with 3 defective items:

$$P(p = .01 \mid r = 3, n = 10) = .005,$$

$$P(p = .05 \mid r = 3, n = 10) = .245,$$

$$P(p = .10 \mid r = 3, n = 10) = .359,$$

and
$$P(p = .25 \mid r = 3, n = 10) = .391.$$

The expected payoffs calculated using the posterior probabilities are

$$ER(\text{major adjustment}) = 380,$$

$$ER(\text{minor adjustment}) = 375.5,$$

and
$$ER(\text{no adjustment}) = 208.1.$$

The optimal act under the prior distribution is "no adjustment," since there is a high probability that p is small (a probability of .90 that p is either .01 or .05). The particular sample observed makes the higher values of p seem more likely (the posterior probability is only .25 that p is either .01 or .05), and the optimal act under the posterior distribution is "major adjustment."

It should be noted that some authors differentiate between what they call "decision making under risk" (decision making when the state of the world is not known but probabilities for the various possible states are known) and "decision making under uncertainty" (decision making when the state of the world is not known and probabilities for the various possible states are also not known). Here, we assume that probabilities are always available (under the subjective interpretation of probability, as discussed in Chapter 2, it is always possible to assess probabilities for the possible states of the world). Therefore, although nonprobabilistic criteria for making decisions were discussed in the previous section, we will not employ the "risk versus uncertainty" dichotomy. In this book any decision-making problem in which the state of the world is not known for certain will be called decision making under uncertainty. The term "risk" will be reserved for discussions relating to utility, which is taken up in the following sections.

In the discussion of decision under uncertainty, we have implicitly assumed that the state of the world is determined by "nature" or by some disinterested party. Suppose that a manufacturer must decide whether or not to produce a new product and that the state of the world represents the demand for the new product. If the manufacturer can express the state of information about the potential demand in the form of a probability distribution, then the problem is one of decision under uncertainty. However, there may be other factors which should be considered! If the new product is produced, perhaps a rival firm will decide to follow suit, and the total demand for the product would then be divided (not necessarily equally) between the firms. The payoff to the manufacturer thus depends not only on the demand for the product, but also on the decision of another firm. The problem is now one of decision making under competitive conditions. Since this type of problem has traditionally been investigated in the context of "games" between individuals, the study of such problems is called **game theory.**

In game theory the decision maker faces not a disinterested opponent, such as "nature," but an opponent who is interested in maximizing his own payoff. The state of the world is thus the action selected by the opponent. Some examples of realistic game-theoretic situations are (1) a bidding situation, in which competing firms bid for a contract and the firm with the lowest bid is awarded the contract; (2) a bargaining situation, in which two parties (for example, union and management) attempt to reach an agreement; and (3) a military situation, in which each side must consider the possible actions that could be taken by the opponents. Much of the study of game theory has not involved the use of probabilistic decision rules. Rules such as the maximin rule have been used with the justification that the decision maker's opponent will try to act so as to maximize his own payoff and hence minimize the decision maker's payoff.

Through the use of probabilities, the game-theory problem can be thought of as a problem of decision making under uncertainty, which can be treated by the methods discussed in this chapter. The decision maker must assess a probability distribution on the set of actions available to the opponent. If the opponent has two actions, and we call these actions θ_1 and θ_2, then the decision maker must assess $P(\theta_1)$ and $P(\theta_2)$. In order to make such assessments, we must take into consideration any information concerning the opponent and concerning the decision problem facing the opponent. Suppose that a manufacturer's decision as to whether or not to introduce a new product depends on the manufacturer's assessment of the chances that a rival firm will follow suit. By considering past actions of the rival firm in similar situations, their general willingness or lack of willingness to introduce new products, the capabilities of their key personnel, and so on, the manufacturer should be able to represent the information by a set of probabilities. It may be even more difficult to assess probabilities in this case than it is when the state of the world is determined by a "disinterested opponent," since it is necessary to take into account the fact that the opponent is also trying to make a "good" decision and is considering the possible actions that our decision maker might take. Nevertheless, given probabilities for the opponent's possible actions, we can use the *ER* and *EL* criteria discussed in this section.

Bear in mind that the *ER* and *EL* criteria take into consideration the probabilities of the various states of nature in a decision-making problem. The uncertainty in the decision-making situation is expressed in terms of probabilities, and the probabilities are used to calculate expectations of payoffs and losses. Even the *ER* and *EL* criteria are not without their drawbacks, however, as we shall see in the next section.

9.5 UTILITY

Although the *ER* and *EL* criteria take probabilities into account, they ignore another factor that may be important: the decision maker's **attitude toward risk.** Therefore, they sometimes lead to decisions that seem intuitively unreasonable. To avoid this difficulty, we shall introduce the concept of utility in this section. First, the difficulty is illustrated with some simple examples.

Suppose that you are offered the following bet on one toss of a coin (you have the opportunity to bet only once—repeated bets are not allowed). You win \$1 if the coin comes up heads, and you lose \$.75 if the coin comes up tails. If you are convinced that the coin is a fair coin, so that $P(\text{heads}) = P(\text{tails}) = \frac{1}{2}$, your expected payoff is $\frac{1}{2}(\$1) + \frac{1}{2}(-\$.75) = \$.125$ if you take the bet and \$0 if you do not take the bet. According to the *ER* criterion, you should take the bet. This is reasonable, since the bet looks intuitively advantageous. Now suppose that the amounts involved are \$10,000 and \$7500 rather than \$1 and \$.75. The expected payoff is now \$1250 if you take the bet and \$0 if you do not take the bet. According to the *ER* rule, you should take the bet. Would you do so? Probably not, unless you are wealthy enough so that a loss of \$7500 would not seriously affect your financial position. The possible gain of \$10,000 is tempting, but there is still a 50–50 chance of losing \$7500. By the way, if you still would take the bet, change the values to \$100,000 and \$75,000 and make the choice again.

If you choose not to take the bet in any of these examples, you are violating the *ER* criterion. Why? The monetary payoffs are unambiguous, so there must be factors involved in this example other than these payoffs. If you lose \$7500, the consequence is not just that loss, but the possible accompanying factors, such as the impact of a sharp reduction in your savings account or the embarrassment when you tell your spouse or your friends that you lost \$7500 on the toss of a coin.

For another example, consider a choice between two bets. In the first bet, you win \$1 million if a coin comes up heads and you win \$1 million if the coin comes up tails. In the second bet, you win \$10 million if the coin comes up heads but you win nothing if the coin comes up tails. The expected payoffs of the two bets are

$$ER(\text{bet 1}) = \tfrac{1}{2}(\$1 \text{ million}) + \tfrac{1}{2}(\$1 \text{ million}) = \$1 \text{ million}$$

and $$ER(\text{bet 2}) = \tfrac{1}{2}(\$10 \text{ million}) + \tfrac{1}{2}(\$0) = \$5 \text{ million.}$$

Bet 2 has a much larger expected payoff than bet 1. Which bet would you choose? In spite of the great difference in expected payoffs, most persons, if given this opportunity on a one-shot basis, would probably choose bet 1. This is because \$1 million is a very large sum of money, clearly enough to allow a person to lead a quite comfortable life (even at today's prices!). If this amount could be invested at 5 percent interest, you would receive \$50,000 per year without touching the principal. It would be even better, of course, to have \$10 million, but most people feel that \$10 million is not so preferable to \$1 million that it is worth a risk of winding up with nothing. Informally, this amounts to saying that to most persons, \$10 million is not worth 10 times as much as \$1 million. If it were, then bet 2 would be strongly preferred to bet 1.

This example merits a few additional comments. The amounts involved are much larger than they would be in most decision-making situations. Unless the decision at hand is a very important decision involving a large corporation, it would be expected that the amounts involved would be much smaller. The large amounts were chosen in order to stress the point, but other examples with smaller amounts could be found that give essentially the same results. An everyday example is the purchase of an insurance policy that has a negative expected payoff or an expected payoff less than that of a savings account in a bank. Also, the bets are to be offered strictly as a one-shot affair, as was pointed out. If the situation were to be repeated several times, bet 2 would surely be preferred to bet 1, for the chances of winding up with nothing would be greatly reduced. If the choice of bets is repeated just 4 times, the probability of winding up with nothing if bet 2 is chosen each time is just $(\tfrac{1}{2})^4$, or $\tfrac{1}{16}$. Of course, if the bets are to be repeated, then you must choose an overall strategy. You might, for instance, choose bet 1 on the first trial to be assured of \$1 million and then choose bet 2 thereafter.

The examples illustrate the fact that the "value" of a dollar may differ from person to person and that for any specific person, the value of x dollars is not necessarily x times the value of a single dollar. As a result, the ER rule may not be satisfactory when the payoffs are expressed in terms of money. If it could somehow be possible to measure the true relative value to the decision maker of the various possible payoffs in a problem of decision making under uncertainty, expectations could be taken in terms of these "true" values. The theory of **utility** prescribes such a decision-making rule: the **maximization of expected utility,** or the EU criterion. In order to understand this criterion, it is first necessary to discuss briefly the concept of utility.

Essentially, the theory of utility makes it possible to measure the *relative* value to a decision maker of the payoffs, or consequences, in a decision problem. Formally, a **utility function** U can be interpreted in terms of a preference relationship. The two basic axioms of utility are as follows.

1. *If payoff R_1 is preferred to payoff R_2, then $U(R_1) > U(R_2)$; if R_2 is preferred to R_1, then $U(R_2) > U(R_1)$; and if neither is preferred to the other, then $U(R_1) = U(R_2)$.*
2. *If you are indifferent between* (a) *receiving payoff R_1 for certain and* (b) *taking a bet or lottery in which you receive payoff R_2 with probability p and payoff R_3 with probability $(1 - p)$, then*

$$U(R_1) = pU(R_2) + (1 - p)U(R_3).$$

The first axiom simply states that a payoff that is preferred to another payoff has a higher utility than the other payoff. The second axiom implies that in decision-making situations involving uncertainty, actions can be compared on the basis of their expected utilities.

It is important to note that the utility function we have defined measures utility on an interval scale (recall the discussion of measurement scales in Section 5.3). Moreover, if U satisfies the two axioms given above, then so does any linear transformation of the form $a + bU$ with $b > 0$. That is, a utility function is only unique up to a positive linear transformation. This means that the values taken on by a utility function are meaningful only in relation to each other. The fact that $U(R) = 100$ or that $U(R)$ is negative means nothing by itself. Therefore, you should be careful not to think that a positive utility is "good" and a negative utility is "bad." It is true that for a given utility function U, positive values are preferred to negative values, but even the positive values may represent "bad" payoffs or the negative values may represent "good" payoffs. The discussion of the assessment of utility functions in the next section should clarify this point somewhat.

The basis for the theory of utility, as well as for a joint theory of utility and subjective probability, is provided by certain **axioms of coherence.** These axioms provide a model of how a "rational man" makes decisions in the face of uncertainty. For example, one such axiom states that preferences can be ordered. That is, given any two payoffs R_1 and R_2, you can decide whether you prefer payoff R_1 to R_2, you prefer R_2 to R_1, or you are indifferent between R_1 and R_2. At first glance, you may not be sure which payoff you prefer, but if you still cannot decide after some careful thought, it should be safe to say that you are, for all practical purposes, indifferent between the two payoffs. Another axiom provides the justification for the elimination of inadmissible actions from consideration in a decision-making problem. This axiom states that if R_1 is preferred to R_2 and if R_3 is some other payoff, then Lottery 1 is preferred to Lottery 2, where the two lotteries are as follows.

Lottery 1:   You receive R_1 with probability p.
              You receive R_3 with probability $1 - p$.

Lottery 2:   You receive R_2 with probability p.
              You receive R_3 with probability $1 - p$.

The two lotteries are identical except for the fact that R_2 is used in place of R_1 in Lottery 2. But R_1 is preferred to R_2, so in the terminology of Section 9.2, Lottery 2 is inadmissible because it is dominated by Lottery 1. These two examples illustrate the axioms of coherence, and two other examples (transitivity and substitutability) were presented in Section 2.16.

If the choices of a person satisfy the axioms of coherence, this implies the existence of a utility function reflecting preferences for various payoffs and a subjective probability distribution reflecting judgments about the random variables of interest. We will not go into further detail with regard to the axioms of coherence. However, for the purposes of decision theory, the most important thing to note is that the axioms of coherence imply that **a person should make decisions in such a manner as to maximize expected utility.** This provides a very powerful argument in favor of the *EU* criterion. Anyone willing to accept the axioms of coherence must also accept the *EU* criterion, since it follows logically from the axioms.

It should be emphasized that the outline of utility theory presented in this section and the following section is only a brief, rough development of some of the important points of this theory, although it will suffice for our purposes. In the next section, we turn to a practical problem of interest: the assessment of utility functions.

9.6   THE ASSESSMENT OF UTILITY FUNCTIONS

Using the axioms of utility, it is possible for a person to assess a utility function. Suppose that there are a number of possible payoffs in a decision-making problem facing an individual. First of all, it is necessary to determine what the decision maker considers to be the most preferable and the least preferable payoffs. Call the most preferable payoff R^* and the least preferable payoff R_*. Since a utility function is unique only up to a positive linear transformation, the decision maker can arbitrarily assign any values to $U(R^*)$ and $U(R_*)$, provided that $U(R^*) > U(R_*)$. Suppose that $U(R^*) = 1$ and $U(R_*) = 0$. The choice of 1 and 0 is arbitrary; we could just as easily choose 3 and -5, or 21.8 and 10.2, or any such pair of values, but 1 and 0 simplify the calculations somewhat, as we shall see.

Now consider any payoff R. From the choice of R^* and R_* and from the first axiom of utility presented in the preceding section, it must be true that

$$U(R_*) \leq U(R) \leq U(R^*),$$

or
$$0 \leq U(R) \leq 1.$$

To determine the value of $U(R)$ more precisely, consider the following choice of lotteries.

Lottery I: Receive R for certain.

Lottery II: Receive R^* with probability p and receive R_* with probability $(1 - p)$.

How should a person choose between these lotteries? According to the *EU* criterion (essentially, according to the second axiom of utility presented in the preceding section), the lottery with the higher expected utility should be selected. The expected utilities are

$$EU(\text{Lottery I}) = U(R)$$

and
$$EU(\text{Lottery II}) = pU(R^*) + (1 - p)U(R_*)$$
$$= p(1) + (1 - p)(0) = p.$$

Thus, if $U(R) > p$, Lottery I should be chosen; if $U(R) < p$, Lottery II should be chosen; and if $U(R) = p$, the person should be indifferent between the two lotteries.

This relationship between $U(R)$ and p can be exploited to determine the utility of R. If the decision maker can determine a probability p that makes him or her indifferent between the two lotteries, then the utility of R must be equal to this value, p. In this manner, the utility of any consequence, or payoff, can be found once the most and least preferable payoffs, R^* and R_*, have been determined. Because of the choice of $U(R^*) = 1$ and $U(R_*) = 0$, the utility for any other payoff can be interpreted in terms of an indifference probability.

For example, consider the oil-drilling example presented in Section 2.17, with the payoff table given in Table 9.6.1 in terms of dollars. Here, $R^* = \$100,000$ and $R_* = -\$30,000$. To find the utility of $\$20,000$, the follow-

Table 9.6.1   Payoff Table for Oil-Drilling Example

STATE OF THE WORLD

		Oil	*No oil*
	Drill	100,000	−30,000
ACTION			
	Do not drill	20,000	10,000

ing lotteries are used.

Lottery I: Receive $20,000 for certain.

Lottery II: Receive $100,000 with probability p and receive −$30,000 with probability $(1 − p)$.

The decision maker might decide that Lottery I is clearly preferable if $p = .30$ and that Lottery II is clearly preferable if $p = .60$. After giving it some thought, the decision maker chooses $p = .45$ as the indifference point.

Next, to find the utility of $10,000, the decision maker looks at these lotteries.

Lottery I: Receive $10,000 for certain.

Lottery II: Receive $100,000 with probability p and receive −$30,000 with probability $(1 − p)$.

Suppose that the indifference point for these lotteries is $p = .35$. The payoff table is given in Table 9.6.2 in terms of utilities. If $P(\text{oil}) = .3$, the expected utilities of the two actions are

$$EU(\text{drill}) = .3(1) + .7(0) = .30$$

and $$EU(\text{do not drill}) = .3(.45) + .7(.35) = .38.$$

According to the EU criterion, the decision maker should not drill.

Table 9.6.2   Utility Table for Oil-Drilling Example

STATE OF THE WORLD

		Oil	*No oil*
	Drill	1	0
ACTION			
	Do not drill	.45	.35

In the oil-drilling example, the problem is sufficiently simple that utilities are needed for only four payoffs. In some cases many different payoffs are of interest. Instead of assessing a utility separately for each payoff, it is often easier to attempt to determine a function relating monetary payoffs to utility over some interval of monetary values. For example, suppose that an individual selects the interval from $-\$100$ to $+\$100$ and lets $U(-\$100) = 0$ and $U(+\$100) = 1$. Next, the individual takes several values between $-\$100$ and $+\$100$ and determines their utilities. For instance, to determine the utility of $0, consider the following two lotteries.

Lottery I: Receive $0 for certain.

Lottery II: Receive $100 with probability p and receive $-\$100$ with probability $(1 - p)$.

After considering several values of p, suppose that the individual decides that $p = .70$ is the indifference point for these two lotteries (that is, for $p < .70$, Lottery I is preferred; for $p > .70$, Lottery II is preferred). From the results of the previous section, this means that $U(\$0) = .70$. Even though $p = .5$ would make the expected payoffs of the two lotteries equal in terms of money, the individual feels that the probability of winning in Lottery II has to be .70 before he or she is indifferent between the sure thing in Lottery I and the risky situation in Lottery II. In a similar fashion, the individual assesses $U(-\$50)$ to be .40 and $U(+\$50)$ to be .90. These values can be plotted on a graph, as in Figure 9.6.1, and a rough curve can be drawn through them. This curve is the individual's **utility function for money.** The utility of any monetary payoff between $-\$100$ and $+\$100$ can be read directly from the curve.

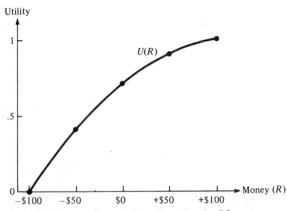

Figure 9.6.1   A Utility Function for Money

It is important to keep in mind that the limits of 0 and 1 that we have chosen for utility functions in this section are arbitrary limits. In general, utility functions can take on any values. Particularly when mathematical functions are used to approximate utility functions (this will be discussed briefly in the next section), values outside of the interval from 0 to 1 are common. The important point to remember is that only the *relative* values of utilities with respect to each other are important and that a utility of 0 has no fixed, special significance.

9.7 SHAPES OF UTILITY FUNCTIONS

It is useful to distinguish among several types, or classes, of utility functions, although there are certainly utility functions not falling into any of the classes to be described. In Figures 9.7.1 through 9.7.3 three utility curves are presented. The curve in Figure 9.7.1 represents the utility function of a *risk-avoider*, the curve in Figure 9.7.2 represents the utility function of a *risk-taker*, and the curve in Figure 9.7.3 represents the utility function of a person who is neither a risk-taker nor a risk-

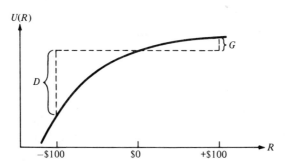

Figure 9.7.1   A Utility Function for a Risk-Avoider

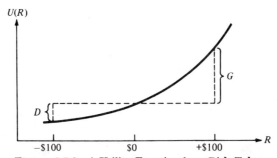

Figure 9.7.2   A Utility Function for a Risk-Taker

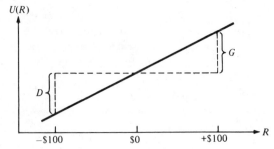

Figure 9.7.3   A Utility Function for a Risk-Neutral Individual

avoider. To see why these terms aptly describe the curves, consider a bet in which you win \$100 with probability $\frac{1}{2}$ and you lose \$100 with probability $\frac{1}{2}$. This can be thought of as a bet of \$100 on the toss of a fair coin. In terms of expected *payoff*, you should be indifferent about the bet since it has an expected payoff of zero. In terms of expected *utility*, however, the decision as to whether or not to take the bet depends on the shape of your utility function.

In Figures 9.7.1 through 9.7.3, the *gain* in utility if the bet is won is

$$G = U(+\$100) - U(\$0), \qquad (9.7.1)$$

and the decrease in utility if the bet is lost is

$$D = U(\$0) - U(-\$100). \qquad (9.7.2)$$

The expected utility of the bet is

$$EU(\text{bet}) = \tfrac{1}{2}U(+\$100) + \tfrac{1}{2}U(-\$100).$$

The alternative action is not to bet, and the expected utility of this is just

$$EU(\text{not bet}) = U(\$0).$$

Under what circumstances would you take the bet? Using the EU rule, you would take the bet if

$$EU(\text{bet}) > EU(\text{not bet});$$

that is, if

$$EU(\text{bet}) - EU(\text{not bet}) > 0. \qquad (9.7.3)$$

But from the above equations,

$$EU(\text{bet}) - EU(\text{not bet}) = \tfrac{1}{2}U(+\$100) + \tfrac{1}{2}U(-\$100) - U(\$0).$$

The right-hand side of this equation can be written as

$$\tfrac{1}{2}U(+\$100) + \tfrac{1}{2}U(-\$100) - \tfrac{1}{2}U(\$0) - \tfrac{1}{2}U(\$0),$$

which is equal to

$$[\tfrac{1}{2}U(+\$100) - \tfrac{1}{2}U(\$0)] - [\tfrac{1}{2}U(\$0) - \tfrac{1}{2}U(-\$100)],$$

or, using Equations (9.7.1) and (9.7.2),

$$\tfrac{1}{2}G - \tfrac{1}{2}D = \tfrac{1}{2}(G - D).$$

From Equation (9.7.3), the decision rule is as follows.

Take the bet if $\tfrac{1}{2}(G - D) > 0$.

Do not take the bet if $\tfrac{1}{2}(G - D) < 0$.

Thus, in order to make the decision, you need only look at the sign of $G - D$.

For the curve in Figure 9.7.1, G is smaller than D, so that $(G - D)$ is negative, and you should not take the bet. Since you will not take a bet with an expected monetary payoff of zero, you are called a **risk–avoider.** In fact, with this curve it is possible to find some bets with expected monetary payoffs *greater than zero* that you would consider unfavorable in terms of expected utility. Mathematically, the utility function graphed in Figure 9.7.1 is called a concave function.

In Figure 9.7.2, G is greater than D, and as a result you *should* take the bet. Furthermore, there are some bets with *negative* expected monetary payoffs that you would consider to be favorable bets. As a result, this curve represents the utility function of a **risk–taker.** Mathematically, this type of function is called a convex function.

Finally, $G = D$ in Figure 9.7.3. In this case you are indifferent between taking the bet and not taking it, and thus you are neither a risk-taker nor a risk-avoider. For a person with a **linear** utility function (that is, the curve is a straight line), maximizing EU is equivalent to maximizing ER. To prove this, note that if the utility curve is linear, it can be written in the form

$$U = a + bR, \tag{9.7.4}$$

where R is expressed in terms of money and a and b are constants, with $b > 0$. The requirement that b be greater than 0 is necessary because the line must have a positive slope. Since a and b are constants,

$$EU = a + bER, \tag{9.7.5}$$

and EU is at a maximum when ER is at a maximum. This is an important result.

If a person's utility function is linear with respect to money, then the ER criterion and the EU criterion are equivalent.

In the above example, the probabilities for winning and losing are equal, and thus the decision depends only on the sign of $G - D$. If the probabilities are unequal, the difference in expected utilities of the two actions, and hence the decision, will depend on the relative magnitudes of G and D, not just on the sign of $G - D$. However, the results regarding the shapes of the curves are unchanged. Furthermore, even if the bet (or the decision-making problem) is a complex problem with numerous possible consequences, the results still hold provided that the expected payoff of the bet is $0. You should keep in mind that a bet with an expected payoff of $0 is a "fair" bet, regardless of the exact nature of the probabilities or the payoffs. For instance, risk-avoiders would not take a bet in which they could win $10 with probability .1, win $9 with probability .2, and lose $4 with probability .7; the expected payoff is $0, and risk-avoiders will not take a "fair" bet. In addition, be careful to note that not all decision-making problems are "fair bets." Thus, it is not true that a risk-avoider will decline all gambles. If a gamble is favorable enough (if the ER of the gamble is large enough), a risk-avoider will take the gamble. Similarly, a risk-taker will find some gambles (those with highly negative expected payoffs) unfavorable. Thus, refusing a gamble does not necessarily mean that the person is a risk-avoider, and taking a gamble does not necessarily imply a risk-taker.

The curves discussed above are by no means the only possible forms for utility functions. For instance, some individuals appear to be risk-takers for some decisions (such as gambling) and risk-avoiders for other decisions (such as purchasing insurance). One possible form for such an individual's utility function is shown in Figure 9.7.4. Also, it should be noted that it is sometimes convenient to approximate a person's utility function by a simple mathematical function. One example is, of course, a linear function, as noted above. Another example is that of an exponential

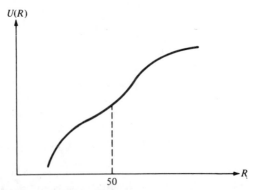

Figure 9.7.4 A Utility Function Exhibiting Both Risk-Aversion and Risk-Preference

utility function:

$$U(R) = a - e^{-bR}, \qquad (9.7.6)$$

where a and b are constants. If the decision maker's utility function can be approximated in this manner, it is necessary only to "plug" values of R into Equation (9.7.6) in order to convert the payoffs in a decision-making problem into utilities. On the other hand, without such a convenient mathematical function, it is necessary to assess the utility for each payoff separately or to attempt to read values from a graph. Attempting to approximate a utility function with some mathematical function is analogous to attempting to approximate a probability distribution with a member of some well-known family of distributions (for example, attempting to approximate a prior distribution with a beta distribution when sampling from a Bernoulli process, as in Section 8.11).

In any given decision-making problem, the utility function of interest is that of the individual (or group, or corporation) who will receive the payoff. Different individuals may have different utility functions, of course. Moreover, a person's utility function may change over time. For instance, if someone gave you $100,000, your preferences for various payoffs might change significantly.

In the examples presented in this chapter, the payoffs have all been expressed in terms of money. This is not always the case; it is easy to think of examples in which the consequences of a decision are nonmonetary. For instance, in many decisions factors such as time, inconvenience, prestige, social approval, and so on, are important. In the development of utility theory, there is no reason to restrict the payoffs to monetary payoffs. In general, payoffs can be thought of as **consequences** that may involve nonmonetary as well as monetary factors. The utility function then represents preferences among such consequences. For example, in a job decision the consequences might be of the form ($15,000 salary, Chicago, good opportunity for promotion), ($18,000 salary, San Francisco, little opportunity for promotion), ($18,000 salary, New York City, moderate opportunity for promotion), and so on. In the examples in this book, we will assume that nonmonetary factors can be expressed in terms of money. In other words, we will replace nonmonetary factors by their *cash equivalents*. You should keep in mind, however, that this is not always an easy thing to do in practice.

9.8 A FORMAL STATEMENT OF THE DECISION PROBLEM

Now that all of the components of a problem of decision under uncertainty have been discussed, some new notation will assist us in exploring their use. In any decision-making problem, we start out with a set of possible actions, or decisions. Call this set A and denote a typical member of this set (that is, an action) by a. In addition, it is presumed that the uncertainty in the problem is uncertainty about the state of the world θ. The set of possible states of the world, or possible values of θ, will be labeled S. We further assume that the uncertainty about θ can be expressed in probabilistic terms, in the form of a probability distribution of θ. This distribution will be called $P(\theta)$ if θ is discrete and $f(\theta)$ if θ is continuous. Note that this distribution may be a posterior distribution $f''(\theta \mid y)$ following an observed sample result y, or it may be a prior distribution $f'(\theta)$ if there is no sample information. Unless it becomes necessary to do so (and it will when optimal sample size is discussed), we will not be concerned about whether the distribution is prior or posterior. There is also a set of consequences, with one consequence for each combination of an action and a state of the world: (a, θ). Finally, there is a utility function U, giving the utility of each consequence. Since the consequence depends on the combination (a, θ), we will write the utility function in the form $U(a, \theta)$.

Recapitulating, we have the following inputs to the decision problem:

1. A, the set of actions, or acts.
2. S, the set of states of the world.
3. $P(\theta)$ or $f(\theta)$, the probability distribution of θ, the state of the world.
4. $U(a, \theta)$, the utility function which associates a utility with each pair (a, θ), where a is an action and θ is a state of the world.

Given these inputs, we must make a decision, which amounts to selecting an act a from the set A. We will select the act which is optimal under the EU rule. In other words, we will select the act which has the highest expected utility of all acts, or of all members of A. The expected utility of any act a can be calculated by using the definition of expectation, where the expectation is taken with regard to the distribution of θ. We will denote the expected utility of act a by $EU(a)$. If θ is discrete, the expected utility of a is

$$EU(a) \;=\; \sum U(a, \theta)P(\theta), \qquad (9.8.1^*)$$

where the summation is taken over all θ in S. If θ is continuous, we have

$$EU(a) \;=\; \int U(a, \theta)f(\theta) \; d\theta, \qquad (9.8.2^*)$$

where the integration is over the set S. An act a^* is **optimal** with respect to the decision problem if

$$EU(a^*) \;\geq\; EU(a) \quad \text{for all acts } a \text{ in } A. \qquad (9.8.3^*)$$

It should be mentioned that it is mathematically possible for such an optimal act not to exist but that in real-world problems this will not be the case.

As we saw in the previous section, the maximization of expected utility is equivalent to the maximization of ER (or the minimization of EL) *if* the utility function is linear with respect to money. To the extent that a utility function in a given situation can be closely approximated by a linear function, then, this provides a strong justification for the ER and EL criteria. How reasonable is the assumption that the utility function is linear with respect to money? For "small" amounts of money, attitude toward risk seems not to be too crucial in decision making, so linearity might not be a bad assumption. Of course, what is a small amount to one person (say, a millionaire) may be a very large amount to another person (say, a student). For a corporation, amounts that would be large to most individuals may be quite small in relation to the total assets of the corporation. In any given situation, it is necessary to investigate carefully the assumption of linearity before working with monetary payoffs. Whenever the assumption seems reasonable, the decision-making problem is simplified in the sense that it is not necessary to convert payoffs into utilities.

In some examples in the remainder of this chapter we will work with payoffs rather than utilities, thus implicitly assuming a linear utility function. In this case, an act a^* is optimal with respect to the decision problem if

$$ER(a^*) \geq ER(a) \quad \text{for all acts } a \text{ in } A, \qquad (9.8.4^*)$$

or, equivalently, if

$$EL(a^*) \leq EL(a) \quad \text{for all acts } a \text{ in } A. \qquad (9.8.5^*)$$

We hope that the student is not bothered by the overly formal nature of this section. The purpose is to take the parts that we have been developing (primarily the prior or posterior distribution and the utility function) and see how they fit together. In the process, the notation is also coordinated. There is nothing in this section that is really new, although we have not yet attacked a decision problem in which the state of the world is a continuous random variable.

9.9   DECISION MAKING UNDER UNCERTAINTY: AN EXAMPLE

To illustrate the application of the framework presented in this chapter for decision making under uncertainty, consider an oil-drilling venture. Assume that the decision maker owns drilling rights at a particular location and that he must decide how best to take advantage of these rights. To begin, he must determine A, the set of actions available to him. After giving this matter some thought, he arrives at five potential actions: (1) drilling with 100 percent interest; (2) finding a partner and drilling with 50 percent interest; (3) "farming out" the drilling rights for a $\frac{1}{8}$ override [this means that he will not have to share in the costs of drilling, but he will receive $\frac{1}{8}$ of the revenue, or net profit (net with respect to all costs except drilling costs), from any oil]; (4) "farming out" the drilling rights for $10,000 and a $\frac{1}{16}$ override; and (5) not drilling. Obviously, many other actions might be considered here; several partners could be found instead of just one, and numerous "farming out" arrangements could be devised. Perhaps the decision maker has chosen these five because he is sure that these particular options are available. The decision maker might not be sure about the availability of other deals that have not yet been proposed, or he may feel that these five actions provide him with a reasonable range of alternatives and that it is not worthwhile to seek out additional alternatives. Sometimes the search for additional actions can be quite costly and time-consuming.

The next input to the problem involves S, the set of states of the world. In this oil-drilling venture, the states of the world correspond to the random variable θ, which is defined as the amount of oil (in barrels) recovered from the well. This variable could be taken to be continuous, but in order to keep the model small and manageable (or because other values are considered somewhat unlikely), the decision maker decides to consider only five possible values: 0 (dry hole), 50,000 barrels, 100,000 barrels, 500,000 barrels, and 1,000,000 barrels.

Taking into consideration information about other oil wells in the general vicinity and about the geological structure at the site in question, the decision maker (or a geologist hired by the decision maker) feels that the odds are in favor of a dry hole. In particular, he feels indifferent between the following two lotteries.

Lottery A: Receive $0 with probability .7 and $100 with probability .3.

Lottery B: Receive $0 if there is no oil (a dry hole) and $100 if there is oil.

This implies that $P(\text{dry hole}) = P(\theta = 0) = .7$. Of the remaining four values of θ, the decision maker judges $\theta = 100,000$ to be 3 times as likely as any of the other three values, which he considers to be approximately equally likely. This implies that

$$P(\theta = 50,000) = P(\theta = 500,000) = P(\theta = 1,000,000)$$

and $$P(\theta = 100,000) = 3P(\theta = 50,000).$$

Since $P(\theta = 0) = .7$, the sum of the probabilities for the remaining four values of θ must be .3. From this fact and from the preceding equations, the decision maker's probability distribution of θ is as follows:

$$P(\theta = 0) = .70,$$

$$P(\theta = 50{,}000) = .05,$$

$$P(\theta = 100{,}000) = .15,$$

$$P(\theta = 500{,}000) = .05,$$

and $$P(\theta = 1{,}000{,}000) = .05.$$

The mean and the standard deviation of θ are 92,500 and 236,000, respectively.

Now the set of potential *actions*, the set of *states of the world* (values of θ), and the *probability distribution* representing the decision maker's uncertainty about θ have been determined. The next step is to consider the *payoffs* associated with each particular combination of an action and a state of the world. With regard to the cost of drilling, the decision maker has a firm estimate of $68,000. He is not absolutely certain about the value of any oil that is recovered, but he is willing to assume that the net profit from the oil will be approximately $.96 per barrel. With this information, he can construct a payoff table for the oil-drilling venture. If he drills with 100 percent interest, his payoff in dollars will be $.96\theta - 68{,}000$ (the total revenue from the oil, less the cost of drilling). If he takes a partner and drills with 50 percent interest, the payoff will be split into two halves, so that his payoff will be $.48\theta - 34{,}000$. If he farms out the drilling rights for a $\frac{1}{8}$ override, he will incur no drilling costs and will receive $\frac{1}{8}$ of the revenue from the oil, which will be $.96\theta/8 = .12\theta$. If he farms out the drilling rights for $10,000 and a $\frac{1}{16}$ override, his payoff will consist of $10,000 plus $\frac{1}{16}$ of the revenue, or $.06\theta + 10{,}000$. Finally, if he does not drill, his payoff will be 0. By calculating the values of the payoffs for each of the five values of θ, the decision maker arrives at a payoff table for his problem, and this is shown in Table 9.9.1.

The last action is clearly dominated by the next-to-last action, so not drilling is inadmissible. However, it is unlikely that the payoff corresponding to "do not drill" is actually $0, since the decision maker might be able to wait and drill at a later date. Of course, actions such as "do nothing now, then drill in a year" could be included in the set of actions. For the purposes of this example, assume that the drilling rights expire shortly. Therefore, it is not possible to postpone drilling, and "do not drill" is an inadmissible action.

Table 9.9.1　Payoff Table for Oil-Drilling Example

STATE OF THE WORLD (VALUE OF θ)

		0	50,000	100,000	500,000	1,000,000
	Drill with 100% interest	−68,000	−20,000	28,000	412,000	892,000
	Drill with 50% interest	−34,000	−10,000	14,000	206,000	446,000
ACTION	Farm out, $\frac{1}{8}$ override	0	6,000	12,000	60,000	120,000
	Farm out, $10,000 and $\frac{1}{16}$ override	10,000	13,000	16,000	40,000	70,000
	Do not drill	0	0	0	0	0

For the remaining four actions, the expected payoffs are

$$ER(\text{drill with 100\% interest}) = (-\$68{,}000)(.70)$$
$$+ (-\$20{,}000)(.05)$$
$$+ (\$28{,}000)(.15)$$
$$+ (\$412{,}000)(.05)$$
$$+ (\$892{,}000)(.05)$$
$$= \$20{,}800,$$

$$ER(\text{drill with 50\% interest}) = \$10{,}400,$$

$$ER(\text{farm out, } \tfrac{1}{8} \text{ override}) = \$11{,}100,$$

and　$ER(\text{farm out, \$10,000 and } \tfrac{1}{16} \text{ override}) = \$15{,}500.$

Under normal conditions, the decision maker feels that his utility function for money is approximately linear. However, he has suffered some unexpected losses recently, and his cash position is dangerously low. As a result, his current utility function for money is quite nonlinear and is shaped like the utility function of a risk-avoider. In order to graph his utility function, he considers various lotteries of the nature discussed in Section 9.6. First, he lets $U(\$900{,}000) = 1$ and $U(-\$100{,}000) = 0$, since all of the payoffs in his oil-drilling venture lie between these two values. Next, he considers the following two lotteries.

Lottery I: Receive $0 for certain.

Lottery II: Receive $900,000 with probability p and $-$100,000 with probability $(1 - p)$.

At first glance, he decides that he would surely prefer Lottery I if $p = .25$ and Lottery II if $p = .40$. After some serious thought, he chooses $p = .32$ as his indifference probability; if $p = .32$, he is indifferent between the two lotteries. It is obvious from this choice that the decision maker is a risk-avoider, for the expected monetary payoff for Lottery II with $p = .32$ is $220,000; he is indifferent between $0 for certain and a risky gamble with $ER = $220,000$.

The decision maker assesses utilities for several other amounts, plots these utilities as points on a graph, and attempts to draw a smooth curve that is a good fit to the assessed points. This is illustrated in Figure 9.9.1. After considering several additional gambles in order to check the "fit" of the utility function to his preferences, the decision maker decides that the smooth curve in Figure 9.9.1 is a very good approximation to his utility function and that he is willing to use this curve as his utility function in the oil-drilling venture. By reading values from the graph, the decision maker can express the payoff table in terms of utilities. For instance,

U(drill with 100 percent interest, dry hole) $= U(-$68,000) = .11$.

The entire table of utilities is shown as Table 9.9.2.

Table 9.9.2   Utility Table for Oil-Drilling Example

STATE OF THE WORLD (VALUE OF θ)

		0	50,000	100,000	500,000	1,000,000
	Drill with 100% interest	.11	.27	.38	.81	1.00
	Drill with 50% interest	.23	.30	.35	.63	.84
ACTION	Farm out, $\frac{1}{8}$ override	.32	.34	.35	.44	.53
	Farm out, $10,000 and $\frac{1}{16}$ override	.34	.35	.36	.40	.45

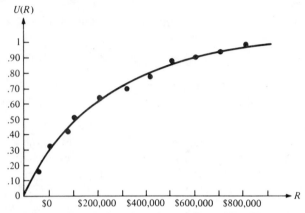

Figure 9.9.1   Utility Function for Oil-Drilling Example

The expected utilities of the four actions to the decision maker are

$$EU(\text{drill with } 100\% \text{ interest}) = (.11)(.70) + (.27)(.05)$$
$$+ (.38)(.15) + (.81)(.05)$$
$$+ 1(.05)$$
$$= .2380,$$
$$EU(\text{drill with } 50\% \text{ interest}) = .3020,$$
$$EU(\text{farm out, } \tfrac{1}{8} \text{ override}) = .3420,$$
$$\text{and } EU(\text{farm out, } \$10,000 \text{ and } \tfrac{1}{16} \text{ override}) = .3520.$$

Thus, the action with the largest expected utility is to farm out the drilling rights for $10,000 and a $\tfrac{1}{16}$ override. Because of the risk-avoiding nature of the decision maker's utility function, the action with the largest expected payoff (drill with 100 percent interest) has the smallest expected utility. It just happens that this is the riskiest action in this oil-drilling venture. Note that "drill with 50 percent interest" has a much higher expected utility than "drill with 100 percent interest," although the expected monetary payoff of the former is only half as large as that of the latter. This illustrates the concept of *risk-sharing*; by finding a partner to take a 50 percent interest, the decision maker is reducing his risk. The same principle is illustrated by insurance companies when they "split up" a large insurance policy among several companies to spread the risk. An insurance company would much rather have a $\tfrac{1}{10}$ share in each of 10 large policies than have a 100 percent share in one of the policies.

Of course, the above results depend on the precise inputs to the formal decision-making model. For example, if the decision maker could not find anyone who would offer him \$10,000 in cash and a $\frac{1}{16}$ override for the drilling rights, then this action would have to be eliminated from consideration. Perhaps someone might offer him \$5000 in cash and a $\frac{1}{16}$ override; he could then include this action in the model, determine the appropriate payoffs and utilities, and compute the expected utility.

Similarly, any change in the set of states of the world, S, could change matters. For instance, the decision maker might consider θ to be continuous rather than discrete. Or even if the set of actions, S, is not changed, the probability distribution for θ could change. By looking at the table of utilities for the oil-drilling venture, we see that if the probabilities for $\theta = 0$, $\theta = 50,000$, or $\theta = 100,000$ are increased, this will lend even more support to "farm out for \$10,000 and a $\frac{1}{16}$ override." Increases in the probabilities for $\theta = 500,000$ and $\theta = 1,000,000$, however, might make one of the other three actions optimal under the expected utility criterion.

Suppose that the decision maker feels that the highest that $P(\theta = 500,000)$ or $P(\theta = 1,000,000)$ could possibly be is .10. In order to see how sensitive the oil-drilling venture is to changes in the probabilities, then, he considers the following probability distribution:

$$P(\theta = 0) = .60,$$

$$P(\theta = 50,000) = .10,$$

$$P(\theta = 100,000) = .10,$$

$$P(\theta = 500,000) = .10,$$

and $$P(\theta = 1,000,000) = .10.$$

Using this distribution, the expected utilities are

$$EU(\text{drill with 100\% interest}) = .3120,$$

$$EU(\text{drill with 50\% interest}) = .3500,$$

$$EU(\text{farm out, } \tfrac{1}{8} \text{ override}) = .3580,$$

and $$EU(\text{farm out, \$10,000 and } \tfrac{1}{16} \text{ override}) = .3600.$$

The optimal action is still to farm out the drilling rights for \$10,000 and a $\frac{1}{16}$ override. Since the decision maker chose this probability distribution as an extreme distribution that would favor the other actions as much as possible (while still remaining somewhat in line with his judgments), it appears that the oil-drilling venture is fairly insensitive with respect to changes in the probability distribution.

Another input to the oil-drilling venture is the set of payoffs, which was determined on the basis of the decision maker's judgment that the costs of drilling are $68,000 and that the revenue per barrel of oil is $.96. The decision maker is virtually certain that the $68,000 figure is correct, but he is somewhat uncertain about the revenue per barrel of oil. To investigate the sensitivity of the problem to changes in the revenue per barrel of oil, he decides to consider two values: a high value, $1.12, which he is virtually certain will *not* be exceeded, and a low value, $.80, which he is virtually certain *will* be exceeded. For each of these values, a new payoff table is determined and converted into a utility table. Using the original set of probabilities, the expected utilities for the four actions turn out to be .2470, .3100, .3435, and .3535, respectively, when the revenue per barrel of oil is $1.12. Thus, farming out the drilling rights for $10,000 and a $\frac{1}{16}$ override is still the optimal action. If the revenue per barrel is $.80, the expected utilities are .2285, .2925, .3370, and .3505, so that the last action is once again optimal. Therefore, the oil-drilling venture appears to be quite insensitive to changes in the revenue per barrel of oil.

Possible changes in the set of actions A and in the set of states of the world S have been informally discussed, and possible variations in the probability distribution of θ and in the payoffs have been formally investigated in terms of a sensitivity analysis. Similarly, variations in the decision maker's utility function could be investigated. Sensitivity analysis, whether conducted informally or formally, is an important part of decision making under uncertainty. It may pinpoint the crucial elements of a problem, and it might suggest new actions. In the oil-drilling venture, the results indicate that maintaining only a part interest in the venture or farming the venture out is better for the decision maker than keeping a 100 percent interest. This result might cause the decision maker to work harder at seeking out partners or "farming out" deals. This could lead to possible new actions to be included in the model. Given the decision maker's aversion to risk, perhaps he should try to sell the drilling rights for a fixed amount of money.

This oil-drilling example illustrates the application of the framework presented in this chapter for decision making under uncertainty. The importance of the determination of the inputs to the general framework has been stressed; this is largely a question of making the model for any particular application as realistic as possible without making it too complex to handle. Once all of the inputs are determined, the process of finding the optimal decision is merely a computational task.

9.10 TERMINAL DECISIONS AND PREPOSTERIOR DECISIONS

It has been assumed in this chapter that a statistician (decision maker) is faced with a problem of decision making under uncertainty and that a decision must be made on the basis of the current "state of information." Such a decision is called a **terminal decision.** If the decision maker's current state of information is represented by a posterior distribution, then the terminal decision should be based on this distribution (and, of course, on a loss function or a payoff function).

Suppose that the decision maker has the option of obtaining more sample information before making a terminal decision. Such sample information might be valuable in reducing the decision maker's uncertainty about the state of the world. For example, in a medical situation, it may be possible to run further diagnostic tests before making a decision regarding surgery; in oil drilling, the decision maker might run geological tests before deciding whether or not to drill; in an investment decision, the investor might be able to obtain more information about the investments under consideration (or about some related factor, such as the chances that taxes will be raised); in a marketing decision, it may be possible to obtain more information about consumers' intentions by conducting a market survey or to gain more information about the future actions of competitors by carefully observing their current actions (building new plants, initiating advertising campaigns, and so on). In these and many other examples, additional sample information could be quite useful to the decision maker.

Of course, there is a cost involved in sampling, so the decision maker must decide if the additional sample information is expected to be useful enough to justify its cost. This type of decision is called a **preposterior decision** because it involves the *potential* posterior distributions following the *proposed* sample. Note that this sample has not been observed yet; it is just being contemplated. The "value" of the sample usually depends on the observed result. Before taking the sample, the decision maker does not know what this result will be, although some probabilistic statements *can* be made about possible results.

It should be stressed that the value of the sample information is to be interpreted with respect to the terminal decision. In the framework presented in Section 9.8 for decision making under uncertainty, the decision of primary interest is the terminal decision. Sample information is valuable to the extent that it may be of some use to the decision maker in terms of the terminal decision.

Before we attack the problem of deciding whether or not to obtain *sample* information, we will investigate a related, but somewhat easier, problem. Suppose that the decision maker has the opportunity to purchase "perfect" information (that is, perfect knowledge of the state of the world), so that the problem of decision under uncertainty could be changed into the easier problem of decision under certainty. How much is this perfect information worth to the decision maker?

9.11 THE EXPECTED VALUE OF PERFECT INFORMATION

Suppose that the decision maker knows the value of θ for certain. Then in terms of the notation of Section 9.8, the optimal action a_θ is the act such that

$$R(a_\theta, \theta) \geq R(a, \theta) \quad \text{for all acts } a \text{ in } A, \tag{9.11.1}$$

or, equivalently,

$$L(a_\theta, \theta) \leq L(a, \theta) \quad \text{for all acts } a \text{ in } A. \tag{9.11.2}$$

In other words, if the payoffs or losses in a particular problem are expressed in the form of a payoff or loss table, then perfect information enables the decision maker to ignore all of the columns of the table except one.

In the advertising example in Section 2.17, the payoff table was as shown in Table 9.11.1. Letting $\theta = 1$ if the new product is introduced and $\theta = 0$ if the new product is not introduced, then $a_\theta =$ "major ad campaign" if $\theta = 1$ and $a_\theta =$ "no advertising" if $\theta = 0$.

For an example using a loss table, consider the stock-purchasing example (Section 9.2), with the loss table presented in Table 9.11.2. If you knew that state I would occur, then you would just consider the first column. The action "buy A" has the smallest loss in this column, so this is the optimal action. Similarly, the optimal action under perfect knowledge indicating state II (state III) will occur is to buy A (buy B). If you are sure that state IV will occur, then it makes no difference which action you select; the losses are the same.

Note from the above example that because of the way in which we defined "losses," the optimal act under certainty will always have a loss

Table 9.11.1 Payoff Table for Advertising Example

STATE OF THE WORLD

		Competitor introduces new product	Competitor does not introduce new product
	No advertising	100,000	700,000
ACTION	*Minor ad campaign*	300,000	600,000
	Major ad campaign	400,000	500,000

Table 9.11.2   Loss Table for Stock-Purchasing Example

STATE OF THE WORLD

		I	II	III	IV
	Buy A	0	0	400	0
ACTION	Buy B	200	400	0	0
	Buy C	400	800	600	0

equal to zero. In every column of a loss table, there is at least one zero, and there are never any negative losses. Therefore, if you have perfect information, your loss is zero.

It would be useful to find out exactly how much perfect information is worth to the decision maker. We will do this under the assumption that the decision maker's utility function is linear with respect to money. (Finding the expected value of perfect information is more complicated when the utility function is not linear, so we will not consider that case.) If a^* is the optimal action under the decision maker's current state of information, then

$$EL(a^*) \leq EL(a) \quad \text{for all actions } a \text{ in } A.$$

Under perfect information, the loss will be zero, as noted above, so the **expected value of perfect information** (EVPI) is equal to

$$\text{EVPI} = EL(a^*) - 0 = EL(a^*). \qquad (9.11.3^*)$$

That is, the expected value of perfect information to the decision maker is equal to the expected loss of the action which is optimal under the current state of information.

In the stock-purchasing example, suppose that the probabilities of the four states of the world are .3, .2, .1, and .4. The expected losses of the three actions are then

$$EL(\text{buy } A) = 0(.3) + 0(.2) + 400(.1) + 0(.4) = 40,$$

$$EL(\text{buy } B) = 200(.3) + 400(.2) + 0(.1) + 0(.4) = 140,$$

and $$EL(\text{buy } C) = 400(.3) + 800(.2) + 600(.1) + 0(.4) = 340.$$

The optimal action, buying stock A, has an expected loss of \$40, so this is the EVPI. *The investor should be willing to pay up to \$40 to obtain perfect information.*

Just as the expected loss of the currently optimal action is the EVPI, the loss in any single column of the payoff table for a^* can be thought of as the value of perfect information for that column, or that value of θ. In the stock-purchasing example, $a^* = $ "buy A," so consider the first row of the loss table. Under perfect information indicating that states I, II, or IV will occur, the decision maker has no reason to shift from buying A. Therefore, such information does not change the decision (that is, $a_\theta = a^*$), so the *value of perfect information* (VPI) is zero:

$$\text{VPI(state I)} = \text{VPI(state II)} = \text{VPI(state IV)} = 0.$$

(If state IV occurs, it makes no difference which action is taken, but the point is simply that the decision maker suffers no loss by sticking with "buy A.") For state III, on the other hand, the decision maker can avoid a loss of 400 by learning that state III will occur. Thus,

$$\text{VPI(state III)} = 400.$$

In general, the value of perfect information associated with a particular value of θ is equal to the difference between the payoff if the action optimal for that value of θ is taken and the payoff if the currently optimal action is taken. This difference is equal to the loss that is incurred if the currently optimal action is taken and the state of the world turns out to be θ:

$$\text{VPI}(\theta) = R(a_\theta, \theta) - R(a^*, \theta) = L(a^*, \theta). \qquad (9.11.4^*)$$

Taking expectations on both sides of Equation (9.11.4), we arrive at Equation (9.11.3), our equation for EVPI.

It is also possible to think of EVPI in terms of payoffs rather than losses. Under the current state of uncertainty, the decision marker's expected payoff is $ER(a^*)$.

With perfect information, the expected payoff (as calculated *before* the perfect information is obtained) is

$$\text{ERPI} = \sum_\theta R(a_\theta, \theta) P(\theta). \qquad (9.11.5^*)$$

Note that $R(a_\theta, \theta)$ is simply the largest payoff in the column corresponding to the state of the world θ in the payoff table, so that ERPI is the expectation of these "largest payoffs."

The EVPI is now simply the difference between the expected payoff under perfect information and the current expected payoff:

$$\text{EVPI} = \text{ERPI} - ER(a^*). \qquad (9.11.6^*)$$

In the advertising example, suppose that the odds are 3 to 2 in favor of the new product being introduced by the competitor. Thus, P(new product is introduced) $= 3/(3 + 2) = .6$ and P(new product is not introduced) $= .4$. The expected payoffs are

$$ER(\text{no advertising}) = \$340,000,$$

$$ER(\text{minor ad campaign}) = \$420,000,$$

and $$ER(\text{major ad campaign}) = \$440,000.$$

Thus, $a^* =$ "major ad campaign" and $ER(a^*) = \$440,000$. From Equations (9.11.5) and (9.11.6), then,

ERPI $= R$(major ad campaign, new product is introduced)P(new product is introduced) $+ R$(no advertising, new product is not introduced)P(new product is not introduced)

$$= \$400,000(.6) + \$700,000(.4) = \$520,000$$

and EVPI $=$ ERPI $- ER$(major ad campaign)

$$= \$520,000 - \$440,000 = \$80,000.$$

The decision maker should be willing to pay up to $80,000 for perfect information about the competitor's decision regarding the new product.

9.12   THE EXPECTED VALUE OF SAMPLE INFORMATION

From the EVPI, we know how much the decision maker should be willing to pay to obtain perfect information. If it is available at a cost less than or equal to the EVPI, then it is worthwhile to purchase it. Unfortunately, perfect information is seldom available, and the decision maker must take a sample if more information is desired. A sample will generally provide imperfect information (that is, some information but not perfect information), and for any proposed sampling procedure, it is useful to determine the **expected value of sample information.** Since sample information can never be any better than perfect information, the expected value of sample information (EVSI) will always be less than or equal to EVPI. Thus, although it is generally not possible to purchase perfect information, EVPI is still useful as an upper bound for EVSI. In this section, we will discuss the determination of the exact value of EVSI. As in the previous section, we assume that the decision maker's utility function is linear with respect to money.

Suppose that the prior distribution for θ is represented by $P'(\theta)$ or $f'(\theta)$ and that the optimal action under the prior distribution is a^*:

$$E'R(a^*) \geq E'R(a) \quad \text{for all acts } a \text{ in } A \qquad (9.12.1)$$

or $\qquad E'L(a^*) \leq E'L(a) \quad \text{for all acts } a \text{ in } A, \qquad (9.12.2)$

where the primes in $E'R$ and $E'L$ indicate that the expectations are taken with regard to the prior distribution. Now, a sample is taken and the sample result y is observed. The prior distribution is revised, yielding a posterior distribution $P''(\theta \mid y)$ or $f''(\theta \mid y)$. Denote the optimal act under this posterior distribution by a_y:

$$E''R(a_y \mid y) \geq E''R(a \mid y) \quad \text{for all acts } a \text{ in } A \qquad (9.12.3)$$

or $\qquad E''L(a_y \mid y) \leq E''L(a \mid y) \quad \text{for all acts } a \text{ in } A, \qquad (9.12.4)$

where the double primes in $E''R$ and $E''L$ indicate that the expectations are taken with regard to the posterior distribution.

Under the posterior distribution the optimal act is a_y, whereas under the prior distribution it is a^*. **The value of this particular sample information y to the decision maker is simply**

$$\text{VSI}(y) = E''R(a_y \mid y) - E''R(a^* \mid y) \qquad (9.12.5^*)$$

or $\qquad \text{VSI}(y) = E''L(a^* \mid y) - E''L(a_y \mid y). \qquad (9.12.6^*)$

In words, this is the posterior expected payoff of the now-optimal act a_y minus the posterior expected payoff of the previously-optimal act a^* (or, equivalently, the posterior expected loss of a^* minus the posterior expected loss of a_y). It should be emphasized that VSI depends on the particular sample result y.

For example, consider the simple oil-drilling example from Section 9.2. The payoff table is given in Table 9.12.1. If the prior probability of oil is .30, the expected payoffs (as calculated in Section 9.4) are $E'R(\text{drill}) = \$9000$ and $E'R(\text{do not drill}) = \$13,000$. Thus, $a^* = $ "do not drill."

Table 9.12.1 Payoff Table for Oil-Drilling Example

| | | STATE OF THE WORLD | |
		Oil	*No oil*
	Drill	100,000	−30,000
ACTION	*Do not drill*	20,000	10,000

Suppose that the decision maker can purchase sample information in the form of a seismic test that will indicate whether the geological structure at the drilling site is "closed" or "open." From past data regarding seismic tests, it is known that

$$P(\text{closed structure} \mid \text{oil}) = .8$$

and $$P(\text{closed structure} \mid \text{no oil}) = .4.$$

Of course, this implies that

$$P(\text{open structure} \mid \text{oil}) = .2$$

and $$P(\text{open structure} \mid \text{no oil}) = .6.$$

For each of the two possible sample outcomes, posterior probabilities can be computed via Bayes' theorem, as in Table 9.12.2. If the seismic test results in a closed structure, the posterior probabilities are 6/13 and 7/13, and the expected payoffs are

$$E''R(\text{drill} \mid \text{closed}) = 100{,}000 \left(\frac{6}{13}\right) + (-30{,}000) \left(\frac{7}{13}\right) = 30{,}000$$

and

$$E''R(\text{do not drill} \mid \text{closed}) = 20{,}000 \left(\frac{6}{13}\right) + 10{,}000 \left(\frac{7}{13}\right) = 14{,}615.$$

The optimal act a_y for $y =$ "closed structure" is thus to drill. In other words, a closed structure is favorable enough (increases the probability of oil enough) to change the action from "do not drill" to "drill." Using

Table 9.12.2 Bayes' Theorem for Oil-Drilling Example

Sample outcome	State of the world	Prior probability	Likelihood	(Prior probability) × (likelihood)	Posterior probability
Closed structure	Oil	.3	.8	.24	24/52 = 6/13
	No oil	.7	.4	.28	28/52 = 7/13
				.52	
Open structure	Oil	.3	.2	.06	6/48 = 1/8
	No oil	.7	.6	.42	42/48 = 7/8
				.48	

Equation (9.12.5), we can compute the VSI:

$$\text{VSI(closed)} = E''R(\text{drill} \mid \text{closed}) - E''R(\text{do not drill} \mid \text{closed})$$
$$= 30,000 - 14,615 = 15,385.$$

If the seismic test results in an open structure, the posterior probabilities are 1/8 and 7/8, and the expected payoffs are

$$E''R(\text{drill} \mid \text{open}) = 100,000 \left(\frac{1}{8}\right) + (-30,000) \left(\frac{7}{8}\right) = -13,750$$

and

$$E''R(\text{do not drill} \mid \text{open}) = 20,000 \left(\frac{1}{8}\right) + 10,000 \left(\frac{7}{8}\right) = 11,250.$$

The optimal act a_y for y = "open structure" is thus not to drill. An open structure reduces the probability of oil and therefore reinforces the previously-optimal action of not drilling. The VSI is

$$\text{VSI(open)} = E''R(\text{do not drill} \mid \text{open}) - E''R(\text{do not drill} \mid \text{open})$$
$$= 11,250 - 11,250 = 0.$$

The VSI depends on the specific sample result y, just as the VPI depends on the specific value of θ.

However, the decision maker must choose whether or not to purchase sample information before seeing the actual sample result. Therefore, what we need is the expected value of sample information, or EVSI:

$$\text{EVSI} = \sum_y \text{VSI}(y)P(y). \qquad (9.12.7^*)$$

In the oil-drilling example, the EVSI for the seismic test is equal to

$$\text{EVSI} = \text{VSI(closed)}P(\text{closed}) + \text{VSI(open)}P(\text{open}).$$

But what are the probabilities $P(\text{closed})$ and $P(\text{open})$? They can be calculated as follows:

$$P(\text{closed}) = P(\text{closed} \mid \text{oil})P(\text{oil}) + P(\text{closed} \mid \text{no oil})P(\text{no oil})$$
$$= .8(.3) + .4(.7) = .52$$

and $P(\text{open}) = P(\text{open} \mid \text{oil})P(\text{oil}) + P(\text{open} \mid \text{no oil})P(\text{no oil})$
$$= .2(.3) + .6(.7) = .48.$$

Actually, these probabilities had already been calculated when we determined the posterior probabilities. Each is equal to a sum in the denominator of Bayes' theorem, or the sum of the column headed (Prior probability) $\times$ (likelihood) in the appropriate part of the table used to illustrate the calculation of posterior probabilities for this example. These probabilities are the predictive probabilities of the possible sample results, as calculated before the sample is taken.

We can now determine the EVSI for the seismic test in the oil-drilling example:

$$\text{EVSI} = \text{VSI}(\text{closed})P(\text{closed}) + \text{VSI}(\text{open})P(\text{open})$$
$$= 15{,}385(.52) + 0(.48) = 8000.$$

Thus, the decision maker should be willing to pay up to $8000 for the seismic test. If the cost of the seismic test is $5000, then the decision maker's expected net gain from sampling is just $8000 − $5000 = $3000.

In general, the expected net gain from sampling (ENGS) is defined as the difference between the expected value of sample information (EVSI) and the cost of sampling (CS):

$$\text{ENGS} = \text{EVSI} - \text{CS.} \qquad (9.12.8^*)$$

If the ENGS of a proposed sample is greater than zero, then it is worthwhile to take the sample.

An alternative procedure for calculating EVSI involves taking the expected value (as calculated before the sample is observed) of the posterior expected payoff (assuming that optimal actions are always chosen) and subtracting from it the prior expected payoff $E'R(a^*)$. For the oil-drilling example,

$$E'R(a^*) = E'R(\text{do not drill}) = 13{,}000.$$

If the seismic test yields closed structure, the optimal act is to drill, and the posterior expected payoff is

$$E''R(a_y \mid \text{closed}) = E''R(\text{drill} \mid \text{closed}) = 30{,}000.$$

On the other hand, if the test yields open structure, the optimal act is not to drill, and the posterior expected payoff is

$$E''R(a_y \mid \text{open}) = E''R(\text{do not drill} \mid \text{open}) = 11{,}250.$$

Since $P(\text{closed}) = .52$ and $P(\text{open}) = .48$, the expected value of the posterior expected payoff of the act that is optimal under the posterior distribution, $E''R(a_y)$ (as calculated before the sample is actually observed), is

$$E''R(a_y) = E''R(a_y \mid \text{closed})P(\text{closed}) + E''R(a_y \mid \text{open})P(\text{open})$$

$$= 30,000(.52) + 11,250(.48) = 21,000.$$

In terms of these overall expected payoffs, the EVSI can be expressed as follows:

$$\text{EVSI} = E''R(a_y) - E'R(a^*). \tag{9.12.9*}$$

For the example,

$$\text{EVSI} = 21,000 - 13,000 = 8000,$$

which is the same value for EVSI as was determined via Equation (9.12.7). Equation (9.12.7) can be thought of as an "incremental" approach in which the value of each possible sample is considered separately by determining how it would change the decision. Equation (9.12.9), on the other hand, determines an overall posterior expected payoff and subtracts from it the optimal prior expected payoff. The two methods will, of course, always yield identical results.

The process of investigating the value of sample information via Equation (9.12.9) can be represented conveniently on a *tree diagram*. Such a tree diagram is presented for the oil-drilling example in Figure 9.12.1. The

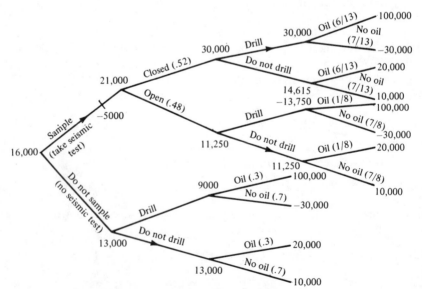

Figure 9.12.1   Tree Diagram for Oil-Drilling Example

first fork is an "action fork" corresponding to the preposterior decision: sample (take seismic test) or do not sample. If the decision is not to sample (the lower branch), then the terminal decision is made, and the action fork at the end of the "do not sample" branch represents the terminal decision, drill or do not drill. Based on the prior probabilities (since no sample is taken), the expected payoffs for the two actions are $9000 and $13,000, as previously calculated. The optimal action is not to drill, and the expected payoff, $E'R(a^*)$, is equal to $13,000.

If the sample is taken, then the upper part of the decision tree is of interest. The "sample" branch is followed by a "chance fork" that represents the two possible sample results, closed structure and open structure. At the end of each of *these* branches is an action fork representing the terminal decision, drill or do not drill. The expected payoffs for these terminal actions are now posterior expected payoffs because the decision forks follow the branches representing sample information. As calculated above, the posterior expected payoffs are $30,000 and $14,615 if the test yields closed structure and $-$13,750 and $11,250 if the test yields open structure. Thus, following closed structure, the terminal decision-making problem has an expected payoff of $E''R(a_y \mid \text{closed})$ = $30,000; following open structure, this expected payoff is $E''R(a_y \mid \text{open})$ = $11,250. Moving backward on the diagram, the overall posterior expected payoff, as computed before the sample is observed, can be determined by looking at the chance fork corresponding to the sample. The predictive probabilities are .52 and .48 for closed and open structures, respectively, and the overall posterior expected payoff is $21,000.

Thus, before the sample is seen, the overall expected payoff is $21,000 if the sample is taken and $13,000 if the sample is not taken, as shown by the ends of the "sample" and "do not sample" branches on the decision tree. The expected value of sample information, then, is $21,000 − $13,000 = $8000. The vertical slash on the "sample" branch represents the cost incurred by taking the seismic test; in evaluating the "sample" branch, it is necessary to subtract the cost of sampling from the overall expected payoff. This yields a *net* expected payoff of $16,000 for the "sample" branch, as compared with $13,000 for the "do not sample" branch, so the decision should be to take the seismic test.

The purpose of analyzing the decision tree in Figure 9.12.1 in such detail is to familiarize you with the use of decision trees in the investigation of the value of information. Actually, this problem is simple enough that it is easy to solve without using a tree diagram. Tree diagrams are particularly useful in complex problems, however, especially in problems involving sequences of actions. The solution of a decision-making problem that is expressed in the form of a decision tree is sometimes called **backward**

induction. In order to make the first decision (the decision regarding sampling), start at the right-hand side of the tree and work backward to determine the expected payoff at each fork in the diagram. The process of backward induction is also referred to as "averaging out and folding back." Starting at the right-hand side of the tree, simply "average out" (take expectations) at chance forks and "fold back" (choose the action with the largest overall expected payoff) at action forks. The arrows on the tree diagram in Figure 9.12.1 indicate the optimal actions at action forks.

9.13   SAMPLE SIZE AND ENGS

Equation (9.12.8) gives the expected net gain of sampling for a particular sample. This may differ for different sample designs or different sample sizes. Suppose that the sample design is fixed and that it is necessary to determine how large a sample to take. One way to determine sample size is to consider the relationship between the sample size and the accuracy of estimation; if a desired degree of accuracy is specified, it is possible to find out how large a sample is necessary to attain this degree of accuracy. This is an informal procedure for determining sample size (although not as informal as the more common "oh, an n of about 20 should be sufficient"). Using decision theory, it is possible to determine an optimal sample size.

If the sample design is fixed and if the statistician wishes to determine n, the sample size, the expected net gain of sampling can be computed for various values of n. Equation (9.12.8) could be written in the form

$$\text{ENGS}(n) = \text{EVSI}(n) - \text{CS}(n). \qquad (9.13.1^*)$$

The optimal sample size is then the value of n for which ENGS is maximized.

Denoting this optimal sample size by n^*,

$$\text{ENGS}(n^*) \geq \text{ENGS}(n) \quad \text{for } n = 0, 1, 2, 3, \ldots. \qquad (9.13.2^*)$$

It is entirely possible for n^* to be 0, in which case the statistician should make the terminal decision without further sampling.

A commonly encountered type of ENGS curve is shown in Figure 9.13.1. The ENGS curve represents the difference between the EVSI curve and the CS curve. In this graph, the expected value of sample information rises rather quickly and then levels off. The leveling off reflects the fact that as the sample size gets larger and larger, the additional value of each trial becomes smaller and smaller, and EVSI approaches EVPI. The cost of sampling, on the other hand, is shown as a linear function of n. This would be reasonable if each trial, or item sampled, costs a given amount and if there are no "economies of scale" associated with large values of n. Because the cost of sampling curve (CS) is continually increasing and the EVSI curve levels off, the ENGS curve rises until $n = n^*$ and then declines, eventually becoming negative. The curve presented in Figure 9.13.1 is by no means the only type of ENGS curve, although it is encountered quite often. In some cases the ENGS curve is never above the horizontal axis, and thus the statistician should take no sample. In yet other cases the curve may be below the axis for small n but may eventually rise above the axis, in which case there is an optimal n^* different from 0.

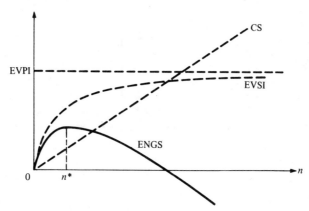

Figure 9.13.1 EVSI, CS, and ENGS as a Function of n

After the optimal sample size n^* is determined and the sample is taken (provided that $n^* > 0$), the decision maker can use Bayes' theorem to revise the prior probability distribution on the basis of the sample results. There are then two options once again: (1) to make a terminal decision on the basis of the newly determined posterior distribution, or (2) to take another sample. The second option involves the calculation of another $ENGS(n)$ curve and the determination of another optimal sample size, n^{**}. Formally, the decision maker could continue this sequential process of taking a sample and conducting preposterior analysis until the point at which the optimal sample size is 0 is reached. A terminal decision should then be made. Incidentally, it is important to remember that sample information has no value by itself. The value in EVSI refers to the expected increase in expected payoff or decrease in expected loss *with respect to the terminal decision.*

9.14   PREPOSTERIOR ANALYSIS: AN EXAMPLE

To illustrate preposterior analysis, we return once again to the quality-control example of Section 8.2. Here, there are four states of the world, corresponding to four values of p, the proportion of defective items: .01, .05, .10, and .25. The three actions are to make a major adjustment, to make a minor adjustment, or to leave the process as is. The payoff table is given in Table 9.14.1. The largest entry in each column is starred to show the optimal action for that particular value of p.

Table 9.14.1   Payoff Table for Quality-Control Example

| | | STATE OF THE WORLD (VALUE OF p) | | | |
		.01	.05	.10	.25
	Major adjustment	380	380	380*	380*
ACTION	*Minor adjustment*	455	375	375	375
	No adjustment	480*	400*	300	0

The prior probabilities are

$$P(p = .01) = .60,$$
$$P(p = .05) = .30,$$
$$P(p = .10) = .08,$$

and

$$P(p = .25) = .02.$$

Under this prior distribution the expected payoffs of the three actions are

$$ER(\text{major adjustment}) = 380,$$
$$ER(\text{minor adjustment}) = 423,$$

and

$$ER(\text{no adjustment}) = 432.$$

The optimal act (a^*) is thus "no adjustment," and the value of perfect information can be calculated:

$$VPI(p = .01) = 480 - 480 = 0,$$
$$VPI(p = .05) = 400 - 400 = 0,$$
$$VPI(p = .10) = 380 - 300 = 80,$$

and

$$VPI(p = .25) = 380 - 0 = 380.$$

The expected value of perfect information is thus

$$\text{EVPI} = \sum \text{VPI}(p)P(p) = 0(.6) + 0(.3) + 80(.08) + 380(.02) = 14,$$

so the production manager should be willing to pay up to \$14 to know p for certain.

Suppose that perfect information is not available but that sample information can be purchased. In this section, we will go through the computations involved in the determination of EVSI for samples of one and three items from the process. We will assume that it costs \$1 to observe an item selected randomly from the process.

For the sample of size 1, the two possible sample results are 0 defectives and 1 defective. The posterior probabilities for each of these two situations are calculated in Table 9.14.2. For each value of p, the likelihood is simply p if a defective item is observed and $(1 - p)$ if a nondefective item is observed. Note that the predictive probability that the item is defective is .0340 and the probability that it is not defective is .9660.

Table 9.14.2 Bayes' Theorem for $n = 1$ in Quality-Control Example

Sample outcome (number of defectives in one trial)	State of the world (value of p)	Prior probability	Likelihood	(Prior probability) × (likelihood)	Posterior probability
One defective	.01	.60	.01	.0060	.177
	.05	.30	.05	.0150	.441
	.10	.08	.10	.0080	.235
	.25	.02	.25	.0050	.147
				.0340	
No defectives	.01	.60	.99	.5940	.615
	.05	.30	.95	.2850	.295
	.10	.08	.90	.0720	.075
	.25	.02	.75	.0150	.015
				.9660	

If a defective item is observed, the posterior probabilities are .177, .441, .235, and .147, and the expected payoffs are

$$E''R(\text{major adjustment} \mid \text{defective}) = 380.00,$$

$$E''R(\text{minor adjustment} \mid \text{defective}) = 389.12,$$

and
$$E''R(\text{no adjustment} \mid \text{defective}) = 331.76.$$

The optimal decision is to make a minor adjustment, and the expected payoff is 389.12. To avoid difficulties that might be caused by rounding errors, all expected payoffs in this example are computed from exact probabilities rather than from probabilities rounded to three decimal places (these rounded values are given in the tables representing the application of Bayes' theorem). For instance, in calculating the expected payoffs following the observance of a defective item, the posterior probabilities used are 6/34, 15/34, 8/34, and 5/34 rather than the rounded figures .177, .441, .235, and .147.

If a nondefective item is observed, the posterior probabilities are .615, .295, .075, and .015, and the expected payoffs are

$$E''R(\text{major adjustment} \mid \text{nondefective}) \ = \ 380.00,$$

$$E''R(\text{minor adjustment} \mid \text{nondefective}) \ = \ 424.19,$$

and $\quad\quad E''R(\text{no adjustment} \mid \text{nondefective}) \ = \ 435.53.$

The optimal decision is to make no adjustment, and the expected payoff is 435.53.

Under the prior distribution, the optimal decision is to make no adjustment. Therefore, the value of sample information for each possible sample result can be computed, using Equation (9.12.5), as follows:

$$\text{VSI(defective)} \ = \ E''R(\text{minor adjustment} \mid \text{defective})$$

$$- \ E''R(\text{no adjustment} \mid \text{defective})$$

$$= \ 389.12 \ - \ 331.76 \ = \ 57.36,$$

and

$$\text{VSI(nondefective)} \ = \ E''R(\text{no adjustment} \mid \text{nondefective})$$

$$- \ E''R(\text{no adjustment} \mid \text{nondefective})$$

$$= \ 435.53 \ - \ 435.53 \ = \ 0.$$

The EVSI for a sample of size 1 is thus

$$\text{EVSI}(n \ = \ 1) \ = \ \text{VSI(defective)} P(\text{defective})$$

$$+ \ \text{VSI(nondefective)} P(\text{nondefective})$$

$$= \ 57.36(.0340) \ + \ 0(.9660)$$

$$= \ 1.95.$$

For the contemplated sample of size 3 there are four possible sample outcomes: 0, 1, 2, or 3 defectives. The likelihoods can be calculated from the binomial distribution with $n = 3$ and p equal to .01, .05, .10, or .25, as the case may be. The posterior probabilities are calculated in Table 9.14.3.

Table 9.14.3   Bayes' Theorem for $n = 3$ in Quality-Control Example

Sample outcome (number of defectives in three trials)	State of the world (value of p)	Prior probability	Likelihood	(Prior probability) × (likelihood)	Posterior probability
No defectives	.01	.60	.9703	.582180	.643
	.05	.30	.8574	.257220	.284
	.10	.08	.7290	.058320	.064
	.25	.02	.4219	.008438	.009
				.906158	
One defective	.01	.60	.0294	.017640	.205
	.05	.30	.1354	.040620	.471
	.10	.08	.2430	.019440	.226
	.25	.02	.4219	.008438	.098
				.086138	
Two defectives	.01	.60	.0003	.000180	.025
	.05	.30	.0071	.002130	.292
	.10	.08	.0270	.002160	.297
	.25	.02	.1406	.002812	.386
				.007282	
Three defectives	.01	.60	.0000	.000000	.000
	.05	.30	.0001	.000030	.071
	.10	.08	.0010	.000080	.190
	.25	.02	.0156	.000312	.739
				.000422	

For each action and each sample result, a posterior expected payoff can be calculated from the relevant posterior probabilities. These expected payoffs are given in Table 9.14.4, and the expected payoff corresponding to the best action for each sample result is starred.

Table 9.14.4   Posterior Expected Payoffs for $n = 3$ in Quality-Control Example

SAMPLE OUTCOME

Number of defectives in sample of size 3

		0	1	2	3
	Major adjustment	380.00	380.00	380.00*	380.00*
ACTION	*Minor adjustment*	426.40	391.38*	376.98	375.00
	No adjustment	441.24*	354.63	217.85	85.31

To calculate the value of sample information, note once again that the act "no adjustment" is optimal under the prior distribution.

$$\text{VSI}(0 \text{ defectives}) = E''R(\text{no adjustment} \mid 0 \text{ defectives})$$
$$- E''R(\text{no adjustment} \mid 0 \text{ defectives}) = 0.$$

$$\text{VSI}(1 \text{ defective}) = E''R(\text{minor adjustment} \mid 1 \text{ defective})$$
$$- E''R(\text{no adjustment} \mid 1 \text{ defective})$$
$$= 391.38 - 354.63$$
$$= 36.75.$$

$$\text{VSI}(2 \text{ defectives}) = E''R(\text{major adjustment} \mid 2 \text{ defectives})$$
$$- E''R(\text{no adjustment} \mid 2 \text{ defectives})$$
$$= 380.00 - 217.85$$
$$= 162.15.$$

$$\text{VSI}(3 \text{ defectives}) = E''R(\text{major adjustment} \mid 3 \text{ defectives})$$
$$- E''R(\text{no adjustment} \mid 3 \text{ defectives})$$
$$= 380.00 - 85.31$$
$$= 294.69.$$

The predictive probabilities of the various sample outcomes are

$$P(0 \text{ defectives}) = .906158,$$
$$P(1 \text{ defective}) = .086138,$$
$$P(2 \text{ defectives}) = .007282,$$

and
$$P(3 \text{ defectives}) = .000422,$$

so that the expected value of sample information for a sample of size 3 is

$$\text{EVSI}(n = 3) = 0(.906158) + 36.75(.086138)$$
$$+ 162.15(.007282) + 294.69(.000422)$$
$$= 4.47.$$

The cost of sampling is \$1 per item, so the ENGS for samples of size 1 and 3 can be computed:

$$\text{ENGS}(n = 1) = \text{EVSI}(n = 1) - \text{CS}(n = 1) = 1.95 - 1.00 = .95$$

and

$$\text{ENGS}(n = 3) = \text{EVSI}(n = 3) - \text{CS}(n = 3) = 4.47 - 3.00 = 1.47.$$

Also, although we will not present the calculations here, the EVSI for a sample of size 5 is 5.80, so that

$$\text{ENGS}(n = 5) = \text{EVSI}(n = 5) - \text{CS}(n = 5) = 5.80 - 5.00 = .80.$$

Of the three samples considered, then, the sample of size 3 has the largest ENGS. Of course, it may be that some sample size other than 1, 3, or 5 has yet a larger ENGS. To carry the problem any further conveniently would require the use of a computer, in which case it would be necessary to take into account the cost of using the computer. The problem of determining optimal sample size can become quite complicated from a computational standpoint, as this example illustrates. As a result, decisions regarding sample size are often made in an informal manner, although for important decisions involving large payoffs it clearly would be worthwhile to apply preposterior analysis formally, hopefully on a high-speed computer.

A tree diagram for the quality-control example is presented in Figure 9.14.1, with the case of $n = 5$ omitted. On the left-hand side of the tree diagram are the potential preposterior decisions, which in this case are $n = 0$ (do not sample), $n = 1$, and $n = 3$. Of course, other values of n could be considered, in which case there would be more "branches" on the tree diagram. If the preposterior decision is not to sample, the terminal decision is made on the basis of the current information by finding the action with the highest expected payoff. If the preposterior decision is to sample, then the sample result is observed and used to determine a posterior distribution; the terminal decision is based on expected payoffs calculated from this posterior distribution. Using the backward induction procedure, the values of $E''R(a_y)$ are 433.95 and 436.47. Thus, using Equation (9.12.9),

$$\text{EVSI}(n = 1) = 433.95 - 432.00 = 1.95$$

and $\quad\quad \text{EVSI}(n = 3) = 436.47 - 432.00 = 4.47.$

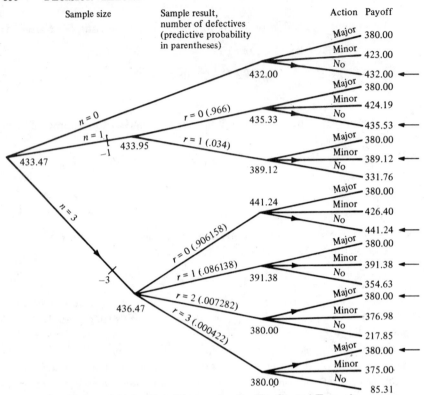

Figure 9.14.1   Tree Diagram for Quality-Control Example

After the costs of sampling are subtracted, the optimal action, as noted above, is to take a sample of size 3, and the net expected payoff is 433.47.

9.15   LINEAR PAYOFF FUNCTIONS: THE TWO-ACTION PROBLEM

In examples concerning the value of information, we have only considered the situation in which there are a finite number of states of the world and a finite number of actions. Under these conditions, it is possible to set up a payoff or loss table and to use this table to calculate EVPI and EVSI. In many decision-making problems, however, the state of the world is represented by a continuous random variable. The production process example would be more realistic if it was assumed that p was continuous and could take on any value between 0 and 1 instead of being limited to .01, .05, .10, and .25. In the stock-purchasing example, the four given states of the world are by no means the only possible states;

it would be more reasonable to assume that the price of a stock 1 year from now is a continuous random variable. In numerous such decision problems, the state of the world can be thought of as continuous.

If the state of the world is continuous, a tabular representation of payoffs is impossible, since an (uncountably) infinite number of columns would be needed. Consequently, an attempt is made to specify the payoffs in functional form. In this section it is assumed that the functional form is *linear in terms of the state of the world*. In other words, if θ is the state of the world, then the **payoff function** can be expressed in the form

$$R(a, \theta) = c + d\theta, \qquad (9.15.1*)$$

where c and d are constants. It is assumed in the following discussion that the decision maker has a utility function that is linear in R.

Payoff functions of the form (9.15.1) greatly simplify the decision-making problem because expected payoffs can be written as

$$ER(a) = E(c + d\theta) = c + dE(\theta). \qquad (9.15.2*)$$

This result implies that in order to make a terminal decision, all the knowledge that is needed about the distribution of θ is the mean, $E(\theta)$. In making a decision, the decision maker can act as though the mean is equal to the true value of θ with certainty; hence, the mean of the distribution is called a **certainty equivalent** in this situation.

There are many situations in which the payoff functions for a given decision-making problem are linear. For instance, the payoff functions in the oil-drilling example presented in Section 9.9 are linear:

$$R(\text{farm out for \$10,000 and } \tfrac{1}{16} \text{ override, } \theta) = .06\theta + 10,000,$$

$$R(\text{farm out for } \tfrac{1}{8} \text{ override, } \theta) = .12\theta,$$

$$R(\text{drill with 50\% interest, } \theta) = .48\theta - 34,000,$$

and $$R(\text{drill with 100\% interest, } \theta) = .96\theta - 68,000,$$

where θ represents the number of barrels of oil that can be recovered if a well is drilled. In producing and selling a product, a firm's profits often can be expressed in the form

$$R(a, \theta) = -c + d\theta,$$

where c represents fixed costs, d represents net profit per unit sold (price per unit minus variable profit per unit), and θ is the number of units sold. The point is simply that linear payoff functions are widely applicable in real-world decision-making problems.

Suppose that there are just two possible actions, a_1 and a_2, and that their payoff functions are

$$R(a_1, \theta) = c_1 + d_1\theta$$

and $$R(a_2, \theta) = c_2 + d_2\theta,$$

(9.15.3)

where $d_2 > d_1$. Under what conditions will a_1 be optimal under a prior distribution $f(\theta)$, and under what conditions will a_2 be optimal? The first act, a_1, will be optimal if

$$ER(a_1) > ER(a_2).$$

(9.15.4)

But all that is needed from $f(\theta)$ in order to compute the expected payoffs is the mean, $E(\theta)$, so Equation (6.6.4) is equivalent to

$$c_1 + d_1E(\theta) > c_2 + d_2E(\theta).$$

Subtracting c_2 and $d_1E(\theta)$ from both sides, we get

$$c_1 - c_2 > E(\theta)(d_2 - d_1).$$

Since $d_2 > d_1$, the term $(d_2 - d_1)$ is greater than 0, so dividing both sides by this term will not affect the inequality:

$$\frac{c_1 - c_2}{d_2 - d_1} > E(\theta).$$

(9.15.5)

Therefore, if Equation (9.15.5) is satisfied, a_1 is the optimal action; if the inequality is reversed, a_2 is the optimal action. For this decision-making problem, θ_b **is called the breakeven value of** θ:

$$\theta_b = \frac{c_1 - c_2}{d_2 - d_1}.$$

(9.15.6*)

If the expected value of θ is lower than θ_b, then a_1 is optimal; if it is greater than θ_b, then a_2 is optimal. If $E(\theta) = \theta_b$, the expected payoffs of a_1 and a_2 are equal.

For example, consider a choice between two investments. Action a_1 corresponds to a half interest in a particular venture, whereas action a_2 corresponds to a full interest in the venture. The half interest costs \$45,000, while the full interest costs \$94,000 (the owner of the second half of the venture is more reluctant to sell, so the cost is higher for that half). Let μ represent the amount of money that the venture will yield. The payoff functions are thus

$$R(a_1, \mu) = -45{,}000 + .5\mu$$

and $$R(a_2, \mu) = -94{,}000 + \mu.$$

The breakeven value of μ is

$$\mu_b = \frac{-45,000 - (-94,000)}{1 - .50} = \frac{49,000}{.5} = 98,000.$$

Suppose that the investor assesses a normal distribution for μ with mean 100,000 and standard deviation 5000. Since the mean is greater than the breakeven value, a_2 is the optimal action; the investor should purchase a full interest in the venture.

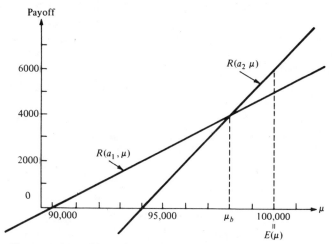

Figure 9.15.1   Linear Payoff Functions in Investment Example

This example, which is illustrated in Figure 9.15.1, is somewhat unrealistic in that it ignores other potential investments. In particular, the possibility of purchasing a half interest and investing the $49,000 (the cost of the second half) in an alternative investment (for example, a savings account) is not considered. Also, note that if $\mu < 90,000$, the payoffs for both actions are negative, in which case the investor would be better off to do nothing. The point of the example, however, is to illustrate a decision-making situation with linear payoff functions, and it seems realistic to assume that alternative actions (including doing nothing) would also have linear payoff functions. Therefore, considering such actions would just generalize the problem from one involving two actions to one involving more than two actions. As long as the payoff functions are all linear, such a generalization is not difficult to handle, although we will not go into details here.

When the payoff functions in a two-action problem are of the form of

Equation (9.15.3), the corresponding **loss functions** are

$$L(a_1, \theta) = \begin{cases} 0 & \text{if } \theta \leq \theta_b, \\ (d_2 - d_1)(\theta - \theta_b) & \text{if } \theta \geq \theta_b; \end{cases} \tag{9.15.7}$$

and

$$L(a_2, \theta) = \begin{cases} (d_2 - d_1)(\theta_b - \theta) & \text{if } \theta \leq \theta_b, \\ 0 & \text{if } \theta \geq \theta_b. \end{cases} \tag{9.15.8}$$

In terms of a graph, the loss suffered at any value of θ is either zero (if the better action at that point was taken) or the vertical difference between the two linear payoff functions. But this vertical difference can be expressed as a product of (1) the difference in the slopes of the lines and (2) the distance of θ from the breakeven value.

The expected losses corresponding to the loss functions in Equations (9.15.7) and (9.15.8) are

$$EL(a_1) = (d_2 - d_1) \int_{\theta_b}^{\infty} (\theta - \theta_b) f(\theta) \, d\theta \tag{9.15.9}$$

and

$$EL(a_2) = (d_2 - d_1) \int_{-\infty}^{\theta_b} (\theta_b - \theta) f(\theta) \, d\theta. \tag{9.15.10}$$

The EVPI is equal to the smaller of the two expected losses. Expressions such as Equations (9.15.9) and (9.15.10) may be difficult to calculate because the integration in each case is over only part of the real line, representing the part for which the loss function is nonzero. These expressions can be simplified for certain families of distributions, however, and we will consider only the case in which the distribution of θ is a normal distribution. Moreover, we will present equations for EVPI and EVSI without proof in order to avoid mathematical details.

Assume that the parameter of interest in a decision-making problem with two actions is μ, the mean of a normal process with known variance σ^2. The payoff functions are of the form of Equation (9.15.3) with $\theta = \mu$, and the prior distribution of μ is a normal distribution with mean m' and standard deviation σ'. The EVPI is then

$$\text{EVPI} = |\, d_2 - d_1 \,|\, \sigma' L_N(u), \tag{9.15.11*}$$

where

$$u = \left| \frac{\mu_b - m'}{\sigma'} \right|, \tag{9.15.12}$$

$$L_N(u) = \int_u^{\infty} (z - u) f_{N*}(z) \, dz, \tag{9.15.13}$$

and $f_{N*}(z)$ is the standard normal density function. A table of values of $L_N(u)$, the unit normal linear loss integral, is presented in the Appendix. Similarly, the EVSI for a sample of size n from the process is given by

$$\text{EVSI} = |\, d_2 - d_1 \,|\, \sigma^* L_N(u^*), \tag{9.15.14*}$$

where

$$\sigma^{*2} = \frac{n\sigma'^4}{\sigma^2 + n\sigma'^2} \tag{9.15.15}$$

and

$$u^* = \left| \frac{\mu_b - m'}{\sigma^*} \right|. \tag{9.15.16}$$

The first term of the expressions for EVPI and EVSI is simply $|\, d_2 - d_1 \,|$, the difference in the slopes of the payoff functions. As this difference increases, EVPI and EVSI increase, since a greater difference implies greater losses for wrong decisions and hence, a greater value for information that reduces or eliminates the chance of a wrong decision. The second term in Equations (9.15.11) and (9.15.14) is a standard deviation. For the EVPI, it is the prior standard deviation, and the effect of this term is that the more current information we have, as measured by a small variance, the less perfect information is worth to us. The effect is similar in the case of the EVSI, where the standard deviation of interest is σ^*, which turns out to be equal to the square root of the reduction in variance due to the sample (that is, the prior variance minus the posterior variance). From Equation (8.8.2),

$$\sigma''^2 = \frac{1}{(1/\sigma'^2) + (n/\sigma^2)} = \frac{1}{(\sigma^2 + n\sigma'^2)/\sigma'^2\sigma^2} = \frac{\sigma'^2\sigma^2}{\sigma^2 + n\sigma'^2}.$$

Therefore,

$$\sigma'^2 - \sigma''^2 = \sigma'^2 - \frac{\sigma'^2\sigma^2}{\sigma^2 + n\sigma'^2} = \frac{\sigma'^2(\sigma^2 + n\sigma'^2) - \sigma'^2\sigma^2}{\sigma^2 + n\sigma'^2}$$

$$= \frac{n\sigma'^4}{\sigma^2 + n\sigma'^2} = \sigma^{*2}.$$

The third term in the expressions for EVPI and EVSI is a unit normal linear loss integral evaluated at a standardized version of the difference between the prior mean and the breakeven value. The larger this difference is, the smaller the linear loss integral is, as we can see by looking at the table of $L_N(u)$ presented in the Appendix. The larger the difference is, the more certain we are that the optimal decision under the prior distribution is the correct decision.

For the investment example presented earlier in the section,

$$u = \left| \frac{98{,}000 - 100{,}000}{5000} \right| = .4,$$

and $\qquad d_2 - d_1 = 1 - .5 = .5,$

so that, from Equation (9.15.11),

$$\text{EVPI} = .5(5000)L_N(.4).$$

From the table of $L_N(u)$,

$$L_N(.4) = .2304,$$

and $\qquad \text{EVPI} = .5(5000)(.2304) = 576.$

Thus, the investor should be willing to pay up to \$576 for perfect information.

Suppose that perfect information is not available, but the investor can take a sample from previous investments that are judged to be very similar to the current investment. For each of the previous investments, the mean payoff is μ and the standard deviation is 10,000. For a sample of size 10,

$$\sigma^* = \sqrt{\frac{10(5000)^4}{(10{,}000)^2 + 10(5000)^2}} = 4226$$

and $\qquad u^* = \left| \dfrac{98{,}000 - 100{,}000}{4226} \right| = .47,$

so that, from Equation (9.15.14),

$$\text{EVSI} = .5(4226)L_N(.47) = .5(4226)(.2072) = 438.$$

The investor should be willing to pay up to \$438 for a sample of size 10. If the cost of looking at previous investments is \$30 per investment, the cost of a sample of size 10 is \$300, so the ENGS is \$438 − \$300 = \$138.

In a similar manner, EVSI and ENGS can be computed for other sample sizes. Some values are given in Table 9.15.1. The optimal sample size is 5, and the general shapes of the EVSI, CS, and ENGS curves are similar to the shapes shown in Figure 9.13.1. The EVSI curve rises rapidly (by $n = 10$, the EVSI is 76 percent of the EVPI) and levels off, while the CS curve continues to rise linearly.

In this section we have demonstrated how to handle problems with a particular structure: two-action problems with linear payoff functions. Methods have been developed to deal with other "special structures," but we do not have the space to discuss such methods here. In each case the general approach to decision theory presented in this chapter is used and special features of the structure of the problem are exploited to yield relatively simple formulas for dealing with problems of that structure.

Table 9.15.1 EVSI, CS, and ENGS as a Function of n for Investment Example

Sample size (n)	$EVSI(n)$	$CS(n)$	$ENGS(n)$
1	115	30	85
2	210	60	150
3	272	90	182
4	319	120	199
5	354	150	204
6	378	180	198
7	396	210	186
8	411	240	171
9	424	270	154
10	438	300	138
11	444	330	114

Now that we have discussed the basic notions of decision theory, we will revisit the inferential procedures of Chapters 6–8, treating them in a decision-theoretic framework. As we shall see, this approach can be considered as an extension of the inferential framework, with the major addition being the formal consideration of payoffs and losses.

9.16 DECISION THEORY AND POINT ESTIMATION

In Chapter 6 we discussed the classical approach to point estimation, which involves certain desirable properties of estimators, such as unbiasedness, consistency, efficiency, and sufficiency. Two methods for determining "good" point estimators are the method of maximum likelihood and the method of moments. In Chapter 8 the Bayesian approach was discussed, and this involves the posterior distribution. Since estimates should reflect all of the available information, according to the Bayesian, they should be based on the posterior distribution. Both of these approaches are informal procedures in the sense that there is no single "best" estimator. For example, S^2 is a maximum-likelihood estimator of the variance of a normal population (or process), but it is biased. The unbiased estimator $\hat{S}^2$, on the other hand, is not a maximum-likelihood estimator of σ^2. Which should be used to estimate σ^2? The choice between the two must be made subjectively. If all that is wanted by the statistician is a "rough and ready" estimate, or a general idea of the value of the parameter of interest, such an informal approach should prove satisfactory. If the estimate is to be used in a situation where errors in estimation may lead to serious consequences, a more formal approach would be desirable.

The choice of an estimate can be thought of as a problem of decision under uncertainty. Here the actions correspond to the possible estimates. If θ is being estimated and S is the set of values of θ, then the set of actions A is identical to the set S. If a payoff function $R(a, \theta)$ or a loss function $L(a, \theta)$ can be determined which gives the payoff or loss associated with an incorrect estimate, then the problem of point estimation can be treated in accordance with the general framework presented in this chapter.

An example should help to clarify the decision-theoretic approach to point estimation. As usual, the example is simplified to emphasize the points of interest. Consider an automobile dealer who sells a line of luxury automobiles. The entire inventory of cars has been sold, and therefore more must be ordered from the manufacturer. Furthermore, next year's model will be introduced shortly and this will be the last chance to order the current model. How many automobiles should be ordered? This is essentially a point-estimation problem, for the dealer must estimate the demand for the cars—the number of customers that will come in and want to purchase a car. The economic factors are quite simple; the dealer will make a profit of \$500 for every car that is sold before the new model comes out. Once the new model comes out, any of the current models that are still in inventory will have to be sold at a large discount, and there will be a loss of \$300 for each car that must be sold at this discount.

The automobile dealer's loss function can be found from the facts given above. Remember that an action, a, corresponds to the number of cars that are ordered from the manufacturer; a state of the world, or value of θ, corresponds to the number of cars that are demanded by the customers. Both a and θ can take on the values 0, 1, 2, ... (all nonnegative integers). If $a = \theta$, then the number ordered is exactly equal to the number demanded, and $L(a, \theta) = 0$ because the losses are "opportunity losses," or "regrets." Given any value of θ, the best possible action is $a = \theta$. If $a > \theta$, then the number of cars ordered is greater than the number demanded. In this case there is a loss associated with the "extra" cars. For each car not sold before the new models are introduced, the loss is \$300. Thus, if $a > \theta$, the loss function is $L(a, \theta) = 300(a - \theta)$. Finally, suppose that $a < \theta$. In this situation more cars are demanded than the dealer has available, and for each car demanded but not available there is an opportunity loss of \$500, the "lost profit." Therefore, if $a < \theta$, the loss function is $L(a, \theta) = 500(\theta - a)$. The entire loss function is thus

$$
L(a, \theta) = \begin{cases} 500(\theta - a) & \text{if } a < \theta, \\[2mm] 0 & \text{if } a = \theta, \\[2mm] 300(a - \theta) & \text{if } a > \theta. \end{cases} \tag{9.16.1}
$$

In order to make a decision, the dealer first must specify the other input to the problem: the probability distribution for demand. Because it is very close to the introduction of next year's model and because the cars are quite expensive, the dealer is certain that demand will be either 0, 1, 2, or 3. The following probability distribution is assessed by the dealer:

$$P(\theta = 0) = .1,$$
$$P(\theta = 1) = .2,$$
$$P(\theta = 2) = .3,$$
and
$$P(\theta = 3) = .4.$$

Since there is only a finite number of possible values of θ, there will be only a finite number of possible values of a (the dealer cannot order a negative number of cars and would not be foolish enough to order more than three as long as it is thought that demand cannot possibly be greater than three). The loss function in Equation (9.16.1) can be used to find the losses in Table 9.16.1.

Table 9.16.1 Loss Table for Automobile-Ordering Example

STATE OF THE WORLD

Number of cars demanded

		0	*1*	*2*	*3*
	0	0	500	1000	1500
ACTION	*1*	300	0	500	1000
Number of cars ordered	*2*	600	300	0	500
	3	900	600	300	0

The expected losses for the four acts are

$$EL(a = 0) = 0(.1) + 500(.2) + 1000(.3) + 1500(.4) = 1000,$$
$$EL(a = 1) = 300(.1) + 0(.2) + 500(.3) + 1000(.4) = 580,$$
$$EL(a = 2) = 600(.1) + 300(.2) + 0(.3) + 500(.4) = 320,$$
and $EL(a = 3) = 900(.1) + 600(.2) + 300(.3) + 0(.4) = 300.$

The optimal act is thus $a = 3$; the dealer should order 3 cars from the manufacturer. In a decision-theoretic sense, then, the "best point estimate" of demand is 3. This point estimate depends both on the loss function and on the probability distribution. If the last two probabilities are switched around, so that $P(\theta = 2) = .4$ and $P(\theta = 3) = .3$, the expected losses are 950, 530, 270, and 330. In this case the optimal point estimate, or number of cars to order, is 2.

This example illustrates the decision-theoretic approach to point estimation. Whenever it is possible to determine a loss function $L(a, \theta)$ or a payoff function $R(a, \theta)$ in a point-estimation situation, where a is the estimate and θ is the true value of the variable of interest, this procedure is applicable. It is then possible to find an optimal, or "best" (in the sense of minimizing EL or maximizing ER) estimate. In the next section we will present a quick way to determine this optimal estimate under certain assumptions regarding the loss function. Of course, not all point-estimation problems involve losses. If a "rough and ready" estimate is all that is wanted, the approaches in Chapters 6 and 8 should prove satisfactory. If losses are involved and the potential losses are important, then the decision-theoretic approach should be used.

9.17   POINT ESTIMATION: LINEAR AND QUADRATIC LOSS FUNCTIONS

The example in the last section illustrated the idea of point estimation as decision making under uncertainty. It is clear, however, that such a problem could be quite difficult to solve if the number of possible values of θ (and hence the number of possible estimates) were large. Instead of automobiles, suppose that the dealer sells radios and that the demand for radios could be as low as 0 or as high as 100. It would appear to be more trouble than it is worth to calculate the expected losses for all 101 possible actions. Fortunately, if certain assumptions can be made about the loss function, it is not necessary to calculate all of these EL's. If the expected loss can be expressed in a certain form, it is possible to use the calculus to find the point at which EL is smallest. This involves taking the derivative of EL with respect to a, setting it equal to zero, and solving the resulting equation. Because EL is expressed as an integral or as a sum, where the limits of integration or limits of summation are related to a, the differentiation of EL is not as elementary as might be thought. As a result, we will present some general statements about the optimal act without proof.

First of all, suppose that the loss function is of the following form:

$$L(a, \theta) = \begin{cases} k_u(\theta - a) & \text{if } a < \theta, \\ 0 & \text{if } a = \theta, \\ k_o(a - \theta) & \text{if } a > \theta. \end{cases} \tag{9.17.1*}$$

Such a loss function is called a linear loss function with respect to a point-estimation problem.

This should not be confused with the concept of a utility function which is linear with respect to money. This particular loss function is *linear with respect to the difference between the estimate a and the true state of the world θ.*

Notice that if $a < θ$, the loss is $k_u (θ - a)$. In this case, a is smaller than $θ$, so we say that we have underestimated $θ$. The *cost per unit of underestimation,* or simply the *cost of underestimation,* is k_u. Similarly, if $a > θ$, we have overestimated $θ$, and the *cost per unit of overestimation* is k_o.

It can be shown mathematically that for the loss function given by Equation (9.17.1), the optimal point estimate is simply a

$$\frac{k_u}{k_u + k_o}$$ **fractile of the decision maker's distribution of $\bar{θ}$.**

Recalling the definition of a fractile (Section 3.12), this would be a value a such that

$$P(θ \leq a) \geq \frac{k_u}{k_u + k_o} \quad \text{and} \quad P(θ \geq a) \geq 1 - \frac{k_u}{k_u + k_o}. \quad (9.17.2^*)$$

If $θ$ is continuous, this reduces to

$$P(θ \leq a) = \frac{k_u}{k_u + k_o}.$$

This greatly simplifies the problem from a computational standpoint. Instead of calculating a number of expected losses, it is only necessary, given k_u, k_o, and the distribution of $θ$, to find a single fractile of the distribution.

In the example presented in the previous section, the loss function is linear with respect to the difference between a and $θ$. The values of k_u and k_o are 500 and 300, and thus

$$\frac{k_u}{k_u + k_o} = \frac{500}{500 + 300} = \frac{5}{8}.$$

To find the $\frac{5}{8}$ fractile of the dealer's probability distribution $P(θ)$, consider the corresponding CDF, presented in Figure 9.17.1. The $\frac{5}{8}$ fractile is 3, so this is the optimal value of a, just as we found out by calculating the expected losses for all of the possible actions.

Equation (9.17.2) is applicable whether the variable $θ$ is discrete or continuous. Suppose that a grocer has to decide how many oranges to order for a particular week. The oranges cost the grocer $.60 per dozen, and they are sold for $.80 per dozen. Any oranges that remain unsold at

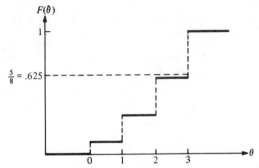

Figure 9.17.1   CDF of Demand in Automobile-Ordering Example

the end of the week will spoil and therefore will be worthless. If the grocer orders more oranges than are demanded, a loss of $.60 per dozen is suffered for all unsold oranges. If the grocer orders fewer than are demanded, an opportunity loss of $.20 per dozen, reflecting lost profits, is suffered for all oranges demanded over and above the amount stocked. The loss function is of the form (9.17.1), with $k_u = .20$ and $k_o = .60$.

Based on past experience, the grocer feels that the weekly demand for oranges is normally distributed with a mean of 80 dozen and a standard deviation of 25 dozen. How many dozen oranges should the grocer order? We need to find the $k_u/(k_u + k_o) = .2/(.2 + .6) = \frac{1}{4}$ fractile of the distribution $f(\theta)$. From tables of the normal distribution, the $\frac{1}{4}$ fractile of the standard normal distribution is approximately $-.675$. This implies that the $\frac{1}{4}$ fractile of a normal distribution with mean μ and standard deviation σ is simply .675 standard deviations to the left of the mean, or $\mu - .675\sigma$. The $\frac{1}{4}$ fractile of the grocer's distribution is then

$$80 - .675(25),$$

or approximately 63 dozen oranges. To minimize expected loss, the grocer should order 63 dozen oranges.

In our two examples, the linear loss function seemed to be realistic. In many cases, especially problems such as these involving inventories, the loss function is linear or almost linear. The result given by Equation (9.17.2) should therefore have wide applicability. Let us see if (9.17.2) is intuitively reasonable. If $k_u = k_o$, the optimal estimate is the .50 fractile, or the median, of the decision maker's distribution. This is reasonable; if the costs of overestimation and underestimation are equal, the optimal act should be somewhere in the "middle" of the distribution. Why specifically the median, rather than the mean? Because if the median is chosen, the probability of overestimating is equal to the probability of under-estimating. If $k_u > k_o$, then $k_u/(k_u + k_o)$ is greater than .50. The estimate will be higher than the median. In this case, the cost of underestimation is greater than the cost of overestimation, so we "hedge" our estimate upward to avoid the greater cost of underestimation. In the automobile-dealer example, the cost of ordering too many cars was only $300 per extra car, as compared with the $500 per car opportunity loss if too few cars were ordered. The dealer would rather overestimate slightly than underestimate, so the optimal estimate is 3 cars. In the limiting case, as k_o becomes very small in relation to k_u, the decision maker will move farther and farther toward the right-hand tail of the distribution in an attempt to avoid suffering the high cost of underestimation. If $k_u < k_o$, just the opposite is true. In the grocer example, the cost of not having enough oranges on hand was just the lost profit of $.20 per dozen, while any extra oranges that spoiled would cost $.60 per dozen. To avoid the higher cost of overestimating, the grocer ordered only 63 dozen oranges, although the mean of the distribution was 80 dozen. For highly perishable goods, k_o is often greater than k_u. For nonperishables, k_o is usually lower than k_u, since the only cost of overestimation is the cost of storing the unsold inventory. At any rate, the result given by Equation (9.17.2) seems to be intuitively reasonable; if $k_o < k_u$, the estimate will be to the right of the median, and if $k_o > k_u$, the estimate will be to the left of the median. In each case, the greater the *relative* difference between k_u and k_o (as represented by the ratio of the larger of the two to the smaller), the farther from the median the estimate will be.

A linear loss function of the form (9.17.1) is often quite realistic, as we have seen. There may be situations in which it is not applicable, however. Suppose that the costs of underestimation and overestimation are equal (let $k_o = k_u = k$) and that an error in estimation of 2 units is *more* than twice as costly as an error of just 1 unit. In particular, we will assume that an error of 2 units is 4 times as costly as an error of just 1 unit and that in general, the loss associated with an error is proportional to the *square* of the error.

The resulting loss function, which is called a quadratic loss function or a squared-error loss function, can be written as follows:

$$L(a, \theta) = k(a - \theta)^2. \qquad (9.17.3^*)$$

For this loss function, the optimal point estimate is the mean of the decision maker's distribution of θ.

That is, the optimal estimate is

$$a = E(\theta). \qquad (9.17.4^*)$$

It is possible to prove the above result without using the calculus. We want to find a so as to minimize

$$EL(a) = E[k(a - \theta)^2] = kE(a - \theta)^2.$$

Clearly, since k is a constant, if we can find a so as to minimize $E(a - \theta)^2$, this a will also minimize $EL(a)$. Adding and subtracting $E(\theta)$ within the parentheses, we get

$$
\begin{aligned}
E(a - \theta)^2 &= E[a - E(\theta) + E(\theta) - \theta]^2 \\
&= E[(a - E(\theta))^2 + 2(a - E(\theta))(E(\theta) - \theta) + (E(\theta) - \theta)^2] \\
&= E[a - E(\theta)]^2 + 2E[(a - E(\theta))(E(\theta) - \theta)] \\
&\quad + E[E(\theta) - \theta]^2.
\end{aligned}
$$

The expectation is being taken with respect to the distribution of θ, so a and $E(\theta)$ are constants. Thus, using the algebra of expectations,

$$
E(a - \theta)^2 = [a - E(\theta)]^2 + 2(a - E(\theta))E[E(\theta) - \theta] + E[E(\theta) - \theta]^2.
$$

But the middle term is zero, since

$$
E[E(\theta) - \theta] = E(\theta) - E(\theta) = 0.
$$

The last term does not involve a, so it is irrelevant to our choice of an optimal a. This leaves the first term, which is

$$
[a - E(\theta)]^2.
$$

Since this is a squared term, it must be either positive or zero. But it is zero only when $a = E(\theta)$, so this is the value of a that minimizes EL. Note the similarity between this proof and the proof in Section 3.17 that the expected squared deviation (that is, the variance) is smallest when calculated from the mean. In fact, this is the reason that $E(\theta)$ is the optimal estimate for a quadratic, or squared-error, loss function, and it is also one of the reasons that the mean is considered a "good" estimator in the classical approach to statistics. *In taking the sample mean as an estimator of the population mean, the statistician is essentially acting as though the loss function is quadratic.* Of course, the estimate is being based solely on the sample information, rather than on the posterior distribution, but the general idea is the same. If the loss function is difficult to define exactly, but appears to be similar to the quadratic loss function in Equation (9.17.3), then the statistician is somewhat justified, in a decision-theoretic sense, in using the mean as an estimator. Similarly, if the loss function appears to be similar to the linear loss function in Equation (9.17.1) with $k_u = k_o$, then there is justification for using the median as an estimator.

In a point-estimation problem with a quadratic loss function of the form (9.17.3), the optimal point estimate is the mean of the distribution of θ. Thus, it is not necessary to know the entire distribution; we just need to know the mean in order to make a decision. We can act as though the mean were equal to the true value of θ with certainty; hence, the mean is called a *certainty equivalent* in this situation. Similarly, if the loss function is linear of the form (9.17.1), the $k_u/(k_u + k_o)$ fractile of the distribution is a *certainty equivalent*.

9.18   DECISION THEORY AND HYPOTHESIS TESTING

As we suggested in Chapter 7, hypothesis testing can be thought of as a decision-making problem, where the two decisions are (1) to accept H_0 and (2) to reject H_0 in favor of H_1. Thus, the losses associated with the two types of errors should be assessed and the rejection region determined so as to minimize expected loss. Suppose that there are two hypotheses of interest,

$$H_0: \theta \in A$$

and $$H_1: \theta \in B. \tag{9 18.1*}$$

Furthermore, assume that A and B are mutually exclusive and exhaustive, so that $A \cap B = \varnothing$ and $A \cup B = S$, where S is the set of all possible values of θ. From the decision maker's distribution of θ, it is possible to compute

$$P(H_0) = P(\theta \in A)$$

and $$P(H_1) = P(\theta \in B). \tag{9.18.2*}$$

Since A and B are mutually exclusive and exhaustive, $P(H_0) + P(H_1) = 1$.

The other input to the decision problem is a loss function. In hypothesis testing, there are two types of errors. A Type I error is committed if H_0 is rejected when in fact it is true, and a Type II error is committed if H_0 is accepted when in fact it is false. Denote the loss associated with a Type I error by $L(I)$, and denote the loss associated with a Type II error by $L(II)$. The loss function can be represented by the loss table in Table 9.18.1, since there are just two actions and two relevant states of the world.

Table 9.18.1   Loss Table in Hypothesis-Testing Situation

		STATE OF THE WORLD	
		H_0 *true* $(\theta \in A)$	H_0 *false* $(\theta \in B)$
ACTION	*Accept* H_0	0	$L(II)$
	Reject H_0	$L(I)$	0

Calculating expected losses in the usual manner, we find that

$$EL(\text{accept } H_0) = 0P(H_0) + L(\text{II})P(H_1) = L(\text{II})P(H_1) \qquad (9.18.3^*)$$

and $EL(\text{reject } H_0) = L(\text{I})P(H_0) + 0P(H_1) = L(\text{I})P(H_0).$ $\qquad (9.18.4^*)$

According to the EL criterion for making decisions, we will reject H_0 if

$$EL(\text{reject } H_0) < EL(\text{accept } H_0). \qquad (9.18.5^*)$$

Using Equations (9.18.3) and (9.18.4), this is equivalent to

$$L(\text{I})P(H_0) < L(\text{II})P(H_1).$$

Dividing both sides by $L(\text{I})$ and by $P(H_1)$, we get

$$\frac{P(H_0)}{P(H_1)} < \frac{L(\text{II})}{L(\text{I})} . \qquad (9.18.6^*)$$

If Equation (9.18.6) is satisfied, the optimal decision is to reject H_0. If the inequality is reversed, the optimal decision is to accept H_0, and if the two sides of the equation are equal, the two expected losses are equal, so neither action is better than the other.

The left-hand side of Equation (9.18.6) is an **odds ratio** and the right-hand side is a ratio of losses, or a **loss ratio.** In order to make a decision in a hypothesis-testing situation, it is necessary only to compare an odds ratio with a loss ratio. In fact, suppose that in Equation (9.18.6), we divide both sides of the equation by the loss ratio $L(\text{II})/L(\text{I})$. The result is

$$\frac{P(H_0)}{P(H_1)} \cdot \frac{L(\text{I})}{L(\text{II})} < 1. \qquad (9.18.7^*)$$

Denoting the loss ratio by $L_{\text{I/II}}$, the decision rule can be written as follows:

$$\text{Reject } H_0 \quad \text{if } \frac{P(H_0)}{P(H_1)} L_{\text{I/II}} < 1. \qquad (9.18.8^*)$$

Let us investigate Equation (9.18.8) to see if it seems reasonable. For a given loss ratio $L_{\text{I/II}}$, we are more likely to reject H_0 when $P(H_0)$ gets smaller relative to $P(H_1)$. This makes sense; as the probability of H_0 gets smaller, H_0 appears to be less likely to be true, so we should be more likely to reject H_0. For a given odds ratio, we are more likely to reject H_0 when $L(\text{I})$ gets smaller relative to $L(\text{II})$. But $L(\text{I})$ is the loss associated with the error of rejecting H_0 when it is in fact true. As this loss decreases relative to $L(\text{II})$, we should be more likely to reject H_0 for the same reason that we are more likely to underestimate in a point-estimation problem as the cost of underestimation decreases relative to the cost of overestimation.

The framework developed in this section is related to both classical *and* Bayesian hypothesis testing, since we have not specified *exactly* what the odds ratio represents. From Equation (8.16.4), the posterior odds ratio is equal to the product of the prior odds ratio and the likelihood ratio:

$$\Omega'' = \Omega'\Omega. \qquad (9.18.9*)$$

If it is not possible to assess the losses $L(\mathrm{I})$ and $L(\mathrm{II})$, the posterior odds ratio will tell us which hypothesis is more likely but will not enable us to make a decision. If we informally decided to reject H_0 if H_1 was more likely (that is, if $\Omega'' < 1$) and to accept H_0 if H_0 itself was more likely (that is, if $\Omega'' > 1$), then our decision rule would resemble Equation (9.18.8) without the loss ratio. This amounts to assuming that $L(\mathrm{I}) = L(\mathrm{II})$, so that $L_{\mathrm{I/II}} = 1$. The inclusion of the loss ratio, $L_{\mathrm{I/II}}$, represents an *extension* of our informal decision rule to a more formal rule which satisfies the *EL* criterion.

How does this relate to classical hypothesis testing? We pointed out in Section 7.6 that classical tests of hypotheses are based on the likelihood ratio, LR. In Equation (9.18.9), if the prior odds ratio is equal to one (that is, if the two hypotheses are equally likely under the prior information), then the posterior odds ratio is equal to the likelihood ratio:

$$\Omega'' = \Omega'\Omega = (1)\Omega = \Omega.$$

In this case, the decision rule becomes

$$\text{Reject } H_0 \quad \text{if } \Omega L_{\mathrm{I/II}} < 1. \qquad (9.18.10*)$$

Dividing both sides by $L_{\mathrm{I/II}}$, the rejection region becomes

$$\Omega < \frac{1}{L_{\mathrm{I/II}}} = \frac{L(\mathrm{II})}{L(\mathrm{I})}. \qquad (9.18.11*)$$

Now, from Equation (7.6.1), the rejection region for a classical test is of the form

$$\Omega \leq k.$$

Thus, the classical "likelihood ratio" test is essentially equivalent to the test given by Equation (9.18.11), with $k = L(\mathrm{II})/L(\mathrm{I})$. The difference is that the classical statistician usually does not formally assess $L(\mathrm{I})$ and $L(\mathrm{II})$. Instead, a convention which implies that $L(\mathrm{I})$ is more serious than, and therefore higher than, $L(\mathrm{II})$ is used. The *formal* assessment of losses and the use of the *EL* criterion is an extension of the informal analysis. It is not even necessary to assess the two losses $L(\mathrm{I})$ and $L(\mathrm{II})$ separately, since all that is needed is their ratio. Sometimes it is convenient to think in terms of loss ratios. For example, it is difficult to quantify the losses due to (1) convicting an innocent man or (2) setting a guilty man free. We may, however, be able to assess the loss ratio of (1) to (2). Some persons might assess this ratio as 100, others might think it to be 10, and still others might choose 1, saying that the two losses are equal. The point is simply that it is often much easier to assess a loss *ratio* than it is to directly assess the two losses comprising the loss ratio.

For a simple (and brief) example of hypothesis testing as a decision-making procedure, consider a manufacturer who is concerned about the potential demand for a new product. For the sake of simplicity, assume that the following two hypotheses are of interest, where p is the proportion of consumers that will purchase the new product:

$$H_0: p = .20$$

and

$$H_1: p = .30.$$

If the manufacturer accepts H_0, then the new product will not be produced; if H_0 is rejected in favor of H_1, then the poduuct will be produced.

The manufacturer's prior probabilities are

$$P'(H_0) = P'(p = .20) = .25$$

and

$$P'(H_1) = P'(p = .30) = .75.$$

In a sample of 20 consumers, only 2 will buy the new product. Thus, the likelihood ratio is (from the binomial tables):

$$\Omega = \frac{P(R = 2 \mid n = 20, p = .20)}{P(R = 2 \mid n = 20, p = .30)} = \frac{.1369}{.0278} = 4.92.$$

Multiplying this by the prior odds ratio of $.25/.75 = \frac{1}{3}$, we get

$$\Omega'' = \Omega'\Omega = \tfrac{1}{3}(4.92) = 1.64.$$

If the manufacturer decided just to accept the most likely hypothesis, then H_0 would be accepted, since the posterior odds ratio is greater than one. Suppose, however, that potential losses are taken into consideration. If H_0 is accepted (and therefore the new product is not produced), the firm might suffer a large opportunity loss in the form of lost profits if H_1 is true. This is $L(\text{II})$. On the other hand, if H_0 is rejected (and the new product is produced), the new product might result in losses to the firm if the demand is not high enough (that is, if H_0 is true). This is $L(\text{I})$. After some serious thought, the manufacturer decides that $L(\text{II})$ is twice as large as $L(\text{I})$. This means that

$$L_{\text{I/II}} = \frac{L(\text{I})}{L(\text{II})} = \frac{1}{2}.$$

Multiplying the posterior odds ratio by the loss ratio, we get

$$\Omega'' L_{\text{I/II}} = (1.64)(\tfrac{1}{2}) = .82.$$

Since this is less than one, the optimal decision is to reject H_0 and to produce the new product. Even though the posterior odds ratio favors H_0, the losses are such that H_1 is the "better" hypothesis from a decision-theory standpoint.

This example was purposely made simple in order to clearly demonstrate the idea of hypothesis testing as a decision-making procedure. In particular, both H_0 and H_1 were taken to be exact hypotheses. It should be mentioned that although the general procedure is still valid when the hypotheses are *not* both exact, there are certain subtleties involved in some situations which we will not consider here.

9.19 THE DIFFERENT APPROACHES TO STATISTICAL PROBLEMS

In this section we shall attempt to recapitulate very briefly the relationship among what we have labeled classical statistics, Bayesian statistics, and decision theory. First of all, classical statistics prescribes inferential techniques which are based on sample information alone. Inferential methods in Bayesian statistics are based on the posterior distribution, which is a combination of the prior distribution, representing the prior state of information, and the likelihood function, representing the information of a sample. Thus, the prior distribution is an input to Bayesian methods but not to classical methods. In either case, the objective of inferential statistics is to make inferences about some population. These inferences may be in the form of point estimates, interval estimates, entire probability distributions, and so on. In decision theory we are interested in taking some *action* rather than just in making some inference about a population.

In the preceding chapter, we mentioned that Bayesian techniques can be thought of as an *extension* of classical techniques, since they utilize both sample information *and* prior information. In the same way, the decision-theoretic approach is yet a further extension, using the sample information, the prior information, and a third input as well: a loss function, or payoff function. The inclusion of the loss function enables the statistician to "optimize" in the sense of minimizing expected loss or maximizing expected payoff. It should be noted at this point that the Bayesian approach has become so closely associated with decision theory that the procedures described in this chapter are called Bayesian methods by some authors. In this book we prefer to reserve the term "Bayesian" for the procedures described in Chapter 8, changing to the term "decision theory" when losses or payoffs are introduced into the analysis. This distinction serves a useful pedagogical function, enabling us to consider the introduction of prior information and the introduction of losses or payoffs as separate extensions of classical inferential procedures. We are thus able to compare inferences based on the posterior distribution (Bayesian inferences) with inferences based on the sampling distribution *without* having to be concerned with losses. In this chapter we introduced losses in order to see how they would affect the analysis.

For important decisions, the formal decision-theoretic approach is the most useful of the three approaches because it takes into account all available information which bears on the decision, both information concerning the state of the world and information concerning the consequences of potential decisions. If the decision is not too important, an informal analysis may be preferred, since applying decision theory may be quite time-consuming. It is in this sense that the classical and Bayesian techniques may be useful, since they are somewhat easier to apply, requiring fewer inputs and not specifying a single "optimizing" criterion. They may be useful in other situations as well, even situations in which the decision is important, if the statistician is unable to determine some of the inputs to the problem. In complex decision-making situations, it may not be clear what the relevant losses or payoffs are. If a big decision problem consists of a combination of several smaller problems, it may be difficult to see exactly how the "smaller" decisions relate to the major decision. In a way, this involves model-building; how can we state the problem in decision-theoretic terms and still have it be as realistic as possible? Decision theory provides a useful framework for making decisions under uncertainty. It is up to the statistician to attempt to express real-world problems in this framework without sacrificing realism. For instance, if the statistician makes assumptions which are unrealistic, then the optimal solution to the decision-theory problem may not be the same as the best solution to the real-world problem. If this is the case, an informal analysis may prove to be just as useful as a formal decision-theoretic approach. Of course, as we pointed out, the formal approach is particularly valuable for important problems, in which case it might be worthwhile to carefully develop a realistic decision model.

Chapters 5–9 have covered the basic theory of statistical inference and decision. In Chapters 10–12, a number of important statistical techniques will be presented. Because many of the Bayesian and decision-theoretic extensions of these techniques are fairly complex, involving such things as the assessment of multivariate prior distributions, most of what follows will be presented in the classical vein. However, we will attempt to indicate briefly how prior information and losses or payoffs could be brought into the analysis. If you prefer the Bayesian interpretation of classical inferential statements, then the classical results can be viewed as being roughly (sometimes *very* roughly, especially with small samples) equivalent to Bayesian results under a diffuse prior distribution.

EXERCISES

1. Explain the difference between decision making under certainty and decision making under uncertainty.

2. A particular product is both manufactured and marketed by two different firms. The total demand for the product is virtually fixed, so neither firm has advertised in the past. However, the owner of Firm A is considering an advertising campaign to woo customers away from Firm B. The ad campaign he has in mind will cost $200,000. He is uncertain about the number of customers that will switch to his firm as a result of the advertising, but he is willing to assume that he will gain either 10 percent, 20 percent, or 30 percent of the market. For each 10 percent gain in market share, the firm's profits will increase by $150,000. Construct the payoff table for this problem and find the corresponding loss table.

3. In Exercise 2, the owner of Firm A is worried that if he proceeds with the ad campaign, Firm B will do likewise, in which case the market shares of the two firms will remain constant. How does this affect the set of possible states of the world? Construct a modified payoff table to allow for this change.

4. You are given the following payoff table.

STATE OF THE WORLD

		A	B	C	D	E
	1	−50	80	20	100	0
	2	30	40	70	20	50
ACTION	*3*	10	30	−30	10	40
	4	−10	−50	−70	−20	200

(a) Are any of the actions inadmissible? If so, eliminate them from further consideration.

(b) Find the loss table corresponding to the above payoff table.

(c) Which action is optimal if it is known that the actual state of the world is C?

5. Consider the following *loss* table.

STATE OF THE WORLD

	I	II	III
1	0	3	6
2	1	1	0
3	4	0	1

ACTION (rows 1, 2, 3)

Given this loss table, complete the corresponding *payoff* table.

STATE OF THE WORLD

	I	II	III
1	12	7	9
2			
3			

ACTION (rows 1, 2, 3)

Are any of the actions inadmissible?

6. The owner of a clothing store must decide how many men's shirts to order for the new season. For a particular type of shirt, he must order in quantities of 100 shirts. If he orders 100 shirts, his cost is $10 per shirt; if he orders 200 shirts, his cost is $9 per shirt; and if he orders 300 or more shirts, his cost is $8.50 per shirt. His selling price for the shirt is $12, but if any are left unsold at the end of the season, they will be sold in his famous "half-price, end-of-the-season sale." For the sake of simplicity, he is willing to assume that the demand for this type of shirt will be either 100, 150, 200, 250, or 300 shirts. Of course, he cannot sell more shirts than he stocks. He is also willing to assume that he will suffer no loss of goodwill among his customers if he understocks and the customers cannot buy all the shirts they want. Furthermore, he must place his order today for the entire season; he cannot wait to see how the demand is running for this type of shirt.

(a) Given these details of the clothing-store owner's problem, list the set of actions available to him (including both admissible and inadmissible actions).

(b) Construct a payoff table and eliminate any inadmissible actions.

(c) Represent the problem in terms of a tree diagram.

(d) If the owner decides that there is a loss of goodwill that is roughly equivalent to $.50 for each person who wants to buy the shirt but cannot because it is out of stock, make the appropriate changes in the payoff table. (Assume that the goodwill cost is $.50 *per shirt* if a customer wants to buy more than one shirt after the stock is exhausted.)

7. In Exercise 6(d), attempt to construct a loss table for the owner's problem *without* referring to the payoff table determined in that exercise. That is, using the concept of opportunity loss, try to express the consequences to the clothing-store owner in terms of losses. Then use the payoff table constructed in Exercise 6(d) to determine a loss table and compare the two loss tables, reconciling any differences.

8. How can the concept of inadmissible actions be used to simplify a decision-making problem?

9. Comment on the following statement: "Whenever the loss in a loss table is zero, the corresponding payoff must be zero or positive."

10. For the payoff and loss tables in Exercise 4, find the actions that are optimal under the following decision-making criteria:
    (a) maximin,   (b) maximax,   (c) minimax loss.

11. For the payoff and loss tables in Exercises 6(d) and 7, find the actions that are optimal under the maximin, maximax, and minimax loss rules.

12. For the payoff table in Exercise 2, find the actions that are optimal under the maximin and maximax rules. Comment on the implications of these particular rules in this example.

13. What is the primary disadvantage of decision-making rules such as maximin, maximax, and minimax loss?

14. One nonprobabilistic decision-making criterion not discussed in the text involves the consideration of a weighted average of the highest and lowest payoffs for each action. The weights, which must sum to 1, can be thought of as an optimism-pessimism index. The action with the highest weighted average of the highest and lowest payoffs is the action chosen by this criterion. Comment on this decision-making criterion and use it for the payoff table in Exercise 4, with the highest payoff in each row receiving a weight of .4 and the lowest payoff receiving a weight of .6.

15. In Exercise 4, find the expected payoff and the expected loss of each action and find the action that has the largest expected payoff and the smallest expected loss, given that the probabilities of the various states of the world are $P(A) = .10$, $P(B) = .20$, $P(C) = .25$, $P(D) = .10$, and $P(E) = .35$.

16. In Exercise 5, if $P(I) = .25$, $P(II) = .45$, and $P(III) = .30$, find the expected payoff and the expected loss of each of the three actions. Using ER (or EL) as a decision-making criterion, which action is optimal?

17. In Exercise 6(d), the owner decides that in comparing possible demands of 200 and 100, the demand is twice as likely to be 200 as it is to be 100. Similarly, 200 is 3 times as likely as 300, $1\frac{1}{2}$ times as likely as 150, and $1\frac{1}{2}$ times as likely as 250. Find the expected payoff of each action. How many shirts should the owner of the store order if the ER criterion is used?

18. In Exercise 2, the owner of Firm A feels that if the ad campaign is initiated, the events "gain 20 percent of the market" and "gain 30 percent of the market" are equally likely, and each of these events is 3 times as likely as the event "gain 10 percent of the market." Using the ER criterion, what should be done?

19. In Exercise 18, suppose that the probabilities given are conditional on the rival firm not advertising. If Firm B also advertises, then the owner of Firm A is certain (for all practical purposes) that there will be no change in the market share of either firm. The chances are 2 in 3 that Firm B will advertise if Firm A does. What should the owner of Firm A do?

20. In Exercises 16, 17, and 18, determine the relationship between the expected payoffs of the various actions and the corresponding expected losses.

21. Prove that the ER and EL criteria always lead to identical decisions. [*Hint*: Express the definition of loss (as given in Section 9.2) in symbols and use this to find an expression for EL.]

22. Explain how game-theory problems (that is, decision-making problems in which there is some "opponent") can be analyzed using the same techniques that are used in decision making under uncertainty. Instead of "states of the world," we must consider "actions of the opponent." Might this make it more difficult to determine the probabilities necessary to calculate expected payoffs and losses? Explain.

23. Consider a situation in which you and a friend both must choose a number from the two numbers 1 and 2. You must make your choices without communicating with each other. The relevant payoffs are as follows.

    If you choose 1 and he chooses 1, you both win $10.
    If you choose 1 and he chooses 2, he wins $15 and you win $0.
    If you choose 2 and he chooses 1, you win $15 and he wins $0.
    If you choose 2 and he chooses 2, you both win $5.

    (a) How would you go about assigning probabilities to your friend's two actions?
    (b) On the basis of the probabilities assigned in (a), what is your optimal action?
    (c) Could you have determined that this was your optimal action without using any probabilities? Why?
    (d) Would you expect your action to be different if you could get together with your friend and make a bargain before the game is played? Explain how this might change matters.

24. In the automobile-salesman example in Section 8.4, the owner must decide (after the sample of 24 days) whether to keep or to fire the salesman. The owner's marginal profit per car is $50 (after subtracting commissions and similar marginal costs), and salesmen are paid $10 per day plus commissions on sales. On the basis of all of the information available to the owner, should the salesman be retained or fired?

25. In a special type of decision-making problem frequently encountered in meteorology, the states of the world are "adverse weather" and "no adverse weather," and the actions are "protect against adverse weather" and "do not protect against adverse weather." The cost of protecting against adverse weather is denoted by c, and the profit that will be obtained unless adverse weather occurs and no protection is purchased is denoted by d.
   (a) Construct a payoff table and a decision tree for this decision-making problem.
   (b) For what values of P(adverse weather) should one protect against adverse weather?
   (c) Given the result in (b), is it necessary to know the absolute magnitudes of c and d?

26. A probabilistic decision-making criterion discussed in Chapter 6 but not in this chapter is the maximum-likelihood criterion. In the context of a decision-making problem, the maximum-likelihood criterion states that for each action, you find the most likely payoff, and then you choose the action with the largest most likely payoff. Use this criterion for the decision-making problems in Exercises 16, 17, and 18. Do you think that it is a good decision-making criterion? Explain.

27. Why is the maximization of expected monetary value (that is, ER with the payoffs expressed in terms of money) not always a reasonable criterion for decision making? What problems does this create for the decision maker?

28. You must choose between three acts, where the payoff table is as follows (in terms of dollars).

STATE OF THE WORLD

		A	B	C
	1	100	70	20
ACTION	2	10	50	120
	3	50	80	30

What is the optimal act according to the expected utility criterion if $P(A) = .3$, $P(B) = .3$, $P(C) = .4$, and the utility function in the relevant range is
   (a) $U(R) = 50 + 2R$,
   (b) $U(R) = 50 + 2R^2$,
   (c) $U(R) = R$,
   (d) $U(R) = R^2 + 5R + 6$.

29. Suppose that you are offered a choice between bets A and B.

   Bet A: You win $1,000,000 with certainty (that is, with probability 1).
   Bet B: You win $5,000,000 with probability .10.
        You win $1,000,000 with probability .89.
        You win $0 with probability .01.

Which bet would you choose? Similarly, choose between bets C and D:

Bet C: You win $1,000,000 with probability .11.
You win $0 with probability .89.
Bet D: You win $5,000,000 with probability .10.
You win $0 with probability .90.

Prove that if you chose bet A, then you should have chosen bet C, and if you chose bet B, then you should have chosen bet D. If you selected A and D or B and C, explain your choices. In light of the proof, would you change your choices?

30. Comment on the following statement from the text: "Some individuals appear to be risk-takers for some decisions (such as gambling) and risk-avoiders for other decisions (such as purchasing insurance)." Can you explain why this phenomenon occurs? Can you draw a utility function which would explain it?

31. Attempt to determine your own utility function for monetary payoffs in the range from $-\$500$ to $+\$500$. If you were given actual decision-making situations, would you act in accordance with this utility function? If you inherited $50,000 tomorrow, would your utility function change? Explain.

32. Prove that if the function U satisfies the two axioms of utility given in Section 9.5, then the function $W = a + bU$ also satisfies these axioms, where a and b are constants with $b > 0$. What are the implications of this result for the application of utility theory?

33. Suppose that a person's utility function for *total* assets (*not* changes in assets) is

$$U(A) = 200A - A^2 \quad \text{for } 0 \le A \le 100,$$

where A represents total assets in thousands of dollars.
(a) Graph this utility function. How would you classify this person with regard to attitude toward risk?
(b) If the person's total assets are currently $10,000, should a bet in which he or she will win $10,000 with probability .6 and lose $10,000 with probability .4 appear favorable?
(c) If the person's total assets are currently $90,000, should the bet given in (b) appear favorable?
(d) Compare your answers to (b) and (c). Does the person's betting behavior seem reasonable to you? How could you intuitively explain such behavior?

34. For each of the following utility functions for changes in assets, graph the function and comment on the attitude toward risk that is implied by the function (R represents dollars). All of the functions are defined for $-1000 < R < 1000$.
(a) $U(R) = (R + 1000)^2$.
(b) $U(R) = -(1000 - R)^2$.
(c) $U(R) = 1000R + 2000$.
(d) $U(R) = \log (R + 1000)$.
(e) $U(R) = R^3$.
(f) $U(R) = 1 - e^{-R/100}$.

35. For each of the utility functions in Exercise 34, find out if the decision maker should take a bet in which $100 will be gained with probability p and $50 will be lost with probability $(1 - p)$, if
   (a) $p = 1/2$,
   (b) $p = 1/3$,
   (c) $p = 1/4$.

36. Two persons, A and B, make the following bet: A wins $40 if it rains tomorrow, and B wins $10 if it does not rain tomorrow.
   (a) If they both agree that the probability of rain tomorrow is .10, what can you say about their utility functions?
   (b) If they both agree that the probability of rain tomorrow is .30, what can you say about their utility functions?
   (c) Given no information about their probabilities, is it possible that their utility functions could be identical? Explain.
   (d) If they both agree that the probability of rain tomorrow is .20, is it possible that their utility functions could be identical? Explain.

37. Suppose that an investor is considering two investments. With each investment, he will double his money with probability .4 and lose half of his money with probability .6. He has $1000 to invest, and he can either put $500 in each investment, put the entire $1000 in one of the two investments, put $500 in one investment and not invest the remaining $500, or not invest at all. Assume that the outcomes of the two investments are independent and that the investor's utilities for *changes* in assets are $U($1000) = 1$, $U($500) = .85$, $U($250) = .75$, $U($0) = .5$, $U(-$250) = .25$, and $U(-$500) = 0$. What should the investor do? What commonly encountered financial concept does this illustrate?

38. In Exercise 17, suppose that $U($2000) = 1$, $U(-$1000) = 0$, and the owner of the clothing store is indifferent between the following lotteries.
   Lottery A: Receive $1000 for certain.
   Lottery B: Receive $2000 with probability .87 and receive $-$1000 with probability .13.
   What is $U($1000)$? By considering lotteries such as these, the owner assesses $U(-$500) = .42$, $U($0) = .62$, $U($500) = .77$, and $U($1500) = .95$. Plot these points on a graph and attempt to fit a smooth curve to them. Using this curve, find the action that maximizes the owner's expected utility.

39. In Exercise 19, suppose that the owner of Firm A assesses $U($500,000) = 1$, $U($400,000) = .70$, $U($300,000) = .43$, $U($200,000) = .28$, $U($100,000) = .18$, $U($0) = .12$, $U(-$100,000) = .06$, and $U(-$200,000) = 0$. Graph these points, fit a smooth utility function to them, and use this utility function to find EU(ad campaign) and EU(no ad campaign).

40. In Exercise 18, suppose that the owner of Firm A is uncertain about some of the elements in his payoff table. In particular, the ad campaign will cost either $150,000, $200,000, or $250,000, with $P($150,000) = .2, P($200,000) = .4$, and $P($250,000) = .4$. In addition, the increase in profits for each 10 percent gain in market share will be either $100,000, $140,000, or $180,000, with $P($100,000) = .3, P($140,000) = .5$, and $P($180,000) = .2$.

(a) Express the problem in terms of a decision tree, including branches for the uncertainties about the payoffs.

(b) What should the owner of Firm A do?

41. Suppose that one person (A) claims that $U(\$100) = .8$ and that $U(\$50) = .3$, whereas a second person (B) claims that $U(\$100) = .8$ and that $U(\$50) = .6$. What, if anything, can you say about the relative preferences of A and B? In general, is it meaningful to make interpersonal comparisons of utility functions? Explain.

42. A contractor must decide whether to buy or rent equipment for a job up for bid. Because of lead-time requirements, the decision must be made before it is known whether the bid is successful. If the equipment is bought, a contract will result in a net profit of $120,000, but failure to win the contract means that the equipment will have to be sold for $40,000 less than the purchase price. If the equipment is rented, the profit from the contract will be only $50,000, but there will be no loss of money if the bid is not successful. The probability that the bid is successful is .4, and the contractor's utility function for monetary payoffs is $U(R) = \sqrt{R + 40,000}$, where R represents dollars. What should the contractor do?

43. Suppose that you could obtain an interest-free $1000 loan with the restriction that it has to be invested in stocks, bonds, or savings accounts (or a combination of these types of investments) and that the loan has to be repaid in exactly 1 year. How could you set up this problem in terms of the framework presented in Section 9.8? First of all, how would you reduce the number of potential actions to a manageable number; second, how would you define the set of states of the world; third, how would you determine the possible payoffs or losses; fourth, how would you introduce a utility function; fifth, how would you quantify your judgments to assess probabilities for the various states of nature?

44. In Exercise 17, suppose that one of the employees in the clothing store assesses the following probabilities for the demand for shirts: $P(50) = .05$, $P(100) = .15$, $P(150) = .25$, $P(200) = .35$, $P(250) = .10$, $P(300) = .05$, and $P(350) = .05$. Note that two additional values for demand, 50 and 350, are being considered, so the payoff table must be expanded. Using the expanded payoff table and the employee's probabilities, which action has the largest expected payoff? Comment on the implications of this result with regard to the sensitivity of the shirt-ordering problem to changes in the set of states of the world or to changes in the probabilities.

45. Discuss the importance of sensitivity analysis in problems of decision making under uncertainty. When might the results of a sensitivity analysis greatly simplify the decision maker's problem? When might the results of a sensitivity analysis make the decision maker's problem extremely difficult?

46. What is the difference between a terminal decision and a preposterior decision? Are they at all related?

47. In Exercises 15 and 16, find the EVPI.

48. In Exercise 17, how much should the store owner be willing to pay for perfect information about the demand for the type of shirts he is about to order?

49. In Exercise 19, what is the value of perfect information that the rival firm, Firm B, will advertise if Firm A advertises? What is the value of perfect information that Firm B will not advertise and the increase in Firm A's market share will be 10 percent? What is the value of perfect information that Firm B will not advertise and the increase in Firm A's market share will be 20 percent? Finally, what is the value of perfect information that Firm B will not advertise and the increase in Firm A's market share will be 30 percent? From these VPI's, calculate the EVPI. Also, calculate the ERPI and use Equation (9.11.6) to determine the EVPI.

50. In Exercise 25, give a general expression for the expected value of perfect information regarding the weather and find the EVPI if
   (a) P(adverse weather) $= .4$, $c = 3.5$, and $d = 10$,
   (b) P(adverse weather) $= .3$, $c = 3.5$, and $d = 10$,
   (c) P(adverse weather) $= .4$, $c = 10$, and $d = 10$,
   (d) P(adverse weather) $= .3$, $c = 2$, and $d = 8$.

51. In Exercise 40, find the expected value of perfect information about (a) the cost of the ad campaign, (b) the increase in profits for each 10 percent gain in market share, and (c) both the cost of the ad campaign *and* the increase in profits for each 10 percent gain in market share. Does the answer to part (c) equal the sum of the answers to parts (a) and (b)? Explain.

52. A store must decide whether or not to stock a new item. The decision depends on the reaction of consumers to the item, and the payoff table (in dollars) is as follows.

PROPORTION OF CONSUMERS PURCHASING
THE ITEM

		.10	.20	.30	.40	.50
	Stock 100	−10	−2	12	22	40
DECISION	Stock 50	−4	6	12	16	16
	Do not stock	0	0	0	0	0

If $P(.10) = .2$, $P(.20) = .3$, $P(.30) = .3$, $P(.40) = .1$, and $P(.50) = .1$, what decision should be made? If perfect information is available, find VPI for each of the five possible states of the world and compute EVPI.

53. In Exercise 52, suppose that sample information is available in the form of a random sample of consumers. For a sample of size *1*,
   (a) find the posterior distribution if the one person sampled will purchase the item, and find the value of this sample information;

(b)   find the posterior distribution if the one person sampled will *not* purchase the item, and find the value of this sample information;

(c)   find the expected value of sample information.

54.   In Exercise 53, suppose that you also want to consider other sample sizes.
   (a)   Find EVSI for a sample of size 2.
   (b)   Find EVSI for a sample of size 5.
   (c)   Find EVSI for a sample of size 10.
   (d)   If the cost of sampling is $.50 per unit sampled, find the expected net gain of sampling (ENGS) for samples of sizes 1, 2, 5, and 10.

55.   Consider a bookbag filled with 100 poker chips. You know that either 70 of the chips are red and the remainder blue, or 70 are blue and the remainder red. You must guess whether the bookbag has 70R–30B or 70B–30R. If you guess correctly, you win $5. If you guess incorrectly, you lose $3. Your prior probability that the bookbag contains 70R–30B is .40.
   (a)   If you had to make your guess on the basis of the prior information, what would you guess?
   (b)   If you could purchase perfect information, what is the most that you should be willing to pay for it?
   (c)   If you could purchase sample information in the form of one draw from the bookbag, how much should you be willing to pay for it?
   (d)   If you could purchase sample information in the form of five draws (with replacement) from the bookbag, how much should you be willing to pay for it?

56.   Do Exercise 55 with the following payoff table (in dollars).

STATE OF THE WORLD

		70R–30B	70B–30R
YOUR GUESS	70R–30B	6	−2
	70B–30R	−6	10

57.   In the automobile-salesman example discussed in Section 8.4, suppose that the owner of the dealership must decide whether or not to hire a new salesman. The payoff table (in terms of dollars) is as follows.

STATE OF THE WORLD

		Great salesman	Good salesman	Poor salesman
ACTION	Hire	20,000	5,000	−10,000
	Do not hire	0	0	0

The prior probabilities for the three states of the world are $P(\text{great}) = .10$' $P(\text{good}) = .50$, and $P(\text{poor}) = .40$. The process of selling cars is assumed to behave according to a Poisson process with $\lambda = \frac{1}{2}$ per day for a great salesman, $\lambda = \frac{1}{4}$ per day for a good salesman, and $\lambda = \frac{1}{8}$ per day for a poor salesman.

(a) Find VPI (great salesman), VPI (good salesman), and VPI (poor salesman).

(b) Find the expected value of perfect information.

(c) Find the ERPI.

(d) Suppose that the owner of the dealership can purchase sample information at the rate of $10 per day by hiring the salesman on a temporary basis. Samples must be taken in units of 4 days since a salesman's work week consists of 4 days. The owner considers hiring the salesman for 1 week (4 days). Find EVSI and ENGS for this proposed sample.

58. In Exercise 53, find $E''R(a_y)$ and use Equation (9.12.9) to determine EVSI.

59. In Exercise 19, suppose that the owner of Firm A can obtain information about Firm B's reaction to advertising by Firm A. In particular, information can be obtained from a contact in Firm B. However, the information from the contact cannot be regarded as perfect information. If Firm B would really advertise if Firm A does, the chances are 4 in 5 that the contact would report this correctly. However, if Firm B would not advertise, the chances are only 2 in 3 that the contact would report this correctly.

(a) If the contact reports that Firm B will advertise if Firm A does, what is the value of this sample information?

(b) If the contact reports that Firm B will not advertise even if Firm A does, what is the value of this sample information?

(c) What is the expected value of sample information?

60. A firm is considering the marketing of a new product. For convenience, suppose that the events of interest are simply $\theta_1 = $ "new product is a success" and $\theta_2 = $ "new product is a failure." The prior probabilities are $P(\theta_1) = .3$ and $P(\theta_2) = .7$. If the product is marketed and is a failure, the firm suffers a loss of $300,000. If the product is not marketed and it would be a success, the firm suffers an opportunity loss of $500,000. The firm is considering two separate surveys, A and B, and the results from each survey can be classified as favorable, neutral, and unfavorable. The conditional probabilities for survey A are

$$P(\text{favorable} \mid \theta_1) = .6, \; P(\text{neutral} \mid \theta_1) = .3, \; P(\text{unfavorable} \mid \theta_1) = .1,$$
$$P(\text{favorable} \mid \theta_2) = .1, \; P(\text{neutral} \mid \theta_2) = .2, \; \text{and} \; P(\text{unfavorable} \mid \theta_2) = .7.$$

The conditional probabilities for survey B are

$$P(\text{favorable} \mid \theta_1) = .8, \; P(\text{neutral} \mid \theta_1) = .1, \; P(\text{unfavorable} \mid \theta_1) = .1,$$
$$P(\text{favorable} \mid \theta_2) = .1, \; P(\text{neutral} \mid \theta_2) = .4, \; \text{and} \; P(\text{unfavorable} \mid \theta_2) = .5.$$

Survey A costs $20,000 and survey B costs $30,000.

(a) Find the expected value of perfect information about θ.

(b) Find the expected net gain from survey A.

(c) Find the expected net gain from survey B.

(d) Suppose that the firm has a choice. It can use no survey, survey A, or survey B, but not both surveys. What is the optimal course of action?

61. In Exercise 60, suppose that the firm also has the option of using *both* surveys. Furthermore, since both surveys would be conducted by the same marketing research firm, the total cost of the two surveys is only $40,000, provided that a decision is made in advance to use both surveys (that is, the firm cannot use one survey and then decide whether or not to use the other). Given any one of the three events θ_1, θ_2, and θ_3, the results of survey B are considered to be independent of the results of survey A. What should the firm do?

62. In Exercise 60, suppose that the firm has the additional option of using survey A and, after seeing the results of survey A, deciding whether or not to use survey B. The reverse procedure, using survey B first and then considering survey A, is not possible. Of course, if the sequential plan is used and both surveys are taken, the total cost of the surveys is $50,000 rather than $40,000. Find the expected net gain of the sequential plan and compare this with the expected net gains of the sampling plans considered in Exercises 60 and 61.

63. Do part (d) of Exercise 55 if
(a) the bag contains only 10 chips, either 7R–3B or 7B–3R,
(b) the bag contains only 10 chips, either 7R–3B or 7B–3R, *and* the sampling is done *without* replacement.

64. Comment on the statement, "Linear payoff functions have wide applicability in real-world decision-making problems." Give some examples of problems in which the payoff functions are linear or approximately linear.

65. Suppose that the payoff functions (in dollars) of two actions are

$$R(a_1, \mu) = 70 - .5\mu$$

and

$$R(a_2, \mu) = 50 + .5\mu,$$

where μ is the mean of a normal process with variance 1200.
(a) Find the breakeven value, μ_b.
(b) If the prior distribution is a normal distribution with mean $m' = 25$ and variance $\sigma'^2 = 400$, which action should you choose?
(c) What is the value of perfect information that $\mu = 15$?
(d) What is the value of perfect information that $\mu = 21$?
(e) What is the expected value of perfect information?
(f) Graph the payoff functions and the associated loss functions.
(g) What would the expected value of perfect information be if the prior mean $m' = 30$? Explain the difference between this answer and the answer to (e).

66. The payoff in a certain decision-making problem depends on p, the parameter of a Bernoulli process. The payoff functions are

$$R(a_1, p) = 50p$$

and

$$R(a_2, p) = -10 + 100p.$$

(a) What is the breakeven value of p?
(b) Graph the payoff functions and the associated loss functions.

(c) If the prior distribution is a beta distribution with $r' = 4$ and $n' = 24$, which action should be chosen?

(d) If a sample of size 15 is taken, with 4 successes, find the posterior distribution and the optimal action under this distribution.

67. In Exercise 65(b), find the EVSI and ENGS for samples of size 1, 2, 3, 4, 5, 6, 7, 8, 9, and 10, assuming that the cost of sampling is $.15 per trial. On a graph, draw curves representing EVSI, ENGS, and CS.

68. A firm is contemplating the purchase of 500 typewriter ribbons. One supplier, supplier A, offers the ribbons at $1.50 each, guarantees each ribbon, and will replace all defective ribbons free. A second supplier, supplier B, offers the ribbons at $1.40 each with no guarantee. However, supplier B will replace defective ribbons with good ribbons for $1.00 per ribbon. Suppose that the proportion of defective ribbons produced by supplier B is denoted by p, and suppose that the prior distribution for p is a beta distribution with parameters $r' = 2$ and $n' = 50$.

(a) What should the firm do on the basis of the prior distribution?

(b) How much is it worth to the firm to know the proportion of defective ribbons for certain?

(c) Suppose that supplier B can provide a randomly chosen sample of 10 ribbons. What is the expected value of this sample?

(d) In the sample of 10 ribbons, 1 is defective. Find the posterior distribution of p and use this distribution to determine which supplier the firm should deal with.

69. A salesman can work on a "straight commission" basis, receiving a 10 percent commission on all sales, or on a "salary plus commission" basis, receiving a fixed salary of $5000 yearly plus a 5 percent commission on all sales. Let θ denote the salesman's total sales in the coming year (it is assumed that the salesman can switch plans at the end of any year, so it is not necessary to consider potential sales in future years). Draw the payoff functions and the loss functions for the salesman's choice of a compensation plan. If the distribution of θ is a normal distribution with $P(\theta < 90{,}000) = .60$, which plan is optimal?

70. What are certainty equivalents, and how can they help simplify problems of decision making under uncertainty?

71. Give some examples of decision-making situations in which the payoff or loss functions might be nonlinear.

72. In Exercise 66, assume that the payoff functions are given in terms of dollars and that the decision maker's utility function for money is of the form

$$U(R) = (R + 100)^2 \qquad \text{for } -100 \le R \le 100,$$

where R represents dollars. Graph the payoff functions and the associated loss functions in terms of utility, determine which of the two actions has the higher expected utility under the prior distribution, and determine which of the two actions has the higher expected utility under the posterior distribution. [*Hint*: The laws of expectation are useful here.]

73. A contractor must decide whether or not to build any speculative houses (houses for which the contractor would have to find a buyer), and if so, how many. The houses are sold for a price of $30,000, and they cost $26,000 to build. Since the contractor cannot afford to have too much cash tied up at once, any houses that remain unsold 3 months after they are completed will have to be sold to a realtor for $25,000. The contractor's prior distribution for θ, the number of houses that will be sold within 3 months of completion, is as follows.

θ	$P(\theta)$
0	.05
1	.10
2	.10
3	.20
4	.25
5	.20
6	.10

If the contractor's utility function is linear with respect to money, how many houses should be built? How much should the contractor be willing to pay to find out for certain how many houses will be sold within 3 months?

74. A hot-dog vendor at a football game must decide in advance how many hot dogs to order. He makes a profit of $.10 on each hot dog that is sold, and he suffers a $.20 loss on hot dogs that are unsold. If his distribution of the number of hot dogs that will be demanded at the football game is a normal distribution with mean 10,000 and standard deviation 2000, how many hot dogs should he order?

75. A wine merchant must decide how many cases of Chambertin to purchase. Chambertin cannot be purchased again for another month. The demand for expensive wine has been increasing, and the merchant feels that the demand for Chambertin during the next month is normally distributed with a mean of 20 cases and a standard deviation of 5 cases. Chambertin costs the merchant $110 a case and retails for $155 a case. Moreover, although Chambertin ages well and any that is left at the end of the month will be sold later, the merchant feels that the cost of storage and having funds tied up in inventory amount to about $5 per case for any cases not sold by the end of the month. If the merchant currently has 5 cases of Chambertin on hand, how many additional cases should be purchased?

76. A book is about to go into its last printing before a new edition comes out. The new edition will appear in 1 year, and the demand for the current edition during this year is denoted by θ, which is expressed in thousands of copies. Suppose that the demand for the book can be treated as a Poisson process

with intensity $\lambda = 1$ per month, where λ is expressed in thousands of copies. The net profit to the publishing firm is \$4 per copy, and it costs \$8 to produce each copy. If books are always printed in multiples of a thousand copies, how many copies should be printed in the last printing of the current edition of the book? Assume that copies of the old edition are worthless once the new edition is available.

77. Comment on the following statement: "In taking the sample mean as an estimator of the population mean, the statistician is acting essentially as though the loss function were quadratic." Explain the difference between the approach to point estimation taken in Chapter 6 and the approach taken in this chapter.

78. If a person faces a point-estimation problem with a linear loss function with $k_u = 4$ and $k_o = 3$, is it necessary to assess an entire probability distribution or can a certainty equivalent be determined? Explain.

79. If a statistician wishes to estimate a parameter θ subject to a loss function which is linear with $k_o = 2k_u$, and the distribution of θ is an exponential distribution with $\lambda = 4$, what is the optimal estimate of θ?

80. If the loss function in a point-estimation problem is of the form

$$ L(a, \theta) = \begin{cases} 0 & \text{if } |a - \theta| < k, \\ 1 & \text{otherwise,} \end{cases} $$

where k is some very small positive number, what the optimal estimate is? Can you think of any realistic situations in which the loss function might be of this form?

81. For a particular individual charged with a crime, a judge must decide between H_0: innocent and H_1: guilty. The judge feels that the loss due to convicting an innocent person is 10 times as great as the loss due to setting a guilty person free. On the basis of the evidence presented up to a particular point at the trial, the probabilities for H_0 and H_1 are .2 and .8, respectively. A new witness is then called to the stand, and the witness states that the accused was at the scene of the crime. The judge feels that the probability of this testimony is only .1 if the person is innocent and .9 if the person is guilty. Revise the judge's probabilities and find out whether H_0 should be accepted or rejected on the basis of the revised probabilities.

82. In Exercise 75, Chapter 8, the loss due to claiming incorrectly that A authored the essay is considered to be 3 times as great as the loss due to claiming incorrectly that B authored the essay. If you must make a claim concerning the authorship, which one should you claim as the author?

83. In Exercise 74, Chapter 8, the marketing manager decides that the loss which will be suffered if the company markets the product and p is in fact only .10 is 3 times as great as the loss which will be suffered if the company fails to market the product and p is actually .20. Should the product be marketed?

84. A statistician is interested in testing the hypotheses

$$H_0: \mu \geq 120$$

and $$H_1: \mu < 120,$$

where μ is the mean of a normally distributed population with variance 144. The prior distribution for μ is a normal distribution with mean $m' = 115$ and variance $\sigma'^2 = 36$. A sample of size 8 is taken, with sample mean 121.
(a) Find the prior odds ratio of H_0 to H_1.
(b) Find the posterior odds ratio of H_0 to H_1.
(c) If $L(I) = 4$ and $L(II) = 6$, which hypothesis should be accepted according to the posterior distribution?

85. Consider the hypotheses $H_0: \mu = 10$ and $H_1: \mu = 12$, where μ is the mean of a normally distributed population with variance 1. The prior distribution is diffuse (take a normal distribution with $n' = 0$), and a sample of size 10 is to be taken.
(a) If $L(I) = 100$ and $L(II) = 50$, find the region of rejection (the sample results for which you would reject H_0 in favor of H_1), using the decision-theoretic approach.
(b) From the region of rejection determined in (a), find $P(\text{Type I error})$ and $P(\text{Type II error})$.
(c) Do (a) and (b) if $L(I) = 50$ and $L(II) = 50$.

86. In Exercise 62, Chapter 7, suppose that from past experience, it is known that the population variance of the contents (in ounces) is 1. The prior judgments of a consumer research group with respect to the mean contents can be represented by a normal distribution with mean 11.80 oz and standard deviation .2. Furthermore, the sample given in Exercise 62, Chapter 7, consists of 16 bottles with a sample mean of 11.6 oz.
(a) Find the posterior odds ratio of $H_0: \mu \geq 12$ versus $H_1: \mu < 12$.
(b) If the loss due to unjustly accusing the bottling firm of short-weighting is judged to be 3 times as great as the loss due to incorrectly deciding that the firm is not guilty of short-weighting, should the bottling firm be accused of short-weighting?

87. Compare and contrast the determination of α and β in hypothesis-testing problems according to (a) the classical "convention" presented in Chapter 7 and (b) the decision-theoretic approach, as illustrated in Exercises 85 and 86.

88. In Exercise 80, Chapter 8, a statement is made about losses. On the basis of that statement, what can you say about $L(I)$ and $L(II)$?

89. Carefully distinguish among "classical" inferential statistics, Bayesian inferential statistics, and decision theory.

10

REGRESSION
AND CORRELATION

Basic concepts of statistical inference and decision were covered in Chapters 5–9. These concepts will be used in Chapters 10–12 as we discuss statistical methods that have been developed to handle some specific types of problems. This chapter deals with problems involving relationships among variables and the use of such relationships in prediction problems. For example, a medical researcher might be interested in the relationship between the incidence of lung cancer and variables such as the average number of cigarettes smoked per day. A college counselor might be interested in the relationship between college grade-point averages and variables such as high-school grade-point averages or scores on college entrance examinations. A security analyst might be interested in the relationship between price changes of stocks and variables such as the previous market performance of the stocks, the earnings per share for the stocks, and changes in economic indicators or stock-market indices. In some cases relationships among variables are of interest in and of themselves (for example, for scientific purposes); in other cases, relationships are of interest because they are useful in the prediction of a particular variable given values of certain other variables.

Investigating relationships between two or more variables and using such relationships in making predictions, then, are the problems of interest in this chapter. These problems are called problems of **correlation** and **regression,** and they embody questions such as the following.

1. Does a statistical relation affording some predictability appear to exist between the random variables of interest?

2. How strong is the apparent degree of the statistical relation, in the sense of the possible predictive ability the relation affords?
3. Can a simple rule be formulated for predicting one variable from the other variable or variables, and if so, how good is this rule?

We will first attempt to answer the first two questions by investigating problems of correlation, and we will then attack the third question, which involves problems of regression. The analysis will be presented both for the case in which there are only two variables and for the case in which there may be more than two variables. Since we will be concerned with relationships between two or more random variables, it might be useful for the reader to review the material on joint distributions that was presented in Sections 3.21–3.23.

10.1 CORRELATION

Two measures involving the relationship between two random variables, the covariance and the correlation coefficient, were defined in Section 3.23. From Equations (3.23.8) and (3.23.9), the covariance is

$$\text{cov}\,(X, Y) = E[(X - \mu_X)(Y - \mu_Y)] = E(XY) - E(X)E(Y). \quad (10.1.1^*)$$

As we pointed out in Section 3.23, the sign of the covariance gives us some idea regarding the **direction** of the relationship between X and Y. Thus, if the covariance is positive, then high values of X tend to be associated with high values of Y, and low values of X with low values of Y. If the covariance is negative then high values of X tend to be associated with low values of Y, and vice versa.

Because the covariance is affected by the variability of X and Y taken individually, it tells us little of the strength of the relationship between the two variables. A better measure is the **correlation coefficient**:

$$\rho_{XY} = \frac{\text{cov}\,(X, Y)}{\sigma_X \sigma_Y}. \quad (10.1.2^*)$$

The correlation coefficient is one of the most commonly used measures of the relationship between two random variables.

More specifically, ρ_{XY} is a measure of the linear relationship between the two variables X and Y.

Of course, it is perfectly possible for two variables to be perfectly related in a nonlinear fashion; for a trivial example, if X can take on only positive values, then X and $Y = X^2$ are perfectly related. Given any value of X, we can with certainty predict the corresponding value of Y, and vice versa. Despite this, the relationship is nonlinear, so that the correlation coefficient will not reflect the perfect functional relationship between the two random variables.

Often the correlation coefficient is not known, but a sample from the bivariate population or process of interest is available. Just as the sample mean can be used to estimate the population mean, the sample correlation coefficient can be used to estimate ρ_{XY}.

The sample correlation coefficient (called the Pearson product-moment correlation coefficient) is defined as follows:

$$r_{XY} = \frac{\sum\limits_{i=1}^{n} (x_i - m_X)(y_i - m_Y)}{n s_X s_Y}, \qquad (10.1.3^*)$$

where the n pairs of values (x_i, y_i) represent a sample of size n from the bivariate population and m_X, m_Y, s_X, and s_Y represent the sample means and sample standard deviations of the two variables.

Comparing the sample correlation coefficient (10.1.3) with the population correlation coefficient (10.1.2), we see that it is an intuitively appealing estimate. The sample standard deviations serve to estimate the population standard deviations, and the remaining term,

$$\frac{\sum\limits_{i=1}^{n} (x_i - m_X)(y_i - m_Y)}{n},$$

is the sample covariance, which is, of course, an estimate of cov (X, Y). There is also a more theoretical justification for the use of r_{XY}, provided that certain assumptions are met. If it is assumed that the joint distribution of X and Y is a bivariate normal distribution (this distribution will be briefly discussed in the following section; suffice it to say for now that this is a bivariate, or two-variable, version of the normal distribution which has proven so useful in previous chapters), then r_{XY} is the maximum-likelihood estimate of ρ_{XY}. In Section 6.7 we pointed out that maximum-likelihood estimators possess certain desirable properties.

The formula presented above for r_{XY} [Equation (10.1.3)] is somewhat difficult to apply computationally. So that we can present some examples of sample correlation coefficients, it will be helpful to develop a "computational formula" for r_{XY}. Working with the numerator of Equation

(10.1.3), we have

$$\sum_i (x_i - m_X)(y_i - m_Y) = \sum_i (x_i y_i - x_i m_Y - y_i m_X + m_X m_Y).$$

But m_X and m_Y are constants with respect to the summation, so this is equal to

$$\sum_i x_i y_i - m_Y \sum_i x_i - m_X \sum_i y_i + n m_X m_Y,$$

or

$$\sum_i x_i y_i - \frac{\sum_i x_i \sum_i y_i}{n} - \frac{\sum_i x_i \sum_i y_i}{n} + \frac{\sum_i x_i \sum_i y_i}{n}.$$

Simplifying, we have

$$\sum_i (x_i - m_X)(y_i - m_Y) = \frac{n \sum_i x_i y_i - \sum_i x_i \sum_i y_i}{n}. \qquad (10.1.4)$$

Similarly, from Equation (5.11.4),

$$s_X = \sqrt{\frac{n \sum_i x_i^2 - (\sum_i x_i)^2}{n^2}} = \frac{1}{n}\sqrt{n \sum_i x_i^2 - (\sum_i x_i)^2} \qquad (10.1.5)$$

and

$$s_Y = \sqrt{\frac{n \sum_i y_i^2 - (\sum_i y_i)^2}{n^2}} = \frac{1}{n}\sqrt{n \sum_i y_i^2 - (\sum_i y_i)^2}. \qquad (10.1.6)$$

Using Equations (10.1.3)–(10.1.6), r_{XY} can be expressed as follows:

$$r_{XY} = \frac{n \sum_i x_i y_i - \sum_i x_i \sum_i y_i}{\sqrt{n \sum_i x_i^2 - (\sum_i x_i)^2}\ \sqrt{n \sum_i y_i^2 - (\sum_i y_i)^2}}. \qquad (10.1.7^*)$$

Although this may look more formidable than Equation (10.1.3), it is easier to apply from a computational standpoint.

To demonstrate the calculation of r_{XY}, consider the relationship between the grades of 7 students on successive examinations. Let x_i and y_i denote the scores earned by Student i ($i = 1, \ldots, 7$) on the first and second examinations in a statistics class. The numbers, which are given in Table 10.1.1, have been scaled down to make the computations simpler; as the scores are presented, the highest possible grade was 15 on each of the two

Table 10.1.1 Grades of 7 Students on Two Examinations

Student	Score on first examination (x_i)	Score on second examination (y_i)	$x_i y_i$	x_i^2	y_i^2
1	12	11	132	144	121
2	11	14	154	121	196
3	5	11	55	25	121
4	10	13	130	100	169
5	13	15	195	169	225
6	13	14	182	169	196
7	12	12	144	144	144
	76	90	992	872	1172

examinations. From these data and Equation (10.1.7),

$$r_{XY} = \frac{7(992) - (76)(90)}{\sqrt{7(872) - (76)^2}\,\sqrt{7(1172) - (90)^2}} = \frac{104}{\sqrt{34,112}} = .563.$$

How can we interpret this sample correlation of .563? Like the population correlation coefficient ρ_{XY}, the sample correlation coefficient measures the strength of the *linear* relationship between X and Y. It is helpful to consider the **scatter diagram** presented in Figure 10.1.1. Scatter diagrams are merely graphs showing, in two-dimensional space, the pairs of values

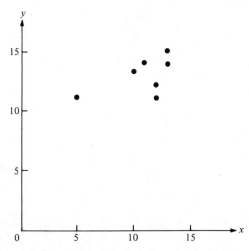

Figure 10.1.1 Scatter Diagram for Examination-Scores Example

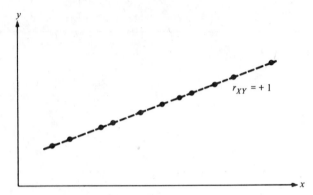

Figure 10.1.2   A Scatter Diagram Showing a Perfect Positive
Linear Relationship

(x_i, y_i). These diagrams are extremely valuable in that they give the statistician some idea about the form of the functional relationship between X and Y, if such a functional relationship in fact exists. In this example, there does seem to be a positive relationship between X and Y; that is, high values of X are associated with high values of Y, and low values of X are associated with low values of Y. This is also indicated by the fact that r_{XY} is greater than zero. A sample correlation coefficient less than zero, on the other hand, indicates a negative relationship. We see from Figure 10.1.1 that the relationship is certainly not a perfect relationship (in a linear sense or in any other sense). In a perfect linear relationship, all of the points would lie on a single straight line. The value of r_{XY} would then be $+1$ if the relationship was positive (if the straight line had a positive slope) and -1 if the relationship was negative. From Equation (3.23.11), ρ must be between -1 and $+1$; this also holds for r_{XY}, which is an estimate of ρ. Figures 10.1.2 and 10.1.3 illustrate scatter diagrams for

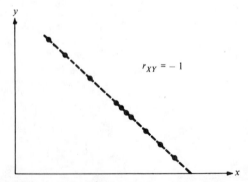

Figure 10.1.3   A Scatter Diagram Showing a Perfect Negative
Linear Relationship

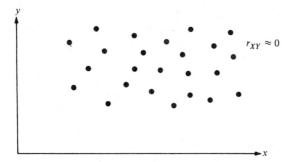

Figure 10.1.4 A Scatter Diagram Showing No Linear Relationship

which r_{XY} is equal to $+1$ and -1, respectively. Figure 10.1.4 is a situation in which r_{XY} is approximately zero; in other words, there is no linear relationship between the two random variables. Finally, in Figure 10.1.5, the points in the scatter diagram all lie on the smooth curve which is shown as a dashed line. This would tend to indicate that there is a perfect functional relationship between the two variables (although we would need more data to investigate this assertion further). The value of r_{XY}, however, is less than 1 because the functional relationship is not linear. This demonstrates clearly that r_{XY} measures the strength of the *linear* relationship between the sample outcomes of X and Y.

Incidentally, it should be noted that the sample correlation coefficient (as well as the theoretical correlation coefficient ρ_{XY}) is a *dimensionless* quantity. That is, if

$$U = cX + g$$

and
$$W = dY + h,$$

where c and d are any positive constants, and g and h are any constants,

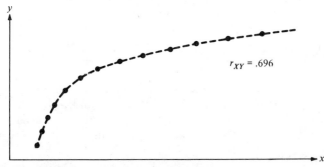

Figure 10.1.5 A Scatter Diagram Showing a Nonlinear Relationship

positive, zero, or negative, then r_{UW} is equal to r_{XY}. The proof of this follows directly from the definition (10.1.3) and is left as an exercise for the reader. In words, if the variables U and W are positive linear functions of X and Y, respectively, then the sample correlation between U and W is the same as the sample correlation between X and Y. This fact makes it possible to simplify the computation of r_{XY} by transforming X and Y into new variables U and W. For instance, if all of the X values observed are between 365 and 385, it will make the computations much simpler if we let $U = X - 375$. The correlation between U and Y will then be the same as the correlation between X and Y.

It should be mentioned that the Pearson product-moment correlation coefficient, which we denote by r_{XY}, is properly used only when the data are measured on an interval scale or a ratio scale, as defined in Section 5.3. If the data are nominal (expressed just in categories) or ordinal (just ordered), then there are other techniques available to study the association between two variables, as we shall see in Chapter 12.

10.2 THE BIVARIATE NORMAL DISTRIBUTION

In order to make inferences about ρ_{XY} or about other measures involving the relationship between two variables, it is necessary to make some assumptions about the joint distribution of the two variables. As noted in Section 3.21, a joint distribution of two variables X and Y can be represented by a joint mass function of the form $P(X = x, Y = y)$ in the discrete case and by a joint density function of the form $f(x, y)$ in the continuous case. Since two variables are involved, distributions of this nature are called **bivariate** distributions.

Although any number of theoretical bivariate distributions are possible in principle, by far the most studied is the **bivariate normal distribution,** an example of which is illustrated in Figure 10.2.1. The density function, unfortunately, has a rather elaborate-looking rule:

$$f(x, y) = \frac{1}{2\pi\sigma_X\sigma_Y\sqrt{1 - \rho_{XY}^2}} \exp\left\{-\frac{(z_X^2 + z_Y^2 - 2\rho_{XY}z_Xz_Y)}{2(1 - \rho_{XY}^2)}\right\}, \quad (10.2.1)$$

where $z_X = (x - \mu_X)/\sigma_X$ and $z_Y = (y - \mu_Y)/\sigma_Y$. The bivariate normal distribution has five parameters: μ_X, μ_Y, σ_X, σ_Y, and ρ_{XY}. Thus, in order to completely specify a bivariate normal distribution, we need to specify the correlation coefficient as well as the means and standard deviations of the two variables.

The bivariate normal distribution has some attractive mathematical properties. For example, if X and Y have a bivariate normal distribution,

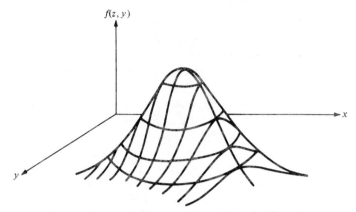

Figure 10.2.1 A Bivariate Normal Density Function

then the marginal distribution of X,

$$f(x) = \int_{-\infty}^{\infty} f(x, y) \, dy,$$

is a normal distribution, as is the marginal distribution of Y. (On the other hand, just because the marginal distributions of X and Y both happen to be normal distributions, this does *not* necessarily mean that the joint distribution of X and Y is bivariate normal.) Furthermore, given any value of X, the *conditional* distribution of Y,

$$f(y \mid x) = \frac{f(x, y)}{f(y)},$$

is a normal distribution; similarly, given any value of Y, the conditional distribution of X is normal.

A particularly important feature of the bivariate normal distribution with respect to the study of the relationship between two variables follows.

Given that X and Y have a bivariate normal distribution, then X and Y are independent if and only if $\rho_{XY} = 0$.

For *any* joint distribution of X and Y, the independence of X and Y implies that $\rho_{XY} = 0$, but it may happen that $\rho_{XY} = 0$ even though X and Y are *not* independent. However, for the bivariate normal distribution, $\rho_{XY} = 0$ both implies and is implied by the statistical independence of X and Y. This result means that any relationship between two variables with a bivariate normal distribution is strictly a *linear* relationship.

Most inferential techniques involving correlation have been developed in terms of the bivariate normal distribution. If we can assume such a joint population distribution, inferences about correlation are equivalent to inferences about independence or dependence between two random variables. As always, of course, by adopting more stringent assumptions about the form of the population distribution, we are able to make stronger statements from sample results.

The general notion of a **multivariate normal distribution** involves the case of more than two variables. For example, there might be K variables, with each observation represented by a K-tuple of the form $(x_1, x_2, \ldots, x_K)$. The parameters of a K-variate normal distribution are the means and standard deviations of the K variables and the correlations between *each pair* of variables. As in the case of the bivariate normal distribution, any marginal or conditional densities determined from a multivariate normal distribution are normal in form.

10.3 INFERENCES IN CORRELATION PROBLEMS

If we assume that the population of interest is bivariate normal in form, then the relationship between X and Y is strictly linear and can be summarized by the parameter ρ_{XY}. Of course, the sample correlation coefficient r_{XY} can be used to estimate ρ_{XY}. Although it is a sufficient and consistent estimator of ρ_{XY}, the sample correlation coefficient is slightly biased. However, the amount of bias involves terms of the order of $1/n$, so that for most practical purposes it can be ignored.

Unfortunately, the sampling distribution of r_{XY} is not always of a convenient form. For very large samples from a bivariate normal population, the distribution of the sample correlation coefficient may be regarded as approximately normal when $\rho_{XY} = 0$. Even for relatively small samples $(n > 4)$, this sampling distribution is unimodal and symmetric. However, when ρ_{XY} is other than zero, the distribution of the sample correlation coefficient tends to be very skewed. When ρ_{XY} is greater than zero, the skewness tends to be toward the left, with intervals of high values of r_{XY} relatively more probable than similar intervals of negative values. When ρ_{XY} is negative, this situation is just reversed, and the distribution is skewed in the opposite direction.

Often we are interested in whether or not two variables are independent. Under the assumption of bivariate normality, independence is equivalent to zero correlation, so the hypotheses of interest might be H_0: $\rho_{XY} = 0$ and H_1: $\rho_{XY} \neq 0$. Even for relatively small n, the test statistic

$$t = \frac{r_{XY}\sqrt{n-2}}{\sqrt{1 - r_{XY}^2}} \tag{10.3.1*}$$

with $n - 2$ degrees of freedom can be used in a test of H_0 versus H_1.

For example, consider the data from Section 10.1 concerning examination scores. In that case $n = 7$ and $r_{XY} = .563$, so that

$$t = \frac{.563\sqrt{5}}{\sqrt{1 - (.563)^2}} = 1.52$$

with $n - 2 = 5$ degrees of freedom. Looking in the table of percentage points of the T distribution, we see that this corresponds to a p-value slightly less than .20. Thus if ρ_{XY} actually equals zero, we would expect to observe an r_{XY} greater than .563 or less than $-.563$ in a sample of size 5 about 20 percent of the time.

Unfortunately, the test statistic given by Equation (10.3.1) is not valid for testing hypotheses other than the hypothesis that $\rho_{XY} = 0$ and it does not provide interval estimates for ρ_{XY}. However, R. A. Fisher showed that tests of hypotheses about ρ_{XY}, as well as confidence intervals, can be made from moderately large samples from a bivariate normal population if we use a particular *function* of r_{XY} rather than the sample correlation coefficient itself. The function, which is known as the **Fisher r to z transformation,** is given by the rule

$$w = \frac{1}{2} \log_e \left(\frac{1 + r_{XY}}{1 - r_{XY}} \right). \qquad (10.3.2)$$

(We use w rather than z in order to avoid confusion with standardized random variables.) The function is of the type called "one to one"; for each possible value of r there can exist one and only one value of w, and for each w, one and only one value of r. This fact makes it possible to convert a sample r value to a w value, to make inferences in terms of w, and then to turn those inferences back into statements about correlation once again.

For virtually any value of ρ_{XY}, the sampling distribution of W is approximately normal for samples of moderate size, with an expectation given approximately by

$$E(W) = \frac{1}{2} \log_e \left(\frac{1 + \rho_{XY}}{1 - \rho_{XY}} \right). \qquad (10.3.3)$$

The sampling variance of W is approximately

$$V(W) = \frac{1}{n - 3}. \qquad (10.3.4)$$

The goodness of these approximations increases the *smaller* the absolute value of ρ_{XY} and the *larger* the sample size. For moderately large samples the hypothesis that ρ_{XY} is equal to any value ρ_0 (not too close to 1 or -1)

can be tested. This is done in terms of the test statistic

$$Z = \frac{W - E(W)}{\sqrt{V(W)}} = \frac{W - E(W)}{\sqrt{1/(n - 3)}} \qquad (10.3.5)$$

referred to a standard normal distribution. The value taken for $E(W)$ depends on the value given for ρ_0 by the null hypothesis:

$$E(W) = \frac{1}{2} \log_e \left(\frac{1 + \rho_0}{1 - \rho_0} \right).$$

It should be emphasized that the use of this r to w transformation *does* require the assumption that X and Y have a bivariate normal distribution in the population. On the surface, this assumption seems to be a very stringent one, which may not be reasonable in some situations, although there is some evidence that the assumption may be relatively innocuous in others. However, the consequences when this assumption is not met seem largely to be unknown. Perhaps the safest course is to require rather large samples in uses of this test when the assumption of a bivariate normal population is very questionable.

Table IX in the Appendix gives the w values corresponding to various values of r. This table is quite easy to use and makes carrying out the test itself extremely simple. Only positive r and w values are shown, since if r is negative, the sign of the w value is taken as negative also.

For example, suppose that $n = 100$ and $r_{XY} = .35$ instead of $n = 7$ and $r_{XY} = .563$ in the case of the examination scores from Section 10.1. The hypotheses of interest are H_0: $\rho_{XY} = .50$ and H_1: $\rho_{XY} \neq .50$. From Table IX, we find that for $r_{XY} = .35$, $w = .3654$; for $\rho_{XY} = .50$, $E(W) = .5493$. Thus,

$$z = \frac{.3654 - .5493}{\sqrt{1/97}} = -1.81.$$

The p-value for a two-tailed test is .07, so H_0 should be rejected only if $\alpha \geq .07$.

The above transformation can also be used to find confidence intervals for ρ_{XY}. An approximate $100(1 - \alpha)$ percent confidence interval for $E(W)$ is given by the limits

$$W \pm a \sqrt{\frac{1}{n - 3}}, \qquad (10.3.6)$$

where a is the $1 - (\alpha/2)$ fractile of the standard normal distribution. On changing the limiting w values back into correlation values, we have a confidence interval for ρ_{XY}. In the example of the preceding paragraph, a

95 percent confidence interval for $E(W)$ is given approximately by the limits

$$.3654 \pm 1.96 \sqrt{\frac{1}{97}},$$

or $(.1694, .5614).$

Transforming via Table IX, the corresponding 95 percent confidence interval for ρ_{XY} has limits of approximately .168 and .510.

10.4   AN EXAMPLE OF A CORRELATION PROBLEM

Consider the relationship between the height of a wife and the height of her husband. If we assume that the joint distribution of heights of married couples can be reasonably approximated by a bivariate normal distribution, then the relationship of interest can be summarized by a correlation coefficient. In order to investigate this relationship, a sample of 15 American couples was drawn at random, and the resulting data are shown in Table 10.4.1 and Figure 10.4.1.

For these data the computations for the correlation coefficient can be simplified by subtracting 60 from the height of each wife and 70 from the height of each husband; this does not alter the value of r_{XY} obtained (Section 10.1). Then the new scores are as shown in Table 10.4.2.

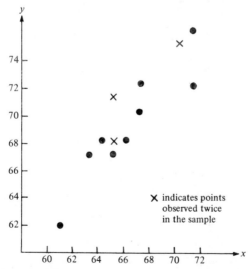

Figure 10.4.1   A Scatter Diagram of Heights of Wives and Husbands

Table 10.4.1   Heights of Wives and Husbands for 15 Couples

	HEIGHTS IN INCHES	
Couple	x (Wife's height)	y (Husband's height)
1	70	75
2	67	72
3	70	75
4	71	76
5	67	70
6	64	68
7	71	72
8	63	67
9	65	67
10	64	68
11	65	68
12	65	71
13	66	68
14	65	71
15	61	62

Table 10.4.2   Transformed Heights for 15 Couples

Couple	u	v	uv	u^2	v^2
1	10	5	50	100	25
2	7	2	14	49	4
3	10	5	50	100	25
4	11	6	66	121	36
5	7	0	0	49	0
6	4	−2	−8	16	4
7	11	2	22	121	4
8	3	−3	−9	9	9
9	5	−3	−15	25	9
10	4	−2	−8	16	4
11	5	−2	−10	25	4
12	5	1	5	25	1
13	6	−2	−12	36	4
14	5	1	5	25	1
15	1	−8	−8	1	64
	94	0	142	718	194

The scatter diagram (Figure 10.4.1) indicates that there is some degree of linear association between the two variables in this example. The sample correlation coefficient computed from Equation (10.1.7) turns out to be

$$r_{XY} = \frac{(15)(142) - (94)(0)}{\sqrt{[(15)(718) - (94)^2][(15)(194) - (0)^2]}}$$

$$= \frac{2130}{\sqrt{(1934)(2910)}} = .89.$$

It appears, then, that there is a strong linear relationship between the heights of wives and husbands. Using the test of independence given by Equation (10.3.1), we get

$$t = \frac{(.89)\sqrt{15 - 2}}{\sqrt{1 - (.89)^2}} = \frac{3.209}{.456} = 7.04.$$

From the table of the T distribution, we see that the p-value for a two-tailed test is much less than .01 in this case. Moreover, from Equation (10.3.6), limits for a 95 percent confidence interval are

$$1.422 \pm 1.96 \sqrt{\tfrac{1}{12}},$$

or $\qquad\qquad .856 \quad$ and $\quad 1.988,$

in terms of $E(W)$. (The w value corresponding to $r_{XY} = .89$ is 1.422.) Transforming back to correlation values, a 95 percent confidence interval for ρ_{XY} is approximately $(.694, .963)$.

We must be careful in interpreting the results of this example. Even though the correlation found is sizable, it makes no sense at all to think of the height of the wife as "causing" the height of the husband, or that of the husband causing that of the wife. These are simply two numerical measurements that happen to occur together in a more or less linear way, according to the evidence of this sample. The reason *why* this linear relation exists is completely out of the realm of statistics, and the correlation coefficient and tests shed absolutely no light on this problem. In this example, it is perfectly obvious that personal preferences and current standards of society cause *some* selection to occur in the process of mating, and these factors in turn underlie our observations that (x, y) pairs occur in a particular kind of relationship. As a description of a population situation, our inferences may very well be valid, but this fact alone gives us no license to talk about the *cause* of the apparent linear relation. *It is most important in correlation problems to distinguish between correlation and causation.* A high correlation in a particular set of data does *not* necessarily imply a *causal* relationship between variables.

10.5   THE REGRESSION CURVE

In correlation problems we were interested in measuring the strength of the statistical relationship between two variables X and Y. In **regression** problems the statistician wants to predict the value of one of the random variables, given a value of the other variable. For example, suppose that we want to predict the sales of a product, given the price of the product. In this case we say that price is the independent variable and that the other variable, sales, is dependent on price. Suppose that we let X represent price and we let Y represent sales. How can we predict Y? If we know the marginal distribution of Y, we might use the mean of the distribution, $E(Y)$. This ignores the information concerning X, however. Since x, the value of X, is known, the distribution of interest is the *conditional* distribution of Y given that $X = x$. This distribution is represented by $P(Y \mid X = x)$ in the discrete case and by $f(y \mid x)$ in the continuous case. An intuitively reasonable estimator (or predictor) of Y is thus the mean of the conditional distribution of Y given that $X = x$.

The conditional mean, $E(Y \mid X = x)$, may vary for different values of x; in other words, it is a function of x, and this function is called the regression curve of Y on X.

A regression curve is illustrated graphically in Figure 10.5.1. For any value x, there is a conditional distribution of Y given $X = x$; three such distributions are shown in the illustration (since X is assumed to be continuous, there is an infinite number of such distributions, corresponding to the infinite number of possible values of X). For each conditional distribution, the mean, $E(Y \mid X = x)$, can be determined. If the set of all such conditional means is traced out, as in Figure 10.5.1, the result is the regression curve of Y on X. Remember that X is the independent variable and Y is the dependent variable.

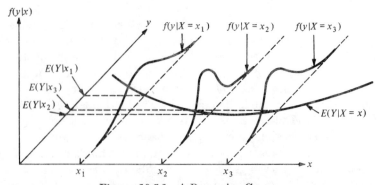

Figure 10.5.1   A Regression Curve

$f(x, y)$

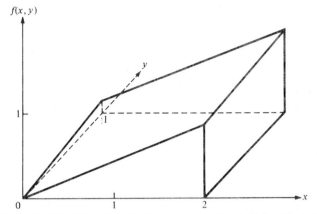

Figure 10.5.2 Joint Density Function of X and Y for Market-Share Example

To demonstrate the determination of a regression curve, suppose that the market share of Brand A of a particular product is currently about $\frac{2}{3}$. The manufacturer of Brand A is concerned that the firm marketing the chief competing brand is about to increase advertising expenditures by up to $2 million. The relationship of interest is that between X, the increase (in millions of dollars) in the competitor's advertising expenditures, and Y, the market share of Brand A.

Suppose that the joint density function of X and Y is

$$f(x, y) = \begin{cases} (3x + y)/7 & \text{for } 0 < x < 2, 0 < y < 1, \\ 0 & \text{elsewhere.} \end{cases}$$

This joint density function is graphed in Figure 10.5.2. We want to find the regression curve, $E(Y \mid x)$. In order to do this, we will first find the marginal density of X, $f(x)$, and then we will divide $f(x, y)$ by $f(x)$ in order to find $f(y \mid x)$, from which $E(Y \mid x)$ can be obtained. The marginal density of X is obtained by integrating $f(x, y)$ with respect to y:

$$f(x) = \int f(x, y) \, dy = \int_0^1 \left(\frac{3x + y}{7} \right) dy$$

$$= \frac{1}{7} \int_0^1 (3x + y) \, dy = \frac{1}{7} \left[3xy + \frac{y^2}{2} \right]_0^1$$

$$= \frac{1}{7} \left[3x + \frac{1}{2} \right] = \frac{(6x + 1)}{14} \quad \text{for } 0 < x < 2.$$

Now, for $0 < x < 2$ and $0 < y < 1$,

$$f(y \mid x) = \frac{f(x, y)}{f(x)} = \frac{(3x + y)/7}{(6x + 1)/14} = \frac{2(3x + y)}{6x + 1}.$$

The regression curve is the conditional expectation

$$E(Y \mid x) = \int yf(y \mid x) \, dy = \int_0^1 y \left[\frac{2(3x + y)}{6x + 1} \right] dy$$

$$= \frac{2}{6x + 1} \int_0^1 y(3x + y) \, dy = \frac{2}{6x + 1} \int_0^1 (3xy + y^2) \, dy$$

$$= \frac{2}{6x + 1} \left[\frac{3xy^2}{2} + \frac{y_3}{3} \right]_0^1 = \frac{2}{6x + 1} \left[\frac{3x}{2} + \frac{1}{3} \right]$$

$$= \frac{2}{6x + 1} \left[\frac{9x + 2}{6} \right] = \frac{9x + 2}{3(6x + 1)}.$$

This is the regression curve of Y on X, which is graphed in Figure 10.5.3. For example, if $X = 1$, the value of Y predicted from the regression curve is

$$E(Y \mid X = 1) = \frac{9(1) + 2}{3[6(1) + 1]} = \frac{11}{21}.$$

Thus, if the competitor increases advertising expenditures by \$1 million, the expected market share for Brand A is 11/21, or about .523, as compared with a current market share of about .67. Of course, this is just an expected value, and we are not at all certain that Y will equal 11/21. We can determine how "confident" we feel about $E(Y \mid x)$ by calculating the variance of the conditional distribution. The calculation of this variance is summarized as follows [the reader can verify that the results are

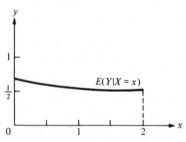

Figure 10.5.3   Regression Curve of Y on X for Market-Share Example

correct by performing the required integration for $E(Y^2 \mid X = x)$]:

$$E(Y^2 \mid x) = \int y^2 f(y \mid x) \, dy = \frac{4x + 1}{2(6x + 1)}$$

and
$$V(Y \mid x) = E(Y^2 \mid x) - [E(Y \mid x)]^2$$

$$= \left[\frac{4x + 1}{2(6x + 1)}\right] - \left[\frac{9x + 2}{3(6x + 1)}\right]^2.$$

Thus, for $X = 1$,

$$V(Y \mid X = 1) = \left[\frac{4 + 1}{2(6 + 1)}\right] - \left[\frac{9 + 2}{3(6 + 1)}\right]^2 = \frac{5}{14} - \left(\frac{11}{21}\right)^2 = \frac{73}{882}.$$

The standard deviation of the conditional distribution of Y given that $X = 1$ is therefore equal to $\sqrt{73/882}$, or .288. This gives us some idea of how accurate the prediction for Y will be.

In using the regression curve to predict Y, then, we are attempting to utilize the information we have concerning X. The motivation here is very similar to the motivation in Chapter 8 when we studied Bayesian inference; we would like to take advantage of all available information that might pertain to the variable of interest, which is the market share of Brand A in the example.

10.6 LINEAR REGRESSION

The regression curve $E(Y \mid x)$ depends on the form of the joint density function $f(x, y)$, and in some cases $E(Y \mid x)$ may be a reasonably complicated mathematical function. Fortunately, in other cases $E(Y \mid x)$ can be expressed in a very simple form. For example, if the joint distribution of X and Y is a bivariate normal distribution, then the regression curve is **linear**:

$$E(Y \mid x) = \beta_1 + \beta_2 x. \qquad (10.6.1^*)$$

Graphically, β_1 is the y-intercept and β_2 is the slope of the line. Of course, the regression curve $E(Y \mid x)$ may be linear or close to linear for other joint distributions as well. Moreover, even if the regression curve is not linear in x, it may be possible to transform X and Y into two new variables, U and W, such that $E(U \mid w)$ is linear in w. We will investigate this sort of transformation when we discuss nonlinear regression later in the chapter. The point we are trying to make here is that linear regression has widespread applicability. Because of this applicability and because linear regression functions are relatively easy to work with, much of the theoretical study of regression analysis and most of the practical applications of regression analysis involve the assumption of linearity. As noted above, nonlinear regression will be studied later in the chapter; the next few sections, however, will be devoted to linear regression.

If the regression curve is of the form (10.6.1), then the values of β_1 and β_2, which are called the coefficients of the linear regression, or simply the **regression coefficients,** can be expressed in terms of the means and standard deviations of X and Y and the correlation coefficient ρ_{XY}:

$$\beta_1 = \mu_Y - \rho_{XY} \frac{\sigma_Y}{\sigma_X} \mu_X \qquad (10.6.2^*)$$

and
$$\beta_2 = \rho_{XY} \frac{\sigma_Y}{\sigma_X}. \qquad (10.6.3^*)$$

We will just briefly sketch the derivation of Equations (10.6.2) and (10.6.3) from Equation (10.6.1), omitting the details. Taking expectations with respect to x on both sides of Equation (10.6.1) yields the equation

$$\mu_Y = \beta_1 + \beta_2 \mu_X. \qquad (10.6.4)$$

Next, multiplying both sides of Equation (10.6.1) by x and *then* integrating with respect to x yields

$$E(XY) = \beta_1 \mu_X + \beta_2 E(X^2). \qquad (10.6.5)$$

Finally, solving Equations (10.6.4) and (10.6.5) simultaneously, we get Equations (10.6.2) and (10.6.3).

When β_1 and β_2 are expressed in terms of μ_X, μ_Y, σ_X, σ_Y, and ρ_{XY}, the regression line [Equation (10.6.1)] is of the form

$$E(Y \mid x) = \beta_1 + \beta_2 x = \mu_Y - \rho_{XY} \frac{\sigma_Y}{\sigma_X} \mu_X + \rho_{XY} \frac{\sigma_Y}{\sigma_X} x,$$

or
$$E(Y \mid x) = \mu_Y + \rho_{XY} \frac{\sigma_Y}{\sigma_X} (x - \mu_X). \qquad (10.6.6^*)$$

Notice the role of the correlation coefficient in Equation (10.6.6). If $\rho_{XY} = 0$, then the regression line is simply $E(Y \mid x) = \mu_Y$. In other words, if there is *no* linear relationship between X and Y, then X is of no value in the prediction of Y via linear regression. The closer ρ_{XY} gets to -1 or $+1$, the more effect the term involving $(x - \mu_X)$ has on the predicted value of Y.

To determine the precise effect that the correlation coefficient has on the prediction of Y, it is helpful to consider the conditional variance of Y, given x. If X and Y have a bivariate normal distribution, this variance is

$$V(Y \mid x) = \sigma_Y^2(1 - \rho_{XY}^2).$$

Let us denote $V(Y \mid x)$ by $\sigma_{Y.x}^2$, so that

$$\sigma_{Y.x}^2 = \sigma_Y^2(1 - \rho_{XY}^2). \tag{10.6.7*}$$

Notice that $\sigma_{Y.x}^2$ is the variance of Y, given a value of X; on the other hand, if we had no knowledge concerning X, the variance of Y would just be σ_Y^2. The knowledge of X therefore reduces the variance from σ_Y^2 to $\sigma_Y^2(1 - \rho_{XY}^2)$.

From Equation (10.6.7), the *square* of the correlation coefficient, which is called the **coefficient of determination,** can be written as

$$\rho_{XY}^2 = \frac{\sigma_Y^2 - \sigma_{Y.x}^2}{\sigma_Y^2}. \tag{10.6.8*}$$

The interpretation of this equation is that the square of the correlation coefficient is the proportion of the variance accounted for by the linear regression.

The original variance is σ_Y^2, and the remaining variance, that is, the variance *not* accounted for by the linear regression, is $\sigma_{Y.x}^2$.

Consider an example in which $\sigma_Y^2 = 100$. If $\rho_{XY} = 0$, then the knowledge of X will not improve the prediction of Y, and $\sigma_{Y.x}^2 = 100$. None of the variance is accounted for by the linear regression. At the other extreme, suppose that ρ_{XY} is equal to $+1$ or to -1. Then $\sigma_{Y.x}^2 = \sigma_Y^2(1 - 1) = 0$. The knowledge of X enables us to predict Y *perfectly*, and therefore *all* of the variance is accounted for by the linear regression. If $\rho_{XY} = .5$, then $\sigma_{Y.x}^2 = 100(1 - .25) = 75$, and 25 percent of the variance is accounted for by the linear regression, thus leaving 75 percent of the variance unaccounted for. Observe that it is the absolute magnitude of ρ_{XY}, and not the sign, that determines the proportion of the variance which is accounted for by the linear regression. If $\rho_{XY} = -.5$, for example, 25 percent of the variance is accounted for by the linear regression, just as when $\rho_{XY} = +.5$.

As this discussion indicates, knowledge that $X = x$ does not allow us to predict Y perfectly unless ρ_{XY} equals $+1$ or -1. In using the regression curve to predict Y, we can consider the difference between the actual value of Y and the predicted value of Y as the error in our prediction. Denoting this error by e, we have

$$e = Y - E(Y \mid x), \tag{10.6.9}$$

so Y can be written as

$$Y = E(Y \mid x) + e. \tag{10.6.10}$$

When $E(Y \mid x)$ is linear in x, this becomes

$$Y = \beta_1 + \beta_2 x + e. \tag{10.6.11*}$$

Given that $X = x$, then, Y can be written as the sum of a linear function of x and an additional term e, which may be thought of as a **random-error term.** This error term includes the variability that keeps $\beta_1 + \beta_2 x$ from being a perfect predictor of Y. With $X = x$, the first two terms on the right-hand side of Equation (10.6.11) are constants. Thus, using the laws of expectation,

$$E(Y \mid x) = E(\beta_1 + \beta_2 x + e \mid x) = \beta_1 + \beta_2 x + E(e \mid x).$$

But, from Equation (10.6.1), $E(Y \mid x) = \beta_1 + \beta_2 x$, so we must have

$$E(e \mid x) = 0. \tag{10.6.12}$$

Also,

$$V(Y \mid x) = V(\beta_1 + \beta_2 x + e \mid x) = V(e \mid x); \tag{10.6.13}$$

the conditional variance of the error term, given that $X = x$, is identical to $\sigma_{Y.x}^2$. In attempting to make inferences about the regression parameters, we will make further assumptions concerning the distribution of e. First, we will attempt to estimate the regression parameters without invoking such distributional assumptions.

10.7 ESTIMATING THE REGRESSION LINE

In most applications, the exact form of the joint distribution of X and Y is not known, and it is thus not possible to determine the theoretical regression curve $E(Y \mid x)$. Suppose, however, that the results of a sample of size n are available. These results are in the form of n pairs of values (x_i, y_i). The problem now becomes one of *estimating* the regression curve, or of fitting a curve to the data. Since we have noted the wide applicability of linear regression, we will attempt to fit a straight line to the data (later in the chapter we will discuss nonlinear regression). In terms of the model presented in the preceding section, we want to estimate β_1 and β_2. Denote the estimates of the regression coefficients by b_1 and b_2, respectively, so that the estimated regression line is of the form

$$\hat{y} = b_1 + b_2 x. \tag{10.7.1*}$$

Note that the left-hand side of Equation (10.7.1) is $\hat{y}$, an estimate, rather than y, since $b_1 + b_2 x$ will obviously not be exactly equal to y. Letting $\hat{e}$ denote the error, just as e denoted the error in Equation (10.6.11), we have

$$y = b_1 + b_2 x + \hat{e}. \qquad (10.7.2^*)$$

The problem, then, is to determine estimates b_1 and b_2 of the theoretical regression parameters β_1 and β_2.

There are a number of criteria that might be chosen to determine estimates of β_1 and β_2. For example, we might want to "hit the actual value on the nose" the largest proportion of times or to obtain the least absolute error on the average. The curve-fitting technique that we will use, however, is called the **least-squares criterion** (due to the mathematicians Legendre and Gauss).

According to the least-squares criterion, we should select b_1 and b_2 so that the sum of the squared errors is as small as possible.

In terms of decision theory, the least-squares criterion implicitly assumes a quadratic, or squared-error, loss function. Given the n pairs of values (x_i, y_i), the error in each case is simply

$$\hat{e}_i = y_i - (b_1 + b_2 x_i). \qquad (10.7.3)$$

According to the least-squares criterion, we want to minimize the sum of the squared errors, or the sum of squared deviations between the actual and predicted values of Y:

$$\text{minimize} \sum_{i=1}^{n} \hat{e}_i^2 = \sum_{i=1}^{n} [y_i - (b_1 + b_2 x_i)]^2. \qquad (10.7.4^*)$$

To minimize this function, we take the partial derivatives of $\sum_{i=1}^{n} \hat{e}_i^2$ with respect to b_1 and b_2 and set them equal to zero (the reader unfamiliar with partial differentiation can safely skip to the results):

$$\frac{\partial \sum_{i=1}^{n} \hat{e}_i^2}{\partial b_1} = \sum_{i=1}^{n} (-2)[y_i - (b_1 + b_2 x_i)] = 0 \qquad (10.7.5)$$

and

$$\frac{\partial \sum_{i=1}^{n} \hat{e}_i^2}{\partial b_2} = \sum_{i=1}^{n} -2x_i[y_i - (b_1 + b_2 x_i)] = 0. \qquad (10.7.6)$$

Simplifying, we get

$$\sum y_i - nb_1 - b_2 \sum x_i = 0 \qquad (10.7.7)$$

and

$$\sum x_i y_i - b_1 \sum x_i - b_2 \sum x_i^2 = 0. \qquad (10.7.8)$$

Now, from Equation (10.7.7),

$$b_1 = \frac{\sum y_i - b_2 \sum x_i}{n}.$$
(10.7.9*)

Inserting this into Equation (10.7.8), we find that

$$\sum x_i y_i - \left(\frac{\sum y_i - b_2 \sum x_i}{n} \right) \sum x_i - b_2 \sum x_i^2 = 0,$$

which can be simplified to

$$n \sum x_i y_i - \sum x_i \sum y_i + b_2 (\sum x_i)^2 - b_2 n \sum x_i^2 = 0,$$

or
$$b_2 = \frac{n \sum x_i y_i - \sum x_i \sum y_i}{n \sum x_i^2 - (\sum x_i)^2}.$$
(10.7.10*)

This is the least-squares estimator of β_2, and if it is inserted into Equation (10.7.9), the result is the least-squares estimator of β_1. It can be shown that the estimators given by Equations (10.7.9) and (10.7.10) minimize (rather than maximize) $\sum \hat{e}_i^2$.

To see that these estimators seem intuitively reasonable, note the similarity between b_2 [Equation (10.7.10)] and r_{XY} [Equation (10.1.7)]:

$$r_{XY} = \frac{n \sum x_i y_i - \sum x_i \sum y_i}{\sqrt{n \sum x_i^2 - (\sum x_i)^2} \sqrt{n \sum y_i^2 - (\sum y_i)^2}}.$$

On multiplying both sides of Equation (10.7.10) by the ratio s_X/s_Y, we see that

$$b_2 \frac{s_X}{s_Y} = r_{XY},$$

or
$$b_2 = r_{XY} \frac{s_Y}{s_X}.$$
(10.7.11*)

Comparing this with Equation (10.6.3), it is clear that b_2 is a reasonable estimator of β_2. Similarly, b_1 can be written in the form

$$b_1 = m_Y - b_2 m_X,$$
(10.7.12*)

where m_X and m_Y are the sample means; this is analogous to the expression for β_1 given in Equation (10.6.2).

The estimated regression line can thus be written as follows:

$$\hat{y} = b_1 + b_2 x = m_Y - b_2 m_X + b_2 x = m_Y + b_2(x - m_X),$$

or $\quad \hat{y} = m_Y + r_{XY} \dfrac{s_Y}{s_X} (x - m_X).$ $\qquad (10.7.13^*)$

But this is simply Equation (10.6.6) with m_X, m_Y, s_X, s_Y, and r_{XY} used to estimate the population parameters μ_X, μ_Y, σ_X, σ_Y, and ρ_{XY}, respectively.

The sample variance of the value of Y which is predicted by Equation (10.7.13) is equal to the sample variance of the error terms:

$$s_{Y.X}^2 = \frac{\sum (\hat{e}_i - m_{\hat{e}})^2}{n} \qquad (10.7.14)$$

[compare with Equation (10.6.13)]. But $m_{\hat{e}}$ is simply

$$m_{\hat{e}} = \frac{\sum \hat{e}_i}{n} = \frac{\sum [y_i - (b_1 + b_2 x_i)]}{n} = \frac{\sum \left[y_i - m_Y - r_{XY} \dfrac{s_Y}{s_X} (x_i - m_X) \right]}{n}$$

$$= \frac{\sum (y_i - m_Y)}{n} - r_{XY} \frac{s_Y}{s_X} \frac{\sum (x_i - m_X)}{n},$$

which equals zero since $\sum (y_i - m_Y) = \sum (x_i - m_X) = 0$, so that

$$s_{Y.X}^2 = \frac{\sum \hat{e}_i^2}{n} = \frac{\sum [y_i - (b_1 + b_2 x_i)]^2}{n}. \qquad (10.7.15^*)$$

It is also possible to express $s_{Y.X}^2$ in terms of s_Y and r_{XY}. From Equations (10.7.13) and (10.7.15),

$$s_{Y.X}^2 = \frac{\sum \left[y_i - m_Y - r_{XY} \dfrac{s_Y}{s_X} (x_i - m_X) \right]^2}{n}$$

$$= \frac{\sum (y_i - m_Y)^2}{n} + r_{XY}^2 \frac{s_Y^2}{s_X^2} \frac{\sum (x_i - m_X)^2}{n}$$

$$- 2r_{XY} \frac{s_Y}{s_X} \frac{\sum (x_i - m_X)(y_i - m_Y)}{n}$$

$$= s_Y^2 + r_{XY}^2 \frac{s_Y^2}{s_X^2} s_X^2 - 2r_{XY} \frac{s_Y}{s_X} \frac{\sum (x_i - m_X)(y_i - m_Y)}{n}.$$

But, from Equation (10.1.3),

$$r_{XY} = \frac{\sum (x_i - m_X)(y_i - m_Y)}{n s_X s_Y},$$

so

$$s_{Y \cdot X}^2 = s_Y^2 + r_{XY}^2 s_Y^2 - 2r_{XY} \frac{s_Y}{s_X}(s_X s_Y r_{XY})$$

$$= s_Y^2 + r_{XY}^2 s_Y^2 - 2r_{XY}^2 s_Y^2,$$

or

$$s_{Y \cdot X}^2 = s_Y^2(1 - r_{XY}^2). \qquad (10.7.16^*)$$

[Compare with Equation (10.6.7).] This is an estimate of $\sigma_{Y \cdot X}^2$. [It is a biased estimate; to get an unbiased estimate, multiply $s_{Y \cdot X}^2$ by $n/(n-2)$.] Just as ρ_{XY}^2 is equal to the proportion of the population variance of Y which is accounted for by the theoretical linear regression, r_{XY}^2 is equal to the proportion of the *sample* variance of Y which is accounted for by the *estimated* regression line.

To illustrate the application of the least-squares criterion to determine an estimated regression line, consider once again the data presented in Section 10.1 concerning the two examinations. Suppose that we want to predict Y, the score on the second examination, given x, the score on the first examination. Using Equations (10.7.10) and (10.7.11),

$$b_2 = \frac{n \sum x_i y_i - \sum x_i \sum y_i}{n \sum x_i^2 - (\sum x_i)^2} = \frac{7(992) - (76)(90)}{7(872) - (76)^2} = .32$$

and

$$b_1 = \frac{\sum y_i - b_2 \sum x_i}{n} = \frac{90 - .32(76)}{7} = 9.38.$$

The estimated regression line is therefore

$$\hat{y} = 9.38 + .32x.$$

The reader can verify that we could have used Equation (10.7.13) to arrive at the same result. Using this alternative approach is more troublesome from a computational standpoint, however.

The sample variance of the predicted value of Y is

$$s_{Y \cdot X}^2 = s_Y^2(1 - r_{XY}^2) = \left[\frac{n \sum y_i^2 - (\sum y_i)^2}{n^2} \right](1 - r_{XY}^2) = 1.45.$$

An unbiased estimate of $\sigma_{Y \cdot X}^2$ is thus

$$\frac{n}{n-2} s_{Y \cdot X}^2 = \frac{7}{5}(1.45) = 2.03.$$

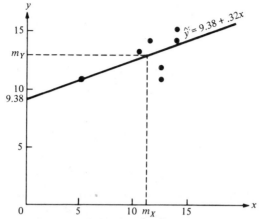

Figure 10.7.1   Scatter Diagram and Estimated Regression Line
for Examination-Scores Example

In Figure 10.7.1, a scatter diagram is presented together with the estimated regression line. In the line $\hat{y} = b_1 + b_2x$, b_2 is the slope of the line and b_1 is the y-intercept of the line. For the example, we see that the line intersects the y-axis at $y = 9.38$ and has a slope of .32, which means that $\hat{y}$ increases .32 units for every unit that x increases. Also, notice that the line goes through the point (m_X, m_Y). This is true for all estimated linear regression curves, as we can see from Equation (10.7.13); when $x = m_X$, the last term on the right-hand side of the equation becomes zero, and thus $\hat{y} = m_Y$. Finally, observe how the least-squares criterion can be interpreted graphically. For each sample point, the deviation, or error in prediction, is the *vertical* distance between the point and the estimated regression line. Least squares amounts to minimizing the sum of the squares of these vertical distances. Figure 10.7.2 shows the regression line and only a single one of the seven sample points in order to demonstrate more clearly how the deviation is equivalent to the vertical distance between the point and the line. Given x_i, the predicted value is $b_1 + b_2x_i$, while the observed value is y_i.

It should be emphasized that the least-squares technique is a curve-fitting technique and that while it is a method for arriving at estimators, it requires no assumptions about the form of the joint distribution of X and Y. This is in contrast to the method of maximum likelihood (Section 6.7), which requires distributional assumptions (so that the likelihood function can be specified). The least-squares criterion is by no means the only available criterion for curve fitting, but it is fairly easy to work with and it often results in estimators that are intuitively as well as theoretically

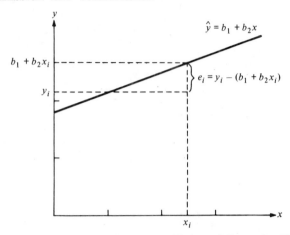

Figure 10.7.2   A Deviation from an Estimated Regression Line

appealing. We have seen that the least-squares estimates of β_1 and β_2 are intuitively reasonable, and when normal regression (regression with the assumption of bivariate normality) is discussed, we will see that the estimates are theoretically attractive. In conjunction with least squares, it should be pointed out that the technique can be used to fit curves other than straight lines to a set of data. This will be discussed later in this chapter under the heading of nonlinear regression.

At this point the reader may be somewhat puzzled by the use of the term "prediction" in the context of regression analysis. In our example, we actually have the two examination scores for each of the 7 students. Why don't we just look at the value of Y, the score on the second examination, for any student? Actually, the regression line is of particular interest in situations where the statistician wants to go beyond the immediate data and to forecast the value of Y for an individual for which this information is not already available. This is more clearly seen in an example from economic forecasting. On the basis of past data, we can estimate the regression line of Y, representing Gross National Product, on X, representing personal income. Then, if we know the current personal income but not the GNP, we could use the estimated regression line to predict GNP (although there might be problems in this example because of the different time periods involved—we will discuss problems of this nature briefly later in the chapter).

10.8   THE IDEA OF REGRESSION TOWARD THE MEAN

The term regression has come to be applied to the general problem of prediction by use of a wide variety of rules, although the original application of this term had a very specific meaning, as we shall show. The term regression is a shortened form of **regression toward the mean in prediction.** The general idea is that given any value x of X, the best linear prediction of Y is one relatively nearer the mean of Y than is x to its mean. By "relatively nearer" we mean nearer in terms of standard deviations. In simpler terms, if we standardize x and the predicted value $\hat{y}$, the resulting standardized value of $\hat{y}$ will be nearer to zero than the standardized value of x.

This can be illustrated quite simply from the regression equation (10.7.13). Subtracting m_Y from both sides of the equation and dividing both sides by s_Y, we arrive at the following expression:

$$\left(\frac{\hat{y} - m_Y}{s_Y}\right) = r_{XY} \left(\frac{x - m_X}{s_X}\right). \tag{10.8.1}$$

Taking the absolute value of both sides of this equation, we get

$$\left|\frac{\hat{y} - m_Y}{s_Y}\right| = | r_{XY} | \left|\frac{x - m_X}{s_X}\right|.$$

We know, however, that $-1 \le r_{XY} \le +1$, which means that

$$0 \le | r_{XY} | \le 1.$$

If $| r_{XY} | = 1$ (the sample values are perfectly correlated), then the standardized values of x and $\hat{y}$ are equal in absolute value (although the signs would be different if $r_{XY} = -1$). Otherwise, from Equation (10.8.1), the standardized value of $\hat{y}$ is closer to zero than the standardized value of x.

The idea of regression, and, indeed, most of the foundations for the theory of correlation and regression equations, came from the work of Sir Francis Galton in the nineteenth century. In his studies of hereditary traits, Galton pointed out the apparent regression toward the mean in the prediction of natural characteristics. For example, he found that unusually tall men tend to have sons shorter than themselves and unusually short men tend to have sons taller than themselves. This suggested a "regression toward mediocrity" in height from generation to generation, hence the term "regression." The interpretation of such results as a "regression toward mediocrity" has been aptly called the regression fallacy. A better explanation of the results is that the unusual, or extreme, values occur by chance, and the odds are that the genetic factors causing the unusual height or lack of height will not be completely passed on to the offspring. Note that we are talking particularly about extreme values; it is still true, for example, that if A is taller than B, then the odds are in favor of A's sons being taller than B's sons.

In some biological traits (such as height), regression toward the mean does seem to occur just as the linear theory implies. Often the same is true for the profits of business firms. Extraordinarily high profits in a particular year may be due to nonrecurring factors (a current fad for a particular product, for example), in which case the following year's profits would be expected to be somewhat lower. In many instances, however, regression toward the mean is not necessarily a feature of the natural world. Nevertheless, it *is* a feature of linear regression because it is built into the statistical assumptions and methods we use for prediction. Given the value of X, the corresponding value of Y can be anything, and regression toward the mean simply describes our best guess about the value of Y, when "best" is defined in terms of minimal squared error.

The idea of regression toward the mean points out an important difference between correlation problems and regression problems. The correlation coefficient is a *symmetric* measure of the linear relationship between two variables. In regression analysis, however, it makes an important difference which variable is the dependent variable and which is the independent variable, for the regression line of X on Y is not the same as the regression line of Y on X unless the two variables are perfectly correlated. The estimated regression line of Y on X is given by Equation (10.8.1), whereas the estimated regression line of X on Y is given by

$$\left(\frac{\hat{x} - m_X}{s_X}\right) = r_{XY}\left(\frac{y - m_Y}{s_Y}\right). \tag{10.8.2}$$

In terms of a graph with x on the horizontal axis and y on the vertical axis, Equation (10.8.1) minimizes the sum of the squared *vertical* deviations from the line and Equation (10.8.2) minimizes the sum of the squared *horizontal* deviations from the line.

10.9   INFERENCES IN REGRESSION PROBLEMS

In Sections 10.7 and 10.8 no assumptions were made about the joint distribution of X and Y. To determine the estimated regression line, we used the least-squares criterion. If we assume a linear model, this determines the straight line that is a "best" fit to the data, where "best" is to be interpreted in terms of minimal squared error.

If it can be assumed that X and Y have a bivariate normal distribution, further justification is provided for the results of the preceding two sections. First, as noted in Section 10.2, **any relationship between two variables with a bivariate normal distribution is strictly a linear relationship.** In terms of regression, the regression curve $E(Y \mid x)$ is linear. Therefore, under bivariate normality, it is not necessary to consider nonlinear regression models.

Second, under bivariate normality it is possible to determine a likelihood function and to utilize the method of maximum likelihood. The results provide support for the estimates given in Equations (10.7.9) and (10.7.10). **If X and Y have a bivariate normal distribution, the maximum-likelihood estimates of the regression coefficients β_1 and β_2 are identical to the least-squares estimates derived in Section 10.7.** It is important to note that although the results are identical, the two methods for arriving at them are quite different. The least-squares technique, as we have pointed out, requires no distributional assumptions and should be regarded more as a curve-fitting technique than as a statistical procedure. The method of maximum likelihood does require distributional assumptions so that the likelihood function can be determined. Note that if the joint distribution is not normal, the maximum-likelihood estimates need not coincide with the least-squares estimates.

In regression problems we are attempting to predict a dependent variable (Y) given a value of an independent variable (X). Since the independent random variable is regarded as "given," it is not necessary to make any distributional assumptions concerning X. In fact, it is not even necessary for the independent variable to be a random variable. For example, a firm may be interested in a regression equation with the price of a product as the independent variable and the sales of the product as the dependent variable. The firm may have complete control over the price of the product, in which case the independent variable is fixed by the firm rather than being a random variable. In general, we will continue to consider X as a random variable, although it must be kept in mind that this is by no means necessary for the purposes of inferences concerning the regression of Y on X.

In the linear regression of Y on X, the assumption of normality can be brought into the picture as follows (remember that linearity has already been assumed).

1. For any given x, the distribution of Y is a normal distribution.
2. The conditional variance of Y, $V(Y \mid x)$, is the same for all x.
3. The process is independent.

Of course, once x is fixed, any variability in Y is due strictly to the random-error term in the regression equation. Thus, the above assumptions can be restated in terms of such random-error terms.

1. For any given x, the random-error term has a normal distribution.
2. The variance of the random-error term is the same for all x.
3. The random-error terms are independent.

These assumptions are represented graphically in Figure 10.9.1, where the distribution $f(y \mid x)$ is shown for three different values of x. The dis-

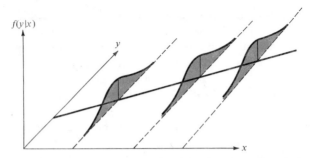

Figure 10.9.1 An Example with Identically Distributed Error
Terms

tributions are normal and they all have the same variance. Figure 10.9.2
demonstrates what the situation might be like if the assumption of equal
variances was violated; in the situation depicted in that figure, the variance
appears to increase as x increases. For instance, suppose that we want to
predict family expenditures for food as a function of family income. It
might be expected that the variability among families in terms of food
expenditures is greater for high-income families than for low-income
families. The assumptions given above must be kept in mind in attempting
to apply the inferential techniques presented in the remainder of this section to any real-world situation, of course.

Estimates for the regression coefficients β_1 and β_2 and for the error
variance $\sigma_{Y \cdot x}^2$ were given in Section 10.7. No distributional assumptions
were used in that section, but in order to go beyond the point estimates
to interval estimates and tests of hypotheses, we will use the assumptions
presented above regarding the error terms.

The error terms are assumed to be normally distributed, but in general
the variance of the error terms is not known. In dealing with a normally
distributed population with unknown variance in previous chapters, we

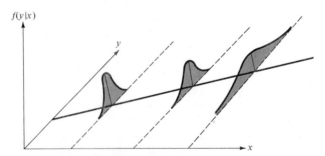

Figure 10.9.2 An Example for which the Variance of the Random-
Error Term Changes as x Changes

encountered T distributions, and the same is true in the linear regression model. Under normality, the sampling distributions of the estimators of the regression coefficients are T distributions with $n - 2$ degrees of freedom. In particular,

$$t = \frac{b_1 - \beta_1}{\text{est. } \sigma_{b_1}} = \frac{b_1 - \beta_1}{s_{Y \cdot X} \sqrt{\sum x_i^2 / n} / s_X \sqrt{n - 2}} \tag{10.9.1*}$$

and

$$t = \frac{b_2 - \beta_2}{\text{est. } \sigma_{b_2}} = \frac{b_2 - \beta_2}{s_{Y \cdot X} / s_X \sqrt{n - 2}} \tag{10.9.2*}$$

can be treated as values drawn from T distributions with $n - 2$ degrees of freedom.

Equations (10.9.1) and (10.9.2) enable us to make inferences about β_1 and β_2. For example, $100(1 - \alpha)$ percent confidence intervals for β_1 and β_2 are given by

$$b_1 \pm a \frac{s_{Y \cdot X} \sqrt{\sum x_i^2 / n}}{s_X \sqrt{n - 2}} \tag{10.9.3}$$

and

$$b_2 \pm a \frac{s_{Y \cdot X}}{s_X \sqrt{n - 2}}, \tag{10.9.4}$$

respectively, where a is the $1 - (\alpha/2)$ fractile of the T distribution with $n - 2$ degrees of freedom. Also, Equations (10.9.1) and (10.9.2) can be used to test hypotheses concerning β_1 and β_2. For example, a test of the hypothesis that $\beta_2 = 0$ is a test of whether any improvement in predictability is afforded by the use of a linear regression equation. [When $\beta_2 = 0$, Equation (10.9.2) can be shown to be equivalent to Equation (10.3.1), so a test of the hypothesis that $\beta_2 = 0$ is equivalent to a test of the hypothesis that $\rho_{XY} = 0$. In view of Equation (10.6.3), this is not at all a surprising result.]

The above results can be interpreted in a Bayesian sense. If the prior distribution of β_1, β_2, and $\sigma_{Y \cdot X}$ is an improper diffuse prior distribution (see Section 8.13) of the form

$$f(\beta_1, \beta_2, \sigma_{Y \cdot X}) = \begin{cases} 1/\sigma_{Y \cdot X} & \text{for} \quad -\infty < \beta_1 < \infty, \\ & \qquad -\infty < \beta_2 < \infty, \\ & \qquad \sigma_{Y \cdot X} > 0, \quad (10.9.5) \\ 0 & \text{elsewhere,} \end{cases}$$

then Equations (10.9.1) and (10.9.2) can be thought of in terms of the marginal posterior distributions of β_1 and β_2, respectively. The intervals given by Equations (10.9.3) and (10.9.4) are then Bayesian credible intervals, and the posterior T distributions can be used to make other inferences as well. Be sure to remember, from Chapter 8, that although the classical and Bayesian results are numerically similar under a diffuse prior distribution, the interpretations are quite different under the two approaches. Also, if the prior distribution is not diffuse, we would expect the Bayesian and classical results to differ numerically as well as in terms of interpretation. For example, if the conditional prior distribution of β_1 and β_2 given $\sigma_{Y \cdot X}{}^2$ is a bivariate normal distribution, then the corresponding conditional posterior distribution of β_1 and β_2 given $\sigma_{Y \cdot X}{}^2$ will also be a bivariate normal distribution. We will not present the details of such an analysis here; this paragraph is intended primarily to indicate that the distributional results given in this section can be interpreted in terms of posterior distributions following a diffuse prior distribution.

Suppose that we are interested in making inferences not just about β_1 and β_2 separately, but about the theoretical linear regression curve,

$$E(Y \mid x) = \beta_1 + \beta_2 x.$$

Because b_1 and b_2 are just estimates of β_1 and β_2, with

$$\hat{y} = b_1 + b_2 x,$$

the predicted value of Y from the estimated regression line for a given x does not necessarily agree with $E(Y \mid x)$. Under the distributional assumptions we have made,

$$t = \frac{\hat{y} - E(Y \mid x)}{\dfrac{s_{Y \cdot X}}{\sqrt{n-2}} \sqrt{1 + \dfrac{(x - m_X)^2}{s_X{}^2}}} \tag{10.9.6}$$

can be treated as a value drawn from a T distribution with $n - 2$ degrees of freedom. This distribution can be used to make inferences about $E(Y \mid x)$; for example,

$$\hat{y} \pm a \, \frac{s_{Y \cdot X}}{\sqrt{n-2}} \sqrt{1 + \frac{(x - m_X)^2}{s_X{}^2}} \tag{10.9.7}$$

represents a $100(1 - \alpha)$ percent confidence interval for $E(Y \mid x)$, where a is the $1 - (\alpha/2)$ fractile of the T distribution with $n - 2$ degrees of freedom.

Notice, from Equations (10.9.6) and (10.9.7), that the term involving $(x - m_X)^2$ implies that the estimated regression line found from a sample is not equally good as an approximation to the theoretical regression line over all of the different values of x. The estimated line is at its best as a substitute for the theoretical line when $x = m_X$, the sample mean, since the confidence interval given by Equation (10.9.7) is shortest at this point. However, as x is increasingly deviant from m_X in either direction, the confidence intervals grow wider. For extreme values of x, we cannot be nearly as certain that the estimated line provides a reasonable approximation for $E(Y \mid x)$ as we might be for "central" values of x.

The inferential procedures presented in this section have involved the parameters of the theoretical regression line $E(Y \mid x)$, the entire line itself. Often in regression problems the primary goal is to use the estimated regression line to predict actual values y of Y. In making inferences about such values, we note that $\hat{y}$ is related to y:

$$t = \frac{\hat{y} - y}{\dfrac{s_{Y \cdot X}}{\sqrt{n-2}} \sqrt{n + 1 + \dfrac{(x - m_X)^2}{s_X{}^2}}} \tag{10.9.8}$$

can be treated as a value drawn from a T distribution with $n - 2$ degrees of freedom. Notice that the denominator of Equation (10.9.8) is larger than that of Equation (10.9.6) because of the n that is added under the square root sign. The denominator of Equation (10.9.6) involves the error variability in attempting to use $\hat{y}$ to predict $E(Y \mid x)$. In attempting to use $\hat{y}$ to predict y, this variability is still present, and the additional term n that appears in the denominator of Equation (10.9.8) represents the effect of the additional variability encountered in going from $E(Y \mid x)$ to the actual value y of Y. From Equation (10.9.8), a $100(1 - \alpha)$ percent confidence interval for Y is given by

$$\hat{y} \pm a \, \frac{s_{Y \cdot X}}{\sqrt{n-2}} \sqrt{n + 1 + \frac{(x - m_X)^2}{s_X{}^2}} \,, \tag{10.9.9}$$

where a is the $1 - (\alpha/2)$ fractile of the T distribution with $n - 2$ degrees of freedom.

The inferential procedures in this section depend, of course, on the distributional assumptions concerning the random-error terms. It is possible to estimate the regression equation and to use the estimated line for prediction without assuming anything except random sampling, but interval estimates and tests of hypotheses depend on distributional assumptions for their validity. Later in the chapter we will discuss (briefly) some situations in which the assumptions appear doubtful. Suffice it to say for now that there may often be cause to think that one or more of the assumptions may be violated and that it is quite useful to check the data for this. For instance, once the estimated regression line is determined, we can calculate the predicted Y, $\hat{y}$, corresponding to each x, and compare it with the actual y. By performing certain tests concerning the error terms $y - \hat{y}$, the assumptions of the linear regression model can be investigated. Some tests of this nature, such as tests of "goodness of fit," will be discussed in Chapter 12.

10.10   AN EXAMPLE OF A REGRESSION PROBLEM

Suppose that a safety expert is interested in the relationship between the number of licensed vehicles in a community and the number of accidents per year in that community. In particular, the expert wishes to use the number of licensed vehicles to predict the number of accidents per year. A random sample of 10 communities yields the results shown in Table 10.10.1. Here, X is defined as the number of licensed vehicles (in thousands) and Y is the number of accidents (in hundreds). In the first community, for example, there were 4000 licensed vehicles and 100 accidents in the year in which the sample was taken. Expressing the data in this form simplifies the calculations. As we saw before, such transformations do not affect the value of r_{XY}. In addition, it is easy to take a predicted value of Y from the estimated regression line and multiply it by 100 to convert it into terms of accidents rather than hundreds of accidents.

Table 10.10.1   Number of Licensed Vehicles and Number of Accidents for 10 Communities

i Community	x_i Licensed vehicles (thousands)	y_i Number o accidents (hundreds)	x_i^2	y_i^2	$x_i y_i$
1	4	1	16	1	4
2	10	4	100	16	40
3	15	5	225	25	75
4	12	4	144	16	48
5	8	3	64	9	24
6	16	4	256	16	64
7	5	2	25	4	10
8	7	1	49	1	7
9	9	4	81	16	36
10	10	2	100	4	20
	96	30	1060	108	328

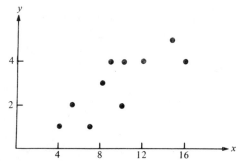

Figure 10.10.1 Scatter Diagram for Accident Example

The scatter diagram (Figure 10.10.1) indicates that there is some positive linear relationship between X and Y, as the safety expert might have suspected. The sample correlation coefficient is

$$r_{XY} = \frac{n \sum x_i y_i - (\sum x_i)(\sum y_i)}{\sqrt{n \sum x_i^2 - (\sum x_i)^2}\sqrt{n \sum y_i^2 - (\sum y_i)^2}}$$

$$= \frac{10(328) - (30)(96)}{\sqrt{10(1060) - (96)^2}\sqrt{10(108) - (30)^2}} = .80.$$

The estimated regression coefficients are

$$b_2 = \frac{n \sum x_i y_i - (\sum x_i)(\sum y_i)}{n \sum x_i^2 - (\sum x_i)^2} = \frac{10(328) - (30)(96)}{10(1060) - (96)^2} = .29$$

and $b_1 = \dfrac{\sum y_i - b_2 \sum x_i}{n} = \dfrac{30 - .29(96)}{10} = .22.$

Thus, the estimated regression line is

$$\hat{y} = .22 + .29x.$$

The sample variances of X and Y are

$$s_X^2 = \frac{n \sum x_i^2 - (\sum x_i)^2}{n^2} = \frac{10(1060) - (96)^2}{100} = 13.84$$

and $s_Y^2 = \dfrac{n \sum y_i^2 - (\sum y_i)^2}{n^2} = \dfrac{10(108) - (30)^2}{100} = 1.80.$

By using the estimated regression line and thereby taking advantage of the positive linear relationship between X and Y, we reduce the sample variance of Y by a factor of r_{XY}^2, the proportion of the variance of Y accounted for by the linear regression. Thus, the variance of Y left unaccounted for by the linear regression is

$$s_{Y \cdot X}^2 = s_Y^2 (1 - r_{XY}^2) = 1.8(1 - .64) = .648.$$

Suppose that the safety expert wants to test the hypothesis that β_2 is equal to zero. From Equation (10.9.2),

$$t = \frac{b_2 - \beta_2}{s_{Y \cdot X}/s_X \sqrt{n-2}} = \frac{.29 - 0}{\sqrt{.648}/\sqrt{13.84}\sqrt{8}} = 3.8$$

with 8 degrees of freedom. The two-tailed p-value is approximately .005, indicating that the sample provides evidence against the hypothesis that $\beta_2 = 0$. Alternatively, suppose that the safety expert is interested in whether an increase in the number of licensed vehicles by 1000 vehicles leads to an expected increase in the number of accidents of more than 50. In terms of β_2, the hypotheses of interest are

$$H_0: \beta_2 \geq .50$$

and
$$H_1: \beta_2 < .50.$$

The value of the T statistic is

$$t = \frac{b_2 - \beta_2}{s_{Y \cdot X}/s_X \sqrt{n-2}} = \frac{.29 - .50}{\sqrt{.648}/\sqrt{13.84}\sqrt{8}} = -2.76$$

with 8 degrees of freedom, and the p-value for the one-tailed test is slightly greater than .1. For a final test, consider the hypotheses $H_0: \beta_1 = 0$ and $H_1: \beta_1 \neq 0$. From Equation (10.9.1),

$$t = \frac{b_1 - \beta_1}{s_{Y \cdot X}\sqrt{\sum x_i^2/n}/s_X \sqrt{n-2}} = \frac{.22 - 0}{\sqrt{.648}\sqrt{1060/10}/\sqrt{13.84}\sqrt{8}} = .28$$

with 8 degrees of freedom, and the p-value is approximately .80.

These tests provide some information concerning β_1 and β_2, but confidence intervals provide a better idea of how accurate our estimates of the regression coefficients are. From Equations (10.9.3) and (10.9.4), 95 percent confidence intervals for β_1 and β_2 are as follows:

$$b_1 \pm 2.306 \frac{s_{Y \cdot X}\sqrt{\sum x_i^2/n}}{s_X \sqrt{n-2}} = .22 \pm 2.306 \frac{\sqrt{.648}\sqrt{1060/10}}{\sqrt{13.84}\sqrt{8}} = .22 \pm .78,$$

or $(-.56, 1.00)$, and

$$b_2 \pm 2.306 \frac{s_{Y \cdot X}}{s_X \sqrt{n-2}} = .29 \pm 2.306 \frac{\sqrt{.648}}{\sqrt{13.84}\sqrt{8}} = .29 \pm .18,$$

or (.11, .47). Notice, from the width of these confidence intervals, that our estimate of the slope of the regression line is much more accurate than our estimate of the y-intercept of the regression line. Considering the situation from a Bayesian standpoint with a diffuse prior distribution of the form (10.9.5), the marginal posterior distributions of β_1 and β_2 are T distributions with 8 degrees of freedom:

$$T = \frac{\beta_1 - b_1}{s_{Y \cdot X} \sqrt{\sum x_i^2/n}/s_X \sqrt{n-2}} = \frac{\beta_1 - .22}{.7876}$$

and

$$T = \frac{\beta_2 - b_2}{s_{Y \cdot X}/s_X \sqrt{n-2}} = \frac{\beta_2 - .29}{.0765}.$$

The interval estimates given above can be interpreted as posterior credible intervals determined from these T distributions.

Next, suppose that another community chosen at random has 12,000 vehicles. The predicted value of Y from the estimated regression equation is

$$\hat{y} = .22 + .29(12) = 3.70.$$

To determine a confidence interval for $E(Y \mid x = 12)$, Equation (10.9.7) is used. For example, 95 percent confidence limits are

$$\hat{y} \pm 2.306 \frac{s_{Y \cdot X}}{\sqrt{n-2}} \sqrt{1 + \frac{(x - m_X)^2}{s_X^2}}$$

$$= 3.70 \pm 2.306 \frac{\sqrt{.648}}{\sqrt{8}} \sqrt{1 + \frac{(12 - 9.6)^2}{13.84}}$$

$$= 3.70 \pm .78,$$

or (2.92, 4.48). A 95 percent confidence interval for the *actual* value of Y can be determined from Equation (10.9.9):

$$\hat{y} \pm 2.306 \frac{s_{Y \cdot X}}{\sqrt{n-2}} \sqrt{n + 1 + \frac{(x - m_X)^2}{s_X^2}}$$

$$= 3.70 \pm 2.306 \frac{\sqrt{.648}}{\sqrt{8}} \sqrt{10 + 1 + \frac{(12 - 9.6)^2}{13.84}}$$

$$= 3.70 \pm 2.22,$$

or (1.48, 5.92). This interval is wider than the corresponding interval for $E(Y \mid x)$. Notice that both of these intervals were computed under the assumption that $x = 12$. It is of some interest to see how the confidence

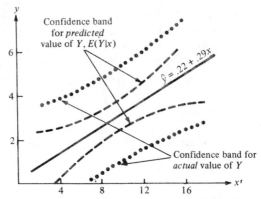

Figure 10.10.2   Estimated Regression Line and Confidence Bands
for Y and $E(Y \mid X)$ in Accident Example

intervals vary as x varies. If we determine the confidence limits as a function of x, the result is a **confidence band** for the regression line. In Figure 10.10.2, the solid line is the estimated regression line; the dashed lines enclose the 95 percent confidence band for $E(Y \mid x)$, the theoretical predicted value of Y; and the dotted lines enclose the 95 percent confidence band for the actual value of Y. In similar fashion, $100(1 - \alpha)$ percent confidence bands could be found for any choice of α.

We must be careful about the inferences drawn from the above results, however. For instance, suppose that the safety expert is particularly interested in a community in which there are 24,000 licensed vehicles. Could the estimated regression line $\hat{y} = .22 + .29x$ be used to predict the number of accidents in this community? From the data used to determine the estimated regression line, we see that no community in the sample had more than 16,000 vehicles. The sample data provide no justification for using the regression line to predict Y when the independent variable is outside the range of values encountered in the sample. In fact, as Figure 10.10.2 illustrates, the confidence bands get wider as the independent variable moves away from its sample mean. For example, we cannot make as precise a prediction of Y when there are 15,000 vehicles as we can when there are 10,000 vehicles (the sample mean was 9600 vehicles). *If the safety expert has some reason to believe that the association of Y and X could still be represented by the estimated regression line for communities with more than 16,000 vehicles, then the line might be used to predict Y for the community with 24,000 vehicles.* Such a prediction is strictly informal, however, and is not justified by the formal statistical analysis presented in this section. Unless some data are collected from communities with more vehicles, the safety expert has no statistical basis upon which to make predictions of Y for such communities. It may well be that the relationship between X and Y becomes extremely nonlinear as the number of vehicles increases; more data would be needed to investigate this possibility.

Just as in a correlation study, a regression study is concerned with the statistical relationship between two variables. Even though the variables are sometimes labeled as "independent" and "dependent," the reader should be careful not to read in implications of causation. It may or may not be that a larger number of vehicles "causes" an increase in the accident rate. All the statistical analysis does is suggest a possible relationship between the two variables; it does not necessarily imply that changes in X "cause" changes in Y, or vice versa.

10.11 MULTIPLE REGRESSION

In practical prediction situations, it is seldom that only one independent variable is of interest in the prediction of a particular dependent variable. In attempting to predict how much steak an individual will buy next week, we might take into account the individual's income, the price of steak, and the prices of other cuts of meat. In predicting the degree of success which will be attained in college by a particular student (as measured, perhaps, by a grade-point average), some relevant independent variables are the student's college entrance examination scores, high-school grade-point average, educational level of parents, and so on. In other words, there are often several variables with known values, each of which may contribute to the prediction about the value of the dependent variable.

When we have K variables and we wish to determine a function rule to predict the value of one of the variables, given values of the other $K - 1$ variables, we call this a **multiple regression** problem. Bivariate regression is a special case of multiple regression with $K = 2$. Because of this, it will be convenient to relate the theory of multiple regression to the theory of bivariate regression.

The notation for multiple regression will be as follows. The K variables will be denoted by $Y, X_2, X_3, \ldots, X_{K-1}$, and X_K, where Y is the dependent variable and the X_k $(k = 2, 3, \ldots, K)$ are the $(K - 1)$ independent variables. If we know the joint distribution of $(Y, X_2, X_3, \ldots, X_K)$, it is possible to find the conditional expectation of Y, given values of $X_2, \ldots, X_K$. This conditional expectation $E(Y \mid x_2, x_3, \ldots, x_K)$ is called the **theoretical regression surface** of Y on the remaining $K - 1$ variables (the term "surface" is used in place of "curve" to distinguish this from the bivariate situation). This is, of course, analogous to the development of $E(Y \mid x)$ in Section 10.5. The regression surface $E(Y \mid x_2, x_3, \ldots, x_K)$ can take on many different forms. However, if the K variables have a multivariate normal distribution, then the regression surface is linear:

$$E(Y \mid x_2, \ldots, x_K) = \beta_1 + \beta_2 x_2 + \beta_3 x_3 + \cdots + \beta_K x_K. \qquad (10.11.1^*)$$

In actual applications, we seldom know the joint distribution of the K variables of interest, and therefore we are faced with the problem of estimating the regression surface. In our discussion of multiple regression in this chapter, we will assume a linear model with theoretical regression coefficients $\beta_1, \beta_2, \ldots$, and β_K. Just as in the bivariate case, we will attempt to estimate these coefficients on the basis of sample information. Denoting the estimates by $b_1, b_2, \ldots$, and b_K, the estimated linear regression function is

$$\hat{y} = b_1 + b_2 x_2 + b_3 x_3 + \cdots + b_K x_K. \qquad (10.11.2^*)$$

Denoting the error of prediction by $\hat{e}$, as in the bivariate case, we have

$$y = b_1 + b_2 x_2 + b_3 x_3 + \cdots + b_K x_K + \hat{e}. \qquad (10.11.3^*)$$

Suppose that we have a sample of size n, where the sample is represented by n K-tuples $(y_i, x_{2i}, x_{3i}, \ldots, x_{Ki})$. For each of these K-tuples, the error of prediction is

$$\hat{e}_i = y_i - (b_1 + b_2 x_{2i} + b_3 x_{3i} + \cdots + b_K x_{Ki}). \qquad (10.11.4)$$

If we use the least-squares criterion to determine the estimated regression equation, we want to find $b_1, b_2, \ldots, b_K$ such as to minimize $\sum \hat{e}_i^2$. Finding the least-squares estimators involves taking K partial derivatives, setting them equal to zero, and solving the resulting set of K equations in K unknowns. This can be quite tedious, especially when K is large. For $K = 3$, we have the following three equations in three unknowns:

$$\sum y_i = nb_1 + b_2 \sum x_{2i} + b_3 \sum x_{3i},$$

$$\sum x_{2i} y_i = b_1 \sum x_{2i} + b_2 \sum x_{2i}^2 + b_3 \sum x_{2i} x_{3i}, \qquad (10.11.5)$$

$$\sum x_{3i} y_i = b_1 \sum x_{3i} + b_2 \sum x_{2i} x_{3i} + b_3 \sum x_{3i}^2.$$

Given n sets of values (y_i, x_{2i}, x_{3i}), $i = 1, \ldots, n$, we can compute the sums required in these equations and then solve for b_1, b_2, and b_3. It should be clear that more and more computations will be necessary as K becomes larger. By noticing the "pattern" to Equations (10.11.5), we can write the set of K equations in K unknowns for the general multivariate linear model. These equations are known as the "normal" equations:

$$\sum y_i = nb_1 + b_2 \sum x_{2i} + b_3 \sum x_{3i} + \cdots + b_K \sum x_{Ki},$$

$$\sum x_{2i} y_i = b_1 \sum x_{2i} + b_2 \sum x_{2i}^2 + b_3 \sum x_{2i} x_{3i} + \cdots + b_K \sum x_{2i} x_{Ki},$$

$$\vdots \qquad \vdots \qquad \vdots \qquad \vdots \qquad \vdots$$

$$\sum x_{Ki} y_i = b_1 \sum x_{Ki} + b_2 \sum x_{Ki} x_{2i} + b_3 \sum x_{Ki} x_{3i} + \cdots + b_K \sum x_{Ki}^2.$$

A procedure for solving a set of K linear equations in K unknowns by using matrix algebra is presented in the Appendix. In Section 10.13, we will see that multiple regression problems are more easily understood if they are expressed in the notation of linear algebra, using the concepts of vectors and matrices. Because of this, we will wait until Section 10.14 to present a numerical example of the application of multiple linear regression. Even the use of linear algebra will not reduce the computational burden, however, and for most multiple regression problems it is necessary to turn to high-speed computers for aid. Conceptually, multiple regression is a straightforward extension of bivariate regression; computationally, it may prove much more troublesome.

10.12　MULTIPLE AND PARTIAL CORRELATION

For a two-variable problem, ρ_{XY}^2 describes the proportion of variance in Y accounted for by the theoretical linear regression of Y on X. But the predicted value of Y, determined from the theoretical linear regression of Y on X, is a linear function with respect to X, and linear transformations do not affect correlation coefficients, as noted in Section 10.1. Therefore, the correlation between the predicted and actual values of Y must be equal to the correlation between X and Y:

$$\rho_{\hat{Y}Y} = \rho_{XY}. \qquad (10.12.1)$$

Thus, $\rho_{\hat{Y}Y}^2$ can be interpreted as describing the proportion of variance in Y which is accounted for by the theoretical linear regression of Y on X:

$$\rho_{\hat{Y}Y}^2 = \frac{\sigma_Y^2 - \sigma_{Y \cdot X}^2}{\sigma_Y^2}. \qquad (10.12.2)$$

In a similar fashion, it can be shown that the sample correlation coefficient $r_{\hat{Y}Y}$ between the predicted and actual scores in a sample is equal to r_{XY}. The term $r_{\hat{Y}Y}^2$ thus represents the proportion of the sample variance of Y which is accounted for by the estimated linear regression of Y on X:

$$r_{\hat{Y}Y}^2 = \frac{s_Y^2 - s_{Y \cdot X}^2}{s_Y^2}. \qquad (10.12.3)$$

It is possible to extend this idea to a multivariate situation in order to define a **multiple correlation coefficient.** To simplify the notation, a subscript will refer to a particular variable, with 1 referring to Y and $2, 3, \ldots, K$ referring to $X_2, X_3, \ldots, X_K$, respectively. Thus, r_{12} is the sample correlation coefficient of Y and X_2; ρ_{24} is the population correlation coefficient of X_2 and X_4; s_1^2 is the sample variance of Y; s_3^2 is the sample variance of X_3; $\sigma^2_{1\cdot234\cdots K}$ is $V(Y \mid x_2, \ldots, x_K)$, the error variance; and so on.

In the two-variable problem, $\rho_{\hat{Y}Y}^2$ describes the proportion of variance in Y accounted for by the theoretical linear regression of Y on X, as we saw in Equation (10.12.2). Extending this to the K-variable problem, we define the square of the multiple correlation coefficient, $\rho^2_{1\cdot23\cdots K}$, to be the proportion of variance in Y accounted for by the theoretical linear regression of Y on X_2, X_3, $\ldots$, X_K:

$$\rho^2_{1\cdot23\cdots K} = \frac{\sigma_1^2 - \sigma^2_{1\cdot23\cdots K}}{\sigma_1^2} . \tag{10.12.4*}$$

The multiple correlation coefficient is simply the square root of this term.

Since it is unlikely that we would know the population variances required to calculate the theoretical multiple correlation coefficient in any actual application, we need an estimate of $\rho_{1\cdot23\cdots K}$. This estimate, which is the multivariate extension of r_{XY}, is denoted by $r_{1\cdot23\cdots K}$, and its square is equal to the proportion of variance in Y which is accounted for by the estimated linear regression of Y on the $(K - 1)$ independent variables:

$$r^2_{1\cdot23\cdots K} = \frac{s_1^2 - s^2_{1\cdot23\cdots K}}{s_1^2} , \tag{10.12.5*}$$

where

$$s^2_{1\cdot23\cdots K} = s_e^2 = \frac{\sum \hat{e}_i^2}{n} , \tag{10.12.6}$$

with $\hat{e}_i$ given by Equation (10.11.4). The estimated linear regression is of course of the form

$$\hat{y} = b_1 + b_2x_2 + b_3x_3 + \cdots + b_Kx_K.$$

For a given set of sample data and the resulting set of estimates $b_1, \ldots, b_K$, we can determine the sample correlation between the predicted values and the observed values [using Equation (10.1.7)]:

$$r_{\hat{Y}Y} = \frac{n \sum \hat{y}_iy_i - \sum \hat{y}_i \sum y_i}{\sqrt{n \sum \hat{y}_i^2 - (\sum \hat{y}_i)^2}\sqrt{n \sum y_i^2 - (\sum y_i)^2}} . \tag{10.12.7}$$

The sample multiple correlation coefficient, by an extension of the argument given above for the two-variable case, is equal to $r_{\hat{Y}Y}$. Thus, we can determine $r_{1\cdot23\cdots K}$ either from Equation (10.12.5) or (10.12.7) [be sure to notice that if Equation (10.12.5) is used, the result is the *square* of the sample multiple correlation coefficient].

Complementary to the notion of multiple correlation is that of the **par-**

tial correlation between two variables, with the effects of some other variables being held constant. For instance, consider three variables Y, X_2, and X_3. When the correlation between, say, Y and X_2 is computed, the value of X_3 is left completely free to vary. If the value of X_3 were fixed at some constant level, what would happen to the correlation between Y and X_2? This correlation might be different because some of the apparent linear predictability of Y from X_2 (and vice versa) may be due to the association of each with X_3. For example, there is undoubtedly some positive correlation between body weight and the ability to read among normal children. However, it is true that the tendency for such variables to be correlated arises in part from another factor strongly related to each, the child's chronological age. If we could hold chronological age constant, then this correlation between weight and reading ability might vanish or be appreciably lowered. For studies centered on the influence of some extraneous variable (or variables) on the tendency of two other variables either to correlate or to fail to correlate, the partial correlation coefficient is a very useful descriptive device. It is particularly useful in situations in which ordinary correlation may be quite misleading, such as the above example involving weight and reading ability.

The partial correlation coefficient $\rho_{12 \cdot 3}$ is the correlation between Y and X_2, adjusted for the linear regression of each on X_3. The partial correlation can be defined as follows. Suppose that Y is predicted from X_3 using a linear regression equation and that X_2 is predicted from X_3 using another linear regression equation. These regressions account for the linear association of X_3 with Y and X_2, respectively. Therefore, the variability remaining, which is

$$\sigma_{1 \cdot 3}{}^2 = \sigma_1{}^2 (1 - \rho_{13}{}^2)$$

when Y is the dependent variable and

$$\sigma_{2 \cdot 3}{}^2 = \sigma_2{}^2 (1 - \rho_{23}{}^2)$$

when X_2 is the dependent variable, must represent the operation of all factors associated with variance in the dependent variable *other than the linear association with X_3*.

The tendency for linear association between Y and X_2, free of the linear association each has with X_3, is represented by the correlation of the predicted values of Y from the regression of Y on X_3 with the predicted values of X_2 from the regression of X_2 on X_3. Thus, we define the partial correlation between Y and X_2, with X_3 held constant, as follows:

$$\rho_{12 \cdot 3} = \frac{\text{cov } (\hat{Y}_{1 \cdot 3}, \hat{X}_{2 \cdot 3})}{\sigma_{1 \cdot 3} \sigma_{2 \cdot 3}}. \tag{10.12.8}$$

It can be shown that

$$\rho_{12\cdot3} = \frac{\rho_{12} - \rho_{13}\rho_{23}}{\sqrt{(1 - \rho_{13}{}^2)(1 - \rho_{23}{}^2)}}, \tag{10.12.9}$$

and this is the usual computational formula for a partial correlation coefficient based on the three original correlations. Note that there is no special significance to our choice of subscripts; we could define $\rho_{13\cdot2}$ and $\rho_{23\cdot1}$ in an analogous way.

Often we may wish to hold several variables constant and determine the partial correlation between two other variables. For instance, if we wanted to find the partial correlation of Y and X_2 with the $(K - 2)$ variables $X_3, X_4, \ldots, X_K$ held constant, the extension of Equation (10.12.8) would give us

$$\rho_{12\cdot34\cdots K} = \frac{\text{cov}\ (\hat{Y}_{1\cdot34\cdots K}, \hat{X}_{2\cdot34\cdots K})}{\sigma_{1\cdot34\cdots K}\sigma_{2\cdot34\cdots K}}, \tag{10.12.10}$$

where $\hat{Y}_{1\cdot34\cdots K}$ is the value of Y predicted from the linear regression of Y on $X_3, X_4, \ldots, X_K$; $\hat{X}_{2\cdot34\cdots K}$ is the value of X_2 predicted from the linear regression of X_2 on $X_3, X_4, \ldots, X_K$; and $\sigma_{1\cdot34\cdots K}$ and $\sigma_{2\cdot34\cdots K}$ are the standard deviations of $\hat{Y}_{1\cdot34\cdots K}$ and $\hat{X}_{2\cdot34\cdots K}$.

There is an intimate connection between partial correlation and multiple regression. The multiple linear regression equation can be written in the form

$$Y = \mu_1 + \rho_{12\cdot34\cdots K}\left(\frac{\sigma_{1\cdot34\cdots K}}{\sigma_{2\cdot34\cdots K}}\right)(x_2 - \mu_2) + \rho_{13\cdot245\cdots K}\left(\frac{\sigma_{1\cdot245\cdots K}}{\sigma_{3\cdot245\cdots K}}\right)(x_3 - \mu_3)$$

$$+ \cdots + \rho_{1K\cdot234\cdots(K-1)}\left(\frac{\sigma_{1\cdot234\cdots(K-1)}}{\sigma_{K\cdot234\cdots(K-1)}}\right)(x_K - \mu_K). \tag{10.12.11}$$

Thus, the coefficient of $(x_K - \mu_K)$ in this equation is the partial correlation coefficient of Y and X_K multiplied by a ratio of standard deviations. Note the similarity between this and the two-variable case, in which β_2 was equal to $\rho_{XY}(\sigma_Y/\sigma_X)$.

The above discussion, of course, gives us the theoretical linear regression surface of Y on the $(K - 1)$ independent variables. If we do not know the theoretical partial correlation coefficients and so on, but have sample data, it is necessary to estimate the linear regression surface. In the three-variable case, we can estimate $\rho_{12\cdot3}$ by using Equation (10.12.9) with each ρ being replaced by the corresponding estimate, r:

$$r_{12\cdot3} = \frac{r_{12} - r_{13}r_{23}}{\sqrt{(1 - r_{13}{}^2)(1 - r_{23}{}^2)}}. \tag{10.12.12}$$

In general, the estimate of the partial correlation coefficient [Equation (10.12.10)] is

$$r_{12 \cdot 34 \cdots K} = \frac{\sum [\hat{y}_i - m_{\hat{Y}}][\hat{x}_{2i} - m_{\hat{X}_2}]}{n s_{1 \cdot 34 \cdots K} s_{2 \cdot 34 \cdots K}}, \qquad (10.12.13)$$

where $\hat{y}_i$ is the value of Y predicted by the estimated regression of Y on X_3, X_4, $\ldots$, X_K; $\hat{x}_{2i}$ is the value of X_2 predicted by the estimated linear regression of X_2 on X_3, X_4, $\ldots$, X_K; and the sample standard deviations of these predicted values are $s_{1 \cdot 34 \cdots K}$ and $s_{2 \cdot 34 \cdots K}$, respectively.

We note that the estimated linear regression surface of Y on X_2, X_3, $\ldots$, X_K is now given by Equation (10.12.11) with μ replaced by m, σ replaced by s, and ρ replaced by r. In other words, each time a population parameter appears, replace it with the value of its usual estimator. Expressing the estimated linear regression equation in this form is of some interest, particularly for comparing it with the estimated regression line in the two-variable case. For example, the partial correlation appearing as a factor in the coefficient of x_j represents the proportion of the variance of Y which is attributable to the linear regression of Y on X_j, with the other independent variables held constant. This can also be interpreted in terms of the addition of X_j to the set of independent variables. The estimated multiple linear regression of Y on the other $K - 2$ independent variables explains a certain proportion of the variance of Y and leaves the remaining variance unexplained. If X_j is then added to the set of independent variables and the estimated multiple linear regression of Y on all $K - 1$ independent variables is considered, then the square of the partial correlation coefficient $r_{1j \cdot 234 \cdots K}$ represents the proportion of the previously unexplained variance (that is, the variance not explained by the estimated multiple regression of Y on the other $K - 2$ independent variables) that is explained by adding X_j to the set of independent variables. This interpretation is quite useful in **stepwise regression** procedures, in which independent variables are added to the regression equation or deleted from the regression equation at each step. In particular, at a given step, the appropriate partial correlation coefficients associated with the independent variables not currently in the regression equation can be computed, and the variable that explains the largest proportion of the remaining unexplained variance can be the next variable to "enter" the equation.

Unfortunately, the form of the estimated multiple linear regression equation represented by Equation (10.12.11) does not lend itself easily to computational techniques. Indeed, the notation used in this section has become quite cumbersome and hard to follow. It is of some value to understand what partial correlation is and how it relates to multiple regression, and this is the purpose of this section; but we will have to appeal to the branch of mathematics which is known as linear algebra, or matrix algebra, in order to find more convenient ways to determine the estimated multiple regression equation.

10.13 MULTIPLE REGRESSION IN MATRIX FORM

A useful tool in multiple regression analysis, as well as in other procedures involving more than two variables, is **linear algebra,** or **matrix algebra.** A brief review of the basic concepts of matrix algebra is presented in Appendix B. The reader not wishing to take the time to learn these concepts can omit this section. It should be pointed out, however, that matrix algebra is absolutely essential to multivariate statistical techniques and that an effort will be made to keep the discussion as simple as possible.

From the linear model specified in Section 10.11, a value y_i of the dependent variable Y, given values $x_{2i}, y_{3i}, \ldots, x_{Ki}$ of the independent variables, is equal to a linear combination of the independent variables plus a random-error term:

$$y_i = \beta_1 + \beta_2 x_{2i} + \beta_3 x_{3i} + \cdots + \beta_K x_{Ki} + e_i. \qquad (10.13.1^*)$$

Suppose that we have a sample of size n; that is, a sample of n sets of values $(y_i, x_{2i}, x_{3i}, \ldots, x_{Ki})$, where $i = 1, 2, \ldots, n$. Then we have n equations of the form (10.13.1), where the subscript i goes from 1 to n:

$$y_1 = \beta_1 + \beta_2 x_{21} + \beta_3 x_{31} + \cdots + \beta_K x_{K1} + e_1,$$

$$y_2 = \beta_1 + \beta_2 x_{22} + \beta_3 x_{32} + \cdots + \beta_K x_{K2} + e_2,$$

$$y_3 = \beta_1 + \beta_2 x_{23} + \beta_3 x_{33} + \cdots + \beta_K x_{K3} + e_3, \qquad (10.13.2)$$

$$\vdots \qquad\quad \vdots \qquad\quad \vdots \qquad\qquad \vdots$$

$$y_n = \beta_1 + \beta_2 x_{2n} + \beta_3 x_{3n} + \cdots + \beta_K x_{Kn} + e_n.$$

By using vectors and matrices, it is possible to express these n equations in much simpler form. We will use boldface letters to denote vectors and matrices. First, let **y** be a row vector with n elements $y_1, y_2, \ldots,$ and y_n. Similarly, let **e** be a row vector with n elements $e_1, e_2, \ldots,$ and e_n. Let **β** be a row vector with K elements $\beta_1, \beta_2, \ldots,$ and β_K. Finally, let **X** denote a matrix with n rows and K columns, in which the element in the ith row and the jth column is equal to x_{ji}, the ith value of the variable x_j. Note that there is no variable X_1; we will set x_{1i} equal to 1 for $i = 1, \ldots, n$.

Thus, we have the following:

$$
\mathbf{y} = \begin{bmatrix} y_1 \\ y_2 \\ y_3 \\ \cdot \\ \cdot \\ \cdot \\ y_n \end{bmatrix}, \quad
\mathbf{X} = \begin{bmatrix} 1 & x_{21} & x_{31} & \cdots & x_{K1} \\ 1 & x_{22} & x_{32} & \cdots & x_{K2} \\ 1 & x_{23} & x_{33} & \cdots & x_{K3} \\ \cdot & \cdot & \cdot & \cdot & \cdot \\ \cdot & \cdot & \cdot & \cdot & \cdot \\ \cdot & \cdot & \cdot & \cdot & \cdot \\ 1 & x_{2n} & x_{3n} & \cdots & x_{Kn} \end{bmatrix}, \quad
\boldsymbol{\beta} = \begin{bmatrix} \beta_1 \\ \beta_2 \\ \beta_3 \\ \cdot \\ \cdot \\ \beta_K \end{bmatrix}, \quad
\mathbf{e} = \begin{bmatrix} e_1 \\ e_2 \\ e_3 \\ \cdot \\ \cdot \\ e_n \end{bmatrix}.
$$

Using this new notation, the set of n equations (10.13.2) can be written

$$\mathbf{y} = \mathbf{X\boldsymbol{\beta}} + \mathbf{e}. \tag{10.13.3*}$$

You can verify for yourself, using the rules of matrix multiplication, that Equations (10.13.2) and (10.13.3) are equivalent. For instance, multiply the first row of $\mathbf{X}$ by the vector $\boldsymbol{\beta}$, add the first element of $\mathbf{e}$, and the result is equal to the first element of $\mathbf{y}$. The reason for including a column of 1's in the $\mathbf{X}$ matrix should be clear now; it allows for the constant term β_1, which is not multiplied by any variable.

In estimating the linear regression surface, our task is to estimate the values $\beta_1, \beta_2, \ldots, \beta_K$, or simply to estimate the vector $\boldsymbol{\beta}$. Suppose that we denote an estimator of $\boldsymbol{\beta}$ by b, which is a vector of K elements $b_1, b_2, \ldots, b_K$. The estimated regression equation is then

$$\mathbf{y} = \mathbf{Xb} + \hat{\mathbf{e}}, \tag{10.13.4*}$$

where $\hat{\mathbf{e}}$ is the vector of error terms when $\mathbf{b}$ is used in place of $\boldsymbol{\beta}$. Now recall that the least-squares criterion requires that we minimize the sum of the squared errors, $\sum_{i=1}^{n} \hat{e}_i^2$. But notice that this sum of squared errors can be represented in matrix notation by $\hat{\mathbf{e}}^t\hat{\mathbf{e}}$, where $\hat{\mathbf{e}}^t$ is the transpose of $\hat{\mathbf{e}}$:

$$\hat{\mathbf{e}}^t\hat{\mathbf{e}} = \begin{bmatrix} \hat{e}_1 & \hat{e}_2 & \cdots & \hat{e}_N \end{bmatrix} \begin{bmatrix} \hat{e}_1 \\ \hat{e}_2 \\ \vdots \\ \hat{e}_N \end{bmatrix} = \sum_{i=1}^{n} \hat{e}_i^2.$$

Therefore, according to the least-squares criterion, we want to choose $\mathbf{b}$ so as to minimize $\hat{\mathbf{e}}^t\hat{\mathbf{e}}$. But, from Equation (10.13.4), $\hat{\mathbf{e}} = \mathbf{y} - \mathbf{Xb}$, so we want to minimize

$$(\mathbf{y} - \mathbf{Xb})^t(\mathbf{y} - \mathbf{Xb}).$$

To minimize this, it is necessary to use vector differentiation, so we will present the results without proof. In matrix notation, the least-squares estimator of $\boldsymbol{\beta}$ is given by

$$\mathbf{b} = (\mathbf{X}^t\mathbf{X})^{-1}\mathbf{X}^t\mathbf{y}, \tag{10.13.5*}$$

where $(\mathbf{X}^t\mathbf{X})^{-1}$ is the inverse of the matrix $\mathbf{X}^t\mathbf{X}$ (see Appendix B). The existence of this solution requires an assumption concerning the matrix $\mathbf{X}$, an assumption involving linear algebra which guarantees that the matrix $\mathbf{X'X}$ has an inverse. We shall not concern ourselves with this assumption, which is satisfied if the determinant of $\mathbf{X'X}$ is not zero. Incidentally, the estimator $\mathbf{b}$, in addition to being a least-squares estimator, is also a maximum-likelihood estimator under the assumption that the error terms are normally distributed with mean zero and fixed variance σ_e^2 (remember that the least-squares criterion requires no distributional assumptions).

A note regarding computations is in order here. The only computational difficulty encountered in Equation (10.13.5) is the determination of the inverse of the matrix $\mathbf{X'X}$. For small K, this problem is not too serious; for larger K, it could be quite serious and would surely require the use of a high-speed computer. With the computers currently available, it is advantageous to use them for problems involving more than two or three variables. We will discuss this at greater length in Section 10.15.

The matrix notation also makes it possible to express the multiple correlation coefficient in a form which does not require the computation of s_1^2 and $s^2{}_{1\cdot23\ldots K}$. The square of the multiple correlation coefficient is equal to

$$r^2{}_{1\cdot23\ldots K} = \frac{\mathbf{b'X'y} - \dfrac{1}{n}(\sum y_i)^2}{\mathbf{y'y} - \dfrac{1}{n}(\sum y_i)^2}, \tag{10.13.6}$$

and the sample error variance is equal to

$$s_e^2 = s^2{}_{1\cdot23\ldots K} = \frac{\mathbf{y'y} - \mathbf{b'X'y}}{n}. \tag{10.13.7}$$

To illustrate the use of vectors and matrices in solving the linear model, we will rework the two-variable example from Section 10.10 in our new notation. In this case, $n = 10$ and $K = 2$, and from Table 10.10.1 we have

$$\mathbf{y} = \begin{bmatrix} 1 \\ 4 \\ 5 \\ 4 \\ 3 \\ 4 \\ 2 \\ 1 \\ 4 \\ 2 \end{bmatrix} \quad \text{and} \quad \mathbf{X} = \begin{bmatrix} 1 & 4 \\ 1 & 10 \\ 1 & 15 \\ 1 & 12 \\ 1 & 8 \\ 1 & 16 \\ 1 & 5 \\ 1 & 7 \\ 1 & 9 \\ 1 & 10 \end{bmatrix}.$$

In order to apply Equation (10.13.5), we must calculate $(X'X)^{-1}$. First,

$$X'X = \begin{bmatrix} 1 & 1 & 1 & \cdots & 1 \\ 4 & 10 & 15 & \cdots & 10 \end{bmatrix} \begin{bmatrix} 1 & 4 \\ 1 & 10 \\ 1 & 15 \\ \vdots & \vdots \\ 1 & 10 \end{bmatrix} = \begin{bmatrix} 10 & 96 \\ 96 & 1060 \end{bmatrix}.$$

Now, using the rules in the Appendix to invert this matrix, we get

$$(X'X)^{-1} = \begin{bmatrix} \dfrac{1060}{1384} & \dfrac{-96}{1384} \\ \\ \dfrac{-96}{1384} & \dfrac{10}{1384} \end{bmatrix} = \begin{bmatrix} .76590 & -.06936 \\ -.06936 & .00723 \end{bmatrix}.$$

We also need to determine $X'y$:

$$X'y = \begin{bmatrix} 1 & 1 & 1 & \cdots & 1 \\ 4 & 10 & 15 & \cdots & 10 \end{bmatrix} \begin{bmatrix} 1 \\ 4 \\ \vdots \\ 2 \end{bmatrix} = \begin{bmatrix} 30 \\ 328 \end{bmatrix}.$$

We are now ready to apply Equation (10.13.5):

$$b = (X'X)^{-1}X'y = \begin{bmatrix} .76590 & -.06936 \\ -.06936 & .00723 \end{bmatrix} \begin{bmatrix} 30 \\ 328 \end{bmatrix} = \begin{bmatrix} .22 \\ .29 \end{bmatrix}.$$

The estimated regression line is $\hat{y} = .22 + .29x$, which is identical to the result obtained in Section 10.10.

We can also use Equation (10.13.6) to calculate r^2, which in the two-variable case is simply r_{XY}^2, the square of the correlation coefficient between X and Y. Using the above results,

$$b'X'y = \begin{bmatrix} .22 & .29 \end{bmatrix} \begin{bmatrix} 30 \\ 328 \end{bmatrix} = 101.72$$

and

$$y'y = \begin{bmatrix} 1 & 4 & 5 & \cdots & 2 \end{bmatrix} \begin{bmatrix} 1 \\ 4 \\ 5 \\ \vdots \\ 2 \end{bmatrix} = 108.$$

Thus,

$$r^2 = \frac{\mathbf{b}^t X^t \mathbf{y} - \dfrac{1}{n}(\sum y_i)^2}{\mathbf{y}^t \mathbf{y} - \dfrac{1}{n}(\sum y_i)^2} \qquad \frac{101.72 - \left(\dfrac{900}{10}\right)}{108 - \left(\dfrac{900}{10}\right)} = .65.$$

Taking the square root of this, we get $r_{XY} = .806$, which differs only slightly (due to rounding error) from the value obtained in Section 10.10.

It is obvious that in the two-variable case it is easier to use the formulas developed in Section 10.7 than to use the methods of linear algebra. With more than two variables, the use of linear algebra does, in general, simplify the understanding (but not necessarily the computations) of the general linear model. In most applications of multiple regression, high-speed computers are used, so the computational burden is of little concern; the important thing is to *understand* the underlying statistical procedures. The purpose of reworking the vehicles-accidents example was to demonstrate, as simply as possible, the use of linear algebra in regression analysis. In Section 10.14, we will attack a problem involving three variables, a true *multiple* regression problem.

In order to make inferences in multiple regression problems, we will assume that the error terms in the linear regression equation are normally distributed with mean zero and variance σ_e^2, where σ_e^2 is not known. Under this assumption, the marginal sampling distributions of individual regression coefficients are T distributions. For a given value of β_i, where i can take on any value from 1 to K,

$$t = \frac{b_i - \beta_i}{s_e \sqrt{n/(n-K)}\sqrt{a_{ii}}} \qquad (10.13.8)$$

is a value drawn from a T distribution with $n - K$ degrees of freedom, where a_{ii} is the element in the ith row and the ith column of the matrix $(X^t X)^{-1}$. Thus, a $100(1 - \alpha)$ percent confidence interval for β_i is given by

$$b_i \pm a s_e \sqrt{a_{ii}} \sqrt{\frac{n}{n-K}}, \qquad (10.13.9)$$

where a is the $1 - (\alpha/2)$ fractile of the T distribution with $n - K$ degrees of freedom. Incidentally, just as in Section 10.9 in the bivariate case, the T distribution represented in Equation (10.13.8) can be interpreted in a Bayesian sense as the marginal posterior distribution of β_i following a diffuse prior distribution. The Bayesian, then, can use this marginal posterior distribution to make inferences about β_i; for example, Equation (10.13.9) may be regarded as a Bayesian credible interval.

Let us drop the subscript $1 \cdot 23 \cdots K$ from the symbol for multiple correlation, since it is understood that we are talking about the prediction of Y, given values of X_2, X_3, ..., X_K. One possible hypothesis of interest is the hypothesis that the square of the multiple correlation coefficient is equal to zero; that is, $\rho^2 = 0$. Under this hypothesis and under the assumption of normality, it can be shown that

$$\left(\frac{r^2}{1 - r^2}\right)\left(\frac{n - K}{K - 1}\right) \tag{10.13.10}$$

has an F distribution with $K - 1$ and $n - K$ degrees of freedom. Therefore, we can use this statistic to test the hypothesis that ρ^2 is equal to zero. Remember, however, that ρ^2 is the proportion of the variance of Y which is accounted for by the linear regression of Y on the $(K - 1)$ independent variables. If $\rho^2 = 0$, therefore, there is no reduction in variance due to the linear regression. In terms of β_2, β_3, ..., and β_K, this means that these regression coefficients are equal to zero, since they are the coefficients of the terms involving the independent variables (be careful to note that nothing is said here about β_1). The above F statistic can thus be used to test the hypothesis that $\beta_2 = \beta_3 = \beta_4 = \cdots = \beta_K = 0$.

There are other hypotheses of yet more interest, such as those involving the increase in r^2 due to the addition of more independent variables or the decrease in r^2 associated with the deletion of one or more independent variables. The methods used to test these hypotheses, however, will not be covered here. If you are interested in pursuing this matter further, consult the references listed at the end of the book.

10.14   AN EXAMPLE OF A MULTIPLE REGRESSION PROBLEM

Consider once again the accidents-vehicles example of Section 10.10, and suppose that the safety expert decides to introduce another independent variable to the linear model: the size of the community's police force. The expert now wants to predict Y [the number of accidents per year (in hundreds)] from the model

$$y = \beta_1 + \beta_2 x_2 + \beta_3 x_3 + e,$$

where X_2 represents the number of licensed vehicles in the community (in thousands) and X_3 represents the size of the police force. Data are obtained regarding X_3 for the 10 communities for which data were already available on Y and X_2; all of the data are presented in Table 10.14.1.

Table 10.14.1. Number of Accidents, Number of Licensed Vehicles, and Size of Police Force for 10 Communities

Community	y_i	x_{2i}	x_{3i}
1	1	4	20
2	4	10	6
3	5	15	2
4	4	12	8
5	3	8	9
6	4	16	8
7	2	5	12
8	1	7	15
9	4	9	10
10	2	10	10

From the data in Table 10.14.1, it is possible to determine the vector $\mathbf{y}$ and the matrix $\mathbf{X}$:

$$\mathbf{y} = \begin{bmatrix} 1 \\ 4 \\ 5 \\ 4 \\ 3 \\ 4 \\ 2 \\ 1 \\ 4 \\ 2 \end{bmatrix} \quad \text{and} \quad \mathbf{X} = \begin{bmatrix} 1 & 4 & 20 \\ 1 & 10 & 6 \\ 1 & 15 & 2 \\ 1 & 12 & 8 \\ 1 & 8 & 9 \\ 1 & 16 & 8 \\ 1 & 5 & 12 \\ 1 & 7 & 15 \\ 1 & 9 & 10 \\ 1 & 10 & 10 \end{bmatrix}.$$

Using the rules for multiplying and inverting matrices which are presented in the Appendix, you can verify that

$$\mathbf{X'X} = \begin{bmatrix} 10 & 96 & 100 \\ 96 & 1060 & 821 \\ 100 & 821 & 1218 \end{bmatrix},$$

$$(\mathbf{X'X})^{-1} = \begin{bmatrix} 5.68689 & -.32099 & -.25054 \\ -.32099 & .02009 & .01281 \\ -.25054 & .01281 & .01276 \end{bmatrix},$$

and

$$\mathbf{X'y} = \begin{bmatrix} 30 \\ 328 \\ 244 \end{bmatrix}.$$

To obtain the least-squares estimate of the vector of regression coefficients $\boldsymbol{\beta}$, we compute

$$\mathbf{b} = (\mathbf{X}^t\mathbf{X})^{-1}\mathbf{X}^t\mathbf{y} = \begin{bmatrix} 4.190 \\ .085 \\ -.201 \end{bmatrix}.$$

The estimated regression line is therefore given by

$$\hat{y} = 4.190 + .085x_2 - .201x_3.$$

The proportion of the variance of Y which is accounted for by this multiple linear regression is

$$r_{1 \cdot 23}{}^2 = \frac{\mathbf{b}^t\mathbf{X}^t\mathbf{y} - \dfrac{1}{n}\left(\sum y_i\right)^2}{\mathbf{y}^t\mathbf{y} - \dfrac{1}{n}\left(\sum y_i\right)^2} = .81.$$

This means that 81 percent of the variance of Y is accounted for by the estimated multiple regression line, leaving 19 percent unaccounted for. The variance that is unaccounted for is the error variance, which is

$$s_e{}^2 = \frac{\mathbf{y}^t\mathbf{y} - \mathbf{b}^t\mathbf{X}^t\mathbf{y}}{n} = .346.$$

From Equation (10.13.9), 95 percent confidence intervals for the regression coefficients β_1, β_2, and β_3 are, respectively,

$$b_1 \pm 2.365 s_e \sqrt{a_{11}} \sqrt{\frac{n}{n-k}} = 4.190 \pm 2.365 \sqrt{.346} \sqrt{5.68689} \sqrt{\frac{10}{7}}$$

$$= 4.190 \pm 3.965 = (.225, 8.155),$$

$$b_2 \pm 2.365 s_e \sqrt{a_{22}} \sqrt{\frac{n}{n-k}} = .085 \pm 2.365 \sqrt{.346} \sqrt{.02009} \sqrt{\frac{10}{7}}$$

$$= .085 \pm .236 = (-.151, .321),$$

and $\quad b_3 \pm 2.365 s_e \sqrt{a_{33}} \sqrt{\dfrac{n}{n-k}} = -.201 \pm 2.365 \sqrt{.346} \sqrt{.01276} \sqrt{\dfrac{10}{7}}$

$$= -.201 \pm .188 = (-.389, -.013).$$

Note that the 95 percent confidence intervals for β_1 and β_3 contain all positive values and all negative values, respectively, whereas the 95 percent confidence interval for β_2 contains both positive and negative values.

In Section 10.10, s_Y^2 (or s_1^2 in multiple regression notation) was calculated and found to be 1.8, and r_{XY} (or r_{12}) was found to be .80. Thus, for a bivariate linear regression of Y on X_2,

$$s_{1.2}^2 = s_1^2(1 - r_{12}^2) = 1.8(.36) = .648.$$

Also, from Table 10.14.1 we can calculate r_{13}:

$$r_{13} = \frac{n \sum x_{3i}y_i - (\sum x_{3i})(\sum y_i)}{\sqrt{n \sum x_{3i}^2 - (\sum x_{3i})^2}\sqrt{n \sum y_i^2 - (\sum y_i)^2}} = -.89.$$

As the safety expert might have suspected, the number of accidents is inversely related to the number of police; furthermore, the linear association is quite strong, indicating that the use of a linear prediction rule with X_3 as an independent variable would improve the expert's predictions of Y. In the bivariate linear regression of Y on X_3, we see that

$$s_{1.3}^2 = s_1^2(1 - r_{13}^2) = 1.8(.208) = .374.$$

How can we interpret these sample variances? First, we can say that if just the single variable Y is considered, the sample variance about m_Y is 1.8. If information regarding X_2 is introduced into the analysis via a linear regression equation, the sample variance about this regression line is $s_{1.2}^2$, which is equal to .648. If, instead of X_2, X_3 is used as the independent variable in a bivariate linear regression, the sample variance about the resulting regression line is $s_{1.3}^2$, which is .374. Finally, if *both* X_2 and X_3 are used as independent variables in a multiple linear model, the sample variance about the resulting regression equation is $s_{1.23}^2$, which is .346. Consider the safety expert, who is interested in predicting Y, the number of accidents per year (in hundreds) for a particular community. The use of a single independent variable, either X_2 or X_3, greatly increases the precision of the prediction. Observe that the precision can be increased more with X_3 than with X_2, because the correlation of Y with X_3 is greater than the correlation of Y with X_2. Furthermore, the precision can be increased even more by using a multiple linear prediction rule, with both X_2 (number of licensed vehicles, in thousands) *and* X_3 (size of community police force) as independent variables. If the safety expert could obtain data on other relevant variables (such as the number of miles of roads in the community), even greater precision could be obtained by using a multiple regression with more than two independent variables.

Suppose that the safety expert is also interested in the partial correlations among the three variables. We already have calculated $r_{12} = .80$ and $r_{13} = -.89$. In order to compute the values of the *partial* correlation coefficients, we also need to determine r_{23}:

$$r_{23} = \frac{n \sum x_{2i} x_{3i} - (\sum x_{2i})(\sum x_{3i})}{\sqrt{n \sum x_{2i}^2 - (\sum x_{2i})^2} \sqrt{n \sum x_{3i}^2 - (\sum x_{3i})^2}} = -.80.$$

Now, using Equation (10.12.9) with the sample estimates (r's) used in place of the population correlation coefficients (ρ's), the safety expert finds that

$$r_{12 \cdot 3} = \frac{r_{12} - r_{13} r_{23}}{\sqrt{(1 - r_{13}^2)(1 - r_{23}^2)}} = .32$$

and

$$r_{13 \cdot 2} = \frac{r_{13} - r_{12} r_{23}}{\sqrt{(1 - r_{12}^2)(1 - r_{23}^2)}} = -.69.$$

Notice that the correlation between Y and X_2 is greatly reduced (from .80 to .32) when X_3 is held constant. This implies that the linear relationship between Y and X_2 is partly due to the strong linear association of each of these variables with X_3. Similarly, the strength of the relationship between Y and X_3 is reduced somewhat when X_2 is held constant, but not as much as is the relationship between Y and X_2. This is also reflected by the fact that the linear regression of Y on X_3 explains almost as much of the variance in Y as does the linear regression of Y on *both* X_2 and X_3. Thus, the high correlation (.80) between Y and X_2 appears as though it might be due at least in part to the relationships of Y and X_2 with the third variable, X_3.

To demonstrate the notion of stepwise regression with this very simple example, we start by considering the two independent variables and their correlations with Y. In a simple bivariate linear regression, X_2 explains the proportion $(.80)^2 = .64$ of the variance of Y, whereas X_3 explains the proportion $(-.89)^2 = .79$ of the variance of Y. Thus, if we want to choose the variable that explains the greater proportion of the variance of Y, X_3 should be chosen to enter the equation as an independent variable. In a typical application of stepwise regression, there would be several other variables contesting for the right to enter the equation at the second step. Here there is only one remaining independent variable, X_2. Of the remaining 21 percent of the variance of Y (the other 79 percent was explained by the use of X_3 as an independent variable), the proportion $r_{12 \cdot 3}^2 = (.32)^2$ is explained by the inclusion of X_2 as another independent variable. But $(.32)^2$ multiplied by 21 percent is only about 2 percent, so the percentage of variance explained by the multiple regression equation is 81 percent, consisting of 79 percent explained by X_3 and 2 percent explained by adding X_2 as a second independent variable. The result that 81 percent of the variance of Y is explained by the estimated multiple regression line was noted previously when the square of the multiple correlation coefficient was computed.

This example should serve to give you some idea of the use of the general linear model in a multiple regression problem and the interpretation of the results. Because most multiple regression problems are too big to be analyzed by hand calculations, a high-speed computer is invaluable for these problems and other problems involving more than two variables, as we point out in the next section. As a result, it is not too important that you become adept at making calculations such as those in this example; it is much more important that you understand the interpretation of the resulting numbers.

10.15 COMPUTERS AND MULTIPLE REGRESSION

It should be clear by now that a multiple regression analysis requires a large number of arithmetic operations. If the number of variables K is greater than 3 or 4, or if the sample size n is quite large, the number of operations required makes it a next-to-impossible task to carry out the calculations by hand or even on a mechanical calculator. The alternative, of course, is to conduct the analysis on a high-speed computer, which can handle a large amount of data in a short time. With reference to the general linear model presented in matrix form in Section 10.13, it is possible to have the computer invert the matrix $\mathbf{X'X}$ as well as perform the other required operations in the estimation of the vector of regression coefficients and the calculation of statistics such as those in Section 10.13 for testing purposes. It is not even necessary to write a computer program for a multiple regression problem, for there are numerous "standard programs" available for multiple regression and other statistical techniques. These programs provide the statistician not only with the estimated linear regression equation, but also with such output as the multiple correlation coefficient, the standard errors of all of the estimated coefficients, means and variances of all of the variables, covariances and correlation coefficients between all pairs of variables, partial correlation coefficients, values of the F and T statistics used to test hypotheses in regression (along with the corresponding p-values), plots of the values of the error term (in order to check the assumptions concerning the error term), and so on. There are also stepwise regression programs available, in which numerous independent variables may be included and the computer proceeds through a series of steps, at each step adding and/or deleting certain independent variables from the regression equation after determining how such changes will affect the proportion of the variance in the dependent variable which is explained by the linear regression. All the statistician must do is put the data in the form required by the particular program that is being used and prepare a few "control statements" to tell the computer a few details about the problem, such as the number of variables and the sample size. Incidentally, specific "standard programs" will not be discussed in this book because they would be outdated much too fast.

The evolution of standard statistical computer programs has greatly eased the computational burden on the statistician and has made it possible to analyze larger and larger problems, problems that would be impossible to handle other than by computer. This development has not been entirely without its drawbacks, however. With some programs, the printed output of results is almost as massive as the data used as input. The statistician is now faced with the task of sorting out meaningful results from relatively unimportant results. Just looking at simple correlations, if we have 10 variables, there are $\binom{10}{2}$, or 45 correlations between pairs of variables. In a multiple regression problem, we are faced with innumerable standard errors, partial correlations, and so on. If we know exactly what we are looking for, then we should have little difficulty, and some experience with multiple regression problems should be helpful in this regard. On the other hand, in performing hand calculations or working on a calculating machine, the statistician is able to follow the analysis step by step. In so doing, a "feel" for the problem (and for the statistical techniques involved) is acquired, a "feel" which is often hard to get from reading printed output from a computer. This weakness is probably counterbalanced by the fact that the computer output, being quite thorough, often suggests ideas for further analysis or research that might not have been noticed otherwise.

In some respects, the standard statistical programs are almost *too* easy to use. By this we mean that it is possible for a person with very little knowledge of statistics to use one of these complex programs even though that person has little idea as to what the program does or what much of the output means. This is a weakness not of the computer programs themselves, but of the users of the programs. Grinding out numerous results on the computer is of little value if the underlying statistical techniques are not understood. That is why the emphasis in this book has been on *understanding* the basis for various statistical procedures rather than on the computational aspects of statistics. In the computer age, we can spend more time concentrating on the concepts and less time on the calculations.

10.16 NONLINEAR REGRESSION

The discussion in this chapter has been centered on correlation, the measurement of the degree of *linear* relationship between two variables, and *linear* regression, the use of a *linear* function rule for the prediction of one variable from another variable or set of variables. However, the theory of regression is much more extensive than the preceding discussion of linear regression might suggest. Indeed, the linear rule for prediction is only the simplest of a large number of rules that might apply to a given statistical relation. Linear regression equations may serve quite well to describe many statistical relations that are roughly like linear functions. One justification for linear regression is the fact that if the variables of interest have a joint distribution which is multivariate normal, then the theoretical regression function will always be linear. Nevertheless, there is no law of nature requiring all important relationships among variables to have a linear form. It thus becomes important to extend the idea of regression equations to the situation where the relation is *not* best described by a linear rule. Now we are going to consider problems of **nonlinear regression**— problems in which the best rule for prediction need not be a simple linear function.

In terms of the theoretical regression curve, it should be clear by now that it is by no means necessary that $E(Y \mid x)$ be a linear function of x in the bivariate case. Indeed, the example in Section 10.5 resulted in a regression curve which was not perfectly linear, although it was nearly linear. *If* we know $f(x, y)$, *and if* it is not too difficult to compute $f(y \mid x)$ and $E(Y \mid x)$, then we can determine the exact form of the regression curve, and we need not worry about whether it is linear or nonlinear. In virtually all applications of regression analysis, however, the exact joint distribution is not known. As a result, it is necessary to estimate a regression curve on the basis of sample data. In order to estimate such a curve, it is first necessary to specify some sort of mathematical model for the curve. Once this has been done, the next task is to estimate the unknown parameters of this mathematical model. For instance, if we specify the usual linear model, then we must estimate the regression coefficients β_1 and β_2 from the sample data.

Of these two problems (specifying a model and estimating the parameters of the model), the first is the most important. If the model chosen is inappropriate for the problem at hand, then the second problem becomes almost meaningless. Of course, there must be some trade-off between the two problems, since the difficulty of the estimation problem is directly related to the complexity of the model. We want our model to be as realistic as possible while maintaining a reasonably simple mathematical form.

How, then, might we specify a model? Obviously, this question is not entirely within the realm of statistics. One possibility is the existence of some underlying theory which indicates that the relationship between the two variables of interest should be of a particular form. This is more common in the physical sciences than it is in business, the social sciences, and

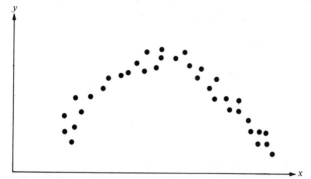

Figure 10.16.1 A Scatter Diagram that Indicates a Nonlinear Relationship

the behavioral sciences, but it still does occur in the latter disciplines. Consider, for example, the concepts of "learning curves" (learning as a function of time) in psychology and "demand curves" (the amount demanded of a product as a function of the price of the product) in economics. In fact, one way to investigate the applicability of a proposed theory is to compare the predictions of a model determined by the theory with the predictions of alternative models.

More often than not, however, the form of the relationship between two variables will not be specified by any theory. If this is the case, perhaps the data themselves will suggest a model. The scatter diagram thus may play an important role in the selection of a mathematical model. Consider, for example, the scatter diagrams presented in Figures 10.16.1–10.16.5. The importance of such informal tools as scatter diagrams is often ignored in statistics texts because of the desire to discuss more "high-powered" statistical techniques. However, it is perfectly possible for the statistician to (1) find that the correlation coefficient and the regression coefficient of

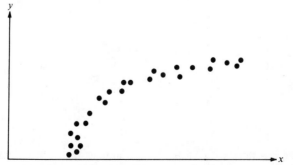

Figure 10.16.2 A Scatter Diagram that Indicates a Nonlinear Relationship

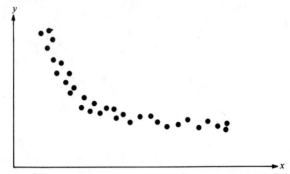

Figure 10.16.3   A Scatter Diagram that Indicates a Nonlinear Relationship

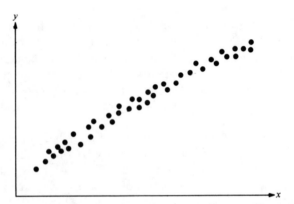

Figure 10.16.4   A Scatter Diagram that Indicates a Linear Relationship

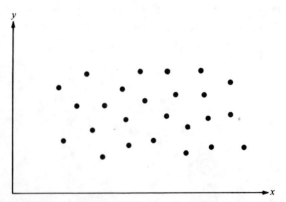

Figure 10.16.5   A Scatter Diagram that Indicates no Readily Discernible Relationship

a linear model are "significantly different from zero," implying that there is some linear relationship between the variables, while (2) a nonlinear model would provide a much better fit to the data. A look at the scatter diagram prior to application of the standard linear model might make this clear.

To see how the scatter diagram could suggest a mathematical model, consider Figures 10.16.1–10.16.5. In Figure 10.16.1 it appears that a parabola might provide a good fit, in which case we might use the following model:

$$y = \beta_0 + \beta_1 x + \beta_2 x^2. \tag{10.16.1}$$

This differs from the linear model in that one term, involving X^2, has been added. This is one of a general class of **polynomial** functions (also called power functions). A polynomial function of degree M may be represented as

$$y = \beta_0 + \beta_1 x + \beta_2 x^2 + \cdots + \beta_M x^M. \tag{10.16.2}$$

The model of Equation (10.16.1) is a polynomial function of degree 2 (a quadratic function).

In Figure 10.16.2, the data appear to follow a **logarithmic** function:

$$y = \beta_0 + \beta_1 \log x. \tag{10.16.3}$$

In Figure 10.16.3, an **exponential** function might be a good fit:

$$y = \beta_0 \beta_1^x. \tag{10.16.4}$$

In Figure 10.16.4 it appears that it will be unnecessary to resort to a nonlinear model, since the points seem to follow roughly a linear path.

The final scatter diagram presented in this section, Figure 10.6.5, illustrates a slightly different, but equally important, use of scatter diagrams: to indicate whether or not it might be worthwhile to attempt to fit *any* function to the data. In Figure 10.16.5, there is no discernible relationship between X and Y, and it is apparent that any attempt to fit a straight line or any nonlinear function would prove to be fruitless.

It should be emphasized that the scatter diagrams encountered in actual situations are not nearly as clear-cut as the above examples. It may not always be easy to "eyeball" the diagram and select a mathematical model or determine whether or not any model is worthwhile. At a minimum, however, the diagram should provide the statistician with some "feeling" for the data and for the relationship between the two variables. If the data do not clearly point to one particular model, perhaps they suggest two or three potential models, in which case the statistician could fit different models to the data and attempt to determine which provides the "best fit."

Once the model (or a set of potential models) is determined, the problem facing the statistician is the estimation of the parameters of the model, which we will discuss in the next two sections.

10.17 APPLYING LEAST SQUARES IN NONLINEAR REGRESSION

The least-squares method of curve-fitting is equally applicable to linear and nonlinear functions. In the case of a quadratic function as given in Equation (10.16.1), the least-squares estimators of β_0, β_1, and β_2 would be the values b_0, b_1, and b_2 resulting in the minimization of the sum of the squared deviations between the actual values and the predicted values:

$$\text{minimize} \sum_{i=1}^{n} \hat{e}_i{}^2 = \sum_{i=1}^{n} [y_i - (b_0 + b_1 x_i + b_2 x_i{}^2)]^2,$$

where the data consist of n pairs of values (x_i, y_i). To find the least-squares estimators, we must use a procedure similar to that presented in Section 10. This involves taking the partial derivatives of $\sum \hat{e}_i{}^2$ with respect to b_0, b_1, and b_2, setting these partial derivatives equal to zero, and solving the resulting equations. Following this procedure, we get the following three equations:

$$\sum y_i = b_0 n + b_1 \sum x_i + b_2 \sum x_i{}^2, \tag{10.17.1}$$

$$\sum x_i y_i = b_0 \sum x_i + b_1 \sum x_i{}^2 + b_2 \sum x_i{}^3, \tag{10.17.2}$$

and $\quad\sum x_i{}^2 y_i = b_0 \sum x_i{}^2 + b_1 \sum x_i{}^3 + b_2 \sum x_i{}^4. \tag{10.17.3}$

Rather than solve these equations in terms of the various sums, it is easier to calculate the sums from the data, insert these values in the equations, and then solve for b_0, b_1, and b_2.

For example, suppose that the dial controlling a machine can be set at any one of 11 possible settings, ranging from -5 to $+5$ in unit increments. Let X represent the setting on the dial. Each of the settings is tried for a 15-minute period, and Y, the output of the machine during that period, is recorded in each case. (In a realistic situation each setting would no doubt be tried more than once in order to obtain some information about the variability of Y given X, but for the purposes of this example each setting is used only once, so that $n = 11$.) The data are given in Table 10.17.1, and Equations (10.17.1)–(10.17.3) are

$$106 = 11b_0 + 0b_1 + 110b_2,$$

$$20 = 0b_0 + 110b_1 + 0b_2,$$

and $\quad 688 = 110b_0 + 0b_1 + 1958b_2.$

Table 10.17.1 Data for Machine-Output Example

x_i	y_i	x_iy_i	x_i^2	x_i^3	x_i^4	$x_i^2y_i$
-5	2	-10	25	-125	625	50
-4	7	-28	16	-64	256	112
-3	9	-27	9	-27	81	81
-2	12	-24	4	-8	16	48
-1	13	-13	1	-1	1	13
0	14	0	0	0	0	0
1	14	14	1	1	1	14
2	13	26	4	8	16	52
3	10	30	9	27	81	90
4	8	32	16	64	256	128
5	4	20	25	125	625	100
0	106	20	110	0	1958	688

Solving these three equations simultaneously, we get

$$b_0 = 13.97,$$

$$b_1 = .18,$$

and $$b_2 = -.43,$$

which are the least-squares estimates of β_0, β_1, and β_2. The estimated regression curve is

$$\hat{y} = 13.97 + .18x - .43x^2.$$

The scatter diagram and the estimated regression curve for this example are presented in Figure 10.17.1. It appears that the output predicted from

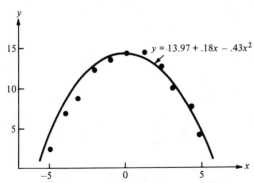

Figure 10.17.1 Scatter Diagram and Estimated Regression Curve with Quadratic Model for Machine-Output Example

the curve is greater for settings near zero than for settings near the extremes, -5 and $+5$.

In a similar manner, the least-squares criterion could be used to fit any nonlinear model to a set of data—that is, to estimate the parameters of the model. If the model can be represented by the function h, so that

$$\hat{y} = h(x),$$

then the least-squares criterion amounts to minimizing

$$\sum_{i=1}^{n} \hat{e}_i^2 = \sum_{i=1}^{n} [y_i - h(x_i)]^2.$$

In addition, least squares could be used to compare different models. For each of the competing models, we could estimate the parameters using least squares and then calculate the corresponding value of the sum of squared errors. The model producing the smallest value of $\sum \hat{e}_i^2$ would be the "best" model, where "best" is to be interpreted solely in the least-squares sense (that is, the sense of minimum squared error). We shall discuss other methods for investigating "goodness of fit" in Chapter 12.

Unfortunately, as the model becomes more complex, it becomes increasingly difficult to determine the least-squares estimators of the parameters. This is particularly true when the model has a large number of parameters. One alternative might be to make some distributional assumptions and attempt to determine maximum-likelihood estimators of the parameters. Making distributional assumptions may also simplify the selection of a mathematical model for the regression curve, just as the assumption of bivariate normality allows us to restrict our attention to linear functions. However, bivariate distributions are not always easy to deal with, and in general it will be just as difficult, if not more difficult, to determine maximum-likelihood estimators as it is to determine least-squares estimators. Obviously some other ways of handling nonlinear regression problems are needed.

10.18   TRANSFORMATION OF VARIABLES IN NONLINEAR REGRESSION

In some cases it may be possible to greatly simplify a nonlinear regression problem by a transformation of variables. For instance, consider the curve given by

$$y = \beta_0 x^{\beta_1}.$$

Taking logarithms, we get

$$\log y = \log \beta_0 + \beta_1 \log x.$$

Now, letting $u = \log x$, $w = \log y$, $\beta_0{}^* = \log \beta_0$, and $\beta_1{}^* = \beta_1$, we have

$$w = \beta_0{}^* + \beta_1{}^* u.$$

By transforming from the variables X and Y to the variables U and W, we get a linear function. We could then use least squares to estimate the parameters of the new equation, $\beta_0{}^*$ and $\beta_1{}^*$. Since

$$\beta_0{}^* = \log \beta_0$$

and

$$\beta_1{}^* = \beta_1,$$

these estimates can be converted back into estimates for the parameters of the original equation.

This approach is applicable for numerous curvilinear models. For the logarithmic model in Equation (10.16.3),

$$y = \beta_0 + \beta_1 \log x,$$

we simply need to use the transformation $u = \log x$, in which case

$$y = \alpha + \beta u.$$

For the exponential model in Equation (10.16.4),

$$y = \beta_0 \beta_1{}^x,$$

we take logarithms on both sides of the equation:

$$\log y = \log \beta_0 + x \log \beta_1.$$

Now, letting $w = \log y$, $\beta_0{}^* = \log \beta_0$, and $\beta_1{}^* = \log \beta_1$,

$$w = \beta_0{}^* + \beta_1{}^* x.$$

It is once again easy to convert estimates of $\beta_0{}^*$ and $\beta_1{}^*$ into estimates of β_0 and β_1.

For an example, suppose that we want to fit an exponential curve to the data in Table 10.18.1. The least-squares estimates of $\beta_0{}^*$ and $\beta_1{}^*$ are

$$b_1{}^* = \frac{n \sum x_i w_i - (\sum x_i)(\sum w_i)}{n \sum x_i^2 - (\sum x_i)^2} = \frac{8(52.42) - 36(8.62)}{8(204) - (36)^2}$$

$$= .325$$

and $$b_0{}^* = \frac{\sum w_i - b_1{}^* \sum x_i}{n} = \frac{8.62 - .325(36)}{8}$$

$$= -.385.$$

Table 10.18.1   A Sample of Size 8

x_i	y_i	$w_i = \log_e y_i$	$x_i w_i$	x_i^2
1	1.00	.00	.00	1
2	1.20	.18	.36	4
3	1.80	.59	1.77	9
4	2.50	.92	3.68	16
5	3.60	1.28	6.40	25
6	4.70	1.55	9.30	36
7	6.60	1.89	13.23	49
8	9.10	2.21	17.68	64
36	30.50	8.62	52.42	204

But
$$\beta_0^* = \log_e \beta_0$$

and
$$\beta_1^* = \log_e \beta_1,$$

so that
$$\beta_0 = e^{\beta_0^*}$$

and
$$\beta_1 = e^{\beta_1^*}.$$

Expressing this in terms of estimates,

$$b_0 = e^{b_0^*} = e^{-.385} = .68$$

and
$$b_1 = e^{b_1^*} = e^{.325} = 1.39.$$

Using the transformation, then, we arrive at the estimated exponential curve

$$\hat{y} = (.68)(1.39)^x.$$

This curve, along with the sample points, is illustrated in Figure 10.18.1.

Using transformations such as these, it is possible to convert a fairly complex nonlinear model into a linear model. The parameters can then be estimated for the linear model and the results transformed back in order to be expressed in terms of the parameters of the original nonlinear model. Unfortunately, there are some difficulties involved. *Minimizing the sum of squared deviations for the transformed model is not necessarily equivalent to minimizing the sum of squared deviations for the original model.* Although the results are least-squares estimators for the transformed model, they do not result in least-squares estimators for the original model. In the example,

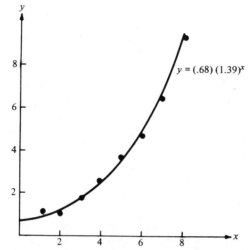

Figure 10.18.1   Scatter Diagram and Estimated Regression Curve for Exponential Model with Data from Table 10.18.1

applying least squares to the model

$$\log y = \beta_0{}^* + \beta_1{}^* x$$

amounts to minimizing the sum of the squared deviations in the $(x, \log y)$ plane, which is not necessarily the same as minimizing the sum of the squared deviations in the (x, y) plane for the original function

$$y = \beta_0 \beta_1{}^x.$$

If we think in terms of regression models with random-error terms, the difficulty lies in the transformation of the error term. Even though the error terms for the original model satisfy the assumptions of being independent and distributed normally with mean zero and constant variance for any given x, the transformed error terms may no longer satisfy the assumptions. Of course, the opposite may be true; the original error terms violating the assumptions, but the transformed error terms satisfying the assumptions. At any rate, the correspondence between the old and new error terms must be taken into account.

Another way to simplify a nonlinear regression problem is to fit a piecewise linear function to the data, as illustrated in Figure 10.18.2. To do this, we just fit a series of linear regression models, with each one being restricted to a certain interval of x values. This is a particularly useful technique when there is no reasonably simple curve which provides a good model or when transformations cause problems with the random-error

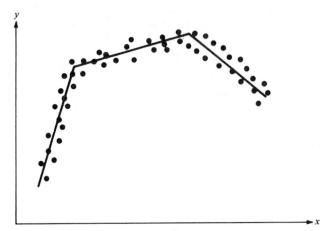

Figure 10.18.2   A Scatter Diagram and a Piecewise Linear Function

terms. Fitting a piecewise linear function requires more data than fitting a single function, because data are needed to fit each individual segment. If there are very few data this procedure degenerates to a simple "connect-the-dots" exercise, which, although it may provide the "best fit" in a least-squares sense for the small set of available data, is of very little value in making predictions.

In summary, we have discussed three ways for dealing with nonlinear regression models. The first method is to apply least squares directly to the nonlinear model; this requires a fairly simple model. The second approach is to somehow transform the nonlinear equation into a linear equation and then to apply least squares to this transformed equation. Finally, the third technique is to fit a piecewise linear function to the data; that is, to approximate the nonlinear model with a piecewise linear model. We emphasize once again that although nonlinear regression may be a more difficult problem than linear regression, there are many cases in which variables are strongly related in some nonlinear fashion. The statistician is then better off (in the sense of having a more accurate model) if a nonlinear model is used. Because of the emphasis placed on linear regression as a predictive tool and because of the ease with which a linear regression curve can be estimated, there is perhaps too much of a tendency among users of statistics to use the linear model in situations where some other model would provide a better fit. In this regard, the scatter diagram is a most useful informal tool, for it may suggest an alternative to the linear model. The statistician should not discount the possibility of important and predictive nonlinear relations between variables. On the other hand, it should be emphasized that functions (linear or nonlinear) fitted to data;

like all other statistical devices, are only as good as the data to which they are applied. In Chapter 11 we will discuss tests for the presence of linear and nonlinear regression; these tests should be quite useful in the development of regression models.

10.19   TIME SERIES AND REGRESSION ANALYSIS

In our study of correlation and regression, we have assumed that the sample upon which our estimates and tests are based is a random sample from the population of interest. In many instances, particularly in problems of prediction, we must deal with **time-series data;** that is, the observation of the values of certain random variables at successive points in time. For instance, variables such as the price of a stock, the Gross National Product, and the temperature at a given place change over time. To obtain a series of observations on variables such as these, the observations must be made at different times. We might observe the price of a stock at the end of each hour, day, month, or year; whatever the interval between observations, the resulting set of data forms a time series. In economics, where the variables of interest are often observed over time, a distinction is made between *cross-section* data and *time-series* data. If we take a sample of individuals at a given point in time and determine their incomes, we have cross-section data; if we consider a single individual and determine his or her income at various points in time, we have time-series data.

If we are dealing with time-series data, then the sample may not be a random sample. This casts doubt on the assumptions used in our linear-regression models. In particular, it is very likely that the error terms are not statistically independent; instead, they may be correlated with each other. This is a problem of **autocorrelation.** If we have a series of observations of a variable Y which is taken over time, denoted by $Y_1, Y_2, \ldots, Y_n$, then the correlation between successive terms of the series (that is, the correlation between Y_t and Y_{t-1}) is called **autocorrelation, or serial correlation.** The autocorrelation in a time series may invalidate the assumptions of a regression model. Recall, however, that these assumptions are needed primarily for the development of tests of hypotheses concerning the model. The least-squares curve-fitting method does not require any assumptions such as independence of the error terms. Thus, it should be possible to follow the standard least-squares procedure to estimate the regression equation, but it may not be possible to use the usual formulas for computing standard errors of the estimators, confidence intervals, and test statistics. Of course, if a check of the actual error terms (following the estimation of the regression equation) indicates that there is little or no autocorrelation, then we may proceed as if the data had not arisen from a time series. There are statistical procedures available for testing for the presence of autocorrelation.

A special type of regression analysis when we have time-series data is **autoregression,** which is the use of regression analysis to predict a dependent variable when the independent variables are merely lagged terms of the dependent variable:

$$\hat{y}_t = \beta_1 + \beta_2 y_{t-1} + \beta_3 y_{t-2} + \cdots + \beta_{K+1} y_{t-K}.$$

Of course, other independent variables could be added to this regression equation.

An important point that should be mentioned with regard to time-series data is the problem of extrapolation over time. If the dependent variable is observed over time, we may want to predict this dependent variable for some future time period. But the regression equation is based on data from past periods, so essentially we are predicting beyond the range of the past data, using the justification that the trend observed in the past will continue in the future.

How do time-series data affect correlation analysis? If two time series both have a similar trend (say, upward) over time, then they will be highly correlated over a long period of time even though they are not highly correlated "in the short run," that is, for smaller samples. Such high correlation is sometimes referred to as **spurious correlation.** If the trend can somehow be removed from the two time series, then such spurious correlation will be eliminated. One way to do this is to correlate *differences* between successive terms of the series rather than to correlate the terms themselves. Another technique is to use multiple regression of one series on another, with time also being included as an independent variable. It should thus be possible to determine how much of the correlation is due to trend (time) and how much is actually attributable to association between the variables. Techniques such as these are of interest in the study of time series. One approach is to consider the time series as being made up of four components: (1) trend, (2) seasonal fluctuations, (3) cyclical fluctuations, and (4) irregular fluctuations. Generally, the statistician is primarily interested in the trend, so methods have been developed to attempt to remove the effect of the other three components from the time series. Two such methods are the use of moving averages and the technique of exponential smoothing. Because much of the traditional time-series analysis is primarily descriptive rather than inferential, because other useful techniques for working with time-series data (for example, spectral analysis) are beyond the level of this book, and because of space limitations, we will not discuss time series in any greater detail. The purpose of this section is merely to point out some of the problem areas involving time series, particularly those relating to regression and correlation, and to note briefly how these problems have been attacked.

10.20   A BRIEF LOOK AT ECONOMETRICS: OTHER PROBLEMS IN REGRESSION

A great deal of the research involving multiple regression has been conducted by economists. This research falls under the general heading of **econometrics,** which is a combination of economics and statistics. Although this section concerns regression in econometrics, the *statistical* problems to be discussed apply equally well to situations in other disciplines.

As we noted in Section 10.16, a major problem in the estimation of a regression curve (or surface) is the specification of the model. The economist must consider the role of economic theory as well as the data at hand. Yet the theory (or the data, or both) may suggest a very complex model that would greatly complicate the estimation procedure. For some models, it is not even theoretically possible, given the data, to determine unique estimators through use of the least-squares criterion. It is necessary, then, to balance off (1) the desire to have a realistic model and (2) the desire to have a model for which the parameters can be estimated from the available data. The problem of the specification of the model includes the choice of variables to be included in the model. The dependent variable which we wish to predict may be related to innumerable independent variables. In order to keep the model as simple as possible, we might include only those variables that are thought to be strongly associated with the dependent variable. Incidentally, although we have been working solely with quantitative variables in this chapter, it is possible that a variable of interest could be qualitative. One variable might simply be the answer to the question, "Is your income over d dollars per year?" In other words, data regarding income may not consist of an individual's income, but simply of whether or not the income is greater than d dollars per year. The usual way to handle this is to set up a "dummy variable," which is equal to 1 if the income is over d and equal to 0 otherwise. In this way, variables which are measured on a nominal or ordinal scale rather than an interval scale can be included in the analysis.

In addition to the problem of specifying the model, there may be some statistical problems facing the economist. For instance, since economic data frequently are in time-series form, the problems discussed in the preceding section (such as autocorrelation) could be encountered. There are also other ways in which the assumptions of the multiple regression model could be violated. Suppose that we wish to use a person's income to predict how much that person will spend on clothing. A sample is taken, consisting of 5 persons in a low-income bracket, 5 in a middle-income bracket, and 5 in a high-income bracket. We would expect that there would be a greater variability in clothing expenditures within the high-income bracket than within the low-income bracket. But the assumption that the variance of the error terms, σ_e^2, is constant over all values of the independent variable implies that this cannot be true. Therefore, this assumption, which is the assumption of constant variance, or **homoscedasticity,** is violated.

Another difficulty often faced in multiple regression studies is that of **multicollinearity,** which arises when the independent variables are highly correlated *with each other*. In the extreme case, if the independent variables are perfectly correlated with each other, it is not possible to determine the least-squares estimates of the regression coefficients. In terms of the matrix model presented in Section 10.13, the inverse of the matrix $\mathbf{X}'\mathbf{X}$ does not exist. In nonmathematical terms, this means that the relationship among the independent variables has obscured the relationship between the independent variables and the dependent variable.

Another problem of interest arises when the model is so complex that it cannot be expressed in terms of just one equation. Instead it must be expressed in terms of two equations which are interrelated. Because of the interrelationship, it is not feasible to solve the two equations separately by using the multiple-regression model. Instead, it becomes necessary to solve (perhaps "estimate" would be a better term here) the two equations simultaneously, and this is what is called a **simultaneous-equation problem.** Several techniques, some involving variations of the standard least-squares approach (two-stage least-squares, indirect least-squares) have been developed to handle such problems.

In this section we have pointed out, in the context of econometrics, some problems that may arise in regression problems. Further discussions may be found in most standard econometrics books. A considerable amount of work has been done regarding these problems, both from a classical standpoint and from a Bayesian standpoint. The major point of interest for the purposes of this book is that in regression and correlation problems (just as in problems involving other statistical techniques), the statistician must be aware of conditions under which the standard techniques may not be applicable. In particular, as we have emphasized so many times, it is important to investigate possible violations of the assumptions underlying any statistical technique and to understand the implications of such violations. Insofar as possible within the scope of this book, we have attempted to discuss the importance of assumptions and the sensitivity of different statistical methods to violations of these assumptions.

EXERCISES

1. For a particular population of married couples under 40 years of age, let X represent the age of the husband when the couple was married, and let Y represent the age of the wife when the couple was married. All ages have been rounded to the nearest age ending in the digit 0 or the digit 5. The joint probability distribution of X and Y is as follows:

AGE OF WIFE

		15	20	25	30	35
	15	.02	.01	.01	.01	.00
	20	.05	.10	.03	.01	.01
AGE OF HUSBAND	25	.04	.15	.13	.02	.01
	30	.02	.06	.09	.07	.01
	35	.02	.03	.04	.04	.02

(a) Find the covariance of X and Y.
(b) Find ρ_{XY}.

2. In Exercise 67, Chapter 3, find ρ_{XY}.

3. Suppose that X and Y are continuous random variables with joint density functions given by the rule

$$f(x, y) = \begin{cases} x + y & \text{if } 0 < x < 1,\, 0 < y < 1, \\ 0 & \text{elsewhere.} \end{cases}$$

(a) Find the marginal density function of X.
(b) Find the conditional density function of Y, given that $X = x$.
(c) Find the conditional expectation, $E(Y \mid X = x)$.

4. For the random variables given in Exercise 3, find
(a) the marginal density function of Y,
(b) the conditional density of X, given that $Y = y$,
(c) the conditional expectation, $E(X \mid Y = y)$.

5. For the random variables given in Exercise 3, find
(a) the covariance, cov (X, Y),
(b) the correlation coefficient, ρ_{XY}.

6. Discuss the following statement: "A correlation coefficient of $+1$ or -1 indicates a perfect functional relationship between two variables, but a correlation coefficient different from $+1$ and -1 does *not* rule out the possibility of a perfect functional relationship."

7. Prove that if $U = cX + g$ and $W = dY + h$, then $\rho_{XY} = \rho_{UW}$ and $r_{XY} = r_{UW}$. How might this fact be useful in calculating correlation coefficients?

8. For the joint distribution in Exercise 69, Chapter 3, find ρ_{XY}.

9. Under what circumstances must uncorrelated variables (that is, variables with $\rho = 0$) also be independent? Is it possible for two independent variables to be correlated?

10. A college administrator is interested in the possible linear relationship between a student's score on a college entrance examination and that student's performance in the freshman year at a given college. Let X be the entrance examination score, and let Y be the grade average during the freshman year. A random sample of 25 students from the freshman class yields the following results:

Student	x	y
1	75	84
2	77	94
3	75	90
4	76	90
5	75	91
6	76	86
7	73	87
8	75	95
9	74	83
10	75	85
11	76	88
12	74	91
13	72	80
14	75	85
15	73	87
16	75	82
17	78	86
18	76	83
19	74	85
20	74	88
21	77	100
22	75	98
23	76	89
24	74	91
25	75	99

(a) Construct a scatter diagram. Does there appear to be any relationship between X and Y? Any linear relationship?

(b) Compute the correlation coefficient, r_{XY}.

11. A marketing manager was interested in the ability of consumers to judge the price of an item from a simple inspection of the item. To investigate this, he randomly selected 12 pieces of material worth $2 per yard, 12 pieces worth $3 per yard, 12 pieces worth $4 per yard, and 12 pieces worth $5 per yard. Each of a panel of consumers guessed the price of each piece of material. The data below show the four prices together with the average price guessed for each piece of material:

ACTUAL PRICE

	2	3	4	5
Average "guessed" prices	4.37	4.46	3.51	3.09
	2.98	2.80	2.53	2.87
	4.33	2.12	3.60	4.25
	3.33	2.13	2.63	2.77
	4.10	3.25	4.25	3.51
	3.68	3.78	2.77	3.69
	3.35	3.59	3.49	3.40
	3.27	2.64	3.69	3.10
	4.29	2.56	3.40	4.40
	2.91	2.66	3.10	2.59
	2.73	4.09	4.40	3.09
	4.12	3.87	2.59	2.87

Find the correlation coefficient between X, actual price, and Y, average "guessed" price.

12. A statistician is interested in the relationship between the height (in inches) and the weight (in pounds) of students from a particular school. A random sample of students is taken, with the following results (X = height in inches, Y = weight in pounds):

x: 70   75   64   67   71   70   68   76   68   69   70

y: 175 198 156 180 178 182 160 204 167 169 162

(a) Construct a scatter diagram.
(b) Find the correlation between X and Y.

13. The officials of a bank are interested in the relationship between X, the income of an individual who has a checking account with the bank, and Y, the balance in the individual's checking account. A random sample of two customers is selected from all of the individuals with checking accounts. The first customer has an income of $15,000 and a checking account balance of $600; the second customer has an income of $12,000 and a checking account balance of $480.
(a) Compute the sample correlation coefficient, r_{XY}.
(b) Compute r_{XY} if the checking account balance of the second customer is $1200 instead of $480.

(c) Compute r_{XY} if the checking account balance of the second customer is $100 instead of $480.

(d) What general result is illustrated by this exercise? Prove that this general result must always be true and give an intuitive explanation of why it is true.

14. In Exercise 10, the college administrator would like to use the sample results to make inferences about the entire freshman class.

(a) Find the p-value for the test of the hypothesis that $\rho_{XY} = 0$ against the alternative that $\rho_{XY} \neq 0$ (do this in two different ways).

(b) Find the p-value for the test of the hypothesis that $\rho_{XY} = .50$ against the alternative that $\rho_{XY} \neq .50$.

(c) Find a 90 percent confidence interval for ρ_{XY}.

(d) Discuss the interpretation of the results in parts (a)–(c) and discuss any assumptions which were made.

15. In Exercise 11, is there a significant linear relation between X and Y at the .05 level? Plot the data on a scatter diagram and discuss the results.

16. In Exercise 12, find 75 percent and 90 percent confidence intervals for ρ_{XY}.

17. Clearly explain the distinction between correlation and causation.

18. Are any assumptions required to compute sample correlation coefficients? Are any assumptions required to make inferences from these sample correlation coefficients to the entire population?

19. A statistician has independent samples from two bivariate normal distributions and wishes to make inferences about the difference between the correlation coefficient for the first population and the correlation coefficient for the second population. How can the statistician determine confidence intervals for this difference and test hypotheses concerning this difference? [*Hint:* Use the transformation given by Equation (10.3.2) for each of the two samples and consider the difference between the transformed values.]

20. Discuss the rationale behind the use of $E(Y \mid X = x)$ as the regression curve of Y on X. Why is it often reasonable to approximate this curve by a straight line?

21. Discuss the situation in which the regression curve $E(Y \mid X = x)$ is a horizontal line.

22. For the distribution in Exercise 1, find $E(Y \mid x)$ for each of the five possible x values. Also, find $V(Y \mid X = 20)$ and $V(Y \mid X = 30)$.

23. In Exercise 2, find the regression curve $E(Y \mid X = x)$.

24. In Exercise 8, find the regression curve of Y on X, $E(Y \mid X = x)$, and find the regression curve of X on Y, $E(X \mid Y = y)$.

25. For the data in Exercise 10, estimate the linear regression of Y on X. What proportion of the variance of Y is accounted for by the linear regression?

26. For the data in Exercise 11, estimate the linear regression of Y on X and plot this line on the scatter diagram. Assuming normality, find 95 percent confidence limits for β_2. If $X = 3$, find 95 percent confidence limits for the actual value of Y.

27. In Exercise 13, consider a random sample of 100 customers with checking accounts. Let X, income, be expressed in units of $10,000, and let Y, checking account balance, be expressed in units of $500. For the sample, $\sum x_i = 120$, $\sum y_i = 10$, $\sum x_i y_i = 16$, $\sum x_i^2 = 169$, and $\sum y_i^2 = 3.25$.

   (a) Find the estimated regression line expressing checking account balance as a function of income.

   (b) Find the estimated regression line expressing income as a function of checking account balance.

   (c) In (a), find the proportion of the variance of checking account balance explained by the linear regression and find the amount of variance explained by the linear regression.

   (d) In (b), find the proportion of the variance of income explained by the linear regression and find the amount of variance explained by the linear regression.

   (e) Compare your answers to (a) and (b) and compare your answers to (c) and (d).

28. If $E(X) = 5$, $E(Y) = 80$, $V(X) = 100$, $V(Y) = 144$, $E(XY) = 460$, and X and Y have a bivariate normal distribution, find

   (a) the regression line of Y on X, $E(Y \mid X = x)$,

   (b) the variance of the predicted value of Y given X, $V(Y \mid X = x)$,

   (c) the proportion of the variance of Y which is *not* accounted for by the linear regression.

29. Discuss the statement, "The interpretation of the result that unusually tall men tend to have sons shorter than themselves and unusually short men tend to have sons taller than themselves as a regression toward mediocrity has been aptly called the regression fallacy."

30. In general, the estimated regression line of Y on X, $\hat{y} = b_1 + b_2 x$, is different from the estimated regression line of X on Y, $\hat{x} = b_1' + b_2' y$. Is there any situation in which the two lines will coincide? Explain.

31. Show that when the coefficients of a regression line in the bivariate case are estimated via the method of least squares, the average value of the error terms, $\sum \hat{e}_i / n$, is equal to zero, and the sample variance of the error terms can be expressed in the form $\sum \hat{e}_i^2 / n$.

32. Use Equation (10.6.7) to show that $-1 \le \rho_{XY} \le 1$.

33. A stereo cartridge manufacturer has conducted a regression analysis to study L, the lifetime of a cartridge in hours, as a function of F, the tracking force (in grams). For a sample of 100 cartridges, the estimated regression line is $\hat{l} = 1300 - 200f$. The standard deviation of values predicted with this estimated line is $s_{L \cdot F} = 100$ hours. (a) By decreasing the tracking force from 3 grams to 1 gram, how much longer can you expect a cartridge to last on the average? (b) If the sample correlation coefficient is $r_{LF} = .5$, what is s_L^2, the sample variance of L?

34. In the linear regression model which includes an error term, what assumptions are generally made about the error term in making inferences about the theoretical regression line? Can you give some situations in which these assumptions would not be reasonable?

35. A statistician is interested in the possible linear relation between the time spent per day in practicing a foreign language and the ability of the person to speak the language at the end of a 6-week period. Fifty language students are divided randomly into 5 groups, each group spending a specific number of hours per day in practice. At the end of 6 weeks, each student is scored for proficiency in the language. The data are as follows, with proficiency scores (values of Y) in the body of the table:

PRACTICE IN HOURS (VALUES OF X)

.25	.50	1	2	3
117	106	76	125	85
85	81	88	113	129
112	74	115	93	90
81	79	113	89	124
105	118	108	117	117
109	110	84	118	121
80	82	83	81	97
73	86	81	86	93
110	111	112	88	122
78	113	120	120	92

(a) Find the linear regression equation for predicting Y from X.
(b) Plot the linear regression equation, along with the data, on a scatter diagram.
(c) What is the appropriate measure of the "scatter" or vertical deviations of the obtained points in the scatter diagram about the regression line?
(d) Find 95 percent confidence bands for the *predicted* value of Y (from the theoretical regression line) and for the *actual* value of Y. How can you explain the differences between these two confidence bands? Graph the two bands and the estimated regression line.

36. A stock-market analyst is interested in using the *change in price* of a stock on any given day to predict the change in price of the same stock on the following day. The analyst chooses a widely traded stock and observes the following daily sequence of *prices* (not price changes):

80, 82, 85, 81, 80, 80, 80, 84, 88, 89, 90, 88, 84.

(a) Construct the appropriate scatter diagram and plot the data.
(b) Find the estimated regression line.
(c) What proportion of the variance is explained by this line?
(d) What do you conclude about the relationship the analyst is interested in? Might this lead to improved predictions? Can you think of any extensions which might improve the model?

37. Consider the data given in Exercise 12.
   (a) Find b_1 and b_2 and draw the estimated regression line of Y on X.
   (b) Find the correlation between X and Y.
   (c) Does the use of a linear regression model improve our ability to predict Y, given X?
   (d) If a student chosen at random is 70 inches tall, use the estimated regression line to predict the student's weight.

38. Suppose that you are interested in using past expenditures on research and development by a firm to predict current expenditures on R&D. You obtain the following data by taking a random sample of firms, where X is the amount spent on R&D (in millions of dollars) 5 years ago and Y is the amount spent on R&D (in millions of dollars) in the current year:

$$x: \quad 3 \quad 5 \quad 2 \quad 8 \quad 1 \quad 2 \quad 2 \quad 4$$

$$y: \quad 5 \quad 8 \quad 3 \quad 11 \quad 2 \quad 2 \quad 4 \quad 5$$

   (a) Find the estimated linear regression of Y on X.
   (b) If another firm is chosen randomly and $X = 4$, use the regression to predict the value of Y.
   (c) If another firm is chosen randomly and $X = 10$, can you use the regression to predict the value of Y? Discuss.

39. In Exercise 26, find 90 percent confidence bands for the predicted value of Y from the theoretical regression line and for the actual value of Y.

40. Why are confidence bands such as those illustrated in Figure 10.10.2 narrower for values of the independent variable that are near the "middle" of the range of observed values of that variable than they are for values that are more "extreme"?

41. In Exercise 38, test the hypothesis that $\beta_2 = 0$ versus the alternative that $\beta_2 \neq 0$, with $\alpha = .10$. What is the p-value?

42. In Exercise 37, find 99 percent confidence limits for β_1 and for β_2.

43. In Exercise 35, suppose that the prior distribution of β_1, β_2, and $\sigma_{Y \cdot X}$ is of the form of Equation (10.9.5). Find 90 percent credible intervals for β_1 and β_2 from the posterior distribution.

44. Discuss how the Bayesian approach to statistical inference can be used in regression problems.

45. If $U = cX + g$ and $W = dY + h$, and you know that the estimated regression line of Y on X is of the form $\hat{y} = b_1 + b_2 x$, what can you say about the estimated regression line of W on U? How might this fact be used in calculating regression coefficients?

46. A spark plug manufacturer conducts a study in which the gas mileage of 100 cars is found for a 50-mile drive around a test track at a fixed speed of 40 miles per hour. The cars are all of the same model, but they differ in terms of the number of miles driven since the last tune-up. Let Y denote the gas mileage

(in miles per gallon) and let X denote the number of miles driven since the last tune-up. From the sample of 100 cars, the estimated regression line is $\hat{y} = 19.6 - .0005x$. The sample correlation coefficient is $r_{XY} = -.80$, and the sample standard deviation of gas mileage is $s_Y = 3.1$. The cars in the sample had been driven from 1000 to 24,000 miles since the last tune-up, and $m_X = 9000$.

(a) Give an interpretation of the coefficients, 19.6 and $-.0005$, of the estimated linear regression.

(b) If a car has been driven 9000 miles since its last tune-up, find the predicted gas mileage from the estimated linear regression and find the variance of this predicted value.

(c) A car that had been driven 4000 miles since its last tune-up obtained 19.8 miles per gallon for the test run. Should this be considered as unusual gas mileage? Briefly explain your answer.

47. If a statistician is attempting to estimate the regression line of Y on X, does it make any difference whether X is actually a random variable or whether the values taken on by this independent variable are fixed by the statistician? Carefully explain your answer.

48. For the data in Exercise 10, use matrix algebra to find the regression line of Y on X and the correlation coefficient.

49. Suppose that you want to use your knowledge about two variables, X_2 and X_3, to predict a third variable, Y. The data are as follows:

y:	3	2	4	1	3	5	2
x_2:	6	8	5	12	7	2	9
x_3:	12	10	16	7	9	20	12

(a) Put the data in matrix form for the multiple linear regression model.
(b) Find $\mathbf{X'X}$ and $(\mathbf{X'X})^{-1}$.
(c) Find $\mathbf{X'y}$.
(d) Find $\mathbf{b}$, the least-squares estimate of the vector of regression coefficients, and write out the estimated regression surface.
(e) Find the multiple correlation coefficient.
(f) What proportion of the variance of Y is accounted for by the multiple linear regression?

50. Consider the data in Exercise 49.
(a) Find r_{12}, r_{13}, and r_{23}.
(b) If X_2 alone is used as an independent variable to predict Y, what proportion of the variance of Y will be accounted for by this simple linear regression?
(c) If X_3 alone is used as an independent variable to predict Y, what proportion of the variance of Y will be accounted for by *this* simple linear regression?
(d) Do your answers to parts (b) and (c) of this exercise, when added together, equal the answer to part (f) in Exercise 49? Why or why not?

51. Consider the data in Exercise 49.
    (a) Find the partial correlation coefficients $r_{12\cdot3}$ and $r_{13\cdot2}$. Compare these partial correlation coefficients with r_{12} and r_{13} and discuss the results.
    (b) If X_2 alone is used as an independent variable to predict Y, a certain proportion of the variance of Y is left unaccounted for by the linear regression of Y on X_2. Of this "unaccounted-for" variance, what proportion is explained by introducing X_3 as a second independent variable?
    (c) If X_3 is the first independent variable considered and X_2 is then added to the regression equation, how much of the unexplained variance (that is, the variance not explained by the linear regression of Y on X_3) is still left unexplained after X_2 is included in the regression equation?

52. For the data in Exercise 49, test the hypothesis that
    (a) $\beta_2 = \beta_3 = 0$,
    (b) $\beta_1 = 0$,
    (c) $\beta_2 = 0$,
    (d) $\beta_3 = 0$.

53. In comparing the two regression equations

    $$y = a + bx$$

    and $$y = a' + b'x + c'w,$$

    the coefficients b and b' are interpreted differently. Explain this difference.

54. A statistician is interested in using the price of a product and the amount of advertising for the product to predict the sales of the product. Several combinations of price and advertising are tried, with the following results (sales are given in ten-thousands, advertising in thousands of dollars):

Sales (y):	12	8	9	14	6	11	10	8
Price (x_2):	4	4	5	5	6	6	7	7
Advertising (x_3):	3	0	5	7	3	8	6	8

    Determine the estimated multiple regression line and the value of r^2.

55. Consider the data in Exercise 54.
    (a) Find the regression line of Y on X_2; what proportion of $V(Y)$ is accounted for by this line?
    (b) Find the regression line of Y on X_3; what proportion of $V(Y)$ is accounted for by this line?
    (c) What proportion of $V(Y)$ is accounted for by the regression of Y on both X_2 and X_3?

56. For the data in Exercise 54,
    (a) find r_{12}, r_{13}, and r_{23},
    (b) find the partial correlations $r_{12\cdot3}$ and $r_{13\cdot2}$,
    (c) explain the difference between r_{12} and $r_{12\cdot3}$ and between r_{13} and $r_{13\cdot2}$.

57. In Exercise 54, find 90 percent confidence intervals for the regression coefficients β_1, β_2, and β_3.

58. In Exercise 38, W, total assets, is found for each of the 8 firms in the sample. The values (in millions of dollars) are, respectively, 140, 160, 100, 180, 140, 60, 120, and 150.
    (a) Find the estimated linear regression of Y on X and W.
    (b) Find the proportion of the variance of Y that is accounted for by this linear regression.
    (c) Find 80 percent confidence intervals for the regression coefficients.
    (d) Another firm chosen randomly spent \$4 million on R&D 5 years ago and currently has total assets of \$170 million. Use the estimated linear regression obtained in (a) to predict the firm's current R&D expenditures, and estimate the variance of Y given these values of the independent variables.

59. Clearly differentiate between simple correlation, partial correlation, and multiple correlation, indicating when each type of correlation is useful.

60. What is stepwise regression? Discuss some possible advantages and disadvantages of the use of a stepwise procedure in regression problems.

61. Explain the interpretation of a partial correlation coefficient in terms of a reduction in the variance of Y in a multiple regression problem.

62. In Exercise 54, a third independent variable, the size of the package in ounces, is considered. This variable is denoted by X_4, and the values (in the same order as the values of Y, X_2, and X_3 in Exercise 54) are 12, 3, 6, 12, 3, 9, 9, and 6, respectively.
    (a) Find the estimated multiple regression line of Y on X_2, X_3, and X_4.
    (b) Find the value of r^2 for the regression line in (a).

63. In Exercise 62, find the correlation coefficients r_{12}, r_{13}, and r_{14}. Which of the three independent variables accounts for the greatest proportion of the variance of Y in a simple bivariate regression line?

64. In Exercise 63, let the variable with the highest squared correlation with Y be the first variable to enter the regression equation. Then compute the partial correlation coefficients of the other two independent variables with Y, holding constant the variable already in the equation. Which of the remaining two independent variables accounts for the greater proportion of the unexplained variance (that is, the variance of Y not accounted for by the regression of Y on the first independent variable in the equation)?

65. Briefly discuss three methods of approaching regression problems for which the relationship between the two variables is clearly nonlinear.

66. Discuss the technique of fitting a curve by the least-squares method and demonstrate the use of least squares to fit an exponential function

$$y = \alpha e^{\beta x}$$

on the basis of a sample of size n from the joint population of X and Y.

67. Consider a nonlinear regression curve of the form

$$y = \alpha x^\beta.$$

Given the following data, estimate the regression curve through the use of a logarithmic transformation (estimate the resulting regression line, then transform back to the original model):

$$x: \quad 1 \quad 2 \quad 3 \quad 4 \quad 5$$

$$y: \quad 12 \quad 38 \quad 80 \quad 170 \quad 252$$

68. From the equations given in Section 10.17, use least squares to find the estimated regression curve

$$\hat{y} = b_0 + b_1 x + b_2 x^2$$

from the following data:

$$x: \quad 8 \quad 10 \quad 6 \quad 4 \quad 8 \quad 9 \quad 7 \quad 9 \quad 2$$

$$y: \quad 5 \quad 14 \quad 4 \quad 12 \quad 6 \quad 8 \quad 4 \quad 9 \quad 26$$

Construct a scatter diagram showing the data and the estimated regression curve.

69. How might certain types of graph paper, such as graph paper with logarithmic axes, be useful in the investigation of suitable models for a particular set of data?

70. In the machine-output example in Section 10.17, suppose that each of the settings is tried for 5 separate 15-minute periods, with the following results (the outputs are given in the body of the table):

					SETTING					
-5	-4	-3	-2	-1	0	1	2	3	4	5
2	7	9	12	13	14	14	13	10	8	4
1	3	15	10	15	18	16	15	11	7	2
4	6	11	15	14	20	13	12	13	8	5
2	9	13	13	15	15	17	14	8	4	4
6	5	10	12	17	16	16	14	12	9	7

Using least squares, fit a polynomial function of the form used in the machine-output example in Section 10.17 to these observations. On a graph, plot the observations and the estimated regression curve.

71. For the data in Exercise 70, try to fit a piecewise linear function to the data. How might you compare the fit of this function with the fit of the polynomial function used in Exercise 70?

72. For the data in Exercise 70, try to fit a linear function to the data. Do you think that this provides a good fit? What proportion of the variance of output is explained by the estimated linear regression of output on the setting of the machine?

73. For a given model, such as a linear model, we can find a single equation that in some sense (for example, in a least-squares sense) is the "best fit" of all of the possible equations satisfying the linear model. Discuss the problem of comparing different models. For example, a statistician might attempt to fit a linear model, a polynomial model, and yet other models to some data; and a procedure for comparing the "fit" of these different models is needed.

74. In fitting a curve to some data, a statistician might try to minimize the sum of the *absolute values* of the errors instead of the sum of the *squares* of the errors. Does this seem like an intuitively reasonable criterion for curve-fitting? Does it have any disadvantages?

75. Suppose that you wish to investigate the relationship between the price *changes* of two stocks. You are given the following data:

Day	Change in price of stock A	Change in price of stock B
1	+3	No change
2	+2	+1
3	−1	−3
4	+2	+1
5	−1	+2

(a) Construct a scatter diagram.
(b) Find the correlation between the price changes of the two stocks.
(c) If you wanted to use the change in price of stock *A* to predict the change in price of stock *B*, what would you do?
(d) Briefly discuss how you might interpret your results.

76. Consider the series of prices of a stock given in Exercise 36.
(a) Find the amount of autocorrelation, or serial correlation, in the series.
(b) Determine a regression line (autoregression), using the price from the previous period as the independent variable.
(c) Determine a multiple linear regression, using the prices from the two previous periods as the two independent variables.
(d) In parts (b) and (c), use the estimated regression to predict the next value in the series (that is, the value following the last value given, 84).
(e) Can you make inferences about the regression coefficients in parts (c) and (d)? Discuss.

77. The prices of 2 stocks on 10 consecutive days are as follows:

X (*stock A*):   60   62   63   62   63   63   63   65   64   64

Y (*stock B*):   22   25   25   24   26   25   26   27   27   27

(a) Find the correlation of X and Y.
(b) Find the amount of autocorrelation in the series for stock A.
(c) Find the amount of autocorrelation in the series for stock B.
(d) Find the correlation between the price *changes* for the two stocks.
(e) Compare your answers to (a) and (d) and discuss briefly.

78. What is "spurious correlation" and how can it be reduced or eliminated?

79. Give a few examples of regression problems from your field of interest in which the assumptions of homoscedasticity and independence are likely to be violated.

SAMPLING THEORY,
EXPERIMENTAL DESIGN,
AND ANALYSIS
OF VARIANCE

As we suggested in Chapter 5, sampling methods other than simple random sampling may be quite valuable, particularly in areas such as survey sampling. We will discuss some of these alternative sampling methods and note their advantages in certain situations. Such sampling plans, which involve taking a sample from a population over which the statistician has no control, can be contrasted with plans used in experiments in which the statistician does exercise control over one or more of the relevant variables. The study of these latter plans is called **experimental design,** and a widely used set of procedures for analyzing the results of an experiment is called the **analysis of variance.** After discussing **sampling theory** and briefly noting its relationship to experimental design, we will present some basic rudiments of the analysis of variance.

11.1 SAMPLING FROM FINITE POPULATIONS

In the preceding chapters we have often assumed that any sample information has been the result of random sampling from some infinitely large population or process. If we are sampling from a population which obviously contains only a *finite* number of items, we can still act as though the population were infinite provided that we sample *with replacement.* However, this introduces the possibility that one or more items from the population could appear more than once in the sample. In many applications, this would appear to be a wasteful procedure, since sampling is not without some cost and it does not seem reasonable to sample the same item a second time, thus duplicating previous information. In a survey to determine voter preferences in a community, we presumably would not want any single voter to be represented twice or more in the sample, since that voter will only be able to vote once on election day. A sociologist conducting a survey regarding the institution of marriage would not want to conduct a long personal interview twice with the same couple; a production manager would not test the same item twice to see if it was defective. In short, there are numerous situations in which the statistician wants to sample *without replacement* from a finite population. We have developed one distribution, the hypergeometric, which deals with sampling without replacement from a finite population. This distribution, which was discussed in Section 4.9, involves a population in which each element can be classified into one of a finite number of sets. In this chapter we will discuss sampling from a finite population in which the variable of interest is a quantitative variable measured at least at the ordinal level, and preferably at the interval or ratio level of measurement (Section 5.3). This contrasts with the hypergeometric distribution, in which the measurement is only at the nominal, or categorical, level.

Because our discussion of sampling from infinite populations concentrated on simple random sampling, and because it is the easiest to comprehend of the various sampling plans, we will use it to begin our discussion of sampling plans for finite populations. Recall the definition of simple random sampling presented in Section 5.1.

A method of drawing samples such that each and every distinct sample of the same size n has exactly the same probability of being selected is called simple random sampling.

This definition holds whether we are sampling with replacement or without replacement. Either way, a random number table can be used to determine which items are to be included in the sample. In sampling *without* replacement, if the numbers from the table indicate that a particular item is to be included in the sample more than once, we will have to throw out the numbers associated with the second (and third, and so on, if necessary) appearance of the item, replacing them with additional numbers drawn from the table.

The first question of interest concerning any sampling plan is whether or not it is a "probabilistic" sampling plan. A sampling plan is called **probabilistic** if it is possible to determine the probability of any particular sample outcome; otherwise, it is a nonprobabilistic plan. Simple random sampling, for example, is probabilistic; if a random sample (without replacement) of n items is taken from a population of w items, then the probability of any particular sample is just $1/\binom{w}{n}$, the reciprocal of the number of ways of choosing n items from w items. As we shall see, simple random sampling is not the only probabilistic sampling plan; it is not necessary for all samples to have the *same* probabilities, just as long as their probabilities can be determined.

It should be obvious that in statistical work we would like to deal with probabilistic samples. In making statistical inferences about a population on the basis of a sample, it is necessary to work with the probabilities of the various outcomes of the sample. Sample information enters the inferential and decision-making procedures in the form of a sampling distribution or likelihood function; these are probabilistic statements. Without such statements, it is dangerous to generalize from the sample to the population. The reader might wonder why nonprobabilistic sampling plans are ever used if this is the case. The answer is that they usually cost less than probabilistic plans, both in terms of money and time, and such costs may be important factors. Because we are interested in making inferences from the sample to the population, we will be primarily concerned with probabilistic sampling plans.

One other caution regarding sampling; it is important to carefully define the population of interest and to insure that the sample is taken from *that population.* A famous blunder in survey work was the *Literary Digest* poll of 1936, which erred by 19 percent in predicting the percentage of votes that Franklin D. Roosevelt would get in the presidential election of that year. The poll was conducted by mailing out postcards, which were to be sent back marked with the individual's voting preferences. The difficulty was partly due to the fact that the mailing lists were made up from readily available lists such as lists of telephone subscribers, automobile owners, and so on. In such lists, those individuals with high incomes were overrepresented in relation to their representation in the overall voting population. As a result, since this misrepresentation was not allowed for in reporting the results of the poll, and since the high-income voters had more of a tendency to vote against Roosevelt than did the low-income voters, the poll seriously underestimated Roosevelt's strength, as the election results subsequently indicated. This illustrates the importance of carefully defining the population of interest and sampling from that population rather than from other, perhaps related, populations.

Once the population of interest is defined and a sample is taken from that population, we can proceed to make inferences on the basis of the sample. For example, suppose that we take a simple random sample of size n without replacement from a finite population of size w. Remember that n is the **sample size** and w is the **population size.** We observe the value x of a variable X for each member of the sample (for example, x might represent the height of a person, the diameter of a part, or the price of a security), and we wish to estimate the *population* mean μ. If the w items in the population are $X_1, X_2, \ldots, X_w$, then the population mean is simply

$$\mu = \frac{\sum_{i=1}^{w} X_i}{w}. \tag{11.1.1}$$

We do not know the values of all of the items in the population, however, so we cannot calculate μ. We *do* know the values of n items which have been randomly selected from the population, and these n items can be used to make inferences about the entire population. Denote the n items in the sample by $x_1, x_2, \ldots, x_n$. The usual estimator of the population mean μ is the sample mean M, just as in the case of sampling from an infinite population. It can be shown that M is an unbiased estimator of μ.

We are also interested in the variance of the sample mean. In sampling from an infinite population, the variance of the sample mean is σ^2/n, where σ^2 is the variance of the infinite population. For a finite population,

$$\sigma^2 = \frac{\sum_{i=1}^{w} (X_i - \mu)^2}{w}, \tag{11.1.2}$$

and it can be shown that the variance of the sample mean is equal to

$$V(M) = \left(\frac{w-n}{w-1}\right)\left(\frac{\sigma^2}{n}\right). \tag{11.1.3*}$$

Thus, the difference between the finite and infinite population situations is the presence of the factor

$$\frac{w-n}{w-1}$$

in Equation (11.1.3). This term is often referred to as the **finite population correction,** as we noted when we encountered it in relation to the hypergeometric distribution in Section 4.9. Sometimes it is written in the form $(1 - f)$, where $f = (n - 1)/(w - 1)$ is approximately the proportion of the finite population which is being sampled. From Equation (11.1.3), then, the effect on $V(M)$ of the finiteness of the population is to reduce it by a factor of $(1 - f)$. If this factor is not included, and the variance of M is written simply as σ^2/n, the true variance is being overstated. The difference is small when f is small, though, and this occurs when only a small proportion of the population is being sampled. This is why the statistician may act as if the population were infinite, even though it is really finite, provided that the sample size n is small in relation to the population size w. One rule of thumb is to ignore the finite population correction if less than 5 percent of the population is being sampled.

To calculate $V(M)$, it is necessary to know the population variance σ^2. But this will generally not be available to the statistician, so it is necessary to estimate it. Just as in sampling from an infinite population, $\hat{s}^2$ is an unbiased estimate of the population variance σ^2. It then follows that

$$\hat{s}_M{}^2 = \left(\frac{w - n}{w - 1}\right)\left(\frac{\hat{s}^2}{n}\right) \qquad (11.1.4)$$

is an unbiased estimate of the variance of the sample mean.

At this point the reader may wonder why simple random sampling is not always used; why do we consider other sampling plans? The relevant factors in determining a sampling plan are essentially (1) cost and (2) precision. We want the sample to cost as little as possible, and yet we want as much precision as possible in our estimates. Ideally, we would like to be able to sample the entire population and obtain perfect estimates, estimates with no sampling variance at all. If we sample the entire population, then the sample mean is obviously equal to the population mean, and the variance of the sample mean is zero. Unfortunately, it is seldom possible to sample the entire population because it is much too costly, both in terms of time and money. Unless it is very important that the statistician have extremely precise estimates, or unless the costs of sampling are very low, only a portion of the population will be sampled. In determining the sampling plan (which includes the choice of sample size as well as type of sampling plan), the statistician must balance off cost and precision. To obtain greater precision will in almost all cases require greater cost; conversely, to reduce costs will necessitate some sacrifice of precision. In some situations, it may be possible to find a plan other than simple random sampling which gives the same precision at less cost or the same cost and greater precision. Simple random sampling is not an easy procedure to implement, particularly when the population size is large. It is difficult to select a sample at random from a long list of items. In terms of survey sampling, a random sample of families will probably result in high interviewing costs because of the cost of going from one house to another house in a different neighborhood, and so on. In the next two sections we will discuss two alternatives to simple random sampling, stratified sampling and cluster sampling, and we will attempt to indicate the conditions under which each of these sampling plans is advantageous.

11.2 STRATIFIED SAMPLING

In **stratified sampling**, the population is divided into a number of sub-populations, or strata, and a sample is taken from each of the subpopulations. If simple random sampling is used within each of the strata, then the term **stratified random sampling** is used to refer to the overall sampling plan. Unless otherwise specified, this is what we refer to as stratified sampling. For example, we may divide a population into age groups and then take a random sample within each age group. Or, instead of stratifying by age, we may stratify by income, or sex, or occupation, or almost anything. There are several possible advantages to stratification. First, in addition to estimates for the entire population, stratified sampling provides us with estimates for each of the subpopulations. It is possible, for instance, to estimate the proportion of consumers in each age group who will buy a particular new product and then to combine these estimates to find an estimate for the entire population of consumers of all ages. Second, it may be more convenient from an administrative standpoint to stratify. For instance, it may be easier to take separate random samples from a number of neighborhoods than to take a single random sample from an entire city. Third, stratification may be used to guarantee a "representative" sample. If simple random sampling is used in a survey, it would be perfectly possible (although perhaps unlikely) to end up with a sample which is not "representative" of the population as a whole. Stratification reduces the chances of a nonrepresentative sample by assuring that a specified number of individuals will be chosen from each stratum. Finally, stratification may result in increased precision for the estimates of interest, as compared with simple random sampling. After we introduce the notation and the estimates used in stratified sampling, we will discuss the conditions under which it is likely to lead to increased precision.

Once again, w is the population size and n is the sample size. The population is divided into L strata of size $w_1, w_2, \ldots,$ and w_L, where

$$w = w_1 + w_2 + \cdots + w_L = \sum_{h=1}^{L} w_h.$$

A sample of size n_h is taken from the hth stratum, and

$$n = n_1 + n_2 + \cdots + n_L = \sum_{h=1}^{L} n_h.$$

Now, within the hth stratum, where $h = 1, 2, \ldots, L$, the population mean of the stratum is μ_h and the n_h sample values which are observed are $x_{h1}, x_{h2}, \ldots, x_{hn_h}$. Since each stratum is a finite population and we are using simple random sampling within each stratum, the results of the preceding section are applicable. The estimate of μ_h is simply the sample mean within that stratum,

$$m_h = \frac{\sum_{i=1}^{n_h} x_{hi}}{n_h}. \tag{11.2.1}$$

The variance of the sample mean is given [from Equation (11.1.3)] by

$$V(M_h) = \left(\frac{w_h - n_h}{w_h - 1}\right)\left(\frac{\sigma_h^2}{n_h}\right), \tag{11.2.2}$$

where σ_h^2 is the population variance of the hth stratum. The term

$$\frac{w_h - n_h}{w_h - 1}$$

is the finite population correction for the hth stratum. If σ_h^2 is not known, then it can be estimated by the unbiased estimate of the variance within the hth stratum, $\hat{s}_h^2$.

Since the L strata are finite populations and simple random sampling is used, the statistician can make inferences about each stratum and can also combine the results of the L samples to make inferences about the overall population. To determine the population mean μ, we take the population total and divide it by w (since the population total is simply the population mean multiplied by the population size). But the population total is the sum of the totals for the L strata. The total within the hth stratum is simply the stratum mean μ_h multiplied by the size of the stratum, w_h, and thus we have

$$\mu = \frac{\sum_{h=1}^{L} w_h \mu_h}{w}. \tag{11.2.3}$$

To estimate μ from the sample results, we notice that w_h and w are given, so that we simply need to estimate each μ_h with the corresponding sample mean from that stratum [Equation (11.2.1)]:

$$\text{est. } \mu = \frac{\sum_{h=1}^{L} w_h m_h}{w}. \tag{11.2.4*}$$

Since each m_h is an unbiased estimate of μ_h, est. μ is obviously an unbiased estimate of μ. Incidentally, notice that est. μ as given by Equation

(11.2.4) is *not* the same as the overall sample mean m, which is equal to

$$m = \frac{\sum_{h=1}^{L} n_h m_h}{n}. \tag{11.2.5}$$

In determining est. μ, the estimated stratum means receive their correct weights according to their representation in the *population*, not in the *sample*. In determining m, they are weighted according to their representation in the overall sample, and m and est. μ will not necessarily be equal unless

$$\frac{n_h}{n} = \frac{w_h}{w} \quad \text{for all } h = 1, 2, \ldots, L,$$

in which case

$$\frac{n_h}{w_h} = \frac{n}{w} \quad \text{for all } h, \tag{11.2.6}$$

implying that the same proportion is sampled from each stratum. This type of stratification is called stratification with **proportional allocation** of the sample size. Of course, it is not at all necessary to allocate the sample size in a proportional manner, so in general m is not equal to est. μ.

To determine the variance of the estimate given by Equation (11.2.4), we apply the rules for determining variances:

$$V(\text{est. } \mu) = V\left(\frac{\sum_{h=1}^{L} w_h M_h}{w}\right) = V\left(\sum_{h=1}^{L} \frac{w_h}{w} M_h\right).$$

The L samples (from the L strata) are independent, so we have

$$V(\text{est. } \mu) = \sum_{h=1}^{L} V\left[\left(\frac{w_h}{w}\right) M_h\right] = \sum_{h=1}^{L} \left(\frac{w_h}{w}\right)^2 V(M_h).$$

Using Equation (11.2.2), we get

$$V(\text{est. } \mu) = \sum_{h=1}^{L} \left(\frac{w_h}{w}\right)^2 \left(\frac{w_h - n_h}{w_h - 1}\right) \left(\frac{\sigma_h^2}{n_h}\right), \tag{11.2.7}$$

or

$$V(\text{est. } \mu) = \sum_{h=1}^{L} \frac{w_h^2 (w_h - n_h) \sigma_h^2}{w^2 (w_h - 1) n_h}. \tag{11.2.8}$$

Now we return to the question posed previously. When is it advantageous to use a stratified sampling plan? The answer: when it

is possible to divide the population into groups (strata) in such a way that the items within each group are very similar but the groups themselves are quite dissimilar.

We would like to select strata such that the variability in the population is primarily reflected in the variability *between strata*, so that the variability *within each stratum* is small. If we can achieve this goal, then it is possible to get precise estimates within each stratum with a small sample size n_h. Combining these precise estimates to get estimates for the entire population, we can get the same precision as we would with simple random sampling, and the total sample size will be smaller, since we only take a small sample within each stratum. Or, if the total sample size is the same, then we can get greater precision with stratified sampling. Remember, though, that this is true only to the extent that the heterogeneous population can be divided into a number of reasonably homogeneous groups.

For a concrete example, suppose that a sociologist is conducting an opinion survey of families and is considering the use of a stratified sampling plan. He must decide what to use as a stratifying factor; in other words, should he stratify by sex, by age, by race, by income, or by something else (or even by a combination of factors)? He feels that the respondent's sex should have little effect on the results; thus, if he stratified by sex, there would not be much variability between strata. Similarly, he thinks that race should have little effect on the results. Sex and race, then, do not appear to be good stratifying factors in this example. The sociologist *does* think that both age and income should have an effect on the responses to the survey. As a result, he decides to divide age into three classes (below 25, 25–50, and above 50) and income into two classes (below \$15,000 and above \$15,000). Taking each possible combination of age class and income class (for example, age below 25 and income below \$15,000; age 25–50 and income below \$15,000; and so on), there are six strata. The sociologist thinks that each stratum is relatively homogeneous (relative to the entire population) with respect to the topics covered by the opinion survey, and therefore he is of the opinion that he can obtain his desired precision with less cost under a stratified sampling plan than under simple random sampling.

Earlier in this section we hinted at the problem of the allocation of the total sample size among the L strata. One possibility is proportional allocation, where n_h is proportional to w_h. If the cost per unit of sampling in stratum h is constant and equal to c_h, then it can be shown that the optimal allocation is to take n_h proportional to

$$\frac{w_h \sigma_h}{\sqrt{c_h}},$$

where "optimal" means maximizing precision for a given cost or minimizing cost for a given precision. Note that if the costs c_h and the variances $\sigma_h{}^2$ are the same for all strata, then n_h will be proportional to w_h, and the resulting allocation of the total sample size among the L strata is what we have called proportional allocation. In general, of course, this will not be the case. The optimal allocation rule tells us to take a larger sample in a particular stratum if (1) the stratum size is larger, (2) the stratum variance is larger, or (3) the cost in the stratum is smaller. These three rules seem intuitively reasonable. If w_h or σ_h is larger, we need a larger sample to attain a given precision within the stratum, as we can see from Equation (11.2.2). If c_h is smaller, we can take a larger sample (and hence get greater precision) for the same cost.

Finally, it should be pointed out that it is possible that the sampling within each stratum is not simple random sampling, in which case we do not have stratified *random* sampling, but some other form of stratified sampling. In fact, a popular nonprobabilistic sampling plan, called **quota sampling,** is a form of stratified sampling. In quota sampling in the context of interviewing, interviewers are sent out and told to obtain a specified number of interviews from each stratum (the strata may be neighborhoods, blocks, income levels, age levels, sex, and so on). They are not given a specific list of persons, or families, to interview, which would be the case with stratified random sampling. Instead, they are free to use their own judgment in selecting persons to interview, and as a result the members of the sample are not chosen in a probabilistic fashion. Instead, they are usually chosen for the convenience of the interviewer, and this may lead to biased estimates of the population values. Some recent studies have indicated, however, that the amount of bias is often quite small and that quota sampling may sometimes be justified because it is much cheaper than probabilistic sampling, particularly in studies requiring personal interviews.

11.3   CLUSTER SAMPLING

In **cluster sampling,** as in stratified sampling, the population is divided into a number of subpopulations, which are called clusters in this case. Unlike stratified sampling, however, samples are not taken from *each* of the subpopulations. Instead, one or more of the clusters are randomly chosen from the entire set of clusters, and sampling is carried out only in these clusters. In the extreme example of cluster sampling, only one cluster is selected, and the entire cluster is sampled. We will not present the estimation formulas for cluster sampling because they depend to a great extent on such things as whether or not the clusters are of equal size, how many clusters are chosen, and whether the entire cluster or just a portion of the cluster is sampled. Instead, we will attempt to explain heuristically the conditions under which cluster sampling is a useful sampling plan.

The main advantage of cluster sampling is a cost advantage. With little loss in precision, it may be possible to reduce costs greatly with a cluster sample. Consider the following simple example. A firm is interested in marketing a new product, but the potential consumer acceptance of the product is quite uncertain. As a result, it is decided to conduct a market test, in which the product will be sold in a sample of stores. Simple random sampling would be prohibitively expensive, for this would mean marketing the product at a random selection of stores across the country. If it is felt that this product is not one for which regional preferences would exist (that is, the item should be equally popular in all areas of the country), cluster sampling would be almost as precise as simple random sampling, and it would be much cheaper. One or more cities could be randomly selected from all cities in the country, and the new product could be test marketed in all of the stores in the chosen city or cities. This would lead to great savings in, for example, distributional costs.

Notice that the key feature of the above example is the assumption that there are no regional preferences for the product. Perhaps it would also be wise to investigate the possibility of urban versus rural preferences, high-income versus low-income preferences, and so on. If just one city is selected, then it is hoped that, with regard to consumer preferences for the new product, the city could be considered as a microcosm of the entire country. What does this mean in statistical terms? The population (of consumers in the country) will have some variability in its feelings toward the new product. Some persons will like it, others will hate it, and yet others will be indifferent. *The population should be divided into clusters* (cities) *in such a manner that there is little or no variability between clusters* (no regional preferences). Then the variability in the population will be reflected in the variability *within* each cluster.

Recapitulating, we would like the clusters to be as similar as possible, with each one being a miniature version of the entire population.

This is just the opposite of what we wanted in stratified sampling; we wanted the strata to be as *different* as possible, with each one being internally homogeneous.

A type of sampling plan which is related to cluster sampling is **systematic sampling.** In systematic sampling, it is assumed that a list of the members of the population is available. Suppose that $w = 100$ and $n = 20$. Then a systematic sample can be drawn as follows. Randomly choose a number between 1 and 5, then take that number and every fifth number thereafter. If the number chosen is 2, then the 2nd, 7th, 12th, 17th, and so on, items from the list will be included in the sample. This is a quick way to take a sample from a long list. It is related to cluster sampling because the population is essentially divided into 5 clusters as follows:

$$Cluster\ 1: \quad \{1, \quad 6, 11, 16, \ldots, \quad 96\}$$
$$Cluster\ 2: \quad \{2, \quad 7, 12, 17, \ldots, \quad 97\}$$
$$Cluster\ 3: \quad \{3, \quad 8, 13, 18, \ldots, \quad 98\}$$
$$Cluster\ 4: \quad \{4, \quad 9, 14, 19, \ldots, \quad 99\}$$
$$Cluster\ 5: \quad \{5, 10, 15, 20, \ldots, 100\}.$$

One cluster is then randomly selected, and the entire cluster is sampled.

The relative advantages of the three basic sampling plans discussed here should be clear by now. If it is possible to divide the population into a number of homogeneous groups, then stratified sampling may provide the greatest precision for a given cost or the least cost for a given precision. This is because the homogeneity of the groups means that precise estimates can be made with a small sample within each group. Stratified sampling is also quite useful if we are interested in separate estimates for each of the groups as well as estimates for the entire population. Cluster sampling is useful if (1) we can divide the population into a number of similar groups, each of which is heterogeneous and is similar to the whole population, and (2) taking large samples in one or two of these groups is much cheaper than taking a random sample from the entire population. Finally, simple random sampling should not be discarded, for there are many situations in which the requirements for stratified or cluster sampling cannot be met. If the type of survey is such that simple random sampling is not much more expensive than the other plans (for example, telephone interviews rather than personal interviews), or if not enough is known about the population to allow the statistician to divide it successfully into strata or clusters, then simple random sampling may be the best plan. As we shall see in the next section, many applications of sampling theory are complex enough to require combinations of the various sampling plans.

11.4 OTHER TOPICS IN SAMPLING THEORY

Entire books have been written on sampling theory, and obviously we do not have the space to do justice to the topic. The sampling plans discussed in Sections 11.1–11.3 are important plans (although other plans exist), but we have discussed them in a limited manner. Often estimates other than estimates of the population mean are desired. The population *total* is sometimes of interest, and estimates of this can be derived directly from estimates of the mean, since it is equal to $w\mu$. It is also possible to develop interval estimates as well as point estimates for μ, $w\mu$, and other population parameters.

Most sampling plans are actually combinations of the simple plans discussed in the previous sections. In a survey, for example, sampling may take place on several levels, and a different plan may be needed for each level. In conducting a nationwide opinion poll, we might first divide the country into several regions (strata), randomly select one subregion (cluster) within each region, randomly select a number of cities or towns within each subregion (simple random sampling), divide each town into neighborhoods (strata), and within each neighborhood randomly select a few blocks (clusters), interviewing everyone living on the chosen blocks. This is a **multistage sampling plan,** with five stages of sampling. At each stage of sampling, we must make a decision as to which sampling plan to use. In doing so, we must keep in mind our overall objective, for the choice of a plan at any stage affects every stage thereafter.

One area of some interest within sampling theory is the concept of ratio estimation. Suppose that we are interested in estimating μ_X, but when we sample an item, we can observe both the value of X and the value of a second variable Y for that item. Furthermore, suppose that we know the population mean of the Y values, which we label μ_Y. It would be nice to take advantage of our knowledge about Y somehow in order to estimate μ_X. We can do this by taking

$$\text{est. } \mu_X = \mu_Y \frac{m_X}{m_Y}.$$

This approach is called **ratio estimation,** since we are dealing with the ratio of X to Y. In particular, we equate the ratio of population means and the ratio of sample means in order to determine an estimate of μ_X. Although ratio estimates are, in general, biased, they are often preferred to the corresponding unbiased estimates (in this case, m_X) because of their small variance. This small variance is made possible by the inclusion of additional information, the information concerning Y. To use the above formula, of course, it is necessary that μ_Y be known. To give a simple example, suppose that we have information regarding last year's income for everyone living in a certain town. We wish to estimate the mean *current* income in the town, μ_X. Suppose that last year's mean income, μ_Y, was $9000. A random sample of 10 persons is taken, their incomes are ascertained, and the sample mean is $10,200. Looking up the incomes in the previous year of these 10 persons, we find that $m_Y = 9800$. The ratio estimate of μ_X is

$$\text{est. } \mu_X = \mu_Y \frac{m_X}{m_Y} = (9000) \frac{10,200}{9800} = \$9367.$$

The unbiased estimate, m_X, is \$10,200. But the average income of the 10 persons chosen was not equal to the population mean last year, so there is no reason to believe that it will be this year; indeed, there is every reason to believe that it will *not* be. The ratio estimate allows for this by assuming that the ratio of this year's to last year's income in the population is roughly the same as in the sample.

Brief mention should be made of the possibility of introducing prior information and using the Bayesian approach in sampling from finite populations. The general analysis is the same as that presented in Chapter 8; a prior distribution is assessed and combined with the sample information. Instead of inferences about some process parameter, however, we are interested in inferences about the finite population. Given a sample of size n, we attempt to determine a probability distribution for the remaining $w - n$ items in the population. Combining what we already know about the n items sampled with our distribution for the remaining $w - n$ items, we can make probability statements about the entire population. We shall not go into the details of this Bayesian procedure here, but it is worth noting that if the prior distribution is diffuse, or "informationless," then some Bayesian results are virtually identical numerically to the classical results (for example, the estimates discussed in the previous few sections are virtually identical to the corresponding posterior means following diffuse prior distributions), although the interpretation is not the same, of course.

One problem faced in sampling is that of *missing data*. In survey sampling this may most likely be due to *nonresponse*. Perhaps an individual is not at home when the interviewer calls, refuses to consent to the interview, or does not reply to a mail survey. In each of these situations, a person has been selected by some probabilistic sampling plan, but the statistician is unable to obtain the data of interest from that person. The most convenient way to handle nonresponse would be to replace the nonrespondent with another randomly selected person. The danger here is that the nonresponse is somehow related to the variable of interest. Perhaps working individuals who are not home when the interviewer calls, tend to have different voting preferences, or brand preferences, or whatever, than those who stay home all day. Perhaps persons with high or low incomes are more likely to refuse to answer questions regarding financial matters than persons in the middle-income bracket. Thus, completely eliminating the nonrespondents might create a bias, and the results could then only be generalized to populations of persons like the respondents. One way to combat nonresponse is through the use of "call-backs," in which the interviewer makes repeated attempts to complete the interview successfully. If we can get some of the original nonrespondents to cooperate, then we can use their responses to make inferences about the responses of the remaining nonrespondents. This approach is based on the idea that the nonrespondents are likely to differ from those responding on the first attempt.

Another serious problem in surveys is that of *questionnaire design*. This refers to the wording of questions, the order of questions, and the responses available to the subject. Since this is a psychological rather than a statistical problem, we shall not concern ourselves with it. Similar difficulties involve interviewer bias, errors in recording answers, errors in tabulating data, and so on. For instance, it has been demonstrated that a person will tend to give different answers to a person of the same sex and/or race than to a person of the opposite sex or of a different race. Again, such problems belong in the realm of psychology and sociology, so we shall return to our discussion of statistical techniques.

In this section, we have attempted to discuss very briefly some topics of interest in sampling theory, topics which we cannot present at more length because of space limitations but which are important enough to deserve at least some mention. In the next section, we attempt to compare the general problem of sample design with the related, yet different, problem of experimental design.

11.5 SAMPLE DESIGN AND EXPERIMENTAL DESIGN

The discussion of sampling theory indicates that there is often more to taking a sample than just determining the sample size. First, there is the problem of carefully defining the population of interest. Once this is done, various sampling plans must be considered. If, as is often the case, a multi-stage sampling plan is needed, then the different plans, such as simple random sampling, stratified sampling, and cluster sampling, must be evaluated for each stage. The specific details of each plan, such as the number of strata and the sample size within each stratum, must be worked out. Once all this has been decided upon, the actual items to be sampled can be determined. This entire procedure falls under the general heading of "sample design." There may be other problems to iron out, particularly in survey sampling (questionnaire design, the mechanics of conducting the survey and analyzing the results, and the like), but these are not statistical problems and hence are not included in the concept of sample design (perhaps a broader term, such as survey design, could be used to incorporate these details).

The overall objective in designing a sample is to get the greatest possible precision for a given cost, or to attain some given level of precision in the least costly manner. In our discussion of stratified sampling, we saw that, in terms of this objective, there is an optimal way to allocate the total sample size n among the L strata. There is a tradeoff between precision and cost in sampling, and some sample designs are more efficient than others in terms of precision versus cost. In any given situation, we would like to determine an efficient sample design. Incidentally, how do we decide when to stop spending money to get more precision? If the sample is conducted with a fixed budget, we will sample until the budget is exhausted. If a specific level of precision is predesignated, then we will spend the amount required to attain that precision. In general, recall the concept of utility from Chapter 9; in terms of utility, we should sample until the increase in expected utility due to further increments in precision is equal to the decrease in expected utility due to the cost associated with these increments.

In sampling theory, it is presumed that the sample consists of observations on an *existing* population; that is, some members of the population are observed and the desired information about these members is recorded. This contrasts with the idea of a **controlled experiment,** which is widely encountered in scientific research. In a controlled experiment, the statistician attempts to control, or to modify, certain factors and observes the effect of these modifications on one or more variables of interest. Instead of merely observing the proportion of defectives in a sample of items from a production process, the experimental statistician might try a modification of the process and observe the proportion defective both with and without the modification. This should not be done in a haphazard manner, however. The statistician should take care to insure that all factors (other than the one being varied) remain constant. In terms of the production example, the conditions under which the process is run without the modification should be as similar as possible to the conditions under which the process is run with the modification. If the modification is used during the night shift but not during the day shift, then any differences which appear may be caused by the different sets of personnel rather than by the modification. Unfortunately, it is often quite difficult, if not impossible, to hold such factors constant in actual problems. This is particularly true in experimentation in the social sciences, the behavioral sciences, and business. In these areas, experiments usually deal in some way or another with people, and as a result there may be numerous factors, psychological, social, environmental, and otherwise, which cannot be completely controlled. The experimeter should try very hard, however, to reduce the effects of such factors; and certain principles of experimental design are useful in this regard, as we shall see. In contrast, the experimenter in the physical sciences finds it easier to identify and to control extraneous factors.

A medical example should serve to illustrate the concept of simple experimental design. Suppose that a medical researcher is interested in the relative effectiveness of three different drugs in the treatment of a particular illness. To study the drugs, he designs an experiment which is to be performed on a large group of patients from a nearby hospital, all of whom are suffering from the disease in question. The researcher decides to divide the patients into four groups of equal size. The members of the first three groups will be given controlled amounts of drugs A, B, and C, respectively. The members of the fourth group will be given a placebo, which is similar in appearance to the three drugs but contains no medication (this makes the conditions in the four groups appear more nearly identical to the patients than if nothing was given to the fourth group). The researcher can then observe how many patients in each group are cured of the disease.

We have so far avoided the question of how the patients are to be assigned to the four groups. One appealing possibility is to assign them randomly to groups, thus invoking the principle of **randomization,** which is an important principle in experimental statistics. The use of randomization permits the researcher to make inferential statements from the sample to the "population." The quotation marks around "population" indicate that this is not the same as the finite population dealt with in sampling theory. The population corresponding to the first group would be a population of patients with the disease who are given drug A. Since there is no existing population which the researcher has in mind, the inferences refer to a hypothetical population rather than to an actual set of individuals. Randomization permits inferences to be made to such hypothetical "as if" populations.

Although randomization is a good procedure, it may in some cases lead to a poor allocation of patients to groups. It may just happen that the oldest patients are all assigned to one group and the youngest patients are assigned to another group. Randomization can be useful in letting chance take care of the unknown, but relevant, factors in the experiment, but it may be rather ineffective in coping with known factors such as "age" in this example. Instead of randomizing, then, the researcher might first divide the population into two age classes to make sure that each age class is represented equally in each of the four groups. Within each age class, patients could be randomly assigned to groups. This is a slightly more complicated design than the first one, but it also provides the experimenter with more information. The effect of age as well as the effect of the different drugs on the cure rate can be investigated. **Interactions** between age and the drugs can be considered. Suppose that for the entire population of patients in the experiment, it is ascertained that drug A has no effect on the cure rate (as compared with the control group, with the placebo) and that age has no effect on the cure rate. However, it is found that among those given drug A, the older patients suffer a great decline in the cure rate and the younger patients enjoy a large increase in the cure rate. This type of result is said to be due to an *interaction* between age and cure rate. In experiments where more than one factor is varied systematically, it is often possible to investigate interactions between factors as well as the effects of the individual factors.

The medical example may give you an idea of what experimental design is all about and how it differs from sample design. In each case, the statistician wants to make some inferences on the basis of the results, and the objective is to balance off precision and cost (experiments, like samples, are costly). The difference between the two situations is straightforward. In sample design, the statistician wants to take observations from an existing population; in experimental design, the statistician more or less creates a given population of interest by varying certain factors and by using principles such as randomization. Next, we shall investigate a body of statistical techniques called the analysis of variance, which is often used to analyze experimental data. Let it be emphasized that the analysis of variance is *not* experimental design; it is merely a commonly used technique for the analysis of data obtained via such designs. Nevertheless, the analysis of variance does serve to illustrate some of the basic features of some commonly used experimental designs.

11.6 ANALYSIS OF VARIANCE: THE ONE-FACTOR MODEL

Ordinarily, the primary objective in an experiment is to compare different, fixed, experimental treatments. For instance, in the medical example of the preceding section, the researcher is interested in differences in cure rates among the four groups receiving four particular treatments—drug A, drug B, drug C, and a placebo. Recall that in Section 7.16 we discussed a procedure for comparing *two* means; this procedure can be used when we have just an experimental group and a control group or perhaps two experimental groups. Most experiments are far more complicated than this, however, employing several groups given different treatments. We might consider looking at each *pair* of treatments separately and applying the procedure of Section 7.16. However, this would be bothersome, and probability statements about the experiment as a whole are difficult to make on the basis of a separate look at each pair of treatments. Thus, we need a simple method for the *simultaneous* comparison of several treatments in order to make inferences about the relationship between the treatments and the variable that is being measured in the experiment (cure rate in the medical example). The **analysis of variance,** which is the topic of the remainder of this chapter, is such a method. Essentially, the analysis of variance allows us to make inferences about differences in treatment effects by estimating how much of the variability in the experimental data is due to such differences and how much is simply "chance variation," or "random error."

First, we consider the simplest analysis of variance model, the **one-factor model.** In this model, only one factor is being varied in the experiment. Every other possibly relevant factor is assumed to be randomized or to be held constant. For example, in the medical example of the preceding section, the only factor being varied was the drug given to the patients. In many situations, it is desirable to attempt to isolate only one factor and to attempt to keep other factors constant.

Suppose that there are J different "treatments" for the one factor of interest (for example, 4 different drugs), so that there are J different samples, or groups, in the experiment, where the sample size in the jth group is n_j. Now we are ready to state a model for the composition of any observed score y_{ij}, corresponding to the ith observation in treatment j, or sample j. The model used is a linear model; it states that the value of any observation in any treatment group is composed of the following simple sum:

$$y_{ij} = \mu + \alpha_j + e_{ij}, \quad \text{where } i = 1, \ldots, n_j$$

$$\text{and } j = 1, \ldots, J. \quad (11.6.1^*)$$

This model asserts that the value of observation i in sample j is based on the sum of three components: the **grand mean** μ of all of the J different treatment populations, the **effect** α_j associated with the particular treatment j, and a **random–error term** e_{ij}. Notice the similarity of Equation (11.6.1) to the linear regression model studied in Chapter 10; later in this chapter we will discuss the regression model in terms of the analysis of variance.

If we think of a treatment population j as the hypothetical set of all possible unit observations that might be made under treatment j, then the experimental group to which treatment j is actually applied represents a sample from this hypothetical population. Let the mean of the potential treatment population be μ_j. Furthermore, suppose that the different treatment populations could somehow be pooled so that the representation of each treatment in the pooled population is proportional to its representation in the total sample. Then the grand mean μ of this pooled population would be

$$\mu = \frac{\sum_j n_j \mu_j}{\sum_j n_j}. \quad (11.6.2)$$

The effect of treatment j is defined as the deviation of μ_j, the

mean of population j, from μ, the grand population mean:

$$\alpha_j = (\mu_j - \mu). \tag{11.6.3*}$$

This symbol α_j (*not* to be confused with the alpha standing for the probability of a Type I error) will stand for the effect of any single treatment j.

Since the grand population mean μ is also the mean of all of the treatment population means, it follows that the weighted average of all of the effects must be zero:

$$\frac{\sum\limits_{j} n_j \alpha_j}{\sum\limits_{j} n_j} = \frac{\sum\limits_{j} n_j \mu_j}{\sum\limits_{j} n_j} - \frac{\sum\limits_{j} n_j \mu}{\sum\limits_{j} n_j} = \mu - \mu = 0. \tag{11.6.4}$$

If there is absolutely no effect associated with any treatment, then $\alpha_j = 0$ for each and every treatment population j. This is equivalent to the statement that

$$\mu_1 = \mu_2 = \cdots = \mu_J = \mu. \tag{11.6.5}$$

The complete absence of effects is equivalent to the absolute equality of all of the population means.

Any individual i observed under treatment j is a random sample of one from the corresponding treatment population j, so that the expectation of an observation in treatment j is

$$E(Y_{ij}) = \mu_j. \tag{11.6.6}$$

By the linear model,

$$E(Y_{ij}) = \mu + \alpha_j + E(e_{ij}),$$

or

$$E(Y_{ij}) = \mu_j + E(e_{ij}). \tag{11.6.7}$$

Thus, for any population j, the expectation of e_{ij} over all observations is zero:

$$E(e_{ij}) = 0. \tag{11.6.8}$$

Similarly, if we take the *sample mean* of the error terms, either over a single treatment or over the entire set of J treatments, we find that its expectation is zero. Then, since both individual errors and mean errors have expectations of zero in this model, an unbiased estimate of the effect of any treatment j can be found by using Equation (11.6.3) and estimating μ_j and μ:

$$\text{est. } \alpha_j = (m_j - m), \tag{11.6.9}$$

where

$$m_j = \frac{\sum\limits_{i} y_{ij}}{n_j} \tag{11.6.10}$$

is the sample mean for the jth treatment and

$$m = \frac{\sum_i \sum_j y_{ij}}{\sum_j n_j} \tag{11.6.11}$$

is the overall sample mean for the entire experiment (note that if the sample sizes for the individual treatments are not equal, m will not necessarily equal the average of the individual sample means).

To gain some understanding of this linear model, suppose that we are interested in comparing the fuel economy (as measured in miles per gallon) of three particular automobile brands: one from Germany, one from Japan, and one from the United States. For each brand, 3 cars are selected randomly and each is driven at a steady 50 miles per hour for 100 miles. First, if each of the 3 brands has a population mean of 20 miles per gallon and if there is *no* variability *within* any of the populations, then our sample results should look like the data in Table 11.6.1. In terms of the model, $\alpha_j = 0$ and $e_{ij} = 0$ for all i and j, so the model becomes simply $y_{ij} = \mu$.

Table 11.6.1   Data Exhibiting No Variability for Gas Mileage Example

German brand	Japanese brand	U. S. brand
20	20	20
20	20	20
20	20	20

Of course, it is quite unrealistic to have no treatment effects in this example. Thus, suppose that there are differences between the brands but that there is still no variability within any single sample. Our results might look like the data in Table 11.6.2. Here the linear model is $y_{ij} = \mu + \alpha_j$, with $\alpha_1 = 1$, $\alpha_2 = 4$, and $\alpha_3 = -5$.

Table 11.6.2   Data Exhibiting Variability Only Between Brands for Gas Mileage Example

German brand	Japanese brand	U. S. brand
$20 + 1 = 21$	$20 + 4 = 24$	$20 - 5 = 15$
$20 + 1 = 21$	$20 + 4 = 24$	$20 - 5 = 15$
$20 + 1 = 21$	$20 + 4 = 24$	$20 - 5 = 15$

Table 11.6.3 Data Exhibiting Variability Both Between and Within Brands for Gas Mileage Example

German brand	Japanese brand	U. S. brand
$20 + 1 + 3 = 24$	$20 + 4 + 0 = 24$	$20 - 5 - 4 = 11$
$20 + 1 - 2 = 19$	$20 + 4 + 1 = 25$	$20 - 5 - 1 = 14$
$20 + 1 + 1 = 22$	$20 + 4 - 3 = 21$	$20 - 5 + 2 = 17$
$m_1 = 21.67$	$m_2 = 23.33$	$m_3 = 14.00$ $m = 19.67$

In actuality it is seldom possible to avoid random error, so the actual data we obtain would undoubtedly look something like the data in Table 11.6.3. Here a random-error component has been added, so the model is now of the form (11.6.1). Not only do differences exist between brands, but also between different cars of the same brand.

If we use Equation (11.6.9) to estimate the treatment effects from the above data, we get

$$\text{est. } \alpha_1 = m_1 - m = 21.67 - 19.67 = 2.00,$$

$$\text{est. } \alpha_2 = m_2 - m = 23.33 - 19.67 = 3.66,$$

and
$$\text{est. } \alpha_3 = m_3 - m = 14.00 - 19.67 = -5.67.$$

The estimates are in error by 1.00, $-.67$, and $-.67$, respectively. Although the errors seem rather slight in this example, there is no guarantee in any given experiment that they will not be very large. Thus, we need to evaluate how much of the apparent effect of any experimental treatment is, in fact, due to error before we can decide that something systematic is actually occurring.

This example should suggest that evidence for experimental effects has something to do with the differences *between* the different groups relative to the differences that exist *within* each group. Next, we will turn to the problem of separating the variability among observations into two parts: the part that should reflect both experimental effects and sampling error and that part that should reflect sampling error alone.

11.7 PARTITIONING THE VARIATION IN AN EXPERIMENT

Any observation y_{ij} in sample j exhibits some deviation from the grand sample mean, m. The extent of deviation is merely

$$(y_{ij} - m).$$

This deviation can be thought of as composed of two parts,

$$(y_{ij} - m) = (y_{ij} - m_j) + (m_j - m),$$ (11.7.1)

the first part being the deviation of y_{ij} from the mean of group j and the second being the deviation of the group mean from the grand mean.

Now suppose that we square the deviation from m for each score in the entire sample and sum these squared deviations across all observations i in all sample groups j:

$$\sum_j \sum_i (y_{ij} - m)^2 = \sum_j \sum_i [(y_{ij} - m_j) + (m_j - m)]^2$$

$$= \sum_j \sum_i (y_{ij} - m_j)^2 + \sum_j \sum_i (m_j - m)^2$$

$$+ 2 \sum_j \sum_i (y_{ij} - m_j)(m_j - m).$$ (11.7.2)

The last term on the right-hand side of Equation (11.7.2) is equal to

$$2 \sum_j (m_j - m) \sum_i (y_{ij} - m_j) = 0,$$

since the value represented by the term $(m_j - m)$ is the same for all i in group j, and the sum of $(y_{ij} - m_j)$ must be zero when taken over all i in any group j.

Furthermore,

$$\sum_j \sum_i (m_j - m)^2 = \sum_j n_j(m_j - m)^2$$

since, once again, $(m_j - m)$ is a constant for each observation i figuring in the sum. Putting these results together, we have

$$\sum_j \sum_i (y_{ij} - m)^2 = \sum_j \sum_i (y_{ij} - m_j)^2 + \sum_j n_j(m_j - m)^2.$$ (11.7.3)

This equality is usually called the **partition of the sum of squares** and is true for any set of J distinct samples. Verbally, this fact can be stated as follows. The total sum of squared deviations from the grand mean can always be separated into two parts, the sum of squared deviations within groups and the weighted sum of squared deviations of group means from the grand mean. It is convenient to call these two parts

$$\text{SS within} = \sum_j \sum_i (y_{ij} - m_j)^2,$$ (11.7.4*)

for **sum of squares within groups,** and

$$\text{SS between} = \sum_j n_j(m_j - m)^2,$$ (11.7.5*)

for **sum of squares between groups.** Thus,

$$\text{SS total} = \sum_j \sum_i (y_{ij} - m)^2 = \text{SS within} + \text{SS between}. \quad (11.7.6^*)$$

The meaning of this partition of the sum of squares into two parts can easily be put into common-sense terms. Individual observations in any sample will differ from each other, or show variability. These obtained differences among observations can be due to two things. Some pairs of observations are in different treatment groups, and their differences are due either to the different treatments, or to chance variation, or to both. The sum of squares between groups reflects the contribution of different treatments, as well as chance, to intergroup differences. On the other hand, observations in the *same* treatment group can differ only because of chance variation, since each observation within the group received exactly the same treatment. The sum of squares within groups reflects these intragroup differences due only to chance variation. Thus, in any sample, two kinds of variability can be isolated: the sum of squares between groups, reflecting variability due to treatments *and* chance, and the sum of squares within groups, reflecting chance variation alone.

From the data in Table 11.6.3, Equations (11.7.4)–(11.7.6) yield the following results:

$$
\begin{aligned}
\text{SS within} = {} & (24 - 21.67)^2 + (19 - 21.67)^2 + (22 - 21.67)^2 \\
& + (24 - 23.33)^2 + (25 - 23.33)^2 + (21 - 23.33)^2 \\
& + (11 - 14)^2 + (14 - 14)^2 + (17 - 14)^2 \\
= {} & 39.33,
\end{aligned}
$$

$$
\begin{aligned}
\text{SS between} = {} & 3(21.67 - 19.67)^2 + 3(23.33 - 19.67)^2 \\
& + 3(14.00 - 19.67)^2 \\
= {} & 148.67,
\end{aligned}
$$

and $\text{SS total} = (24 - 19.67)^2 + (19 - 19.67)^2 + \cdots + (14 - 19.67)^2$

$$= 188.00.$$

Some computational formulas that are easier to apply than Equations (11.7.4)–(11.7.6) will be presented in Section 11.9.

From Equation (11.6.8), each error term in the one-factor model has a mean of zero. In addition, suppose we assume that the error terms are independent and that each error term has the same variance, which we will denote by σ_e^2. An unbiased estimate of σ_e^2 is given by the **mean square**

within groups, which is defined as

$$\text{MS within} = \frac{\text{SS within}}{\sum_j n_j - J} = \frac{\text{SS within}}{n - J}, \qquad (11.7.7^*)$$

where $n = \sum_j n_j$ is the total sample size. Recall that for a *single* sample, an unbiased estimate of the population variance is given by a sum of squared deviations divided by $n - 1$, where n is the sample size. Here we have J samples, and 1 degree of freedom is lost in determining the estimated variance within each sample, so we subtract J from the total sample size in the denominator of Equation (11.7.7).

To show that MS within is an unbiased estimate of σ_e^2, we consider the expectation, as taken before the sample is observed, of SS within:

$$E(\text{SS within}) = E\left[\sum_j \sum_i (Y_{ij} - M_j)^2\right].$$

For any given sample j,

$$E\left(\frac{\sum_i (Y_{ij} - M_j)^2}{n_j - 1}\right) = \sigma_e^2,$$

since for any sample j this value is an unbiased estimate of the population error variance, σ_e^2. Thus,

$$E(\text{SS within}) = \sum_j E\left[\sum_i (Y_{ij} - M_j)^2\right]$$

$$= \sum_j (n_j - 1)\sigma_e^2 = (n - J)\sigma_e^2,$$

so that

$$E(\text{MS within}) = \frac{E(\text{SS within})}{n - J} = \sigma_e^2. \qquad (11.7.8^*)$$

Next, we consider the **mean square between groups:**

$$\text{MS between} = \frac{\text{SS between}}{J - 1}. \qquad (11.7.9^*)$$

In order to investigate MS between, we will consider the expectation, as taken before the sample is observed, of SS between:

$$E(\text{SS between}) = E\left[\sum_j n_j(M_j - M)^2\right] = \sum_j n_j E(M_j - M)^2.$$

But, for a given j,

$$E(M_j - M)^2 = V(M_j - M) + [E(M_j - M)]^2,$$

where

$$E(M_j - M) = \mu_j - \mu = \alpha_j$$

and it can be shown that

$$V(M_j - M) = \frac{(n - n_j)\sigma_e^2}{nn_j}.$$

Thus,

$$E(\text{SS between}) = \sum_j n_j \left\{ \left[\frac{(n - n_j)\sigma_e^2}{nn_j} \right] + \alpha_j^2 \right\},$$

which simplifies to

$$E(\text{SS between}) = (J - 1)\sigma_e^2 + \sum_j n_j\alpha_j^2.$$

Therefore, from Equation (11.7.9),

$$E(\text{MS between}) = \frac{E(\text{SS between})}{J - 1} = \sigma_e^2 + \frac{\sum_j n_j\alpha_j^2}{J - 1}. \quad (11.7.10^*)$$

As we stated earlier, the variation between samples is due to both random error and treatment effects. If there are no treatment effects, so that $\alpha_j = 0$ for all j, then $E(\text{MS between})$ is equal to σ_e^2; however, if *any* nonzero treatment effects exist, then $E(\text{MS between})$ is greater than σ_e^2.

11.8 THE *F* TEST IN THE ONE-FACTOR MODEL

The partition of the sum of squares given in the preceding section requires no special assumptions beyond the basic formulation of the one-factor model, as given in Section 11.6. The calculation of MS between and MS within also requires no assumptions, although the determination of Equations (11.7.8) and (11.7.10) utilized the assumption that the error terms are independent and have the same variance, σ_e^2. In order to make further inferences, we will make the additional assumption that the error terms are normally distributed. Our assumptions about the error terms in the one-factor analysis of variance model are now identical to the assumptions about the error terms in the linear regression model (Section 10.9).

Given the assumption of normality, we can investigate the sampling distributions of MS between and MS within. First, when there are no treatment effects, MS between is an unbiased estimate of σ_e^2, and in Chapter 6 we found that for normal parent populations,

$$\frac{(\text{est. } \sigma_e^2)}{\sigma_e^2} = \frac{\chi^2_{(\nu)}}{\nu}.$$

The ratio of MS between to σ_e^2 must be a chi-square variable divided by degrees of freedom, *when* there are no treatment effects *and* the parent populations are normal. Since SS between involves J squared deviations between individual sample means and the grand sample mean, there are $J - 1$ degrees of freedom for MS between.

Looking at MS within, we see that it is a *pooled* estimate of σ_e^2, which was assumed to be the same for each group. Under normality, then, the ratio of MS within to σ_e^2 must be a chi-square variable divided by degrees of freedom. (Note that this is true whether or not the treatment effects are zero.) Here, the chi-square variable is actually a sum of independent chi-square variables with $n_j - 1$ degrees of freedom for $j = 1, \ldots, J$. Therefore, by the additivity property of chi-square variables (Section 6.18), MS within has $\sum_j (n_j - 1) = n - J$ degrees of freedom.

The usual hypothesis tested using the analysis of variance is

$$H_0: \mu_1 = \cdots = \mu_j = \cdots = \mu_J,$$

the hypothesis that all treatment population means are equal. The alternative is just "not H_0," implying that some of the population means are different from others. As we have seen, these two hypotheses are equivalent to the hypothesis of no treatment effects and its contrary:

$$H_0: \alpha_j = 0 \quad \text{for all } j$$

and

$$H_1: \alpha_j \neq 0 \quad \text{for some } j.$$

The argument in the two preceding sections has shown that *when H_0 is true,*

$$E(\text{MS within}) = E(\text{MS between}) = \sigma_e^2. \tag{11.8.1}$$

On the other hand, *when the null hypothesis is false,* then

$$E(\text{MS within}) < E(\text{MS between}). \tag{11.8.2}$$

Since both of these mean squares divided by σ_e^2 are distributed as chi-square variables divided by their respective degrees of freedom when H_0 is true, it follows that their ratio should have an F distribution provided that MS between and MS within are *independent* estimates of σ_e^2. By an extension of the principle given in Section 5.15, MS between, based on the J values of m_j, must be independent of MS within, based on the several $\hat{s}_j^2$ values; each piece of information making up MS between is independent of the information making up MS within, given normal parent distributions.

Thus, we have all the justification needed in order to say that the ratio

$$F = \frac{(\text{MS between}/\sigma_e^2)}{(\text{MS within}/\sigma_e^2)} = \frac{\text{MS between}}{\text{MS within}} \tag{11.8.3}$$

is distributed as F with $J - 1$ and $n - J$ degrees of freedom *when the null hypothesis is true.* This statistic is the ratio of two independent chi-square variables, each divided by its degrees of freedom, and thus it is exactly distributed as F when H_0 is true (from Section 6.21).

The F ratio used in the analysis of variance always provides a *one-tailed* test of H_0 in terms of the sampling distribution of F. If H_1 is true, then from Equation (11.8.2), we would expect the numerator of Equation (11.8.3) to be greater than the denominator. If we compare Equations (11.7.8) and (11.7.10), we see that if there are nonzero treatment effects, $E(\text{MS between})$ is expected to be larger than $E(\text{MS within})$ because of the term $\sum n_j \alpha_j^2 / (J - 1)$. Therefore, for the analysis of variance, the F ratio obtained can be compared directly with the one-tailed values given in Table IV.

For the example of Section 11.6 involving the three automobile brands, the sums of squares were calculated in Section 11.7, and the mean squares are

$$\text{MS within} = \frac{\text{SS within}}{n - J} = \frac{39.33}{9 - 3} = 6.55$$

and

$$\text{MS between} = \frac{\text{SS between}}{J - 1} = \frac{148.67}{3 - 1} = 74.33.$$

Therefore, from Equation (11.8.3), the value of F is $74.33/6.55 = 11.35$. From Table IV, we see that $P(F \geq 10.92) = .01$ for 2 and 6 degrees of freedom, so the p-value is less than .01. In other words, it appears that there is some difference among the three brands in terms of fuel economy.

In the F test given here, then, we are concerned primarily with large values of F. Under the null hypothesis, the expected value of F is $\nu_2 / (\nu_2 - 2)$, where ν_2 represents the degrees of freedom for the denominator. Of course, by chance it is possible to obtain a value of F less than 1. However, a very small value of F can also serve as a signal to the experimenter. In such situations, a close analysis of the experimental situation often shows a systematic, but uncontrolled, factor to be in operation, resulting in an MS within that reflects something other than error variation alone. The existence of such an uncontrolled but nonrandom factor is, of course, a failure of the assumptions, but before the experimenter appeals to chance or to some less obvious failure of assumptions, the possibility that the experiment itself is open to suspicion should be entertained.

The results presented in this section require the assumption that the error terms are independent and normally distributed with identical variances. Violations of the normality assumption are not too serious for relatively large samples because the central limit theorem, discussed in

Section 5.17, comes to our aid. For this reason, the more severely the distribution of the error terms is suspected to depart from normality, the larger the n_j's should be. With regard to the assumption of homogeneous variances, violations are not too serious provided that the sample sizes are equal, but they can be quite serious when the sample sizes differ. Thus, whenever possible, an experiment should be planned so that the sample size is the same in each experimental group. The remaining assumption, that of independence of errors, is most important for the justification of the F test in the analysis of variance, and dependence among error terms makes inferences invalid. In general, great care should be taken to see that data treated by analysis of variance procedures are based on independent observations, both within and across groups.

Finally, a word about the general assumption embodied in the linear model. Each observation is assumed to be a *sum*, consisting of a grand mean, plus a treatment effect, plus an independent error component. The appealing simplicity of this model nonwithstanding, many situations exist where we know that an additive model such as this is not realistic. For example, in some experimental problems it may be far more reasonable to suppose that random errors serve to multiply treatment effects rather than add to them. In such instances, it is often possible to transform the original values by, for example, taking their logarithms, so as to make the transformed values correspond to the linear model. This is similar to the general problem of nonlinear regression discussed in Sections 10.16–10.18.

Inferences other than the F test presented in this section are of interest in the analysis of variance. Quite a few new concepts have been introduced in Sections 11.6–11.8, however, so we will present an example (and introduce some computational forms for the one-factor analysis of variance in the example) before introducing yet more concepts.

11.9   AN EXAMPLE OF A ONE-FACTOR ANALYSIS OF VARIANCE

A marketing manager is interested in the effect of different promotional schemes for a particular product. One such scheme is currently being used, and two new schemes have been proposed. The manager draws a random sample of 18 stores from the population of stores that stock the product. One of the new schemes seems particularly promising, and that scheme (treatment 1) is used at 8 stores selected randomly from the 18 stores. The second new scheme (treatment 2) is used at 5 stores that are chosen randomly from the remaining 10 stores, and the final 5 stores constitute a control group using the current promotional scheme (treatment 3). The sales of the product at each of the 18 stores are carefully recorded for a period of 1 month, and the results are presented in Table 11.9.1. For example, the first store in Group I sold 43, the next store sold 49, and so on down to the last store in Group III, which sold 29. In this example, there are three treatments, or groups, so $J = 3$. Furthermore, the group sizes are $n_1 = 8$, $n_2 = 5$, and $n_3 = 5$.

Table 11.9.1   Total Sales During Experiment for Each Store, According to Promotional Scheme

Group I	Group II	Group III
43	35	34
49	32	37
42	36	25
38	40	32
46	38	29
34	—	—
37	181	157
47		
336		
$m_1 = 42.1$	$m_2 = 36.2$	$m_3 = 31.4$

$$m = \frac{336 + 181 + 157}{18} = 37.4$$

To apply the one-factor analysis of variance model to this example, we could use the formulas developed in Sections 11.7 and 11.8. However, it is possible to develop some formulas for the sums of squares that are easier to use computationally than formulas such as Equations (11.7.4) and (11.7.5). First, we consider SS total:

$$\text{SS total} = \sum_j \sum_i (y_{ij} - m)^2 = \sum_j \sum_i (y_{ij}^2 - 2y_{ij}m + m^2)$$

$$= \sum_j \sum_i y_{ij}^2 - 2m \sum_j \sum_i y_{ij} + \sum_j \sum_i m^2,$$

which reduces further to

$$\sum_j \sum_i y_{ij}^2 - 2m(nm) + nm^2,$$

or

$$\sum_j \sum_i y_{ij}^2 - nm^2,$$

by the definition of the sample grand mean, $m = \sum_j \sum_i y_{ij}/n$. Making one last substitution for m gives the computing formula

$$\text{SS total} = \sum_j \sum_i y_{ij}^2 - \frac{(\sum_j \sum_i y_{ij})^2}{n}. \tag{11.9.1*}$$

The computing formula for the sum of squares between groups can be worked out in a similar way:

$$\text{SS between} = \sum_j n_j(m_j - m)^2 = \sum_j n_j(m_j^2 - 2mm_j + m^2)$$

$$= \sum_j n_j m_j^2 - 2m \sum_j n_j m_j + m^2 \sum_j n_j$$

$$= \sum_j \frac{(\sum_i y_{ij})^2}{n_j} - 2nm^2 + nm^2,$$

or $$\text{SS between} = \sum_j \frac{(\sum_i y_{ij})^2}{n_j} - \frac{(\sum_j \sum_i y_{ij})^2}{n}. \tag{11.9.2*}$$

Finally, the computing formula for the sum of squares within groups is found by subtracting Equation (11.9.2) from Equation (11.9.1):

$$\text{SS within} = \text{SS total} - \text{SS between}$$

$$= \sum_j \sum_i y_{ij}^2 - \sum_j \frac{(\sum_i y_{ij})^2}{n_j}. \tag{11.9.3*}$$

Ordinarily, the simplest computational procedure is to calculate both the sum of squares total and the sum of squares between directly and then to subtract SS between from SS total in order to find the SS within.

It is natural for the beginner in statistics to be a little staggered by all of the arithmetic that the analysis of variance involves. However, with a bit of organization and with the aid of a desk calculator simple analyses can be done quite quickly (more complicated analyses usually are handled on a high-speed computer). The important thing is to form a clear mental picture of the different sample quantities you will need to compute. In this book we make every effort to avoid presenting "recipes," and you should always keep the underlying assumptions of the model in mind, but the following outline may be helpful in terms of the computational steps in the one-factor analysis of variance.

1. Start with a listing of the data separated by columns into the treatment groups to which they belong.
2. Square each observation y_{ij} and then add these squared values over all individuals in all groups. The result is $\sum_j \sum_i y_{ij}^2$. Call this quantity A.
3. Now sum the observations themselves over all individuals in all groups to find $\sum_j \sum_i y_{ij}$. Call the resulting value B.
4. Now for a single group, say group j, sum all of the observations in that group and square the sum to find $(\sum_i y_{ij})^2$, then divide by the number in that group: $(\sum_i y_{ij})^2/n_j$.
5. Repeat step 4 for each group, and then sum the results across the several groups to find $\sum_j (\sum_i y_{ij})^2/n_j$. Call this quantity C.
6. The **sum of squares total** is $A - (B^2/n)$.
7. The **sum of squares between** is $C - (B^2/n)$.
8. The **sum of squares within** is $A - C$.
9. Divide SS between by $J - 1$ to give MS **between.**
10. Divide SS within by $n - J$ to give MS **within.**
11. Divide MS between by MS within to find the F **ratio.**

It is customary to display the results of an analysis of variance in a table similar to Table 11.9.2, and one should form the habit of arranging the results of an analysis of variance in this way. Not only is it a good way to display the results for maximum clarity, but it also forms a convenient device for organizing and remembering the computational steps.

Table 11.9.2 An Analysis of Variance Table

Source	SS	df	MS	F
Treatments (between groups)	$\sum_j \dfrac{(\sum_i y_{ij})^2}{n_j} - \dfrac{(\sum_j \sum_i y_{ij})^2}{n}$	$J - 1$	$\dfrac{\text{SS between}}{J - 1}$	$\dfrac{MS \text{ between}}{MS \text{ within}}$
Error (within groups)	$\sum_j \sum_i y_{ij}^2 - \sum_j \dfrac{(\sum_i y_{ij})^2}{n_j}$	$n - J$	$\dfrac{\text{SS within}}{n - J}$	
Totals	$\sum_j \sum_i y_{ij}^2 - \dfrac{(\sum_j \sum_i y_{ij})^2}{n}$	$n - 1$		

For the marketing example, we calculate the following from the data in Table 11.9.1:

$$\sum_j \sum_i y_{ij}^2 = 25{,}912 = A,$$

$$\sum_j \sum_i y_{ij} = 674 = B,$$

$$\sum_j \frac{\left(\sum_i y_{ij}\right)^2}{20} = \frac{(336)^2}{8} + \frac{(181)^2}{5} + \frac{(157)^2}{5} = 25{,}594 = C,$$

$$\text{SS total} = A - \frac{B^2}{18} = 674.4,$$

$$\text{SS between} = C - \frac{B^2}{18} = 356.4,$$

and

$$\text{SS within} = A - C = 318.0.$$

The analysis of variance is summarized in Table 11.9.3. The p-value for the F test of the hypothesis of no treatment effects is less than .01, so we can feel confident in asserting that the type of packaging does have *some* effect on the sales of the item in question. Moreover, the estimated treatment effects are

$$\text{est. } \alpha_1 = m_1 - m = 42.1 - 37.4 = 4.7,$$

$$\text{est. } \alpha_2 = m_2 - m = 36.2 - 37.4 = -1.2,$$

and

$$\text{est. } \alpha_3 = m_3 - m = 31.4 - 37.4 = -6.0.$$

From the estimated treatment effects, it appears that the promotional scheme used in Group I produces greater sales than the other two schemes. If all of the schemes have identical costs, then the scheme used in Group I appears to be preferable to the other two, although the choice among the

Table 11.9.3   Analysis of Variance Table for Marketing Example

Source	SS	df	MS	F
Between groups	356.4	$3 - 1 = 2$	178.2	8.41
Within groups	318.0	$18 - 3 = 15$	21.2	
Totals	674.4	$18 - 1 = 17$		

three schemes may not be so simple if, say, the first scheme is more costly than the other two.

The F test indicates that there are differences among the schemes with respect to sales. In addition, the marketing manager might be interested in making pairwise comparisons. If the difference between two particular treatment effects α_j and α_k is of interest, then a $100(1 - \alpha)$ percent confidence interval for this difference is given by

$$(\text{est. } \alpha_j - \text{est. } \alpha_k) \pm a \sqrt{\text{MS within} \left(\frac{1}{n_j} + \frac{1}{n_k}\right)}, \qquad (11.9.4)$$

where a is the $1 - (\alpha/2)$ fractile of the T distribution with $n - J$ degrees of freedom. Here MS within is being used as an estimate of σ_e^2 [compare Equation (11.9.4) with Equations (6.16.6) and (6.16.9)].

Unfortunately, the use of Equation (11.9.4) to find confidence intervals for differences between all possible pairs of treatments is invalid because the confidence coefficient applies to a single interval and not to the simultaneous realization of several intervals, and multiple intervals of this nature are not independent. However, methods have been developed to look simultaneously at all possible pairwise differences. The result presented here (without proof) is based on the S-method, which is a general procedure for making **multiple comparisons.** Using this method, the intervals

$$(\text{est. } \alpha_j - \text{est. } \alpha_k) \pm \sqrt{(J - 1)a(\text{MS within}) \left(\frac{1}{n_j} + \frac{1}{n_k}\right)} \qquad (11.9.5)$$

for all $j \neq k$ form a simultaneous set of confidence intervals at the $1 - \alpha$ level of confidence, where a is the $1 - \alpha$ fractile of the F distribution with $J - 1$ and $n - J$ degrees of freedom. In other words, the probability is $1 - \alpha$ that all of these intervals *simultaneously* cover the appropriate differences in treatment effects.

For the marketing example, the following three intervals hold simultaneously with a 95 percent level of confidence:

$$[4.7 - (-1.2)] \pm \sqrt{2(3.68)(21.2)(\tfrac{1}{8} + \tfrac{1}{5})}$$
$$= (-1.2, 13.0) \quad \text{for } \alpha_1 - \alpha_2,$$

$$[4.7 - (-6.0)] \pm \sqrt{2(3.68)(21.2)(\tfrac{1}{8} + \tfrac{1}{5})}$$
$$= (3.6, 17.8) \quad \text{for } \alpha_1 - \alpha_3,$$

and

$$[-1.2 - (-6.0)] \pm \sqrt{2(3.68)(21.2)(\tfrac{1}{5} + \tfrac{1}{5})}$$
$$= (-3.1, 12.7) \quad \text{for } \alpha_2 - \alpha_3.$$

The F test provides us with information concerning the existence of treatment effects, and multiple comparisons procedures enable us to look at pairwise differences simultaneously.

11.10 ESTIMATING THE STRENGTH OF A STATISTICAL RELATION IN THE ONE-FACTOR MODEL

It may appear that the main purpose of the analysis of variance is to generate an F test and that the partition of the sum of squares is only a means to this end. However, this is really a very narrow view of the role of this form of analysis in experimentation. The really important feature of the analysis of variance is that it permits the separation of all of the potential information in the data into distinct and nonoverlapping portions, each reflecting only certain aspects of the experiment. For example, in the one-factor analysis of variance, the mean square between groups reflects both the systematic differences among observations that are attributable to the experimental manipulations and the chance, unsystematic differences attributable to all of the other circumstances of the experiment. On the other hand, the mean square within the groups reflects only these latter, unsystematic, features. Under the linear model, these two statistics are independent, completely nonoverlapping, ways of summarizing the data. The information contained in one is nonredundant with the information contained in the other. Estimates of the effects of the treatments are independent of estimates of error variability.

In short, it is useful to think of the analysis of variance as a device for "sorting" the information in an experiment into nonoverlapping and meaningful portions. By way of comparison, multiple T tests carried out on the same data do not provide this feature; the various differences between means do overlap in the information they provide, as we pointed out in the preceding section, and it is not easy to assess the evidence for overall existence or importance of treatment effects from a complete set of such differences. On the other hand, the analysis of variance packages the information in the data into neat, distinct "bundles," permitting a relatively simple judgment to be made about the effects of the experimental treatments. The real importance of the analysis of variance lies in the fact that it routinely provides such succinct overall "packaging" of the data, and this feature becomes even more important as the experiment becomes more complex (for example, as more factors are added to the experiment).

If the statistician is interested in testing to see if there are any differences among treatment effects, then the F test provides some information. Although the F test indicates the existence or nonexistence of treatment effects, it does not tell us about the overall *strength* of association between the independent variable (the treatments) and the dependent variable (the variable being measured in the experiment). In the marketing example, the F test does not tell us about the strength of the association between promotional schemes and sales. In order to measure association in the one-factor model, we will introduce an index which is analogous to the square of the correlation coefficient in regression problems.

This index, which will be denoted by ω^2 (omega, squared), represents the proportion of variance in the dependent variable Y which is accounted for by knowledge about the independent variable X:

$$\omega^2 = \frac{\sigma_Y^2 - \sigma_{Y|X}^2}{\sigma_Y^2}. \tag{11.10.1*}$$

Just as in the regression model, σ_Y^2 is the variance of the marginal distribution of Y, and $\sigma_{Y|X}^2$ is the variance of the conditional distribution of Y given a value of X. One of the assumptions of the linear one-way analysis of variance model is that for any given treatment (value of X), the variability in Y is due solely to the random-error term, which has variance σ_e^2 no matter what value of X is given. Thus, $\sigma_{Y|X}^2$ is equal to σ_e^2.

Suppose that the probability that any observation unit falls into the treatment category x_j is $P(x_j)$. Then

$$\sigma_Y^2 = E(Y_{ij} - \mu)^2 = E(Y_{ij} - \mu_j + \alpha_j)^2$$
$$= E(Y_{ij} - \mu_j)^2 + E(\alpha_j^2) + 2E[\alpha_j(Y_{ij} - \mu_j)]$$
$$= \sigma_e^2 + \sum_j \alpha_j^2 P(x_j),$$

and, since $\sigma_{Y|X}^2 = \sigma_e^2$, Equation (11.10.1) becomes

$$\omega^2 = \frac{\sum_j \alpha_j^2 P(x_j)}{\sigma_e^2 + \sum_j \alpha_j^2 P(x_j)}. \tag{11.10.2*}$$

This population index should reflect how much knowledge of the particular treatment (x_j) represents in terms of an increased ability to predict the value of Y. Notice that if α_j is zero for each j, then ω^2 must be zero. That is, if there are no treatment effects whatsoever, then knowledge about the treatment will not improve our ability to predict Y. In other words, there is no reduction in the variance due to this information. At the other extreme, suppose that σ_e^2 is equal to 0. This means that if we know the treatment, we can predict Y perfectly, since there is no error variance. In this case, notice that ω^2 is equal to 1; *all* of the variance in Y can be accounted for by knowledge about X.

In order to estimate the value of ω^2 from sample data, we recall from Section 11.7 that

$$E(\text{MS between}) = \sigma_e^2 + \sum_j \frac{n_j \alpha_j^2}{J - 1}$$

and

$$E(\text{MS within}) = \sigma_e^2.$$

When the probability $P(x_j) = n_j/n$, so that the proportional representation of cases in the J samples is the same as the proportions in the respective populations,

$$\omega^2 = \frac{\sum_j \alpha_j^2 n_j/n}{\sigma_e^2 + (\sum_j \alpha_j^2 n_j/n)} = \frac{\sum_j \alpha_j^2 n_j}{n\sigma_e^2 + \sum_j \alpha_j^2 n_j}, \qquad (11.10.3)$$

and a reasonable (though rough) estimate of ω^2 is given by

$$\text{est. } \omega^2 = \frac{\text{SS between} - (J - 1) \text{ MS within}}{\text{SS total} + \text{MS within}}. \qquad (11.10.4)$$

This estimate is reasonable in the sense that the expectation of the numerator equals the numerator of ω^2 as given in Equation (11.10.3), and the same is true for the denominator. [Be careful to note, however, that this does *not* mean that the expectation of the *ratio* of the numerator to the denominator in Equation (11.10.4) is equal to ω^2. The estimate given in Equation (11.10.4) is biased, and it may not be a good estimator in some other respects as well.] It is possible for the estimate to be *negative*, in which case the estimate of ω^2 is set equal to 0; if $F > 1$, however, a nonnegative estimate is guaranteed.

For example, let us apply the estimate of ω^2 to the data of Section 11.9. Here, the result was very significant; how much does knowing the promotional scheme let us reduce our uncertainty about sales? For this example,

$$\text{est. } \omega^2 = \frac{356.4 - 2(21.2)}{674.4 + 21.2} = .45.$$

The independent variable (promotional scheme) is estimated to account for 45 percent of the variance in the Y values. This shows how the estimate of ω^2 can reinforce the meaning of a significant finding. Not only is there evidence for *some* association between independent and dependent variables; our rough estimate suggests that the association is reasonably strong.

Table 11.10.1 Analysis of Variance Table for Example

Source	SS	df	MS	F
Between groups	1500	5	300	4.5
Within groups	11,605.8	174	66.7	
Totals	13,105.8	179		

By way of contrast, suppose that another study employed 6 groups of 30 observations each and that the analysis of variance turned out as in Table 11.10.1. For 5 and 174 degrees of freedom, the p-value is slightly less than .01. Now we will see how much statistical association is apparently represented by this finding.

$$\text{est. } \omega^2 = \frac{1500 - 333.5}{13,105.8 + 66.7} = .089.$$

In this instance, slightly less than 9 percent of the variance in the dependent variable seems to be accounted for by the independent variable. This still may be enough to make the statistical association an important one from the experimenter's point of view, but the example shows, nevertheless, that a significant result need not correspond to a *very* strong association. A result that is both significant at some small α level *and* that gives an estimate of relatively strong association is usually far more informative than a significant result taken alone. Furthermore, even though a result is not significant, estimating whether or not a fairly high degree of association may in fact be present may help the experimenter decide whether to conduct further experimentation or to forget the whole business. If ω^2 is relatively high, an increase in sample size or a refinement in the experimental procedure (to reduce error variance) may lead to more "significant" results in terms of the F test. In summary, it is helpful to consider both the F test *and* the value of an index such as ω^2 in analysis of variance problems.

11.11 ANALYSIS OF VARIANCE: THE TWO-FACTOR MODEL

We will now discuss how the simple one-factor analysis of variance can be extended to cover a more complicated experimental setup in which there are two different sets of treatments. Returning to the example of Section 11.6, suppose that an experimenter is interested in the fuel economy of a car, as measured in terms of miles per gallon, as the dependent variable, and that the brand of car is an independent variable, with 3 brands (one German, one Japanese, and one American) being used. In addition, the experimenter decides to control a second independent variable, the transmission of the car. For this variable there will be two "values," standard transmission and automatic transmission.

In this example, interest is focused on **two distinct experimental factors:** the brand of car and the transmission. Either or both of these factors might possibly influence fuel economy. For each of the 3 brands, 4 cars with standard transmissions and 4 cars with automatic transmissions are selected randomly from a large shipment of cars. Here there are 6 possible **treatment combinations** (German car with standard transmission, German car with automatic transmission, Japanese car with standard transmission, and so on). Since all 6 possible combinations are used in the experiment, the treatments are said to be **completely crossed;** this means that each category or level of one factor (brand of car) occurs with each level of the other factor (transmission). Furthermore, this experiment is said to be **balanced,** since each of these 6 groups occurs the same number of times. The experiment we have described, then, is completely crossed and balanced.

In the following sections, we will assume that there are two sets of treatments, or two variables, of interest, and that the experiment is completely crossed and balanced. Returning to the example, there are three questions of interest.

1. Are there systematic effects due to the brand of car alone (irrespective of transmission)?
2. Are there systematic effects due to transmission alone (irrespective of the brand of car)?
3. Are there systematic effects due neither to the brand of car alone, nor to transmission alone, but attributable only to the *combination* of particular brands of cars with particular transmissions?

Notice that the study could be viewed as two separate experiments carried out on the same set of cars: (a) there are 3 groups of 8 cars each, differing in the brand of car; and (b) there are 2 groups of 12 cars each, differing in the transmission. The third question above cannot, however, be answered by the comparison of brands alone or transmissions alone. This is a question of **interaction,** the unique effects of combinations of treatments. This is an important feature of the two-way analysis of variance; we will be able to examine **main effects** of the separate experimental variables or factors just as in the one-way analysis, as well as **interaction effects,** differences apparently due to the unique combinations of treatments.

Just as in the linear model for the one-way analysis of variance, in the corresponding model for a two-way analysis it is assumed that each observed value of the dependent variable is a sum of systematic effects associated with experimental treatments, plus random error:

$$y_{ijk} = \mu + \alpha_j + \beta_k + \gamma_{jk} + e_{ijk}, \qquad (11.11.1^*)$$

where α_j is the effect of treatment j of the first factor,

$$\alpha_j = \mu_j - \mu, \qquad (11.11.2^*)$$

and β_k is the effect of treatment k of the second factor,

$$\beta_k = \mu_k - \mu. \qquad (11.11.3^*)$$

Here, μ_j is the mean of the population given treatment j of factor 1 and pooled over all of the K different treatments k of factor 2, and μ_k is the mean of the population given treatment k of factor 2 and pooled over all of the J different treatments j of factor 1. The grand mean μ is the mean of the population formed by pooling all of the different populations given the possible treatment combinations j and k.

The new feature of Equation (11.11.1) is the inclusion of a term representing the **interaction effect,** γ_{jk} (small Greek gamma). The interaction effect is the experimental effect created by the combination of treatments j and k of factors 1 and 2, respectively, over and above any effects associated with treatments j and k considered separately:

$$\gamma_{jk} = \mu_{jk} - \mu - \alpha_j - \beta_k$$
$$= \mu_{jk} - \mu_j - \mu_k + \mu. \qquad (11.11.4^*)$$

The interaction effect γ_{jk} is thus equal to the mean of the population given both of the treatments j of factor 1 and k of factor 2, minus the mean of the treatment population j of factor 1, minus the mean of the population given treatment k of factor 2, plus the grand mean.

The following equalities are assumed true of the effects:

$$\sum_j \alpha_j = 0,$$

$$\sum_k \beta_k = 0,$$

$$\sum_j \gamma_{jk} = 0,$$

and
$$\sum_k \gamma_{jk} = 0.$$

The effects, being deviations from a grand mean μ, sum to zero over all the different "levels" of a given kind of treatment.

Some intuition about the meaning of interaction effects may be gained by examining another set of artificial data, just as we did in Section 11.6. In the experiment outlined earlier in this section, suppose that the only nonzero effects are associated with the brand of car and that the effects are $+1$ for the German brand, $+4$ for the Japanese brand, and -5 for the U. S. brand. Additionally, for convenience assume that there is no random error, so that all observations in a given treatment combination are identical. Therefore, the data look like the data in Table 11.11.1. (Since all observations in a given treatment combination are identical, we just write 1 observation for each combination instead of 4.) Here each observation is of the form $y_{ijk} = 20 + \alpha_j$.

Table 11.11.1   Data Exhibiting Column Effects but No Row Effects for Gas Mileage Example

BRAND OF CAR

		German brand	Japanese brand	U. S. brand
TRANSMISSION	Standard	$20 + 1 = 21$	$20 + 4 = 24$	$20 - 5 = 15$
	Automatic	$20 + 1 = 21$	$20 + 4 = 24$	$20 - 5 = 15$

On the other hand, it might turn out that nonzero effects exist only for the transmission, so that no effects are associated with the brand of car or with interactions. Once again, if we have no random error, we should observe something like Table 11.11.2. Here the effect associated with standard transmission is $\beta_1 = 2$, and the effect associated with automatic transmission is $\beta_2 = -2$. Each observation is simply of the form $y_{ijk} = 20 + \beta_k$.

Table 11.11.2   Data Exhibiting Row Effects but No Column Effects for Gas Mileage Example

BRAND OF CAR

		German brand	Japanese brand	U. S. brand
TRANSMISSION	Standard	$20 + 2 = 22$	$20 + 2 = 22$	$20 + 2 = 22$
	Automatic	$20 - 2 = 18$	$20 - 2 = 18$	$20 - 2 = 18$

Now suppose that there are *both* column and row effects, but no interaction effects or random error. In this case, $y_{ijk} = 20 + \alpha_j + \beta_k$, and

Table 11.11.3   Data Exhibiting Row Effects *and* Column Effects for Gas Mileage Example

BRAND OF CAR

	German brand	Japanese brand	U. S. brand
TRANSMISSION Standard	$20 + 1 + 2 = 23$	$20 + 4 + 2 = 26$	$20 - 5 + 2 = 17$
Automatic	$20 + 1 - 2 = 19$	$20 + 4 - 2 = 22$	$20 - 5 - 2 = 13$

the data are given in Table 11.11.3. In this instance the 6 treatment combinations yield means differing across the different *cells* of the table. However, the effect of a combination, $\mu_{jk} - \mu$, associated with cell jk, is exactly equal to the effect associated with its row, β_k, plus the effect associated with its column, α_j, so that $y_{ijk} = \mu + \alpha_j + \beta_k$. When there is no interaction, effects are said to be additive, since the effect of a combination is the sum of the effects of the treatments involved. Notice that the difference between a particular pair of columns is the same over the rows and that any row difference is constant over columns.

Finally, let us add interaction effects, giving us a table something like Table 11.11.4. The effect associated with a combination of treatments is now no longer the simple sum of the effects of its row and its column, an indication that interaction effects are present. **Notice that columns are different in different ways within rows, and vice versa, when interaction is present.** In the example, the brand of car and transmission interact in such a way that, for instance, the transmission makes a big difference in the fuel economy of the Japanese brand but no difference whatsoever in the fuel economy of the U. S. brand.

Naturally, for any real data there will be random error as well. Thus, the observations within any particular treatment combination will not

Table 11.11.4   Data Exhibiting Row Effects, Column Effects, and Interaction Effects for Gas Mileage Example

BRAND OF CAR

	German brand	Japanese brand	U. S. brand
TRANSMISSION Standard	$20 + 1 + 2 - 1$ $= 22$	$20 + 4 + 2 + 3$ $= 29$	$20 - 5 + 2 - 2$ $= 15$
Automatic	$20 + 1 - 2 + 1$ $= 20$	$20 + 4 - 2 - 3$ $= 19$	$20 - 5 - 2 + 2$ $= 15$

Table 11.11.5   Data Exhibiting Row Effects, Column Effects, Interaction Effects, and Random Error for Gas Mileage Example

BRAND OF CAR

		German brand	Japanese brand	U. S. brand
TRANSMISSION	Standard	21 20 $m_{11} = 21.5$ 23 22	32 28 $m_{21} = 30.5$ 32 30	16 14 $m_{31} = 15.5$ 17 15
	Automatic	20 18 $m_{12} = 19.5$ 19 21	20 21 $m_{22} = 19.5$ 18 19	14 16 $m_{32} = 14.5$ 13 15

necessarily be equal, and the actual data may look something like Table 11.11.5. The sample means for each of the treatment combinations are given in the table. Moreover, with respect to the brand of car, the sample mean is 20.5 for the German brand, 25.0 for the Japanese brand, and 15.0 for the U. S. brand; with respect to the transmission, the sample mean is 22.5 for the standard transmission and 17.8 for the automatic transmission. The grand sample mean is 20.2. Estimates of the main effects and interaction effects can be obtained by subtracting the grand sample mean from the row means, column means, and treatment combination means. For example, the effects due to the two transmissions are estimated as follows:

$$\text{est. } \beta_1 = 22.5 - 20.2 = 2.3,$$

$$\text{est. } \beta_2 = 17.8 - 20.2 = -2.4;$$

an estimate of the interaction effect for the combination of a Japanese car and a standard transmission is

$$\text{est. } \gamma_{21} = 30.5 - 25.0 - 22.5 + 20.2 = 3.2.$$

In order to investigate the effects further, we will take the variability in the dependent variable and separate it into different parts, just as we did in the one-factor model.

11.12   PARTITIONING THE VARIATION IN THE TWO-FACTOR MODEL

Once more we start off by looking at how the total sum of squares can be partitioned for a set of data. The experiment is assumed to be completely crossed and balanced, and we will denote the common sample size for each treatment combination by n. For simplicity, we let R and C represent the number of rows and columns, respectively, so that C is the number of levels of the first factor (brand of car in the example of the previous section) and R is the number of levels of the second factor (transmission in the example). For any individual i in any treatment combination jk, the deviation of y_{ijk} from the sample grand mean m can be written as

$$y_{ijk} - m = (y_{ijk} - m_{jk}) + (m_j - m) + (m_k - m)$$
$$+ (m_{jk} - m_j - m_k + m), \tag{11.12.1}$$

where
$$m_{jk} = \frac{\sum\limits_{i} y_{ijk}}{n}, \tag{11.12.2}$$

$$m_j = \frac{\sum\limits_{k}\sum\limits_{i} y_{ijk}}{Rn}, \tag{11.12.3}$$

$$m_k = \frac{\sum\limits_{j}\sum\limits_{i} y_{ijk}}{Cn}, \tag{11.12.4}$$

and
$$m = \frac{\sum\limits_{k}\sum\limits_{j}\sum\limits_{i} y_{ijk}}{RCn}. \tag{11.12.5}$$

If the deviation for each score is squared, and these squares are summed over all individuals in all combinations j and k, we have

$$\sum\limits_{j}\sum\limits_{k}\sum\limits_{i} (y_{ijk} - m)^2 = \sum\limits_{j}\sum\limits_{k}\sum\limits_{i} (y_{ijk} - m_{jk})^2$$
$$+ \sum\limits_{j} Rn(m_j - m)^2 + \sum\limits_{k} Cn(m_k - m)^2$$
$$+ \sum\limits_{j}\sum\limits_{k} n(m_{jk} - m_j - m_k + m)^2. \tag{11.12.6}$$

The proof of this equation is similar to the development of Equation (11.7.3), so we will not present it here, although you may find it profitable to try to derive it for yourself.

Now let us examine the various individual terms on the right of expression (11.12.6). We call

$$\sum\limits_{j}\sum\limits_{k}\sum\limits_{i} (y_{ijk} - m_{jk})^2 = \text{SS error} \tag{11.12.7}$$

the **sum of squares for error,** since it is based on deviations from a cell mean for individuals treated in exactly the same way; the only possible contribution to this sum of squares should be error variation.

Next, consider

$$\sum_j Rn(m_j - m)^2 = \text{SS columns,} \qquad (11.12.8)$$

which is the **sum of squares between columns.** Here the deviations of column treatment means from the grand mean make up this sum of squares. This sum of squares reflects two things: the treatment effects of the columns *and* error. Notice that this sum of squares is identical to the sum of squares between groups found in the one-way analysis if the different experimental groups are regarded as columns in the table.

The third term is

$$\sum_k Cn(m_k - m)^2 = \text{SS rows,} \qquad (11.12.9)$$

which is the **sum of squares between rows.** It is based upon deviations of the row means from the grand mean and thus reflects both row-treatment effects and error. This is the same as the sum of squares between groups if data were regarded as coming only from experimental groups corresponding to the rows.

Finally, the fourth term is

$$\sum_j \sum_k n\ (m_{jk} - m_j - m_k + m)^2 = \text{SS interaction,} \qquad (11.12.10)$$

the **sum of squares for interaction.** This sum of squares involves only *interaction effects and error.*

The partition of the sum of squares for a two-way analysis can be written in the following schematic form:

$$\text{SS total} = \text{SS error} + \text{SS columns} + \text{SS rows} + \text{SS interaction.} \qquad (11.12.11)$$

Whereas in the one-factor analysis the total sum of squares can be broken into only two parts, a sum of squares between groups and a sum of squares within groups (error), in the two-factor analysis with replication the total sum of squares can be broken into *four* distinct parts. The principle generalizes to experimental layouts with any number of treatments and treatment combinations, but we shall stop with the two-factor situation.

As in Section 11.7, suppose that we assume that the error terms are independent with zero mean and constant variance σ_e^2. An unbiased estimate of σ_e^2 is given by the **mean square for error,** which is defined as

$$\text{MS error} = \frac{\text{SS error}}{RC(n-1)}. \qquad (11.12.12)$$

The other mean squares of interest in the two-factor model are the **mean square between columns,**

$$\text{MS columns} = \frac{\text{SS columns}}{C-1} ; \tag{11.12.13}$$

the **mean square between rows,**

$$\text{MS rows} = \frac{\text{SS rows}}{R-1} ; \tag{11.12.14}$$

and the **mean square for interaction,**

$$\text{MS interaction} = \frac{\text{SS interaction}}{(R-1)(C-1)} . \tag{11.12.15}$$

We state the expectations of these mean squares without proof:

$$E(\text{MS error}) = \sigma_e^2, \tag{11.12.16}$$

$$E(\text{MS columns}) = \sigma_e^2 + \frac{Rn \sum_j \alpha_j^2}{C-1}, \tag{11.12.17}$$

$$E(\text{MS rows}) = \sigma_e^2 + \frac{Cn \sum_k \beta_k^2}{R-1}, \tag{11.12.18}$$

and

$$E(\text{MS interaction}) = \sigma_e^2 + \frac{n \sum_j \sum_k \gamma_{jk}^2}{(R-1)(C-1)} . \tag{11.12.19}$$

If there are no column effects, row effects, or interaction effects, the expectation of each mean square will simply be σ_e^2. However, nonzero column effects imply that $E(\text{MS columns}) > \sigma_e^2$, nonzero row effects imply that $E(\text{MS rows}) > \sigma_e^2$, and nonzero interaction effects imply that $E(\text{MS interaction}) > \sigma_e^2$. Just as in the case of the one-factor model, we will use these results concerning the expected mean squares to make inferences about the effects. In the two-factor model, of course, there are three different types of effect to investigate: column effects, row effects, and interaction effects.

11.13 *F* TESTS IN THE TWO-FACTOR MODEL

As in the case of the one-factor model, the calculation of sums of squares and mean squares requires no assumptions, and the determination of the expectations of the mean squares utilizes only the assumption that the error terms are independent with zero mean and constant variance σ_e^2. In order to make inferences, we will add the assumption that the error terms are normally distributed. Thus, the assumptions concerning the error terms are identical to the assumptions in Section 11.8. The development of the inferential procedures discussed in this section is precisely analogous to the development in Section 11.8, so we will present the procedures briefly and then attempt to discuss their implications.

First, consider the column effects. If the null hypothesis of no column effects is true, then $\alpha_j = 0$ for all j and $E(\text{MS columns}) = \sigma_e^2$. Under the assumption of normality, MS columns/σ_e^2 is a chi-square variable divided by its degrees of freedom, $C - 1$. Similarly, whether or not there are any treatment effects, MS error/σ_e^2 is a chi-square variable divided by its degrees of freedom, $RC(n - 1)$. Furthermore, MS columns and MS error are *independent* estimates of σ_e^2 when the hypothesis of no column effects is true. Thus, their ratio has an F distribution,

$$F = \frac{\text{MS columns}}{\text{MS error}}, \qquad (11.13.1)$$

with $C - 1$ and $RC(n - 1)$ degrees of freedom. As in the test of Section 11.8, high values of F support the hypothesis of nonzero effects.

For row effects, an identical argument to that in the preceding paragraph can be used to show that the ratio of MS rows to MS error has an F distribution,

$$F = \frac{\text{MS rows}}{\text{MS error}}, \qquad (11.13.2)$$

with $R - 1$ and $RC(n - 1)$ degrees of freedom *if* there are no row effects. Thus, this ratio can be used to test the hypothesis that $\beta_k = 0$ for all k.

Finally, $E(\text{MS interaction}) = \sigma_e^2$ if $\gamma_{jk} = 0$ for all j and k, and the ratio of MS interaction to MS error has an F distribution,

$$F = \frac{\text{MS interaction}}{\text{MS error}}, \qquad (11.13.3)$$

with $(R - 1)(C - 1)$ and $RC(n - 1)$ degrees of freedom. This ratio can be used to test the hypothesis of no interaction effects. Thus, we see that it is possible to make separate tests of the hypothesis of no row effects, the hypothesis of no column effects, and the hypothesis of no interaction effects, all from the same data. The "impact" of row and column effects is much the same as the "impact" of treatment effects in the one-factor model, so we need not dwell on that subject. However, the notion of nonzero interaction effects will be discussed in more detail.

The presence or absence of interaction effects, as inferred from the *F* test for interaction, can have a very important bearing on how we interpret and use the results of an experiment. Significant interaction effects usually reflect a situation in which overall estimates of differences due to one factor are fine as predictors of average differences over *all possible levels of the other factor*, but it will not necessarily be true that these are good estimates of the differences to be expected when information about the specific level of the other factor is given. Significant interaction serves as a warning that treatment differences *do* exist, but to specify exactly *how* the treatments differ, and especially to make good individual predictions, we must look *within* levels of the *other* factor. The presence of interaction effects is a signal that in any predictive use of the experimental results, effects attributed to particular treatments representing one factor are best qualified by specifying the level of the *other* factor. This is extremely important if we are going to try to use estimated effects in forecasting the result of applying a treatment to an individual; when interaction effects are present, the best forecast can be made only if the individual's status on *both* factors is known.

For example, suppose that an experimenter is comparing two methods of instruction in golf. Let us represent these two methods as the column treatments in the data table. The other factor considered is the sex of the student; the study employs a group of 50 boys and a group of 50 girls, with 25 subjects in each group taught by Method I and the remainder by Method II. After a fixed period of instruction by one or the other method, each member of the sample is given a proficiency test. Table 11.13.1 gives the sample means for the four subgroups. For a small enough estimated error variance, such data would lead to the conclusion that no difference exists between boys and girls in terms of performance on the proficiency test but that *both column effects and interaction effects do exist*. Now suppose that the experimenter wants to decide which method to use for the instruction of an individual student. If *low* scores indicate good performance, but the sex of the student is not known, then Method II clearly is

Table 11.13.1   Sample Means for Golf-Instruction Example

METHOD

		I	II	
	Girls	55	65	60
SEX				
	Boys	75	45	60
		65	55	

called for, since the experimenter's best estimate is that Method I gives a higher overall mean than Method II. However, suppose that the experimenter knows that the individual to be instructed is a *girl*; in this case, the experimenter does much better to choose Method I, since there is evidence that *within the population of girls, mean II is higher than mean I*.

Although it is necessary to consider possible interaction effects even in fairly simple experiments, the subject of interaction and of the interpretation that should be given to significant tests for interaction is neither elementary nor fully explored. To a very large extent, the presence or absence of interactions in an experiment is governed by the scale of measurement used for the dependent variable. Thus, in terms of the original scale of measurement, interaction may be present, but if, for example, the values are transformed into their respective logarithms, interaction effects may vanish. It is clear that in many circumstances evidence for interaction reflects not so much a state of nature as our own inability to find the proper measurement scales for the phenomena we study. Since simple additive models are so much more tractable theoretically and practically than models including all the qualifications introduced by interaction, it is often desirable to transform the original data to eliminate interactions. Such considerations are, however, far beyond our limited scope.

Interaction effects can be studied separately only in a two-way (or higher) analysis of variance with crossed factors, where the experiment is carried out **with replication.** Furthermore, the procedures we develop here apply only to the situation where the experimental design is **orthogonal.** We need to have a look at what these terms "with replication" and "orthogonal" imply about the way the experiment is designed.

The discussion of the two-factor analysis of variance is limited here to replicated experiments. For our purposes this means that **within each treatment combination there are at least two independent observations made under identical experimental circumstances.** The requirement that the experiment be replicated is introduced here so that an error sum of squares will be available, permitting the study of tests both for treatment effects and for interaction. If there were only one observation for each treatment combination, we would not be able to test separately for interaction effects, since in this situation there is no direct way to estimate error variance apart from interaction effects. Occasionally, experiments are carried out where only one observation is made per treatment combination; under our two-factor model this makes it necessary to know or to assume that no interaction effects exist if a test for main treatment effects is to be carried out. This assumption is often very questionable, and most circumstances requiring a nonreplicated experiment will fit into a model slightly different from those discussed in this chapter.

We now turn to the term "orthogonal design." An **orthogonal design** for an experiment can be defined as a way of collecting observations that will permit us to estimate and test for the various treatment effects and for interaction effects separately. The potential information in the experiment can be "pulled apart" for study in an orthogonal design. Any experimental layout can be regarded as an orthogonal design provided that: (1) the observations within a given treatment *combination* are sampled at random and independently from a normal population, and (2) the number of observations in *each possible combination* of treatments is the same. Thus, the usual procedure in setting up an experiment to be analyzed by the two-factor (or higher order) analysis of variance is to assign experimental units at random and independently to each combination of treatments so as to have an equal number in each combination. If at all possible, experiments should be set up in this way, not only to insure orthogonality, but also to minimize the effect of nonhomogeneous population variances should they exist.

In applying the procedures developed earlier in this section, we must keep the assumptions in mind. With regard to violations of the assumptions, the comments in Section 11.8 are relevant in the two-factor case as well: nonnormality is generally a serious problem only for small samples and nonhomogeneous variances are not too serious if the sample sizes are equal for all treatment combinations, but lack of independence of the error terms creates very serious difficulties.

11.14 COMPUTING FORMS FOR THE TWO-FACTOR MODEL

In carrying out an analysis of variance the following computing forms are generally used. These sums of squares are algebraically equivalent to those given in Section 11.12.

$$\text{SS total} = \sum_j \sum_k \sum_i y_{ijk}^2 - \frac{\left(\sum_j \sum_k \sum_i y_{ijk}\right)^2}{RCn}. \tag{11.14.1}$$

$$\text{SS rows} = \frac{\sum_k \left(\sum_j \sum_i y_{ijk}\right)^2}{Cn} - \frac{\left(\sum_j \sum_k \sum_i y_{ijk}\right)^2}{RCn}. \tag{11.14.2}$$

$$\text{SS columns} = \frac{\sum_j \left(\sum_k \sum_i y_{ijk}\right)^2}{Rn} - \frac{\left(\sum_j \sum_k \sum_i y_{ijk}\right)^2}{RCn}. \tag{11.14.3}$$

$$\text{SS error} = \sum_j \sum_k \sum_i y_{ijk}^2 - \frac{\sum_j \sum_k (\sum_i y_{ijk})^2}{n}. \qquad (11.14.4)$$

$$\text{SS interaction} = \frac{\sum_j \sum_k (\sum_i y_{ijk})^2}{n} - \frac{\sum_k (\sum_j \sum_i y_{ijk})^2}{Cn}$$

$$- \frac{\sum_j (\sum_k \sum_i y_{ijk})^2}{Rn} + \frac{(\sum_j \sum_k \sum_i y_{ijk})^2}{RCn}$$

$$= \text{SS total} - \text{SS rows} - \text{SS columns} - \text{SS error}. \qquad (\text{11.14.5})$$

Notice that the sum of squares for columns is calculated just as for a one-factor analysis of data arranged into columns. Furthermore, the sum of squares for rows is identical to the sum of squares between groups when the data are arranged into a table where the experimental groups are designated by rows. The total sum of squares is also calculated in exactly the same way as for a one-factor analysis. The only new features here are the computations for error and for interaction. Generally, the error term is calculated directly, and then the interaction term is found by subtracting the sums of squares for rows, columns, and error from the total sum of squares.

As might be expected, the computational burden is even greater in the two-way model than in the one-way model. A brief outline of the steps to follow is presented below. It should be emphasized that this outline is merely for convenience, and by no means implies that you can learn the two-way analysis of variance by simply memorizing the successive steps. Computer methods are replacing hand computation in applications of the analysis of variance and other computationally cumbersome statistical methods. In view of these new technological aids, the important thing is to *understand* the underlying theory and the development of the various formulas.

1. Arrange the data into an $R \times C$ table, in which the R rows represent the R different treatments of one kind, and the C columns the C different treatments of the other kind. Each cell in the table should contain the same number n of observations. There are RCn distinct observations in all.

2. Square each raw score and sum over all individuals in all cells to find $\sum_j \sum_k \sum_i y_{ijk}^2$. Call this quantity A.

3. Sum the raw scores in a given *cell jk* to find $\sum_i y_{ijk}$. Do this for *each cell*, and reserve these values for use in later steps.

4. Now sum the resulting values (step 3) over *all cells* to find $\sum_j \sum_k \sum_i y_{ijk}$. Call this quantity B. Find the **sum of squares total** by $A - (B^2/RCn)$.

5. Next, take the RC different values found in step 3 and sum the cell totals for a *given row across columns* to find $\sum_j \sum_i y_{ijk}$. The result for any row k will be designated by D_k.

6. Having carried out step 5 for *each row*, square each D_k and then sum over all of the various rows to find $\sum_k D_k^2$. Divide this quantity by Cn, the number of observations per row. Then

$$\frac{\sum\limits_{k} D_k^2}{Cn} - \frac{B^2}{RCn}$$

is the **sum of squares for rows.**

7. Now return to the quantities found in step 3. This time sum the cell totals for *a given column across rows* to find $\sum_k \sum_i y_{ijk}$ and, for column j, call this value G_j.

8. Having carried out step 7 for each column, square each G_j and then sum across the various columns to find $\sum_j G_j^2$. Divide this quantity by Rn, the number of observations per column. Then

$$\frac{\sum\limits_{j} G_j^2}{Rn} - \frac{B^2}{RCn}$$

is the **sum of squares for columns.**

9. Once again return to the cell totals found in step 3. For a given cell jk, call the total H_{jk}. Now square H_{jk} for each cell and sum across *all cells* to find $\sum_j \sum_k H_{jk}^2$. Divide this by n, the number of observations per cell. Then

$$A - \frac{\sum\limits_{j} \sum\limits_{k} H_{jk}^2}{n}$$

is the **sum of squares for error.**

10. Find the **sum of squares for interaction** by taking

$$\text{SS total} - \text{SS rows} - \text{SS columns} - \text{SS error,}$$

or

$$\frac{\sum\limits_{j} \sum\limits_{k} H_{jk}^2}{n} - \frac{\sum\limits_{k} D_k^2}{Cn} - \frac{\sum\limits_{j} G_j^2}{Rn} + \frac{B^2}{RCn}.$$

11. Enter these sums of squares in the summary table.

Table 11.14.1   Analysis of Variance Table for the Two-Factor Model

Source	SS	df	MS	F
Rows	$\displaystyle \frac{\sum_k \left(\sum_j \sum_i y_{ijk}\right)^2}{Cn} - \frac{\left(\sum_j \sum_k \sum_i y_{ijk}\right)^2}{RCn}$	$R-1$	$\dfrac{\text{SS rows}}{R-1}$	$\dfrac{\text{MS rows}}{\text{MS error}}$
Columns	$\displaystyle \frac{\sum_j \left(\sum_k \sum_i y_{ijk}\right)^2}{Rn} - \frac{\left(\sum_j \sum_k \sum_i y_{ijk}\right)^2}{RCn}$	$C-1$	$\dfrac{\text{SS col.}}{C-1}$	$\dfrac{\text{MS col.}}{\text{MS error}}$
Interaction	$\displaystyle \frac{\sum_j \sum_k \left(\sum_i y_{ijk}\right)^2}{n} - \frac{\sum_k \left(\sum_j \sum_i y_{ijk}\right)^2}{Cn} - \frac{\sum_j \left(\sum_k \sum_i y_{ijk}\right)^2}{Rn} + \frac{\left(\sum_j \sum_k \sum_i y_{ijk}\right)^2}{RCn}$	$(R-1)(C-1)$	$\dfrac{\text{SS int.}}{(R-1)(C-1)}$	$\dfrac{\text{MS int.}}{\text{MS error}}$
Error (within cells)	$\displaystyle \sum_j \sum_k \sum_i y_{ijk}{}^2 - \frac{\sum_j \sum_k \left(\sum_i y_{ijk}\right)^2}{n}$	$RC(n-1)$	$\dfrac{\text{SS error}}{RC(n-1)}$	—
Totals	$\displaystyle \sum_j \sum_k \sum_i y_{ijk}{}^2 - \frac{\left(\sum_j \sum_k \sum_i y_{ijk}\right)^2}{RCn}$	$RCn-1$	—	—

12. Divide the SS rows by $R - 1$ to find MS rows.
13. Divide the SS columns by $C - 1$ to find MS columns.
14. Divide the SS interaction by $(R - 1)(C - 1)$ to find MS interaction.
15. Divide the SS error by $RC(n - 1)$ to find MS error.
16. Conduct the three F tests presented in Section 11.13.

The results of a two-factor analysis of variance can be displayed in a summary table; such a table is illustrated in Table 11.14.1.

11.15   AN EXAMPLE OF A TWO-FACTOR ANALYSIS OF VARIANCE

In the example of Section 11.9, suppose that the marketing manager is interested not only in different promotional schemes, but also in different models of the product in question. The product is a household appliance, and the different models are identical in every respect except the placement of the controls. There are just 2 models and 3 promotional schemes, so there are 6 treatment combinations. A random sample of 60 stores is selected, and these stores are randomly assigned to the treatment combinations in such a way as to give 10 stores in each treatment combination. The sales at each store are recorded for a 1-month period, and the results are presented in Table 11.15.1. The marketing manager wants to examine the effects due to promotional schemes, the effects due to models, and the interaction effects.

Table 11.15.1   Data for the Marketing Example

PROMOTIONAL SCHEME

		I		*II*		*III*	
		52	44	28	27	15	20
		48	46	35	31	14	21
	A	43	46	34	27	23	16
		50	43	32	29	21	20
		43	49	34	25	14	14
MODEL							
		38	38	43	37	23	26
		42	39	34	37	25	20
	B	42	34	33	40	18	19
		35	33	42	36	26	22
		33	34	41	35	18	17

Following the computational outline given in Section 11.14, we first find the square of each of the scores, and sum:

$$A = \sum_j \sum_k \sum_i y_{ijk}{}^2 = (52)^2 + (48)^2 + \cdots + (22)^2 + (17)^2 = 66{,}872.$$

Next, taking the sum of the cell sums gives the total sum

$$B = \sum_j \sum_k \sum_i y_{ijk} = 464 + 302 + \cdots + 214 = 1904.$$

Hence, the total sum of squares is

$$A - \frac{B^2}{RCn} = 66{,}872 - \frac{(1904)^2}{60} = 6451.7.$$

Now the cell totals are summed for each row:

$$D_1 = \sum_j \sum_i y_{ij1} = 464 + 302 + 178 = 944,$$

$$D_2 = \sum_j \sum_i y_{ij2} = 368 + 378 + 214 = 960.$$

The sum of squares for rows is found from

$$\frac{\sum_k D_k{}^2}{Cn} - \frac{B^2}{RCn} = \frac{(944)^2 + (960)^2}{30} - \frac{(1904)^2}{60} = 4.2.$$

In a similar way, we find the sum of squares for columns by first summing cell totals for each column:

$$G_1 = \sum_k \sum_i y_{i1k} = 464 + 368 = 832,$$

$$G_2 = \sum_k \sum_i y_{i2k} = 302 + 378 = 680,$$

and $\qquad G_3 = \sum_k \sum_i y_{i3k} = 178 + 214 = 392.$

The sum of squares for columns is found from

$$\frac{\sum_j G_j{}^2}{Rn} - \frac{B^2}{RCn} = \frac{(832)^2 + (680)^2 + (392)^2}{20} - \frac{(1904)^2}{60}$$

$$= 4994.1.$$

Next, the sum of squares for error will be calculated. We begin by squaring and summing the *cell totals*:

$$\sum_j \sum_k H_{jk}{}^2 = (464)^2 + (302)^2 + \cdots + (214)^2 = 662{,}288.$$

The sum of squares for error is

$$A - \frac{\sum_j \sum_k H_{jk}^2}{n} = 66{,}872 - \frac{662{,}288}{10} = 643.2.$$

The only remaining value to be calculated is the sum of squares for interaction; this is done by subtraction, as follows:

SS total − SS rows − SS cols. − SS error = 6451.7 − 4.2 − 4994.1 − 643.2,

or SS interaction = 810.2.

Table 11.15.2 is the summary table for this analysis of variance. The hypothesis of no row effects cannot be rejected, since the F value is less than unity. For the hypothesis of no column effects, an F of approximately 3.15 is required for rejection at the 5 percent level; the obtained F of 209 far exceeds this, and so we may conclude with considerable confidence that column effects exist. In the same way, the F for interaction effects greatly exceeds that required for rejecting the null hypothesis, and so there seems to be reliable evidence for such interaction effects.

Table 11.15.2 Analysis of Variance Table for Marketing Example

Source	SS	df	MS	F
Rows (models)	4.2	1	4.2	.35
Columns (promotional schemes)	4994.1	2	2497.05	209.8
Interaction	810.2	2	405.1	34.0
Error (within cells)	643.2	54	11.9	
Totals	6451.7	59		

Our conclusions from this analysis of variance make it reasonably safe to make the following assertions.

1. There is apparently little or no effect of models alone on sales.
2. The promotional schemes do appear to affect sales when considered over the two different models.
3. There is apparently an interaction between models and promotional schemes, implying that the magnitudes and the directions of the effects of promotional schemes differ for the two different models.

In short, the promotional scheme used makes a difference in sales, but the kind and extent of the difference depends upon the model.

The different column effects can be estimated from the column means and the overall mean:

$$\text{est. } \alpha_1 = 41.6 - 31.7 = 9.9,$$

$$\text{est. } \alpha_2 = 34.0 - 31.7 = 2.3,$$

and
$$\text{est. } \alpha_3 = 19.6 - 31.7 = -12.1.$$

(Because of rounding error these do not quite total zero, as they should.) In a similar way, interaction effects may be estimated from the means of the cells, the rows, and the columns:

$$\text{est. } \gamma_{11} = 46.4 - 31.5 - 41.6 + 31.7 = 5.0,$$

$$\text{est. } \gamma_{21} = 30.2 - 31.5 - 34.0 + 31.7 = -3.6,$$

and so on. The estimated total effect of promotional scheme I combined with model A is thus

$$\text{est. } (\alpha_1 + \gamma_{11}) = 9.9 + 5.0 = 14.9.$$

Note that for a store selected at random from those with scheme I, the best guess we can make about the effect of scheme I is 9.9 units. However, if in addition we are told that the store is in the group with model A, our best bet of the effect is 14.9. In the same way, the estimated effect of any column treatment j within a row-treatment population k is est. $(\alpha_j + \gamma_{jk})$. Observe that we are ignoring the β_k terms because of assertion (1) above. Obviously, if we had to make a decision, we would probably want to include estimates of β_k, whether or not the row effects were found to be significant. In this case our best bet of the effect of the combination of Type I packaging and advertising is

$$\text{est. } (\alpha_1 + \beta_1 + \gamma_{11}) = 9.9 - .2 + 5.0 = 14.7.$$

11.16 ESTIMATING STRENGTH OF ASSOCIATION IN THE TWO-FACTOR MODEL

The analysis for a two-factor experiment yields a mean square for one treatment factor and a separate mean square for the other. These two mean squares reflect nonredundant aspects of the experiment, even though they are each based on the same basic data; the first sum of squares reflects only the effects attributable to the first experimental factor (plus error), and the second reflects effects attributable to the other (plus error). Under the statistical assumptions we make, these two mean squares are independent of each other. Furthermore, the mean squares for interaction and for error are independent of each other and of the treatment mean squares. The analysis of variance lets the experimenter "pull apart" the factors that contribute to variation in the experiment and identify them exclusively with particular summary statistics. For experiments of the orthogonal, balanced type considered here, the analysis of variance is a routine method for finding the statistics that reflect particular, meaningful, aspects of the data.

Let us now consider the several F tests obtained from a two- or multi-factor experiment. The sums of squares and mean squares for columns, for rows, for interaction, and for error are all, under the assumptions made, independent of each other. However, are the three or more F tests themselves independent? Does the level of significance shown by any one of the tests in any way predicate the level of significance shown by the others? Unfortunately, it can be shown that such F tests are *not* independent. Some connection exists among the various F values and significance levels. This is due to the fact that each of the F ratios involves the same mean square for error in the denominator; the presence of this same value in each of the ratios creates some statistical dependency among them. If three F tests are carried out, and these tests actually are independent, the probability is $1 - (.95)^3$, or about .14, that *at least one* of the tests will show spurious significance at the .05 level. However, for the usual situation where the tests are not independent, the probability that at least one is spuriously significant is not necessarily .14.

For really complicated analyses of variance, the problem becomes much more serious, since a fairly large number of F tests may be carried out, and the probability may be quite large that one or more tests gives spuriously significant results. The matter is further complicated by the fact that the F tests are not independent, and the number of "significant" results to be expected by chance is quite difficult to calculate exactly. For this reason, when large numbers of F tests are performed, the experimenter should not pay too much attention to isolated results that happen to be significant. Rather, the pattern and interpretability of results, as well as the strength of association represented by the findings, form a more reasonable basis for the overall evaluation of the experiment. When the number of degrees of freedom for the mean square error is very large, then the various F tests may be regarded as approximately independent, and the number of significant results at the .05 level to be expected should be close to 5 percent. Even here, however, the importance of a particular result is very difficult to interpret on the basis of significance level alone. A great deal of thought must go into the interpretation of a complicated experiment, quite over and above the information provided by the significance tests.

In some circumstances, a statistician applying the two-factor analysis of variance model may be primarily interested in the estimates of the effects. Even when the F tests are of interest, it is informative to supplement the F tests by attempting to assess the strength of association represented by either main effects or by interaction effects. In Section 11.10 we developed a measure of association, denoted by ω^2, for the one-factor model. In the two-factor model, there are three such measures, corresponding to the two main effects and the interaction effect.

Suppose that X represents the column variable in a two-factor situation, W represents the row variable, and Y represents the dependent variable. If the probability of an observation in column j is $1/C$ for all j, the probability of an observation in row k is $1/R$ for all k, and the probability of an observation in cell jk is $1/RC$ for all j and k, then the variance of the *marginal* distribution of Y is

$$\sigma_Y^2 = \sigma_e^2 + \frac{\sum_j \alpha_j^2}{C} + \frac{\sum_k \beta_k^2}{R} + \frac{\sum_j \sum_k \gamma_{jk}^2}{RC}. \tag{11.16.1}$$

The definition of $\omega_{Y|X}^2$, *the proportion of the variance of Y accounted for by X alone in the population, is*

$$\omega_{Y|X}^2 = \frac{(\sum_j \alpha_j^2)/C}{\sigma_Y^2}. \tag{11.16.2}$$

Similarly, we can define

$$\omega_{Y|W}^2 = \frac{(\sum_k \beta_k^2)/R}{\sigma_Y^2} \tag{11.16.3}$$

and

$$\omega_{Y|XW}^2 = \frac{(\sum_j \sum_k \gamma_{jk}^2)/RC}{\sigma_Y^2}. \tag{11.16.4}$$

This last index is the proportion of the variance of Y accounted for uniquely by the combination of *both X and W.*

Given these definitions, and given our results about the expectations of mean squares for the two-way analysis of variance (Section 11.12), we can estimate these values of ω^2 by taking

$$\text{est. } \omega_{Y|X}^2 = \frac{\text{SS columns} - (C-1)\ \text{MS error}}{\text{MS error} + \text{SS total}}, \tag{11.16.5}$$

$$\text{est. } \omega_{Y|W}^2 = \frac{\text{SS rows} - (R-1)\ \text{MS error}}{\text{MS error} + \text{SS total}}, \tag{11.16.6}$$

$$\text{est. } \omega_{Y|XW}^2 = \frac{\text{SS interaction} - (R-1)(C-1)\ \text{MS error}}{\text{MS error} + \text{SS total}}. \tag{11.16.7}$$

For the example in Section 11.15, these estimated values are

$$\text{est. } \omega_{Y|X}^2 = \frac{4994.1 - (2)(11.9)}{11.9 + 6451.7} = .77$$

and $\text{est. } \omega_{Y|XW}^2 = \dfrac{810.2 - (2)(11.9)}{11.9 + 6451.7} = .12.$

Since the F ratio shows a value less than 1.00 in the test for row differences, the estimate of $\omega_{Y|W}^2$ is set equal to zero. These estimates suggest that a very strong association exists between the treatments symbolized by X and the dependent variable Y. Knowing X alone tends to reduce our "uncertainty" about Y by about 77 percent. Moreover, there is apparently a further accounting for around 12 percent of the variance of Y if we know *both* of the categories represented by X and W, the treatment combination. In other words, we may safely conclude not only that association exists between independent variables, but also that this association is quite sizable in a predictive sense for any population situation corresponding to our experiment.

As always, we cannot be sure that any association at all exists; the validity of this statement depends upon the assumptions being correct and on these data not representing a chance result. However, the significance level assures us that the probability of error in such a statement is rather small, and our estimates of the strength of association are the best guesses we are able to make about the association's magnitude. Estimates of the size of effects and of strength of association are aids to the experimenter in trying to figure out what went on in the experiment and the meaning of the results. The F test per se is capable of indicating merely that something systematic seems to have happened. Only a careful examination of the data can make the meaning of the experiment clear, and this is why estimation of effects or of strength of association can form an informative part of an experimental analysis.

11.17   THE FIXED-EFFECTS MODEL AND THE RANDOM-EFFECTS MODEL

In this chapter we have discussed the analysis of variance only for the **fixed-effects model.** This model is appropriate when the experimental treatments actually administered are thought of as exhausting all treatments of interest. That is, given any experimental factor, all "levels" or categories of that factor which are of interest are observed. The only inferences to be drawn from the experiment concern the effects of those levels actually represented. In the example involving promotional schemes, this means that the three schemes investigated in the experiment were the only types of interest to the marketing manager. Similarly, in an experiment with two or more crossed factors, each combination of factor levels ordinarily is represented in the experiment, and the only inferences drawn concern those observed levels and their combinations. Thus, experiments to which the fixed-effects model applies are distinguished by the fact that inferences are to be made only about differences among the treatments actually administered, and about no other treatments that might have been included but were not. The effect of any treatment is "fixed," in the sense that it must appear in any complete repetition of the experiment.

On the other hand, suppose that the marketing manager is interested in the difference (with respect to sales) among a large number of promotional schemes. Because of cost considerations, three schemes are randomly selected and the experiment discussed in Section 11.9 is conducted. Here inferences are to be drawn about an entire set of distinct treatments or factor levels, including some not actually observed. The experimenter is interested in the *whole range* of possible levels, and the observed factor levels or experimental treatments only represent a random sample of the potential set that might have been observed. Before the experiment a sample is drawn from among all possible levels of a particular experimental factor, and then inferences are made about the effects of all such levels from the sample of factor levels. In our two-factor example, there are many models of the product that may be of interest, even though only two models are used in the experiment. If these two models were randomly chosen from the set of all possible models, then the experiment provides data for only a few of the many possible combinations of a promotional scheme and a model.

In the preceding paragraph, the fixed-effects model is no longer appropriate; instead, what is called the **random-effects model** now applies. The random-effects model applies when the experiment involves only a random sample of the set of treatments about which the experimenter wants to make inferences. The various treatments actually applied do not exhaust the set of all treatments of interest. Here, the effect of a treatment is not regarded as fixed, since any particular treatment itself need not be included each time the experiment is carried out; on each repetition of the entire experiment a new sample of treatments is to be taken. The experimenter may not actually plan to repeat the experiment, but conceptually each repetition involves a fresh sample of treatments.

In the simple one-way random-effects model, we can write the linear model as

$$y_{ij} = \mu + a_j + e_{ij}.$$ (11.17.1)

Observe that this is identical to the model in the fixed-effects situation, Equation (11.6.1), except that the term involving the *fixed* effect of treatment j, α_j, has been replaced by a_j. The term a_j represents a *random* effect. A random sample of treatments has been taken, and one of these is labeled the jth treatment in the experiment. Since the jth treatment is a randomly chosen member of a larger set of treatments, its effect, a_j, is a random variable. The assumptions concerning e_{ij} in the random-effects model are the same as in the fixed-effects model, and in addition some assumptions are made about a_j.

1. The possible values a_j represent a random variable having a distribution with a mean of zero and a variance $\sigma_A{}^2$.
2. The J values of the random variable a_j occurring in the experiment are completely independent of each other.
3. Each pair of random variables a_j and e_{ij} is completely independent.

The variance $\sigma_A{}^2$ represents the variance of the treatment effects. The hypothesis of no treatment effects is true when, and only when, the value of $\sigma_A{}^2$ is zero. Thus, in the random-effects model the test of no treatment effects concerns a variance, $\sigma_A{}^2$ (for all practical purposes, so did the test of no treatment effects in the fixed-effects model, but in that case the variance was given by $\sum_j \alpha_j{}^2/J$).

Before you begin to feel that our detailed study of the fixed-effects model is of no value in a random-effects situation, let us put in a word of reassurance: *computationally*, analyses using the two models are identical, although the inferences drawn are different. The F test provides inferences about the entire population of treatment effects in the random-effects model, even though only some of them are used in the experiment.

Instead of estimating effects directly by taking differences of the treatment means from the grand mean, as in the fixed-effects model, we wish to estimate $\sigma_A{}^2$ in the random-effects model. It can be shown that if the same number of observations, n, is made under each treatment,

$$E(\text{MS between}) = n\sigma_A{}^2 + \sigma_e{}^2 \qquad (11.17.2)$$

and
$$E(\text{MS within}) = \sigma_e{}^2. \qquad (11.17.3)$$

Thus, an unbiased estimate of $\sigma_A{}^2$ may be found by taking

$$\text{est. } \sigma_A{}^2 = \frac{\text{MS between} - \text{MS within}}{n}. \qquad (11.17.4)$$

This concludes our discussion of the random-effects model. We have attempted to present the basic conceptual differences between the fixed-effects model and the random-effects model. Fortunately, the computations are similar (differing only in the choice of quantities to estimate after the F test has been conducted) in the two models. It should be pointed out that in an experiment involving two or more factors, it is possible that one or more of the factors have fixed levels and the remaining factors are sampled as in the random-effects model. This situation calls for a third model, in which each observation is a sum of *both* fixed and random effects. This model, which we shall not discuss in this book, is called a **mixed model**.

11.18   REGRESSION AND ANALYSIS OF VARIANCE: THE GENERAL LINEAR MODEL

At various points in this chapter we have noted the similarity between the linear analysis of variance model and the linear regression model. The bivariate linear regression model is similar to the simple one-factor analysis of variance model, and the multiple regression model is similar to a two (or more)-factor analysis of variance model without the interaction terms. Even the assumptions concerning the error terms are the same in the two models. Let us see, then, if we can apply the technique of analysis of variance to a regression problem.

Suppose that in a simple one-way analysis of variance, the J treatment categories correspond to J values of the independent variable X and that we want to investigate the linear regression of a second variable Y on X. We can write the model in the form

$$y_{ij} = \beta_1 + \beta_2 x_j + e_{ij}.$$

There are J levels of X, and a sample of size n_j is taken in the jth level. On the basis of this sample, the parameters β_1 and β_2 are estimated by using the least-squares criterion, and the estimated linear regression is

$$y_{ij} = b_1 + b_2 x_j + \hat{e}_{ij},$$

or $$y_{ij} = \hat{y}_j + \hat{e}_{ij}.$$

Here, y_{ij} is the observed value and $\hat{y}_j$ the value predicted by the estimated linear regression.

Using the above model, the deviation of an observed value y_{ij} from the grand mean m_Y can be thought of as the sum of three parts:

$$(y_{ij} - m_Y) = (y_{ij} - m_{Yj}) + (m_{Yj} - \hat{y}_j) + (\hat{y}_j - m_Y). \quad (11.18.1)$$

The first term on the right-hand side of Equation (11.18.1) is simply the deviation of the particular observation from the mean of its group, or treatment level. The second term is the deviation of the mean of group j from the *predicted* value from the estimated linear regression for that group. The third term is the deviation of the predicted value itself from the grand mean.

By an argument like that in Section 11.7, it can be shown that over all observations in all groups, the total sum of squares can be partitioned into

$$\sum_j \sum_i (y_{ij} - m_Y)^2 = \sum_j \sum_i (y_{ij} - m_{Yj})^2 + \sum_j n_j(m_{Yj}^2 - \hat{y}_j^2)$$

$$+ \sum_j n_j(\hat{y}_j - m_Y)^2. \tag{11.18.2}$$

The first of these parts is

$$\text{SS error} = \sum_j \sum_i (y_{ij} - m_{Yj})^2, \tag{11.18.3}$$

which is just the ordinary SS *within* found as for a one-way analysis of variance. The second sum of squares is

$$\text{SS deviations from linear regression} = \sum_j n_j(m_{Yj}^2 - \hat{y}_j^2), \tag{11.18.4}$$

and the third part is

$$\text{SS linear regression} = \sum_j n_j(\hat{y}_j - m_Y)^2. \tag{11.18.5}$$

In short, for any regression problem it is true that

SS total = (SS error) + (SS deviations from lin. reg.) + (SS lin. reg.).

$$\tag{11.18.6}$$

Given the linear model, the first two terms on the right-hand side of Equation (11.18.6) reflect only error. Therefore, we combine these terms to arrive at

SS deviations and error = SS error + SS deviations from linear regression.

$$\tag{11.18.7}$$

The appropriate number of degrees of freedom associated with SS deviations and error is $n - 2$, and it can be shown that

$$E(\text{MS deviations and error}) = E\left(\frac{\text{SS deviations and error}}{n - 2}\right) = \sigma_e^2. \tag{11.18.8}$$

The expectation of MS linear regression, on the other hand, is

$$E(\text{MS linear regression}) = E\left(\frac{\text{SS linear regression}}{1}\right) = \sigma_e^2 + n\beta_2^2 s_X^2. \tag{11.18.9}$$

The analysis of variance table for this situation is presented in Table 11.18.1. SS total is calculated in the same way as in the one-factor analysis of variance (Section 11.9), and SS deviations and error is found by subtracting SS linear regression from SS total. The resulting F test, by the way, is equivalent to the T test which was discussed in Section 10.9 for the hypothesis that the slope of the regression line is zero.

Table 11.18.1   Analysis of Variance Table for Linear Regression

Source	SS	df	MS	F
Linear regression	$nr_{XY}^2 s_Y^2$	1	$\dfrac{\text{SS linear regression}}{1}$	$\dfrac{\text{MS linear regression}}{\text{MS deviations and error}}$
Deviations and error	—	$n - 2$	$\dfrac{\text{SS deviations and error}}{n - 2}$	
Total	—	$n - 1$		

This section demonstrates the fact that a bivariate linear regression model can be thought of as a simple analysis of variance model. Similarly, a multiple regression model can be thought of as an analysis of variance model with several factors. Both linear regression analysis and the analysis of variance fall under the heading of the **general linear model.** In multivariate statistical problems (problems involving several variables), the general linear model is of great value, both because it is often a realistic model and because nonlinear models are very difficult to work with when there are many variables. In the bivariate case, it is less difficult to work with nonlinear models, so the choice between linear and nonlinear models is not always obvious. We turn to this problem in the next section.

11.19   TESTING FOR LINEAR AND NONLINEAR REGRESSION

In the last section we mentioned that for any regression problem,

SS total $=$ (SS error) $+$ (SS deviations from linear regression)

$+$ (SS linear regression).   (11.19.1)

Under a linear model, the first two terms on the right-hand side of this equation were combined and used to estimate σ_e^2, since nothing could contribute to these sums of squares except random errors. If we admit the possibility of a nonlinear relationship between X and Y, however, the SS deviations from linear regression reflects not only random error, but also the possible nonlinear relationship between the two variables. The other two terms do not change: SS error just reflects error variance, and SS linear regression reflects both the effect of linear regression *and* error terms.

Under the simple one-factor analysis of variance model, SS total can be partitioned into the sum of SS within and SS between (Section 11.7). But we know that SS within is due only to error variance, so SS within corresponds to what we have called SS error in this section. From Equation (11.19.1), then, SS between must be equal to the sum of SS linear regression and SS deviations from linear regression:

SS between = SS linear regression + SS deviations from linear regression.

As usual, $E(\text{MS error})$ is simply σ_e^2, and $E(\text{MS linear regression})$ is given by Equation (11.18.9). The third mean square of interest is MS deviations from linear regression, and its expectation is given by

$$E(\text{MS deviations from linear regression}) = \sigma_e^2 + \frac{(\omega^2 - \rho_{XY}^2)n\sigma_Y^2}{J - 2}.$$

$$(11.19.2)$$

The summary table for this situation is given in Table 11.19.1. The

Table 11.19.1 Analysis of Variance Table for Linear and Nonlinear Regression

Source	SS	df	MS	F
Between groups	—	$(J - 1)$		
Linear regression	—	1	$\dfrac{\text{SS linear regression}}{1}$	$\dfrac{\text{MS linear regression}}{\text{MS error}}$
Deviations from linear regression	—	$J - 2$	$\dfrac{\text{SS deviations from linear regression}}{J - 2}$	$\dfrac{\text{MS deviations from linear regression}}{\text{MS error}}$
Within groups (error)	—	$n - J$	$\dfrac{\text{SS error}}{n - J}$	
Total	—	$n - 1$		

formulas for SS are not given in the table. SS total is computed in the usual manner [Equation (11.9.1)], as are SS between [Equation (11.9.2)] and SS within [Equation (11.9.3)]. The only problem is to partition SS between into its two components, SS linear regression and SS deviations from linear regression. We noted in the last section that

$$\text{SS linear regression} = n r_{XY}^2 s_Y^2. \tag{11.19.3}$$

A more convenient formula for finding SS linear regression directly from the data is

$$\text{SS linear regression} = \frac{n\left[\sum_j \sum_i x_j y_{ij} - \left(\sum_j n_j x_j\right)\left(\sum_j \sum_i y_{ij}\right)/n\right]^2}{n\left(\sum_j n_j x_j^2\right) - \left(\sum_j n_j x_j\right)^2}. \tag{11.19.4}$$

Once SS linear regression is computed,

SS deviations from linear regression = (SS between)

$$- \text{(SS linear regression)}. \tag{11.19.5}$$

This term reflects only *systematic differences between group means which are not due to linear regression*, as can be seen from Equation (11.19.2).

We can now discuss the two F tests in the above summary table. The first F test, given by

$$F = \frac{\text{MS linear regression}}{\text{MS error}}, \tag{11.19.6}$$

is a test for the existence of linear regression. This test is similar to the test for linear regression which was presented in the last section, so we shall not elaborate on it. The second F test is a test for the existence of curvilinear regression:

$$F = \frac{\text{MS deviations from linear regression}}{\text{MS error}}. \tag{11.19.7}$$

Here the hypothesis being tested is that there is no curvilinear regression.

At this point, let us present a simple example to illustrate these tests for linear and curvilinear regression. Suppose that a statistician is interested in the effect of intensity of background noise on the output of production line workers in a particular plant. An experiment is designed in which 6 different levels of noise intensity are employed, each representing a one-step interval in a scale of intensity. The dependent variable Y is the observed output of a worker under noise intensity x_j. The data are shown in Table 11.19.2.

Table 11.19.2   Data for Study of Noise and Output

		Noise intensity levels, x_j			
1	2	3	4	5	6
18	34	39	37	15	14
24	36	41	32	18	19
20	39	35	25	27	5
26	43	48	28	22	25
23	48	44	29	28	7
29	28	38	31	24	13
27	30	42	34	21	10
33	33	47	38	19	16
32	37	53	43	13	20
38	42	33	23	33	11
270	370	420	320	220	140

The usual computations for a one-way analysis of variance are carried out first:

$$\text{SS total} = (18)^2 + (24)^2 + \cdots + (11)^2 - \frac{(1740)^2}{60}$$

$$= 7252;$$

$$\text{SS between} = \frac{(270)^2 + \cdots + (140)^2}{10} - \frac{(1740)^2}{60}$$

$$= 5200;$$

and $\text{SS error} = \text{SS total} - \text{SS between} = 7252 - 5200 = 2052.$

Next the SS for linear regression is found from Equation (11.19.4). Since $n_j = 10$ for each group, this becomes

$$\text{SS linear regression} = \frac{60[\sum_j \sum_i x_j y_{ij} - (\sum_j 10x_j)(\sum_j \sum_i y_{ij})/60]^2}{60(\sum_j 10x_j{}^2) - (\sum_j 10x_j)^2}.$$

Here

$$\sum_j \sum_i x_j y_{ij} = \sum_j x_j (\sum_i y_{ij}) = 1(270) + 2(370)$$

$$+ \cdots + 6(140) = 5490,$$

$$\sum_j 10x_j = 10(1) + \cdots + (10)(6) = 210,$$

and

$$\sum_j 10x_j^2 = 10(1 + 4 \cdots + 36) = 910,$$

so that SS linear regression $= \dfrac{60[5490 - (210)(1740)/60]^2}{60(910) - (210)^2} = 2057.1.$

Then

SS deviations from linear regression = SS between − SS linear regression

$$= 5200 - 2057.1 = 3142.9.$$

The completed summary table is given in Table 11.19.3.

Table 11.19.3   Analysis of Variance Table for Study of Noise and Output

Source	SS	df	MS	F
Between groups	5200	5	—	—
Linear reg.	2057.1	1	2057.1	54.1
Dev. from lin. reg.	3142.9	4	785.7	20.7
Error	2052	54	38	
Totals	7252	59		

On evaluating these two F tests, we find that each is significant far beyond the .01 level. In short, we can reject both the hypothesis that there is no linear regression and the hypothesis that there is no curvilinear regression. We may say with some confidence that both linear and curvilinear regression exist. This contention is supported by a look at Figure 11.19.1, which shows the scatter diagram together with linear and quadratic functions fitted to the data.

The next question is this: how can we *estimate* the strength of both the linear and curvilinear relationship in these data? Recall that ρ_{XY}^2 represents the strength of the *linear* relationship between X and Y, and that ω^2 represents the strength of the *total* relationship between X and Y. Therefore, since the total relationship consists of the linear relationship and the

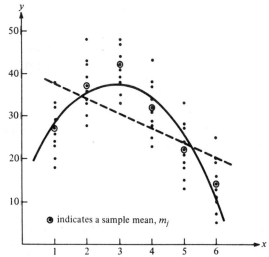

Figure 11.19.1 Scatter Diagram with Linear and Quadratic Functions for Study of Noise and Output

curvilinear relationship, the *strength of the curvilinear relationship between X and Y must be represented by the difference*

$$\omega^2 - \rho_{XY}^2. \tag{11.19.8}$$

It can be shown that the F test for the existence of curvilinear regression is actually a test of the hypothesis that $\omega^2 - \rho_{XY}^2$ is equal to zero.

Our immediate problem is to estimate ρ_{XY}^2 and $\omega^2 - \rho_{XY}^2$ from the data. In terms of the language of the analysis of variance, estimates can be found as follows:

$$\text{est. } \rho_{XY}^2 = \frac{\text{SS linear regression} - \text{MS error}}{\text{SS total} + \text{MS error}}, \tag{11.19.9}$$

$$\text{est. } (\omega^2 - \rho_{XY}^2) = \frac{\text{SS deviations from lin. reg.} - (J - 2) \text{ MS error}}{\text{SS total} + \text{MS error}}.$$

$$\tag{11.19.10}$$

Applying Equation (11.19.9) to the data from our example, we find that

$$\text{est. } \rho_{XY}^2 = \frac{2057.1 - 38}{7252 + 38} = \frac{2019.1}{7290} = .28.$$

From the evidence at hand, it appears that about 28 percent of the variance in Y may be attributable to a *linear* relationship with X, the noise levels.

Furthermore, using Equation (11.19.10), we have

$$\text{est. } (\omega^2 - \rho_{XY}{}^2) = \frac{3142.9 - (4)(38)}{7290} = \frac{2990.9}{7290} = .41.$$

Something on the order of 41 percent of the variance of Y is attributable to *curvilinear* relationship with X. Notice that we estimate a considerably stronger curvilinear than linear relationship in the population, even though the F tests were both very significant.

Finally, the total contribution of X to the variance of Y, or ω^2, is estimated to be

$$\text{est. } \omega^2 = \text{est. } \rho_{XY}{}^2 + \text{est. } (\omega^2 - \rho_{XY}{}^2) = .69,$$

so that we infer that quite a strong statistical association exists between X and Y.

11.20 THE ANALYSIS OF VARIANCE AND THE GENERAL PROBLEM OF EXPERIMENTAL DESIGN

Although the methods and examples discussed in this chapter are a small subset of the potential forms and applications of the analysis of variance, these methods and examples illustrate the essential features of the analysis of variance. The basic reasoning underlying our discussion of the simple one- and two-factor situations extends quite easily to more complicated situations. In experiments involving three or more different experimental factors, the total sum of squares is partitioned into even more parts, but the basic ideas of the partition, the mean squares, and the F tests are the same. In a three-factor experiment, not only are there mean squares representing the interactions of particular *pairs* of the experimental factors, but also a mean square representing the simultaneous interaction of *all three* of the factors.

The experiments discussed in this chapter are examples of **factorial designs,** which involve several experimental "factors" each represented at several "levels." In a complete factorial experiment, the set of experimental factors is completely crossed, so that every possible combination of factor levels is observed. It is also possible to consider experimental designs in which only some of the possible combinations are observed. For example, in some experiments categories or levels of one factor occur only *within* levels of another factor. Thus, we might be comparing three different teaching methods, and another factor of interest might be the particular classrooms in which the methods were tried. It is impractical to cross the factors of methods and classrooms, and so it is decided to apply the same method to each of two different classrooms, for a total of 6 classrooms, 2 per method. Here, the factor of "classroom" is said to be **nested** within the factor of "teaching method." A particular classroom occurs in the experiment only in association with one particular method. In a nested experiment such as this, any comparison of methods that we make is also a comparison of sets of classrooms; the only meaningful evidence for the possible effects of classrooms, over and above the effects of the methods, must come from comparison of classrooms *within* the particular methods. Many experiments calling for a simple one-way analysis of variance can be thought of as nested designs, where a factor corresponding to "individual subjects" is nested within the main treatments factor of the experiment.

If there are several factors, each with a number of treatment levels, a very large sample would be needed to include all possible combinations of treatment levels. In addition, to consider interactions would require replication, so the sample size would have to be at least twice as large as the number of combinations. For instance, if there are 4 factors and each has 4 levels, then there are 4^4, or 256 possible combinations, and with replication this would require at least 512 observations. Since observations are not without some cost, this could easily get out of hand. This is where the question of experimental design becomes important. Many specialized designs have been worked out for different purposes. For example, in a three-factor situation with a specialized design that is called a **Latin Square,** only one level of the third factor is used with each possible combination of levels of the first two factors. Two other types of designs that lead to reductions in the overall sample size are **incomplete block designs** and **fractional factorial designs.** For detailed discussions of these and other designs, you should consult a textbook on experimental design (see the references given in the Appendix).

The actual design of an experiment involves many considerations. We must consider the situation of interest carefully and make sure that the data collected will contain exactly the information needed to make inferences regarding this situation. For example, what are the factors of interest? How many levels of a factor should be observed? Will these levels be sampled or regarded as fixed? Must all combinations of factors be observed, or is there interest in only some of the treatment combinations? If only part of the possible set of combinations can be observed, is it possible to make inferences separately about the various factors and combinations? Is the contemplated sample size large enough to give the precision that we feel is necessary for our inferences?

In answering questions such as these, we need to consider the potential "costs" associated with the experimental designs as well as the "benefits" (the potential advantages of the experimental design in terms of making inferences). For example, the more treatments there are, the more the experiment will cost in time, effort, and other expenses. The experimenter would like to get by with as few assumptions and as few observations as possible. The more kinds of information the experimenter wants to gain, however, the more necessary it is to make additional assumptions and to take larger samples. Moreover, if many significance tests are carried out on the same data, the probability of spuriously significant results may be large. Thus, complicated experiments with many factors are somewhat uneconomical to perform, since they require a large number of observations as a rule, and a complete analysis of the data yields so many statistical results that the experiment as a whole is often very difficult to interpret in a statistical light. In planning an experiment, it is a temptation to throw in many experimental treatments, especially if the data are inexpensive and the experimenter is adventuresome. However, this is *not* always good policy if the experimenter is interested in finding meaning in the results; other things being equal, the simpler the experiment the better will be its execution, and the more likely will the experimenter be able to decide what actually happened and what the results actually mean.

All in all, there are several things the experimenter wants from the experiment and several ways to get them. Each desideratum has its price, however, and the experimenter must somehow decide if the gain in designing the experiment in a particular way is offset by the cost of so doing. Thus, the problem of experimental design has strong economic overtones. In this sense, the design of a complex experiment is very much like the design of a complicated, multistage sampling plan. As we pointed out, sample design and experimental design differ in that the latter is concerned with carefully controlled experiments (implying that the population of interest is more or less "created" by the experimenter), while the former deals with existing populations. In terms of economics, however, the problems of sample design and experimental design are very similar.

Texts in experimental design present ways of laying out the experiment so as to get "the most for the least" in a given situation. Various designs emphasize one aspect or another of the considerations and costs involved in obtaining and analyzing experimental data. Texts in design can give only a few standard types or layouts that study and experience have shown optimal in one or more ways, with, nevertheless, some price paid for using each design. Obviously, the best design for every conceivable experiment does not exist "canned" somewhere in a book, and experienced researchers and statisticians often come up with novel ways of designing an experiment for a special purpose. On the other hand, a study of the standard experimental designs enables us to appreciate the strengths and weaknesses of different ways of laying out an experiment.

The design and the analysis of experiments are intimately related, of course. The analysis of variance is a statistical technique which is encountered often in experimental work. Its advantages are many; the general technique is extremely flexible and applies to a wide variety of experimental arrangements. Indeed, the availability of a statistical technique such as the analysis of variance has done much to stimulate inquiry into the logic and economics of the *planning* of experiments. Statistically, the F test in the analysis of variance is relatively *robust*; as we have seen, the failure of at least two of the underlying statistical assumptions does not necessarily disqualify the application of this method in practical situations. Computationally, the analysis is relatively simple and routine and provides a condensation of the main statistical results of an experiment into an easily understood form.

However, the application of the analysis of variance never transforms a sloppy experiment into a good one, no matter how elegant the experimental design appears on paper, nor how neat and informative the summary table appears to the reader. Quite often, the statistician is able to tailor the experimental problem to fit the available methodology *without losing the essence of the problem in so doing*. However, this is not always the case. In general, experiments should be planned so as to capture the phenomena under study in their clearest, most easily understood forms, and this does not necessarily mean that one of the "textbook" experimental designs, nor a treatment by the analysis of variance, will best clarify matters. The experimental *problem* must come first in planning, and not the requirements of some particular form of analysis, even though, ideally, both should be considered together from the outset. If it should come to a choice between preserving the essential character of the experimental problem or using a relatively elegant technique such as the analysis of variance, then the problem should come first.

The point has been repeated several times, but it bears repetition. Statistics should aid in the clarification of meaning in an experimental situation, but the production of a statistical summary in some impressive and elegant form should never be the primary goal of an investigation. If the analysis of variance does not fit the problem, do not use it. If *no* inferential statistical techniques are available to fit the problem, do not alter the problem in essential ways to make some pet or fashionable technique apply. Above all, do not "jam" the data into some wildly inappropriate statistical analysis simply to get a significance test; there is little or nothing to gain by doing this. Thoughtless application of statistical techniques makes the reader wonder about the care that went into the experiment itself. *A really good experiment, carefully planned and controlled, often speaks for itself with little or no aid from inferential statistics.*

EXERCISES

1. Discuss the differences between sampling with replacement and without replacement, both from an infinite population and a finite population. If we wish to make inferences about the population on the basis of a sample, does it make any difference whether we sample with replacement or without replacement? Explain.

2. Distinguish between probabilistic and nonprobabilistic sampling plans, and comment on the following statement: "Any sampling plan can be thought of as probabilistic if we follow the subjective interpretation of probability."

3. We have a population of 4 items, divided into 2 groups of 2 items each. The items in the first group are A and B, and the items in the second group are C and D. A sampling plan is formulated as follows. Two fair coins are to be tossed. If the first coin comes up heads, A will be in the sample; if tails, B will be in the sample. If the second coin comes up heads, C will be in the sample; if tails, D will be in the sample. Thus, the sample will consist of 1 item from each group. For each item, find the probability that it will be included in the sample. Is this a simple random sample? Explain your answer, and discuss the general principle which is demonstrated by this exercise.

4. In the formula for the variance of the sample mean of a simple random sample without replacement from a finite population, can you explain why it is necessary to include the finite population correction? Why should the variance be reduced by this factor?

5. A population consists of 4 members, with values 1, 2, 3, and 4. A sample of size 2 is to be taken *without* replacement from this population.
   (a) Enumerate the possible sample outcomes, and for each calculate the sample mean.
   (b) Use the results of (a) to determine the sampling distribution of M, the sample mean.
   (c) From the sampling distribution of M, find $E(M)$ and $V(M)$.

6. In Exercise 5, do (a), (b), and (c) under the assumption that the sample of size 2 is to be taken *with* replacement, and then compare the variance of the sample mean with that obtained in Exercise 5. Why are the variances different?

7. In Exercise 5, suppose that the sample to be taken is of size 4. Find $V(M)$ if the sample is taken
   (a) with replacement,
   (b) without replacement.

8. In Exercise 14, Chapter 5, estimate the variance of the sample mean if the firm is actually quite small and the total number of long-distance telephone calls made by employees during the past year was only 200, if the sample is taken
   (a) with replacement,
   (b) without replacement.

9. In Exercises 5 and 6, let the sample size be 3, and consider the sample median instead of the sample mean. Find the variance of the sample median in each case.

10. Under what conditions might stratified sampling or cluster sampling be preferred to simple random sampling? Explain your answer.

11. What type of sampling plan is the one described in Exercise 3? If numerical values are assigned to the 4 items in the population as follows,

$$A = 4, \qquad B = 6, \qquad C = 18, \qquad D = 20,$$

(a) enumerate all possible samples and calculate the sample mean for each sample,
(b) find the sampling distribution of M and calculate its variance, $V(M)$.
(c) Suppose that two items were selected from the given population by simple random sampling, and repeat parts (a) and (b). With simple random sampling, is $V(M)$ less than, equal to, or greater than it is under the original sampling plan? Carefully discuss your results.

12. For the population given in Exercise 11, the following sampling plan is used. Toss a fair coin; if it comes up heads, then the sample consists of A and C; if it comes up tails, then the sample consists of B and D.
(a) Calculate the two possible values of M and determine $V(M)$.
(b) What type of sampling plan is this? Compare $V(M)$ with the variances obtained in Exercise 11.

13. Do Exercises 11 and 12 with the numerical values as follows:

$$A = 4, \qquad B = 18, \qquad C = 6, \qquad D = 20.$$

Explain how the change in values affects your answers. What does this suggest about the choice of sampling plans?

14. A marketing manager is interested in the number of trips per month that a person takes to a nearby shopping center. Denote the number of trips per month by X. The manager feels that the variance of X is 16 for women and 9 for men. He decides to stratify by sex, and there are 100 males and 200 females in the population of interest. He takes a random sample of 10 males and another random sample of 15 females, with the following results:

Males: 4   2   1   1   4   7   10   5   7   3

Females: 6   9   12   13   4   2   10   9   7   5   8   16   14   10   8

(a) Estimate the mean of each of the two strata (males and females), and determine the variance of each of the sample means.
(b) Estimate the mean of the entire population and determine the variance of your estimate.
(c) Pool the two samples and find the sample mean for all 25 observations. Is this the same as the estimate you found in (b)? Explain.

15. Is the sampling plan in Exercise 14 stratified sampling with proportional allocation? When might proportional allocation *not* be the best form of allocation in a stratified sampling plan?

16. Briefly discuss some of the problems that might arise in survey sampling; can you think of some problems that are not mentioned in the text?

17. Suppose you want to conduct a survey regarding voting preferences for the entire United States. This would obviously require a multistage sampling plan which might well be quite complex. Briefly outline two or three possible plans, indicating why you would choose a certain type of sampling plan at a particular level in the multistage plan.

18. The number of cars licensed in a particular state last year was 5 million, and the number of cars licensed in a neighboring state was 4 million. For the current year, the officials of the neighboring state estimate that there will be 4.6 million cars registered. Can you use this information to determine an estimate of the number of cars that will be registered in the first state in the current year?

19. In Exercise 14, suppose that the marketing manager is interested in the *total* number of trips to the shopping center in a given month, not the average number of trips per person for that month.
    (a) Estimate the population totals for the two strata (males and females), and determine the sample variance of estimation in each case.
    (b) Estimate the total for the entire population of males *and* females and determine the variance of estimation.

20. In Exercise 69, Chapter 7, the two towns form a congressional district. The population of town A is 45,000 and the population of town B is 60,000. From the data given in Exercise 69, Chapter 7, estimate the mean income for the congressional district and find the sample variance of estimation. (Ignore the problem of individual income versus family income by assuming that the random sample is from all *individuals* in each case.)

21. In Exercise 20, how would your answer change if the population of town A were 90,000 instead of 45,000? Explain.

22. The occurrence of defective and nondefective items among items from a particular production process is regarded as a Bernoulli process. A lot of 10 items from the process is considered, and 9 of these items are tested for defects. Exactly 1 of the 9 is defective, and *after* this information is obtained, the probability distribution of p, the probability that an item from the process is defective, can be represented by $P(p = .10) = .8$ and $P(p = .20) = .2$.
    (a) Find the conditional probability that the tenth item in the lot is defective, given that $p = .10$.
    (b) Find the conditional probability that the tenth item in the lot is defective, given that $p = .20$.
    (c) Find the marginal (unconditional) probability that the tenth item in the lot is defective after seeing that 1 of the other 9 was defective.

(d)   From the probability obtained in (c) and from your knowledge of the condition of the other 9 items in the lot, find the distribution of the proportion of defectives *in the lot of* 10 *items.*

23.   In Exercise 22, suppose that the lot contains 12 items but the other details of the problem are unchanged.
   (a)   Find the probability distribution of the number of defective items among the 3 untested items in the lot, given that $p = .10$.
   (b)   Find the probability distribution of the number of defective items among the 3 untested items in the lot, given that $p = .20$.
   (c)   Find the marginal (unconditional) probability distribution of the number of defective items among the 3 untested items in the lot after seeing that 1 of the first 9 items was defective.
   (d)   From the probability distribution obtained in (c) and from your knowledge of the condition of the first 9 items in the lot, find the distribution of the proportion of defectives *in the lot of* 12 *items.*

24.   Discuss the Bayesian approach to sampling without replacement from a finite population, as illustrated by Exercises 22 and 23.

25.   Carefully explain the difference between the terms "sample design" and "experimental design."

26.   Discuss the principle of randomization in the context of experimentation. Consider a situation in which a psychologist randomly assigns the participants in an experiment to groups and finds out that just by chance, all of the male subjects are in one group and all of the female subjects are in the other group. What options are available to the psychologist?

27.   The mean age of the population of one country (A) is 28.2 years, the mean age of the population of a second country (B) is 25.4 years, and the mean age of the population of a third country (C) is 30.4 years. The populations of the 3 countries are close enough in size that for all practical purposes, they can be regarded as equal. A sample of 5 individuals is to be taken at random within each of the 3 countries, and the age of each individual is recorded.
   (a)   If there is no variation within each country (a highly unrealistic assumption, to say the least), what can you say about the observations?
   (b)   What is the grand mean?
   (c)   What are the treatment effects for the 3 countries?
   (d)   If the random-error terms are 22.4, -10.1, 17.6, 2.4, and -8.7 in the sample from country A, what are the observed ages for country A?

28.   In Exercise 27, the sampling process can be simulated as follows. For each observation, start at a randomly chosen spot in a table of random numbers and record 4 digits from the table. The sum of these digits is then taken as the absolute value of the random-error term. Next, toss a coin; if it comes up heads, let the random-error term be positive, and if it comes up tails, let the random-error term be negative. The random-error term is then added to the population mean of the country from which the observation is assumed to

arise. (If a negative age results, discard it and start over again for that observation!)

(a) Follow this procedure 5 times for each country and display the resulting observations in a table.

(b) Find the sample mean for each country and the grand sample mean.

(c) Estimate the three treatment effects.

29. In Exercise 28, what are the implications of the sampling procedure with respect to the distribution of ages within the countries? Do you feel that these implications are realistic? Explain.

30. Prove, without consulting the text, that for the simple one-factor analysis of variance model given in this chapter,

$$\text{SS total} = \text{SS between} + \text{SS within}.$$

31. Why should a statistician want to partition the total sum of squares in a set of experimental data? Why is the approach discussed in Sections 11.6–11.20 called the analysis of variance?

32. Discuss the proposition that "If one defines MS total as SS total divided by $n - 1$ in the simple one-factor analysis of variance, then MS total is equal to the sum of MS between and MS within."

33. An experiment concerning the output per hour of four machines gave the following results:

	MACHINE		
A	B	C	D
160	134	104	86
155	139	175	71
170	144	96	112
175	150	83	110
152	156	89	87
167	159	79	100
180	170	84	105
154	133	86	93
141	128	83	85

(a) Construct an analysis of variance table.

(b) Test the hypothesis of equality among the four population means, using $\alpha = .01$.

(c) Estimate ω^2, the proportion of the total variance explained by the machines in this experiment.

(d) How would you interpret the results of this experiment?

34. A number of company presidents were sampled at random from each of 6 large geographical areas of the United States, and the annual income of each president sampled was recorded. The data are as follows (incomes in thousands of dollars):

Southeast	Southwest	Northeast	Northwest	Midwest	Far West
27	29	34	44	32	45
43	49	43	36	28	50
40	27	30	30	54	30
30	46	44	28	50	33
42	26	32	42	46	35
29	48	42		36	47
30	28	41		41	
41	30	33			
28	47	31			
	50	40			

Construct an analysis of variance table and test the hypothesis that the mean income of company presidents is the same in different parts of the United States.

35. In Exercise 34, estimate the effect associated with each geographical region. What is the estimate of ω^2, the proportion of the variance of income which is accounted for by geographical region? Given the estimated effects associated with the geographical regions and the estimate of the grand mean μ, compute an estimated value of the error term e_{ij} for each observation in the Southeast and Southwest groups.

36. Consider the following data:

Group 1	Group 2	Group 3	Group 4
1.69	1.82	1.71	1.69
1.53	1.93	1.82	1.82
1.91	1.94	1.75	1.86
1.82	1.60	1.64	1.90
1.57	1.78	1.52	1.39
1.77	1.85	1.73	1.56
1.94	1.98	1.86	1.74
1.60	1.72	1.68	1.83
1.74	1.83	1.54	1.47
1.74	1.75	1.75	1.64

You wish to carry out an analysis of variance and test for equality of means. However, you would like to simplify the computations by subtracting 1.00

from each observed value and then multiplying by 100. Complete the analysis and carry out the F test. Should the transformation of the numbers $[X' = 100(X - 1)]$ affect the results of the F test? Does it affect the values of MS between and MS within? Explain.

37. Estimate ω^2 for the data in Exercise 36 and estimate the effects associated with the 4 groups.

38. Suppose that you were given the sample mean, the sample standard deviation, and the sample size for each of 5 groups. Could you carry out an analysis of variance for these 5 groups, or would you need more information? How would you compute SS between and SS within?

39. Suppose that 4 randomly selected groups of 5 observations each were used in an experiment. Furthermore, imagine that for all 20 cases, the grand sample mean was 60. What would the data be like if the F test resulted in $F = 0$? What would the data be like if $F \to \infty$? If the hypothesis of equality of the means for the 4 groups were true, how large should we expect F to be?

40. Explain the advantage, if any, of a comparison of J means by an analysis of variance and an F test over the practice of carrying out a T test separately for each pair of means.

41. In Exercise 33,
    (a) find 95 percent confidence intervals *separately* for the difference between each possible pair of treatment effects,
    (b) find 95 percent confidence intervals *simultaneously* for the difference between each possible pair of treatment effects.
    (c) Compare your answers to (a) and (b) and explain any differences.

42. In Exercise 36, there are 6 possible pairs of treatments. Find 6 intervals (one for each of the 6 pairwise differences between treatment effects) that hold simultaneously with a 99 percent level of confidence.

43. Explain the term "multiple comparisons" and explain why it is desirable to consider all possible pairs of treatment effects simultaneously.

44. Consider the following data:

Group 1	Group 2
8	7
7	9
9	14
8	6
6	8
10	10

   (a) Test the hypothesis that the means are equal, using the T test developed in Chapter 7. (Assume that the variances are equal.)

(b) Test the hypothesis that the means are equal, using the F test from the analysis of variance.

(c) Is there any relationship between the value of T obtained in (a) and the value of F obtained in (b)? Explain.

45. In Exercise 73, Chapter 7, perform an analysis of variance on the data concerning the two detergents. Estimate the treatment effects and test the hypothesis that the population means are equal at the .025 level. Finally, estimate ω_2, the proportion of the total variance explained by the detergents.

46. In Exercise 73, Chapter 7, suppose that the experiment is conducted with 18 dirty sheets instead of 12 and with 3 detergents instead of 2. The results are as follows:

Bright detergent		Pure detergent		Clean detergent	
8	8	7	10	4	3
7	6	5	6	6	2
9	10	8	6	3	5

(a) At the .05 level, is there evidence that the three detergents differ in their mean ratings?

(b) Estimate the treatment effect for each detergent.

(c) Estimate ω^2.

47. A random sample of supermarkets was selected from a very large chain, and within each supermarket the number of items purchased on a single day by each of 10 customers selected at random was noted. The entries in the following table are the numbers of items purchased, arranged by store:

STORE											
1		*2*		*3*		*4*		*5*		*6*	
13	22	11	26	18	18	45	27	26	32	18	31
16	19	14	24	27	19	48	36	29	18	36	14
19	19	17	26	51	28	27	40	32	21	14	25
16	13	17	17	57	30	33	28	38	12	14	28
19	6	25	23	9	34	18	10	41	25	28	30

(a) Estimate the mean number of items purchased for each store and estimate the grand mean.

(b) Estimate the treatment effects.

(c) Estimate ω^2.

(d) Do the stores differ at the .01 level in terms of mean number of items purchased?

48. Discuss the claim that the main purpose of the analysis of variance is to generate one or more F tests.

49. Why is ω^2 a useful index in the one-factor analysis of variance?

50. Show that if $F > 1$, the estimate of ω^2 given by Equation (11.10.4) must be positive.

51. Construct an example, similar to that in the text, of a two-factor experiment in which
    (a) row, column, interaction, and error effects are all absent,
    (b) row effects only are present,
    (c) row and column effects, but no interaction effects, are present,
    (d) row, column, and interaction effects are all present, but there is no error.

52. Explain in your own words what an interaction effect is.

53. Construct an original example of a situation in which effective prediction from the results of an experiment makes it necessary to take account of interaction effects.

54. Given the following data, carry out an analysis of variance and the appropriate F tests.

20	−43	−25	−12
12	−10	7	14
−22	20	−32	−16
2	38	−6	19
−34	−29	−13	−15
−2	5	47	−9

55. For the data in Exercise 54, estimate the column effects, the row effects, and the interaction effects. Also, estimate ω^2 for rows, columns, and interaction. How would you interpret the results?

56. In an experiment, 36 patients with the same type of illness were divided into groups according to the hospital at which they were treated (A, B, or C) and according to sex. The number of days spent in the hospital by each patient was recorded, and the results follow:

	Hospital A		Hospital B		Hospital C	
	29	36	14	5	22	25
Women patients	35	33	8	7	20	30
	28	38	10	16	23	32
	25	35	3	5	18	7
Men patients	31	32	8	9	15	11
	26	34	4	6	8	10

Carry out an analysis of variance, testing for

(a) effects of the hospitals,

(b) effects of sex,

(c) interaction effects.

57. For the data of Exercise 56, estimate the ω^2 values for rows, columns, and interaction. Furthermore, estimate row, column, and interaction effects. Using these estimates and an estimate of the grand mean μ, estimate the error terms e_{ijk} corresponding to each patient.

58. In Exercise 36, suppose that the first 5 observations in each column are associated with level 1 of a second factor and the last 5 observations in each column are associated with level 2 of a second factor. Carry out a two-factor analysis of variance, testing for row effects, column effects, and interaction effects.

59. In Exercise 58, estimate ω^2 for rows, columns, and interaction. Comment on these results in light of the arbitrary fashion in which the second "factor" was created in Exercise 58.

60. Show that the sum of the estimates of $\omega_{Y|X}^2$, $\omega_{Y|W}^2$, and $\omega_{Y|XW}^2$ given by Equations (11.16.5)–(11.16.7) must be less than or equal to 1. Under what conditions will this sum equal 1 exactly?

61. Discuss the basic conceptual differences between the fixed-effects model and the random-effects model in analysis of variance, and make up two realistic examples involving analysis of variance, one using the fixed-effects model and the other using the random-effects model.

62. Analyze the data of Exercise 11, Chapter 10, by considering the analysis of variance model of Section 11.18, and carry out the appropriate F test for the existence of linear regression.

63. In Exercise 62, analyze the data by means of the analysis of variance model of Section 11.19, and carry out the appropriate F tests for the existence of linear regression and curvilinear regression. Also, estimate the strength of the linear relationship, the strength of the total relationship, and the strength of the curvilinear relationship.

64. Prove Equation (11.18.2).

65. Analyze the data of Exercise 35, Chapter 10, by considering an analysis of variance model, and carry out F tests for the existence of linear regression and curvilinear regression.

66. In Exercise 65, estimate the strength of the linear relationship and the strength of the curvilinear relationship.

67. Analyze the data of Exercise 70, Chapter 10, by considering an analysis of variance model, and carry out the appropriate F tests for the existence of linear regression and curvilinear regression. Also, estimate the strength of the linear relationship, the strength of the total relationship, and the strength of the curvilinear relationship.

68. In Exercise 67, plot the observations on a scatter diagram. Does the scatter diagram appear to be consistent with the results of Exercise 58? Explain.

69. Discuss the following concepts, which are important in experimental design and the analysis of variance: completely crossed, nested, balanced, randomization, replication, orthogonal designs.

70. In the design of an experiment, what desiderata should the statistician keep in mind? Discuss the general problem of experimental design.

12

NONPARAMETRIC
METHODS

In the discussion of statistical inference in this book, we have been primarily concerned with procedures that require distributional assumptions, such as normality or homogeneity of variances, concerning the population of interest. Once such distributional assumptions are made, the population can be expressed in terms of the parameters of the assumed distribution. For example, if it is assumed that a certain variable is normally distributed, then its distribution can be expressed in terms of μ and σ; if it is assumed that two variables have a bivariate normal distribution, then the theoretical regression curve of one variable on the other is a straight line, and that line can be expressed in terms of β_1 and β_2, the intercept and the slope. Thus, the problem of statistical inference is reduced to making inferences about parameters, and the methods used to make such inferences are often called **parametric methods.**

In many circumstances there may be considerable doubt about some of the parametric assumptions, such as that of normality. In the preceding chapters, we have tried to discuss the *robustness* of parametric procedures with respect to possible violations of the assumptions; for instance, we pointed out that the normality assumption is more vital for inferences concerning variances than for inferences concerning means. Even for relatively robust procedures, however, there are obviously situations in which the violations of the assumptions are serious enough to render the procedures inapplicable.

In addition to the problem of violations of distributional assumptions, we must consider a more basic problem involving levels of measurement. Most of the parametric methods discussed in this book have involved quantitative variables, implying measurement on either an interval scale or a ratio scale (see Section 5.3 for a discussion of the different levels of measurement). In simpler terms, we have been dealing with numerical values. In many situations, it is not possible or feasible to obtain interval-scale measurement, and it is very seldom possible to obtain ratio-scale measurement. If this is the case, then it is necessary to assign numerical values to different qualitative classes in order to use parametric techniques. Such assignments are somewhat arbitrary, and there is considerable doubt concerning the interpretation of the results.

If parametric methods are not applicable (either because the level of measurement is inappropriate or because the distributional assumptions are violated), how can the statistician make inferences and decisions? There is a class of statistical procedures which do not require stringent assumptions such as that of normality and which can be used with nominal or ordinal levels of measurement. Such procedures are generally called **nonparametric methods,** or **distribution-free methods.** These methods will be the subject matter of this chapter. First, we will discuss techniques requiring only nominal measurement, including procedures for comparing entire distributions (parametric methods usually just involve comparisons of certain parameters of distributions, such as means, rather than entire distributions). In the second half of the chapter we will deal with "order statistics," a set of methods requiring ordinal measurement. Some of the nonparametric methods will deal with problems which have not been discussed before; for example, we will discuss tests of the hypothesis that a particular distribution is normal (such a test might be used to see whether or not a parametric method is applicable). Other nonparametric methods will serve as alternatives to certain parametric methods, in which case we will briefly attempt to compare the competing methods.

12.1 METHODS INVOLVING CATEGORICAL DATA

Quite often the variables of interest are qualitative in nature rather than quantitative. That is, the level of measurement is nominal, so that each item observed can be placed in a particular category, hence the expression **categorical data.** Examples of such variables are the sex of a person, the color of a person's hair, the brand of a particular product that is purchased by a consumer, the industry to which a firm belongs, the type of drug that is administered to a patient, and so on. There are many instances in which the level of measurement is nominal. Furthermore, sometimes it is convenient to group quantitative data into categories and to use procedures applicable to categorical data. For instance, instead of recording a person's height in inches, we may simply label the person as "very tall" (say, taller than 6'4''), "tall" (between 6' and 6'4''), and so on.

In the next few sections we will discuss methods for dealing with categorical data. The first topic to be discussed is the comparison of a sample with a hypothetical population distribution. We would like to infer whether or not the sample result actually does represent some particular population distribution. We will deal only with discrete or grouped population distributions, and our inferences will be made through an *approximation* to the exact multinomial probabilities. Such problems are said to involve **goodness of fit** between a single sample and a single population distribution.

Next, we will extend this idea to the simultaneous comparison of several discrete distributions. Ordinarily, the reason for comparing such distributions in the first place is to find evidence for **association** between two qualitative attributes. In short, we are going to employ a test for independence between attributes, a test which is based on the comparison of *sample* distributions.

Finally, we will take up the problem of measuring the strength of association between two attributes from sample data. Tests and measures of association for qualitative data are very important for the social and behavioral sciences and business, where many of the variables of interest are essentially qualitative or categorical in nature. Because of this and because of their computational simplicity, the methods to be discussed in the following sections are widely used. However, the underlying theory is not simple, and misapplication of these methods is common. For this reason we will make a special attempt to discuss some of the basic ideas underlying these methods and to emphasize their inherent limitations.

12.2 COMPARING SAMPLE AND POPULATION DISTRIBUTIONS: GOODNESS OF FIT

There are problems in which we want to make direct inferences about two or more distributions, either by asking if a population distribution has some particular specifiable form or by asking if two or more population distributions are identical. Remember that population distributions may have some random variable as the domain, as in the examples in previous chapters, or the distribution may consist of a probability assigned to each of a set of mutually exclusive and exhaustive qualitative classes. Such a set of mutually exclusive and exhaustive *qualitative* events is often called an **attribute.** The methods of the first half of this chapter were originally developed for the study of theoretical distributions of attributes and especially for the problem of independence or association of attributes. However, as we shall see, the distribution of a random variable may be studied by these methods if the domain of the distribution is thought of as divided into a set of distinct class intervals.

It stands to reason that the best evidence we have about a population distribution grouped into qualitative classes is the sample distribution, grouped in the same way. Presumably the discrepancy between sample and theoretical distributions should have some bearing on the "goodness" of the theory in light of the evidence.

Suppose that a study of educational achievement of American men is being carried on. The population sampled is the set of all American males who are 25 years old at the time of the study. Each subject observed can be put into one and only one of the following categories, based on his *maximum* formal educational achievement:

1. college graduate
2. some college
3. high-school or preparatory-school graduate
4. some high school or preparatory school
5. finished eighth grade
6. did not finish eighth grade.

These categories are mutually exclusive and exhaustive; each man observed must fall into one and only one classification.

We happen to know that 10 years ago the distribution of educational achievement on this scale for 25-year-old men was as given in Table 12.2.1. We would like to know if the present population distribution on this scale is exactly like that of 10 years ago. Therefore, the hypothesis of "no change" in the distribution for the present population specifies the exact distribution given above. The alternative hypothesis is that the present population differs from the distribution given above in some unspecified way.

Table 12.2.1 Distribution of Educational Achievement of 25-Year-Old Men 10 Years Ago

Category	Relative frequency
1	.18
2	.17
3	.32
4	.13
5	.17
6	.03

Table 12.2.2 Frequency Distribution of Educational Achievement

Category (j)	Observed frequency (f_{o_j})	Expected frequency (f_{e_j})
1	35	36
2	40	34
3	83	64
4	16	26
5	26	34
6	0	6
	200	200

A random sample of 200 subjects is drawn from the current population of 25-year-old males, and the frequency distribution presented in Table 12.2.2 is obtained. The last column on the right gives the *expected* frequencies under the hypothesis that the population has the same distribution as 10 years ago. For category j, the expected frequency is

$$f_{e_j} = np_j,$$

where p_j is the relative frequency dictated by the hypothesis for category j. For example, under the "no change" hypothesis, the expected *relative* frequency of college graduates among 25-year-old males is .18, so in a sample of size 200, the expected frequency of college graduates is $200(.18) = 36$.

How well do these two distributions, the observed and the expected, agree? At first glance, you might think that the difference in observed frequency and expected frequency across the categories, or

$$\sum_j (f_{o_j} - f_{e_j}),$$

would describe the difference in the two distributions. However, it must be true that

$$\sum_j (f_{o_j} - f_{e_j}) = \sum_j f_{o_j} - \sum_j f_{e_j} = n - n = 0,$$

so this is definitely not a satisfactory index of disagreement.

On the other hand, the sum of the *squared* differences in observed and expected frequencies does begin to reflect the extent of disagreement:

$$\sum_j (f_{o_j} - f_{e_j})^2. \tag{12.2.1}$$

This quantity can be zero only when the fit between the observed and expected distributions is perfect, and it must be large when the two distributions are quite different.

Remember, however, that the real purpose in the comparison of these distributions is to investigate the hypothesis that the expectations are correct and that the current distribution actually is the same as 10 years ago. We might proceed in this way; given the probabilities shown as relative frequencies for the hypothetical population distribution, the exact probability of any sample distribution can be found. That is, given the hypothesis and the assumption of independent random sampling of individuals (with replacement), the exact probability of a particular sample distribution can be found from the **multinomial** rule (Section 4.8). Thus, in terms of the hypothetical population distribution, the probability of a sample distribution exactly like the one observed is, from Equation (4.8.1),

$$P(\text{observed distribution} \mid H_0)$$

$$= \frac{200!}{35!40!.83!16!26!0!} (.18)^{35}(.17)^{40}(.32)^{83}(.13)^{16}(.17)^{26}(.03)^{0}.$$

With some effort we could work this value out exactly. However, the idea of working out a multinomial probability for each possible such sample result is ridiculous for this large a sample and this many categories, because an absolutely staggering amount of calculation would be involved. In this kind of impasse, the theoretical statistician usually begins looking around for an approximation device. In this particular instance, it turns out that the multivariate normal distribution provides an approximation to the multinomial distribution for very large n, and thus the problem can be solved. We will not go into this derivation here; suffice it to say that the basic rationale for this test does depend on the possibility of this approximation and that the approximation itself is really good only for very large n. Then the following procedure is justified: we form the statistic

$$\chi^2 = \sum_j \frac{(f_{o_j} - f_{e_j})^2}{f_{e_j}}, \qquad (12.2.2^*)$$

which is known as the **Pearson χ^2 statistic** (after its inventor, Karl Pearson). Given that the exact probabilities for samples follow a multinomial distribution, and given a very large n, **when H_0 is true this statistic χ^2 is distributed approximately as chi-square with $J - 1$ degrees of freedom,** where J represents the number of categories.

Note that Equation (12.2.2) differs from Equation (12.2.1) in that each squared difference in frequency is weighted inversely by the frequency expected in that category. This weighting makes sense if we consider that a departure from expectation should get relatively more weight if we expect rather few individuals in that category than if we expect a great many. Also, note that the number of degrees of freedom here is $J - 1$. You may have anticipated this from the fact that the sum of the differences between observed and expected frequencies is zero; given any $J - 1$ such differences, the remaining difference is fixed. This is very similar to the situation for deviations from a sample mean, and the mathematical argument for degrees of freedom here would be much the same as for degrees of freedom in a variance estimate.

To return to our example, the value of the χ^2 statistic is

$$\chi^2 = \frac{(35 - 36)^2}{36} + \frac{(40 - 34)^2}{34} + \frac{(83 - 64)^2}{64}$$

$$+ \frac{(16 - 26)^2}{26} + \frac{(26 - 34)^2}{34} + \frac{(0 - 6)^2}{6} = 18.46.$$

This value is referred to the chi-square table (Table III) for $J - 1 = 6 - 1 = 5$ degrees of freedom. We are interested only in the upper tail of the chi-square distribution in such problems, because the only reasonable alternative hypothesis (disagreement between the observed and hypothetical distributions) must be reflected in *large* values of χ^2. Looking at Equation (12.2.2), it is obvious that small values of χ^2 (that is, values near zero) reflect *agreement* between the observed and expected frequencies. Table III shows that for 5 degrees of freedom, the p-value corresponding to $\chi^2 = 18.46$ is between .001 and .005. This strongly indicates that the current distribution of educational achievement is *not* exactly like that of 10 years ago.

Tests such as that in the example, based on a single sample distribution, are called **goodness-of-fit tests.** Chi-square tests of goodness of fit may be carried out for any hypothetical population distribution we might specify, provided that the population distribution is discrete or is thought of as grouped into some relatively small set of class intervals. However, in the use of the Pearson χ^2 statistic to approximate multinomial probabilities, it *must* be true that:

1. each and every sample observation falls into one and only one category or class interval;
2. the outcomes for the n respective observations in the sample are independent;
3. n is large.

These assumptions apply to the χ^2 tests presented in the next two sections as well as to the test discussed in this section.

The first two requirements stem from the multinomial sampling distribution itself; the multinomial rule for probability holds only for mutually exclusive and exhaustive categories and for independent observations in a sample (random sampling with replacement). The third requirement comes from the use of the chi-square distribution to approximate these exact multinomial probabilities; this approximation is good only for large sample size. Even when H_0 is true, the Pearson χ^2 statistic for goodness of fit is not distributed *exactly* as the random chi-square variable. The expected value and the variance of the Pearson χ^2, where ν symbolizes the degrees of freedom, $J - 1$, are

$$E(\chi^2) = \nu \qquad (12.2.3)$$

and

$$V(\chi^2) = 2\nu + \frac{1}{n}\left(\sum_j \frac{1}{p_j} - \nu^2 - 4\nu - 1\right). \qquad (12.2.4)$$

Recall that we learned in Chapter 6 that the expected value of a chi-square variable with ν degrees of freedom is ν and the variance is 2ν. Although the expected value of the Pearson χ^2 statistic is also ν, the variance of this statistic need not equal the variance of the random variable chi-square, unless n is infinitely large.

How large should sample size be in order to permit the use of the Pearson χ^2 goodness-of-fit tests? Opinions vary on this question, and some fairly sharp debate has been raised by this issue over the years. Many rules of thumb exist, but as a conservative rule one is usually safe in using this chi-square test for goodness of fit if each *expected* frequency, f_{e_i}, is 10 or more when the number of degrees of freedom is 1 (that is, two categories), or if the expected frequencies are each 5 or more where the number of degrees of freedom is greater than 1 (more than two categories). Be sure to notice that *these rules of thumb apply to expected, not observed, frequencies per category.*

12.3   A SPECIAL PROBLEM: A GOODNESS-OF-FIT TEST FOR A NORMAL DISTRIBUTION

One use of the Pearson goodness-of-fit χ^2 test is in deciding if a continuous population distribution has a particular form. That is, we might be interested in seeing if a sample distribution might have arisen from some theoretical form such as the normal distribution. It is important to recall that like most theoretical distributions useful in statistics, the normal is a family of distributions, particular distributions differing in the parameters entering into the rule as constants; for the normal distribution, these parameters are μ and σ, of course. It is also important to remember that although such a distribution is continuous, it is necessary to think of the population as grouped into a finite number of distinct intervals if the Pearson χ^2 test is to be applied.

For example, suppose that the president of a large chain of stores hypothesizes that for a given week, total sales per store at the member stores of the chain is normally distributed. This hypothesis refers not to the mean, nor to the variance of the distribution of sales, but rather to the form of the distribution. The president decides to draw a random sample of 400 stores from the chain to test the hypothesis of a normal distribution of sales.

The population distribution must be thought of as grouped into intervals. Furthermore, it is necessary that the number expected in each interval be relatively large. Therefore, the president must first decide on the number of intervals. It is desirable to have enough intervals to insure a fairly "fine" description, but not so many that the expected frequency for any interval is very small. Suppose that the president decides to use 8 intervals in such a way that each interval would be expected to include exactly $\frac{1}{8}$ of the population. Table 12.3.1 presents the 8 intervals in terms of standardized values, as determined from Table I.

Table 12.3.1 Eight Equally Likely Intervals for a Standard Normal Random Variable

Class limits in terms of z for class interval j	Approx. p_j	$f_{e_i} = np_j$
below -1.15	1/8	50
-1.15 to $-.68$	1/8	50
$-.68$ to $-.32$	1/8	50
$-.32$ to $\quad 0$	1/8	50
$0\quad$ to $\quad.32$	1/8	50
$.32$ to $\quad.68$	1/8	50
$.68$ to $\quad1.15$	1/8	50
above 1.15	1/8	50
	1	400

Note that this arrangement into intervals refers to the *population* distribution, assumed to be normal under the null hypothesis, and that this choice of intervals is made before the data are seen. This arrangement is quite arbitrary, and some other number of intervals might have been chosen; in fact, with a sample size this large, we would be quite safe in taking many more intervals with a much smaller probability associated with each. Notice also that the population is thought of as divided into intervals of unequal size, in order to give equal probabilities for the intervals. On the other hand, it is perfectly possible to decide on some arbitrary interval size in terms of z values and then to allow the various probabilities to be unequal. Using equal probabilities has two possible advantages: departures from normality in the middle of the score range are relatively more likely to be detected in this way than otherwise, and computations are simplified by having equal expectations for each interval. It must be emphasized, however, that the choice of intervals is arbitrary and that the χ^2 test is sensitive to this choice. Thus, we need to give considerable prior thought to how we want to group the population.

Now to proceed with the example. Before the sample distribution can be grouped into the same categories as the population, something must be known about the population mean and standard deviation. Our best evidence comes from the sample estimates m and $\hat{s}$, and so these are used in place of the unknown μ and σ. In the actual data for this example, suppose that it turns out that the sample mean m is 980 and the estimate $\hat{s}$ is 8.4. Using these estimates, we can convert the intervals of standardized values into intervals of nonstandardized values. This conversion, together with the sample results, is presented in Table 12.3.2.

Now the χ^2 test for goodness of fit is carried out:

$$\chi^2 = \frac{(20-50)^2}{50} + \frac{(48-50)^2}{50} + \frac{(70-50)^2}{50} + \frac{(56-50)^2}{50}$$

$$+ \frac{(46-50)^2}{50} + \frac{(55-50)^2}{50} + \frac{(60-50)^2}{50} + \frac{(45-50)^2}{50} = 30.12.$$

For this statistic, an adjustment must be made in the degrees of freedom.

Table 12.3.2 Observed and Expected Frequencies for Sales Example

Standardized limits	Nonstandardized limits	f_{o_i}	f_{e_i}
below −1.15	below 970.34	20	50
−1.15 to −.68	970.34 to 974.29	48	50
−.68 to −.32	974.29 to 977.31	70	50
−.32 to 0	977.31 to 980	56	50
0 to .32	980 to 982.69	46	50
.32 to .68	982.69 to 985.71	55	50
.68 to 1.15	985.71 to 989.66	60	50
above 1.15	above 989.66	45	50

Two parameters had to be estimated in order to carry out this test: the mean and the standard deviation of the population. *One degree of freedom is subtracted from J — 1 for each separate parameter estimated in such a test.* Therefore, the correct number of degrees of freedom here is

$$\nu = J - 1 - 2 = 5.$$

For 5 degrees of freedom, Table III shows that this obtained value exceeds that required for significance at the .001 level, so the p-value is less than .001. The president can feel quite confident in saying that the variable of interest (total weekly sales) is not normally distributed. By looking at Table 12.3.2, we can see that there are too few observations in the extreme left tail and too many observations in the interval from 974.29 to 977.31 as compared with what would be expected if the population were normally distributed with the same mean and variance.

12.4 PEARSON χ^2 TESTS OF ASSOCIATION

The general rationale for testing goodness of fit also extends to tests of independence (or lack of statistical association) between categorical attributes. Quite often situations arise where n independent observations are made and each and every observation is classified in two or more qualitative ways. For example, a person might be classified according to sex, hair color, and eye color; a firm might be classified according to industry, total assets (above a certain prespecified level or below that level), and earnings (again, above or below a certain level).

In this section we will consider the case in which there are two attributes of interest, A and B. Suppose that there are C different categories A_1, A_2, ..., A_C for attribute A and R different categories B_1, B_2, ..., B_R for attribute B. The data can then be put in the form of a table, with the C classes of attribute A making up the columns and the R classes of attribute B making up the rows. Such a table is called a **contingency table,** and each and every possible combination (A_j, B_k) is represented by one cell in the table. For example, when $R = C = 2$, we have the 2×2 table shown as Table 12.4.1.

Table 12.4.1 A 2×2 Contingency Table

	A_1	A_2
B_1	(A_1, B_1)	(A_2, B_1)
B_2	(A_1, B_2)	(A_2, B_2)

Consider a firm that sells materials to other firms. The president of the firm in question is concerned about overdue accounts and the relationship between the size of a customer firm (classified as large or small) and the state of the firm's account (overdue or not). Let A_1 and A_2 represent "large firm" and "small firm," and let B_1 and B_2 represent "account overdue" and "account not overdue." A random sample of 100 customers is taken from the firm's 10,000 customers; although the sample is taken without replacement, the sample size is small enough relative to the population size that it can be treated as a sample taken with replacement (see Section 11.1). The results are given in Table 12.4.2. Thus, the sample included 19 large firms with overdue accounts, 32 small firms with overdue accounts, 29 large firms without overdue accounts, and 20 small firms without overdue accounts.

Table 12.4.2 Contingency Table for Overdue-Accounts Example

	A_1	A_2	
B_1	19	32	51
B_2	29	20	49
	48	52	$100 = n$

Suppose that we had some hypothesis specifying the probability of each possible combination (A_j, B_k). We might want to ask how this hypothetical *joint* distribution actually fits the data. Just as for the simple frequency distribution considered in Section 12.2, the hypothesis about the joint distribution tells us what frequency to expect for each cell in the table. That is, for cell (A_1, B_1) we should expect exactly $nP(A_1, B_1)$ cases to occur in a random sample; for cell (A_1, B_2) our expectation is $nP(A_1, B_2)$, and so on for the other cells. Given this *complete specification* of the population joint distribution, for sufficiently large n we could apply the Pearson χ^2 test of goodness of fit. Just as for any goodness-of-fit test where no parameters are estimated, the number of degrees of freedom would be the number of distinct event classes minus 1. Since there are RC *joint* events in this instance, $RC - 1$ is the degrees of freedom for a goodness-of-fit test for such a joint distribution.

However, exact hypotheses about joint distributions are quite rare in applications. Very seldom would we want to carry out such a test even though it is possible. Instead, the usual null hypothesis is that the two attributes A and B are independent. If the hypothesis of independence can be rejected, then we say that the attributes A and B are statistically related or associated.

In Chapter 3 it was stated that two discrete attributes are considered independent *if and only if*

$$P(A_j, B_k) = P(A_j)P(B_k)$$

for *all possible* joint events (A_j, B_k). Given that the hypothesis of independence is true, and given the *marginal* distributions showing $P(A_j)$ and $P(B_k)$, we know what the joint probabilities *must* be.

On the other hand, this fact alone does us little good, because independence is defined in terms of the *population* probabilities, $P(A_j)$ and $P(B_k)$, and we do not know these. What can we use instead? *The best estimates we can make of the unknown marginal probabilities are the sample marginal proportions,*

$$\text{est. } P(A_j) = \frac{\text{freq. of } A_j}{n}$$

and

$$\text{est. } P(B_k) = \frac{\text{freq. of } B_k}{n},$$

for each A_j and B_k. Given these estimates of the true probabilities, then we expect that the frequency of the joint event (A_j, B_k) will be

$f_{e_{jk}} = $ expected frequency of $(A_j, B_k) = n[\text{est. } P(A_j)][\text{est. } P(B_k)]$.

Thus,

$$f_{e_{jk}} = \frac{(\text{freq. } A_j)(\text{freq. } B_k)}{n}. \tag{12.4.1}$$

In tests for independence, the expected frequency in any cell is taken to be the product of the frequency in the column times the frequency in the row, divided by n**.** Using these expected frequencies, the Pearson χ^2 statistic in a test for association is simply

$$\chi^2 = \sum_j \sum_k \frac{(f_{o_{jk}} - f_{e_{jk}})^2}{f_{e_{jk}}}, \tag{12.4.2*}$$

where $f_{o_{jk}}$ is the frequency actually observed in cell (A_j, B_k), $f_{e_{jk}}$ is the expected frequency for that cell under the hypothesis of independence, and the sum is taken over all of the RC cells.

The number of degrees of freedom for such a test, where sample estimates of the marginal probabilities are made, differs from the degrees of freedom for a goodness-of-fit test. As we saw, if each joint probability actually were completely specified by the hypothesis, then a goodness-of-fit χ^2 could be carried out over the cells; this would have $RC - 1$ degrees of freedom.

However, in the last section there was an instance of the principle that *1 degree of freedom is subtracted for each estimate made.* How many different estimates must we actually make in order to carry out the χ^2 test for association? Since there are C categories for attribute A, we must actually estimate $C - 1$ probabilities for this attribute, since given the first $C - 1$ probabilities, the last value is determined by the fact that $\sum_j P(A_j) = 1.00$. Furthermore, we also must estimate $R - 1$ probabilities for attribute B. In all, $(C - 1) + (R - 1)$ estimates are made, and this number must be subtracted out of the total degrees of freedom. Therefore, the degrees of freedom for a Pearson χ^2 test of association is

$$\nu = RC - 1 - (C - 1) - (R - 1) = (R - 1)(C - 1).$$

For our example, the expected frequency for cell (A_1, B_1) is found to be

$$f_{e11} = \frac{(\text{freq. } A_1)(\text{freq. } B_1)}{n} = \frac{(48)(51)}{100} = 24.48,$$

that for cell (A_2, B_1) is

$$f_{e21} = \frac{(\text{freq. } A_2)(\text{freq. } B_1)}{n} = \frac{(52)(51)}{100} = 26.52,$$

and so on, until the set of expected frequencies in Table 12.4.3 is found. Notice that in any row the sum of the expected frequencies must equal the observed marginal frequency for that row, and the sum of the expected frequencies in any column must also equal the observed frequency for that column.

Table 12.4.3   Expected Frequencies for Overdue-Accounts Example

	A_1	A_2	
B_1	24.48	26.52	51
B_2	23.52	25.48	49
	48	52	$100 = n$

Given the expected frequencies, the χ^2 test for this example is based on

$$\chi^2 = \frac{(19 - 24.48)^2}{24.48} + \frac{(32 - 26.52)^2}{26.52} + \frac{(29 - 23.52)^2}{23.52} + \frac{(20 - 25.48)^2}{25.48}$$

$$= 4.81.$$

In this 2×2 table, the degrees of freedom are $(2 - 1)(2 - 1) = 1$. For 1 degree of freedom, the p-value is between .025 and .05. This means, for instance, that we would reject the hypothesis of independence at the .05 level of significance, but not at the .025 level of significance. It appears that small firms are more likely to have overdue accounts than are large firms.

Two-by-two contingency or joint frequency tables are especially common. For such fourfold tables, computations for the Pearson χ^2 test can be put into a very simple form. Consider the observed frequencies given in Table 12.4.4. For this table,

$$\chi^2 = \frac{n(ad - bc)^2}{(a + b)(c + d)(a + c)(b + d)}, \tag{12.4.3}$$

and an even better approximation to a chi-square variable with 1 degree of freedom can be obtained from

$$\chi^2 = \frac{n[|\,ad - bc\,| - (n/2)]^2}{(a + b)(c + d)(a + c)(b + d)}. \tag{12.4.4}$$

Equation (12.4.4) differs from Equation (12.4.3) in that it includes a continuity correction, called **Yates' correction for continuity.**

Table 12.4.4   A 2×2 Contingency Table

	A_1	A_2	
B_1	a	b	$a + b$
B_2	c	d	$c + d$
	$a + c$	$b + d$	n

Although the 2×2 case is of some interest, there is no reason at all why the number of categories in A or B must be only 2. Furthermore, one or both of the attributes may represent numerical measurements grouped into intervals. To illustrate these two points, suppose that we are interested in voting preferences in a political race in which there are 3 candidates. In particular, we are interested in the association between the age of a voter and the voter's preference among the 3 candidates. Based on past experience in studying voting trends, we decide to set up 4 "age groups." A random sample of 100 voters is taken, with the results given in Table 12.4.5.

Table 12.4.5   Contingency Table for Voting Example

AGE GROUP

		1 (below 30)	2 (30–40)	3 (41–50)	4 (over 50)	
	1	5	8	5	4	22
PREFERRED CANDIDATE	2	10	7	8	6	31
	3	12	10	11	14	47
		27	25	24	24	100

From Equation (12.4.1), the corresponding expected frequencies under the assumption of independence of the attributes are as given in Table 12.4.6. Thus,

$$\chi^2 = \frac{(5 - 5.94)^2}{5.94} + \frac{(8 - 5.50)^2}{5.50} + \cdots + \frac{(14 - 11.28)^2}{11.28} = 3.28.$$

Here there are $(3 - 1)(4 - 1)$ degrees of freedom, and the p-value is between .75 and .90. This indicates that the association between age and voting preference in this case is not very strong.

This example illustrates the fact that the χ^2 test can be used if one or both of the variables of interest is quantitative rather than qualitative, provided that the quantitative variable is divided into intervals. In the example, the variable "age" was broken up into 4 intervals. Of course, a procedure like this involves "throwing away" some relevant information, since we are unable to distinguish between different ages within any age group. On the other hand, as long as at least one of the variables is strictly

Table 12.4.6   Expected Frequencies for Voting Example

AGE GROUP

		1 (below 30)	2 (30–40)	3 (41–50)	4 (over 50)	
	1	5.94	5.50	5.28	5.28	22
PREFERRED CANDIDATE	2	8.37	7.75	7.44	7.44	31
	3	12.69	11.75	11.28	11.28	47
		27	25	24	24	100

qualitative in nature, it is improper to make inferences such as those made in Chapter 10 concerning correlation coefficients—such procedures require assumptions such as normality, assumptions that are even more restrictive than assumptions concerning levels of measurement. When the parametric assumptions are applicable, procedures such as those in Chapter 10 are more powerful than the procedures discussed here in Chapter 12; when the assumptions are not applicable, the latter procedures can be applied whereas the former cannot.

Before moving on to a discussion of measures of association in·contingency tables, we should mention that some alternatives (including exact tests) to the Pearson χ^2 tests are available. For example, likelihood-ratio tests based on the multinomial rule have been developed for problems of goodness of fit and of association for categorical data (recall that the notion of likelihood-ratio tests was introduced in Section 7.6). We will not cover such methods in detail, but for tests of association, the statistic of interest is

$$\chi^2 = 2n \log n + 2 \sum_j \sum_k f_{o_i k} \log f_{o_i k} - 2 \sum_j f_{o_i .} \log f_{o_i .} - 2 \sum_k f_{o . k} \log f_{o . k},$$

where $f_{o_i .}$ is the observed frequency in column j, $f_{o . k}$ is the observed frequency in row k, and the logarithms are taken to the base e. This statistic, which is referred to the chi-square distribution with $(R - 1)(C - 1)$ degrees of freedom, becomes equivalent to the Pearson χ^2 test given by Equation (12.4.2) for very large n. For small samples, on the other hand, there is some reason to believe that the likelihood ratio procedures are preferable to the Pearson χ^2 procedures. Of course, the assumptions having to do with the completeness of the data and of the independence of observations are just as essential for likelihood-ratio procedures as for the Pearson χ^2 tests.

12.5 MEASURES OF ASSOCIATION IN CONTINGENCY TABLES

The problems to which the Pearson χ^2 tests apply are basically those of comparing distributions. In tests of goodness of fit, the null hypothesis states that some theoretical distribution exists, and the question itself is one of "fit" between the hypothetical and the observed distributions. Tests for association may be regarded in a similar way; the hypothesis of independence between two attributes dictates a particular relationship we should expect to hold between the cell frequencies and the marginal frequencies in the observed joint distribution. Divergence between the expected and observed frequencies is regarded as evidence against independence.

On the face of the matter, χ^2 tests are simple and appealing. They have been widely used in many areas and in studies of the most varied kinds of problems. In light of the statistical requirements for these tests, a large proportion of these applications are doubtless unjustified. However, even granting that the test is justified in the first place, what does the use of a χ^2 test for association actually tell the statistician? If n is very large, as it should be for the best application of the test, virtually any "degree" of true statistical relationship between attributes will show up as a significant result. The test detects virtually any departure from strict independence between the attributes for these large sample sizes. Given the significant result, the statistician can say that the two attributes are not independent, but is that of primary interest? It has been said before, but it bears saying again: surely nothing on earth is completely independent of anything else. Given a large enough sample size, the chances are very good that the statistician can demonstrate the association of any two qualitative attributes via a χ^2 test.

It seems that the really important thing is some measure of the *strength* of association between the attributes studied. Such a measure could tell us much more about the possible importance and meaning of a given relationship than can any χ^2 test alone. Thus, perhaps the emphasis in many studies of contingency tables should be shifted from the sheer significance of the test to an appraisal of the strength of relationship represented. You may recall that in Section 11.10, a similar comment was made with respect to studies using analysis of variance models.

The problem of determining measures of association between categorical attributes is somewhat different than the situation faced when we are dealing with quantitative variables. As we have seen in Chapters 10 and 11, most of our notions of the strength of a statistical association rest on the concept of the variance of a random variable. Thus, indices such as ρ^2 and ω^2 rest on the idea of a proportional reduction in variance in the dependent variable afforded by specifying the value of the independent variable. However, when the independent and dependent variables are categorical in nature, the variance per se is not defined. Something else must be used in specifying how knowledge of the A category to which an observation belongs increases our ability to predict the B category.

Three somewhat different approaches to this problem will be discussed here. The first rests directly on the notion of **statistical independence** between two attributes, defined by $P(A_j, B_k) = P(A_j)P(B_k)$. In this approach, the strength of association is measured basically in terms of a comparison of $P(A_j, B_k)$ and $P(A_j)P(B_k)$. As we shall see, this conception is adequate from a statistical point of view but seems to lack a simple interpretation in terms of how we *use* the statistical relation.

Another and much more recent approach deals with **predictive association.** Association between categorical attributes is indexed by the reduction in the probability of error in prediction afforded by knowing the status of the individual on one of the attributes. This way of defining association makes intuitive sense but is not as directly tied to tests of association as the first approach.

Finally, a brief mention will be made of still another point of view on this problem, based on concepts from **information theory.** Here an analog to the variance does exist for categorical data. This makes it possible to define the strength of association in contingency tables in a way very similar to the usual indices for numerical data.

12.6 THE PHI COEFFICIENT AND INDICES OF CONTINGENCY

Before we go into the problem of describing statistical association in a sample, a general way of viewing statistical association in a population will be introduced. This is the **index of mean square contingency,** originally suggested by Karl Pearson, the originator of the χ^2 test for association. Imagine *a discrete joint probability distribution* represented in a table with C columns and R rows. The columns represent the qualitative attribute A and the rows represent B. The mean square contingency is defined as

$$\varphi^2 = \sum_j \sum_k \frac{P(A_j, B_k)^2}{P(A_j)P(B_k)} - 1. \qquad (12.6.1^*)$$

This population index φ^2 (Greek phi, squared) can be zero only when there is complete independence, so that

$$P(A_j, B_k) = P(A_j)P(B_k)$$

for each joint event (A_j, B_k). However, when there is *complete association* in the table, the value of φ^2 is given by

$$\text{max. } \varphi^2 = L - 1,$$

where L is the *smaller* of the two numbers R or C (the number of rows or columns in the table). Thus, a convenient index of strength of association in a population is provided by

$$\varphi' = \sqrt{\frac{\varphi^2}{L - 1}}, \qquad (12.6.2^*)$$

which will always lie between the values 0 and 1 (the sign is taken as positive).

For the special case in which $R = C = 2$ (a 2×2 table), $L = 2$ and φ is itself an index of association, since $\varphi = \varphi'$. Unfortunately, in order to compute φ, we need to know the joint probabilities. Suppose that instead of these probabilities, we have a sample, and the observed frequencies are arranged into a 2×2 table of the form of Table 12.4.1. Estimates of the four joint probabilities are easy to determine; for example, a/n is an estimate of $P(A_1, B_1)$, b/n is an estimate of $P(A_2, B_1)$, and so on. Similarly, estimates of the four marginal probabilities can be found; $(a + c)/n$ is an estimate of $P(A_1)$, for example. Replacing the probabilities with their estimates in Equation (12.6.1), we get the following estimate of φ^2:

$$\hat{\varphi}^2 = \frac{a^2}{(a + c)(a + b)} + \frac{b^2}{(b + d)(a + b)}$$

$$+ \frac{c^2}{(a + c)(c + d)} + \frac{d^2}{(b + d)(c + d)} - 1.$$

Fortunately, this formidable-looking expression simplifies with the use of some algebra, and the square root of it, the sample value of the phi coefficient, can be expressed as follows:

$$\hat{\varphi} = \frac{|bc - ad|}{\sqrt{(a + b)(c + d)(a + c)(b + d)}}. \tag{12.6.3}$$

Notice that this is almost exactly the square root of the expression for χ^2 in a 2×2 table, given by Equation (12.4.3). In fact,

$$\hat{\varphi} = \sqrt{\frac{\chi^2}{n}}. \tag{12.6.4}$$

Since both χ^2 and $\hat{\varphi}$ reflect the degree to which there is nonindependence between A and B, a test for the hypothesis that $\varphi = 0$ is provided by the ordinary χ^2 test for association in a 2×2 table.

The idea of φ^2 extends to samples in larger contingency tables as well. For a set of data arranged into an $R \times C$ table, the *sample* value of φ^2 is simply

$$\hat{\varphi}^2 = \frac{\chi^2}{n},$$

where χ^2 is computed using Equation (12.4.2). A convenient way to describe the apparent strength of association in a sample is to find

$$\hat{\varphi}' = \sqrt{\frac{\hat{\varphi}^2}{L - 1}} = \sqrt{\frac{\chi^2}{n(L - 1)}}, \tag{12.6.5}$$

which must lie between 0, reflecting complete independence, and 1, showing complete dependence, of the attributes. For the example involving overdue accounts presented in Section 12.4, $\chi^2 = 4.81$, $n = 100$, and $L = 2$. Therefore,

$$\hat{\varphi}' = \hat{\varphi} = \sqrt{\frac{4.81}{100}} = .22.$$

For the example from Section 12.4 involving voting preferences,

$$\hat{\varphi}' = \sqrt{\frac{3.28}{100(2)}} = .13.$$

The index $\hat{\varphi}'$ (Cramér's statistic) is not to be confused with the ordinary *coefficient of contingency*, sometimes used for the same purpose. The coefficient of contingency is defined by

$$C_{AB} = \sqrt{\frac{\chi^2}{n + \chi^2}}. \tag{12.6.6}$$

The coefficient of contingency has the disadvantage that it cannot attain an upper limit of 1. Obviously, this limits the usefulness of C_{AB} as a descriptive statistic.

The sample coefficient $\hat{\varphi}'$ gives a way to discuss the apparent strength of statistical association in any contingency table, and there is a direct connection with χ^2 tests, making it possible to test the significance of any observed $\hat{\varphi}'$ value from a sufficiently large sample. However, it is rather hard to put the meaning of $\hat{\varphi}'$ in common-sense terms, particularly for larger tables. Other indices of association such as ω^2 do have such an interpretation in terms of reduction in variance, or variance accounted for, but this idea is not directly applicable to $\hat{\varphi}'$.

Now we turn to an index of association in the *predictive* sense. How much does knowing the classification A improve our ability to predict the classification B?

12.7 A MEASURE OF PREDICTIVE ASSOCIATION FOR CATEGORICAL DATA

We will introduce the notion of predictive association for categorical data by means of an example. Suppose that we are interested in the association between a student's performance in a particular course and whether or not that student has taken another course that is listed as a recommended but optional prerequisite to the course of interest. Let A_1 and A_2

Table 12.7.1 Joint Probability Distribution for Prerequisite-
Course Example

	A_1	A_2	
B_1	.25	.10	.35
B_2	.10	.30	.40
B_3	.05	.20	.25
	.40	.60	1.00

represent "has taken the prerequisite course" and "has not taken the prerequisite course," respectively. The B category will represent the student's grade in the course of interest (not the prerequisite course), with B_1 = high pass, B_2 = pass, and B_3 = fail. Table 12.7.1 gives the joint probability distribution of the two attributes. For instance, the probability that a student chosen at random from the students completing the course of interest has earned a "high pass" in the course *and* has taken the prerequisite course is .25.

Now suppose that, knowing these probabilities, you were asked to predict the grade in the course (the B category) for a student drawn at random. You know *nothing* about whether the student has taken the prerequisite course (the A category). Which grade should you bet on? Your probability of being correct is largest if you bet on B_2 (pass), since this grade has the largest *marginal* probability. In general, the largest probability in the marginal distribution of B is $\max_k P(B_k)$, which in this case is .40. Thus, if we must predict without knowing the A classification, the probability of an *error* in prediction is

$$P(\text{error} \mid A \text{ unknown}) = 1 - \max_k P(B_k).$$

For this particular example,

$$P(\text{error} \mid A \text{ unknown}) = .60.$$

Now, however, suppose that a student is drawn at random and you are told that the student has taken the prerequisite course (that is, the student falls in category A_1). Given A_1, the largest *conditional* probability, $\max_k P(B_k \mid A_1)$, occurs for category B_1(high pass):

$$\max_k P(B_k \mid A_1) = P(B_1 \mid A_1) = \frac{.25}{.40} = .625.$$

Thus, given A_1, category B_1 should be predicted, and the probability of an error in this instance is

$$P(\text{error} \mid A_1) = 1 - \max_k P(B_k \mid A_1) = .375.$$

On the other hand, if the information were that the student belongs to A_2 (has *not* taken the prerequisite course), then category B_2 (pass) would be predicted, since

$$\max_k P(B_k \mid A_2) = P(B_2 \mid A_2) = \frac{.30}{.60} = .50.$$

An error in this prediction has probability of

$$P(\text{error} \mid A_2) = 1 - \max_k P(B_k \mid A_2) = .50.$$

Under this model of prediction, *on the average*, over all cases, the probability of error is then

$$P(\text{error} \mid A) = P(\text{error} \mid A_1)P(A_1) + P(\text{error} \mid A_2)P(A_2)$$
$$= .375(.40) + .50(.60) = .45.$$

Notice that when A is not specified, then the probability of an error in prediction is .60, but when A is specified, this average probability of an error is only .45. This shows that there is *predictive* association between A and B; knowing whether or not a student has taken the recommended prerequisite course enables us to reduce the probability of error in attempting to predict the student's grade in the course of interest.

This idea is the basis for an **index of predictive association**. This index, which was developed by Goodman and Kruskal, will be called λ_B:

$$\lambda_B = \frac{P(\text{error} \mid A \text{ unknown}) - P(\text{error} \mid A \text{ known})}{P(\text{error} \mid A \text{ unknown})}. \quad (12.7.1^*)$$

This index shows the proportional reduction in the *probability* of error afforded by specifying A. If the information about the A category does not reduce the probability of error at all, the index is 0, and we can say that there is no predictive association. On the other hand, if the index is 1.00, no error is made given the A_j classification, and there is complete predictive association. For our example,

$$\lambda_B = \frac{.60 - .45}{.60} = \frac{.15}{.60} = .25,$$

so that knowing whether or not a student has taken the prerequisite course reduces the probability of error by 25 percent on the average.

It must be emphasized that this idea is not completely equivalent to independence and association as reflected in χ^2 and φ'. It is quite possible for some statistical association to exist even though the value of λ_B is 0. In this situation, A and B are not independent, but the relationship is not such that knowing A_j causes us to change our bet about B_k; the index λ_B is other than 0 only when *different* B_k categories would be predicted for different A_j information.

On the other hand, if there is complete proportionality throughout the table, so that φ' is 0, then λ_B must be 0. Furthermore, when there is complete association, so that perfect prediction is possible, both λ_B and φ' must be 1.00.

Sample values of λ_B can be calculated quite easily from a contingency table. Here, the sample is regarded as though it were the population, and probabilities are estimated by the relative frequencies in the sample. Thus, we interpret $\hat{\lambda}_B$ as the proportional reduction in the probability of error in prediction for cases drawn at random from *this* sample, or, if you will, a population exactly like this sample in its joint distribution.

In terms of the frequencies in the sample, we find that

$$\hat{\lambda}_B = \frac{\sum_j \max_k f_{jk} - \max_k f_{.k}}{n - \max_k f_{.k}}, \qquad (12.7.2)$$

where

f_{jk} is the frequency observed in cell (A_j, B_k),

$\max_k f_{jk}$ is the *largest* frequency in column A_j,

and $\max_k f_{.k}$ is the largest *marginal* frequency among the rows B_k.

To illustrate this index of predictive association, consider the example from Section 12.4 involving overdue accounts. In attempting to predict B (whether or not an account is overdue), how useful will knowledge of A (the size of the firm) be? Using Equation (12.7.2),

$$\hat{\lambda}_B = \frac{29 + 32 - 51}{100 - 51} = \frac{10}{49} = .20.$$

This value is very close to $\hat{\varphi}'$ for this example, which was shown in Section 12.6 to be .22. On the other hand, the voting preferences example of Section 12.4 provides a case in which the two measures are not similar. For that example, $\hat{\varphi}' = .13$, but

$$\hat{\lambda}_B = \frac{12 + 10 + 11 + 14 - 47}{100 - 47} = 0.$$

In all age groups, candidate 3 was the most preferred candidate, so knowing the age group a voter falls into will not change our prediction concerning the voter's preferred candidate. (Recall, from Chapter 9, that if there is no way that a particular piece of information can change our decision, the expected value of that information with respect to our decision-making problem is zero.). Thus, since the prediction is never changed, there is no reduction, on the average, in the probability of error.

The index λ_B is an *asymmetric* measure, much like ω^2. It applies when A is the independent variable, or the variable ordinarily known first, and B is the variable predicted. However, for the same set of data, it is entirely possible to reverse the roles of A and B and obtain the index

$$\lambda_A = \frac{P(\text{error} \mid B_k \text{ unknown}) - P(\text{error} \mid B_k \text{ known})}{P(\text{error} \mid B_k \text{ unknown})}, \qquad (12.7.3)$$

which is suitable for predictions of A from B. In terms of frequencies,

$$\hat{\lambda}_A = \frac{\sum_k \max_j f_{jk} - \max_j f_j.}{n - \max_j f_j.}, \qquad (12.7.4)$$

where $\max_{\cdot j} f_j.$ is the largest marginal frequency among the columns A_j. In general, the two indices λ_B and λ_A will not be identical; it is entirely possible to have situations where B may be quite predictable from A, but not A from B.

Finally, in some contexts it may be desirable to have a *symmetric* measure of the power to predict, where neither A nor B is specially designated as the variable predicted from or known first. Rather, we act as though sometimes the A and sometimes the B information is given beforehand. In this circumstance the index $\hat{\lambda}_{AB}$ can be computed from

$$\hat{\lambda}_{AB} = \frac{\sum_j \max_k f_{jk} + \sum_k \max_j f_{jk} - \max_k f_{\cdot k} - \max_j f_j.}{2n - \max_k f_{\cdot k} - \max_j f_j.}. \qquad (12.7.5)$$

The value of $\hat{\lambda}_{AB}$ will always lie between $\hat{\lambda}_A$ and $\hat{\lambda}_B$.

These measures of *predictive* association form a valuable adjunct to the tests given by χ^2 methods. When the value of χ^2 turns out significant, we can say with confidence that the attributes A and B are not independent. Nevertheless, the significance level alone tells almost nothing about the strength of association. Usually we want to say something about the predictive strength of the relation as well. If there is the remotest interest in actual predictions using the relation studied, then the λ measures are worthwhile. Statistical relations so small as to be almost nonexistent can show up as highly significant χ^2 results, and this is especially likely to occur when sample size is large. All too often the researcher then tends to believe that some relationship which is important and applicable in some real-world situation has been discovered. Plainly, this is not necessarily true. The λ indices do, however, suggest just how much the relationship found implise about real predictions and how much one attribute actually does tell us about the other. Such indices are a most important correction of the tendency to confuse statistical significance with the importance of results for actual prediction. Virtually any statistical relation will show up as highly significant given a sufficient sample size, but it takes a relation of considerable strength to enhance our ability to predict in real, uncontrolled, situations. It can happen that even though a χ^2 test is significant, the predictor's *behavior* is not changed one whit by this new information. The λ measures show how we are led to predict *differentially* in the light of the relationship.

12.8   INFORMATION THEORY AND THE ANALYSIS OF CONTINGENCY TABLES

A few words must be said about still another development for the description of the relation exhibited in a contingency table. These ideas, which come from information theory and suggest analogs to the classical statistics based on variance, are applicable to the qualitative situation. Unfortunately, space does not permit a thorough discussion of these notions, and so only the barest sketch will be given.

Consider a contingency table with R rows and C columns, as in the preceding sections. For the A attribute, there is a marginal probability distribution with probabilities $P(A_1)$, $P(A_2)$, ..., $P(A_C)$, and for the B attribute, there is a marginal probability distribution with probabilities $P(B_1)$, $P(B_2)$, ..., $P(B_R)$. In information theory, the average amount of uncertainty in a probability distribution of a categorical variable is the negative of the sum of the product of the probabilities and their logarithms. Thus, the average amount of uncertainty with respect to the attribute A is

$$H(A) = - \sum_j P(A_j) \log P(A_j), \qquad (12.8.1)$$

and the average amount of uncertainty with respect to the attribute B is

$$H(B) = - \sum_k P(B_k) \log P(B_k). \qquad (12.8.2)$$

From a purely statistical point of view, the interesting thing about this formulation is that the index H is a very close analog to the *variance* of a distribution. When one of the probabilities is 1 and the rest are 0, then H is 0; similarly, if there is only one possible value for a random variable, then $\sigma^2 = 0$. On the other hand, when there is a wide range of possible events, H tends to be large; indeed, the more evenly spread are the probabilities over the various possible events, the larger H will be. In the same way, σ^2 is large when the distribution of a random variable is "spread out." In short, H is a "variance-like" index that can be computed for any discrete distribution, even though the event-classes are purely qualitative.

Furthermore, for any joint probability distribution, we can define the average *joint* uncertainty

$$H(A, B) = -\sum_j \sum_k P(A_j, B_k) \log P(A_j, B_k) \qquad (12.8.3)$$

and the average conditional uncertainties

$$H(A \mid B) = -\sum_k \sum_j P(A_j, B_k) \log P(A_j \mid B_k) \qquad (12.8.4)$$

and

$$H(B \mid A) = -\sum_j \sum_k P(A_j, B_k) \log P(B_k \mid A_j). \qquad (12.8.5)$$

Now given these variance-like measures for *qualitative* data, indices of strength of association much like ρ^2 and ω^2 may be formed:

$$\text{relative reduction in uncertainty in } B \text{ given } A = \frac{H(B) - H(B \mid A)}{H(B)}.$$

$$(12.8.6)$$

Notice the similarity between this index and ω^2. If B is thought of as the dependent variable, this index gives the proportion by which knowing the A category reduces uncertainty about B, just as ω^2 tells the extent to which fixing X reduces the variance in Y.

A symmetric measure of association highly analogous to ρ^2 is

$$\frac{H(A) + H(B) - H(A, B)}{\text{minimum } [H(A), H(B)]}, \qquad (12.8.7)$$

indicating the relative strength of association between both variables.

On the basis of a sample, these indices can be estimated by replacing the probabilities with the relative frequencies from the sample. Furthermore, a sampling theory exists for these statistics, and tests for zero association are possible. Their main contribution to statistical methods lies, however, in the possibility of extending the familiar notions having to do with variance and factors accounting for variance to qualitative data situations.

This completes our discussion of methods for dealing with categorical data. We have discussed tests of goodness of fit and association and measures of association. Of course, although data at any level of measurement can be put into categories, the methods discussed here are of particular value when we are only able to attain the nominal level of measurement. Now we turn to methods that require at least ordinal measurement, a level of measurement one step up from nominal measurement.

12.9 ORDER STATISTICS

In the remainder of this chapter we are going to discuss methods which are based primarily on the **order relations** among observations in a set of data. An example of a statistic which depends only on the ordering of the observations is the median; this is one of the order statistics that we shall utilize in developing nonparametric methods based on order relations. The reasoning behind these methods involves relatively simple applications of probability theory; it happens that discrete sampling distributions often can be found in particularly simple ways if only the order features of the data are considered.

Why should we be interested in analyzing data in terms of ordinal properties? In the first place, the only relevant information in a set of data may be ordinal. That is, it may be that the statistician is able to attain only an ordinal level of measurement rather than an interval level. If this is the case, the numerical values obtained give information only about *relative* magnitudes of the underlying variable, and arithmetic differences between values have no particular meaning.

An equally common reason for using order methods is that one or more assumptions about the population distributions, strictly necessary for a parametric method to apply, may be quite unreasonable. Rather than use the parametric method anyway and wonder about the validity of the conclusion, the statistician prefers to change the question in such a way that another method applies.

A great deal of study has been given to relative merits of parametric and nonparametric tests in situations where both types of methods apply. These studies show clearly that advantages and drawbacks exist in the use of any of these methods. However, statistical practitioners sometimes gain the impression that they "get away with something" by using an order method or some other nonparametric technique in preference to one of the parametric methods. There are both potential gains and losses in the decision to use a nonparametric technique, and the choice among methods can be evaluated only in light of what we want to do and the price we are willing to pay.

Clearly, a word of warning is called for in the use of order methods as stand-ins for the parametric methods in situations where both kinds of methods are appropriate. At least two things must be kept in mind.

First, the actual hypothesis tested by a given order method is seldom exactly equivalent to the hypothesis tested by a parametric technique. For example, when the usual assumptions for the simple analysis of variance are true, then the hypothesis that the means of the populations are equal *both implies and is implied by* the absolute identity of the population distributions. Under these assumptions, the test statistic is distributed as F when and only when the null hypothesis is true. On the other hand, in order methods designed for the comparison of J experimental groups, the actual null hypothesis ordinarily is that all possible orderings of observations by their scores in the data are equally likely. This is implied by the hypothesis of identical populations, so that if the equal-likelihood hypothesis is rejected, the hypothesis of identical populations can also be rejected. Thus, these order tests can be regarded essentially as testing the hypothesis of identical population distributions. Regardless of how well the actual test statistic agrees with expectation, however, the population distributions still *might* be different in particular ways. If we are willing to make only minimal assumptions about the population distributions, then the kinds of true differences among populations that the test fails to detect may be quite unknown.

Second, when both the order method and a parametric method actually do apply (that is, when the parametric assumptions are true), the power of the two kinds of tests may be compared, given α, n, and the true situation. Order techniques share with other nonparametric methods the disadvantage of being relatively low-powered as compared with parametric tests. This means that, other things such as α and n being equal, we are taking more risk of a Type II error in using the order method. If Type II errors are to be avoided, then a relatively larger sample size (or a larger α value) is required in the use of the order technique as compared to the parametric method. This is due at least in part to the fact that order techniques use only ordinal features of the raw numerical data, thereby "throwing away" some information when the original level of measurement is interval or ratio rather than just ordinal.

In connection with the second point, a useful concept in the comparison of tests of hypotheses is that of "power-efficiency." The general idea may be given in this way. Suppose that there is some null hypothesis to be tested and that either of two tests (methods of testing H_0), test U and test V, might appropriately be applied. The power of test U and of test V against any alternative to H_0 will depend upon several things, of course, one of which is sample size. For a given degree of power against a specific true alternative, test U may require n_U cases, whereas test V may require

n_V cases. In general, for different tests, n_U and n_V will be different for the same power level.

Now suppose that U is the more powerful test, in the sense that it requires fewer cases to detect a true alternative to H_0 for some fixed α and $1 - \beta$ probabilities. Then the **power-efficiency** of test V relative to test U is

$$\text{power-efficiency of } V = \frac{(100)n_U}{n_V} \text{ percent.}$$

The more cases n_V that test V requires to attain the same power as test U with n_U cases, the smaller is the power-efficiency of test V relative to test U.

For instance, if test U requires 20 cases to reach power of .95 for a given true situation and $\alpha = .05$, and test V requires 40 cases to reach the same power under the same conditions, then the power efficiency of V relative to U is

$$\frac{100(20)}{40} = 50 \text{ percent.}$$

In this way nonparametric statistics (such as order methods) may be compared with parametric methods such as the T test and the analysis of variance. Results of such comparisons show that **order methods generally have less than 100 percent power-efficiency when used in situations where the most powerful parametric tests such as T and F apply.** Order methods require more evidence than parametric methods to yield comparable conclusions.

Of course, it must be emphasized that these comments on power hold only where the appropriate "high-powered" methods can be applied. If the data are collected as order data in their own right, perhaps because no higher-level measurement operation is available, then this objection does not necessarily hold; the concept of comparative power between parametric and nonparametric tests is useless here since the experimenter really has no choice to make. Similarly, comparative power is difficult or impossible to study when the assumptions underlying parametric tests are not true. For data that are essentially ordinal, or when assumptions are manifestly untrue, some of the tests described here may be about as powerful as can be devised.

The decision to use or not to use order methods in a given problem cannot be given a simple prescription. The statistician must learn to pick and choose among all of the various methods available, finding the one that most clearly, economically, and reasonably sheds light on the particular question to be answered. This is not a simple task, and a brief discussion such as this can only begin to suggest some of the issues that are involved.

In the sections to follow, several types of order statistics will be discussed, including tests and correlational methods for ordinal data. Although we will not discuss any of these procedures at great length, we will attempt to explain the rationale underlying the procedures.

12.10   THE KOLMOGOROV-SMIRNOV TEST FOR GOODNESS OF FIT

The first technique involving order statistics which we will discuss is the **Kolmogorov–Smirnov test.** This test is applicable when the statistician wants a comparison between distributions, either sample and theoretical distributions or two sample distributions. In the former situation, we have a problem of goodness of fit, which can also be handled by the χ^2 test given in Section 12.2. In the latter situation, the comparison of two sample distributions, the Kolmogorov-Smirnov test is one of a number of alternative nonparametric procedures; others (the median test, the Mann-Whitney test, and the Wald-Wolfowitz runs test) will be discussed later in this chapter.

Let us consider the goodness-of-fit problem, in which we wish to determine whether or not a population distribution has a certain form on the basis of some sample data. If the data are categorical (or if they are grouped into categories by the statistician), the χ^2 test is applicable. If the measurement is at least on an ordinal level, so that the distributions considered can be put into cumulative form, then the Kolmogorov-Smirnov test can be used. This test involves a comparison of a theoretical CDF with the sample CDF. If the theory provides a good fit to the data, we would expect these two CDF's to be very similar; any differences between them tend to indicate that the hypothesis of goodness of fit might not be reasonable. In particular, the statistic used to measure the difference between the CDF's is

$$D = \operatorname*{maximum}_{x} |\, F_S(x) - F_T(x) \,|, \qquad (12.10.1)$$

where $F_S(x)$ is the sample CDF and $F_T(x)$ is the theoretical CDF. Thus, D is the *largest absolute vertical deviation* between the two cumulative functions. The sampling distribution of D under the hypothesis that the two distributions are identical is known. The test is a one-tailed test, since only large values of D cast doubt on the hypothesis. Tabled values of the distribution of D are presented in Table X in the Appendix.

To illustrate the Kolmogorov-Smirnov test of goodness of fit, consider once again the example presented in Section 12.3: a goodness-of-fit test for

Table 12.10.1   Observed and Expected Frequencies for Sales Example

Interval	f_{o_i}	Observed relative frequencies	f_{e_i}	Expected relative frequencies
below 970.34	20	.05	50	.125
970.34–974.29	48	.12	50	.125
974.29–977.31	70	.175	50	.125
977.31–980.00	56	.14	50	.125
980.00–982.69	46	.115	50	.125
982.69–985.71	55	.1375	50	.125
985.71–989.66	60	.15	50	.125
above 989.66	45	.1125	50	.125

a normal distribution. The variable involved is sales, and for the purpose of the example, the variable was standardized by subtracting the sample mean and dividing the result by the estimate of the standard deviation, $\hat{s}$. Since sales is a quantitative rather than a qualitative variable, it is possible to determine cumulative distributions, and the Kolmogorov-Smirnov test can thus be applied here. From Table 12.3.2, we have the observed and expected frequencies for a sample of size 400. In Table 12.10.1 we present these frequencies again and also convert them to relative frequencies by dividing each frequency by 400. These two sets of relative frequencies represent two probability mass functions, one for the sample distribution and one for the theoretical distribution (by grouping, the theoretical normal density function has been converted into a discrete mass function). Finally, it is simple to convert the relative frequencies into *cumulative* relative frequencies, which are presented in Table 12.10.2.

The value of the statistic D for this example is simply the largest value in the last column of Table 12.10.2. This value is .08, and it occurs at

Table 12.10.2   Cumulative Relative Frequencies for Sales Example

x	$F_S(x)$	$F_T(x)$	$\mid F_S(x) - F_T(x) \mid$
970.34	.05	.125	.075
974.29	.17	.25	.08
977.31	.345	.375	.03
980.00	.485	.50	.015
982.69	.60	.625	.025
985.71	.7375	.75	.0125
989.66	.8875	.875	.0125

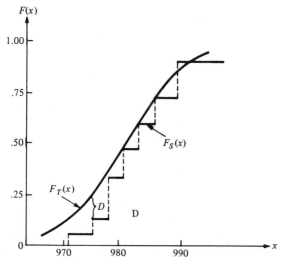

Figure 12.10.1 Theoretical and Sample Cumulative Distributions for Sales Example

$x = 974.29$. From the table of the distribution of D, it can be seen that with $n = 400$, the critical value of D for a significance level of .05 is $1.36/\sqrt{400} = .068$ and the critical value of D for a significance level of .01 is $1.63/\sqrt{400} = .0815$. Therefore, the p-value is between .01 and .05 and is much closer to .01 than to .05.

The above procedure is shown graphically in Figure 12.10.1. Note that the value of D is dependent upon the choice of intervals, that is, upon the particular grouping used. As we emphasized in Section 12.3, the arrangement into class intervals is arbitrary. With the Kolmogorov-Smirnov test, it is not *really* necessary to group the data at all. If no grouping is done, the CDF's may look something like those presented in Figure 12.10.2. Of course, it is more difficult computationally to find the largest vertical difference between the two functions in this case.

In almost all cases, the Kolmogorov-Smirnov test of goodness of fit is a more powerful test than the χ^2 test. An added advantage is that it is not necessary to have a certain minimum expected frequency in each interval, as it is with the χ^2 test. Finally, it is generally easier to compute D than it is to compute χ^2. Of course, the χ^2 test can be used with categorical data, whereas the Kolmogorov-Smirnov test cannot. Technically, the Kolmogorov-Smirnov test requires that the underlying variable be continuous, but it has been shown that the violation of this assumption leads only to very slight errors on the conservative side; obviously, though, we must have at least ordinal measurement or it is meaningless to talk of a cumulative function.

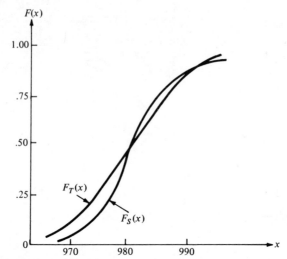

Figure 12.10.2   An Example of Theoretical and Sample Cumulative Distributions

At the beginning of the section we noted that the Kolmogorov-Smirnov test could also be used to compare two sample distributions. The general rationale is exactly the same as for the comparison of a sample distribution and a theoretical distribution. The statistic D is now defined as

$$D = \underset{x}{\text{maximum}} \mid F_{S_1}(x) - F_{S_2}(x) \mid, \qquad (12.10.2)$$

where $F_{S_1}(x)$ is the sample CDF from the first sample and $F_{S_2}(x)$ is the sample CDF from the second sample. The distribution of D is slightly different in this case from the case where a sample CDF and a theoretical CDF are being compared, and tables are available for this two-sample D statistic.

12.11   THE WALD-WOLFOWITZ "RUNS" TEST FOR TWO SAMPLES

The **runs test** applies to the situation where two unmatched samples are to be compared and each observation consists of a numerical score. The underlying variable that these scores represent is assumed to be continuously distributed.

Suppose that the numbers of observations in the two samples are n_1 and n_2, respectively. All of the $n_1 + n_2$ observations in these samples are drawn independently and at random. For convenience, we will call any observation appearing in sample 1 an A and any observation in sample 2 a B.

Now suppose that all these sample observations, irrespective of group, are arranged in order according to the magnitude of the scores. Then there will be some *arrangement* or pattern of A's and B's in order. In particular, there will be *runs* or "clusterings" of the A's and B's. This is easily illustrated by an example.

Suppose that we are interested in comparing the distributions of incomes in two communities. A random sample is taken from each community and yearly incomes (in thousands of dollars) are recorded, with the results shown in Table 12.11.1. When these incomes are combined into a single set and arranged in order of magnitude, we have the following ordering:

6	8	9	10	12	13	15	18	20	22	25
A	A	A	B	B	A	B	A	B	B	B

Below each observed income is an A or a B, denoting the community to which the observation belongs. Now notice that there are runs of A's and B's. That is, the ordering starts out with a run of 3 A's; the second run is a run of 2 B's; and so on. Proceeding in this way, with a new run beginning whenever an A is succeeded by a B or vice versa, we find that there are 6 runs in this set of data.

Table 12.11.1   Incomes for Two Communities

Community A	Community B
8	20
18	10
9	12
6	25
13	22
	15

Now suppose that the two groups are random samples from *absolutely identical population distributions*. In this instance we should expect many runs, since the values for the two samples should be well "mixed up" when put in order. On the other hand, if the populations differ, particularly in central tendency, we should expect there to be less tendency for runs to occur in the sample ordering. In the extreme case, when there is a large difference in central tendency, all of the observations from one sample might be below all of the observations from the other sample, in which case there are only two runs.

Let R symbolize the total number of runs appearing for the samples.

Then it can be shown that for any *odd* number, $2g + 1$,

$$P(R = 2g + 1) = \frac{\binom{n_1 - 1}{g - 1}\binom{n_2 - 1}{g} + \binom{n_1 - 1}{g}\binom{n_2 - 1}{g - 1}}{\binom{n_1 + n_2}{n_1}} \quad (12.11.1)$$

when all arrangements in order are equally likely. For an *even* number, $2g$,

$$P(R = 2g) = \frac{2\binom{n_1 - 1}{g - 1}\binom{n_2 - 1}{g - 1}}{\binom{n_1 + n_2}{n_1}}. \quad (12.11.2)$$

On this basis, the exact sampling distribution of R can be worked out, given equal probability for all possible arrangements of A and B observations. The alternative to the hypothesis of identical population distributions entails "too few" runs, so the test of interest is ordinarily one-tailed.

For the income example, 6 runs were observed, so the p-value is $P(R \leq 6)$. Using Equations (12.11.1) and (12.11.2), exact probabilities can be calculated. For example, since 5 is an odd number, Equation (12.11.1) can be used with $g = 2$ to determine

$$P(R = 5) = \frac{\binom{5 - 1}{2 - 1}\binom{6 - 1}{2} + \binom{5 - 1}{2}\binom{6 - 1}{2 - 1}}{\binom{11}{5}} = \frac{\binom{4}{1}\binom{5}{2} + \binom{4}{2}\binom{5}{1}}{\binom{11}{5}}$$

$$= \frac{4(10) + 6(5)}{462} = \frac{70}{462} = .15.$$

Calculating the probabilities for the values 2 (the smallest possible number of runs), 3, 4, and 6 in a similar fashion, we can find the p-value, $P(R \leq 6)$, which turns out to be .52.

For sample sizes less than or equal to 20 in *either* sample, exact values of R required for significance have been tabled for this test, which is known as the **Wald-Wolfowitz runs test.** For larger samples, the distribution of R can be approximated by a normal distribution with

$$E(R) = \frac{2n_1 n_2}{n_1 + n_2} + 1 \quad (12.11.3)$$

and

$$\sigma_R{}^2 = \frac{2n_1 n_2 (2n_1 n_2 - n_1 - n_2)}{(n_1 + n_2)^2 (n_1 + n_2 - 1)}. \quad (12.11.4)$$

Thus, an approximate large sample test is given by

$$Z = \frac{R - E(R)}{\sigma_R} , \qquad (12.11.5)$$

where Z is normally distributed.

There is some reason to believe that the runs test generally has rather low power-efficiency, compared either with a T test for means or with other order tests for identical populations. However, it is mentioned here because of its utility in various problems where other methods may not apply. Actually, A and B may designate any dichotomy within a *single* sample, and any principle at all that gives an ordering to A's and B's may be used. For example, it may be of interest to see if there is a time-related trend such as learning in a *single* set of data. In this instance, we might find the overall median for the set of scores, calling values above the median A and values below the median B. Here time is used as the ordering principle, and the occurrence of few runs is treated as evidence that time trends do exist. In this regard, a runs test could be used to investigate the randomness of a sample. For example, suppose that a football team plays 10 games and wins 5 of these games, losing the other 5. There are $\binom{10}{5} = 252$ possible orderings of 5 wins and 5 losses. Consider the following three orderings:

$$W \quad W \quad W \quad W \quad W \quad L \quad L \quad L \quad L \quad L,$$

$$W \quad L \quad W \quad L \quad W \quad L \quad W \quad L \quad W \quad L,$$

and $\quad\quad W \quad W \quad L \quad W \quad L \quad L \quad L \quad W \quad W \quad L.$

In the first ordering there are only two runs, and the sample does not appear to be random (too few runs). In the second ordering, there are 10 runs, and this also casts doubt on the randomness of the sample, because it appears that there is a systematic trend (alternating wins and losses). When the runs test is used to investigate randomness in a single sample, then, the test is *two-tailed*, with either too few runs *or* too many runs leading to the rejection of the hypothesis of randomness. In the third ordering presented above, there are 6 runs; this is an intermediate value that would tend to support the randomness hypothesis. The runs test is valuable in a situation such as this, both because it is easy to apply and because other tests do *not* apply directly.

A major technical problem with any runs test is the treatment of ties. In principle, ties should not occur if the scores themselves represent a continuous random variable. But, of course, ties do occur in actual practice, since we seldom represent the underlying variable directly or precisely in the data. If all tied scores occur among observations in the *same* group (as in the example above), then there is no problem; the value of R is unchanged by any method of breaking these ties. However, if members of *different* groups show tied scores, the number of runs depends on how these ties are resolved in the final ordering. One way of meeting this problem is to break all ties in a way *least* conducive to rejecting the null hypothesis (so that the number of runs is made as large as possible). This at least makes for a conservative test of H_0. However, if cross-group ties are very numerous, this test is really inapplicable.

12.12 THE MANN-WHITNEY TEST FOR TWO INDEPENDENT SAMPLES

Unlike the runs test, the Mann-Whitney test employs the actual *ranks* of the various observations as a device for testing hypotheses about the identity of two population distributions. It is a good, relatively powerful alternative to the usual T test for equality of means.

We assume that the underlying variable on which two samples are to be compared is continuously distributed. The null hypothesis to be tested is that the two population distributions are identical. Then we proceed as follows.

The scores from the combined samples are arranged in order (much as in the runs test). However, now we assign a *rank* to each of the observations, in terms of the magnitude of the original score. That is, the lowest score gets rank 1, the next lowest 2, and so on. Now choose one of the samples, say sample 1, and find the *sum* of the ranks associated with observations in that sample. Call this W. Then the statistic of interest is

$$U = n_1 n_2 + \frac{n_1(n_1 + 1)}{2} - W. \qquad (12.12.1)$$

For example, consider the data from Section 12.11 involving incomes. The data are arranged in order and ranked in Table 12.12.1. The sum of

Table 12.12.1 Ordered and Ranked Data for Income Example

Community	A	A	A	B	B	A	B	A	B	B	B
Income	6	8	9	10	12	13	15	18	20	22	25
Rank	1	2	3	4	5	6	7	8	9	10	11

the ranks for the observations in community A is

$$W = 1 + 2 + 3 + 6 + 8 = 20,$$

so that
$$U = 5(6) + \frac{5(6)}{2} - 20 = 25.$$

Now notice that given these 11 incomes, the value of U depends only on how the A's and B's happen to be arranged over the rank order. The number of possible random arrangements is just

$$\binom{n_1 + n_2}{n_1},$$

and if the hypothesis of completely identical populations is true, the random assignment of individuals to groups should be the only factor causing variation in U. Under the null hypothesis, all arrangements should be equally likely, and for large samples, the sampling distribution of U is approximately normal, with

$$E(U) = \frac{n_1 n_2}{2} \tag{12.12.2}$$

and
$$\sigma_U{}^2 = \frac{n_1 n_2 (n_1 + n_2 + 1)}{12}. \tag{12.12.3}$$

Thus, for large samples, the hypothesis of no difference in the population distributions is tested by

$$Z = \frac{U - E(U)}{\sigma_U}.$$

This test, called the **Mann-Whitney test,** is a two-tailed test, since both unusually large values of U and unusually small values of U cast doubt on the hypothesis of identical population distributions.

For situations where the larger of the two samples is 20 or more and the samples are not too different in size, the normal approximation given above should suffice. However, when the larger sample contains fewer than 20 observations, tables are available to evaluate the significance of U.

For the income example, the exact p-value for the Mann-Whitney test turns out to be .082; this casts some doubt on the null hypothesis of identical populations. Note that for the same data, the runs test of the previous section yielded a p-value of .52. For a sample this small (and for this particular situation and this particular sample), the runs test does not appear to be as sensitive to the differences between the populations as is the Mann-Whitney test. Of course, this might be expected, since the Mann-Whitney test uses the *ranks* of the observations, information that is ignored by the runs test.

The Mann-Whitney test is one of the best of the nonparametric techniques with respect to power and power-efficiency. It compares quite well with the T test for differences in means when the assumptions for the T test are met. For the income example, the T test of Section 7.16 yields a p-value between .05 and .10 of approximately .08 (approximately because of interpolation in the T table), which is about the same as the p-value from the Mann-Whitney test. Of course, as noted in the next paragraph, the T test is addressed to a slightly different question, since it is limited to differences in means.

We have now discussed three different nonparametric tests for the comparison of two independent samples: the Kolmogorov-Smirnov two-sample test (discussed only briefly), the Wald-Wolfowitz runs test, and the Mann-Whitney test. Another nonparametric test for this situation is the median test, which we will discuss in Section 12.15 in the context of a situation involving J samples. It is important to note that with all of these tests, rejection of the null hypothesis indicates that the two samples being compared are from different populations; it does not, however, indicate in what way the populations differ. By way of comparison, the parametric test for comparing two independent samples, the T test (Section 7.16), is concerned only with differences in *means*.

Ordinarily, ties are treated in the Mann-Whitney test by giving each of a set of tied scores the *average* rank for that set. Thus, if three scores are tied for fourth, fifth, and sixth place in order, each of the scores gets rank 5. If two scores are tied for ninth and tenth place in order, each gets rank 9.5, and so on. This introduces no particular problem for large sample size when the normal approximation is used and ties are relatively infrequent. However, when ties exist σ_U^2 becomes

$$\sigma_U^2 = \frac{n_1 n_2}{12} \left[n_1 + n_2 + 1 - \frac{\sum\limits_{i=1}^{G} (t_i^3 - t_i)}{(n_1 + n_2)(n_1 + n_2 - 1)} \right],$$

where there are some G distinct *sets* of tied observations, i represents any *one* such set, and t_i is the number of observations tied in set i. For a small number of ties and for large $n_1 + n_2$, this correction for σ_U^2 can safely be ignored.

12.13 THE SIGN TEST FOR MATCHED PAIRS

When the number of samples (or experimental treatments) is only two, and *pairs* of observations are matched, a simple test based on the binomial distribution can be used. Let n be the number of pairs of observations, where one member of each pair belongs to experimental (or natural) treatment 1 and the other to treatment 2. Here, the only relevant information given by the two scores for a pair is taken to be the **sign of the difference** between them. If the two treatments actually represent identical populations and chance is the only determiner of which member of a pair falls into which treatment, we should expect an equal number of differences of plus and of minus signs. The theoretical probability of a "plus" sign is .5, and so the probability of a particular number of plus (or minus) signs can be found by the binomial rule with $p = .5$ and n. Notice, however, that we must assume either that the population distributions are continuous so that exact equality between scores has probability zero, or that ties are otherwise impossible. Ordinarily, pairs showing zero differences are simply dropped from the sample, although this makes the final conclusion have a conditional character, "In a population of untied pairs . . . ," and so on. The test is carried out as follows.

First, the direction of the difference (that is, the sign of the difference) between the two observations in each pair is noted, with the same order of subtraction always maintained. Thus, each *pair* is given a classification of plus or minus, according to the sign of the difference between scores. If the null hypothesis of no difference between the two matched populations were true, we would expect half the nonzero differences to show a positive sign, and half to show a negative sign. Thus, we may simply take the *proportion* of plus differences and test the hypothesis that the sample arose from a true proportion of .50.

If the sample size is less than 10 (that is, if there are fewer than 10 pairs), the binomial distribution should be used to give an exact probability for the sign test. For larger sample sizes, the normal approximation to the binomial distribution can be used (see Section 4.16). The test can be either one-tailed or two-tailed, depending upon the situation.

As an example, suppose that a production manager is interested in comparing the items produced by two different machines. Twenty items produced by machine A are randomly paired with 20 items produced by machine B, and the manager evaluates each pair of items to see which of the two items is "better" than the other. Perhaps the items are subjected to some sort of a test, for example, to find their breaking point; the item with the higher breaking point is judged to be "better" than the other. Of the 20 pairs, the item from machine A was better in 11 pairs, the item from machine B was better in 7 pairs, and the two items were judged to be equal in the remaining 2 pairs. If we exclude the 2 equal pairs, we have an effective n of 18, with 11 positive signs and 7 negative signs (or vice versa, depending on which way was considered "positive"). Using the normal approximation with a continuity correction (see Section 4.16), the observed value of the test statistic is

$$z = \frac{11 - 18(.50) - .50}{\sqrt{18(.50)(.50)}} = .71.$$

From the tables of the normal distribution, the p-value is .48, indicating that the differences in the number of plus signs and minus signs observed could well have occurred by chance.

Observe that since all we know about each pair in this example is the *sign* of the difference, the parametric T test for matched pairs (Section 7.17) is not applicable in this situation. The T test requires knowledge of the *magnitude* of the difference as well as the sign. If the magnitude is known, then the sign test is not very powerful, since it does not consider all of the relevant information. A better nonparametric test in this situation (magnitudes known as well as signs) would be the Wilcoxon test, which we discuss in the next section.

12.14   THE WILCOXON TEST FOR TWO MATCHED SAMPLES

As we saw in Section 12.13, the problem of comparing two matched samples can be treated by the sign test if the only feature of the data considered is the sign of the difference between each pair. If the magnitudes of the differences can be ranked, the sign test can still be used, but it ignores the ranks. The **Wilcoxon test** takes account of both features in the data and thus uses somewhat more of the available information in paired scores than does the sign test. This procedure has very close ties with the Mann-Whitney test; both are based essentially on the idea of randomization. The Wilcoxon test also has high power-efficiency compared to other methods designed specifically for the matched-pair situation.

The mechanics of the test are quite simple. The signed differences between the pairs of observations are rank-ordered in terms of their *absolute* size, and the sign of each difference is attached to the rank associated with that difference. The test statistic is then W, the sum of the ranks with the less-frequently-occurring sign.

In the production example of the previous section, suppose that n is only 10 and that the breaking point of an item can be measured in terms of the maximum amount of weight (in pounds) the item can support

Table 12.14.1 Raw Data, Differences, Ranks of Differences, and Signed Ranks for Production Example

Pair	Machine A	Machine B	Difference	Rank	Signed rank
1	83	75	8	8	8
2	80	78	2	2	2
3	81	66	15	10	10
4	74	77	−3	3.5	−3.5
5	79	80	−1	1	−1
6	78	68	10	9	9
7	72	75	−3	3.5	−3.5
8	84	90	−6	7	−7
9	85	81	4	5	5
10	88	83	5	6	6

without breaking. The observed data are given in Table 12.14.1 along with the differences, the ranks of the differences in absolute size, and the signed ranks. The less frequent sign is minus, and the sum of the ranks of the differences with minus signs is

$$W = 3.5 + 1 + 3.5 + 7 = 15.$$

The hypothesis tested by the Wilcoxon test is that the two populations represented by the respective members of matched pairs are identical. When this hypothesis is true, then each of the 2^n possible sets of *signed* ranks obtained by arbitrarily assigning $+$ or $-$ signs to the ranks 1 through n is *equally* likely. On this basis, the exact distribution of W over all possible randomizations can be worked out, and tables of this distribution are available for small samples. For our production example, $n = 10$ and $W = 15$ yield a p-value of .23 for a test of the hypothesis that the two populations are identical.

For large n, the sampling distribution of W is approximately normal with

$$E(W) = \frac{n(n + 1)}{4} \tag{12.14.1}$$

and

$$\sigma_W{}^2 = \frac{n(n + 1)(2n + 1)}{24}, \tag{12.14.2}$$

so that a large sample test is given by

$$Z = \frac{W - E(W)}{\sigma_W}. \tag{12.14.3}$$

The normal approximation seems reasonably adequate for samples larger than about $n = 8$. For the production example,

$$E(W) = \frac{10(11)}{4} = 27.5$$

and

$$\sigma_W{}^2 = \frac{10(11)(21)}{24} = 96.25,$$

so that

$$z = \frac{15 - 27.5}{9.81} = -1.27.$$

The p-value is .21, which is quite close to the p-value for the exact test, .23. (For these data, the p-value is between .10 and .20 for a T test, as given in Section 7.17, of the hypothesis that the *means* are equal, and the p-value for a sign test is .37.)

Since only one set of differences is ranked in the Wilcoxon test, ties present no special problem unless they occur for zero differences. If an *even* number of zero differences occurs, each zero difference is assigned the average rank for the set (zero differences, of course, rank lowest in absolute size), and then half are arbitrarily given positive signs and half negative signs. If an odd number of zeros occurs, one randomly chosen zero difference is discarded from the data and the procedure for an even number of zeros is followed, except that n is reduced by 1, of course. For other kinds of tied differences, the method used in the example may be followed. Be sure to notice that even when several pairs are tied in absolute size so that they all receive the midrank for that set, the sign given to that midrank for different pairs may be different. For fairly large samples with relatively few ties, this procedure of assigning average ranks introduces negligible error.

All in all, the Mann-Whitney and the Wilcoxon tests are generally regarded as the best of the order tests for two samples. They both compare favorably with T in the appropriate circumstances, and when the assumptions for T are not met they may even be superior to this classical method. However, each is fully equivalent to a classical test of the hypothesis that the *means* of two groups are equal *only* when the assumptions appropriate to T are true. Unless additional assumptions are made, these tests refer to the hypothesis that two population distributions of unspecified form are *exactly* alike. In many instances, this is the hypothesis that we wish to test, especially if we are interested only in the possibility of statistical independence or of association between experimental and dependent variables. However, if we want to make particular kinds of inferences, particularly about population *means*, then other assumptions become necessary.

Without these assumptions, the rejection of H_0 implies only that the populations differ in *some* way, but the test need not be equally sensitive to all ways that population distributions might differ.

12.15 COMPARING J INDEPENDENT GROUPS: THE MEDIAN TEST

One of the simplest of the order methods involves the comparison of several samples on the basis of deviations from the median rather than from the mean. We assume that the underlying variable on which the populations are to be compared is continuous and that the probability of a tie between two observations in actual value of the underlying variable is, in effect, zero.

The null hypothesis to be tested is that the J different populations are absolutely identical in terms of their distributions. The alternative is simply the contrary of H_0. A feature common to many nonparametric tests is that the alternative hypothesis is somewhat unclear; the alternative that different distributions are not identical covers a lot of territory and cannot be expressed in a simple form such as $\mu_1 \neq \mu_2$.

The method is as follows. The J different sample groups are combined into a single distribution and the grand median for the sample is obtained. Now each score in each group is compared with the grand median. If the particular score is above the grand median, the observation is assigned to a "plus" category; if the particular score is not above the grand median, the observation is assigned to a "minus" category. Let a_j be the number of "plus" observations in group j and let $n_j - a_j$ represent the number of "minus" observations. Then the data can be arranged into a $2 \times J$ table such as Table 12.15.1. Notice that this is simply a joint-frequency or contingency table, where one attribute is "group" and the other is "plus or minus."

Now, if the grand median actually divides each of the populations in exactly the same way, then the probability of exactly a_j "plus" observa-

Table 12.15.1   A $2 \times J$ Contingency Table

GROUP

	1	$\cdots$	j	$\cdots$	J	
Plus	a_1	$\cdots$	a_j	$\cdots$	a_J	a
Minus	$n_1 - a_1$		$n_j - a_j$		$n_J - a_J$	$n - a$
	n_1	$\cdots$	n_j	$\cdots$	n_J	n

tions in group j is a binomial probability,

$$\binom{n_j}{a_j} p^{a_j} q^{n_j - a_j},$$

where p is the probability of an observation falling into the "plus" category and $q = 1 - p$. Under the null hypothesis, the probability p should be the same for each and every population.

However, we want the probability of the obtained *sample* result, conditional upon the fact that the *marginal* frequency of "plus" observations is a. Under the null hypothesis, the probability of exactly a observations above the grand median is $\binom{n}{a} p^a q^{n-a}$. Then the conditional probability for a particular arrangement in the table works out to be

$$\frac{\binom{n_1}{a_1} \binom{n_2}{a_2} \cdots \binom{n_j}{a_j} \cdots \binom{n_J}{a_J}}{\binom{n}{a}}.$$

This is the probability of this *particular* sample result conditional upon this particular value of a. Since this probability can be found for any possible sample result, a p-value can be found by exact methods, involving finding all possible sample results differing this much or more among the J groups. Such exact probabilities may be very laborious to work out, of course, unless J is only 2 or 3.

For large samples, a fairly good approximation to the exact significance level is found from the statistic

$$\chi^2 = \frac{(n-1)}{a(n-a)} \sum_{j=1}^{J} \frac{(na_j - n_j a)^2}{n n_j}. \tag{12.15.1}$$

For reasonably large samples ($n \geq 20$, $n_j \geq 5$ for each j), this statistic is distributed *approximately* as chi-square with $J - 1$ degrees of freedom.

One difficulty with this test is that it is based on the assumption that ties in the data will not occur. Naturally, this is most unreasonable since tied scores ordinarily will occur. This is really a problem only if several scores are tied at the overall median, since other ties have no effect on the test statistic itself. When ties occur at the median, several things may be done to remove this difficulty, but the safest general procedure seems to be to allot the tied scores (those tied with the grand median) within each group in such a way that a_j is as close as possible to $n_j - a_j$. This at least makes the test relatively conservative. Remember that this χ^2 method gives an approximate test and really should be used only when the sample size within each group is fairly large. In principle, exact probabilities may be computed, as suggested above, when n is small.

Table 12.15.2 Data for Examination-Score Example

Score	GROUP (SECTION)			
	I	II	III	IV
12	0	0	0	0
11	3	2	1	0
10	7	3	4	1
9	5	5	4	4
8	5	5	1	0
7	3	6	3	10
6	2	2	5	8
5	0	2	4	1
4	0	0	3	1
3	0	0	0	0
	25	25	25	25	$n = 100$

As an example, consider the following situation. An instructor is interested in whether the four sections of a particular course all have the same distribution of scores on a particular examination. The highest possible score on the examination is 12 points, and there are 25 students in each section of the course. The data are shown in Table 12.15.2.

For the combined population (all four sections), the median score is 7.5, since exactly 50 scores are 7 or lower and the other 50 are 8 or higher. When the scores within each section are compared to the median, we get the results given in Table 12.15.3. Thus,

$$\chi^2 = \frac{(99)}{(50)(50)} \left\{ \frac{[(100)(20) - (25)(50)]^2}{2500} + \cdots \right.$$

$$\left. + \frac{[(100)(5) - (25)(50)]^2}{2500} \right\} = 19.8.$$

Table 12.15.3 Observations Above and Below the Overall Median for 4 Different Groups for Examination-Score Example

	I	II	III	IV	
Plus	20	15	10	5	$50 = a$
Minus	5	10	15	20	$50 = n - a$

Looking at the χ^2 table under 3 degrees of freedom, we see that the p-value is less than .001, and so the instructor can say that the populations (sections) are not identical. Notice that the hypothesis tested is not about any particular characteristic of the populations considered (such as central tendency, dispersion, skewness), but rather about the absolute identity of their distributions. Some departure from identity of distributions is observed, indicating that there are differences between sections with regard to distributions of scores.

When $J = 2$, the median test is an alternative to the procedures discussed earlier in this chapter for dealing with 2 independent samples. As such, it is generally less powerful than tests such as the Mann-Whitney test, discussed in Section 12.12. For the more general case of J independent samples (the case considered in this section), we will now consider a procedure related to the Mann-Whitney test.

12.16   THE KRUSKAL-WALLIS "ANALYSIS OF VARIANCE" BY RANKS

The same general argument for the Mann-Whitney test may be extended to the situation where J independent groups are being compared. The version of a J-sample rank test given here is the **Kruskal-Wallis test.** This test has very close ties to the Mann-Whitney and Wilcoxon tests and can properly be regarded as a generalized version of the Mann-Whitney method.

Imagine some J experimental groups in which each observation is associated with a numerical score. As usual, we assume that the underlying variable is continuously distributed. Now, just as in the Mann-Whitney test, the scores from all groups are pooled, arranged in order of size, and ranked. Then the rank-sum attached to *each separate group* is found. Let us denote this sum of ranks for group j by the symbol W_j:

$$W_j = \text{sum of ranks for group } j.$$

For example, suppose that we are interested in comparing the distributions of checking account balances for personal checking accounts at three different branches of a particular bank. A random sample of 12 accounts is chosen from each branch ($n_j = 12$ for $j = 1$, 2, and 3, so that $n = 36$). For each account the current balance (in hundreds of dollars) is recorded. The data are shown in Table 12.16.1, where the numbers in parentheses are the ranks assigned to the various observations in the entire set of 36 cases. Then the sum of the ranks for each particular group j is found and designated W_j. The value of W is the sum of these rank sums; if the ranking has been done correctly, it will be true that

$$W = \sum_j W_j = \frac{n(n+1)}{2}. \tag{12.16.1}$$

For the example, $n = 36$ and $W = 36(37)/2 = 666$.

Table 12.16.1 Data for Checking-Account Example

	BRANCH BANK		
	I	*II*	*III*
	6 (1)	31 (34.5)	13 (10)
	11 (7)	7 (2)	32 (36)
	12 (9)	9 (4)	31 (34.5)
	20 (19)	11 (7)	30 (33)
	24 (23)	16 (14)	28 (31)
	21 (20)	19 (17.5)	29 (32)
	18 (16)	17 (15)	25 (24)
	15 (13)	11 (7)	26 (26.5)
	14 (11.5)	22 (21)	26 (26.5)
	10 (5)	23 (22)	27 (29.5)
	8 (3)	27 (29.5)	26 (26.5)
	14 (11.5)	26 (26.5)	19 (17.5)
W_j	139.0	200.0	327.0
$W = 666$			

For large samples, a fairly good approximate test for identical populations is given by

$$\chi^2 = \frac{12}{n(n+1)}\left[\sum_j \frac{W_j^2}{n_j}\right] - 3(n+1). \tag{12.16.2}$$

This value of χ^2 can be referred to the chi-square distribution with $J - 1$ degrees of freedom for a test of the hypothesis that all J population distributions are identical.

For the example,

$$\chi^2 = \frac{12}{36(37)}\left[\frac{(139)^2 + (200)^2 + (327)^2}{12}\right] - 3(37) = 13.81.$$

However, since there were ties involved in the ranking, this value of χ^2

really should be corrected by dividing by a value found from

$$C = 1 - \left(\frac{\sum\limits_{i=1}^{G} (t_i{}^3 - t_i)}{n^3 - n} \right), \qquad (12.16.3)$$

where G is the number of sets of tied observations and t_i is the number tied in the set i. For the example, there are 4 sets of 2 tied observations, 1 set of 3 ties, and 1 set of 4 tied observations. Thus,

$$C = 1 - \left(\frac{4[(2)^3 - 2] + [(3)^3 - 3] + [(4)^3 - 4]}{(36)^3 - 36} \right) = .998,$$

and the corrected value of χ^2 is

$$\chi_c{}^2 = \frac{\chi^2}{C} = \frac{13.81}{.998} = 13.84.$$

Unless n is small, or unless the number of tied observations is very large relative to n, this correction will make very little difference in the value of χ^2. Certainly this was true here. Furthermore, when each set of tied observations lies within the same experimental group, the correction becomes unnecessary.

From Table III with 2 degrees of freedom, the p-value is approximately .001, indicating that we can be quite confident in saying that the population distributions of checking account balances are not identical at the three branch banks.

It is somewhat difficult to specify the class of alternative hypotheses approprate to the J-sample rank test, and so the question of power is somewhat more obscure than for the Mann-Whitney test. However, there is reason to believe that this test is about the best of the J-sample order methods. Certainly it should be superior in most situations to the median test discussed in Section 12.15. In comparisons with F from the analysis of variance, the Kruskal-Wallis test shows up extremely well.

12.17 THE FRIEDMAN TEST FOR J MATCHED GROUPS

Much as the Kruskal-Wallis test represents an extension of the Mann-Whitney test, so the **Friedman test** is related to the Wilcoxon matched pairs procedure. The Friedman test is appropriate when we are dealing with J different samples, or experimental treatments, and we have K sets of observations that are matched. For example, each of K individuals might be observed under each of J treatments. The data are set up in a table as for a two-way analysis of variance with one observation per cell. The experimental treatments are shown by the respective columns and the matched sets of individuals by the rows. Within each row (matched group) a rank order of the J scores is found. Then the resulting ranks are summed by columns to give values of W_j.

For example, suppose that each of 11 students takes 4 tests, with the order in which the tests are taken being randomized for each student. We are interested in differences among the 4 tests; in particular, any differences in the distributions of scores are of interest. Since there are 4 tests, $J = 4$, and the data are presented in Table 12.17.1 with ranks given in parentheses. It is important to remember that the ranks are given to the scores within *rows* and the W_j values are simply the sums of those ranks within *columns*.

Table 12.17.1   Data for Test-Score Example

		TEST		
Student	I	II	III	IV
1	1 (2)	4 (3)	8 (4)	0 (1)
2	2 (2)	3 (3)	13 (4)	1 (1)
3	10 (3)	0 (1)	11 (4)	3 (2)
4	12 (3)	11 (2)	13 (4)	10 (1)
5	1 (2)	3 (3)	10 (4)	0 (1)
6	10 (3)	3 (1)	11 (4)	9 (2)
7	4 (1)	12 (4)	10 (2)	11 (3)
8	10 (4)	4 (2)	5 (3)	3 (1)
9	10 (4)	4 (2)	9 (3)	3 (1)
10	14 (4)	4 (2)	7 (3)	2 (1)
11	3 (2)	2 (1)	4 (3)	13 (4)
W_j	30	24	38	18

$$W = \frac{K(J)(J+1)}{2} = 110$$

The rationale for this test is really very simple. Suppose that within the population represented by a row, the distribution of values for all the *treatment* populations is identical. Then, under random sampling and randomization of observations, the probability for any given permutation of the ranks 1 through J within a given row should be the same as for any other permutation. Furthermore, across rows, each and every one of the $J!$ possible permutations of ranks across columns should be equally probable. This implies that we should expect the column sums of ranks to be identical under the null hypothesis. However, if there tend to be pile-ups of high or low ranks in particular columns, this is evidence against equal probability for the various permutations and thus against the null hypothesis. The test statistic for large samples is given by

$$\chi^2 = \frac{12}{KJ(J+1)} \left[\sum_j W_j^2\right] - 3K(J+1), \qquad (12.17.1)$$

distributed approximately as chi-square with $J - 1$ degrees of freedom.

For the example, we would take

$$\chi^2 = \frac{12}{11(4)(5)} \left[(30)^2 + (24)^2 + (38)^2 + (18)^2\right] - 3(11)(5)$$

$$= 11.79.$$

For 3 degrees of freedom, this is just significant at the .01 level, and so the experimenter may say with some assurance that the treatment populations differ. If ties in ranks within rows should occur, a conservative procedure is to break the ties so that the W_j values are as close together as possible.

This test may well be the best alternative to the ordinary two-way (matched groups) analysis of variance. Once again, there is every reason to believe that this test, much like the Kruskal-Wallis test for one-factor experiments, is superior to the corresponding median test, and from the little evidence available, the result should compare well with F when both the classical and order methods apply.

The chi-square approximation given above is good only for fairly large K. However, the test should be satisfactory when $J \geq 4$ and $K \geq 10$. As usual, tables are available giving significance levels for small samples.

12.18 RANK-ORDER CORRELATION METHODS

In the remainder of the chapter, we will discuss measures of association which are applicable to ordinal data. It is customary to call some of these rank-order statistics "correlations," but this usage deserves some qualification. The Spearman rank correlation actually *is* a correlation coefficient computed for numerical values that happen to be ranks. However, another index to be considered, Kendall's tau, is not a correlation coefficient at all. The Spearman and the Kendall indices can be thought of as showing "concordance" or "agreement," the tendency of two rank orders to be similar. As descriptive statistics, both indices serve this purpose very well, although the definition of "disagree" is somewhat different for these two statistics.

The size of either the Spearman or the Kendall coefficient *does* tell something about the tendency of the underlying variables to relate in a **monotone** way. High absolute values of either index give evidence that the basic form of the relation between Y and X is monotone. Positive values suggest that the relation tends to be monotone-increasing (an increase in one variable tends to be accompanied by an increase in the other variable), and negative values indicate that the relation tends to be monotone-decreasing (an increase in one variable tends to be accompanied by a decrease in the other variable). On the other hand, small absolute values or zero values of these indices suggest either that the two variables are not related *at all* or that the form of the relation is nonmonotone. In short, rank-order measures of association between variables do not reflect *exactly* the same characteristics as r_{XY} (or more properly, r_{XY}^2). They do not necessarily show the tendency toward linear regression per se for either of the numerical variables underlying the ranks, but rather show a more general characteristic, which includes linear regression as a special case. These indices reflect the **tendency toward monotonicity** and the **direction of relationship** that appears to exist. Thus, rank-order measures of association stand somewhere between indices such as φ (Section 12.6) and r_{XY} (Section 10.1); φ gives evidence for *some unspecified departure from independence*, rank-order measures tell something of the *monotonicity of the relationship*, and r_{XY} reflects *the tendency toward linear relationship* between variables. Unfortunately, values given by these different statistics are not directly comparable with each other, but this difference among the statistics is a point to bear in mind when thinking about what the different rank-order indices actually "tell" about the underlying relationship. These various indices of "correlation" imply different things about the relation between variables and are seldom interchangeable.

12.19   THE SPEARMAN RANK CORRELATION COEFFICIENT

Imagine a group of n cases drawn at random for a problem of correlation. However, instead of having a value of X for each observation, we have only its *rank* in the group on variable X (say for low to high, the ranks 1 through n). In the same way, for each observation we have its rank in the group on variable Y. The question to be asked is "How much does the ranking on variable X tend to agree with the ranking on variable Y?" Suppose that we call the items ranked "individuals." We can take the point of view that if rank orders agree, the ranks assigned to individuals should *correlate* positively with each other, whereas disagreement should be reflected by a negative correlation. A zero correlation represents an intermediate condition: no particular connection between the ranks of an individual on the two variables.

For a descriptive index of agreement between ranks, the ordinary correlation coefficient can be computed on the *ranks* just as for any numerical scores, and this is how the **Spearman rank correlation** for a sample is defined:

$$r_S = \text{correlation between ranks over individuals.}$$

That is, r_S is given by Equation (10.1.7) with x_i and y_i representing the *ranks* of individual i on the two variables of interest:

$$r_S = \frac{n \sum_i x_i y_i - \left(\sum_i x_i\right)\left(\sum_i y_i\right)}{\sqrt{n \sum_i x_i^2 - \left(\sum_i x_i\right)^2}\sqrt{n \sum_i y_i^2 - \left(\sum_i y_i\right)^2}}. \quad (12.19.1)$$

But since we are working with ranks, $(x_1, \ldots, x_n)$ and $(y_1, \ldots, y_n)$ are simply permutations of $(1, \ldots, n)$ provided that no ties in rank exist, and from algebra it is known that the sum of the whole numbers from 1 to n is equal to $n(n + 1)/2$ and the sum of the squares of these numbers is $n(n + 1)(2n + 1)/6$. Therefore,

$$\sum_i x_i = \sum_i y_i = \frac{n(n + 1)}{2} \quad (12.19.2)$$

and

$$\sum_i x_i^2 = \sum_i y_i^2 = \frac{n(n + 1)(2n + 1)}{6}. \quad (12.19.3)$$

Also, if we let $d_i = x_i - y_i$, the *difference* between the ranks associated with individual i, then

$$\sum_i d_i^2 = \sum_i (x_i - y_i)^2$$

$$= \sum_i x_i^2 + \sum_i y_i^2 - 2 \sum_i x_i y_i,$$

so that

$$\sum_i x_i y_i = \frac{\sum_i x_i^2 + \sum_i y_i^2 - \sum_i d_i^2}{2}. \quad (12.19.4)$$

Using Equations (12.19.2)–(12.19.4) in Equation (12.19.1) and simplifying, we arrive at the following expression for r_S:

$$r_S = 1 - \left[\frac{6\left(\sum_i d_i^2\right)}{n(n^2 - 1)}\right]. \quad (12.19.5)$$

The Spearman rank correlation is thus very simple to compute when ranks are untied. All we need to know is n, the number of individuals ranked, and d_i, the difference in ranking for each individual. Despite the different computations involved, r_S is only an ordinary correlation coefficient calculated on ranks.

The computation of r_S will be illustrated in a problem dealing with agreement between judges' rankings of objects. In a test of weight discrimination, two judges each ranked 10 small objects in order of their judged heaviness. The results are shown in Table 12.19.1. Did the judges tend to agree?

$$r_S = 1 - \frac{6(128)}{10(10^2 - 1)} = .224.$$

The Spearman rank correlation is only .224, so that agreement between the two judges was not very high, although there was some slight tendency for similar ranks to be given to the same objects by the judges.

Incidentally, notice that if the relative true weights of the objects are known, we could also find the agreement of each judge with the true ranking. When some criterion ranking is known, it may be useful to compare a judged ranking with this criterion to evaluate the accuracy or "goodness" of the judgments—this is not the same as the agreement of judges with each other, of course.

Formula (12.19.1) may not be used if there are ties in either or both rankings, since the means and variances of the ranks then no longer have the simple relationship to n that is present in the no-tie case. When ties

Table 12.19.1 Observations, Differences, and Squared Differences for Weight-Discrimination Example

Object	Judge I	Judge II	d_i	d_i^2
1	6	4	2	4
2	4	1	3	9
3	3	6	3	9
4	1	7	6	36
5	2	5	3	9
6	7	8	1	1
7	9	10	1	1
8	8	9	1	1
9	10	3	7	49
10	5	2	3	9
				128

exist, perhaps the simplest procedure is to assign mean ranks to sets of tied individuals; that is, when two or more individuals are tied in order, each is assigned the mean of the ranks they would otherwise occupy. Next, an ordinary correlation coefficient is computed, using the ranks as though they were simply numerical scores. The result is a Spearman rank correlation that can be regarded as corrected for ties.

When no ties in rank exist, the exact sampling distribution of r_S can be worked out for small samples. Exact tests of significance for r_S are based on the idea that if one of the two rank orders is known, and the two underlying variables are independent, then each and every permutation in order of the individuals is equally likely for the other ranking. On this basis, the exact distribution of the sum of the squared differences in ranks can be found, and this can be converted into a distribution of r_S. The exact distribution of r_S is rather peculiar. Although the distribution is symmetric when the rankings are independent, the plot of the distribution has a curious jagged or serrated appearance due to particular constraints on the possible values of the sum of squared differences for a given n. With n large, the distribution does approach a normal form, but relatively slowly, so that for samples of small to moderate size the normal approximation is not very good.

The hypothesis of the independence of the two variables represented by rankings can also be given an approximate large sample test in terms of r_S. This test has a form very similar to that for r_{XY} [Equation (10.3.1)]:

$$t = \frac{r_S \sqrt{n-2}}{\sqrt{1 - r_S^2}}, \qquad (12.19.6)$$

with $n - 2$ degrees of freedom. This test is really satisfactory, however, only when n is fairly large; n should be greater than or equal to 10.

Remember that if inferences are to be made about the form of relation holding between two continuous underlying variables, we can fail to reject the null hypothesis either because the variables are independent or because the relation is nonmonotone. Furthermore, the rejection of the hypothesis does not necessarily let us conclude that linear association exists between the underlying variables, but only that some more or less monotone relation holds.

Under particular assumptions, especially that of a bivariate normal distribution, the value of r_S from a large sample can be treated as an estimate of the value of ρ for the variables underlying the ranks. When the population is bivariate normal with $\rho = 0$, values of r_S and r_{XY} correlate very highly over samples. On the other hand, these assumptions are rather special, and the status of r_S as an estimator of ρ *in general* is open to considerable question.

12.20   THE KENDALL TAU COEFFICIENT

A somewhat different approach to the problem of agreement between two rankings is given by the τ coefficient (small Greek tau) due to M. G. Kendall. Instead of treating the ranks themselves as though they were scores and finding a correlation coefficient, as in r_S, in the computation of τ we depend only on the number of *inversions* in order for pairs of individuals in the two rankings. A single inversion in order exists between *any pair* of individuals b and c when $b > c$ in one ranking and $c > b$ in the other. When two rankings are *identical*, no inversions in order exist. On the other hand, when one ranking is exactly the reverse of the other, an inversion exists for *each pair* of individuals; this means that complete disagreement corresponds to $\binom{n}{2}$ inversions. If the two rankings *agree* (show noninversion) for as many pairs as they disagree about (show inversion), the tendency for the two rank orders to agree or disagree should be exactly zero.

This leads to the following definition of the τ **statistic:**

$$\tau = 1 - \left[\frac{2(\text{number of inversions})}{\text{number of pairs of objects}} \right]. \qquad (12.20.1)$$

If the sample size is n, the number of pairs of objects is $\binom{n}{2}$, and

$$\tau = 1 - \frac{2(\text{number of inversions})}{\binom{n}{2}}. \qquad (12.20.2)$$

This is equivalent to

$$\tau = \frac{\begin{array}{c}(\text{number of times rankings agree about a pair}) \\ - (\text{number of times rankings disagree})\end{array}}{\text{total number of pairs}}.$$

It follows that the τ statistic is essentially a difference between two proportions: the proportion of pairs having the *same* relative order in both rankings minus the proportion of pairs showing *different* relative order in the two rankings.

Viewed as coefficients of agreement, r_S and τ thus rest on somewhat different conceptions of "disagree." In the computation of r_S, a disagreement in ranking appears as the *squared* difference between the ranks themselves over the individuals. In τ, an inversion in order for any *pair* of objects is treated in the same way as evidence for disagreement. Although these two conceptions are related, they are not identical; the process of squaring differences between rank values in r_S places somewhat different weight on *particular* inversions in order, whereas in τ all inversions are weighted equally by a simple frequency count. Values of the statistics r_S and τ are correlated over successive random samples from the same population, but the extent of the correlation depends on a number of things, including sample size and the character of the relation between the underlying variables in the population. Nevertheless, the two statistics are closely connected, and a number of mathematical inequalities must be satisfied by the values of the two statistics. For example,

$$-1 \leq 3\tau - 2r_S \leq 1.$$

Various methods exist for the computation of τ, but the simplest is a graphic method. In this method, all we do is list the individuals or objects ranked, once in the order given by the first ranking, and again in the order given by the second. For example, suppose that two television critics rank 7 programs $[a, b, c, d, e, f, g]$ as in Table 12.20.1. Now straight lines are drawn connecting the same objects in the two parallel rankings, thus:

$$
\begin{array}{ccccccc}
1 & 2 & 3 & 4 & 5 & 6 & 7
\end{array}
$$

Then *the number of times that pairs of lines cross is the number of inversions in order.* Here, the number of crossings is 4, and, from Equation (12.20.2),

$$\tau = 1 - \frac{2(4)}{\binom{7}{2}} = 1 - \frac{8}{21} = \frac{13}{21} = .62.$$

Although r_S from a sample has a rather artificial interpretation as a correlation coefficient, the interpretation of the obtained value of τ is quite straightforward. If a *pair* of objects is drawn at random from among those ranked, the probability that these two objects show the *same* relative order in both rankings is .62 *more* than the probability that they would show

Table 12.20.1   Rankings of 7 Television Programs

	RANK						
	1	*2*	*3*	*4*	*5*	*6*	*7*
Critic 1	c	a	b	e	d	g	f
Critic 2	a	c	e	b	f	d	g

different order. In other words, from the evidence at hand it is a considerably better bet that the two judges will tend to order a randomly selected pair in the *same* way than in a different way.

This graphic method of computing τ is satisfactory only when no ties in ranking exist. For nontied rankings, the method is very simple to carry out, even when moderately large numbers of individuals have been ranked. Notice that although both the examples of computations for r_S and τ featured judges' rankings of objects, exactly the same methods apply when the ranking principle is provided by scores shown by individuals on each of two variables.

An exact test of significance for τ is based on the assumption that the variables underlying the ranks are continuously distributed, so that ties are impossible. Alternatively, we can imagine random sampling of complete rank orders of n things from a potential population of such rankings. In either instance, we assume that if there is no true relationship between the pairs of rank orders observed, then given the first rank order, all possible permutations in order are equally likely to occur as the second rank order. This makes finding the exact sampling distribution of τ for fixed n relatively simple.

However, for our purposes, the fact of importance is that the exact sampling distribution of τ approaches a normal distribution very quickly with successive increases in the size of n. Even for fairly small values of n, the distribution of τ is approximated relatively well by the normal distribution. Of course, this is true only when H_0 is true, so that the rankings are each equally likely to show any of the $n!$ permutations in order, and $E(\tau) = 0$. The distribution of τ is not simple to discuss when other conditions hold, and so we will test only the hypothesis of independence between rankings (implying equal probability of occurrence for each and every possible ordering of n observations on the second variable given their ordering on the first).

For n of about 10 or more, the test is given by

$$z = \frac{\tau}{\sigma_\tau}, \qquad (12.20.3)$$

referred to a normal distribution, where

$$\sigma_\tau^2 = \frac{2(2n + 5)}{9n(n - 1)}. \qquad (12.20.4)$$

This approximate test can be improved if a correction for continuity is made. The correction for continuity involves the subtraction of $1/\binom{n}{2}$ from the absolute value of τ. For the example involving the television critics,

$$\sigma_\tau^2 = \frac{2(14 + 5)}{9(7)(6)} = \frac{38}{378} = .10$$

and
$$z = \frac{.62 - (1/21)}{.316} = 1.81.$$

To conclude this section, it should be useful to compare briefly r_S and τ. First, τ can be given a very simple interpretation as a descriptive statistic, since it is a difference between two proportions. On the other hand, the Spearman coefficient is meaningful, at least at an elementary level, only by analogy with the ordinary correlation coefficient. The properties of r_S as an estimator of ρ vary considerably with the form of the population distribution and with the true value of ρ. Of course, both r_S and τ provide useful information about the monotonicity of the relationship between two variables, as noted in Section 12.18. Finally, with respect to tests for independence, an advantage of τ is the fairly rapid convergence of its sampling distribution to a normal form, as opposed to the somewhat slower convergence of the distribution of r_S.

12.21   KENDALL'S COEFFICIENT OF CONCORDANCE

Sometimes we want to know the extent to which members of a set of m distinct rank orderings of n things tend to be similar. For example, in a beauty contest, each of 7 judges ($m = 7$) gives a simple rank order of the 10 contestants ($n = 10$). How much do these rank orders tend to agree, or show "concordance"? This problem is usually handled by application of Kendall's **coefficient of concordance.** As we shall see, the coefficient of concordance, denoted by W, is closely related to the average r_S among the m rank orders.

The coefficient W is computed by putting the data into a table with m rows and n columns. In the cell for column j and row k appears the rank number assigned to individual object j by judge k. Table 12.21.1 shows the data for the judges and the beauty contestants. It is quite clear that the judges did not agree perfectly in their rankings of these contestants. However, what would the column totals of ranks, W_j, have been if the judges had agreed exactly? If each judge had given exactly the same rank to the same girl, then one column should total to $7(1)$, or simply 7, another to $7(2)$, or 14, and so on, until the largest sum should be $7(10)$, or 70. On the other hand, suppose that there was complete disagreement among the

Table 12.21.1 Data for Beauty-Contest Example

					CONTESTANTS					
Judges	*1*	*2*	*3*	*4*	*5*	*6*	*7*	*8*	*9*	*10*
1	8	7	5	6	1	3	2	4	10	9
2	7	6	8	3	2	1	5	4	9	10
3	5	4	7	6	3	2	1	8	10	9
4	8	6	7	4	1	3	5	2	10	9
5	5	4	3	2	6	1	9	10	7	8
6	4	5	6	3	2	1	9	10	8	7
7	8	6	7	5	1	2	3	4	10	9
W_j	45	38	43	29	16	13	34	42	64	61

judges, so that there was no tendency for high or low rankings to pile up in particular columns. Then we should expect each column sum to be about the same. In this example, the column sums of ranks are not identical, so that apparently some agreement exists, but neither are the sums as different as they should be when absolutely perfect agreement exists.

This idea of the extent of variability among the respective sums of ranks is the basis for Kendall's W statistic. Basically,

$$W = \frac{\text{variance of rank sums}}{\text{maximum possible variance of rank sums}},$$

which can be expressed in the form

$$W = \left(\frac{12 \sum_j W_j^2}{m^2 n (n^2 - 1)} \right) - \frac{3(n + 1)}{n - 1}. \qquad (12.21.1)$$

For the example, we find

$$W = \left(\frac{12[(45)^2 + \cdots + (61)^2]}{49(10)(99)} \right) - \frac{3(11)}{9}$$

$$= 4.28 - 3.66 = .62.$$

There is apparently a moderately high degree of "concordance" among the judges, since the variance of the rank sums is 62 percent of the maximum possible. Note that by its definition, W cannot be negative, and its maximum value is 1.

The value of the concordance coefficient is somewhat hard to interpret directly in terms of the tendency for the rankings to agree, but an interpretation can be given in terms of the average value of r_S over all possible *pairs* of rank orders. That is,

$$\text{average } r_S = \frac{mW - 1}{m - 1}. \tag{12.21.2}$$

For the example,

$$\text{average } r_S = \frac{7(.62) - 1}{7 - 1} = .56.$$

If we took all of the possible $\binom{7}{2}$ or 21 *pairs* of judges and found r_S for each such pair, the average rank correlation would be about .56. Thus, on the average, judge-pairs do tend to give relatively similar rankings. An advantage of reporting this finding in terms of W rather than average r_S is that

$$\frac{-1}{m - 1} \leq \text{average } r_S \leq 1,$$

whereas, regardless of the values for n or m,

$$0 \leq W \leq 1.$$

This makes W values more immediately comparable across different sets of data. Nevertheless, the clearest interpretation of W seems to be in terms of average r_S.

EXERCISES

1. Briefly discuss the advantages and disadvantages of nonparametric methods as compared with parametric methods in statistics.

2. In what sense is a chi-square test for independence a "comparison of entire distributions"?

3. Why is sample size a problem in the use of chi-square tests? Why is there some disagreement about when a sample is "large enough" for a chi-square test to be appropriate?

4. Why is the assumption of independent observations important in a chi-square test?

5. An instructor gives a 6-item multiple-choice test, with each item having 4 possible answers. Suppose that when a student is simply guessing, the probability of getting the right answer on any given item is exactly $\frac{1}{4}$. Furthermore, suppose that the answer guessed on any item is independent of the answer

guessed on any other item. The test is given to 40 students, with the following results:

Number correct	Frequency
6	3
5	1
4	3
3	6
2	10
1	12
0	5
	—
	40

Test the hypothesis that each student was guessing independently on each item (find the p-value).

6. In a particular problem in Mendelian genetics, when a parent having the two dominant characteristics A and B is mated with another parent having the two recessive characteristics a and b, the offspring should show combinations of dominant and recessive characteristics with the following relative frequencies:

Type	Relative frequency
AB	9/18
aB	4/18
Ab	4/18
ab	1/18

In an actual experiment, the following frequency distribution resulted:

Type	Frequency
AB	39
aB	19
Ab	16
ab	1
	—
	$75 = n$

Test the hypothesis that the Mendelian theory holds for these dominant and recessive characteristics (find the p-value).

7. A marketing manager claims that 20 percent of the consumers purchasing a particular product will buy Brand A, 30 percent will buy Brand B, and 50 percent will buy Brand C. A series of purchases is observed: out of 200 consumers, 45 bought A, 57 bought B, and 98 bought C. In light of these results, what do you think of the marketing manager's claim?

8. A firm produces razors and razor blades. The president of the firm hypothesizes that in the population of males over 18 in the United States, 40 percent use both the firm's razor and the firm's blades, 20 percent use the razor but not the blades, 10 percent use the blades but not the razor, and 30 percent use neither. A sample of 100 males over 18 yields the following results:

| | | USE RAZOR | |
		Yes	No
USE BLADES	Yes	36	13
	No	19	32

On the basis of these data, comment on the president's claim.

9. A random sample from a certain population of children yields the following weights (in pounds):

66 78 82 75 94 77 69 74 68 60 96 78 89 61 75 95 60

79 83 71 79 62 67 97 78 85 76 65 71 75 86 84 75 81

68 63 62 75 76 77 73 65 88 87 60 62 71 78 85 72

Use a chi-square test to investigate the hypothesis that the distribution of weights in the population is a normal distribution. Use 10 class intervals, each of which has probability 1/10 in a normal distribution.

10. For the sales example in Section 12.3, use a chi-square test for normality with only 4 equally likely intervals. Compare the results with those of the test for normally with 8 intervals.

11. In a chi-square goodness-of-fit test for a normal distribution, distinguish between the case in which the statistician specifies the mean and variance in advance and the case in which the sample mean and sample variance are used as the parameters of the normal distribution from which the expected frequencies are obtained. Is there any difference in the tests in the two cases?

12. Does a normal distribution provide a good fit to the observations in Exercise 17, Chapter 5?

13. In Exercise 16, Chapter 5, an economist wants to assume normality in order to use a T statistic for inferences about mean income and a χ^2 statistic for inferences about the variance of income. Comment on the normality assumption.

14. A production manager thinks that the number of defective items in a lot of fixed size from a production process has a binomial distribution. To investigate this supposition, random samples of 20 items are drawn repeatedly, and each time the number of defective items is recorded. This is done 80 times, with the following results.

Number of defective items	Frequency
0	33
1	25
2	13
3	7
4	2

Comment on the applicability of a binomial model in this situation.

15. A safety expert would like to find a convenient probability distribution to represent the occurrence of fatal accidents along a particular stretch of highway. The information available to the safety expert consists of the numbers of fatal accidents for 100 weeks. During this period 45 weeks were free of fatal accidents, 29 weeks had one fatal accident, 17 weeks had two fatal accidents, and 9 weeks had three fatal accidents. Use a chi-square test to check the goodness-of-fit of a Poisson model in this situation.

16. In Exercise 15, from what you know about the occurrence of fatal accidents, does a Poisson model seem reasonable as a representation of the "accident process?" How might you modify the Poisson model to make it even more suitable for this situation? Can you think of any other models that might be considered here?

17. Show that when the probabilities are equal in a χ^2 goodness-of-fit test with $\nu = J - 1$, the variance of χ^2 is equal to $2\nu(n - 1)/n$, where ν represents the number of degrees of freedom. [*Hint:* See Equation (12.2.4).]

18. A college administrator needs to predict the size of the student body for the coming academic year. The number of students accepted by the college is known, but not all of these students will actually enroll for classes. A random sample from the records of the past few years yields the following data: out of 200 individuals accepted as new students by the university, 140 actually enrolled. On the other hand, a colleague of the administrator claims that 80 percent of those accepted actually enroll in the college. Test the colleague's claim
    (a) with a chi-square test of goodness of fit,

(b)   with a normal approximation,

$$z = \frac{r - np}{\sqrt{np(1-p)}},$$

used to test the hypothesis that p, the proportion of accepted students who actually enroll, is equal to .80.

19.   In Exercise 18, what is the relationship between the χ^2 value determined in (a) and the z value determined in (b)? Will this relationship always hold? Explain.

20.   Many of the procedures discussed in Chapters 6–11 require the assumption of normality. Often there is good reason to believe that this assumption is quite unrealistic but that it may be possible to apply a transformation to achieve normality. For example, the distribution of incomes in a population is frequently skewed positively, and the assumption of normality is questionable. However, the distribution of the logarithms of incomes might be closely approximated by a normal distribution. Here, a logarithmic transformation is used in an attempt to achieve normality. Using natural logarithms, apply a logarithmic transformation to the data of Exercise 16, Chapter 5. Does a normal distribution provide a good fit to the transformed data? (To check the fit, consider eight equally likely intervals.)

21.   The amount of time between calls arriving at a certain switchboard is recorded. For a sample of 100 observations, 32 calls arrived within 30 seconds, 22 calls arrived in between 31 and 60 seconds, 16 in between 61 and 90 seconds, 14 in between 91 and 120 seconds, 10 in between 121 and 150 seconds, and 6 took more than 150 seconds to arrive. It is also known that the average amount of time between calls for this sample was 60 seconds. Does the exponential distribution provide a good fit to these observations?

22.   In a comparison of child-rearing practices within two cultures, a researcher drew a random sample of 100 families representing culture I and another sample of 100 families representing culture II. Each family was classified according to whether the family was father-dominant or mother-dominant in terms of administration of discipline. The results follow:

	Culture I	*Culture II*
Father-dominant	53	37
Mother-dominant	47	63

What is the p-value for the test of the hypothesis that culture and the dominant parent in a family are independent?

23. Four large midwestern universities were compared with respect to the fields in which graduate degrees are given. The graduation rolls for last year from each university were taken and the results put into the following contingency table:

FIELD

		Law	Medicine	Science	Humanities	Other
	A	29	43	81	87	73
UNIVERSITY	B	31	59	128	100	87
	C	35	51	167	112	252
	D	30	49	152	98	215

Is there significant association (at the .05 level) between the university and the fields in which it awards graduate degrees? What are we assuming when we carry out this test?

24. In a study of the possible relationship between the amount of training given to production line workers and their ratings (as given by their superiors) after working for 6 months, the following results were obtained:

RATINGS

	Excellent	Good	Poor
1 day	6	12	0
3 days	12	25	6
1 week	14	31	12
2 weeks	2	23	7

AMOUNT OF TRAINING

Is there significant association (at the .01 level) between the amount of training and the ratings, according to these data?

25. Two independent large samples are drawn, and each is arranged into the same form of contingency table. How could we test the hypothesis that the attributes are independent in *both* of the populations represented by the samples? [*Hint*: Recall the discussion of chi-square variables in Chapter 6.] How might we test the hypothesis that the frequency in any given cell of the contingency table should be the *same* for the two populations?

26. A researcher was interested in the stability of political preference among American women voters. A random sample of 80 women who voted in the elections of 1964 and 1968 showed the following results:

1964 VOTE

		Republican	Democrat
1968 VOTE	Republican	34	11
	Democrat	5	30

Do these data indicate that the true proportion of women who voted Republican in 1968 was different from the true proportion who voted Republican in 1964? Find the p-value.

27. In Exercise 58, Chapter 2, determine if there is a significant association (at the .10 level)
    (a) between sex and attitude toward liberalized sale of beer,
    (b) between attitude toward liberalized sale of beer and student-nonstudent status.

28. In a random sample of 600 men between the ages of 60 and 65, each man was classified according to his drinking habits (heavy drinker, light drinker, not a drinker), his smoking habits (heavy smoker, light smoker, not a smoker), and whether or not he suffers from heart disease. (Assume that for the purposes of the study, all of these classifications were carefully defined; to save space, we ignore the exact definitions, such as the definition of a heavy smoker.) The data are as follows:

	HEAVY DRINKER			LIGHT DRINKER			NOT A DRINKER		
	Heavy smoker	Light smoker	Not a smoker	Heavy smoker	Light smoker	Not a smoker	Heavy smoker	Light smoker	Not a smoker
Heart disease	21	22	16	17	37	14	28	35	12
No heart disease	8	20	33	12	85	47	31	97	65

(a) Is there significant association (at the .05 level) between drinking habits and the existence of heart disease?
(b) Is there significant association (at the .05 level) between smoking habits and the existence of heart disease?
(c) Is there significant association (at the .05 level) between smoking habits and drinking habits?

29. Can the chi-square test for association discussed in Section 12.4 be generalized to a situation in which there are *three* categorical attributes? Explain. If there are R classes of the first attribute, C classes of the second attribute, and D classes of the third attribute, how many degrees of freedom would the chi-square statistic have?

30. Use the generalization in Exercise 29 and the data in Exercise 28 to test for association (at the .05 level) among drinking habits, smoking habits, and heart disease.

31. Prove Equation (12.4.3).

32. In Exercise 26, apply the likelihood ratio test for association based on the multinomial rule. From this test, is there significant association between 1968 votes and 1964 votes at the .10 level of significance?

33. For the data of Exercise 22, find the coefficient of predictive association ($\hat{\lambda}$) for predicting parent-dominance of a family from the culture. Do the data suggest the presence of a very strong predictive association here? Explain.

34. For the data of Exercise 22, find $\hat{\varphi}$, $\hat{\varphi}'$, and C_{AB}, three descriptive measures of association in a contingency table.

35. For the data of Exercise 23, find the value of Cramer's statistic and the coefficient of contingency. Also, find the coefficient of predictive association for predicting the field in which a graduate degree is taken, given the university.

36. Why are the descriptive measures of association (φ, φ', C_{AB}) distinguished from a predictive measure of association such as λ? Briefly discuss both types of measures of association.

37. For the data of Exercise 24, find $\hat{\varphi}$, $\hat{\varphi}'$, and C_{AB}.

38. For the data of Exercise 26, find $\hat{\varphi}$, $\hat{\varphi}'$, C_{AB}, and the coefficient of predictive association for predicting a woman's 1968 vote from her 1964 vote.

39. For the data of Exercise 28, find the coefficient of predictive association for predicting smoking habits from drinking habits, the coefficient of predictive association for predicting drinking habits from smoking habits, and the symmetric index of predictive association with respect to drinking habits and smoking habits.

40. Suppose that the categories in a 2×2 contingency table can be thought of as ordered, with 1 representing the higher category for each attribute. Furthermore, let the contingency table be set up as follows:

		ATTRIBUTE A	
		0	1
ATTRIBUTE B	1	a	b
	0	c	d

For each observation, let $X_i = 0$ if the observation falls in the lower category of A and $X_i = 1$ if it falls in the higher category of A. Similarly, let $Y_i = 0$ if the observation falls in the lower category of B and $Y_i = 1$ if the observation falls in the higher category of B. Prove that the absolute value of the sample correlation coefficient r_{XY} is equal to the sample value of $\hat{\phi}$ computed from the contingency table. (Because of this relationship, $\hat{\phi}$ is often called the *fourfold point correlation*.)

41. For each observation in Exercise 11, Chapter 10, let $X_i = 0$ if the actual price is 2 or 3, $X_i = 1$ if the actual price is 4 or 5, $Y_i = 0$ if the average "guessed" price is less than 3.5, and $Y_i = 1$ if the average "guessed" price is 3.5 or more. With the data dichotomized on each variable in this fashion, calculate the sample correlation coefficient (see Exercise 40). Discuss the relative merits of this sample correlation coefficient and the sample correlation coefficient computed from the raw data.

42. Suppose that you are interested in the proportion of defectives produced by a certain process. Call this proportion p, and assume that your prior distribution for p is as follows:

p	$P(p)$
.01	.40
.02	.30
.03	.20
.04	.10

You observe a random sample of 10 items from the process, which behaves like a Bernoulli process, and there are no defectives in the sample.
(a) Find your posterior distribution for p.
(b) Find $H(A)$, the average prior amount of uncertainty.
(c) Find the posterior $H(A)$.
(d) What is the amount of information in the sample, in terms of information theory?

43. In Exercise 9, test the same hypothesis by using the Kolmogorov-Smirnov test instead of the chi-square test.

44. Repeat Exercises 9 and 43, using 6 class intervals instead of 10. Is it possible for the number of class intervals to have an effect on the result of the tests? Explain.

45. Use the Kolmogorov-Smirnov test for the data given in Exercise 5. Strictly speaking, is the test applicable in this situation? Discuss.

46. Could the Kolmogorov-Smirnov test be used in a situation such as that given in Exercise 6? Explain your answer carefully.

47. Use the Kolmogorov-Smirnov test for the data given in Exercise 12.

48. In Exercise 15, test the same hypothesis by using the Kolmogorov-Smirnov test instead of the chi-square test.

49. Use the Kolmogorov-Smirnov test for the data given in Exercise 21.

50. Whenever parametric tests such as T and F tests are applicable, they are generally more powerful than the corresponding tests based on order statistics. Do you see any reason why they *should* be more powerful?

51. Explain the concept of power-efficiency. Why is this concept relevant to the study of order methods and other nonparametric methods?

52. For the data of Exercise 69, Chapter 7, use the Wald-Wolfowitz runs test to decide if there is any difference between Town A and Town B with regard to the distribution of incomes of residents.

53. In Exercise 71, Chapter 7, use the Wald-Wolfowitz runs test to test for any differences in the outputs of the two machines.

54. In Exercise 52, use the Mann-Whitney test instead of the Wald-Wolfowitz test.

55. In Exercise 52, use the median test instead of the Wald-Wolfowitz test.

56. In Exercise 53, use the Mann-Whitney test and the median test and compare these tests with the Wald-Wolfowitz test used in Exercise 53. Also, compare all three of these nonparametric tests with a T test for this situation.

57. Random samples of size 20 are taken from two production processes, and the weight of each item sampled is recorded. The results are as follows:

Process 1:   124 126 125 131 125 124 127 128 123 130

              132 125 124 127 129 125 125 130 129 128

Process 2:   131 130 128 126 127 129 134 126 123 127

              125 128 130 131 126 125 129 128 128 130

Test to see if there are any differences between the two processes with respect to the weight of the items from the processes, (a) using the median test, (b) using the Wald-Wolfowitz runs test, (c) using the Mann-Whitney test.

58. In Exercise 57, use a T test to test for differences in the mean weights. Does this test deal with the same hypotheses as the tests considered in Exercise 57?

59. In Exercise 13, Chapter 5, use a runs test to investigate the randomness of the sample with regard to the sex of the patient.

60. For the income example in Section 12.11, find the sampling distribution of R, the number of runs, (a) exactly, (b) using the normal approximation given in Section 12.11. Does the normal approximation provide a good fit in this situation? Why or why not?

61. In a study of the preferences of newly married couples with respect to size of family, some 26 couples were asked, independently, the ideal number of children they would like to have. Responses of husbands and wives are listed below:

Husband: 3 0 1 2 0 2 1 2 2 1 2 0 3 5 7 1 0 2 10 5 2 0 1 3 5 2

Wife: 2 1 0 2 3 3 2 3 3 3 4 1 4 2 5 2 3 4   3 3 4 2 3 2 2 1

Use the sign test to determine if there is a significant difference (at the .05 level) between husbands and wives in the size of family desired.

62. For the data in Exercise 61, use the Wilcoxon test to determine if there is a significant difference (at the .05 level) between husbands and wives in the size of family desired.

63. Under what circumstances can the order methods, such as the Mann-Whitney test, be compared in power to, say, an ordinary T test? Are tests based on order methods *always* less powerful than their parametric counterparts? Explain.

64. Two expert wine tasters served as judges at a tasting for which 12 different bottles of burgundy wine were served. Each of the experts ranked the 12 bottles according to the overall quality of the wine, and the rankings are as follows:

Wine	Ranking of Judge 1	Ranking of Judge 2
1	10	7
2	8	9
3	4	2
4	11	8
5	2	3
6	5	6
7	3	10
8	12	12
9	1	4
10	9	11
11	7	5
12	6	1

Use the sign test to investigate the difference between the two rankings.

65. In Exercise 64, analyze the data by means of the Wilcoxon test. Is the T test applicable in this situation? Explain.

66. In Exercise 75, Chapter 7, use the sign test and the Wilcoxon test to test for differences between the old test and the new test. How do these tests differ from the T test used in Exercise 75, Chapter 7?

67. In Exercise 76, Chapter 7, use the sign test and the Wilcoxon test to test for differences between 1960 expenditures on research and development and 1970 expenditures on research and development.

68. Samples of size 10 are chosen randomly from each of five populations, with the following results:

		POPULATION		
1	*2*	*3*	*4*	*5*
87	41	31	60	55
35	19	18	8	67
67	70	64	14	46
44	62	7	49	95
51	43	13	16	79
49	46	22	28	63
18	6	38	16	66
98	13	31	70	82
97	22	18	64	56
84	38	64	30	67

Use the median test to decide if there are differences among the five populations. Also, use the median test to decide if there is any difference between Population 1 and Population 2.

69. In Exercise 68, use the Wald-Wolfowitz runs test to decide if there is any difference between Population 1 and Population 3.

70. In Exercise 68, use the Mann-Whitney test to decide if there is any difference between Population 2 and Population 5.

71. In Exercise 68, use the Kruskal-Wallis "analysis of variance" by ranks to test for differences among the five populations.

72. In Exercise 57, suppose that a sample of size 20 is taken from a *third* production process:

Process 3: 135 120 118 129 126 128 138 120 124 132

123 138 130 127 129 125 135 131 ⁻128 130

Use the median test and the Kruskal-Wallis test to investigate differences among the three processes with respect to the weight of the items from the processes.

73. Analyze the data of Exercise 33, Chapter 11 by means of (a) the median test and (b) the Kruskal-Wallis test. Do these procedures test the same hypotheses that are tested by the F test for a simple one-factor analysis of variance model? Explain.

74. In Exercise 46, Chapter 11, use (a) the median test and (b) the Kruskal-Wallis test to investigate differences among the ratings of the three detergents.

75. In Exercise 64, two additional judges ranked the wines in the following order:

    Judge 3:   6, 8, 3,   9, 1, 4, 11, 10, 2, 12, 5, 7.

    Judge 4:   7, 8, 5, 10, 2, 6, 12,   9, 3, 11, 4, 1.

    Use the Friedman test to investigate the differences among the four rankings.

76. Use the Friedman test to analyze the following data regarding scores on four examinations for 15 students:

	EXAMINATION			
Student	*1*	*2*	*3*	*4*
1	20	25	23	32
2	28	24	30	33
3	26	26	26	28
4	19	16	22	25
5	29	24	25	27
6	31	35	30	32
7	26	20	21	28
8	36	30	26	33
9	25	24	24	24
10	27	28	24	30
11	23	18	19	22
12	19	30	27	28
13	28	17	25	20
14	16	16	31	19
15	25	27	24	29

77. The Kruskal-Wallis and the Friedman tests can be used in circumstances where a simple analysis of variance might ordinarily be applicable. Make up original examples from your field of interest of situations in which it would be desirable to substitute one of these two order methods for the parametric analysis of variance.

78. Explain the difference in interpretation among
    (a)   the measures of association in contingency tables,
    (b)   the measures of rank correlation,
    (c)   the product moment correlation coefficient, r.

79. For the data in Exercise 64, compute r_S and τ.

80. Compare and contrast the Spearman rank correlation r_S and the Kendall tau coefficient τ.

81. In Exercise 75, Chapter 7, find how each IQ test ranks the 5 students and compute r_S and τ for these rankings.

82. For the data in Exercises 64 and 75, compute the coefficient of concordance.

83. In Exercise 76, find out how each of the four examinations ranks the 15 students (in case of ties, use the numbers of the students to break the ties, with the lower-numbered student receiving the lower rank). From the four rankings, find the coefficient of concordance.

84. In Exercise 82, find r_S for each possible pair of judges. Next, compute the average value of r_S from the individual values of r_S and compare this result with the average r_S computed from Equation (12.21.2).

85. In a study of attitudes toward international affairs, each of a group of 15 subjects ranked 10 countries according to their "responsibility" in international affairs. The data follow:

COUNTRY

		A	B	C	D	E	F	G	H	I	J
	1	2	1	3	5	4	6	7	8	9	10
	2	8	7	6	4	9	1	2	10	5	3
	3	3	4	2	1	6	5	7	8	9	10
	4	9	10	8	6	7	1	3	4	5	2
	5	5	2	3	4	6	1	7	8	9	10
	6	2	4	3	1	5	8	9	10	7	6
	7	4	3	2	5	1	6	9	10	8	7
SUBJECT	8	2	4	1	6	5	3	8	7	9	10
	9	9	10	6	7	8	1	4	3	5	2
	10	5	4	1	10	2	3	7	8	6	9
	11	2	4	1	5	3	6	7	9	10	8
	12	3	4	2	5	1	10	9	6	8	7
	13	4	3	5	2	10	1	8	6	9	7
	14	5	3	4	2	1	8	9	6	10	7
	15	4	5	10	1	3	2	8	9	6	7

Find the coefficient of concordance and the average rank correlation for these data. Assuming that the 15 subjects constitute a random sample, test the hypothesis of zero true concordance among subjects in terms of their rankings of countries.

APPENDIX A: SOME COMMON DIFFERENTIATION AND INTEGRATION FORMULAS

The **derivative** of y with respect to x is simply the rate of change of y as x changes. In other words, if y can be shown on a graph as a function of x, $y = f(x)$, then the derivative of y with respect to x at the point x_0 is simply the slope of the line which is tangent to the curve $y = f(x)$ at the point x_0. The derivative of y with respect to x is denoted by dy/dx, or $df(x)/dx$, or $f'(x)$, and it is formally defined as follows:

$$\frac{dy}{dx} = \lim_{h \to 0} \frac{f(x + h) - f(x)}{h}.$$

We are not concerned with deriving differentiation formulas here; for derivations the student can refer to a textbook on calculus. For the student's convenience, several common differentiation formulas are presented.

y	dy/dx
c	0
x^k	kx^{k-1}
cx	c
$cf(x)$	$c\,\dfrac{df(x)}{dx}$
$[f(x)]^k$	$k[f(x)]^{k-1}\,\dfrac{df(x)}{dx}$
$f(x) + g(x)$	$\dfrac{df(x)}{dx} + \dfrac{dg(x)}{dx}$
$f(x) - g(x)$	$\dfrac{df(x)}{dx} - \dfrac{dg(x)}{dx}$
$f(x)g(x)$	$f(x)\,\dfrac{dg(x)}{dx} + g(x)\,\dfrac{df(x)}{dx}$
$f(x)/g(x)$	$\left[g(x)\,\dfrac{df(x)}{dx} - f(x)\,\dfrac{dg(x)}{dx}\right]\Big/ [g(x)]^2$
e^x	e^x
$e^{f(x)}$	$e^{f(x)}\,\dfrac{df(x)}{dx}$
a^x	$a^x \log_e a$
$\log_e x$	$1/x$
$\log_e f(x)$	$\dfrac{1}{f(x)}\,\dfrac{df(x)}{dx}$

An **integral** is simply an antiderivative; that is, integration is essentially the inverse of differentiation. If the derivative of $f(x)$ is $f'(x)$, then the integral of $f'(x)$ with respect to x, written $\int f'(x)\,dx$, is equal to $f(x) + c$, where c is a constant with respect to x. In general, the integral of a function is not unique, for the constant c could be just about anything, provided that it is not a function of x. However, if we specify limits of integration, then the integral is a definite integral, and it is unique. The definite integral of a function over a certain interval can be interpreted graphically as the area between the curve and the x-axis in this interval, where the area is taken to be positive if the curve is above the axis and negative if it is below the axis. This area can be approximated by the sum of the areas of a finite set of rectangles, and this is the basis for the formal definition of a definite integral. Several common integration formulas are presented below (once again the student is referred to a calculus textbook for the derivation of these formulas).

$$\int a\, dx = ax + c$$

$$\int af(x)\, dx = a \int f(x)\, dx$$

$$\int x^k\, dx = \frac{x^{k+1}}{k+1} + c$$

$$\int [f(x) + g(x)]\, dx = \int f(x)\, dx + \int g(x)\, dx$$

$$\int [f(x) - g(x)]\, dx = \int f(x)\, dx - \int g(x)\, dx$$

$$\int e^x\, dx = e^x + c$$

$$\int e^{f(x)} \frac{df(x)}{dx}\, dx = e^{f(x)} + c$$

$$\int \frac{1}{x}\, dx = \log_e |x| + c$$

$$\int \frac{1}{f(x)} \frac{df(x)}{dx}\, dx = \log_e |f(x)| + c$$

$$\int f(x) \frac{dg(x)}{dx}\, dx = f(x)g(x) - \int g(x) \frac{df(x)}{dx}\, dx$$

If

$$\int f(x)\, dx = F(x) + c, \quad \text{then} \quad \int_a^b f(x)\, dx = F(b) - F(a).$$

APPENDIX B:
MATRIX ALGEBRA

A **matrix** is a rectangular array of elements, examples of which are as follows:

$$\begin{bmatrix} 2 & 4 \\ 1 & 6 \end{bmatrix}, \quad \begin{bmatrix} 2 & 6 & 1 \\ 3 & 2 & 4 \\ 1 & 3 & -2 \\ 0 & 4 & 2 \end{bmatrix}, \quad \begin{bmatrix} 3 & 2 \\ 4 & -8 \\ 6 & 1 \end{bmatrix}, \quad \begin{bmatrix} 1 & 5 & 7 \\ 2 & 9 & 4 \end{bmatrix}.$$

A matrix with r rows and c columns is of **order** $r \times c$ and is called an $r \times c$ matrix. A matrix is **square** if $r = c$; that is, if the number of rows equals the number of columns. Of the four matrices shown above, only the first matrix is square (it is 2×2); the other matrices are 4×3, 3×2, and 2×3. A matrix with only one row is called a **row vector,** and a matrix with only one column is called a **column vector.**

Often a matrix is written with symbols as elements. In the following matrix, the element in the ith row and the jth column is denoted by a_{ij}:

$$\mathbf{A} = \begin{bmatrix} a_{11} & a_{12} & a_{13} \\ a_{21} & a_{22} & a_{23} \\ a_{31} & a_{32} & a_{33} \end{bmatrix}.$$

Two matrices which are of the same order are **equal** if their corresponding elements are equal. For instance, if

$$\mathbf{B} = \begin{bmatrix} b_{11} & b_{12} & b_{13} \\ b_{21} & b_{22} & b_{23} \\ b_{31} & b_{32} & b_{33} \end{bmatrix},$$

then **A** and **B** are equal if and only if

$$a_{ij} = b_{ij}$$

for all $i = 1, 2, 3$ and $j = 1, 2, 3$. Be sure to notice that two matrices can be equal only if they have the same number of rows and columns.

One matrix of particular interest is the **identity matrix.** The identity matrix of order n is a square $n \times n$ matrix with the elements along the main diagonal of the matrix equaling 1 and the other elements of the matrix equaling 0. The identity matrices of orders 1, 2, 3, and 4 are

$$[1], \quad \begin{bmatrix} 1 & 0 \\ 0 & 1 \end{bmatrix}, \quad \begin{bmatrix} 1 & 0 & 0 \\ 0 & 1 & 0 \\ 0 & 0 & 1 \end{bmatrix}, \quad \text{and} \quad \begin{bmatrix} 1 & 0 & 0 & 0 \\ 0 & 1 & 0 & 0 \\ 0 & 0 & 1 & 0 \\ 0 & 0 & 0 & 1 \end{bmatrix}.$$

The identity matrix, which is denoted by **I**, will be of particular interest when we discuss the determination of the inverse of a matrix.

The **transpose** of a matrix **A** is the matrix $\mathbf{A}^t$ whose rows are the columns of **A** (or, equivalently, whose columns are the rows of **A**). For example, if we have the matrix

$$\begin{bmatrix} 3 & 4 & 8 & 1 \\ 2 & 3 & 7 & 9 \\ 8 & 5 & 4 & 0 \end{bmatrix},$$

then its transpose is

$$\begin{bmatrix} 3 & 2 & 8 \\ 4 & 3 & 5 \\ 8 & 7 & 4 \\ 1 & 9 & 0 \end{bmatrix}.$$

Thus, the element in the ith row and jth column of the matrix **A** is in the jth row and ith column of the transpose $\mathbf{A}^t$. Clearly, if we take the transpose of $\mathbf{A}^t$, we simply get the original matrix **A**. Thus,

$$(\mathbf{A}^t)^t = \mathbf{A}.$$

It is possible to define certain operations on matrices, just as we define operations such as addition or subtraction for ordinary numbers. First, we consider **matrix addition.** If two matrices **A** and **B** are of the same order, then their sum is defined as the matrix $\mathbf{A} + \mathbf{B}$ consisting of the sums of the corresponding elements of **A** and **B**. If **A** and **B** are of the form given above, then

$$\mathbf{A} + \mathbf{B} = \begin{bmatrix} a_{11} + b_{11} & a_{12} + b_{12} & a_{13} + b_{13} \\ a_{21} + b_{21} & a_{22} + b_{22} & a_{23} + b_{23} \\ a_{31} + b_{31} & a_{32} + b_{32} & a_{33} + b_{33} \end{bmatrix}.$$

For example,

$$\begin{bmatrix} 2 & 6 & 5 \\ 1 & 8 & 8 \\ 8 & 0 & 3 \end{bmatrix} + \begin{bmatrix} 1 & 1 & 2 \\ 7 & 0 & 5 \\ 4 & 5 & 8 \end{bmatrix} = \begin{bmatrix} 3 & 7 & 7 \\ 8 & 8 & 13 \\ 12 & 5 & 11 \end{bmatrix},$$

and

$$\begin{bmatrix} 2 & 1 \\ 4 & 7 \\ 3 & 3 \\ 6 & 8 \end{bmatrix} + \begin{bmatrix} 4 & 5 \\ -2 & 2 \\ 1 & 6 \\ 4 & -6 \end{bmatrix} = \begin{bmatrix} 6 & 6 \\ 2 & 9 \\ 4 & 9 \\ 10 & 2 \end{bmatrix}.$$

Obviously, two matrices can be added only if they have the same number of rows and columns. Similarly, we can define **matrix subtraction**: the difference $\mathbf{A} - \mathbf{B}$ consists of the differences of the corresponding elements of $\mathbf{A}$ and $\mathbf{B}$. For example,

$$\begin{bmatrix} 3 & 1 & 6 & 8 & 5 \\ 2 & 4 & 3 & 1 & 0 \\ 3 & 5 & 2 & 5 & 1 \end{bmatrix} - \begin{bmatrix} 1 & 2 & 6 & 8 & 4 \\ 5 & 7 & 1 & 2 & 2 \\ 2 & 2 & 1 & 3 & 0 \end{bmatrix} = \begin{bmatrix} 2 & -1 & 0 & 0 & 1 \\ -3 & -3 & 2 & -1 & -2 \\ 1 & 3 & 1 & 2 & 1 \end{bmatrix}.$$

Another operation which can be performed is the multiplication of a matrix by a scalar (a single element, or number). To perform this operation, simply multiply *each element* of the matrix by the given scalar. For the matrix $\mathbf{A}$ given above, $k\mathbf{A}$ is thus defined as

$$k\mathbf{A} = \begin{bmatrix} ka_{11} & ka_{12} & ka_{13} \\ ka_{21} & ka_{22} & ka_{23} \\ ka_{31} & ka_{32} & ka_{33} \end{bmatrix}.$$

For instance, let $k = 3$ and

$$\mathbf{A} = \begin{bmatrix} 6 & 2 & 1 \\ 0 & 3 & 3 \\ 4 & 1 & 3 \end{bmatrix}.$$

Then

$$k\mathbf{A} = \begin{bmatrix} 18 & 6 & 3 \\ 0 & 9 & 9 \\ 12 & 3 & 9 \end{bmatrix}.$$

It is also possible, under certain conditions, to multiply two matrices. This operation, however, is more complicated than matrix addition or the multiplication by a scalar. Consider the matrices

$$\mathbf{A} = \begin{bmatrix} a_{11} & a_{12} \\ a_{21} & a_{22} \end{bmatrix} \quad \text{and} \quad \mathbf{B} = \begin{bmatrix} b_{11} & b_{12} & b_{13} \\ b_{21} & b_{22} & b_{23} \end{bmatrix}.$$

The **product AB** is defined as

$$\mathbf{AB} = \begin{bmatrix} a_{11}b_{11} + a_{12}b_{21} & a_{11}b_{12} + a_{12}b_{22} & a_{11}b_{13} + a_{12}b_{23} \\ a_{21}b_{11} + a_{22}b_{21} & a_{21}b_{12} + a_{22}b_{22} & a_{21}b_{13} + a_{22}b_{23} \end{bmatrix}.$$

In other words, the element in the first row and first column of **AB** is obtained by multiplying the elements of the first *row* of **A** by the elements of the first *column* of **B** and summing. This is the multiplication of two vectors, as follows:

$$[a_{11} \quad a_{12}] \times \begin{bmatrix} b_{11} \\ b_{21} \end{bmatrix} = [a_{11}b_{11} + a_{12}b_{21}].$$

Similarly, the element in the first row and the *second* column of **AB** is obtained by multiplying the first row of **A** by the second column of **B**. In general, the element in the *i*th row and *j*th column of **AB** is obtained by multiplying the *i*th row of **A** by the *j*th column of **B**. For an example using numbers rather than symbols, consider the following:

$$\begin{bmatrix} 3 & 6 & 0 \\ 2 & 1 & 4 \end{bmatrix} \times \begin{bmatrix} 2 & 4 & 5 \\ 3 & 1 & 3 \\ 6 & 1 & 5 \end{bmatrix}$$

$$= \begin{bmatrix} 3 \cdot 2 + 6 \cdot 3 + 0 \cdot 6 & 3 \cdot 4 + 6 \cdot 1 + 0 \cdot 1 & 3 \cdot 5 + 6 \cdot 3 + 0 \cdot 5 \\ 2 \cdot 2 + 1 \cdot 3 + 4 \cdot 6 & 2 \cdot 4 + 1 \cdot 1 + 4 \cdot 1 & 2 \cdot 5 + 1 \cdot 3 + 4 \cdot 5 \end{bmatrix}$$

$$= \begin{bmatrix} 24 & 18 & 33 \\ 31 & 13 & 33 \end{bmatrix}.$$

Notice that when we multiplied **A** and **B**, **A** was 2×2, **B** was 2×3, and the product **AB** was 2×3. In the numerical example, the first matrix was 2×3, the second was 3×3, and the product was 2×3. In general, suppose that **A** is $m \times n$ and **B** is $s \times t$. Then the product **AB** is defined only if $n = s$; that is, if the number of columns in the first matrix equals the number of rows in the second matrix. Then the two matrices are $m \times n$ and $n \times t$, and the product matrix is $m \times t$.

When we multiply numbers, it makes no difference in which order they are taken. For instance, 2 multiplied by 4 is the same as 4 multiplied by 2. With matrices, however, this is not true, as the following example demonstrates:

$$\begin{bmatrix} 3 & 1 & 2 \\ 4 & 0 & 2 \end{bmatrix} \times \begin{bmatrix} 1 & 2 \\ 1 & 1 \\ 3 & 0 \end{bmatrix} = \begin{bmatrix} 10 & 7 \\ 10 & 8 \end{bmatrix},$$

but

$$\begin{bmatrix} 1 & 2 \\ 1 & 1 \\ 3 & 0 \end{bmatrix} \times \begin{bmatrix} 3 & 1 & 2 \\ 4 & 0 & 2 \end{bmatrix} = \begin{bmatrix} 11 & 1 & 6 \\ 7 & 1 & 4 \\ 9 & 3 & 6 \end{bmatrix}.$$

It should be noted that it is entirely possible for **AB** to exist and for **BA** not to exist. If **A** is 2×3 and **B** is 3×4, then **AB** exists and is 2×4, but **BA** does not exist because the number of columns in **B**(4) is not equal to the number of rows in $\mathbf{A}(2)$. It should be clear by now that it is very important to specify the order in which two matrices are to be multiplied.

The **inverse** of a square matrix A is defined as the matrix A^{-1} such that

$$\mathbf{AA^{-1} = A^{-1}A = I}.$$

In other words, the product of a matrix and its inverse is equal to the identity matrix. For instance, if

$$\mathbf{A} = \begin{bmatrix} 2 & 1 \\ 1 & 1 \end{bmatrix}, \qquad \text{then} \qquad \mathbf{A^{-1}} = \begin{bmatrix} 1 & -1 \\ -1 & 2 \end{bmatrix},$$

as you can verify by multiplying the two matrices. The inverse is defined only for a square matrix, and even for a square matrix it may not exist.

Although it is quite easy to define what is meant by the inverse of a matrix, it is not quite so easy to find $\mathbf{A^{-1}}$, given $\mathbf{A}$. One way to proceed is illustrated by the following example. Suppose that

$$\mathbf{A} = \begin{bmatrix} 2 & 3 \\ 4 & 1 \end{bmatrix},$$

and you want to find $\mathbf{A^{-1}}$. First, write the identity matrix $\mathbf{I}$ next to $\mathbf{A}$:

$$\begin{bmatrix} 2 & 3 \\ 4 & 1 \end{bmatrix} \begin{bmatrix} 1 & 0 \\ 0 & 1 \end{bmatrix}.$$

Now, by performing certain operations, we want to change this pair of matrices in such a manner that the matrix on the *left* is the identity matrix. If this can be done, then the matrix we end up with on the right is simply $\mathbf{A^{-1}}$. Whatever operations we perform must be performed on *both* matrices, and the following operations are permissible.

1. multiply any row by a constant,
2. add (or subtract) any multiple of a row from another row.

In our example, suppose that we first multiply the first row by $\frac{1}{2}$:

$$\begin{bmatrix} 1 & \frac{3}{2} \\ 4 & 1 \end{bmatrix} \begin{bmatrix} \frac{1}{2} & 0 \\ 0 & 1 \end{bmatrix}.$$

Now, subtract 4 times the first row from the second row (notice that the first row is not changed by this operation):

$$\begin{bmatrix} 1 & \frac{3}{2} \\ 4 - 4(1) & 1 - 4(\frac{3}{2}) \end{bmatrix} \begin{bmatrix} \frac{1}{2} & 0 \\ 0 - 4(\frac{1}{2}) & 1 - 4(0) \end{bmatrix},$$

or

$$\begin{bmatrix} 1 & \frac{3}{2} \\ 0 & -5 \end{bmatrix} \begin{bmatrix} \frac{1}{2} & 0 \\ -2 & 1 \end{bmatrix}.$$

Next, multiply the second row by $-\frac{1}{5}$:

$$\begin{bmatrix} 1 & \frac{3}{2} \\ 0 & 1 \end{bmatrix} \begin{bmatrix} \frac{1}{2} & 0 \\ \frac{2}{5} & -\frac{1}{5} \end{bmatrix}.$$

Finally, subtract $\frac{3}{2}$ times the second row from the first row:

$$\begin{bmatrix} 1 - \frac{3}{2}(0) & \frac{3}{2} - \frac{3}{2}(1) \\ 0 & 1 \end{bmatrix} \begin{bmatrix} \frac{1}{2} - \frac{3}{2}(\frac{2}{5}) & 0 - \frac{3}{2}(-\frac{1}{5}) \\ \frac{2}{5} & -\frac{1}{5} \end{bmatrix},$$

or

$$\begin{bmatrix} 1 & 0 \\ 0 & 1 \end{bmatrix} \begin{bmatrix} -\frac{1}{10} & \frac{3}{10} \\ \frac{2}{5} & -\frac{1}{5} \end{bmatrix}.$$

Therefore,

$$\mathbf{A}^{-1} = \begin{bmatrix} -\frac{1}{10} & \frac{3}{10} \\ \frac{2}{5} & -\frac{1}{5} \end{bmatrix}.$$

You can verify this by multiplying $\mathbf{A}$ and $\mathbf{A}^{-1}$.

There are no hard-and-fast rules for applying the operations to the two matrices in order to change the left matrix into the identity matrix. However, the above example illustrates the usual sequence of operations. First, we multiplied the first row by a constant so that the first element in that row would equal 1. Then, we subtracted an appropriate multiple of row 1 from row 2 so that the first element in row 2 would be equal to 0. Next, we multiplied the second row by a constant so the second element in the row would equal 1. Finally, we subtracted an appropriate multiple of row 2 from row 1 so that the second element in row 1 would equal 0.

To demonstrate the determination of an inverse of a 3 × 3 matrix, the following sequence of operations is presented. The specific operations are not explained, but it should be helpful if the reader tries to determine which operations were used and *why* each one was used.

$$\begin{bmatrix} 3 & 0 & 2 \\ 1 & 4 & 2 \\ 2 & 1 & 5 \end{bmatrix} \begin{bmatrix} 1 & 0 & 0 \\ 0 & 1 & 0 \\ 0 & 0 & 1 \end{bmatrix}$$

$$\begin{bmatrix} 1 & 0 & \frac{2}{3} \\ 1 & 4 & 2 \\ 2 & 1 & 5 \end{bmatrix} \begin{bmatrix} \frac{1}{3} & 0 & 0 \\ 0 & 1 & 0 \\ 0 & 0 & 1 \end{bmatrix}$$

$$\begin{bmatrix} 1 & 0 & \frac{2}{3} \\ 0 & 4 & \frac{4}{3} \\ 2 & 1 & 5 \end{bmatrix} \begin{bmatrix} \frac{1}{3} & 0 & 0 \\ -\frac{1}{3} & 1 & 0 \\ 0 & 0 & 1 \end{bmatrix}$$

$$\begin{bmatrix} 1 & 0 & \frac{2}{3} \\ 0 & 4 & \frac{4}{3} \\ 0 & 1 & \frac{11}{3} \end{bmatrix} \begin{bmatrix} \frac{1}{3} & 0 & 0 \\ -\frac{1}{3} & 1 & 0 \\ -\frac{2}{3} & 0 & 1 \end{bmatrix}$$

$$\begin{bmatrix} 1 & 0 & \frac{2}{3} \\ 0 & 1 & \frac{1}{3} \\ 0 & 1 & \frac{11}{3} \end{bmatrix} \begin{bmatrix} \frac{1}{3} & 0 & 0 \\ -\frac{1}{12} & \frac{1}{4} & 0 \\ -\frac{2}{3} & 0 & 1 \end{bmatrix}$$

$$\begin{bmatrix} 1 & 0 & \frac{2}{3} \\ 0 & 1 & \frac{1}{3} \\ 0 & 0 & \frac{10}{3} \end{bmatrix} \begin{bmatrix} \frac{1}{3} & 0 & 0 \\ -\frac{1}{12} & \frac{1}{4} & 0 \\ -\frac{7}{12} & -\frac{1}{4} & 1 \end{bmatrix}$$

$$\begin{bmatrix} 1 & 0 & \frac{2}{3} \\ 0 & 1 & \frac{1}{3} \\ 0 & 0 & 1 \end{bmatrix} \begin{bmatrix} \frac{1}{3} & 0 & 0 \\ -\frac{1}{12} & \frac{1}{4} & 0 \\ -\frac{7}{40} & -\frac{3}{40} & \frac{3}{10} \end{bmatrix}$$

$$\begin{bmatrix} 1 & 0 & 0 \\ 0 & 1 & \frac{1}{3} \\ 0 & 0 & 1 \end{bmatrix} \begin{bmatrix} \frac{9}{20} & \frac{1}{20} & -\frac{1}{5} \\ -\frac{1}{12} & \frac{1}{4} & 0 \\ -\frac{7}{40} & -\frac{3}{40} & \frac{3}{10} \end{bmatrix}$$

$$\begin{bmatrix} 1 & 0 & 0 \\ 0 & 1 & 0 \\ 0 & 0 & 1 \end{bmatrix} \begin{bmatrix} \frac{9}{20} & \frac{1}{20} & -\frac{1}{5} \\ -\frac{1}{40} & \frac{11}{40} & -\frac{1}{10} \\ -\frac{7}{40} & -\frac{3}{40} & \frac{3}{10} \end{bmatrix}.$$

To check this, you can verify that

$$\begin{bmatrix} 3 & 0 & 2 \\ 1 & 4 & 2 \\ 2 & 1 & 5 \end{bmatrix} \times \begin{bmatrix} \frac{9}{20} & \frac{1}{20} & -\frac{1}{5} \\ -\frac{1}{40} & \frac{11}{40} & -\frac{1}{10} \\ -\frac{7}{40} & -\frac{3}{40} & \frac{3}{10} \end{bmatrix} = \begin{bmatrix} 1 & 0 & 0 \\ 0 & 1 & 0 \\ 0 & 0 & 1 \end{bmatrix}.$$

Using the procedure we have demonstrated, it is possible to find the inverse of any square matrix, provided that the inverse exists. Of course, as the number of rows and columns increases, the computations become increasingly more difficult and it may be necessary to use the computer to find the inverse of a matrix. For 2×2 and 3×3 matrices, it is not too hard to find the inverse without resorting to the use of a computer.

One important use of the inverse of a matrix is in the solution of a set of linear equations. For example, suppose that we have k equations in k unknowns $x_1, x_2, \ldots, x_k$:

$$a_{11}x_1 + a_{12}x_2 + \cdots + a_{1k}x_k = b_1$$

$$a_{21}x_1 + a_{22}x_2 + \cdots + a_{2k}x_k = b_2$$

$$\vdots \qquad \vdots \qquad \qquad \vdots \qquad \vdots$$

$$a_{k1}x_1 + a_{k2}x_2 + \cdots + a_{kk}x_k = b_k.$$

Let

$$\mathbf{A} = \begin{bmatrix} a_{11} & a_{12} & \cdots & a_{1k} \\ a_{21} & a_{22} & \cdots & a_{2k} \\ \vdots & \vdots & & \vdots \\ a_{k1} & a_{k2} & \cdots & a_{kk} \end{bmatrix}, \quad \mathbf{X} = \begin{bmatrix} x_1 \\ x_2 \\ \vdots \\ x_k \end{bmatrix}, \quad \text{and} \quad \mathbf{B} = \begin{bmatrix} b_1 \\ b_2 \\ \vdots \\ b_k \end{bmatrix}.$$

Thus, **A** is $k \times k$, **X** is $k \times 1$, and **B** is $k \times 1$, and the set of k linear equations in k unknown variables can be written in matrix form:

$$\mathbf{AX} = \mathbf{B}.$$

To solve the set of equations, we need to find values of $x_1, x_2, \ldots, x_k$ which satisfy all k equations. Suppose that we multiply both sides of the matrix equation by $\mathbf{A}^{-1}$:

$$\mathbf{A}^{-1}(\mathbf{AX}) = \mathbf{A}^{-1}\mathbf{B}.$$

But $\mathbf{A}^{-1}\mathbf{A} = \mathbf{I}$, so we have

$$\mathbf{IX} = \mathbf{A}^{-1}\mathbf{B}.$$

It is easy to verify that $\mathbf{IX} = \mathbf{X}$, which gives us

$$\mathbf{X} = \mathbf{A}^{-1}\mathbf{B},$$

which is the solution of the set of equations.

For example, consider the set of two equations in two unknowns given by

$$2x_1 + x_2 = 4$$

and

$$x_1 + x_2 = 3.$$

The matrix **A** is

$$\mathbf{A} = \begin{bmatrix} 2 & 1 \\ 1 & 1 \end{bmatrix}, \quad \text{and its inverse is} \quad \mathbf{A}^{-1} = \begin{bmatrix} 1 & -1 \\ -1 & 2 \end{bmatrix}.$$

The solution to the set of two equations is then

$$\mathbf{X} = \mathbf{A}^{-1}\mathbf{B} = \begin{bmatrix} 1 & -1 \\ -1 & 2 \end{bmatrix} \times \begin{bmatrix} 4 \\ 3 \end{bmatrix} = \begin{bmatrix} 1 \\ 2 \end{bmatrix},$$

or simply $x_1 = 1$ and $x_2 = 2$.

Therefore, if a set of n linear equations in n unknowns is expressed in the matrix form $\mathbf{AX} = \mathbf{B}$, the solution is simply $\mathbf{X} = \mathbf{A}^{-1}\mathbf{B}$. As we have noted, not all square matrices possess inverses. If $\mathbf{A}^{-1}$ does not exist, then there is no unique solution to the set of equations.

EXERCISES

1. Find the transpose of

$$\text{(a)} \quad \begin{bmatrix} 2 & 3 & 9 \\ 1 & 4 & 6 \\ 5 & 4 & 8 \end{bmatrix},$$

(b) $\begin{bmatrix} 6 \\ 4 \\ 7 \end{bmatrix}$,

(c) $\begin{bmatrix} 3 & 6 & 9 & 2 \\ 0 & -3 & 7 & 15 \end{bmatrix}$.

2. Perform the following operations:

(a) $\begin{bmatrix} 3 & -2 & 6 \\ 4 & 8 & -3 \\ -1 & 1 & 2 \end{bmatrix} + \begin{bmatrix} 4 & 14 & -4 \\ -5 & -3 & 14 \\ 8 & 3 & 6 \end{bmatrix} - \begin{bmatrix} 2 & 5 & 9 \\ 4 & 6 & 1 \\ -5 & 8 & 3 \end{bmatrix}$,

(b) $\begin{bmatrix} 6 & 8 & -2 \end{bmatrix} \times \begin{bmatrix} 1 \\ 3 \\ -4 \end{bmatrix}$,

(c) $\begin{bmatrix} 2 & 7 & 13 \\ 4 & 8 & 6 \\ 1 & 1 & 1 \end{bmatrix} \times \begin{bmatrix} 4 & 9 & 1 \\ -2 & 0 & 3 \\ 0 & 1 & 10 \end{bmatrix}$,

(d) $4 \begin{bmatrix} 12 & 3 & 4 & 5 \\ 0 & -1 & 3 & -7 \\ -5 & 4 & 9 & -3 \end{bmatrix}$.

3. Find the inverse of each of the following matrices:

(a) $\begin{bmatrix} 4 & 6 \\ -3 & 1 \end{bmatrix}$,

(b) $\begin{bmatrix} 3 & 5 & -2 \\ 5 & 2 & 8 \\ -4 & 0 & 6 \end{bmatrix}$,

(c) $\begin{bmatrix} 2 & 1 & 1 & 1 \\ 1 & 2 & 1 & 1 \\ 1 & 1 & 2 & 1 \\ 1 & 1 & 1 & 2 \end{bmatrix}$.

4. Solve the following sets of linear equations:

(a) $\quad 4x_1 + 5x_2 = \quad 3,$

$\quad -3x_1 + 9x_2 = -2;$

(b) $\quad 2x_1 - 5x_2 + 10x_3 = 30,$

$\quad x_1 \quad\quad + 2x_3 = 15,$

$\quad 3x_1 + x_2 - \quad x_3 = \quad 5.$

TABLES

Table   I.   Cumulative Standard Normal Probabilities

This table gives values of the cumulative distribution function of the standard normal distribution,

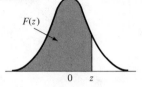

$$F(z) = \int_{-\infty}^{z} \frac{1}{\sqrt{2\pi}}\, e^{-z^2/2}\, dz,$$

for $z = 0(.01)\,2.60(.10)\,3.00(.20)\,4.00(.50)\,5.50$. For $z < 0$, take $F(z) = 1-F(-z)$.

Examples: $F(1.12) = .8686431$,
and $F(-.67) = 1 - F(.67) = 1 - .7485711 = .2514289$.

z	$F(z)$	z	$F(z)$	z	$F(z)$	z	$F(z)$
.00	.5000000	.36	.6405764	.72	.7642375	1.08	.8599289
.01	.5039894	.37	.6443088	.73	.7673049	1.09	.8621434
.02	.5079783	.38	.6480273	.74	.7703500	1.10	.8643339
.03	.5119665	.39	.6517317	.75	.7733726	1.11	.8665005
.04	.5159534	.40	.6554217	.76	.7763727	1.12	.8686431
.05	.5199388	.41	.6590970	.77	.7793501	1.13	.8707619
.06	.5239222	.42	.6627573	.78	.7823046	1.14	.8728568
.07	.5279032	.43	.6664022	.79	.7852361	1.15	.8749281
.08	.5318814	.44	.6700314	.80	.7881446	1.16	.8769756
.09	.5358564	.45	.6736448	.81	.7910299	1.17	.8789995
.10	.5398278	.46	.6772419	.82	.7938919	1.18	.8809999
.11	.5437953	.47	.6808225	.83	.7967306	1.19	.8829768
.12	.5477584	.48	.6843863	.84	.7995458	1.20	.8849303
.13	.5517168	.49	.6879331	.85	.8023375	1.21	.8868606
.14	.5556700	.50	.6914625	.86	.8051055	1.22	.8887676
.15	.5596177	.51	.6949743	.87	.8078498	1.23	.8906514
.16	.5635595	.52	.6984682	.88	.8105703	1.24	.8925123
.17	.5674949	.53	.7019440	.89	.8132671	1.25	.8943502
.18	.5714237	.54	.7054015	.90	.8159399	1.26	.8961653
.19	.5753454	.55	.7088403	.91	.8185887	1.27	.8979577
.20	.5792597	.56	.7122603	.92	.8212136	1.28	.8997274
.21	.5831662	.57	.7156612	.93	.8238145	1.29	.9014747
.22	.5870604	.58	.7190427	.94	.8263912	1.30	.9031995
.23	.5909541	.59	.7224047	.95	.8289439	1.31	.9049021
.24	.5948349	.60	.7257469	.96	.8314724	1.32	.9065825
.25	.5987063	.61	.7290691	.97	.8339768	1.33	.9082409
.26	.6025681	.62	.7323711	.98	.8364569	1.34	.9098773
.27	.6064199	.63	.7356527	.99	.8389129	1.35	.9114920
.28	.6102612	.64	.7389137	1.00	.8413447	1.36	.9130850
.29	.6140919	.65	.7421539	1.01	.8437524	1.37	.9146565
.30	.6179114	.66	.7453731	1.02	.8461358	1.38	.9162067
.31	.6217195	.67	.7485711	1.03	.8484950	1.39	.9177356
.32	.6255158	.68	.7517478	1.04	.8508300	1.40	.9192433
.33	.6293000	.69	.7549029	1.05	.8531409	1.41	.9207302
.34	.6330717	.70	.7580363	1.06	.8554277	1.42	.9221962
.35	.6368307	.71	.7611479	1.07	.8576903	1.43	.9236415

Table I   (continued)

z	F(z)	z	F(z)	z	F(z)	z	F(z)
1.44	.9250663	1.77	.9616364	2.10	.9821356	2.43	.9924506
1.45	.9264707	1.78	.9624620	2.11	.9825708	2.44	.9926564
1.46	.9278550	1.79	.9632730	2.12	.9829970	2.45	.9928572
1.47	.9292191	1.80	.9640697	2.13	.9834142	2.46	.9930531
1.48	.9305634	1.81	.9648521	2.14	.9838226	2.47	.9932443
1.49	.9318879	1.82	.9656205	2.15	.9842224	2.48	.9934309
1.50	.9331928	1.83	.9663750	2.16	.9846137	2.49	.9936128
1.51	.9344783	1.84	.9671159	2.17	.9849966	2.50	.9937903
1.52	.9357445	1.85	.9678432	2.18	.9853713	2.51	.9939634
1.53	.9369916	1.86	.9685572	2.19	.9857379	2.52	.9941323
1.54	.9382198	1.87	.9692581	2.20	.9860966	2.53	.9942969
1.55	.9394292	1.88	.9699460	2.21	.9864474	2.54	.9944574
1.56	.9406201	1.89	.9706210	2.22	.9867906	2.55	.9946139
1.57	.9417924	1.90	.9712834	2.23	.9871263	2.56	.9947664
1.58	.9429466	1.91	.9719334	2.24	.9874545	2.57	.9949151
1.59	.9440826	1.92	.9725711	2.25	.9877755	2.58	.9950600
1.60	.9452007	1.93	.9731966	2.26	.9880894	2.59	.9952012
1.61	.9463011	1.94	.9738102	2.27	.9883962	2.60	.9953388
1.62	.9473839	1.95	.9744119	2.28	.9886962	2.70	.9965330
1.63	.9484493	1.96	.9750021	2.29	.9889893	2.80	.9974449
1.64	.9494974	1.97	.9755808	2.30	.9892759	2.90	.9981342
1.65	.9505285	1.98	.9761482	2.31	.9895559	3.00	.9986501
1.66	.9515428	1.99	.9767045	2.32	.9898296	3.20	.9993129
1.67	.9525403	2.00	.9772499	2.33	.9900969	3.40	.9996631
1.68	.9535213	2.01	.9777844	2.34	.9903581	3.60	.9998409
1.69	.9544860	2.02	.9783083	2.35	.9906133	3.80	.9999277
1.70	.9554345	2.03	.9788217	2.36	.9908625	4.00	.9999683
1.71	.9563671	2.04	.9793248	2.37	.9911060	4.50	.9999966
1.72	.9572838	2.05	.9798178	2.38	.9913437	5.00	.9999997
1.73	.9581849	2.06	.9803007	2.39	.9915758	5.50	.9999999
1.74	.9590705	2.07	.9807738	2.40	.9918025		
1.75	.9599408	2.08	.9812372	2.41	.9920237		
1.76	.9607961	2.09	.9816911	2.42	.9922397		

Table II.   Fractiles of the T Distribution

This table gives the .60, .75, .90, .95, .975, .99, .995, and .999 fractiles of the T distribution with ν degrees of freedom for $\nu = 1(1)30,\ 40,\ 60,\ 120,$ and ∞. For $F(t) < .50$, the $F(t)$ fractile is the negative of the $1 - F(t)$ fractile.

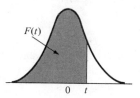

$F(t)$

Examples: For $\nu = 12$, the .90 fractile is 1.356, and for $\nu = 5$, the .025 fractile is -2.571.

ν	F(t)							
	.60	.75	.90	.95	.975	.99	.995	.999
1	0.325	1.000	3.078	6.314	12.706	31.821	63.657	318.31
2	.289	0.816	1.886	2.920	4.303	6.965	9.925	22.326
3	.277	.765	1.638	2.353	3.182	4.541	5.841	10.213
4	.271	.741	1.533	2.132	2.776	3.747	4.604	7.173
5	0.267	0.727	1.476	2.015	2.571	3.365	4.032	5.893
6	.265	.718	1.440	1.943	2.447	3.143	3.707	5.208
7	.263	.711	1.415	1.895	2.365	2.998	3.499	4.785
8	.262	.706	1.397	1.860	2.306	2.896	3.355	4.501
9	.261	.703	1.383	1.833	2.262	2.821	3.250	4.297
10	0.260	0.700	1.372	1.812	2.228	2.764	3.169	4.144
11	.260	.697	1.363	1.796	2.201	2.718	3.106	4.025
12	.259	.695	1.356	1.782	2.179	2.681	3.055	3.930
13	.259	.694	1.350	1.771	2.160	2.650	3.012	3.852
14	.258	.692	1.345	1.761	2.145	2.624	2.977	3.787
15	0.258	0.691	1.341	1.753	2.131	2.602	2.947	3.733
16	.258	.690	1.337	1.746	2.120	2.583	2.921	3.686
17	.257	.689	1.333	1.740	2.110	2.567	2.898	3.646
18	.257	.688	1.330	1.734	2.101	2.552	2.878	3.610
19	.257	.688	1.328	1.729	2.093	2.539	2.861	3.579
20	0.257	0.687	1.325	1.725	2.086	2.528	2.845	3.552
21	.257	.686	1.323	1.721	2.080	2.518	2.831	3.527
22	.256	.686	1.321	1.717	2.074	2.508	2.819	3.505
23	.256	.685	1.319	1.714	2.069	2.500	2.807	3.485
24	.256	.685	1.318	1.711	2.064	2.492	2.797	3.467
25	0.256	0.684	1.316	1.708	2.060	2.485	2.787	3.450
26	.256	.684	1.315	1.706	2.056	2.479	2.779	3.435
27	.256	.684	1.314	1.703	2.052	2.473	2.771	3.421
28	.256	.683	1.313	1.701	2.048	2.467	2.763	3.408
29	.256	.683	1.311	1.699	2.045	2.462	2.756	3.396
30	0.256	0.683	1.310	1.697	2.042	2.457	2.750	3.385
40	.255	.681	1.303	1.684	2.021	2.423	2.704	3.307
60	.254	.679	1.296	1.671	2.000	2.390	2.660	3.232
120	.254	.677	1.289	1.658	1.980	2.358	2.617	3.160
∞	.253	.674	1.282	1.645	1.960	2.326	2.576	3.090

Table III.   Fractiles of the χ^2 Distribution

This table gives the .005, .01, .025, .05, .10, .25, .50, .75, .90, .95, .975, .99, .995, and .999 fractiles of the χ^2 distribution with ν degrees of freedom for $\nu = 1(1)30(10)100$.

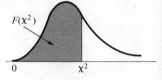

Example: For $\nu = 14$, the .025 fractile is 5.62872.

ν	.005	.01	.025	.05	.10	.25	.50
1	$392704 \cdot 10^{-10}$	$157088 \cdot 10^{-9}$	$982069 \cdot 10^{-9}$	$393214 \cdot 10^{-8}$	0.0157908	0.1015308	0.454937
2	0.0100251	0.0201007	0.0506356	0.102587	0.210720	0.575364	1.38629
3	0.0717212	0.114832	0.215795	0.351846	0.584375	1.212534	2.36597
4	0.206990	0.297110	0.484419	0.710721	1.063623	1.92255	3.35670
5	0.411740	0.554300	0.831211	1.145476	1.61031	2.67460	4.35146
6	0.675727	0.872085	1.237347	1.63539	2.20413	3.45460	5.34812
7	0.989265	1.239043	1.68987	2.16735	2.83311	4.25485	6.34581
8	1.344419	1.646482	2.17973	2.73264	3.48954	5.07064	7.34412
9	1.734926	2.087912	2.70039	3.32511	4.16816	5.89883	8.34283
10	2.15585	2.55821	3.24697	3.94030	4.86518	6.73720	9.34182
11	2.60321	3.05347	3.81575	4.57481	5.57779	7.58412	10.3410
12	3.07382	3.57056	4.40379	5.22603	6.30380	8.43842	11.3403
13	3.56503	4.10691	5.00874	5.89186	7.04150	9.29906	12.3398
14	4.07468	4.66043	5.62872	6.57063	7.78953	10.1653	13.3393
15	4.60094	5.22935	6.26214	7.26094	8.54675	11.0365	14.3389
16	5.14224	5.81221	6.90766	7.96164	9.31223	11.9122	15.3385
17	5.69724	6.40776	7.56418	8.67176	10.0852	12.7919	16.3381
18	6.26481	7.01491	8.23075	9.39046	10.8649	13.6753	17.3379
19	6.84398	7.63273	8.90655	10.1170	11.6509	14.5620	18.3376
20	7.43386	8.26040	9.59083	10.8508	12.4426	15.4518	19.3374
21	8.03366	8.89720	10.28293	11.5913	13.2396	16.3444	20.3372
22	8.64272	9.54249	10.9823	12.3380	14.0415	17.2396	21.3370
23	9.26042	10.19567	11.6885	13.0905	14.8479	18.1373	22.3369
24	9.88623	10.8564	12.4011	13.8484	15.6587	19.0372	23.3367
25	10.5197	11.5240	13.1197	14.6114	16.4734	19.9393	24.3366
26	11.1603	12.1981	13.8439	15.3791	17.2919	20.8434	25.3364
27	11.8076	12.8786	14.5733	16.1513	18.1138	21.7494	26.3363
28	12.4613	13.5648	15.3079	16.9279	18.9392	22.6572	27.3363
29	13.1211	14.2565	16.0471	17.7083	19.7677	23.5666	28.3362
30	13.7867	14.9535	16.7908	18.4926	20.5992	24.4776	29.3360
40	20.7065	22.1643	24.4331	26.5093	29.0505	33.6603	39.3354
50	27.9907	29.7067	32.3574	34.7642	37.6886	42.9421	49.3349
60	35.5346	37.4848	40.4817	43.1879	46.4589	52.2938	59.3347
70	43.2752	45.4418	48.7576	51.7393	55.3290	61.6983	69.3344
80	51.1720	53.5400	57.1532	60.3915	64.2778	71.1445	79.3343
90	59.1963	61.7541	65.6466	69.1260	73.2912	80.6247	89.3342
100	67.3276	70.0648	74.2219	77.9295	82.3581	90.1332	99.3341

Table III   (continued)

ν	$F(\chi^2)$						
	.75	.90	.95	.975	.99	.995	.999
1	1.32330	2.70554	3.84146	5.02389	6.63490	7.87944	10.828
2	2.77259	4.60517	5.99147	7.37776	9.21034	10.5966	13.816
3	4.10835	6.25139	7.81473	9.34840	11.3449	12.8381	16.266
4	5.38527	7.77944	9.48773	11.1433	13.2767	14.8602	18.467
5	6.62568	9.23635	11.0705	12.8325	15.0863	16.7496	20.515
6	7.84080	10.6446	12.5916	14.4494	16.8119	18.5476	22.458
7	9.03715	12.0170	14.0671	16.0128	18.4753	20.2777	24.322
8	10.2188	13.3616	15.5073	17.5346	20.0902	21.9550	26.125
9	11.3887	14.6837	16.9190	19.0228	21.6660	23.5893	27.877
10	12.5489	15.9871	18.3070	20.4831	23.2093	25.1882	29.588
11	13.7007	17.2750	19.6751	21.9200	24.7250	26.7569	31.264
12	14.8454	18.5494	21.0261	23.3367	26.2170	28.2995	32.909
13	15.9839	19.8119	22.3621	24.7356	27.6883	29.8194	34.528
14	17.1170	21.0642	23.6848	26.1190	29.1413	31.3193	36.123
15	18.2451	22.3072	24.9958	27.4884	30.5779	32.8013	37.697
16	19.3688	23.5418	26.2962	28.8454	31.9999	34.2672	39.252
17	20.4887	24.7690	27.5871	30.1910	33.4087	35.7185	40.790
18	21.6049	25.9894	28.8693	31.5264	34.8053	37.1564	42.312
19	22.7178	27.2036	30.1435	32.8523	36.1908	38.5822	43.820
20	23.8277	28.4120	31.4104	34.1696	37.5662	39.9968	45.315
21	24.9348	29.6151	32.6705	35.4789	38.9321	41.4010	46.797
22	26.0393	30.8133	33.9244	36.7807	40.2894	42.7956	48.268
23	27.1413	32.0069	35.1725	38.0757	41.6384	44.1813	49.728
24	28.2412	33.1963	36.4151	39.3641	42.9798	45.5585	51.179
25	29.3389	34.3816	37.6525	40.6465	44.3141	46.9278	52.620
26	30.4345	35.5631	38.8852	41.9232	45.6417	48.2899	54.052
27	31.5284	36.7412	40.1133	43.1944	46.9630	49.6449	55.476
28	32.6205	37.9159	41.3372	44.4607	48.2782	50.9933	56.892
29	33.7109	39.0875	42.5569	45.7222	49.5879	52.3356	58.302
30	34.7998	40.2560	43.7729	46.9792	50.8922	53.6720	59.703
40	45.6160	51.8050	55.7585	59.3417	63.6907	66.7659	73.402
50	56.3336	63.1671	67.5048	71.4202	76.1539	79.4900	86.661
60	66.9814	74.3970	79.0819	83.2976	88.3794	91.9517	99.607
70	77.5766	85.5271	90.5312	95.0231	100.425	104.215	112.317
80	88.1303	96.5782	101.879	106.629	112.329	116.321	124.839
90	98.6499	107.565	113.145	118.136	124.116	128.299	137.208
100	109.141	118.498	124.342	129.561	135.807	140.169	149.449

Table IV.   Fractiles of the F distribution

This table gives the .95, .975, and .99 fractiles of the F distribution with ν_1 and ν_2 degrees of freedom for $\nu_1 = 1(1)\,10,12,15,20,24,30,$ $40,60,120,\infty$ and $\nu_2 = 1(1)\,30,40,60,120,\infty$.

Example: For $\nu_1 = 8$ and $\nu_2 = 13$, the .95 fractile is 2.77.

.95 Fractiles

ν_2\ν_1	1	2	3	4	5	6	7	8	9	10	12	15	20	24	30	40	60	120	∞
1	161.4	199.5	215.7	224.6	230.2	234.0	236.8	238.9	240.5	241.9	243.9	245.9	248.0	249.1	250.1	251.1	252.2	253.3	243.3
2	18.51	19.00	19.16	19.25	19.30	19.33	19.35	19.37	19.38	19.40	19.41	19.43	19.45	19.45	19.46	19.47	19.48	19.49	19.50
3	10.13	9.55	9.28	9.12	9.01	8.94	8.89	8.85	8.81	8.79	8.74	8.70	8.66	8.64	8.62	8.59	8.57	8.55	8.53
4	7.71	6.94	6.59	6.39	6.26	6.16	6.09	6.04	6.00	5.96	5.91	5.86	5.80	5.77	5.75	5.72	5.69	5.66	5.63
5	6.61	5.79	5.41	5.19	5.05	4.95	4.88	4.82	4.77	4.74	4.68	4.62	4.56	4.53	4.50	4.46	4.43	4.40	4.36
6	5.99	5.14	4.76	4.53	4.39	4.28	4.21	4.15	4.10	4.06	4.00	3.94	3.87	3.84	3.81	3.77	3.74	3.70	3.67
7	5.59	4.74	4.35	4.12	3.97	3.87	3.79	3.73	3.68	3.64	3.57	3.51	3.44	3.41	3.38	3.34	3.30	3.27	3.23
8	5.32	4.46	4.07	3.84	3.69	3.58	3.50	3.44	3.39	3.35	3.28	3.22	3.15	3.12	3.08	3.04	3.01	2.97	2.93
9	5.12	4.26	3.86	3.63	3.48	3.37	3.29	3.23	3.18	3.14	3.07	3.01	2.94	2.90	2.86	2.83	2.79	2.75	2.71
10	4.96	4.10	3.71	3.48	3.33	3.22	3.14	3.07	3.02	2.98	2.91	2.85	2.77	2.74	2.70	2.66	2.62	2.58	2.54
11	4.84	3.98	3.59	3.36	3.20	3.09	3.01	2.95	2.90	2.85	2.79	2.72	2.65	2.61	2.57	2.53	2.49	2.45	2.40
12	4.75	3.89	3.49	3.26	3.11	3.00	2.91	2.85	2.80	2.75	2.69	2.62	2.54	2.51	2.47	2.43	2.38	2.34	2.30
13	4.67	3.81	3.41	3.18	3.03	2.92	2.83	2.77	2.71	2.67	2.60	2.53	2.46	2.42	2.38	2.34	2.30	2.25	2.21
14	4.60	3.74	3.34	3.11	2.96	2.85	2.76	2.70	2.65	2.60	2.53	2.46	2.39	2.35	2.31	2.27	2.22	2.18	2.13
15	4.54	3.68	3.29	3.06	2.90	2.79	2.71	2.64	2.59	2.54	2.48	2.40	2.33	2.29	2.25	2.20	2.16	2.11	2.07
16	4.49	3.63	3.24	3.01	2.85	2.74	2.66	2.59	2.54	2.49	2.42	2.35	2.28	2.24	2.19	2.15	2.11	2.06	2.01
17	4.45	3.59	3.20	2.96	2.81	2.70	2.61	2.55	2.49	2.45	2.38	2.31	2.23	2.19	2.15	2.10	2.06	2.01	1.96
18	4.41	3.55	3.16	2.93	2.77	2.66	2.58	2.51	2.46	2.41	2.34	2.27	2.19	2.15	2.11	2.06	2.02	1.97	1.92
19	4.38	3.52	3.13	2.90	2.74	2.63	2.54	2.48	2.42	2.38	2.31	2.23	2.16	2.11	2.07	2.03	1.98	1.93	1.88
20	4.35	3.49	3.10	2.87	2.71	2.60	2.51	2.45	2.39	2.35	2.28	2.20	2.12	2.08	2.04	1.99	1.95	1.90	1.84
21	4.32	3.47	3.07	2.84	2.68	2.57	2.49	2.42	2.37	2.32	2.25	2.18	2.10	2.05	2.01	1.96	1.92	1.87	1.81
22	4.30	3.44	3.05	2.82	2.66	2.55	2.46	2.40	2.34	2.30	2.23	2.15	2.07	2.03	1.98	1.94	1.89	1.84	1.78
23	4.28	3.42	3.03	2.80	2.64	2.53	2.44	2.37	2.32	2.27	2.20	2.13	2.05	2.01	1.96	1.91	1.86	1.81	1.76
24	4.26	3.40	3.01	2.78	2.62	2.51	2.42	2.36	2.30	2.25	2.18	2.11	2.03	1.98	1.94	1.89	1.84	1.79	1.73
25	4.24	3.39	2.99	2.76	2.60	2.49	2.40	2.34	2.28	2.24	2.16	2.09	2.01	1.96	1.92	1.87	1.82	1.77	1.71
26	4.23	3.37	2.98	2.74	2.59	2.47	2.39	2.32	2.27	2.22	2.15	2.07	1.99	1.95	1.90	1.85	1.80	1.75	1.69
27	4.21	3.35	2.96	2.73	2.57	2.46	2.37	2.31	2.25	2.20	2.13	2.06	1.97	1.93	1.88	1.84	1.79	1.73	1.67
28	4.20	3.34	2.95	2.71	2.56	2.45	2.36	2.29	2.24	2.19	2.12	2.04	1.96	1.91	1.87	1.82	1.77	1.71	1.65
29	4.18	3.33	2.93	2.70	2.55	2.43	2.35	2.28	2.22	2.18	2.10	2.03	1.94	1.90	1.85	1.81	1.75	1.70	1.64
30	4.17	3.32	2.92	2.69	2.53	2.42	2.33	2.27	2.21	2.16	2.09	2.01	1.93	1.89	1.84	1.79	1.74	1.68	1.62
40	4.08	3.23	2.84	2.61	2.45	2.34	2.25	2.18	2.12	2.08	2.00	1.92	1.84	1.79	1.74	1.69	1.64	1.58	1.51
60	4.00	3.15	2.76	2.53	2.37	2.25	2.17	2.10	2.04	1.99	1.92	1.84	1.75	1.70	1.65	1.59	1.53	1.47	1.39
120	3.92	3.07	2.68	2.45	2.29	2.17	2.09	2.02	1.96	1.91	1.83	1.75	1.66	1.61	1.55	1.50	1.43	1.35	1.25
∞	3.84	3.00	2.60	2.37	2.21	2.10	2.01	1.94	1.88	1.83	1.75	1.67	1.57	1.52	1.46	1.39	1.32	1.22	1.00

8

Table IV (continued) .975 Fractiles

ν_2 \ ν_1	1	2	3	4	5	6	7	8	9	10	12	15	20	24	30	40	60	120	∞
1	647.8	799.5	864.2	899.6	921.8	937.1	948.2	956.7	963.3	968.6	976.7	984.9	993.1	997.2	1001	1006	1010	1014	1018
2	38.51	39.00	39.17	39.25	39.30	39.33	39.36	39.37	39.39	39.40	39.41	39.43	39.45	39.46	39.46	39.47	39.48	39.49	39.50
3	17.44	16.04	15.44	15.10	14.88	14.73	14.62	14.54	14.47	14.42	14.34	14.25	14.17	14.12	14.08	14.04	13.99	13.95	13.90
4	12.22	10.65	9.98	9.60	9.36	9.20	9.07	8.98	8.90	8.84	8.75	8.66	8.56	8.51	8.46	8.41	8.36	8.31	8.26
5	10.01	8.43	7.76	7.39	7.15	6.98	6.85	6.76	6.68	6.62	6.52	6.43	6.33	6.28	6.23	6.18	6.12	6.07	6.02
6	8.81	7.26	6.60	6.23	5.99	5.82	5.70	5.60	5.52	5.46	5.37	5.27	5.17	5.12	5.07	5.01	4.96	4.90	4.85
7	8.07	6.54	5.89	5.52	5.29	5.12	4.99	4.90	4.82	4.76	4.67	4.57	4.47	4.42	4.36	4.31	4.25	4.20	4.14
8	7.57	6.06	5.42	5.05	4.82	4.65	4.53	4.43	4.36	4.30	4.20	4.10	4.00	3.95	3.89	3.84	3.78	3.73	3.67
9	7.21	5.71	5.08	4.72	4.48	4.32	4.20	4.10	4.03	3.96	3.87	3.77	3.67	3.61	3.56	3.51	3.45	3.39	3.33
10	6.94	5.46	4.83	4.47	4.24	4.07	3.95	3.85	3.78	3.72	3.62	3.52	3.42	3.37	3.31	3.26	3.20	3.14	3.08
11	6.72	5.26	4.63	4.28	4.04	3.88	3.76	3.66	3.59	3.53	3.43	3.33	3.23	3.17	3.12	3.06	3.00	2.94	2.88
12	6.55	5.10	4.47	4.12	3.89	3.73	3.61	3.51	3.44	3.37	3.28	3.18	3.07	3.02	2.96	2.91	2.85	2.79	2.72
13	6.41	4.97	4.35	4.00	3.77	3.60	3.48	3.39	3.31	3.25	3.15	3.05	2.95	2.89	2.84	2.78	2.72	2.66	2.60
14	6.30	4.86	4.24	3.89	3.66	3.50	3.38	3.29	3.21	3.15	3.05	2.95	2.84	2.79	2.73	2.67	2.61	2.55	2.49
15	6.20	4.77	4.15	3.80	3.58	3.41	3.29	3.20	3.12	3.06	2.96	2.86	2.76	2.70	2.64	2.59	2.52	2.46	2.40
16	6.12	4.69	4.08	3.73	3.50	3.34	3.22	3.12	3.05	2.99	2.89	2.79	2.68	2.63	2.57	2.51	2.45	2.38	2.32
17	6.04	4.62	4.01	3.66	3.44	3.28	3.16	3.06	2.98	2.92	2.82	2.72	2.62	2.56	2.50	2.44	2.38	2.32	2.25
18	5.98	4.56	3.95	3.61	3.38	3.22	3.10	3.01	2.93	2.87	2.77	2.67	2.56	2.50	2.44	2.38	2.32	2.26	2.19
19	5.92	4.51	3.90	3.56	3.33	3.17	3.05	2.96	2.88	2.82	2.72	2.62	2.51	2.45	2.39	2.33	2.27	2.20	2.13
20	5.87	4.46	3.86	3.51	3.29	3.13	3.01	2.91	2.84	2.77	2.68	2.57	2.46	2.41	2.35	2.29	2.22	2.16	2.09
21	5.83	4.42	3.82	3.48	3.25	3.09	2.97	2.87	2.80	2.73	2.64	2.53	2.42	2.37	2.31	2.25	2.18	2.11	2.04
22	5.79	4.38	3.78	3.44	3.22	3.05	2.93	2.84	2.76	2.70	2.60	2.50	2.39	2.33	2.27	2.21	2.14	2.08	2.00
23	5.75	4.35	3.75	3.41	3.18	3.02	2.90	2.81	2.73	2.67	2.57	2.47	2.36	2.30	2.24	2.18	2.11	2.04	1.97
24	5.72	4.32	3.72	3.38	3.15	2.99	2.87	2.78	2.70	2.64	2.54	2.44	2.33	2.27	2.21	2.15	2.08	2.01	1.94
25	5.69	4.29	3.69	3.35	3.13	2.97	2.85	2.75	2.68	2.61	2.51	2.41	2.30	2.24	2.18	2.12	2.05	1.98	1.91
26	5.66	4.27	3.67	3.33	3.10	2.94	2.82	2.73	2.65	2.59	2.49	2.39	2.28	2.22	2.16	2.09	2.03	1.95	1.88
27	5.63	4.24	3.65	3.31	3.08	2.92	2.80	2.71	2.63	2.57	2.47	2.36	2.25	2.19	2.13	2.07	2.00	1.93	1.85
28	5.61	4.22	3.63	3.29	3.06	2.90	2.78	2.69	2.61	2.55	2.45	2.34	2.23	2.17	2.11	2.05	1.98	1.91	1.83
29	5.59	4.20	3.61	3.27	3.04	2.88	2.76	2.67	2.59	2.53	2.43	2.32	2.21	2.15	2.09	2.03	1.96	1.89	1.81
30	5.57	4.18	3.59	3.25	3.03	2.87	2.75	2.65	2.57	2.51	2.41	2.31	2.20	2.14	2.07	2.01	1.94	1.87	1.79
40	5.42	4.05	3.46	3.13	2.90	2.74	2.62	2.53	2.45	2.39	2.29	2.18	2.07	2.01	1.94	1.88	1.80	1.72	1.64
60	5.29	3.93	3.34	3.01	2.79	2.63	2.51	2.41	2.33	2.27	2.17	2.06	1.94	1.88	1.82	1.74	1.67	1.58	1.48
120	5.15	3.80	3.23	2.89	2.67	2.52	2.39	2.30	2.22	2.16	2.05	1.94	1.82	1.76	1.69	1.61	1.53	1.43	1.31
∞	5.02	3.69	3.12	2.79	2.57	2.41	2.29	2.19	2.11	2.05	1.94	1.83	1.71	1.64	1.57	1.48	1.39	1.27	1.00

Table IV   (continued)   .99 Fractiles

$v_2 \backslash v_1$	1	2	3	4	5	6	7	8	9	10	12	15	20	24	30	40	60	120	∞
1	4052	4999.5	5403	5625	5764	5859	5928	5982	6022	6056	6106	6157	6209	6235	6261	6287	6313	6339	6366
2	98.50	99.00	99.17	99.25	99.30	99.33	99.36	99.37	99.39	99.40	99.42	99.43	99.45	99.46	99.47	99.47	99.48	99.49	99.50
3	34.12	30.82	29.46	28.71	28.24	27.91	27.67	27.49	27.35	27.23	27.05	26.87	26.69	26.60	26.50	26.41	26.32	26.22	26.13
4	21.20	18.00	16.69	15.98	15.52	15.21	14.98	14.80	14.66	14.55	14.37	14.20	14.02	13.93	13.84	13.75	13.65	13.56	13.46
5	16.26	13.27	12.06	11.39	10.97	10.67	10.46	10.29	10.16	10.05	9.89	9.72	9.55	9.47	9.38	9.29	9.20	9.11	9.06
6	13.75	10.92	9.78	9.15	8.75	8.47	8.26	8.10	7.98	7.87	7.72	7.56	7.40	7.31	7.23	7.14	7.06	6.97	6.88
7	12.25	9.55	8.45	7.85	7.46	7.19	6.99	6.84	6.72	6.62	6.47	6.31	6.16	6.07	5.99	5.91	5.82	5.74	5.65
8	11.26	8.65	7.59	7.01	6.63	6.37	6.18	6.03	5.91	5.81	5.67	5.52	5.36	5.28	5.20	5.12	5.03	4.95	4.86
9	10.56	8.02	6.99	6.42	6.06	5.80	5.61	5.47	5.35	5.26	5.11	4.96	4.81	4.73	4.65	4.57	4.48	4.40	4.31
10	10.04	7.56	6.55	5.99	5.64	5.39	5.20	5.06	4.94	4.85	4.71	4.56	4.41	4.33	4.25	4.17	4.08	4.00	3.91
11	9.65	7.21	6.22	5.67	5.32	5.07	4.89	4.74	4.63	4.54	4.40	4.25	4.10	4.02	3.94	3.86	3.78	3.69	3.60
12	9.33	6.93	5.95	5.41	5.06	4.82	4.64	4.50	4.39	4.30	4.16	4.01	3.86	3.78	3.70	3.62	3.54	3.45	3.36
13	9.07	6.70	5.74	5.21	4.86	4.62	4.44	4.30	4.19	4.10	3.96	3.82	3.66	3.59	3.51	3.43	3.34	3.25	3.17
14	8.86	6.51	5.56	5.04	4.69	4.46	4.28	4.14	4.03	3.94	3.80	3.66	3.51	3.43	3.35	3.27	3.18	3.09	3.00
15	8.68	6.36	5.42	4.89	4.56	4.32	4.14	4.00	3.89	3.80	3.67	3.52	3.37	3.29	3.21	3.13	3.05	2.96	2.87
16	8.53	6.23	5.29	4.77	4.44	4.20	4.03	3.89	3.78	3.69	3.55	3.41	3.26	3.18	3.10	3.02	2.93	2.84	2.75
17	8.40	6.11	5.18	4.67	4.34	4.10	3.93	3.79	3.68	3.59	3.46	3.31	3.16	3.08	3.00	2.92	2.83	2.75	2.65
18	8.29	6.01	5.09	4.58	4.25	4.01	3.84	3.71	3.60	3.51	3.37	3.23	3.08	3.00	2.92	2.84	2.75	2.66	2.57
19	8.18	5.93	5.01	4.50	4.17	3.94	3.77	3.63	3.52	3.43	3.30	3.15	3.00	2.92	2.84	2.76	2.67	2.58	2.49
20	8.10	5.85	4.94	4.43	4.10	3.87	3.70	3.56	3.46	3.37	3.23	3.09	2.94	2.86	2.78	2.69	2.61	2.52	2.42
21	8.02	5.78	4.87	4.37	4.04	3.81	3.64	3.51	3.40	3.31	3.17	3.03	2.88	2.80	2.72	2.64	2.55	2.46	2.36
22	7.95	5.72	4.82	4.31	3.99	3.76	3.59	3.45	3.35	3.26	3.12	2.98	2.83	2.75	2.67	2.58	2.50	2.40	2.31
23	7.88	5.66	4.76	4.26	3.94	3.71	3.54	3.41	3.30	3.21	3.07	2.93	2.78	2.70	2.62	2.54	2.45	2.35	2.26
24	7.82	5.61	4.72	4.22	3.90	3.67	3.50	3.36	3.26	3.17	3.03	2.89	2.74	2.66	2.58	2.49	2.40	2.31	2.21
25	7.77	5.57	4.68	4.18	3.85	3.63	3.46	3.32	3.22	3.13	2.99	2.85	2.70	2.62	2.54	2.45	2.36	2.27	2.17
26	7.72	5.53	4.64	4.14	3.82	3.59	3.42	3.29	3.18	3.09	2.96	2.81	2.66	2.58	2.50	2.42	2.33	2.23	2.13
27	7.68	5.49	4.60	4.11	3.78	3.56	3.39	3.26	3.15	3.06	2.93	2.78	2.63	2.55	2.47	2.38	2.29	2.20	2.10
28	7.64	5.45	4.57	4.07	3.75	3.53	3.36	3.23	3.12	3.03	2.90	2.75	2.60	2.52	2.44	2.35	2.26	2.17	2.06
29	7.60	5.42	4.54	4.04	3.73	3.50	3.33	3.20	3.09	3.00	2.87	2.73	2.57	2.49	2.41	2.33	2.23	2.14	2.03
30	7.56	5.39	4.51	4.02	3.70	3.47	3.30	3.17	3.07	2.98	2.84	2.70	2.55	2.47	2.39	2.30	2.21	2.11	2.01
40	7.31	5.18	4.31	3.83	3.51	3.29	3.12	2.99	2.89	2.80	2.66	2.52	2.37	2.29	2.20	2.11	2.02	1.92	1.80
60	7.08	4.98	4.13	3.65	3.34	3.12	2.95	2.82	2.72	2.63	2.50	2.35	2.20	2.12	2.03	1.94	1.84	1.73	1.60
120	6.85	4.79	3.95	3.48	3.17	2.96	2.79	2.66	2.56	2.47	2.34	2.19	2.03	1.95	1.86	1.76	1.66	1.53	1.38
∞	6.63	4.61	3.78	3.32	3.02	2.80	2.64	2.51	2.41	2.32	2.18	2.04	1.88	1.79	1.70	1.59	1.47	1.32	1.00

Table V.   Binomial Probabilities

This table gives binomial probabilities,

$$P(R = r \mid n, p) = \binom{n}{r} p^r (1 - p)^{n-r},$$

for $n = 1(1)20$, $r = 0(1)n$, and $p = .05(.05).50$. For $p > .50$, take $P(r \mid n, p) = P(n - r \mid n, 1 - p)$.

Examples: $P(R = 3 \mid n = 8, p = .25) = .2076$,
$\longrightarrow$ and $P(R = 2 \mid n = 5, p = .60) = P(R = 3 \mid n = 5, p = .40) = .2304$.

n	r	.05	.10	.15	.20	.25	.30	.35	.40	.45	.50
1	0	.9500	.9000	.8500	.8000	.7500	.7000	.6500	.6000	.5500	.5000
	1	.0500	.1000	.1500	.2000	.2500	.3000	.3500	.4000	.4500	.5000
2	0	.9025	.8100	.7225	.6400	.5625	.4900	.4225	.3600	.3025	.2500
	1	.0950	.1800	.2550	.3200	.3750	.4200	.4550	.4800	.4950	.5000
	2	.0025	.0100	.0225	.0400	.0625	.0900	.1225	.1600	.2025	.2500
3	0	.8574	.7290	.6141	.5120	.4219	.3430	.2746	.2160	.1664	.1250
	1	.1354	.2430	.3251	.3840	.4219	.4410	.4436	.4320	.4084	.3750
	2	.0071	.0270	.0574	.0960	.1406	.1890	.2389	.2880	.3341	.3750
	3	.0001	.0010	.0034	.0080	.0156	.0270	.0429	.0640	.0911	.1250
4	0	.8145	.6561	.5220	.4096	.3164	.2401	.1785	.1296	.0915	.0625
	1	.1715	.2916	.3685	.4096	.4219	.4116	.3845	.3456	.2995	.2500
	2	.0135	.0486	.0975	.1536	.2109	.2646	.3105	.3456	.3675	.3750
	3	.0005	.0036	.0115	.0256	.0469	.0756	.1115	.1536	.2005	.2500
	4	.0000	.0001	.0005	.0016	.0039	.0081	.0150	.0256	.0410	.0625
5	0	.7738	.5905	.4437	.3277	.2373	.1681	.1160	.0778	.0503	.0312
	1	.2036	.3280	.3915	.4096	.3955	.3602	.3124	.2592	.2059	.1562
	2	.0214	.0729	.1382	.2048	.2637	.3087	.3364	.3456	.3369	.3125
	3	.0011	.0081	.0244	.0512	.0879	.1323	.1811	.2304	.2757	.3125
	4	.0000	.0004	.0022	.0064	.0146	.0284	.0488	.0768	.1128	.1562
	5	.0000	.0000	.0001	.0003	.0010	.0024	.0053	.0102	.0185	.0312
6	0	.7351	.5314	.3771	.2621	.1780	.1176	.0754	.0467	.0277	.0156
	1	.2321	.3543	.3993	.3932	.3560	.3025	.2437	.1866	.1359	.0938
	2	.0305	.0984	.1762	.2458	.2966	.3241	.3280	.3110	.2780	.2344
	3	.0021	.0146	.0415	.0819	.1318	.1852	.2355	.2765	.3032	.3125
	4	.0001	.0012	.0055	.0154	.0330	.0595	.0951	.1382	.1861	.2344
	5	.0000	.0001	.0004	.0015	.0044	.0102	.0205	.0369	.0609	.0938
	6	.0000	.0000	.0000	.0001	.0002	.0007	.0018	.0041	.0083	.0156
7	0	.6983	.4783	.3206	.2097	.1335	.0824	.0490	.0280	.0152	.0078
	1	.2573	.3720	.3960	.3670	.3115	.2471	.1848	.1306	.0872	.0547
	2	.0406	.1240	.2097	.2753	.3115	.3177	.2985	.2613	.2140	.1641
	3	.0036	.0230	.0617	.1147	.1730	.2269	.2679	.2903	.2918	.2734
	4	.0002	.0026	.0109	.0287	.0577	.0972	.1442	.1935	.2388	.2734
	5	.0000	.0002	.0012	.0043	.0115	.0250	.0466	.0774	.1172	.1641
	6	.0000	.0000	.0001	.0004	.0013	.0036	.0084	.0172	.0320	.0547
	7	.0000	.0000	.0000	.0000	.0001	.0002	.0006	.0016	.0037	.0078

Table V   (continued)

						p					
n	r	**.05**	**.10**	**.15**	**.20**	**.25**	**.30**	**.35**	**.40**	**.45**	**.50**
8	0	.6634	.4305	.2725	.1678	.1001	.0576	.0319	.0168	.0084	.0039
	1	.2793	.3826	.3847	.3355	.2760	.1977	.1373	.0896	.0548	.0312
	2	.0515	.1488	.2376	.2936	.3115	.2965	.2587	.2090	.1569	.1094
	3	.0054	.0331	.0839	.1468	.2076	.2541	.2786	.2787	.2568	.2188
	4	.0004	.0046	.0185	.0459	.0865	.1361	.1875	.2322	.2627	.2734
	5	.0000	.0004	.0026	.0092	.0231	.0467	.0808	.1239	.1719	.2188
	6	.0000	.0000	.0002	.0011	.0038	.0100	.0217	.0413	.0703	.1094
	7	.0000	.0000	.0000	.0001	.0004	.0012	.0033	.0079	.0164	.0312
	8	.0000	.0000	.0000	.0000	.0000	.0001	.0002	.0007	.0017	.0039
9	0	.6302	.3874	.2316	.1342	.0751	.0404	.0277	.0101	.0046	.0020
	1	.2985	.3874	.3679	.3020	.2253	.1556	.1004	.0605	.0339	.0176
	2	.0629	.1722	.2597	.3020	.3003	.2668	.2162	.1612	.1110	.0703
	3	.0077	.0446	.1069	.1762	.2336	.2668	.2716	.2508	.2119	.1641
	4	.0006	.0074	.0283	.0661	.1168	.1715	.2194	.2508	.2600	.2461
	5	.0000	.0008	.0050	.0165	.0389	.0735	.1181	.1672	.2128	.2461
	6	.0000	.0001	.0006	.0028	.0087	.0210	.0424	.0743	.1160	.1641
	7	.0000	.0000	.0000	.0003	.0012	.0039	.0098	.0212	.0407	.0703
	8	.0000	.0000	.0000	.0000	.0001	.0004	.0013	.0035	.0083	.0176
	9	.0000	.0000	.0000	.0000	.0000	.0000	.0001	.0003	.0008	.0020
10	0	.5987	.3487	.1969	.1074	.0563	.0282	.0135	.0060	.0025	.0010
	1	.3151	.3874	.3474	.2684	.1877	.1211	.0725	.0403	.0207	.0098
	2	.0746	.1937	.2759	.3020	.2816	.2335	.1757	.1209	.0763	.0439
	3	.0105	.0574	.1298	.2013	.2503	.2668	.2522	.2150	.1665	.1172
	4	.0010	.0112	.0401	.0881	.1460	.2001	.2377	.2508	.2384	.2051
	5	.0001	.0015	.0085	.0264	.0584	.1029	.1536	.2007	.2340	.2461
	6	.0000	.0001	.0012	.0055	.0162	.0368	.0689	.1115	.1596	.2051
	7	.0000	.0000	.0001	.0008	.0031	.0090	.0212	.0425	.0746	.1172
	8	.0000	.0000	.0000	.0001	.0004	.0014	.0043	.0106	.0229	.0439
	9	.0000	.0000	.0000	.0000	.0000	.0001	.0005	.0016	.0042	.0098
	10	.0000	.0000	.0000	.0000	.0000	.0000	.0000	.0001	.0003	.0010
11	0	.5688	.3138	.1673	.0859	.0422	.0198	.0088	.0036	.0014	.0005
	1	.3293	.3835	.3248	.2362	.1549	.0932	.0518	.0266	.0125	.0054
	2	.0867	.2131	.2866	.2953	.2581	.1998	.1395	.0887	.0513	.0269
	3	.0137	.0710	.1517	.2215	.2581	.2568	.2254	.1774	.1259	.0806
	4	.0014	.0158	.0536	.1107	.1721	.2201	.2428	.2365	.2060	.1611
	5	.0001	.0025	.0132	.0388	.0803	.1231	.1830	.2207	.2360	.2256
	6	.0000	.0003	.0023	.0097	.0268	.0566	.0985	.1471	.1931	.2256
	7	.0000	.0000	.0003	.0017	.0064	.0173	.0379	.0701	.1128	.1611
	8	.0000	.0000	.0000	.0002	.0011	.0037	.0102	.0234	.0462	.0806
	9	.0000	.0000	.0000	.0000	.0001	.0005	.0018	.0052	.0126	.0269
	10	.0000	.0000	.0000	.0000	.0000	.0000	.0002	.0007	.0021	.0054
	11	.0000	.0000	.0000	.0000	.0000	.0000	.0000	.0000	.0002	.0005
12	0	.5404	.2824	.1422	.0687	.0317	.0138	.0057	.0022	.0008	.0002
	1	.3413	.3766	.3012	.2062	.1267	.0712	.0368	.0174	.0075	.0029
	2	.0988	.2301	.2924	.2835	.2323	.1678	.1088	.0639	.0339	.0161
	3	.0173	.0852	.1720	.2362	.2581	.2397	.1954	.1419	.0923	.0537
	4	.0021	.0213	.0683	.1329	.1936	.2311	.2367	.2128	.1700	.1208
	5	.0002	.0038	.0193	.0532	.1032	.1585	.2039	.2270	.2225	.1934
	6	.0000	.0005	.0040	.0155	.0401	.0792	.1281	.1766	.2124	.2256

Table V   (continued)

n	r	.05	.10	.15	.20	.25	.30	.35	.40	.45	.50
12	7	.0000	.0000	.0006	.0033	.0115	.0291	.0591	.1009	.1489	.1934
	8	.0000	.0000	.0001	.0005	.0024	.0078	.0199	.0420	.0762	.1208
	9	.0000	.0000	.0000	.0001	.0004	.0015	.0048	.0125	.0277	.0537
	10	.0000	.0000	.0000	.0000	.0000	.0002	.0008	.0025	.0068	.0161
	11	.0000	.0000	.0000	.0000	.0000	.0000	.0001	.0003	.0010	.0029
	12	.0000	.0000	.0000	.0000	.0000	.0000	.0000	.0000	.0001	.0002
13	0	.5133	.2542	.1209	.0550	.0238	.0097	.0037	.0013	.0004	.0001
	1	.3512	.3672	.2774	.1787	.1029	.0540	.0259	.0113	.0045	.0016
	2	.1109	.2448	.2937	.2680	.2059	.1388	.0836	.0453	.0220	.0095
	3	.0214	.0997	.1900	.2457	.2517	.2181	.1651	.1107	.0660	.0349
	4	.0028	.0277	.0838	.1535	.2097	.2337	.2222	.1845	.1350	.0873
	5	.0003	.0055	.0266	.0691	.1258	.1803	.2154	.2214	.1989	.1571
	6	.0000	.0008	.0063	.0230	.0559	.1030	.1546	.1968	.2169	.2095
	7	.0000	.0001	.0011	.0058	.0186	.0442	.0833	.1312	.1775	.2095
	8	.0000	.0000	.0001	.0011	.0047	.0142	.0336	.0656	.1089	.1571
	9	.0000	.0000	.0000	.0001	.0009	.0034	.0101	.0243	.0495	.0873
	10	.0000	.0000	.0000	.0000	.0001	.0006	.0022	.0065	.0162	.0349
	11	.0000	.0000	.0000	.0000	.0000	.0001	.0003	.0012	.0036	.0095
	12	.0000	.0000	.0000	.0000	.0000	.0000	.0000	.0001	.0005	.0016
	13	.0000	.0000	.0000	.0000	.0000	.0000	.0000	.0000	.0000	.0001
14	0	.4877	.2288	.1028	.0440	.0178	.0068	.0024	.0008	.0002	.0001
	1	.3593	.3559	.2539	.1539	.0832	.0407	.0181	.0073	.0027	.0009
	2	.1229	.2570	.2912	.2501	.1802	.1134	.0634	.0317	.0141	.0056
	3	.0259	.1142	.2056	.2501	.2402	.1943	.1366	.0845	.0462	.0222
	4	.0037	.0349	.0998	.1720	.2202	.2290	.2022	.1549	.1040	.0611
	5	.0004	.0078	.0352	.0860	.1468	.1963	.2178	.2066	.1701	.1222
	6	.0000	.0013	.0093	.0322	.0734	.1262	.1759	.2066	.2088	.1833
	7	.0000	.0002	.0019	.0092	.0280	.0618	.1082	.1574	.1952	.2095
	8	.0000	.0000	.0003	.0020	.0082	.0232	.0510	.0918	.1398	.1833
	9	.0000	.0000	.0000	.0003	.0018	.0066	.0183	.0408	.0762	.1222
	10	.0000	.0000	.0000	.0000	.0003	.0014	.0049	.0136	.0312	.0611
	11	.0000	.0000	.0000	.0000	.0000	.0002	.0010	.0033	.0093	.0222
	12	.0000	.0000	.0000	.0000	.0000	.0000	.0001	.0005	.0019	.0056
	13	.0000	.0000	.0000	.0000	.0000	.0000	.0000	.0001	.0002	.0009
	14	.0000	.0000	.0000	.0000	.0000	.0000	.0000	.0000	.0000	.0001
15	0	.4633	.2059	.0874	.0352	.0134	.0047	.0016	.0005	.0001	.0000
	1	.3658	.3432	.2312	.1319	.0668	.0305	.0126	.0047	.0016	.0005
	2	.1348	.2669	.2856	.2309	.1559	.0916	.0476	.0219	.0090	.0032
	3	.0307	.1285	.2184	.2501	.2252	.1700	.1110	.0634	.0318	.0139
	4	.0049	.0428	.1156	.1876	.2252	.2186	.1792	.1268	.0780	.0417
	5	.0006	.0105	.0449	.1032	.1651	.2061	.2123	.1859	.1404	.0916
	6	.0000	.0019	.0132	.0430	.0917	.1472	.1906	.2066	.1914	.1527
	7	.0000	.0003	.0030	.0138	.0393	.0811	.1319	.1771	.2013	.1964
	8	.0000	.0000	.0005	.0035	.0131	.0348	.0710	.1181	.1647	.1964
	9	.0000	.0000	.0001	.0007	.0034	.0116	.0298	.0612	.1048	.1527
	10	.0000	.0000	.0000	.0001	.0007	.0030	.0096	.0245	.0515	.0916
	11	.0000	.0000	.0000	.0000	.0001	.0006	.0024	.0074	.0191	.0417
	12	.0000	.0000	.0000	.0000	.0000	.0001	.0004	.0016	.0052	.0139
	13	.0000	.0000	.0000	.0000	.0000	.0000	.0001	.0003	.0010	.0032
	14	.0000	.0000	.0000	.0000	.0000	.0000	.0000	.0000	.0001	.0005
	15	.0000	.0000	.0000	.0000	.0000	.0000	.0000	.0000	.0000	.0000

Table V   (continued)

n	r	.05	.10	.15	.20	.25	.30	.35	.40	.45	.50
							p				
16	0	.4401	.1853	.0743	.0281	.0100	.0033	.0010	.0003	.0001	.0000
	1	.3706	.3294	.2097	.1126	.0535	.0228	.0087	.0030	.0009	.0002
	2	.1463	.2745	.2775	.2111	.1336	.0732	.0353	.0150	.0056	.0018
	3	.0359	.1423	.2285	.2463	.2079	.1465	.0888	.0468	.0215	.0085
	4	.0061	.0514	.1311	.2001	.2252	.2040	.1553	.1014	.0572	.0278
	5	.0008	.0137	.0555	.1201	.1802	.2099	.2008	.1623	.1123	.0667
	6	.0001	.0028	.0180	.0550	.1101	.1649	.1982	.1983	.1684	.1222
	7	.0000	.0004	.0045	.0197	.0524	.1010	.1524	.1889	.1969	.1746
	8	.0000	.0001	.0009	.0055	.0197	.0487	.0923	.1417	.1812	.1964
	9	.0000	.0000	.0001	.0012	.0058	.0185	.0442	.0840	.1318	.1746
	10	.0000	.0000	.0000	.0002	.0014	.0056	.0167	.0392	.0755	.1222
	11	.0000	.0000	.0000	.0000	.0002	.0013	.0049	.0142	.0337	.0667
	12	.0000	.0000	.0000	.0000	.0000	.0002	.0011	.0040	.0115	.0278
	13	.0000	.0000	.0000	.0000	.0000	.0000	.0002	.0008	.0029	.0085
	14	.0000	.0000	.0000	.0000	.0000	.0000	.0000	.0001	.0005	.0018
	15	.0000	.0000	.0000	.0000	.0000	.0000	.0000	.0000	.0001	.0002
	16	.0000	.0000	.0000	.0000	.0000	.0000	.0000	.0000	.0000	.0000
17	0	.4181	.1668	.0631	.0225	.0075	.0023	.0007	.0002	.0000	.0000
	1	.3741	.3150	.1893	.0957	.0426	.0169	.0060	.0019	.0005	.0001
	2	.1575	.2800	.2673	.1914	.1136	.0581	.0260	.0102	.0035	.0010
	3	.0415	.1556	.2359	.2393	.1893	.1245	.0701	.0341	.0144	.0052
	4	.0076	.0605	.1457	.2093	.2209	.1868	.1320	.0796	.0411	.0182
	5	.0010	.0175	.0668	.1361	.1914	.2081	.1849	.1379	.0875	.0472
	6	.0001	.0039	.0236	.0680	.1276	.1784	.1991	.1839	.1432	.0944
	7	.0000	.0007	.0065	.0267	.0668	.1201	.1685	.1927	.1841	.1484
	8	.0000	.0001	.0014	.0084	.0279	.0644	.1143	.1606	.1883	.1855
	9	.0000	.0000	.0003	.0021	.0093	.0276	.0611	.1070	.1540	.1855
	10	.0000	.0000	.0000	.0004	.0025	.0095	.0263	.0571	.1008	.1484
	11	.0000	.0000	.0000	.0001	.0005	.0026	.0090	.0242	.0525	.0944
	12	.0000	.0000	.0000	.0000	.0001	.0006	.0024	.0081	.0215	.0472
	13	.0000	.0000	.0000	.0000	.0000	.0001	.0005	.0021	.0068	.0182
	14	.0000	.0000	.0000	.0000	.0000	.0000	.0001	.0004	.0016	.0052
	15	.0000	.0000	.0000	.0000	.0000	.0000	.0000	.0001	.0003	.0010
	16	.0000	.0000	.0000	.0000	.0000	.0000	.0000	.0000	.0000	.0001
	17	.0000	.0000	.0000	.0000	.0000	.0000	.0000	.0000	.0000	.0000
18	0	.3972	.1501	.0536	.0180	.0056	.0016	.0004	.0001	.0000	.0000
	1	.3763	.3002	.1704	.0811	.0338	.0126	.0042	.0012	.0003	.0001
	2	.1683	.2835	.2556	.1723	.0958	.0458	.0190	.0069	.0022	.0006
	3	.0473	.1680	.2406	.2297	.1704	.1046	.0547	.0246	.0095	.0031
	4	.0093	.0700	.1592	.2153	.2130	.1681	.1104	.0614	.0291	.0117
	5	.0014	.0218	.0787	.1507	.1988	.2017	.1664	.1146	.0666	.0327
	6	.0002	.0052	.0310	.0816	.1436	.1873	.1941	.1655	.1181	.0708
	7	.0000	.0010	.0091	.0350	.0820	.1376	.1792	.1892	.1657	.1214
	8	.0000	.0002	.0022	.0120	.0376	.0811	.1327	.1734	.1864	.1669
	9	.0000	.0000	.0004	.0033	.0139	.0386	.0794	.1284	.1694	.1855
	10	.0000	.0000	.0001	.0008	.0042	.0149	.0385	.0771	.1248	.1669
	11	.0000	.0000	.0000	.0001	.0010	.0046	.0151	.0374	.0742	.1214
	12	.0000	.0000	.0000	.0000	.0002	.0012	.0047	.0145	.0354	.0708

Table V   (continued)

n	r	.05	.10	.15	.20	.25	.30	.35	.40	.45	.50
18	13	.0000	.0000	.0000	.0000	.0000	.0002	.0012	.0045	.0134	.0327
	14	.0000	.0000	.0000	.0000	.0000	.0000	.0002	.0011	.0039	.0117
	15	.0000	.0000	.0000	.0000	.0000	.0000	.0000	.0002	.0009	.0031
	16	.0000	.0000	.0000	.0000	.0000	.0000	.0000	.0000	.0001	.0006
	17	.0000	.0000	.0000	.0000	.0000	.0000	.0000	.0000	.0000	.0001
	18	.0000	.0000	.0000	.0000	.0000	.0000	.0000	.0000	.0000	.0000
19	0	.3774	.1351	.0456	.0144	.0042	.0011	.0003	.0001	.0000	.0000
	1	.3774	.2852	.1529	.0685	.0268	.0093	.0029	.0008	.0002	.0000
	2	.1787	.2852	.2428	.1540	.0803	.0358	.0138	.0046	.0013	.0003
	3	.0533	.1796	.2428	.2182	.1517	.0869	.0422	.0175	.0062	.0018
	4	.0112	.0798	.1714	.2182	.2023	.1491	.0909	.0467	.0203	.0074
	5	.0018	.0266	.0907	.1636	.2023	.1916	.1468	.0933	.0497	.0222
	6	.0002	.0069	.0374	.0955	.1574	.1916	.1844	.1451	.0949	.0518
	7	.0000	.0014	.0122	.0443	.0974	.1525	.1844	.1797	.1443	.0961
	8	.0000	.0002	.0032	.0166	.0487	.0981	.1489	.1797	.1771	.1442
	9	.0000	.0000	.0007	.0051	.0198	.0514	.0980	.1464	.1771	.1762
	10	.0000	.0000	.0001	.0013	.0066	.0220	.0528	.0976	.1449	.1762
	11	.0000	.0000	.0000	.0003	.0018	.0077	.0233	.0532	.0970	.1442
	12	.0000	.0000	.0000	.0000	.0004	.0022	.0083	.0237	.0529	.0961
	13	.0000	.0000	.0000	.0000	.0001	.0005	.0024	.0085	.0233	.0518
	14	.0000	.0000	.0000	.0000	.0000	.0001	.0006	.0024	.0082	.0222
	15	.0000	.0000	.0000	.0000	.0000	.0000	.0001	.0005	.0022	.0074
	16	.0000	.0000	.0000	.0000	.0000	.0000	.0000	.0001	.0005	.0018
	17	.0000	.0000	.0000	.0000	.0000	.0000	.0000	.0000	.0001	.0003
	18	.0000	.0000	.0000	.0000	.0000	.0000	.0000	.0000	.0000	.0000
	19	.0000	.0000	.0000	.0000	.0000	.0000	.0000	.0000	.0000	.0000
20	0	.3585	.1216	.0388	.0115	.0032	.0008	.0002	.0000	.0000	.0000
	1	.3774	.2702	.1368	.0576	.0211	.0068	.0020	.0005	.0001	.0000
	2	.1887	.2852	.2293	.1369	.0669	.0278	.0100	.0031	.0008	.0002
	3	.0596	.1901	.2428	.2054	.1339	.0716	.0323	.0123	.0040	.0011
	4	.0133	.0898	.1821	.2182	.1897	.1304	.0738	.0350	.0139	.0046
	5	.0022	.0319	.1028	.1746	.2023	.1789	.1272	.0746	.0365	.0148
	6	.0003	.0089	.0454	.1091	.1686	.1916	.1712	.1244	.0746	.0370
	7	.0000	.0020	.0160	.0545	.1124	.1643	.1844	.1659	.1221	.0739
	8	.0000	.0004	.0046	.0222	.0609	.1144	.1614	.1797	.1623	.1201
	9	.0000	.0001	.0011	.0074	.0271	.0654	.1158	.1597	.1771	.1602
	10	.0000	.0000	.0002	.0020	.0099	.0308	.0686	.1171	.1593	.1762
	11	.0000	.0000	.0000	.0005	.0030	.0120	.0336	.0710	.1185	.1602
	12	.0000	.0000	.0000	.0001	.0008	.0039	.0136	.0355	.0727	.1201
	13	.0000	.0000	.0000	.0000	.0002	.0010	.0045	.0146	.0366	.0739
	14	.0000	.0000	.0000	.0000	.0000	.0002	.0012	.0049	.0150	.0370
	15	.0000	.0000	.0000	.0000	.0000	.0000	.0003	.0013	.0049	.0148
	16	.0000	.0000	.0000	.0000	.0000	.0000	.0000	.0003	.0013	.0046
	17	.0000	.0000	.0000	.0000	.0000	.0000	.0000	.0000	.0002	.0011
	18	.0000	.0000	.0000	.0000	.0000	.0000	.0000	.0000	.0000	.0002
	19	.0000	.0000	.0000	.0000	.0000	.0000	.0000	.0000	.0000	.0000
	20	.0000	.0000	.0000	.0000	.0000	.0000	.0000	.0000	.0000	.0000

This table is reproduced by permission from R. S. Burington and D. C. May, *Handbook of Probability and Statistics with Tables.* McGraw-Hill Book Company, 1953.

Table VI.   Poisson Probabilities

This table gives Poisson probabilities,

$$P(R = r \mid \lambda, t) = \frac{e^{-\lambda t}(\lambda t)^r}{r!},$$

for $\lambda t = .1(.1)10(1)20$ and suitable values of r.

Example: $P(R = 2 \mid \lambda = 1.5, t = 5) = .0156.$

					λt					
r	0.1	0.2	0.3	0.4	0.5	0.6	0.7	0.8	0.9	1.0
0	.9048	.8187	.7408	.6703	.6065	.5488	.4966	.4493	.4066	.3679
1	.0905	.1637	.2222	.2681	.3033	.3293	.3476	.3595	.3659	.3679
2	.0045	.0164	.0333	.0536	.0758	.0988	.1217	.1438	.1647	.1839
3	.0002	.0011	.0033	.0072	.0126	.0198	.0284	.0383	.0494	.0613
4	.0000	.0001	.0002	.0007	.0016	.0030	.0050	.0077	.0111	.0153
5	.0000	.0000	.0000	.0001	.0002	.0004	.0007	.0012	.0020	.0031
6	.0000	.0000	.0000	.0000	.0000	.0000	.0001	.0002	.0003	.0005
7	.0000	.0000	.0000	.0000	.0000	.0000	.0000	.0000	.0000	.0001

					λt					
r	1.1	1.2	1.3	1.4	1.5	1.6	1.7	1.8	1.9	2.0
0	.3329	.3012	.2725	.2466	.2231	.2019	.1827	.1653	.1496	.1353
1	.3662	.3614	.3543	.3452	.3347	.3230	.3106	.2975	.2842	.2707
2	.2014	.2169	.2303	.2417	.2510	.2584	.2640	.2678	.2700	.2707
3	.0738	.0867	.0998	.1128	.1255	.1378	.1496	.1607	.1710	.1804
4	.0203	.0260	.0324	.0395	.0471	.0551	.0636	.0723	.0812	.0902
5	.0045	.0062	.0084	.0111	.0141	.0176	.0216	.0260	.0309	.0361
6	.0008	.0012	.0018	.0026	.0035	.0047	.0061	.0078	.0098	.0120
7	.0001	.0002	.0003	.0005	.0008	.0011	.0015	.0020	.0027	.0034
8	.0000	.0000	.0001	.0001	.0001	.0002	.0003	.0005	.0006	.0009
9	.0000	.0000	.0000	.0000	.0000	.0000	.0001	.0001	.0001	.0002

					λt					
r	2.1	2.2	2.3	2.4	2.5	2.6	2.7	2.8	2.9	3.0
0	.1225	.1108	.1003	.0907	.0821	.0743	.0672	.0608	.0550	.0498
1	.2572	.2438	.2306	.2177	.2052	.1931	.1815	.1703	.1596	.1494
2	.2700	.2681	.2652	.2613	.2565	.2510	.2450	.2384	.2314	.2240
3	.1890	.1966	.2033	.2090	.2138	.2176	.2205	.2225	.2237	.2240
4	.0992	.1082	.1169	.1254	.1336	.1414	.1488	.1557	.1622	.1680
5	.0417	.0476	.0538	.0602	.0668	.0735	.0804	.0872	.0940	.1008
6	.0146	.0174	.0206	.0241	.0278	.0319	.0362	.0407	.0455	.0540
7	.0044	.0055	.0068	.0083	.0099	.0118	.0139	.0163	.0188	.0216
8	.0011	.0015	.0019	.0025	.0031	.0038	.0047	.0057	.0068	.0081
9	.0003	.0004	.0005	.0007	.0009	.0011	.0014	.0018	.0022	.0027
10	.0001	.0001	.0001	.0002	.0002	.0003	.0004	.0005	.0006	.0008
11	.0000	.0000	.0000	.0000	.0000	.0001	.0001	.0001	.0002	.0002
12	.0000	.0000	.0000	.0000	.0000	.0000	.0000	.0000	.0000	.0001

Table VI   (continued)

λt

r	3.1	3.2	3.3	3.4	3.5	3.6	3.7	3.8	3.9	4.0
0	.0450	.0408	.0369	.0344	.0302	.0273	.0247	.0224	.0202	.0183
1	.1397	.1304	.1217	.1135	.1057	.0984	.0915	.0850	.0789	.0733
2	.2165	.2087	.2008	.1929	.1850	.1771	.1692	.1615	.1539	.1465
3	.2237	.2226	.2209	.2186	.2158	.2125	.2087	.2046	.2001	.1954
4	.1734	.1781	.1823	.1858	.1888	.1912	.1931	.1944	.1951	.1954
5	.1075	.1140	.1203	.1264	.1322	.1377	.1429	.1477	.1522	.1563
6	.0555	.0608	.0662	.0716	.0771	.0826	.0881	.0936	.0989	.1042
7	.0246	.0278	.0312	.0348	.0385	.0425	.0466	.0508	.0551	.0595
8	.0095	.0111	.0129	.0148	.0169	.0191	.0215	.0241	.0269	.0298
9	.0033	.0040	.0047	.0056	.0066	.0076	.0089	.0102	.0116	.0132
10	.0010	.0013	.0016	.0019	.0023	.0028	.0033	.0039	.0045	.0053
11	.0003	.0004	.0005	.0006	.0007	.0009	.0011	.0013	.0016	.0019
12	.0001	.0001	.0001	.0002	.0002	.0003	.0003	.0004	.0005	.0006
13	.0000	.0000	.0000	.0000	.0001	.0001	.0001	.0001	.0002	.0002
14	.0000	.0000	.0000	.0000	.0000	.0000	.0000	.0000	.0000	.0001

λt

r	4.1	4.2	4.3	4.4	4.5	4.6	4.7	4.8	4.9	5.0
0	.0166	.0150	.0136	.0123	.0111	.0101	.0091	.0082	.0074	.0067
1	.0679	.0630	.0583	.0540	.0500	.0462	.0427	.0395	.0365	.0337
2	.1393	.1323	.1254	.1188	.1125	.1063	.1005	.0948	.0894	.0842
3	.1904	.1852	.1798	.1743	.1687	.1631	.1574	.1517	.1460	.1404
4	.1951	.1944	.1933	.1917	.1898	.1875	.1849	.1820	.1789	.1755
5	.1600	.1633	.1662	.1687	.1708	.1725	.1738	.1747	.1753	.1755
6	.1093	.1143	.1191	.1237	.1281	.1323	.1362	.1398	.1432	.1462
7	.0640	.0686	.0732	.0778	.0824	.0869	.0914	.0959	.1002	.1044
8	.0328	.0360	.0393	.0428	.0463	.0500	.0537	.0575	.0614	.0653
9	.0150	.0168	.0188	.0209	.0232	.0255	.0280	.0307	.0334	.0363
10	.0061	.0071	.0081	.0092	.0104	.0118	.0132	.0147	.0164	.0181
11	.0023	.0027	.0032	.0037	.0043	.0049	.0056	.0064	.0073	.0082
12	.0008	.0009	.0011	.0014	.0016	.0019	.0022	.0026	.0030	.0034
13	.0002	.0003	.0004	.0005	.0006	.0007	.0008	.0009	.0011	.0013
14	.0001	.0001	.0001	.0001	.0002	.0002	.0003	.0003	.0004	.0005
15	.0000	.0000	.0000	.0000	.0001	.0001	.0001	.0001	.0001	.0002

λt

r	5.1	5.2	5.3	5.4	5.5	5.6	5.7	5.8	5.9	6.0
0	.0061	.0055	.0050	.0045	.0041	.0037	.0033	.0030	.0027	.0025
1	.0311	.0287	.0265	.0244	.0225	.0207	.0191	.0176	.0162	.0149
2	.0793	.0746	.0701	.0659	.0618	.0580	.0544	.0509	.0477	.0446
3	.1348	.1293	.1239	.1185	.1133	.1082	.1033	.0985	.0938	.0892
4	.1719	.1681	.1641	.1600	.1558	.1515	.1472	.1428	.1383	.1339
5	.1753	.1748	.1740	.1728	.1714	.1697	.1678	.1656	.1632	.1606
6	.1490	.1515	.1537	.1555	.1571	.1584	.1594	.1601	.1605	.1606
7	.1086	.1125	.1163	.1200	.1234	.1267	.1298	.1326	.1353	.1377
8	.0692	.0731	.0771	.0810	.0849	.0887	.0925	.0962	.0998	.1033
9	.0392	.0423	.0454	.0486	.0519	.0552	.0586	.0620	.0654	.0688

Table   VI   (continued)

λt

r	5.1	5.2	5.3	5.4	5.5	5.6	5.7	5.8	5.9	6.0
10	.0200	.0220	.0241	.0262	.0285	.0309	.0334	.0359	.0386	.0413
11	.0093	.0104	.0116	.0129	.0143	.0157	.0173	.0190	.0207	.0225
12	.0039	.0045	.0051	.0058	.0065	.0073	.0082	.0092	.0102	.0113
13	.0015	.0018	.0021	.0024	.0028	.0032	.0036	.0041	.0046	.0052
14	.0006	.0007	.0008	.0009	.0011	.0013	.0015	.0017	.0019	.0022
15	.0002	.0002	.0003	.0003	.0004	.0005	.0006	.0007	.0008	.0009
16	.0001	.0001	.0001	.0001	.0001	.0002	.0002	.0002	.0003	.0003
17	.0000	.0000	.0000	.0000	.0000	.0001	.0001	.0001	.0001	.0001

λt

r	6.1	6.2	6.3	6.4	6.5	6.6	6.7	6.8	6.9	7.0
0	.0022	.0020	.0018	.0017	.0015	.0014	.0012	.0011	.0010	.0009
1	.0137	.0126	.0116	.0106	.0098	.0090	.0082	.0076	.0070	.0064
2	.0417	.0390	.0364	.0340	.0318	.0296	.0276	.0258	.0240	.0223
3	.0848	.0806	.0765	.0726	.0688	.0652	.0617	.0584	.0552	.0521
4	.1294	.1249	.1205	.1162	.1118	.1076	.1034	.0992	.0952	.0912
5	.1579	.1549	.1519	.1487	.1454	.1420	.1385	.1349	.1314	.1277
6	.1605	.1601	.1595	.1586	.1575	.1562	.1546	.1529	.1511	.1490
7	.1399	.1418	.1435	.1450	.1462	.1472	.1480	.1486	.1489	.1490
8	.1066	.1099	.1130	.1160	.1188	.1215	.1240	.1263	.1284	.1304
9	.0723	.0757	.0791	.0825	.0858	.0891	.0923	.0954	.0985	.1014
10	.0441	.0469	.0498	.0528	.0558	.0588	.0618	.0649	.0679	.0710
11	.0245	.0265	.0285	.0307	.0330	.0353	.0377	.0401	.0426	.0452
12	.0124	.0137	.0150	.0164	.0179	.0194	.0210	.0227	.0245	.0264
13	.0058	.0065	.0073	.0081	.0089	.0098	.0108	.0119	.0130	.0142
14	.0025	.0029	.0033	.0037	.0041	.0046	.0052	.0058	.0064	.0071
15	.0010	.0012	.0014	.0016	.0018	.0020	.0023	.0026	.0029	.0033
16	.0004	.0005	.0005	.0006	.0007	.0008	.0010	.0011	.0013	.0014
17	.0001	.0002	.0002	.0002	.0003	.0003	.0004	.0004	.0005	.0006
18	.0000	.0001	.0001	.0001	.0001	.0001	.0001	.0002	.0002	.0002
19	.0000	.0000	.0000	.0000	.0000	.0000	.0000	.0001	.0001	.0001

λt

r	7.1	7.2	7.3	7.4	7.5	7.6	7.7	7.8	7.9	8.0
0	.0008	.0007	.0007	.0006	.0006	.0005	.0005	.0004	.0004	.0003
1	.0059	.0054	.0049	.0045	.0041	.0038	.0035	.0032	.0029	.0027
2	.0208	.0194	.0180	.0167	.0156	.0145	.0134	.0125	.0116	.0107
3	.0492	.0464	.0438	.0413	.0389	.0366	.0345	.0324	.0305	.0286
4	.0874	.0836	.0799	.0764	.0729	.0696	.0663	.0632	.0602	.0573
5	.1241	.1204	.1167	.1130	.1094	.1057	.1021	.0986	.0951	.0916
6	.1468	.1445	.1420	.1394	.1367	.1339	.1311	.1282	.1252	.1221
7	.1489	.1486	.1481	.1474	.1465	.1454	.1442	.1428	.1413	.1396
8	.1321	.1337	.1351	.1363	.1373	.1382	.1388	.1392	.1395	.1396
9	.1042	.1070	.1096	.1121	.1144	.1167	.1187	.1207	.1224	.1241
10	.0740	.0770	.0800	.0829	.0858	.0887	.0914	.0941	.0967	.0993
11	.0478	.0504	.0531	.0558	.0585	.0613	.0640	.0667	.0695	.0722

Table VI (continued)

λt

r	7.1	7.2	7.3	7.4	7.5	7.6	7.7	7.8	7.9	8.0
12	.0283	.0303	.0323	.0344	.0366	.0388	.0411	.0434	.0457	.0481
13	.0154	.0168	.0181	.0196	.0211	.0227	.0243	.0260	.0278	.0296
14	.0078	.0086	.0095	.0104	.0113	.0123	.0134	.0145	.0157	.0169
15	.0037	.0041	.0046	.0051	.0057	.0062	.0069	.0075	.0083	.0090
16	.0016	.0019	.0021	.0024	.0026	.0030	.0033	.0037	.0041	.0045
17	.0007	.0008	.0009	.0010	.0012	.0013	.0015	.0017	.0019	.0021
18	.0003	.0003	.0004	.0004	.0005	.0006	.0006	.0007	.0008	.0009
19	.0001	.0001	.0001	.0002	.0002	.0002	.0003	.0003	.0003	.0004
20	.0000	.0000	.0001	.0001	.0001	.0000	.0001	.0001	.0001	.0002
21	.0000	.0000	.0000	.0000	.0000	.0000	.0000	.0000	.0001	.0001

λt

r	8.1	8.2	8.3	8.4	8.5	8.6	8.7	8.8	8.9	9.0
0	.0003	.0003	.0002	.0002	.0002	.0002	.0002	.0002	.0001	.0001
1	.0025	.0023	.0021	.0019	.0017	.0016	.0014	.0013	.0012	.0011
2	.0100	.0092	.0086	.0079	.0074	.0068	.0063	.0058	.0054	.0050
3	.0269	.0252	.0237	.0222	.0208	.0195	.0183	.0171	.0160	.0150
4	.0544	.0517	.0491	.0466	.0443	.0420	.0398	.0377	.0357	.0337
5	.0882	.0849	.0816	.0784	.0752	.0722	.0692	.0663	.0635	.0607
6	.1191	.1160	.1128	.1097	.1066	.1034	.1003	.0972	.0941	.0911
7	.1378	.1358	.1338	.1317	.1294	.1271	.1247	.1222	.1197	.1171
8	.1395	.1392	.1388	.1382	.1375	.1366	.1356	.1344	.1332	.1318
9	.1256	.1269	.1280	.1290	.1299	.1306	.1311	.1315	.1317	.1318
10	.1017	.1040	.1063	.1084	.1104	.1123	.1140	.1157	.1172	.1186
11	.0749	.0776	.0802	.0828	.0853	.0878	.0902	.0925	.0948	.0970
12	.0505	.0530	.0555	.0579	.0604	.0629	.0654	.0679	.0703	.0728
13	.0315	.0334	.0354	.0374	.0395	.0416	.0438	.0459	.0481	.0504
14	.0182	.0196	.0210	.0225	.0240	.0256	.0272	.0289	.0306	.0324
15	.0098	.0107	.0116	.0126	.0136	.0147	.0158	.0169	.0182	.0194
16	.0050	.0055	.0060	.0066	.0072	.0079	.0086	.0093	.0101	.0109
17	.0024	.0026	.0029	.0033	.0036	.0040	.0044	.0048	.0053	.0058
18	.0011	.0012	.0014	.0015	.0017	.0019	.0021	.0024	.0026	.0029
19	.0005	.0005	.0006	.0007	.0008	.0009	.0010	.0011	.0012	.0014
20	.0002	.0002	.0002	.0003	.0003	.0004	.0004	.0005	.0005	.0006
21	.0001	.0001	.0001	.0001	.0001	.0002	.0002	.0002	.0002	.0003
22	.0000	.0000	.0000	.0000	.0001	.0001	.0001	.0001	.0001	.0001

λt

r	9.1	9.2	9.3	9.4	9.5	9.6	9.7	9.8	9.9	10
0	.0001	.0001	.0001	.0001	.0001	.0001	.0001	.0001	.0001	.0000
1	.0010	.0009	.0009	.0008	.0007	.0007	.0006	.0005	.0005	.0005
2	.0046	.0043	.0040	.0037	.0034	.0031	.0029	.0027	.0025	.0023
3	.0140	.0131	.0123	.0115	.0107	.0100	.0093	.0087	.0081	.0076
4	.0319	.0302	.0285	.0269	.0254	.0240	.0226	.0213	.0201	.0189

Table VI   (continued)

λt

r	9.1	9.2	9.3	9.4	9.5	9.6	9.7	9.8	9.9	10
5	.0581	.0555	.0530	.0506	.0483	.0460	.0439	.0418	.0398	.0378
6	.0881	.0851	.0822	.0793	.0764	.0736	.0709	.0682	.0656	.0631
7	.1145	.1118	.1091	.1064	.1037	.1010	.0982	.0955	.0928	.0901
8	.1302	.1286	.1269	.1251	.1232	.1212	.1191	.1170	.1148	.1126
9	.1317	.1315	.1311	.1306	.1300	.1293	.1284	.1274	.1263	.1251
10	.1198	.1210	.1219	.1228	.1235	.1241	.1245	.1249	.1250	.1251
11	.0991	.1012	.1031	.1049	.1067	.1083	.1098	.1112	.1125	.1137
12	.0752	.0776	.0799	.0822	.0844	.0866	.0888	.0908	.0928	.0948
13	.0526	.0549	.0572	.0594	.0617	.0640	.0662	.0685	.0707	.0729
14	.0342	.0361	.0380	.0399	.0419	.0439	.0459	.0479	.0500	.0521
15	.0208	.0221	.0235	.0250	.0265	.0281	.0297	.0313	.0330	.0347
16	.0118	.0127	.0137	.0147	.0157	.0168	.0180	.0192	.0204	.0217
17	.0063	.0069	.0075	.0081	.0088	.0095	.0103	.0111	.0119	.0128
18	.0032	.0035	.0039	.0042	.0046	.0051	.0055	.0060	.0065	.0071
19	.0015	.0017	.0019	.0021	.0023	.0026	.0028	.0031	.0034	.0037
20	.0007	.0008	.0009	.0010	.0011	.0012	.0014	.0015	.0017	.0019
21	.0003	.0003	.0004	.0004	.0005	.0006	.0006	.0007	.0008	.0009
22	.0001	.0001	.0002	.0002	.0002	.0002	.0003	.0003	.0004	.0004
23	.0000	.0001	.0001	.0001	.0001	.0001	.0001	.0001	.0002	.0002
24	.0000	.0000	.0000	.0000	.0000	.0000	.0000	.0001	.0001	.0001

λt

r	11	12	13	14	15	16	17	18	19	20
0	.0000	.0000	.0000	.0000	.0000	.0000	.0000	.0000	.0000	.0000
1	.0002	.0001	.0000	.0000	.0000	.0000	.0000	.0000	.0000	.0000
2	.0010	.0004	.0002	.0001	.0000	.0000	.0000	.0000	.0000	.0000
3	.0037	.0018	.0008	.0004	.0002	.0001	.0000	.0000	.0000	.0000
4	.0102	.0053	.0027	.0013	.0006	.0003	.0001	.0001	.0000	.0000
5	.0224	.0127	.0070	.0037	.0019	.0010	.0005	.0002	.0001	.0001
6	.0411	.0255	.0152	.0087	.0048	.0026	.0014	.0007	.0004	.0002
7	.0646	.0437	.0281	.0174	.0104	.0060	.0034	.0018	.0010	.0005
8	.0888	.0655	.0457	.0304	.0194	.0120	.0072	.0042	.0024	.0013
9	.1085	.0874	.0661	.0473	.0324	.0213	.0135	.0083	.0050	.0029
10	.1194	.1048	.0859	.0663	.0486	.0341	.0230	.0150	.0095	.0058
11	.1194	.1144	.1015	.0844	.0663	.0496	.0355	.0245	.0164	.0106
12	.1094	.1144	.1099	.0984	.0829	.0661	.0504	.0368	.0259	.0176
13	.0926	.1056	.1099	.1060	.0956	.0814	.0658	.0509	.0378	.0271
14	.0728	.0905	.1021	.1060	.1024	.0930	.0800	.0655	.0541	.0387
15	.0534	.0724	.0885	.0989	.1024	.0992	.0906	.0786	.0650	.0516
16	.0367	.0543	.0719	.0866	.0960	.0992	.0963	.0884	.0772	.0646
17	.0237	.0383	.0550	.0713	.0847	.0934	.0963	.0936	.0863	.0760
18	.0145	.0256	.0397	.0554	.0706	.0830	.0909	.0936	.0911	.0844
19	.0084	.0161	.0272	.0409	.0557	.0699	.0814	.0887	.0911	.0888
20	.0046	.0097	.0177	.0286	.0418	.0559	.0692	.0798	.0866	.0888
21	.0024	.0055	.0109	.0191	.0299	.0426	.0560	.0684	.0783	.0846
22	.0012	.0030	.0065	.0121	.0204	.0310	.0433	.0560	.0676	.0769
23	.0006	.0016	.0037	.0074	.0133	.0216	.0320	.0438	.0559	.0669
24	.0003	.0008	.0020	.0043	.0083	.0144	.0226	.0328	.0442	.0557

Table   VI   (continued)

λt

r	11	12	13	14	15	16	17	18	19	20
25	.0001	.0004	.0010	.0024	.0050	.0092	.0154	.0237	.0336	.0446
26	.0000	.0002	.0005	.0013	.0029	.0057	.0101	.0164	.0246	.0343
27	.0000	.0001	.0002	.0007	.0016	.0034	.0063	.0109	.0173	.0254
28	.0000	.0000	.0001	.0003	.0009	.0019	.0038	.0070	.0117	.0181
29	.0000	.0000	.0001	.0002	.0004	.0011	.0023	.0044	.0077	.0125
30	.0000	.0000	.0000	.0001	.0002	.0006	.0013	.0026	.0049	.0083
31	.0000	.0000	.0000	.0000	.0001	.0003	.0007	.0015	.0030	.0054
32	.0000	.0000	.0000	.0000	.0001	.0001	.0004	.0009	.0018	.0034
33	.0000	.0000	.0000	.0000	.0000	.0001	.0002	.0005	.0010	.0020
34	.0000	.0000	.0000	.0000	.0000	.0000	.0001	.0002	.0006	.0012
35	.0000	.0000	.0000	.0000	.0000	.0000	.0000	.0001	.0003	.0007
36	.0000	.0000	.0000	.0000	.0000	.0000	.0000	.0001	.0002	.0004
37	.0000	.0000	.0000	.0000	.0000	.0000	.0000	.0000	.0001	.0002
38	.0000	.0000	.0000	.0000	.0000	.0000	.0000	.0000	.0000	.0001
39	.0000	.0000	.0000	.0000	.0000	.0000	.0000	.0000	.0000	.0001

This table is reproduced by permission from R. S. Burington and D. C. May, *Handbook of Probability and Statistics with Tables*. McGraw-Hill Book Company, Inc., 1953.

Table   VII.   Standard Normal Density Function

This table gives values of the standard normal density function,

$$f(z) = \frac{1}{\sqrt{2\pi}} e^{-z^2/2},$$

for $z = 0(.01)4.29$. For $z < 0$, take $f(z) = f(-z)$.

Examples: $f(2.16) = .0387$, and $f(-1.57) = f(1.57) = .1163$.

z	.00	.01	.02	.03	.04	.05	.06	.07	.08	.09
0.0	.3989	.3989	.3989	.3988	.3986	.3984	.3982	.3980	.3977	.3973
0.1	.3970	.3965	.3961	.3956	.3951	.3945	.3939	.3932	.3925	.3918
0.2	.3910	.3902	.3894	.3885	.3876	.3867	.3857	.3847	.3836	.3825
0.3	.3814	.3802	.3790	.3778	.3765	.3752	.3739	.3725	.3712	.3697
0.4	.3683	.3668	.3653	.3637	.3621	.3605	.3589	.3572	.3555	.3538

Table VII   (continued)

z	.00	.01	.02	.03	.04	.05	.06	.07	.08	.09
0.5	.3521	.3503	.3485	.3467	.3448	.3429	.3410	.3391	.3372	.3352
0.6	.3332	.3312	.3292	.3271	.3251	.3230	.3209	.3187	.3166	.3144
0.7	.3123	.3101	.3079	.3056	.3034	.3011	.2989	.2966	.2943	.2920
0.8	.2897	.2874	.2850	.2827	.2803	.2780	.2756	.2732	.2709	.2685
0.9	.2661	.2637	.2613	.2589	.2565	.2541	.2516	.2492	.2468	.2444
1.0	.2420	.2396	.2371	.2347	.2323	.2299	.2275	.2251	.2227	.2203
1.1	.2179	.2155	.2131	.2107	.2083	.2059	.2036	.2012	.1989	.1965
1.2	.1942	.1919	.1895	.1872	.1849	.1826	.1804	.1781	.1758	.1736
1.3	.1714	.1691	.1669	.1647	.1626	.1604	.1582	.1561	.1539	.1518
1.4	.1497	.1476	.1456	.1435	.1415	.1394	.1374	.1354	.1334	.1315
1.5	.1295	.1276	.1257	.1238	.1219	.1200	.1182	.1163	.1145	.1127
1.6	.1109	.1092	.1074	.1057	.1040	.1023	.1006	.0989	.0973	.0957
1.7	.0940	.0925	.0909	.0893	.0878	.0863	.0848	.0833	.0818	.0804
1.8	.0790	.0775	.0761	.0748	.0734	.0721	.0707	.0694	.0681	.0669
1.9	.0656	.0644	.0632	.0620	.0608	.0596	.0584	.0573	.0562	.0551
2.0	.0540	.0529	.0519	.0508	.0498	.0488	.0478	.0468	.0459	.0449
2.1	.0440	.0431	.0422	.0413	.0404	.0396	.0387	.0379	.0371	.0363
2.2	.0355	.0347	.0339	.0332	.0325	.0317	.0310	.0303	.0297	.0290
2.3	.0283	.0277	.0270	.0264	.0258	.0252	.0246	.0241	.0235	.0229
2.4	.0224	.0219	.0213	.0208	.0203	.0198	.0194	.0189	.0184	.0180
2.5	.0175	.0171	.0167	.0163	.0158	.0154	.0151	.0147	.0143	.0139
2.6	.0136	.0132	.0129	.0126	.0122	.0119	.0116	.0113	.0110	.0107
2.7	.0104	.0101	.0099	.0096	.0093	.0091	.0088	.0086	.0084	.0081
2.8	.0079	.0077	.0075	.0073	.0071	.0069	.0067	.0065	.0063	.0061
2.9	.0060	.0058	.0056	.0055	.0053	.0051	.0050	.0048	.0047	.0046
3.0	.0044	.0043	.0042	.0040	.0039	.0038	.0037	.0036	.0035	.0034
3.1	.0033	.0032	.0031	.0030	.0029	.0028	.0027	.0026	.0025	.0025
3.2	.0024	.0023	.0022	.0022	.0021	.0020	.0020	.0019	.0018	.0018
3.3	.0017	.0017	.0016	.0016	.0015	.0015	.0014	.0014	.0013	.0013
3.4	.0012	.0012	.0012	.0011	.0011	.0010	.0010	.0010	.0009	.0009
3.5	.0009	.0008	.0008	.0008	.0008	.0007	.0007	.0007	.0007	.0006
3.6	.0006	.0006	.0006	.0005	.0005	.0005	.0005	.0005	.0005	.0004
3.7	.0004	.0004	.0004	.0004	.0004	.0004	.0003	.0003	.0003	.0003
3.8	.0003	.0003	.0003	.0003	.0003	.0002	.0002	.0002	.0002	.0002
3.9	.0002	.0002	.0002	.0002	.0002	.0002	.0002	.0002	.0001	.0001
4.0	.0001	.0001	.0001	.0001	.0001	.0001	.0001	.0001	.0001	.0001
4.1	.0001	.0001	.0001	.0001	.0001	.0001	.0001	.0001	.0001	.0001
4.2	.0001	.0001	.0001	.0001	.0000	.0000	.0000	.0000	.0000	.0000

Table VIII.   Random Digits

10	09	73	25	33	76	52	01	35	86	34	67	35	48	76	80	95	90	91	17	39	29	27	49	45
37	54	20	48	05	64	89	47	42	96	24	80	52	40	37	20	63	61	04	02	00	82	29	16	65
08	42	26	89	53	19	64	50	93	03	23	20	90	25	60	15	95	33	47	64	35	08	03	36	06
99	01	90	25	29	09	37	67	07	15	38	31	13	11	65	88	67	67	43	97	04	43	62	76	59
12	80	79	99	70	80	15	73	61	47	64	03	23	66	53	98	95	11	68	77	12	17	17	68	33
66	06	57	47	17	34	07	27	68	50	36	69	73	61	70	65	81	33	98	85	11	19	92	91	70
31	06	01	08	05	45	57	18	24	06	35	30	34	26	14	86	79	90	74	39	23	40	30	97	32
85	26	97	76	02	02	05	16	56	92	68	66	57	48	18	73	05	38	52	47	18	62	38	85	79
63	57	33	21	35	05	32	54	70	48	90	55	35	75	48	28	46	82	87	09	83	49	12	56	24
73	79	64	57	53	03	52	96	47	78	35	80	83	42	82	60	93	52	03	44	35	27	38	84	35
98	52	01	77	67	14	90	56	86	07	22	10	94	05	58	60	97	09	34	33	50	50	07	39	98
11	80	50	54	31	39	80	82	77	32	50	72	56	82	48	29	40	52	42	01	52	77	56	78	51
83	45	29	96	34	06	28	89	80	83	13	74	67	00	78	18	47	54	06	10	68	71	17	78	17
88	68	54	02	00	86	50	75	84	01	36	76	66	79	51	90	36	47	64	93	29	60	91	10	62
99	59	46	73	48	87	51	76	49	69	91	82	60	89	28	93	78	56	13	68	23	47	83	41	13
65	48	11	76	74	17	46	85	09	50	58	04	77	69	74	73	03	95	71	86	40	21	81	65	44
80	12	43	56	35	17	72	70	80	15	45	31	82	23	74	21	11	57	82	53	14	38	55	37	63
74	35	09	98	17	77	40	27	72	14	43	23	60	02	10	45	52	16	42	37	96	28	60	26	55
69	91	62	68	03	66	25	22	91	48	36	93	68	72	03	76	62	11	39	90	94	40	05	64	18
09	89	32	05	05	14	22	56	85	14	46	42	75	67	88	96	29	77	88	22	54	38	21	45	98
91	49	91	45	23	68	47	92	76	86	46	16	28	35	54	94	75	08	99	23	37	08	92	00	48
80	33	69	45	98	26	94	03	68	58	70	29	73	41	35	53	14	03	33	40	42	05	08	23	41
44	10	48	19	49	85	15	74	79	54	32	97	92	65	75	57	60	04	08	81	22	22	20	64	13
12	55	07	37	42	11	10	00	20	40	12	86	07	46	97	96	64	48	94	39	28	70	72	58	15
63	60	64	93	29	16	50	53	44	84	40	21	95	25	63	43	65	17	70	82	07	20	73	17	90
61	19	69	04	46	26	45	74	77	74	51	92	43	37	29	65	39	45	95	93	42	58	26	05	27
15	47	44	52	66	95	27	07	99	53	59	36	78	38	48	82	39	61	01	18	33	21	15	94	66
94	55	72	85	73	67	89	75	43	87	54	62	24	44	31	91	19	04	25	92	92	92	74	59	73
42	48	11	62	13	97	34	40	87	21	16	86	84	87	67	03	07	11	20	29	25	70	14	66	70
23	52	37	83	17	73	20	88	98	37	68	93	59	14	16	26	25	22	96	63	05	52	28	25	62
04	49	35	24	94	75	24	63	38	24	45	86	25	10	25	61	96	27	93	35	65	33	71	24	72
00	54	99	76	54	64	05	18	81	59	96	11	96	38	96	54	69	28	23	91	23	28	72	95	29
35	96	31	53	07	26	89	80	93	54	33	35	13	54	62	77	97	45	00	24	90	10	33	93	33
59	80	80	83	91	45	42	72	68	42	83	60	94	97	00	13	02	12	48	92	78	56	52	01	06
46	05	88	52	36	01	39	09	22	86	77	28	14	40	77	93	91	08	36	47	70	61	74	29	41
32	17	90	05	97	87	37	92	52	41	05	56	70	70	07	86	74	31	71	57	85	39	41	18	38
69	23	46	14	06	20	11	74	52	04	15	95	66	00	00	18	74	39	24	23	97	11	89	63	38
19	56	54	14	30	01	75	87	53	79	40	41	92	15	85	66	67	43	68	06	84	96	28	52	07
45	15	51	49	38	19	47	60	72	46	43	66	79	45	43	59	04	79	00	33	20	82	66	95	41
94	86	43	19	94	36	16	81	08	51	34	88	88	15	53	01	54	03	54	56	05	01	45	11	76
98	08	62	48	26	45	24	02	84	04	44	99	90	88	96	39	09	47	34	07	35	44	13	18	80
33	18	51	62	32	41	94	15	09	49	89	43	54	85	81	88	69	54	19	94	37	54	87	30	43
80	95	10	04	06	96	38	27	07	74	20	15	12	33	87	25	01	62	52	98	94	62	46	11	71
79	75	24	91	40	71	96	12	82	96	69	86	10	25	91	74	85	22	05	39	00	38	75	95	79
18	63	33	25	37	98	14	50	65	71	31	01	02	46	74	05	45	56	14	27	77	93	89	19	36
74	02	94	39	02	77	55	73	22	70	97	79	01	71	19	52	52	75	80	21	80	81	45	17	48
54	17	84	56	11	80	99	33	71	43	05	33	51	29	69	56	12	71	92	55	36	04	09	03	24
11	66	44	98	83	52	07	98	48	27	59	38	17	15	39	09	97	33	34	40	88	46	12	33	56
48	32	47	79	28	31	24	96	47	10	02	29	53	68	70	32	30	75	75	46	15	02	00	99	94
69	07	49	41	38	87	63	79	19	76	35	58	40	44	01	10	51	82	16	15	01	84	87	69	38

This table is reproduced here by permission from The RAND Corporation, *A Million Random Digits*. The Free Press, New York. 1955.

Table   IX.   The Transformation of r to $w = \frac{1}{2} \log \left[(1 + r)/(1 - r)\right]$

r	r (3rd decimal)					r	r (3rd decimal)				
	.000	**.002**	**.004**	**.006**	**.008**		**.000**	**.002**	**.004**	**.006**	**.008**
.00	.0000	.0020	.0040	.0060	.0080	**.35**	.3654	.3677	.3700	.3723	.3746
1	.0100	.0120	.0140	.0160	.0180	6	.3769	.3792	.3815	.3838	.3861
2	.0200	.0220	.0240	.0260	.0280	7	.3884	.3907	.3931	.3954	.3977
3	.0300	.0320	.0340	.0360	.0380	8	.4001	.4024	.4047	.4071	.4094
4	.0400	.0420	.0440	.0460	.0480	9	.4118	.4142	.4165	.4189	.4213
.05	.0500	.0520	.0541	.0561	.0581	**.40**	.4236	.4260	.4284	.4308	.4332
6	.0601	.0621	.0641	.0661	.0681	1	.4356	.4380	.4404	.4428	.4453
7	.0701	.0721	.0741	.0761	.0782	2	.4477	.4501	.4526	.4550	.4574
8	.0802	.0822	.0842	.0862	.0882	3	.4599	.4624	.4648	.4673	.4698
9	.0902	.0923	.0943	.0963	.0983	4	.4722	.4747	.4772	.4797	.4822
.10	.1003	.1024	.1044	.1064	.1084	**.45**	.4847	.4872	.4897	.4922	.4948
1	.1104	.1125	.1145	.1165	.1186	6	.4973	.4999	.5024	.5049	.4075
2	.1206	.1226	.1246	.1267	.1287	7	.5101	.5126	.5152	.5178	.5204
3	.1307	.1328	.1348	.1368	.1389	8	.5230	.5256	.5282	.5308	.5334
4	.1409	.1430	.1450	.1471	.1491	9	.5361	.5387	.5413	.5440	.5466
.15	.1511	.1532	.1552	.1573	.1593	**.50**	.5493	.5520	.5547	.5573	.5600
6	.1614	.1634	.1655	.1676	.1696	1	.5627	.5654	.5682	.5709	.5736
7	.1717	.1737	.1758	.1779	.1799	2	.5763	.5791	.5818	.5846	.5874
8	.1820	.1841	.1861	.1882	.1903	3	.5901	.5929	.5957	.5985	.6013
9	.1923	.1944	.1965	.1986	.2007	4	.6042	.6070	.6098	.6127	.6155
.20	.2027	.2048	.2069	.2090	.2111	**.55**	.6194	.6213	.6241	.6270	.6299
1	.2132	.2153	.2174	.2195	.2216	6	.6328	.6358	.6387	.6416	.6446
2	.2237	.2258	.2279	.2300	.2321	7	.6475	.6505	.6535	.6565	.6595
3	.2342	.2363	.2384	.2405	.2427	8	.6625	.6655	.6685	.6716	.6746
4	.2448	.2469	.2490	.2512	.2533	9	.6777	.6807	.6838	.6869	.6900
.25	.2554	.2575	.2597	.2618	.2640	**.60**	.6931	.6963	.6994	.7026	.7057
6	.2661	.2683	.2704	.2726	.2747	1	.7089	.7121	.7153	.7185	.7218
7	.2769	.2790	.2812	.2833	.2855	2	.7250	.7283	.7315	.7348	.7381
8	.2877	.2899	.2920	.2942	.2964	3	.7414	.7447	.7481	.7514	.7548
9	.2986	.3008	.3029	.3051	.3073	4	.7582	.7616	.7650	.7684	.7718
.30	.3095	.3117	.3139	.3161	.3183	**.65**	.7753	.7788	.7823	.7858	.7893
1	.3205	.3228	.3250	.3272	.3294	6	.7928	.7964	.7999	.8035	.8071
2	.3316	.3339	.3361	.3383	.3406	7	.8107	.8144	.8180	.8217	.8254
3	.3428	.3451	.3473	.3496	.3518	8	.8291	.8328	.8366	.8404	.8441
4	.3541	.3564	.3586	.3609	.3632	9	.8480	.8518	.8556	.8595	.8634

Table IX   (continued)

r	r (3rd decimal)					r	r (3rd decimal)				
	.000	.002	.004	.006	.008		.000	.002	.004	.006	.008
.70	.8673	.8712	.8752	.8792	.8832	.85	1.256	1.263	1.271	1.278	1.286
1	.8872	.8912	.8953	.8994	.9035	6	1.293	1.301	1.309	1.317	1.325
2	.9076	.9118	.9160	.9202	.9245	7	1.333	1.341	1.350	1.358	1.367
3	.9287	.9330	.9373	.9417	.9461	8	1.376	1.385	1.394	1.403	1.412
4	.9505	.9549	.9549	.9639	.9684	9	1.422	1.432	1.442	1.452	1.462
.75	0.973	0.978	0.982	0.987	0.991	.90	1.472	1.483	1.494	1.505	1.516
6	0.996	1.001	1.006	1.011	1.015	1	1.528	1.539	1.551	1.564	1.576
7	1.020	1.025	1.030	1.035	1.040	2	1.589	1.602	1.616	1.630	1.644
8	1.045	1.050	1.056	1.061	1.066	3	1.658	1.673	1.689	1.705	1.721
9	1.071	1.077	1.082	1.088	1.093	4	1.738	1.756	1.774	1.792	1.812
.80	1.099	1.104	1.110	1.116	1.121	.95	1.832	1.853	1.874	1.897	1.921
1	1.127	1.133	1.139	1.145	1.151	6	1.946	1.972	2.000	2.029	2.060
2	1.157	1.163	1.169	1.175	1.182	7	2.092	2.127	2.165	2.205	2.249
3	1.188	1.195	1.201	1.208	1.214	8	2.298	2.351	2.410	2.477	2.555
4	1.221	1.228	1.235	1.242	1.249	9	2.647	2.759	2.903	3.106	3.453

This table is abridged from Table 14 of the *Biometrika Tables for Statisticians*, Vol. 1, edited by Pearson and Hartley. Used with the permission of E. S. Pearson and the trustees of *Biometrika*.

Table X.   Unit Normal Linear Loss Integral

This table gives values of the unit normal linear loss integral,

$$L_N(u) = \int_u^\infty (z - u)\, f(z)\, dz,$$

for $u = 0(.01)4.99$, where $f(z)$ is a standard normal density function. For $u < 0$, take $L_N(u) = -u + L_N(-u)$.

Examples: $L_N(3.57) = .0^4 44417 = .00004417$,
and $L_N(-1.04) = 1.04 + L_N(1.04) = 1.04 + .07716 = 1.11716.$

u	.00	.01	.02	.03	.04	.05	.06	.07	.08	.09
.0	.3989	.3940	.3890	.3841	.3793	.3744	.3697	.3649	.3602	.3556
.1	.3509	.3464	.3418	.3373	.3328	.3284	.3240	.3197	.3154	.3111
.2	.3069	.3027	.2986	.2944	.2904	.2863	.2824	.2784	.2745	.2706
.3	.2668	.2630	.2592	.2555	.2518	.2481	.2445	.2409	.2374	.2339
.4	.2304	.2270	.2236	.2203	.2169	.2137	.2104	.2072	.2040	.2009
.5	.1978	.1947	.1917	.1887	.1857	.1828	.1799	.1771	.1742	.1714
.6	.1687	.1659	.1633	.1606	.1580	.1554	.1528	.1503	.1478	.1453
.7	.1429	.1405	.1381	.1358	.1334	.1312	.1289	.1267	.1245	.1223
.8	.1202	.1181	.1160	.1140	.1120	.1100	.1080	.1061	.1042	.1023
.9	.1004	.09860	.09680	.09503	.09328	.09156	.08986	.08819	.08654	.08491
1.0	.08332	.08174	.08019	.07866	.07716	.07568	.07422	.07279	.07138	.06999
1.1	.06862	.06727	.06595	.06465	.06336	.06210	.06086	.05964	.05844	.05726
1.2	.05610	.05496	.05384	.05274	.05165	.05059	.04954	.04851	.04750	.04650
1.3	.04553	.04457	.04363	.04270	.04179	.04090	.04002	.03916	.03831	.03748
1.4	.03667	.03587	.03508	.03431	.03356	.03281	.03208	.03137	.03067	.02998
1.5	.02931	.02865	.02800	.02736	.02674	.02612	.02552	.02494	.02436	.02380
1.6	.02324	.02270	.02217	.02165	.02114	.02064	.02015	.01967	.01920	.01874
1.7	.01829	.01785	.01742	.01699	.01658	.01617	.01578	.01539	.01501	.01464
1.8	.01428	.01392	.01357	.01323	.01290	.01257	.01226	.01195	.01164	.01134
1.9	.01105	.01077	.01049	.01022	$.0^2 9957$	$.0^2 9698$	$.0^2 9445$	$.0^2 9198$	$.0^2 8957$	$.0^2 8721$

Table X   (continued)

u	.00	.01	.02	.03	.04	.05	.06	.07	.08	.09
2.0	$.0^{2}8491$	$.0^{2}8266$	$.0^{2}8046$	$.0^{2}7832$	$.0^{2}7623$	$.0^{2}7418$	$.0^{2}7219$	$.0^{2}7024$	$.0^{2}6835$	$.0^{2}6649$
2.1	$.0^{2}6468$	$.0^{2}6292$	$.0^{2}6120$	$.0^{2}5952$	$.0^{2}5788$	$.0^{2}5628$	$.0^{2}5472$	$.0^{2}5320$	$.0^{2}5172$	$.0^{2}5028$
2.2	$.0^{2}4887$	$.0^{2}4750$	$.0^{2}4616$	$.0^{2}4486$	$.0^{2}4358$	$.0^{2}4235$	$.0^{2}4114$	$.0^{2}3996$	$.0^{2}3882$	$.0^{2}3770$
2.3	$.0^{2}3662$	$.0^{2}3556$	$.0^{2}3453$	$.0^{2}3352$	$.0^{2}3255$	$.0^{2}3159$	$.0^{2}3067$	$.0^{2}2977$	$.0^{2}2889$	$.0^{2}2804$
2.4	$.0^{2}2720$	$.0^{2}2640$	$.0^{2}2561$	$.0^{2}2484$	$.0^{2}2410$	$.0^{2}2337$	$.0^{2}2267$	$.0^{2}2199$	$.0^{2}2132$	$.0^{2}2067$
2.5	$.0^{2}2004$	$.0^{2}1943$	$.0^{2}1883$	$.0^{2}1826$	$.0^{2}1769$	$.0^{2}1715$	$.0^{2}1662$	$.0^{2}1610$	$.0^{2}1560$	$.0^{2}1511$
2.6	$.0^{2}1464$	$.0^{2}1418$	$.0^{2}1373$	$.0^{2}1330$	$.0^{2}1288$	$.0^{2}1247$	$.0^{2}1207$	$.0^{2}1169$	$.0^{2}1132$	$.0^{2}1095$
2.7	$.0^{2}1060$	$.0^{2}1026$	$.0^{3}9928$	$.0^{3}9607$	$.0^{3}9295$	$.0^{3}8992$	$.0^{3}8699$	$.0^{3}8414$	$.0^{3}8138$	$.0^{3}7870$
2.8	$.0^{3}7611$	$.0^{3}7359$	$.0^{3}7115$	$.0^{3}6879$	$.0^{3}6650$	$.0^{3}6428$	$.0^{3}6213$	$.0^{3}6004$	$.0^{3}5802$	$.0^{3}5606$
2.9	$.0^{3}5417$	$.0^{3}5233$	$.0^{3}5055$	$.0^{3}4883$	$.0^{3}4716$	$.0^{3}4555$	$.0^{3}4398$	$.0^{3}4247$	$.0^{3}4101$	$.0^{3}3959$
3.0	$.0^{3}3822$	$.0^{3}3689$	$.0^{3}3560$	$.0^{3}3436$	$.0^{3}3316$	$.0^{3}3199$	$.0^{3}3087$	$.0^{3}2978$	$.0^{3}2873$	$.0^{3}2771$
3.1	$.0^{3}2673$	$.0^{3}2577$	$.0^{3}2485$	$.0^{3}2396$	$.0^{3}2311$	$.0^{3}2227$	$.0^{3}2147$	$.0^{3}2070$	$.0^{3}1995$	$.0^{3}1922$
3.2	$.0^{3}1852$	$.0^{3}1785$	$.0^{3}1720$	$.0^{3}1657$	$.0^{3}1596$	$.0^{3}1537$	$.0^{3}1480$	$.0^{3}1426$	$.0^{3}1373$	$.0^{3}1322$
3.3	$.0^{3}1273$	$.0^{3}1225$	$.0^{3}1179$	$.0^{3}1135$	$.0^{3}1093$	$.0^{3}1051$	$.0^{3}1012$	$.0^{4}9734$	$.0^{4}9365$	$.0^{4}9009$
3.4	$.0^{4}8666$	$.0^{4}8335$	$.0^{4}8016$	$.0^{4}7709$	$.0^{4}7413$	$.0^{4}7127$	$.0^{4}6852$	$.0^{4}6587$	$.0^{4}6331$	$.0^{4}6085$
3.5	$.0^{4}5848$	$.0^{4}5620$	$.0^{4}5400$	$.0^{4}5188$	$.0^{4}4984$	$.0^{4}4788$	$.0^{4}4599$	$.0^{4}4417$	$.0^{4}4242$	$.0^{4}4073$
3.6	$.0^{4}3911$	$.0^{4}3755$	$.0^{4}3605$	$.0^{4}3460$	$.0^{4}3321$	$.0^{4}3188$	$.0^{4}3059$	$.0^{4}2935$	$.0^{4}2816$	$.0^{4}2702$
3.7	$.0^{4}2592$	$.0^{4}2486$	$.0^{4}2385$	$.0^{4}2287$	$.0^{4}2193$	$.0^{4}2103$	$.0^{4}2016$	$.0^{4}1933$	$.0^{4}1853$	$.0^{4}1776$
3.8	$.0^{4}1702$	$.0^{4}1632$	$.0^{4}1563$	$.0^{4}1498$	$.0^{4}1435$	$.0^{4}1375$	$.0^{4}1317$	$.0^{4}1262$	$.0^{4}1208$	$.0^{4}1157$
3.9	$.0^{4}1108$	$.0^{4}1061$	$.0^{4}1016$	$.0^{5}9723$	$.0^{5}9307$	$.0^{5}8908$	$.0^{5}8525$	$.0^{5}8158$	$.0^{5}7806$	$.0^{5}7469$
4.0	$.0^{5}7145$	$.0^{5}6835$	$.0^{5}6538$	$.0^{5}6253$	$.0^{5}5980$	$.0^{5}5718$	$.0^{5}5468$	$.0^{5}5227$	$.0^{5}4997$	$.0^{5}4777$
4.1	$.0^{5}4566$	$.0^{5}4364$	$.0^{5}4170$	$.0^{5}3985$	$.0^{5}3807$	$.0^{5}3637$	$.0^{5}3475$	$.0^{5}3319$	$.0^{5}3170$	$.0^{5}3027$
4.2	$.0^{5}2891$	$.0^{5}2760$	$.0^{5}2635$	$.0^{5}2516$	$.0^{5}2402$	$.0^{5}2292$	$.0^{5}2188$	$.0^{5}2088$	$.0^{5}1992$	$.0^{5}1901$
4.3	$.0^{5}1814$	$.0^{5}1730$	$.0^{5}1650$	$.0^{5}1574$	$.0^{5}1501$	$.0^{5}1431$	$.0^{5}1365$	$.0^{5}1301$	$.0^{5}1241$	$.0^{5}1183$
4.4	$.0^{5}1127$	$.0^{5}1074$	$.0^{5}1024$	$.0^{6}9756$	$.0^{6}9296$	$.0^{6}8857$	$.0^{6}8437$	$.0^{6}8037$	$.0^{6}7655$	$.0^{6}7290$
4.5	$.0^{6}6942$	$.0^{6}6610$	$.0^{6}6294$	$.0^{6}5992$	$.0^{6}5704$	$.0^{6}5429$	$.0^{6}5167$	$.0^{6}4917$	$.0^{6}4679$	$.0^{6}4452$
4.6	$.0^{6}4236$	$.0^{6}4029$	$.0^{6}3833$	$.0^{6}3645$	$.0^{6}3467$	$.0^{6}3297$	$.0^{6}3135$	$.0^{6}2981$	$.0^{6}2834$	$.0^{6}2694$
4.7	$.0^{6}2560$	$.0^{6}2433$	$.0^{6}2313$	$.0^{6}2197$	$.0^{6}2088$	$.0^{6}1984$	$.0^{6}1884$	$.0^{6}1790$	$.0^{6}1700$	$.0^{6}1615$
4.8	$.0^{6}1533$	$.0^{6}1456$	$.0^{6}1382$	$.0^{6}1312$	$.0^{6}1246$	$.0^{6}1182$	$.0^{6}1122$	$.0^{6}1065$	$.0^{6}1011$	$.0^{7}9588$
4.9	$.0^{7}9096$	$.0^{7}8629$	$.0^{7}8185$	$.0^{7}7763$	$.0^{7}7362$	$.0^{7}6982$	$.0^{7}6620$	$.0^{7}6276$	$.0^{7}5950$	$.0^{7}5640$

Table XI.   Fractiles of D in the Kolmogorov-Smirnov One-Sample Test

n	$F(D)$				
	.80	.85	.90	.95	.99
1	.900	.925	.950	.975	.995
2	.684	.726	.776	.842	.929
3	.565	.597	.642	.708	.828
4	.494	.525	.564	.624	.733
5	.446	.474	.510	.565	.669
6	.410	.436	.470	.521	.618
7	.381	.405	.438	.486	.577
8	.358	.381	.411	.457	.543
9	.339	.360	.388	.432	.514
10	.322	.342	.368	.410	.490
11	.307	.326	.352	.391	.468
12	.295	.313	.338	.375	.450
13	.284	.302	.325	.361	.433
14	.274	.292	.314	.349	.418
15	.266	.283	.304	.338	.404
16	.258	.274	.295	.328	.392
17	.250	.266	.280	.318	.381
18	.244	.259	.278	.309	.371
19	.237	.252	.272	.301	.363
20	.231	.246	.264	.294	.356
25	.21	.22	.24	.27	.32
30	.19	.20	.22	.24	.29
35	.18	.19	.21	.23	.27
Over 35	$1.07/\sqrt{n}$	$1.14/\sqrt{n}$	$1.22/\sqrt{n}$	$1.36/\sqrt{n}$	$1.63/\sqrt{n}$

This table is reproduced by permission, adapted from F. J. Massey, "The Kolmogorov-Smirnov Test for Goodness of Fit," *Journal of the American Statistical Association*, **46**, 70 (1951).

Table XII.   Binomial Coefficients, $\binom{n}{r}$

r \\ n	0	1	2	3	4	5	6	7	8	9	10
1	1	1									
2	1	2	1								
3	1	3	3	1							
4	1	4	6	4	1						
5	1	5	10	10	5	1					
6	1	6	15	20	15	6	1				
7	1	7	21	35	35	21	7	1			
8	1	8	28	56	70	56	28	8	1		
9	1	9	36	84	126	126	84	36	9	1	
10	1	10	45	120	210	252	210	120	45	10	1
11	1	11	55	165	330	462	462	330	165	55	11
12	1	12	66	220	495	792	924	792	495	220	66
13	1	13	78	286	715	1287	1716	1716	1287	715	286
14	1	14	91	364	1001	2002	3003	3432	3003	2002	1001
15	1	15	105	455	1365	3003	5005	6435	6435	5005	3003
16	1	16	120	560	1820	4368	8008	11440	12870	11440	8008
17	1	17	136	680	2380	6188	12376	19448	24310	24310	19448
18	1	18	153	816	3060	8568	18564	31824	43758	48620	43758
19	1	19	171	969	3876	11628	27132	50388	75582	92378	92378
20	1	20	190	1140	4845	15504	38760	77520	125970	167960	184756

Table XIII.   Factorials of Integers

n	$n!$	n	$n!$
1	1	26	4.03291×10^{26}
2	2	27	1.08889×10^{28}
3	6	28	3.04888×10^{29}
4	24	29	8.84176×10^{30}
5	120	30	2.65253×10^{32}
6	720	31	8.22284×10^{33}
7	5040	32	2.63131×10^{35}
8	40320	33	8.68332×10^{36}
9	362880	34	2.95233×10^{38}
10	3.62880×10^{6}	35	1.03331×10^{40}
11	3.99168×10^{7}	36	3.71993×10^{41}
12	4.79002×10^{8}	37	1.37638×10^{43}
13	6.22702×10^{9}	38	5.23023×10^{44}
14	8.71783×10^{10}	39	2.03979×10^{46}
15	1.30767×10^{12}	40	8.15915×10^{47}
16	2.09228×10^{13}	41	3.34525×10^{49}
17	3.55687×10^{14}	42	1.40501×10^{51}
18	6.40327×10^{15}	43	6.04153×10^{52}
19	1.21645×10^{17}	44	2.65827×10^{54}
20	2.43290×10^{18}	45	1.19622×10^{56}
21	5.10909×10^{19}	46	5.50262×10^{57}
22	1.12400×10^{21}	47	2.58623×10^{59}
23	2.58520×10^{22}	48	1.24139×10^{61}
24	6.20448×10^{23}	49	6.08282×10^{62}
25	1.55112×10^{25}	50	3.04141×10^{64}

Table XIV. Powers and Roots

n	n^2	$\sqrt{n}$	$\sqrt{10n}$
1	1	1.00 000	3.16 228
2	4	1.41 421	4.47 214
3	9	1.73 205	5.47 723
4	16	2.00 000	6.32 456
5	25	2.23 607	7.07 107
6	36	2.44 949	7.74 597
7	49	2.64 575	8.36 660
8	64	2.82 843	8.94 427
9	81	3.00 000	9.48 683
10	100	3.16 228	10.00 00
11	121	3.31 662	10.48 81
12	144	3.46 410	10.95 45
13	169	3.60 555	11.40 18
14	196	3.74 166	11.83 22
15	225	3.87 298	12.24 74
16	256	4.00 000	12.64 91
17	289	4.12 311	13.03 84
18	324	4.24 264	13.41 64
19	361	4.35 890	13.78 40
20	400	4.47 214	14.14 21
21	441	4.58 258	14.49 14
22	484	4.69 042	14.83 24
23	529	4.79 583	15.16 58
24	576	4.89 898	15.49 19
25	625	5.00 000	15.81 14
26	676	5.09 902	16.12 45
27	729	5.19 615	16.43 17
28	784	5.29 150	16.73 32
29	841	5.38 516	17.02 94
30	900	5.47 723	17.32 05
31	961	5.56 776	17.60 68
32	1 024	5.65 685	17.88 85
33	1 089	5.74 456	18.16 59
34	1 156	5.83 095	18.43 91
35	1 225	5.91 608	18.70 83
36	1 296	6.00 000	18.97 37
37	1 369	6.08 276	19.23 54
38	1 444	6.16 441	19.49 36
39	1 521	6.24 500	19.74 84
40	1 600	6.32 456	20.00 00
41	1 681	6.40 312	20.24 85
42	1 764	6.48 074	20.49 39
43	1 849	6.55 744	20.73 64
44	1 936	6.63 325	20.97 62
45	2 025	6.70 820	21.21 32
46	2 116	6.78 233	21.44 76
47	2 209	6.85 565	21.67 95
48	2 304	6.92 820	21.90 89
49	2 401	7.00 000	22.13 59
50	2 500	7.07 107	22.36 07
n	n^2	$\sqrt{n}$	$\sqrt{10n}$

n	n^2	$\sqrt{n}$	$\sqrt{10n}$
50	2 500	7.07 107	22.36 07
51	2 601	7.14 143	22.58 32
52	2 704	7.21 110	22.80 35
53	2 809	7.28 011	23.02 17
54	2 916	7.34 847	23.23 79
55	3 025	7.41 620	23.45 21
56	3 136	7.48 331	23.66 43
57	3 249	7.54 983	23.87 47
58	3 364	7.61 577	24.08 32
59	3 481	7.68 115	24.28 99
60	3 600	7.74 597	24.49 49
61	3 721	7.81 025	24.69 82
62	3 844	7.87 401	24.89 98
63	3 969	7.93 725	25.09 98
64	4 096	8.00 000	25.29 82
65	4 225	8.06 226	25.49 51
66	4 356	8.12 404	25.69 05
67	4 489	8.18 535	25.88 44
68	4 624	8.24 621	26.07 68
69	4 761	8.30 662	26.26 79
70	4 900	8.36 660	26.45 75
71	5 041	8.42 615	26.64 58
72	5 184	8.48 528	26.83 28
73	5 329	8.54 400	27.01 85
74	5 476	8.60 233	27.20 29
75	5 625	8.66 025	27.38 61
76	5 776	8.71 780	27.56 81
77	5 929	8.77 496	27.74 89
78	6 084	8.83 176	27.92 85
79	6 241	8.88 819	28.10 69
80	6 400	8.94 427	28.28 43
81	6 561	9.00 000	28.46 05
82	6 724	9.05 539	28.63 56
83	6 889	9.11 043	28.80 97
84	7 056	9.16 515	28.98 28
85	7 225	9.21 954	29.15 48
86	7 396	9.27 362	29.32 58
87	7 569	9.32 738	29.49 58
88	7 744	9.38 083	29.66 48
89	7 921	9.43 398	29.83 29
90	8 100	9.48 683	30.00 00
91	8 281	9.53 939	30.16 62
92	8 464	9.59 166	30.33 15
93	8 649	9.64 365	30.49 59
94	8 836	9.69 536	30.65 94
95	9 025	9.74 679	30.82 21
96	9 216	9.79 796	30.98 39
97	9 409	9.84 886	31.14 48
98	9 604	9.89 949	31.30 50
99	9 801	9.94 987	31.46 43
100	10 000	10.00 000	31.62 28
n	n^2	$\sqrt{n}$	$\sqrt{10n}$

Table XIV (continued)

n	n^2	$\sqrt{n}$	$\sqrt{10n}$	n	n^2	$\sqrt{n}$	$\sqrt{10n}$
100	10 000	10.00 00	31.62 28	**150**	22 500	12.24 74	38.72 98
101	10 201	10.04 99	31.78 05	151	22 801	12.28 82	38.85 87
102	10 404	10.09 95	31.93 74	152	23 104	12.32 88	38.98 72
103	10 609	10.14 89	32.09 36	153	23 409	12.36 93	39.11 52
104	10 816	10.19 80	32.24 90	154	23 716	12.40 97	39.24 28
105	11 025	10.24 70	32.40 37	155	24 025	12.44 99	39.37 00
106	11 236	10.29 56	32.55 76	**156**	24 336	12.49 00	39.49 68
107	11 449	10.34 41	32.71 09	157	24 649	12.53 00	39.62 32
108	11 664	10.39 23	32.86 34	158	24 964	12.56 98	39.74 92
109	11 881	10.44 03	33.01 51	159	25 281	12.60 95	39.87 48
110	12 100	10.48 81	33.16 62	160	25 600	12.64 91	40.00 00
111	12 321	10.53 57	33.31 67	**161**	25 921	12.68 86	40.12 48
112	12 544	10.58 30	33.46 64	162	26 244	12.72 79	40.24 92
113	12 769	10.63 01	33.61 55	163	26 569	12.76 71	40.37 33
114	12 996	10.67 71	33.76 39	164	26 896	12.80 62	40.49 69
115	13 225	10.72 38	33.91 16	165	27 225	12.84 52	40.62 02
116	13 456	10.77 03	34.05 88	**166**	27 556	12.88 41	40.74 31
117	13 689	10.81 67	34.20 53	167	27 889	12.92 28	40.86 56
118	13 924	10.86 28	34.35 11	168	28 224	12.96 15	40.98 78
119	14 161	10.90 87	34.49 64	169	28 561	13.00 00	41.10 96
120	14 400	10.95 45	34.64 10	170	28 900	13.03 84	41.23 11
121	14 641	11.00 00	34.78 51	**171**	29 241	13.07 67	41.35 21
122	14 884	11.04 54	34.92 85	172	29 584	13.11 49	41.47 29
123	15 129	11.09 05	35.07 14	173	29 929	13.15 29	41.59 33
124	15 376	11.13 55	35.21 36	174	30 276	13.19 09	41.71 33
125	15 625	11.18 03	35.35 53	175	30 625	13.22 88	41.83 30
126	15 876	11.22 50	35.49 65	**176**	30 976	13.26 65	41.95 24
127	16 129	11.26 94	35.63 71	177	31 329	13.30 41	42.07 14
128	16 384	11.31 37	35.77 71	178	31 684	13.34 17	42.19 00
129	16 641	11.35 78	35.91 66	179	32 041	13.37 91	42.30 84
130	16 900	11.40 18	36.05 55	180	32 400	13.41 64	42.42 64
131	17 161	11.44 55	36.19 39	**181**	32 761	13.45 36	42.54 41
132	17 424	11.48 91	36.33 18	182	33 124	13.49 07	42.66 15
133	17 689	11.53 26	36.46 92	183	33 489	13.52 77	42.77 85
134	17 956	11.57 58	36.60 60	184	33 856	13.56 47	42.89 52
135	18 225	11.61 90	36.74 23	185	34 225	13.60 15	43.01 16
136	18 496	11.66 19	36.87 82	**186**	34 596	13.63 82	43.12 77
137	18 769	11.70 47	37.01 35	187	34 969	13.67 48	43.24 35
138	19 044	11.74 73	37.14 84	188	35 344	13.71 13	43.35 90
139	19 321	11.78 98	37.28 27	189	35 721	13.74 77	43.47 41
140	19 600	11.83 22	37.41 66	190	36 100	13.78 40	43.58 90
141	19 881	11.87 43	37.55 00	**191**	36 481	13.82 03	43.70 35
142	20 164	11.91 64	37.68 29	192	36 864	13.85 64	43.81 78
143	20 449	11.95 83	37.81 53	193	37 249	13.89 24	43.93 18
144	20 736	12.00 00	37.94 73	194	37 636	13.92 84	44.04 54
145	21 025	12.04 16	38.07 89	195	38 025	13.96 42	44.15 88
146	21 316	12.08 30	38.20 99	**196**	38 416	14.00 00	44.27 19
147	21 609	12.12 44	38.34 06	197	38 809	14.03 57	44.38 47
148	21 904	12.16 55	38.47 08	198	39 204	14.07 12	44.49 72
149	22 201	12.20 66	38.60 05	199	39 601	14.10 67	44.60 94
150	22 500	12.24 74	38.72 98	200	40 000	14.14 21	44.72 14
n	n^2	$\sqrt{n}$	$\sqrt{10n}$	n	n^2	$\sqrt{n}$	$\sqrt{10n}$

Table XIV   (continued)

n	n^2	$\sqrt{n}$	$\sqrt{10n}$	n	n^2	$\sqrt{n}$	$\sqrt{10n}$
200	40 000	14.14 21	44.72 14	**250**	62 500	15.81 14	50.00 00
201	40 401	14.17 74	44.83 30	251	63 001	15.84 30	50.09 99
202	40 804	14.21 27	44.94 44	252	63 504	15.87 45	50.19 96
203	41 209	14.24 78	45.05 55	253	64 009	15.90 60	50.29 91
204	41 616	14.28 29	45.16 64	254	64 516	15.93 74	50.39 84
205	42 025	14.31 78	45.27 69	255	65 025	15.96 87	50.49 75
206	42 436	14.35 27	45.38 72	**256**	65 536	16.00 00	50.59 64
207	42 849	14.38 75	45.49 73	257	66 049	16.03 12	50.69 52
208	43 264	14.42 22	45.60 70	258	66 564	16.06 24	50.79 37
209	43 681	14.45 68	45.71 65	259	67 081	16.09 35	50.89 20
210	44 100	14.49 14	45.82 58	260	67 600	16.12 45	50.99 02
211	44 521	14.52 58	45.93 47	**261**	68 121	16.15 55	51.08 82
212	44 944	14.56 02	46.04 35	262	68 644	16.18 64	51.18 59
213	45 369	14.59 45	46.15 19	263	69 169	16.21 73	51.28 35
214	45 796	14.62 87	46.26 01	264	69 696	16.24 81	51.38 09
215	46 225	14.66 29	46.36 81	265	70 225	16.27 88	51.47 82
216	46 656	14.69 69	46.47 58	**266**	70 756	16.30 95	51.57 52
217	47 089	14.73 09	46.58 33	267	71 289	16.34 01	51.67 20
218	47 524	14.76 48	46.69 05	268	71 824	16.37 07	51.76 87
219	47 961	14.79 86	46.79 74	269	72 361	16.40 12	51.86 52
220	48 400	14.83 24	46.90 42	270	72 900	16.43 17	51.96 15
221	48 841	14.86 61	47.01 06	**271**	73 441	16.46 21	52.05 77
222	49 284	14.89 97	47.11 69	272	73 984	16.49 24	52.15 36
223	49 729	14.93 32	47.22 29	273	74 529	16.52 27	52.24 94
224	50 176	14.96 66	47.32 86	274	75 076	16.55 29	52.34 50
225	50 625	15.00 00	47.43 42	275	75 625	16.58 31	52.44 04
226	51 076	15.03 33	47.53 95	**276**	76 176	16.61 32	52.53 57
227	51 529	15.06 65	47.64 45	277	76 729	16.64 33	52.63 08
228	51 984	15.09 97	47.74 93	278	77 284	16.67 33	52.72 57
229	52 441	15.13 27	47.85 39	279	77 841	16.70 33	52.82 05
230	52 900	15.16 58	47.95 83	280	78 400	16.73 32	52.91 50
231	53 361	15.19 87	48.06 25	**281**	78 961	16.76 31	53.00 94
232	53 824	15.23 15	48.16 64	282	79 524	16.79 29	53.10 37
233	54 289	15.26 43	48.27 01	283	80 089	16.82 26	53.19 77
234	54 756	15.29 71	48.37 35	284	80 656	16.85 23	53.29 17
235	55 225	15.32 97	48.47 68	285	81 225	16.88 19	53.38 54
236	55 696	15.36 23	48.57 98	**286**	81 796	16.91 15	53.47 90
237	56 169	15.39 48	48.68 26	287	82 369	16.94 11	53.57 24
238	56 644	15.42 72	48.78 52	288	82 944	16.97 06	53.66 56
239	57 121	15.45 96	48.88 76	289	83 521	17.00 00	53.75 87
240	57 600	15.49 19	48.98 98	290	84 100	17.02 94	53.85 16
241	58 081	15.52 42	49.09 18	**291**	84 681	17.05 87	53.94 44
242	58 564	15.55 63	49.19 35	292	85 264	17.08 80	54.03 70
243	59 049	15.58 85	49.29 50	293	85 849	17.11 72	54.12 95
244	59 536	15.62 05	49.39 64	294	86 436	17.14 64	54.22 18
245	60 025	15.65 25	49.49 75	295	87 025	17.17 56	54.31 39
246	60 516	15.68 44	49.59 84	**296**	87 616	17.20 47	54.40 59
247	61 009	15.71 62	49.69 91	297	88 209	17.23 37	54.49 77
248	61 504	15.74 80	49.79 96	298	88 804	17.26 27	54.58 94
249	62 001	15.77 97	49.89 99	299	89 401	17.29 16	54.68 09
250	62 500	15.81 14	50.00 00	300	90 000	17.32 05	54.77 23
n	n^2	$\sqrt{n}$	$\sqrt{10n}$	n	n^2	$\sqrt{n}$	$\sqrt{10n}$

Table   XIV   (continued)

n	n²	√n	√10n
300	90 000	17.32 05	54.77 23
301	90 601	17.34 94	54.86 35
302	91 204	17.37 81	54.95 45
303	91 809	17.40 69	55.04 54
304	92 416	17.43 56	55.13 62
305	93 025	17.46 42	55.22 68
306	93 636	17.49 29	55.31 73
307	94 249	17.52 14	55.40 76
308	94 864	17.54 99	55.49 77
309	95 481	17.57 84	55.58 78
310	96 100	17.60 68	55.67 76
311	96 721	17.63 52	55.76 74
312	97 344	17.66 35	55.85 70
313	97 969	17.69 18	55.94 64
314	98 596	17.72 00	56.03 57
315	99 225	17.74 82	56.12 49
316	99 856	17.77 64	56.21 39
317	100 489	17.80 45	56.30 28
318	101 124	17.83 26	56.39 15
319	101 761	17.86 06	56.48 01
320	102 400	17.88 85	56.56 85
321	103 041	17.91 65	56.65 69
322	103 684	17.94 44	56.74 50
323	104 329	17.97 22	56.83 31
324	104 976	18.00 00	56.92 10
325	105 625	18.02 78	57.00 88
326	106 276	18.05 55	57.09 64
327	106 929	18.08 31	57.18 39
328	107 584	18.11 08	57.27 13
329	108 241	18.13 84	57.35 85
330	108 900	18.16 59	57.44 56
331	109 561	18.19 34	57.53 26
332	110 224	18.22 09	57.61 94
333	110 889	18.24 83	57.70 62
334	111 556	18.27 57	57.79 27
335	112 225	18.30 30	57.87 92
336	112 896	18.33 03	57.96 55
337	113 569	18.35 76	58.05 17
338	114 244	18.38 48	58.13 78
339	114 921	18.41 20	58.22 37
340	115 600	18 43 91	58.30 95
341	116 281	18.46 62	58.39 52
342	116 964	18.49 32	58.48 08
343	117 649	18.52 03	58.56 62
344	118 336	18.54 72	58.65 15
345	119 025	18.57 42	58.73 67
346	119 716	18.60 11	58.82 18
347	120 409	18.62 79	58.90 67
348	121 104	18.65 48	58.99 15
349	121 801	18.68 15	59.07 62
350	122 500	18.70 83	59.16 08
n	n²	√n	√10n

n	n²	√n	√10n
350	122 500	18.70 83	59.16 08
351	123 201	18.73 50	59.24 53
352	123 904	18.76 17	59.32 96
353	124 609	18.78 83	59.41 38
354	125 316	18.81 49	59.49 79
355	126 025	18.84 14	59.58 19
356	126 736	18.86 80	59.66 57
357	127 449	18.89 44	59.74 95
358	128 164	18.92 09	59.83 31
359	128 881	18.94 73	59.91 66
360	129 600	18.97 37	60.00 00
361	130 321	19.00 00	60.08 33
362	131 044	19.02 63	60.16 64
363	131 769	19.05 26	60.24 95
364	132 496	19.07 88	60.33 24
365	133 225	19.10 50	60.41 52
366	133 956	19.13 11	60.49 79
367	134 689	19.15 72	60.58 05
368	135 424	19.18 33	60.66 30
369	136 161	19.20 94	60.74 54
370	136 900	19.23 54	60.82 76
371	137 641	19.26 14	60.90 98
372	138 384	19.28 73	60.99 18
373	139 129	19.31 32	61.07 37
374	139 876	19.33 91	61.15 55
375	140 625	19.36 49	61.23 72
376	141 376	19.39 07	61.31 88
377	142 129	19.41 65	61.40 03
378	142 884	19.44 22	61.48 17
379	143 641	19.46 79	61.56 30
380	144 400	19.49 36	61.64 41
381	145 161	19.51 92	61.72 52
382	145 924	19.54 48	61.80 61
383	146 689	19.57 04	61.88 70
384	147 456	19.59 59	61.96 77
385	148 225	19.62 14	62.04 84
386	148 996	19.64 69	62.12 89
387	149 769	19.67 23	62.20 93
388	150 544	19.69 77	62.28 96
389	151 321	19.72 31	62.36 99
390	152 100	19.74 84	62.45 00
391	152 881	19.77 37	62.53 00
392	153 664	19.79 90	62.60 99
393	154 449	19.82 42	62.68 97
394	155 236	19.84 94	62.76 94
395	156 025	19.87 46	62.84 90
396	156 816	19.89 97	62.92 85
397	157 609	19.92 49	63.00 79
398	158 404	19.94 99	63.08 72
399	159 201	19.97 50	63.16 64
400	160 000	20.00 00	63.24 56
n	n²	√n	√10n

Table XIV (continued)

n	n^2	$\sqrt{n}$	$\sqrt{10n}$	n	n^2	$\sqrt{n}$	$\sqrt{10n}$
400	160 000	20.00 00	63.24 56	**450**	202 500	21.21 32	67.08 20
401	160 801	20.02 50	63.32 46	451	203 401	21.23 68	67.15 65
402	161 604	20.04 99	63.40 35	452	204 304	21.26 03	67.23 09
403	162 409	20.07 49	63.48 23	453	205 209	21.28 38	67.30 53
404	163 216	20.09 98	63.56 10	454	206 116	21.30 73	67.37 95
405	164 025	20.12 46	63.63 96	455	207 025	21.33 07	67.45 37
406	164 836	20.14 94	63.71 81	**456**	207 936	21.35 42	67.52 78
407	165 649	20.17 42	63.79 66	457	208 849	21.37 76	67.60 18
408	166 464	20.19 90	63.87 49	458	209 764	21.40 09	67.67 57
409	167 281	20.22 37	63.95 31	459	210 681	21.42 43	67.74 95
410	168 100	20.24 85	64.03 12	460	211 600	21.44 76	67.82 33
411	168 921	20.27 31	64.10 93	**461**	212 521	21.47 09	67.89 70
412	169 744	20.29 78	64.18 72	462	213 444	21.49 42	67.97 06
413	170 569	20.32 24	64.26 51	463	214 369	21.51 74	68.04 41
414	171 396	20.34 70	64.34 28	464	215 296	21.54 07	68.11 75
415	172 225	20.37 15	64.42 05	465	216 225	21.56 39	68.19 09
416	173 056	20.39 61	64.49 81	**466**	217 156	21.58 70	68.26 42
417	173 889	20.42 06	64.57 55	467	218 089	21.61 02	68.33 74
418	174 724	20.44 50	64.65 29	468	219 024	21.63 33	68.41 05
419	175 561	20.46 95	64.73 02	469	219 961	21.65 64	68.48 36
420	176 400	20.49 39	64.80 74	470	220 900	21.67 95	68.55 65
421	177 241	20.51 83	64.88 45	**471**	221 841	21.70 25	68.62 94
422	178 084	20.54 26	64.96 15	472	222 784	21.72 56	68.70 23
423	178 929	20.56 70	65.03 85	473	223 729	21.74 86	68.77 50
424	179 776	20.59 13	65.11 53	474	224 676	21.77 15	68.84 77
425	180 625	20.61 55	65.19 20	475	225 625	21.79 45	68.92 02
426	181 476	20.63 98	65.26 87	**476**	226 576	21.81 74	68.99 28
427	182 329	20.66 40	65.34 52	477	227 529	21.84 03	69.06 52
428	183 184	20.68 82	65.42 17	478	228 484	21.86 32	69.13 75
429	184 041	20.71 23	65.49 81	479	229 441	21.88 61	69.20 98
430	184 900	20.73 64	65.57 44	480	230 400	21.90 89	69.28 20
431	185 761	20.76 05	65.65 06	**481**	231 361	21.93 17	69.35 42
432	186 624	20.78 46	65.72 67	482	232 324	21.95 45	69.42 62
433	187 489	20.80 87	65.80 27	483	233 289	21.97 73	69.49 82
434	188 356	20.83 27	65.87 87	484	234 256	22.00 00	69.57 01
435	189 225	20.85 67	65.95 45	485	235 225	22.02 27	69.64 19
436	190 096	20.88 06	66.03 03	**486**	236 196	22.04 54	69.71 37
437	190 969	20.90 45	66.10 60	487	237 169	22.06 81	69.78 54
438	191 844	20.92 84	66.18 16	488	238 144	22.09 07	69.85 70
439	192 721	20.95 23	66.25 71	489	239 121	22.11 33	69.92 85
440	193 600	20.97 62	66.33 25	490	240 100	22.13 59	70.00 00
441	194 481	21.00 00	66.40 78	**491**	241 081	22.15 85	70.07 14
442	195 364	21.02 38	66.48 31	492	242 064	22.18 11	70.14 27
443	196 249	21.04 76	66.55 82	493	243 049	22.20 36	70.21 40
444	197 136	21.07 13	66.63 33	494	244 036	22.22 61	70.28 51
445	198 025	21.09 50	66.70 83	495	245 025	22.24 86	70.35 62
446	198 916	21.11 87	66.78 32	**496**	246 016	22.27 11	70.42 73
447	199 809	21.14 24	66.85 81	497	247 009	22.29 35	70.49 82
448	200 704	21.16 60	66.93 28	498	248 004	22.31 59	70.56 91
449	201 601	21.18 96	67.00 75	499	249 001	22.33 83	70.63 99
450	202 500	21.21 32	67.08 20	500	250 000	22.36 07	70.71 07
n	n^2	$\sqrt{n}$	$\sqrt{10n}$	n	n^2	$\sqrt{n}$	$\sqrt{10n}$

Table XIV   (continued)

n	n^2	$\sqrt{n}$	$\sqrt{10n}$	n	n^2	$\sqrt{n}$	$\sqrt{10n}$
500	250 000	22.36 07	70.71 07	**550**	302 500	23.45 21	74.16 20
501	251 001	22.38 30	70.78 14	551	303 601	23.47 34	74.22 94
502	252 004	22.40 54	70.85 20	552	304 704	23.49 47	74.29 67
503	253 009	22.42 77	70.92 25	553	305 809	23.51 60	74.36 40
504	254 016	22.44 99	70.99 30	554	306 916	23.53 72	74.43 12
505	255 025	22.47 22	71.06 34	555	308 025	23.55 84	74.49 83
506	256 036	22.49 44	71.13 37	**556**	309 136	23.57 97	74.56 54
507	257 049	22.51 67	71.20 39	557	310 249	23.60 08	74.63 24
508	258 064	22.53 89	71.27 41	558	311 364	23.62 20	74.69 94
509	259 081	22.56 10	71.34 42	559	312 481	23.64 32	74.76 63
510	260 100	22.58 32	71.41 43	560	313 600	23.66 43	74.83 31
511	261 121	22.60 53	71.48 43	**561**	314 721	23.68 54	74.89 99
512	262 144	22.62 74	71.55 42	562	315 844	23.70 65	74.96 67
513	263 169	22.64 95	71.62 40	563	316 969	23.72 76	75.03 33
514	264 196	22.67 16	71.69 38	564	318 096	23.74 87	75.09 99
515	265 225	22.69 36	71.76 35	565	319 225	23.76 97	75.16 65
516	266 256	22.71 56	71.83 31	**566**	320 356	23.79 08	75.23 30
517	267 289	22.73 76	71.90 27	567	321 489	23.81 18	75.29 94
518	268 324	22.75 96	71.97 22	568	322 624	23.83 28	75.36 58
519	269 361	22.78 16	72.04 17	569	323 761	23.85 37	75.43 21
520	270 400	22.80 35	72.11 10	570	324 900	23.87 47	75.49 83
521	271 441	22.82 54	72.18 03	**571**	326 041	23.89 56	75.56 45
522	272 484	22.84 73	72.24 96	572	327 184	23.91 65	75.63 07
523	273 529	22.86 92	72.31 87	573	328 329	23.93 74	75.69 68
524	274 576	22.89 10	72.38 78	574	329 476	23.95 83	75.76 28
525	275 625	22.91 29	72.45 69	575	330 625	23.97 92	75.82 88
526	276 676	22.93 47	72.52 59	**576**	331 776	24.00 00	75.89 47
527	277 729	22.95 65	72.59 48	577	332 929	24.02 08	75.96 05
528	278 784	22.97 83	72.66 36	578	334 084	24.04 16	76.02 63
529	279 841	23.00 00	72.73 24	579	335 241	24.06 24	76.09 20
530	280 900	23.02 17	72.80 11	580	336 400	24.08 32	76.15 77
531	281 961	23.04 34	72.86 97	**581**	337 561	24.10 39	76.22 34
532	283 024	23.06 51	72.93 83	582	338 724	24.12 47	76.28 89
533	284 089	23.08 68	73.00 68	583	339 889	24.14 54	76.35 44
534	285 156	23.10 84	73.07 53	584	341 056	24.16 61	76.41 99
535	286 225	23.13 01	73.14 37	585	342 225	24.18 68	76.48 53
536	287 296	23.15 17	73.21 20	**586**	343 396	24.20 74	76.55 06
537	288 369	23.17 33	73.28 03	587	344 569	24.22 81	76.61 59
538	289 444	23.19 48	73.34 85	588	345 744	24.24 87	76.68 12
539	290 521	23.21 64	73.41 66	589	346 921	24.26 93	76.74 63
540	291 600	23.23 79	73.48 47	590	348 100	24.28 99	76.81 15
541	292 681	23.25 94	73.55 27	**591**	349 281	24.31 05	76.87 65
542	293 764	23.28 09	73.62 06	592	350 464	24.33 11	76.94 15
543	294 849	23.30 24	73.68 85	593	351 649	24.35 16	77.00 65
544	295 936	23.32 38	73.75 64	594	352 836	24.37 21	77.07 14
545	297 025	23.34 52	73.82 41	595	354 025	24.39 26	77.13 62
546	298 116	23.36 66	73.89 18	**596**	355 216	24.41 31	77.20 10
547	299 209	23.38 80	73.95 94	597	356 409	24.43 36	77.26 58
548	300 304	23.40 94	74.02 70	598	357 604	24.45 40	77.33 05
549	301 401	23.43 07	74.09 45	599	358 801	24.47 45	77.39 51
550	302 500	23.45 21	74.16 20	600	360 000	24.49 49	77.45 97
n	n^2	$\sqrt{n}$	$\sqrt{10n}$	n	n^2	$\sqrt{n}$	$\sqrt{10n}$

Table XIV (continued)

n	n^2	$\sqrt{n}$	$\sqrt{10n}$		n	n^2	$\sqrt{n}$	$\sqrt{10n}$
600	360 000	24.49 49	77.45 97		**650**	422 500	25.49 51	80.62 26
601	361 201	24.51 53	77.52 42		651	423 801	25.51 47	80.68 46
602	362 404	24.53 57	77.58 87		652	425 104	25.53 43	80.74 65
603	363 609	24.55 61	77.65 31		653	426 409	25.55 39	80.80 84
604	364 816	24.57 64	77.71 74		654	427 716	25.57 34	80.87 03
605	366 025	24.59 67	77.78 17		655	429 025	25.59 30	80.93 21
606	367 236	24.61 71	77.84 60		**656**	430 336	25.61 25	80.99 38
607	368 449	24.63 74	77.91 02		657	431 649	25.63 20	81.05 55
608	369 664	24.65 77	77.97 44		658	432 964	25.65 15	81.11 72
609	370 881	24.67 79	78.03 85		659	434 281	25.67 10	81.17 88
610	372 100	24.69 82	78.10 25		660	435 600	25.69 05	81.24 04
611	373 321	24.71 84	78.16 65		**661**	436 921	25.70 99	81.30 19
612	374 544	24.73 86	78.23 04		662	438 244	25.72 94	81.36 34
613	375 769	24.75 88	78.29 43		663	439 569	25.74 88	81.42 48
614	376 996	24.77 90	78.35 82		664	440 896	25.76 82	81.48 62
615	378 225	24.79 92	78.42 19		665	442 225	25.78 76	81.54 75
616	379 456	24.81 93	78.48 57		**666**	443 556	25.80 70	81.60 88
617	380 689	24.83 95	78.54 93		667	444 889	25.82 63	81.67 01
618	381 924	24.85 96	78.61 30		668	446 224	25.84 57	81.73 13
619	383 161	24.87 97	78.67 66		669	447 561	25.86 50	81.79 24
620	384 400	24.89 98	78.74 01		670	448 900	25.88 44	81.85 35
621	385 641	24.91 99	78.80 36		**671**	450 241	25.90 37	81.91 46
622	386 884	24.93 99	78.86 70		672	451 584	25.92 30	81.97 56
623	388 129	24.96 00	78.93 03		673	452 929	25.94 22	82.03 66
624	389 376	24.98 00	78.99 37		674	454 276	25.96 15	82.09 75
625	390 625	25.00 00	79.05 69		675	455 625	25.98 08	82.15 84
626	391 876	25.02 00	79.12 02		**676**	456 976	26.00 00	82.21 92
627	393 129	25.04 00	79.18 33		677	458 329	26.01 92	82.28 00
628	394 384	25.05 99	79.24 65		678	459 684	26.03 84	82.34 08
629	395 641	25.07 99	79.30 95		679	461 041	26.05 76	82.40 15
630	396 900	25.09 98	79.37 25		680	462 400	26.07 68	82.46 21
631	398 161	25.11 97	79.43 55		**681**	463 761	26.09 60	82.52 27
632	399 424	25.13 96	79.49 84		682	465 124	26.11 51	82.58 33
633	400 689	25.15 95	79.56 13		683	466 489	26.13 43	82.64 38
634	401 956	25.17 94	79.62 41		684	467 856	26.15 34	82.70 43
635	403 225	25.19 92	79.68 69		685	469 225	26.17 25	82.76 47
636	404 496	25.21 90	79.74 96		**686**	470 596	26.19 16	82.82 51
637	405 769	25.23 89	79.81 23		687	471 969	26.21 07	82.88 55
638	407 044	25.25 87	79.87 49		688	473 344	26.22 98	82.94 58
639	408 321	25.27 84	79.93 75		689	474 721	26.24 88	83.00 60
640	409 600	25.29 82	80.00 00		690	476 100	26.26 79	83.06 62
641	410 881	25.31 80	80.06 25		**691**	477 481	26.28 69	83.12 64
642	412 164	25.33 77	80.12 49		692	478 864	26.30 59	83.18 65
643	413 449	25.35 74	80.18 73		693	480 249	26.32 49	83.24 66
644	414 736	25.37 72	80.24 96		694	481 636	26.34 39	83.30 67
645	416 025	25.39 69	80.31 19		695	483 025	26.36 29	83.36 67
646	417 316	25.41 65	80.37 41		**696**	484 416	26.38 18	83.42 66
647	418 609	25.43 62	80.43 63		697	485 809	26.40 08	83.48 65
648	419 904	25.45 58	80.49 84		698	487 204	26.41 97	83.54 64
649	421 201	25.47 55	80.56 05		699	488 601	26.43 86	83.60 62
650	422 500	25.49 51	80.62 26		700	490 000	26.45 75	83.66 60
n	n^2	$\sqrt{n}$	$\sqrt{10n}$		n	n^2	$\sqrt{n}$	$\sqrt{10n}$

Table   XIV   (continued)

n	n^2	$\sqrt{n}$	$\sqrt{10n}$		n	n^2	$\sqrt{n}$	$\sqrt{10n}$
700	490 000	26.45 75	83.66 60		**750**	562 500	27.38 61	86.60 25
701	491 401	26.47 64	83.72 57		751	564 001	27.40 44	86.66 03
702	492 804	26.49 53	83.78 54		752	565 504	27.42 26	86.71 79
703	494 209	26.51 41	83.84 51		753	567 009	27.44 08	86.77 56
704	495 616	26.53 30	83.90 47		754	568 516	27.45 91	86.83 32
705	497 025	26.55 18	83.96 43		755	570 025	27.47 73	86.89 07
706	498 436	26.57 07	84.02 38		**756**	571 536	27.49 55	86.94 83
707	499 849	26.58 95	84.08 33		757	573 049	27.51 36	87.00 57
708	501 264	26.60 83	84.14 27		758	574 564	27.53 18	87.06 32
709	502 681	26.62 71	84.20 21		759	576 081	27.55 00	87.12 06
710	504 100	26.64 58	84.26 15		760	577 600	27.56 81	87.17 80
711	505 521	26.66 46	84.32 08		**761**	579 121	27.58 62	87.23 53
712	506 944	26.68 33	84.38 01		762	580 644	27.60 43	87.29 26
713	508 369	26.70 21	84.43 93		763	582 169	27.62 25	87.34 99
714	509 796	26.72 08	84.49 85		764	583 696	27.64 05	87.40 71
715	511 225	26.73 95	84.55 77		765	585 225	27.65 86	87.46 43
716	512 656	26.75 82	84.61 68		**766**	586 756	27.67 67	87.52 14
717	514 089	26.77 69	84.67 59		767	588 289	27.69 48	87.57 85
718	515 524	26.79 55	84.73 49		768	589 824	27.71 28	87.63 56
719	516 961	26.81 42	84.79 39		769	591 361	27.73 08	87.69 26
720	518 400	26.83 28	84.85 28		770	592 900	27.74 89	87.74 96
721	519 841	26.85 14	84.91 17		**771**	594 441	27.76 69	87.80 66
722	521 284	26.87 01	84.97 06		772	595 984	27.78 49	87.86 35
723	522 729	26.88 87	85.02 94		773	597 529	27.80 29	87.92 04
724	524 176	26.90 72	85.08 82		774	599 076	27.82 09	87.97 73
725	525 625	26.92 58	85.14 69		775	600 625	27.83 88	88.03 41
726	527 076	26.94 44	85.20 56		**776**	602 176	27.85 68	88.09 09
727	528 529	26.96 29	85.26 43		777	603 729	27.87 47	88.14 76
728	529 984	26.98 15	85.32 29		778	605 284	27.89 27	88.20 43
729	531 441	27.00 00	85.38 15		779	606 841	27.91 06	88.26 10
730	532 900	27.01 85	85.44 00		780	608 400	27.92 85	88.31 76
731	534 361	27.03 70	85.49 85		**781**	609 961	27.94 64	88.37 42
732	535 824	27.05 55	85.55 70		782	611 524	27.96 43	88.43 08
733	537 289	27.07 40	85.61 54		783	613 089	27.98 21	88.48 73
734	538 756	27.09 24	85.67 38		784	614 656	28.00 00	88.54 38
735	540 225	27.11 09	85.73 21		785	616 225	28.01 79	88.60 02
736	541 696	27.12 93	85.79 04		**786**	617 796	28.03 57	88.65 66
737	543 169	27.14 77	85.84 87		787	619 369	28.05 35	88.71 30
738	544 644	27.16 62	85.90 69		788	620 944	28.07 13	88.76 94
739	546 121	27.18 46	85.96 51		789	622 521	28.08 91	88.82 57
740	547 600	27.20 29	86.02 33		790	624 100	28.10 69	88.88 19
741	549 081	27.22 13	86.08 14		**791**	625 681	28.12 47	88.93 82
742	550 564	27.23 97	86.13 94		792	627 264	28.14 25	88.99 44
743	552 049	27.25 80	86.19 74		793	628 849	28.16 03	89.05 05
744	553 536	27.27 64	86.25 54		794	630 436	28.17 80	89.10 67
745	555 025	27.29 47	86.31 34		795	632 025	28.19 57	89.16 28
746	556 516	27.31 30	86.37 13		**796**	633 616	28.21 35	89.21 88
747	558 009	27.33 13	86.42 92		797	635 209	28.23 12	89.27 49
748	559 504	27.34 96	86.48 70		798	636 804	28.24 89	89.33 08
749	561 001	27.36 79	86.54 48		799	638 401	28.26 66	89.38 68
750	562 500	27.38 61	86.60 25		800	640 000	28.28 43	89.44 27
n	n^2	$\sqrt{n}$	$\sqrt{10n}$		n	n^2	$\sqrt{n}$	$\sqrt{10n}$

Table XIV (continued)

n	n^2	$\sqrt{n}$	$\sqrt{10n}$		n	n^2	$\sqrt{n}$	$\sqrt{10n}$
800	640 000	28.28 43	89.44 27		**850**	722 500	29.15 48	92.19 54
801	641 601	28.30 19	89.49 86		851	724 201	29.17 19	92.24 97
802	643 204	28.31 96	89.55 45		852	725 904	29.18 90	92.30 38
803	644 809	28.33 73	89.61 03		853	727 609	29.20 62	92.35 80
804	646 416	28.35 49	89.66 60		854	729 316	29.22 33	92.41 21
805	648 025	28.37 25	89.72 18		855	731 025	29.24 04	92.46 62
806	649 636	28.39 01	89.77 75		**856**	732 736	29.25 75	92.52 03
807	651 249	28.40 77	89.83 32		857	734 449	29.27 46	92.57 43
808	652 864	28.42 53	89.88 88		858	736 164	29.29 16	92.62 83
809	654 481	28.44 29	89.94 44		859	737 881	29.30 87	92.68 23
810	656 100	28.46 05	90.00 00		860	739 600	29.32 58	92.73 62
811	657 721	28.47 81	90.05 55		**861**	741 321	29.34 28	92.79 01
812	659 344	28.49 56	90.11 10		862	743 044	29.35 98	92.84 40
813	660 969	28.51 32	90.16 65		863	744 769	29.37 69	92.89 78
814	662 596	28.53 07	90.22 19		864	746 496	29.39 39	92.95 16
815	664 225	28.54 82	90.27 74		865	748 225	29.41 09	93.00 54
816	665 856	28.56 57	90.33 27		**866**	749 956	29.42 79	93.05 91
817	667 489	28.58 32	90.38 81		867	751 689	29.44 49	93.11 28
818	669 124	28.60 07	90.44 34		868	753 424	29.46 18	93.16 65
819	670 761	28.61 82	90.49 86		869	755 161	29.47 88	93.22 02
820	672 400	28.63 56	90.55 39		870	756 900	29.49 58	93.27 38
821	674 041	28.65 31	90.60 91		**871**	758 641	29.51 27	93.32 74
822	675 684	28.67 05	90.66 42		872	760 384	29.52 96	93.38 09
823	677 329	28.68 80	90.71 93		873	762 129	29.54 66	93.43 45
824	678 976	28.70 54	90.77 44		874	763 876	29.56 35	93.48 80
825	680 625	28.72 28	90.82 95		875	765 625	29.58 04	93.54 14
826	682 276	28.74 02	90.88 45		**876**	767 376	29.59 73	93.59 49
827	683 929	28.75 76	90.93 95		877	769 129	29.61 42	93.64 83
828	685 584	28.77 50	90.99 45		878	770 884	29.63 11	93.70 17
829	687 241	28.79 24	91.04 94		879	772 641	29.64 79	93.75 50
830	688 900	28.80 97	91.10 43		880	774 400	29.66 48	93.80 83
831	690 561	28.82 71	91.15 92		**881**	776 161	29.68 16	93.86 16
832	692 224	28.84 44	91.21 40		882	777 924	29.69 85	93.91 49
833	693 889	28.86 17	91.26 88		883	779 689	29.71 53	93.96 81
834	695 556	28.87 91	91.32 36		884	781 456	29.73 21	94.02 13
835	697 225	28.89 64	91.37 83		885	783 225	29.74 89	94.07 44
836	698 896	28.91 37	91.43 30		**886**	784 996	29.76 58	94.12 76
837	700 569	28.93 10	91.48 77		887	786 769	29.78 25	94.18 07
838	702 244	28.94 82	91.54 23		888	788 544	29.79 93	94.23 38
839	703 921	28.96 55	91.59 69		889	790 321	29.81 61	94.28 68
840	705 600	28.98 28	91.65 15		890	792 100	29.83 29	94.33 98
841	707 281	29.00 00	91.70 61		**891**	793 881	29.84 96	94.39 28
842	708 964	29.01 72	91.76 06		892	795 664	29.86 64	94.44 58
843	710 649	29.03 45	91.81 50		893	797 449	29.88 31	94.49 87
844	712 336	29.05 17	91.86 95		894	799 236	29.89 98	94.55 16
845	714 025	29.06 89	91.92 39		895	801 025	29.91 66	94.60 44
846	715 716	29.08 61	91.97 83		**896**	802 816	29.93 33	94.65 73
847	717 409	29.10 33	92.03 26		897	804 609	29.95 00	94.71 01
848	719 104	29.12 04	92.08 69		898	806 404	29.96 66	94.76 29
849	720 801	29.13 76	92.14 12		899	808 201	29.98 33	94.81 56
850	722 500	29.15 48	92.19 54		900	810 000	30.00 00	94.86 83
n	n^2	$\sqrt{n}$	$\sqrt{10n}$		n	n^2	$\sqrt{n}$	$\sqrt{10n}$

Table XIV (continued)

n	n^2	$\sqrt{n}$	$\sqrt{10n}$
900	810 000	30.00 00	94.86 83
901	811 801	30.01 67	94.92 10
902	813 604	30.03 33	94.97 37
903	815 409	30.05 00	95.02 63
904	817 216	30.06 66	95.07 89
905	819 025	30.08 32	95.13 15
906	820 836	30.09 98	95.18 40
907	822 649	30.11 64	95.23 65
908	824 464	30.13 30	95.28 90
909	826 281	30.14 96	95.34 15
910	828 100	30.16 62	95.39 39
911	829 921	30.18 28	95.44 63
912	831 744	30.19 93	95.49 87
913	833 569	30.21 59	95.55 10
914	835 396	30.23 24	95.60 33
915	837 225	30.24 90	95.65 56
916	839 056	30.26 55	95.70 79
917	840 889	30.28 20	95.76 01
918	842 724	30.29 85	95.81 23
919	844 561	30.31 50	95.86 45
920	846 400	30.33 15	95.91 66
921	848 241	30.34 80	95.96 87
922	850 084	30.36 45	96.02 08
923	851 929	30.38 09	96.07 29
924	853 776	30.39 74	96.12 49
925	855 625	30.41 38	96.17 69
926	857 476	30.43 02	96.22 89
927	859 329	30.44 67	96.28 08
928	861 184	30.46 31	96.33 28
929	863 041	30.47 95	96.38 46
930	864 900	30.49 59	96.43 65
931	866 761	30.51 23	96.48 83
932	868 624	30.52 87	96.54 01
933	870 489	30.54 50	96.59 19
934	872 356	30.56 14	96.64 37
935	874 225	30.57 78	96.69 54
936	876 096	30.59 41	96.74 71
937	877 969	30.61 05	96.79 88
938	879 844	30.62 68	96.85 04
939	881 721	30.64 31	96.90 20
940	883 600	30.65 94	96.95 36
941	885 481	30.67 57	97.00 52
942	887 364	30.69 20	97.05 67
943	889 249	30.70 83	97.10 82
944	891 136	30.72 46	97.15 97
945	893 025	30.74 09	97.21 11
946	894 916	30.75 71	97.26 25
947	896 809	30.77 34	97.31 39
948	898 704	30.78 96	97.36 53
949	900 601	30.80 58	97.41 66
950	902 500	30.82 21	97.46 79

n	n^2	$\sqrt{n}$	$\sqrt{10n}$
950	902 500	30.82 21	97.46 79
951	904 401	30.83 83	97.51 92
952	906 304	30.85 45	97.57 05
953	908 209	30.87 07	97.62 17
954	910 116	30.88 69	97.67 29
955	912 025	30.90 31	97.72 41
956	913 936	30.91 92	97.77 53
957	915 849	30.93 54	97.82 64
958	917 764	30.95 16	97.87 75
959	919 681	30.96 77	97.92 85
960	921 600	30.98 39	97.97 96
961	923 521	31.00 00	98.03 06
962	925 444	31.01 61	98.08 16
963	927 369	31.03 22	98.13 26
964	929 296	31.04 83	98.18 35
965	931 225	31.06 44	98.23 44
966	933 156	31.08 05	98.28 53
967	935 089	31.09 66	98.33 62
968	937 024	31.11 27	98.38 70
969	938 961	31.12 88	98.43 78
970	940 900	31.14 48	98.48 86
971	942 841	31.16 09	98.53 93
972	944 784	31.17 69	98.59 01
973	946 729	31.19 29	98.64 08
974	948 676	31.20 90	98.69 14
975	950 625	31.22 50	98.74 21
976	952 576	31.24 10	98.79 27
977	954 529	31.25 70	98.84 33
978	956 484	31.27 30	98.89 39
979	958 441	31.28 90	98.94 44
980	960 400	31.30 50	98.99 49
981	962 361	31.32 09	99.04 54
982	964 324	31.33 69	99.09 59
983	966 289	31.35 28	99.14 64
984	968 256	31.36 88	99.19 68
985	970 225	31.38 47	99.24 72
986	972 196	31.40 06	99.29 75
987	974 169	31.41 66	99.34 79
988	976 144	31.43 25	99.39 82
989	978 121	31.44 84	99.44 85
990	980 100	31.46 43	99.49 87
991	982 081	31.48 02	99.54 90
992	984 064	31.49 60	99.59 92
993	986 049	31.51 19	99.64 94
994	988 036	31.52 78	99.69 95
995	990 025	31.54 36	99.74 97
996	992 016	31.55 95	99.79 98
997	994 009	31.57 53	99.84 99
998	996 004	31.59 11	99.89 99
999	998 001	31.60 70	99.95 00
1000	1000 000	31.62 28	100.00 00

REFERENCES AND SUGGESTIONS FOR FURTHER READING

PROBABILITY THEORY

Derman, C., Gleser, L. J., and Olkin, I., *A Guide to Probability Theory and Application*. New York: Holt, Rinehart and Winston, Inc., 1973.

Feller, W., *An Introduction to Probability Theory and Its Applications*, Vol. I, 3rd Ed. New York: John Wiley & Sons, Inc., 1968.

de Finetti, B., *Theory of Probability: A Critical Introductory Treatment*, Vol. I. London: John Wiley & Sons, Inc., 1974.

Kyburg, H. E., and Smokler, H. E., *Studies in Subjective Probability*. New York: John Wiley & Sons, Inc., 1964.

Parzen, E., *Modern Probability Theory and Its Applications*. New York: John Wiley & Sons, Inc., 1960.

Ross, S., *Introduction to Probability Models*. New York: Academic Press, Inc., 1972.

Thompson, W. A., *Applied Probability*. New York: Holt, Rinehart and Winston, Inc., 1969.

STATISTICAL INFERENCE AND DECISION

1. Primarily Classical Inference

Brunk, H. D., *An Introduction to Mathematical Statistics*, 2d Ed. Boston: Ginn & Company, 1965.

Cramér, H., *Mathematical Methods of Statistics*. Princeton, N. J.: Princeton University Press, 1946.

Freund, J. E., *Mathematical Statistics*, 2d Ed. Englewood Cliffs, N.J.: Prentice-Hall, Inc., 1971.

Hodges, J. L., and Lehmann, E. L., *Basic Concepts of Probability and Statistics*, 2d Ed. San Francisco: Holden-Day, Inc., 1970.

Hoel, P. G., *Introduction to Mathematical Statistics*, 4th Ed. New York: John Wiley & Sons, Inc., 1971.

Hogg, R. V., and Craig, A. T., *Introduction to Mathematical Statistics*, 3rd Ed. New York: The Macmillan Company, 1970.

Kendall, M. G., and Stuart, A., *The Advanced Theory of Statistics*, Vols. I, II, and III. London: Charles Griffin & Co., Ltd., 1958, 1961, and 1966.

Lindgren, B. W., *Statistical Theory*. New York: The Macmillan Company, 1960.

Lippman, S. A., *Elements of Probability and Statistics*. New York: Holt, Rinehart and Winston, Inc., 1971.

Mood, A. M., Graybill, F. A., and Boes, D.C., *Introduction to the Theory of Statistics*. 3rd Ed. New York: McGraw-Hill, Inc., 1974.

Wallis, W. A., and Roberts, H. V., *Statistics: A New Approach*. Glencoe, Ill.: Free Press, 1956.

Wilks, S. S., *Mathematical Statistics*. New York: John Wiley & Sons, Inc., 1962.

2. Primarily Bayesian Inference and/or Decision Theory

Box, G. E. P., and Tiao, G. C., *Bayesian Inference in Statistical Analysis*. Reading, Mass.: Addison-Wesley Publishing Company, Inc., 1973.

Brown, R. V., Kahr, A. S., and Peterson, C., *Decision Analysis for the Manager*. New York: Holt, Rinehart and Winston, Inc., 1974.

Chernoff, H., and Moses, L. E., *Elementary Decision Theory*. New York: John Wiley & Sons, Inc., 1959.

DeGroot, M. H., *Optimal Statistical Decisions*. New York: McGraw-Hill, Inc., 1970.

Fishburn, P. C., *Decision and Value Theory*. New York: John Wiley & Sons, Inc., 1964.

LaValle, I. H., *An Introduction to Probability, Decision, and Inference*. New York: Holt, Rinehart and Winston, Inc., 1970.

Lindley, D. V., *Introduction to Probability and Statistics from a Bayesian Viewpoint* (2 Vols.). Cambridge: Cambridge University Press, 1965.

Lindley, D. V., *Making Decisions*. London: John Wiley & Sons, Inc., 1971.

Luce, R. D., and Raiffa, H., *Games and Decisions*. New York: John Wiley & Sons, Inc., 1957.

Novick, M. R., and Jackson, P. H., *Statistical Methods for Educational and Psychological Research*. New York: McGraw-Hill, Inc., 1974.

Pratt, J. W., Raiffa, H., and Schlaifer, R., *Introduction to Statistical Decision Theory*. New York: McGraw-Hill, Inc., 1965.

Raiffa, H., *Decision Analysis*. Reading, Mass.: Addison-Wesley Publishing Company, Inc., 1968.

Schlaifer, R., *Probability and Statistics for Business Decisions*. New York: McGraw-Hill, Inc., 1959.

Schlaifer, R., *Analysis of Decisions Under Uncertainty*. New York: McGraw-Hill, Inc., 1969.

Schmitt, S. A., *Measuring Uncertainty: An Elementary Introduction to Bayesian Statistics*. Reading, Mass.: Addison-Wesley Publishing Company, Inc., 1969.

Winkler, R. L., *An Introduction to Bayesian Inference and Decision*. New York: Holt, Rinehart and Winston, Inc., 1972.

3. General

Clelland, R. C., deCani, J. S., and Brown, F. E., *Basic Statistics With Business Applications*, 2d Ed. New York: John Wiley & Sons, Inc., 1973.

Dyckman, T. R., Smidt, S., and McAdams, A. K., *Management Decision Making Under Uncertainty*. New York: The Macmillan Company, 1969.

Hadley, G., *Introduction to Probability and Statistical Decision Theory*. San Francisco: Holden-Day, Inc., 1967.

Hamburg, M., *Statistical Analysis for Decision Making*. New York: Harcourt, Brace & World, Inc., 1970.

Harnett, D. L., *Introduction to Statistical Methods*, 2d Ed. Reading, Mass.: Addison-Wesley Publishing Company, Inc., 1975.

Hughes, A., and Grawoig, D., *Statistics: A Foundation for Analysis*. Reading, Mass.: Addison-Wesley Publishing Company, Inc., 1971.

Larson, H. J., *Introduction to Probability Theory and Statistical Inference*, 2d Ed. New York: John Wiley & Sons, Inc., 1974.

Spurr, W. A., and Bonini, C. P., *Statistical Analysis for Business Decisions*. Homewood, Ill.: Richard D. Irwin, Inc., 1967.

Wonnacott, T. H., and Wonnacott, R. J., *Introductory Statistics*, 2d Ed. New York: John Wiley & Sons, Inc., 1972.

REGRESSION AND CORRELATION, SAMPLING THEORY, EXPERIMENTAL DESIGN, ANALYSIS OF VARIANCE, MULTIVARIATE ANALYSIS, NONPARAMETRIC METHODS

Bradley, J. V., *Distribution-free Statistical Tests*. Englewood Cliffs, N. J.: Prentice-Hall, Inc., 1968.

Christ, C. F., *Econometric Models and Methods*. New York: John Wiley & Sons, Inc., 1966.

Cochran, W. G., *Sampling Techniques*, 2d Ed. New York: John Wiley & Sons, Inc., 1963.

Cochran, W. G., and Cox, G. M., *Experimental Designs*, 2d Ed. New York: John Wiley & Sons, Inc., 1957.

Cox, D. R., *Planning of Experiments*. New York: John Wiley & Sons, Inc., 1958.

Deming, W. E., *Sample Design in Business Research*. New York: John Wiley & Sons, Inc., 1960.

Draper, N., and Smith, H., *Applied Regression Analysis*. New York: John Wiley & Sons, Inc., 1966.

Fisher, R. A., *The Design of Experiments*, 3d Ed. Edinburgh: Oliver and Boyd, Ltd., 1942.

Gibbons, J. D., *Nonparametric Statistical Inference*. New York: McGraw-Hill, Inc., 1971.

Goldberger, A. S., *Econometric Theory*. New York: John Wiley & Sons, Inc., 1964.

Graybill, F. A., *An Introduction to Linear Statistical Models*. Vol. I. New York: McGraw-Hill, Inc., 1961.

Guenther, W. C., *Analysis of Variance*. Englewood Cliffs, N. J.: Prentice-Hall, Inc., 1964.

Hansen, M. H., Hurwitz, W. N., and Madow, W. G., *Sample Survey Methods and Theory*, Vols. I and II. New York: John Wiley & Sons, Inc., 1953.

Hollander, M., and Wolfe, D. A., *Nonparametric Statistical Methods*. New York: John Wiley & Sons, Inc., 1973.

Johnston, J., *Econometric Methods*, 2d Ed. New York: McGraw-Hill, Inc., 1972.

Kish, L., *Survey Sampling*. New York: John Wiley & Sons, Inc., 1965.

Malinvaud, E., *Statistical Methods of Econometrics*, 2d Ed. Chicago: Rand McNally & Company, 1970.

Mendenhall, W., *Introduction to Linear Models and the Design and Analysis of Experiments*. Belmont, Cal.: Wadsworth Publishing Company, Inc., 1968.

Morrison, D. F., *Multivariate Statistical Methods*. New York: McGraw-Hill, Inc., 1967.

Neter, J., and Wasserman, W., *Applied Linear Statistical Models*. Homewood, Ill.: Richard D. Irwin, Inc., 1974.

Siegel, S., *Nonparametric Statistics*. New York: McGraw-Hill, Inc., 1956.

Snedecor, G. W., and Cochran, W. G., *Statistical Methods*, 6th Ed. Ames, Iowa: Iowa State University Press, 1967.

Tatsuoka, M. M., *Multivariate Analysis*, New York: John Wiley & Sons, Inc., 1971.

Theil, H., *Principles of Econometrics*. New York: John Wiley & Sons, Inc., 1971.

Wonnacott, R. J., and Wonnacott, T. H., *Econometrics*. New York: John Wiley & Sons, Inc., 1970.

Zellner, A., *An Introduction to Bayesian Inference in Econometrics*. New York: John Wiley & Sons, Inc., 1971.

TABLES

Beyer, W. H., *Handbook of Tables for Probability and Statistics*, 2d Ed. Cleveland, Ohio: The Chemical Rubber Co., 1968.

Burington, R. S., and May, D. C., *Handbook of Probability and Statistics with Tables*, 2d Ed. New York: McGraw-Hill, Inc., 1969.

Fisher, R. A., and Yates, F., *Statistical Tables for Biological, Agricultural and Medical Research*, 6th Ed. Edinburgh: Oliver and Boyd, 1963.

Owen, D. B., *Handbook of Statistical Tables*. Reading, Mass.: Addison-Wesley Publishing Company, Inc., 1962.

Pearson, E. S., and Hartley, H. O., *Biometrika Tables for Statisticians*, Vol. I, 3rd Ed. and Vol. II, 2d Ed. Cambridge: Cambridge University Press, 1967 and 1972.

ANSWERS TO
SELECTED EXERCISES

CHAPTER 1

3. (b) $\{2, 4, 6, 8\}$.
4. (a) finite   (d) uncountably infinite   (e) countably infinite.
6. (a) $\{x \mid x^2 \geq 1\}$ (c) $\{x \mid -1 < x < 1\}$ (e) $\{x \mid x > 5\}$.
8. (b) $\bar{A} \cap B \cap \bar{C}$, 0   (d) $(A \cap B \cap \bar{C}) \cup (A \cap \bar{B} \cap C)$, 20
    (h) $(A \cap B \cap \bar{C}) \cup (A \cap \bar{B} \cap C) \cup (\bar{A} \cap B \cap C) \cup (A \cap B \cap C)$, 32.
9. (a) 8 percent   (c) 83 percent.
10. (b) 120   (d) 60.
11. (a) 6   (c) 25   (e) 50.
14. (b) $\{-3, -1, 1, 3\}$ (d) $\{-4, -2, 0, 2, 4\}$ (h) $\{-3, -1\}$.
15. (b) $\{0, 1, 2, 3\}$, $\{1, 2, 3, 4, 5\}$, $\{1, 2, 3\}$, $\{x \mid x = -3, x = -2, \text{ or } x \geq -1\}$, $\{4, 5\}$.
18. (a) true   (c) false.
21. 16 subsets.
25. (a) domain = $\{2, 4, 8, 9\}$, range = $\{9\}$ (c) domain = $\{2, 3\}$, range = $\{4, 9\}$ (g) domain = $\{3\}$, range = $\{9\}$.
26. (b) domain = U, range = $\{7\}$ (d) domain = $\{1, 2, 3, 4, 5\}$, range = $\{2, 4, 6, 8, 10\}$.
27. (e) $\{(4, 2), (5, 3), (6, 4)\}$ (g) the relations in (d) and (e).
29. (a) a, e (c) $b, c, d, u, v, w, x, y, z$ (e) b, c, d.
33. (b) $(6, 5), (5, 6)$ (3) $(1, 1), (6, 6)$ (f) 6/36, 2/36, 6/36, 10/36, 2/36.

CHAPTER 2

5. all positive integers, countably infinite; all states plus "no state," finite.
7. (c) 7/12   (e) 13/18   (g) 1/12.

8.   (a) 5/26   (c) 5/13.
9.   25/52, 1/2.
10. (a) 11/16   (c) 5/16.
11. (b) 61/900   (d) 0.
15. (i) (a) 9/32   (b) 10/32   (c) 31/32.
16. (b) 4/13   (d) 3/13.
17. (a) .52   (b) .49.
19. 125/216.
21. P(at least 2 conservatives) $= 1 - [\binom{60}{10}/\binom{98}{10}] - [\binom{38}{1}\binom{60}{9}/\binom{98}{10}]$.
23. (a) $13(48)/\binom{52}{5}$.
24. 2520.
28. (b) 7 (2!)(6!)   (d) 4! (2^4).
29. 7!
31. (b) .2.
32. (a) 24.
34. (b) 3! (2!) (4!) (3!).
41. (a) 2/3   (c) 3/10.
42. 2/5.
43. (b) 1 to 999   (d) 7 to 1.
48. (b) 10/17.
55. (b) 3/7, 12/13, 4/7, 1/13   (e) no.
56. (b) .82, .42   (d) .76.
57. (c) 1/12, 1/3, .70, .65.
58. (a) .60   (c) .10   (e) .175.
61. 2/3.
64. 8/23, 5/23, 10/23; 8/133, 25/133, 100/133.
65. 5/9, 3/9, 1/9.
67. 2/3.
69. 5/29.
70. (a) .06   (c) .30   (e) .506   (g) .284   (i) .042.
72. (a) .276, .048, .168.
73. (a) 5/18.
76. (b) 19/40   (d) no.
77. (b) .512.
78. (a) 1/32   (c) no   (e) no.
80. (b) no.
81. (b) .68   (d) yes.

CHAPTER 3

1.   (a) yes   (c) no   (e) yes.
2.   (a) discrete   (f) continuous.
3.   (c) .5   (e) .9   (g) 0.
4.   (b) 3/4.
6.   (a) 6/13   (c) 10/13   (e) 11/13.
7.   (b) 5.

9. (c) 53/900.
10. (b) .30.
11. (c) 7/18.
12. (b) .21875   (c) 0.
13. (c) 9/15.
14. (b) .756.
17. (b) .16.
18. (a) 1/9.
19. (c) 5/32.
21. no.
22. (b) .83   (c) .453.
24. (d) 1/3   (e)1/2.
27. $5.18.
29. (a) $E(X) = 100$.
30. mean $= 1.8$, mode $= 0$, median $= 1.5$.
31. (a) 50/16,   (c) 150/16.
33. $E(X) = 101/900$, midrange $= 1.5$.
38. (a) 0   (c) -1.
39. 6.25, 5.36.
40. $E(R) = 2.17$, mode $= 3$, median $= 2.24$.
42. (b) 4.30   (d) 1.13.
43. 1.86 years.
45. $E(X) = 8205$, $V(X) = 445,475$, $E(W) = 3,025$, $V(W) = 11,136,875$.
47. (a) 50/16   (c) 24.5   (e) 13.74.
48. 0.75, 1.00.
52. $a = 1/3, b = 2$.
54. 1.39, 1.18.
56. 3, .775.
58. (a) 1.31   (c) 0.65.
59. (b) 30, 70, 69.5, 24.2, 66.5, 53.3.
62. (a) .22   (b) .59   (c) .26.
64. 1.14.
66. $E(X^2) = 10.625$, $E[(X - \mu)]^3 = -.574$.
69. (a) $P(X = x, Y = y) = (6 - x)(3 + xy - 2y)/15(3x)$ for $x = 1, 2, 3, 4, 5$ and $y = 1, 2$.
71. (a) 1 (c) $f(x \mid y) = 2(x + y)/(2y + 1)$ if $0 \leq x \leq 1, 0 \leq y \leq 1$.
72. (b) 0, 1.29, 1.29.
73. .13, .332.
75. $5/3, 5/3, -1/36, -1/11$.
78. $-.418$.
82. 10, 10.

CHAPTER 4

2. .0584.
3. (b) .0407.

5.   (a) .933   (b) .3828   (c) .074   (d) .1184, 6.
7.   (b) 4, .1746, .4958.
9.   .2639.
10.   .0154, .9838, 2/3.
11.   .2668, .1502, .0015.
15.   (a) .6415   (c) .1652.
16.   (a) .4096   (b) .8493.
17.   (c) .5606.
19.   20.
21.   34,000.
22.   (b) .0823.
23.   .0353, .0014.
24.   .000685, .0154.
27.   .35625.
29.   .0324, .0220.
33.   (a) .0046   (c) .00005.
36.   .0067, .0681, .1251, 10.
38.   1/2, .1353.
43.   (a) .1848   (b) 3.
44.   (b) .0025.
47.   .347.
51.   .0083.
54.   (a) .3679   (c) .3418.
55.   (b) .8164   (d) .01.
56.   (b) .097, .326, .58   (d) .15, .35, .51, .65.
58.   109.1, 10.5.
60.   (b) 629   (f) 885.
61.   (a) .2138   (c) .3793.
63.   (b) .2785   (d) 3.968 years.
64.   (a) .016.
67.   2058.
68.   (b) .8817.
72.   (a) .30   (c) .1897.
73. $a = 3, b = 9$.
76.   .6875.

CHAPTER 5

1.   (a) descriptive   (c) inferential.
3.   (b) no.
5.   no.
8.   (a) interval   (d) ratio   (g) nominal.
25.   (a) .316   (c) .322.
26.   (c) median = .304, mode = .187.
29.   (a) 58.09, 58   (c) 58.22, 58.47.

30. 17.92, 17.5, 6.6, 5.08.
36. (a) 24.3   (c) 28.0.
38. .085.
39. (a) $18,375   (c) $19,600.
44. $P(M = 3) = P(M = 5) = 1/9, P(M = 3.5) = P(M = 4.5) = 2/9,$
    $P(M = 4) = 3/9.$
46. 4, .333.
47. $P(S^2 = 0) = 3/9, P(S^2 = .25) = 4/9, P(S^2 = 1) = 2/9.$
49. $P(M = 3) = 1/27, P(M = 4) = 1/6, P(M = 5) = 11/36, P(M = 6) =$
    $7/24, P(M = 7) = 11/72, P(M = 8) = 1/24, P(M = 9) = 1/216.$
50. standard errors = 1.19, 1.66.
52. $P(M = 0.5) = P(M = 2.0) = 1/6, P(M = 1.0) = P(M = 1.5) = 2/6.$
54. (b) 2.48.
55. (b) 2.20.
56. (b) .8262.
59. (a) .3707   (c) .0418.
60. (a) .2514   (c) .0003.
62. (b) normal, mean = 108,000, variance = 2,430,000   (d) normal, mean = 0,
    variance = 1,620,000.
63. (a) normal, mean = 3.2, variance = .04   (c) .0367.
64. (c) normal, mean = −1600, variance = 29,284.
68. (a) .50   (c) 150, 25.
70. (c) .1898   (h) .875.
73. .8944.

CHAPTER 6

2. (a) yes   (c) $\sigma^2/4, 3\sigma^2/10.$
3. (b) all are unbiased   (d) $\sigma^2, \sigma^2/2, \sigma^2/n.$
9. $-\sigma^2/n.$
11. $\sum a_i = 1.$
13. (a) no   (c) no.
15. (a) no   (b) .25, .56, .59.
17. (a) no   (b) 7.65, 12.69, 4.69.
24. 10/31, 5.75.
27. 1.5.
29. $m - s\sqrt{3}, m + s\sqrt{3}.$
34. (a) 54, 130.89   (d) (47.6, 60.4).
35. (165.32, 170.68), (158.68, 177.32).
37. 52.1, 1302.
39. (a) no   (e) no.
41. (a) 100   (b) 4268, 42.4.
43. $9n.$
47. (48.35, 59.65), (44.67, 63.33).
50. (a) −1.356, .259, 3.055   (d) .875.

53. (a) -11.5 (b) $(-20.55, -2.45)$, $(-29.74, 6.74)$.
57. $(-.025, .425)$, $(-.002, .378)$.
58. $(-7.56, 23.56)$.
60. $(248.81, 251.19)$.
62. $(.150, .178)$, $(.137, .191)$.
66. (a) .01 (c) .975.
67. (b) .74 (d) 384.
68. $(88.17, 337.54)$, $(67.20, 540.37)$.
69. $(17.40, 28.21)$.
74. $(10,833.19, 14,329.31)$ for μ, $(15,105,531$ to $41,503,632)$ for σ^2.
75. (a) .148, 6.63 (c) .95.
76. $(.343, 14.42)$.
79. $(.476, 1.296)$, $(.386, 1.604)$.

CHAPTER 7

2. (a) $\alpha = .16$, $\beta = .16$ (c) $\alpha = .006$, $\beta = .69$ (e) $\alpha = .84$, $\beta = .07$.
4. (a) .1557, .4667 (c) .9626, .2578.
8. (b) .001, .16 (d) .977, .977.
9. (a) reject H_0 if $R \leq 7$.
10. (a) reject H_0 if $M > 141.29$ (b) reject H_0 if $M > 138.71$.
17. 103.84, .16.
18. (a) 62.8, .61 (b) 62.8, .043.
20. $\alpha = .1014$, $\beta = .5659$.
21. .056, 12.365.
23. .16.
24. (a) reject manufacturer's claim if $M \leq 74.88$.
25. reject manufacturer's claim if $M \leq 65.12$.
28. (b) .749.
30. (a) 75.49 (c) .96.
31. 75.25.
33. (a) .1209 (c) .0128.
36. (a) reject if $M \leq 134.72$ or $M \geq 141.28$ (b) .272.
38. (a) $R \leq 8$ (b) .1846.
41. return if $M \leq 2.897$ or $M \geq 3.103$.
44. return if $M \leq 2.907$.
48. no.
50. reject H_0 if $R/n > .521$.
53. .62, no, no.
54. (a) .16 (c) .953.
56. (a) $2.742 < M < 3.258$ (c) $2.974 < M < 3.026$.
59. (a) $.005 < p\text{-value} < .10$ (b) market the drug.
63. .12.
64. $p\text{-value} < .001$.
66. accept H_0, $(44.4, 51.0)$, $.20 < p\text{-value} < .50$.

69. $.25 < p\text{-value} < .40$; ignore the claim.
71. .25.
73. $.10 < p\text{-value} < .25$.
75. $.10 < p\text{-value} < .20$.
76. $.10 < p\text{-value} < .25$, (1.14, 4.61).
78. 11,925.
80. yes, $p\text{-value} < .005$.
82. $.20 < p\text{-value} < .50$, accept H_0, accept H_0.
84. no.
86. no.
87. yes.
89. no.

CHAPTER 8

3. .1933, .3844, .3097, .1126.
5. $P(p = .4) = 2/15$, $P(p = .5) = 10/15$, $P(p = .6) = 3/15$.
6. $E''(p) = .1568$, $V''(p) = .000699$.
8. .530, .391, .079.
9. $P(5\%) = .070$, $P(10\%) = .786$, $P(15\%) = .144$.
14. .109, .561, .330.
16. $P(2) = .304$, $P(2.5) = .550$, $P(3) = .146$.
18. $E''(p) = .299$, no.
25. $P(\mu = 109.4) = .0734$, $P(\mu = 109.7) = .2572$, $P(\mu = 110.0) = .5063$, $P(\mu = 110.3) = .1412$, $P(\mu = 110.6) = .0219$.
27. $P(\theta = 2) = 3/99$, $P(\theta = 1) = 24/99$, $P(\theta = 0) = 72/99$.
30. $P(R = 0) = .66879$, $P(R = 3) = .01217$.
31. $f(p \mid y) = 3(1 - p)^2$ if $0 \le p \le 1$.
35. $r = 10$, $n = 15$.
36. (a) .5122   (c) 2.6144.
38. $E''(p) = .4$, $V''(p) = .0218$.
40. beta, $r'' = 3$, $n'' = 26$.
43. normal, $m'' = 109.73$, $\sigma''^2 = .267$.
44. $m' = 174.6$, $\sigma' = 21.7$.
48. normal, $m'' = 51.4$, $\sigma''^2 = 8.97$.
51. normal, $m'' = 2.32$, $\sigma''^2 = .072$.
52. $n = 66$, $n = 714$.
55. exponential, $\lambda = .55$.
58. beta, $r' = 450$, $n' = 1000$.
68. (a) $E'(p) = .40$ (b) $r/n = .30$ (c) $E''(p) = .35$.
70. (a) $E'(\lambda) = 3.05$ (b) $r/t = 3.33$ (c) $E''(\lambda) = 3.22$.
71. (d) (43.83, 56.17)   (f) (46.50, 56.38).
74. (a) 5.667   (c) 1.275.
75. (b) $H_0: \lambda = 4$, $H_1: \lambda = 2$ (e) .005.
77. (b) 29.64.

78.   (b) 1.45   (c) .36.
79. $\Omega'' = .1689$.
80. $\Omega'' = 2.33$.
84.   0, 10.60.
85.   (b) .3625   (d) .325.
87.   (a) .674, .322, .004   (c) H_3.
88.   (c) normal, mean $= 18.62$, variance $= 6.9$.

CHAPTER 9

 4.   (a) action 3   (c) action 2.
10.   (a) action 2   (c) action 4.
11.   order 100, order 300, order 300.
15. $ER(1) = 26.0$, $ER(2) = 48.0$, $ER(3) = 39.5$.
16. $EL(1) = 3.15$, $EL(2) = 0.70$, $EL(3) = 1.30$.
17.   200 shirts.
19.   not advertise.
24. ER(keep salesman) $= 8.725$, ER(don't keep salesman) $= 0$.
25.   (b) P(adverse weather) $> c/d$.
28.   (a) action 2   (d) action 2.
33.   (b) yes   (c) no.
35.   for $p = 1/3$: (a) yes   (b) no   (d) no   (e) yes   (f) no.
36.   (c) yes.
37. $500 in each investment.
40.   (b) advertise.
42.   rent.
47.   68.5, 0.70.
49.   EVPI $= 50,000$.
50.   (b) 1.95   (c) 1.40.
51.   (a) 0   (c) 2571.48.
53.   (c) .84.
54.   (d) .34, .09, $-.82$, -2.95.
56.   (a) $70B-30R$ (b) 4.80   (c) 1.20   (d) 2.84.
57.   (b) 4,000   (c) 4,500   (d) 1236, 1196.
59.   (a) 0   (b) 14,286   (d) 5,079.
60.   (b) 52,000
61.   ENGS (A and B) $= 62,300$.
62.   ENGS (sequential plan) $= 73,900$.
63.   (a) 1.90   (b) 2.53.
65.   (a) 20   (c) 5   (e) 5.726   (g) 3.956.
67.   EVSI: 1.98, 2.91, 3.51, 3.87, 4.10, 4.30, 4.46, 4.62, 4.69, 4.82.
69.   salary plus commission.
72. a_1, a_2.
73.   5 houses, EVPI $= 2,000$.

75. 21.4 additional cases.
76. 10,000 copies.
79. '.10.
81. $\Omega'' = 1/36$, reject H_0.
83. no.
84. (a) .2552   (b) .6284   (c) H_1.
85. (a) $M \geq 11.0346$ (b) .0005, .0011.
88. $L(\text{II}) = 3L(\text{I})$.

CHAPTER 10

1. (a) 9.00   (b) .309.
3. (b) $2(x+y)/(2x+1)$ if $0 < y < 1$ (c) $(3x+2)/3(2x+1)$.
5. (a) $-1/144$ (b) $-1/11$.
8. .203.
10. (b) .324.
12. (b) .86.
14. (a) .114   (b) .32   (c) $(-.013, .595)$.
16. $(.710, .936)$, $(.612, .954)$.
22. $E(Y \mid X = 15) = 21.00$, $E(Y \mid X = 30) = 24.80$, $V(Y \mid X = 20) = 25.69$.
24. $E(Y \mid x) = 1.277 + .083x$, $E(X \mid y) = 1.596 + .499y$.
26. $\hat{y} = 3.63 - .079x$, $(-.25, .09)$, $(2.06, 4.72)$.
28. (a) $E(Y/x) = 77 + 6x$ (b) 108   (c) .75.
30. if $r_{XY}^2 = 1$.
33. (a) 400 hours   (b) 13,333.
35. (a) $\hat{y} = 93.80 + 4.44x$ (c) $S_{Y \cdot X}^2 = 262$.
36. (b) $\hat{y} = -.095 + .38x$ (c) .11.
38. (b) 5.836.
39. for $X = 3$: $(3.21, 3.57)$, $(2.27, 4.51)$.
42. $(-265.05, 79.89)$, $(1.37, 6.31)$.
45. if $\hat{w} = b_1^* + b_2^* u$, then $b_1^* = db_1 + h - (db_2 g/c)$ and $b_2^* = db_2/c$.
46. (b) 15.1, 3.57.
49. (d) $\hat{y} = 4.85370 - .36199x_2 + .04374x_3$ (e) .9816   (f) .9636.
50. (b) .9594   (c) .8176   (d) no.
51. (a) $-.8947, .3207$ (b) .103.
54. $\hat{y} = 16.625 - 2.10526x_2 + .94079x_3$, .6705.
57. $(9.977, 23.273)$, $(-3.604, -.606)$, $(.306, 1.574)$.
58. (b) .9645   (c) 6.35.
63. $-.2813, .142, .879$; X_4.
64. X_2.
67. $\hat{y} = 11.09x^{1.91}$.
68. $\hat{y} = 48.6 - 13.1x + .97x^2$.
72. $r^2 = .0055$.
75. (b) .278.
76. (a) .747   (d) from (b), 84.28; from (c), 82.04.
77. (b) .594   (d) .728.

CHAPTER 11

3. $1/2$ for each item; not a simple random sample.
5. (c) 2.5, 0.42.
7. (a) 0.3125   (b) 0.
8. (a) .252   (b) .203.
11. (b) $P(M = 11) = P(M = 13) = 1/4$, $P(M = 12) = 1/2$; $V(M) = 0.5$.
12. (a) $P(M = 11) = P(M = 13) = 1/2$; $V(M) = 1$.
14. (b) 7.38, .532   (c) 7.08.
18. 5.75 million.
20. est. $\mu = 13,352$.
22. (a) .10   (b) .20   (c) .12   (d) $P(.10) = .88$, $P(.20) = .12$.
23. (c) $P(0) = .6856$, $P(1) = .2712$, $P(2) = .0408$, $P(3) = .0024$.
27. (b) 28   (c) .2, -2.6, 2.4.
33. (a) SS(between) $= 31,141.2$, SS(within) $= 11,449.1$, MS(between) $= 10,380$, MS(within) $= 358$, $F = 29.0$ (c) .70.
37. est. $\omega^2 = .0625$, est. α_i: -4.25, 8.475, -3.525, -4.525.
41. for $\alpha_1 - \alpha_2$: (a) $(-2.53, 33.87)$ (b) $(-10.64, 41.98)$.
42. for $\alpha_2 - \alpha_4$: $(-9.11, 35.11)$.
46. (a) $F = 11.54$, p-value $< .01$ (b) 1.72, .72, -2.45 (c) .540.
47. (c) .173   (d) yes.
56. (a) $F = 107.12$, p-value $< .01$ (c) $F = 6.84$, p-value $< .01$.
57. est. $\omega_{\text{rows}}^2 = .09$, est. $\omega_{\text{col.}}^2 = .74$, est. $\omega_{\text{int.}}^2 = .04$; estimated column effects: 12.4, -11.5, -1.0; estimated row effects: 3.4, -3.5.
65. $F = 3.568$ (p-value $> .05$), $F = .002$.
66. .052, 0.
67. for linear regression, $F = 1.82$ (p-value $> .05$); for deviations from linear regression, $F = 32.28$ (p-value $< .01$); .002, .837, .835.

CHAPTER 12

5. $\chi^2 = 6.99$ with 4 d.f.; $.05 < p$-value $< .10$.
6. $\chi^2 = 2.84$ with 3 d.f.; $.25 < p$-value $< .50$.
8. $\chi^2 = 1.48$ with 3 d.f.; $.50 < p$-value $< .75$.
9. $\chi^2 = 4.0$ with 7 d.f.; $.75 < p$-value $< .90$.
10. $\chi^2 = 17.26$ with 1 d.f.; p-value $< .001$.
15. grouping "3 or more," $\chi^2 = 3.23$ with 2 d.f.; $.10 < p$-value $< .25$.
18. (a) $\chi^2 = 12.5$ with 1 d.f.; p-value $< .001$.
21. using 60 to estimate $1/\lambda$, $\chi^2 = 9.49$ with 4 d.f.; p-value $= .05$.
23. $\chi^2 = 78.78$ with 12 d.f.; p-value $< .001$.
24. no.
27. (a) yes   (b) yes.
28. (a) yes   (c) yes.
32. $\chi^2 = 32.09$ with 1 d.f.; p-value $< .001$.

33. .067.
35. .119, .201, .044.
37. .270, .191, .261.
39. .023, 0, .011.
42. (b) 1.280   (c) 1.234.
43. $D = .04$, p-value $> .20$.
45. $D = .1556$, p-value $> .20$.
49. $D = .092$, p-value $> .20$.
52. 12 runs.
54. for Town A, $W = 112$ and $U = 41$.
57. (a) $\chi^2 = 1.56$ with 1 d.f., $.10 < p$-value $< .25$ (c) $z = 1.47$, p-value $= .144$.
58. $.1 < p$-value $< .2$.
59. 17 runs, $z = .37$.
60. (a) $P(R = 4) = .0866$, $P(R = 7) = .2165$ (b) $P(R = 5) = .1653$.
62. $W = 129$, $z = -.90$, p-value $= .37$.
65. $W = 36$, $z = .27$, p-value $= .78$.
67. for the sign test, p-value $= .0702$.
68. $\chi^2 = 13.33$ with 4 d.f.; $.005 < p$-value $< .01$.
70. $U = 89.5$, $z = 2.99$, p-value $= .003$.
71. $\chi_c^2 = 17.34$ with 4 d.f.; $.001 < p$-value $< .005$.
73. (a) $\chi^2 = 28.09$ with 3 d.f.; p-value $< .001$ (b) $\chi_c^2 = 22.43$ with 3 d.f.; p-value $< .001$.
76. $\chi^2 = 9.94$ with 3 d.f.; $.01 < p$-value $< .25$.
79. .594, .424.
82. .79, .779.
84. average $r_S = .705$.
85. .2634, .21, $\chi^2 = 35.56$ with 9 d.f. (p-value $< .001$).

INDEX